Edited by
Nicholas E. Geacintov
and Suse Broyde

**The Chemical Biology of
DNA Damage**

Further Reading

Nakamoto, K., Tsuboi, M., Strahan, G. D.

Drug-DNA Interactions

Structures and Spectra

2009
Hardcover
ISBN: 978-0-471-78626-9

Singleton, P.

Dictionary of DNA and Genome Technology

2008
Hardcover
ISBN: 978-1-4051-5607-3

Müller, S. (ed.)

Nucleic Acids from A to Z

A Concise Encyclopedia

2008
Hardcover
ISBN: 978-3-527-31211-5

Matta, C. F. (ed.)

Quantum Biochemistry

2 Volumes

2010
Hardcover
ISBN: 978-3-527-32322-7

Ekins, S. (ed.)

Computational Toxicology

Risk Assessment for Pharmaceutical and Environmental Chemicals

2007
Hardcover
ISBN: 978-0-470-04962-4

O'Brien, P. J., Bruce, W. R. (eds.)

Endogenous Toxins

Targets for Disease Treatment and Prevention

2010
Hardcover
ISBN: 978-3-527-32363-0

Edited by Nicholas E. Geacintov and Suse Broyde

The Chemical Biology of DNA Damage

WILEY-VCH Verlag GmbH & Co. KGaA

The Editors

Prof. Nicholas E. Geacintov
New York University
Chemistry Department
31 Washington Place
New York, NY 10003
USA

Prof. Suse Broyde
New York University
Department of Biology
100 Washington Square
New York, NY 10003
USA

Cover
The cover art is based on the modeling work of Dr. Lei Jia (Nucleic Acids Research, Volume 36, pages 6571-6584 (2008), and with assistance in rendering from Dr. Lihua Wang and Dr. Yuqin Cai, as well as Dr. Martin Graf (Wiley-VCH).

It shows DNA damaged by the environmental chemical carcinogen benzo[a]pyrene in the active site of the human DNA bypass polymerase κ.

■ All books published by Wiley-VCH are carefully produced. Nevertheless, authors, editors, and publisher do not warrant the information contained in these books, including this book, to be free of errors. Readers are advised to keep in mind that statements, data, illustrations, procedural details or other items may inadvertently be inaccurate.

Library of Congress Card No.: applied for

British Library Cataloguing-in-Publication Data
A catalogue record for this book is available from the British Library.

Bibliographic information published by the Deutsche Nationalbibliothek
The Deutsche Nationalbibliothek lists this publication in the Deutsche Nationalbibliografie; detailed bibliographic data are available on the Internet at http://dnb.d-nb.de.

© 2010 WILEY-VCH Verlag GmbH & Co. KGaA, Weinheim

All rights reserved (including those of translation into other languages). No part of this book may be reproduced in any form – by photoprinting, microfilm, or any other means – nor transmitted or translated into a machine language without written permission from the publishers. Registered names, trademarks, etc. used in this book, even when not specifically marked as such, are not to be considered unprotected by law.

Cover Design Formgeber, Eppelheim
Typesetting Toppan Best-set Premedia Limited
Printing and Binding Strauss GmbH, Mörlenbach

Printed in the Federal Republic of Germany
Printed on acid-free paper

ISBN: 978-3-527-32295-4

Contents

Preface *XV*
List of Contributors *XVII*

Part One Chemistry and Biology of DNA Lesions *1*

1   **Introduction and Perspectives on the Chemistry and Biology of DNA Damage** *3*
      Nicholas E. Geacintov and Suse Broyde
1.1   Overview of the Field *3*
1.2   DNA Damage – A Constant Threat *4*
1.3   DNA Damage and Disease *5*
1.3.1 The Inflammatory Response *5*
1.3.2 Reactive Oxygen and Nitrogen Species *5*
1.3.3 Early Recognition of Environmentally Related Cancers: Polycyclic Aromatic Hydrocarbons *6*
1.3.4 Exposure to Environmental Cancer-Causing Substances *6*
1.3.5 Aflatoxins *7*
1.3.6 Aristolochic Acid *7*
1.3.7 Estrogens *8*
1.4   DNA Damage and Chemotherapeutic Applications *8*
1.5   The Cellular DNA Damage Response (DDR) *9*
1.6   Repair Mechanisms that Remove DNA Lesions *10*
1.6.1 Repair of Single- and Double-Strand Breaks *10*
1.6.2 Alkylating Agents *10*
1.6.3 Base Excision Repair *11*
1.6.4 Mismatch Excision Repair *11*
1.6.5 Nucleotide Excision Repair *11*
1.6.6 Translesion Bypass of Unrepaired Lesions by Specialized DNA Polymerases and RNA Polymerases *12*
1.7   Relationships between the Chemical, Structural, and Biological Features of DNA Lesions *12*
      Acknowledgements *15*
      References *15*

The Chemical Biology of DNA Damage. Edited by Nicholas E. Geacintov and Suse Broyde
© 2010 WILEY-VCH Verlag GmbH & Co. KGaA, Weinheim
ISBN: 978-3-527-32295-4

2	**Chemistry of Inflammation and DNA Damage: Biological Impact of Reactive Nitrogen Species** *21*
	Michael S. DeMott and Peter C. Dedon
2.1	Introduction *21*
2.2	DNA Oxidation and Nitration *23*
2.2.1	Spectrum of Guanine Oxidation Products Caused by ONOO$^-$, ONOOCO$_2^-$, and NO$_2^{\bullet}$ *23*
2.2.2	Base Oxidation Products as Biomarkers of Inflammation and Oxidative Stress *25*
2.2.3	Charge Transfer as a Determinant of the Location of G Oxidation Products in DNA *25*
2.3	DNA Deamination *26*
2.3.1	Problem of Oxanine *29*
2.3.2	Analytical Methods and Artifacts *29*
2.4	2′-Deoxyribose Oxidation *30*
2.4.1	Variation of 2′-Deoxyribose Oxidation Chemistry as a Function of the Oxidant *34*
2.5	Indirect Base Damage Caused by RNS *35*
2.5.1	Malondialdehyde and Related Adducts *37*
2.6	Conclusions *38*
	Acknowledgements *38*
	References *38*
3	**Oxidatively Generated Damage to Isolated and Cellular DNA** *53*
	Jean Cadet, Thierry Douki, and Jean-Luc Ravanat
3.1	Introduction *53*
3.1.1	Overview and Summary *53*
3.1.2	Overview of Oxidatively Generated DNA Damage *53*
3.2	Single Base Damage *55*
3.2.1	Singlet Oxygen Oxidation of Guanine *55*
3.2.2	Hydroxyl Radical Reactions *58*
3.2.2.1	Thymine *58*
3.2.2.2	Guanine *60*
3.2.2.3	Adenine *62*
3.2.3	One-Electron Oxidation of Nucleobases *63*
3.2.4	HOCl Acid-Mediated Halogenation of Pyrimidine and Purine Bases *65*
3.3	Tandem Base Lesions *66*
3.4	Hydroxyl Radical-Mediated 2-Deoxyribose Oxidation Reactions *67*
3.4.1	Hydrogen Abstraction at C4′: Formation of Cytosine Adducts *67*
3.4.2	Hydrogen Atom Abstraction at C5′: Formation of Purine 5′,8-Cyclonucleosides *68*
3.5	Secondary Oxidation Reactions of Bases *70*
3.6	Conclusions and Perspectives *71*
	Acknowledgements *71*
	References *72*

4	**Role of Free Radical Reactions in the Formation of DNA Damage** *81*
	Vladimir Shafirovich and Nicholas E. Geacintov
4.1	Introduction *81*
4.2	Importance of Free Radical Reactions with DNA *82*
4.2.1	Free Radical Mechanisms: General Considerations *82*
4.2.2	Types of Free Radicals and their Reactions with Nucleic Acids *83*
4.2.3	Methods for Studying Free Radical Reactions: Laser Flash Photolysis *84*
4.2.4	Types of Radical Reactions and Kinetics *85*
4.2.5	Examples of DNA Radical Reactions *86*
4.2.6	Lifetimes of Free Radicals and Environmental Considerations *88*
4.2.7	Reactions of Free Radicals *89*
4.3	Mechanisms of Product Formation *91*
4.3.1	Reactions of G(-H)· Radicals with Nucleophiles *91*
4.3.2	Combinations of G(-H)· and Oxyl Radicals *93*
4.3.3	Oxidation of 8-oxoG *97*
4.4	Biological Implications *99*
	Acknowledgements *100*
	References *101*

5	**DNA Damage Caused by Endogenously Generated Products of Oxidative Stress** *105*
	Charles G. Knutson and Lawrence J. Marnett
5.1	Lipid Peroxidation *105*
5.2	2′-Deoxyribose Peroxidation *107*
5.3	Reactions of MDA and β-Substituted Acroleins with DNA Bases *109*
5.4	Stability of M_1dG: Hydrolytic Ring-Opening and Reaction with Nucleophiles *112*
5.5	Propano Adducts *114*
5.6	Etheno Adducts *114*
5.7	Mutagenicity of Peroxidation-Derived Adducts *117*
5.8	Repair of DNA Damage *121*
5.9	Assessment of DNA Damage *123*
5.10	Conclusions *126*
	Acknowledgements *126*
	References *126*

6	**Polycyclic Aromatic Hydrocarbons: Multiple Metabolic Pathways and the DNA Lesions Formed** *131*
	Trevor M. Penning
6.1	Introduction *131*
6.2	Radical Cation Pathway *134*
6.2.1	Metabolic Activation of PAHs *134*
6.2.2	Radical Cation DNA Adducts *135*
6.2.3	Limitations of the Radical Cation Pathway *136*
6.3	Diol Epoxides *137*

6.3.1	Metabolic Activation of PAHs	137
6.3.2	Diol Epoxide-DNA Adducts	138
6.3.3	Limitations of the Diol Epoxide Pathway	140
6.4	PAH o-Quinones	141
6.4.1	Metabolic Activation of PAH trans-Dihydrodiols by AKRs	141
6.4.2	PAH o-Quinone-Derived DNA Adducts	142
6.4.2.1	Covalent PAH o-Quinone-DNA Adducts	142
6.4.2.2	Oxidative DNA Lesions from PAH o-Quinones	144
6.4.3	Limitations of the PAH o-Quinone Pathway	146
6.5	Future Directions	147
	Acknowledgements	148
	References	148

7 Aromatic Amines and Heterocyclic Aromatic Amines: From Tobacco Smoke to Food Mutagens 157

Robert J. Turesky

7.1	Introduction	157
7.2	Exposure and Cancer Epidemiology	157
7.3	Enzymes of Metabolic Activation and Genetic Polymorphisms	159
7.4	Reactivity of N-Hydroxy-AAs and N-Hydroxy-HAAs with DNA	161
7.5	Syntheses of AA-DNA and HAA-DNA Adducts	162
7.6	Biological Effects of AA-DNA and HAA-DNA Adducts	162
7.7	Bacterial Mutagenesis	164
7.8	Mammalian Mutagenesis	165
7.9	Mutagenesis in Transgenic Rodents	166
7.10	Genetic Alterations in Oncogenes and Tumor Suppressor Genes	167
7.11	AA-DNA and HAA-DNA Adduct Formation in Experimental Animals and Methods of Detection	168
7.12	AA-DNA and HAA-DNA Adduct Formation in Humans	171
7.13	Future Directions	173
	Acknowledgements	173
	References	173

8 Genotoxic Estrogen Pathway: Endogenous and Equine Estrogen Hormone Replacement Therapy 185

Judy L. Bolton and Gregory R.J. Thatcher

8.1	Risks of Estrogen Exposure	185
8.2	Mechanisms of Estrogen Carcinogenesis	187
8.2.1	Hormonal Mechanism	187
8.2.2	Chemical Mechanism	188
8.2.2.1	Oxidative DNA Damage	188
8.2.2.2	DNA Adducts	189
8.2.2.3	Protection against DNA Damage	192
8.3	Estrogen Receptor as a Trojan Horse (Combined Hormonal/Chemical Mechanism)	193

8.4 Conclusions and Future Directions *194*
Acknowledgements *194*
References *194*

Part Two   New Frontiers and Challenges: Understanding Structure–Function Relationships and Biological Activity *201*

9   Interstrand DNA Cross-Linking 1,N^2-Deoxyguanosine Adducts Derived from α,β-Unsaturated Aldehydes: Structure–Function Relationships *203*
Michael P. Stone, Hai Huang, Young-Jin Cho, Hye-Young Kim, Ivan D. Kozekov, Albena Kozekova, Hao Wang, Irina G. Minko, R. Stephen Lloyd, Thomas M. Harris, and Carmelo J. Rizzo

9.1 Introduction *203*
9.2 Interstrand Cross-Linking Chemistry of the γ-OH-PdG Adduct (9) *205*
9.3 Interstrand Cross-Linking by the α-CH$_3$-γ-OH-PdG Adducts Derived from Crotonaldehyde *207*
9.4 Interstrand Cross-Linking by 4-HNE *207*
9.5 Carbinolamine Cross-Links Maintain Watson–Crick Base-Pairing *209*
9.6 Role of DNA Sequence *210*
9.7 Role of Stereochemistry in Modulating Cross-Linking *210*
9.8 Biological Significance *212*
9.9 Conclusions *213*
Acknowledgements *213*
References *213*

10   Structure–Function Characteristics of Aromatic Amine-DNA Adducts *217*
Bongsup Cho

10.1 Introduction *217*
10.2 Major Conformational Motifs *219*
10.2.1 Fully Complementary DNA Duplexes *219*
10.2.2 Other Sequence Contexts *220*
10.3 Conformational Heterogeneity *221*
10.3.1 Sequence Effects on the S/B Conformational Heterogeneity *222*
10.3.2 Conformational Dynamics of the S/B Heterogeneity *224*
10.3.3 Base Sequence Context and Mutagenesis *224*
10.3.4 Dependence of Nucleotide Excision Repair by *E. coli* UvrABC Proteins on Adduct Conformation *225*
10.3.5 Conformational Heterogeneity in Translesion Synthesis *227*
10.3.6 Sequence Effects on the Conformational Stability of SMIs *230*
10.4 Structures of DNA Lesion–DNA Polymerase Complexes *231*
10.5 Conclusions *232*
Acknowledgements *233*
References *233*

11	**Mechanisms of Base Excision Repair and Nucleotide Excision Repair** *239*
	Orlando D. Schärer and Arthur J. Campbell
11.1	General Features of Base Excision and Nucleotide Excision Repair *239*
11.2	BER *241*
11.2.1	BER Overview–Short-Patch and Long-Patch BER *241*
11.2.2	Lesion Recognition by DNA Glycosylases *242*
11.2.3	Passing the Baton–Abasic Site Removal and Repair *247*
11.3	NER *248*
11.3.1	Subpathways of NER: Global Genome and Transcription-Coupled NER *248*
11.3.2	Damage Recognition in GG-NER *248*
11.3.3	Damage Verification and Lesion Demarcation in NER *251*
11.3.4	Dual-Incision and Repair Synthesis in NER *252*
11.3.5	Damage Recognition in TC-NER *252*
11.4	Conclusions *254*
	References *254*

12	**Recognition and Removal of Bulky DNA Lesions by the Nucleotide Excision Repair System** *261*
	Yuqin Cai, Konstantin Kropachev, Marina Kolbanovskiy, Alexander Kolbanovskiy, Suse Broyde, Dinshaw J. Patel, and Nicholas E. Geacintov
12.1	Introduction *261*
12.2	Overview of Mammalian NER *261*
12.3	Prokaryotic NER *263*
12.4	Recognition of Bulky Lesions by Mammalian NER Factors *263*
12.5	Bipartite Model of Mammalian NER and the Multipartite Model of Lesion Recognition *264*
12.6	DNA Lesions Derived from the Reactions of PAH Diol Epoxides with DNA are Excellent Substrates for Probing the Mechanisms of NER *265*
12.7	Multidisciplinary Approach Towards Investigating Structure–Function Relationships in the NER of Bulky PAH-DNA Adducts *268*
12.8	Dependence of DNA Adduct Conformations and NER on PAH Topology and Stereochemistry *269*
12.8.1	Guanine B[*a*]P Adducts (Figure 12.3a): Minor Groove and Base-Displaced/Intercalative Conformations *270*
12.8.2	Bay Region B[*a*]P-N^6-Adenine Adducts (Figure 12.3b): Distorting Intercalative Insertions from the Major Groove *271*
12.8.3	Fjord Region PAH N^6-Adenine Adducts (Figure 12.3c and d): Minimally Distorting Intercalation from the Major Groove *272*
12.8.4	Dependence of NER Efficiencies on the Conformations of the Bay Region B[*a*]P-N^2-dG Adducts *272*
12.8.5	NER Efficiencies: Bay and Fjord Region PAH Diol Epoxide-N^6-dA Adducts *278*
12.8.6	Why the *trans-anti*-B[*c*]Ph-N^6-dA and Related Fjord Region N^6-dA Adducts do not Destabilize DNA and are Resistant to NER *280*

12.9	Dependence of NER of the 10S (+)-trans-anti-B[a]P-N^2-dG Adduct on Base Sequence Context *280*
12.9.1	Structural Characteristics of the Identical 10S (+)-trans-anti-B[a]P-N^2-dG Adduct in Different Sequence Contexts *281*
12.9.1.1	CG*C and TG*T Sequences *282*
12.9.1.2	G6*G7, G6G7*, and I6G7* Sequences *282*
12.9.2	Hierarchies of Mammalian NER Recognition Signals *286*
12.10	Conclusions *287*
	Acknowledgements *289*
	References *289*

13 Impact of Chemical Adducts on Translesion Synthesis in Replicative and Bypass DNA Polymerases: From Structure to Function *299*

Robert L. Eoff, Martin Egli, and F. Peter Guengerich

13.1	Introduction *299*
13.2	Bypass of Abasic Sites *302*
13.3	Lesions Generated by Oxidative Damage to DNA *305*
13.4	Exocyclic DNA Adduct Bypass *308*
13.5	Alkylated DNA *310*
13.6	Polycyclic Aromatic Hydrocarbons and the Effect of Adduct Size upon Polymerase Catalysis *313*
13.7	Cyclobutane Pyrimidine Dimers and UV Photoproducts *316*
13.8	Inter- and Intrastrand DNA Cross-Links *316*
13.9	Conclusions *318*
	References *319*

14 Elucidating Structure–Function Relationships in Bulky DNA Lesions: From Solution Structures to Polymerases *331*

Suse Broyde, Lihua Wang, Dinshaw J. Patel, and Nicholas E. Geacintov

14.1	Introduction *331*
14.2	Benzo[a]pyrene-Derived DNA Lesions as a Useful Model *331*
14.3	Computational Elucidation of the Structural Properties of B[a]P-Derived DNA Lesions in Solution *333*
14.4	DNA Polymerase Structure–Function Relationships Elucidated with B[a]P-Derived Lesions *335*
14.5	Mechanism of the Nucleotidyl Transfer Reaction *343*
14.6	Conclusions and Future Perspectives *345*
	Acknowledgements *345*
	References *346*

15 Translesion Synthesis and Mutagenic Pathways in *Escherichia coli* Cells *353*

Sushil Chandani and Edward L. Loechler

15.1	Introduction *353*
15.2	Mutagenesis in *E. coli* has Illuminated Our Understanding of Mutagenesis in General *354*

15.3	Why Does *E. coli* have Three Translesion Synthesis DNA Polymerases? *356*	
15.4	Overview of the Steps Leading to Translesion Synthesis *358*	
15.5	Case Studies: AAF-C8-dG and N^2-dG Adducts, Such as +BP *360*	
15.6	Structure–Function Analysis of Y-Family Pols IV and V of *E. coli* *362*	
15.6.1	Structural Basis for a Large versus Small Chimney Opening *366*	
15.6.2	Roof-Amino Acids and Roof-Neighbor-Amino Acids *368*	
15.6.3	Interconnected Architecture of the Chimney and Roof Regions *368*	
15.6.4	dCTP Insertion by Pol IV *369*	
15.6.5	How Does UmuC(V) Insert dATP? *370*	
15.6.6	A Cautionary Note about Dpo4 *371*	
15.6.7	Why is Pol IV Efficient at Extension with −BP, but Inefficient with +BP? *372*	
15.7	Y-Family DNA Polymerase Mechanistic Steps *373*	
15.8	Structure of B-Family Pol II of *E. coli* *373*	
	References *374*	

16 **Insight into the Molecular Mechanism of Translesion DNA Synthesis in Human Cells using Probes with Chemically Defined DNA Lesions** *381*
Zvi Livneh

16.1	Introduction *381*	
16.2	Overview of TLS *382*	
16.3	Plasmid Model Systems with Defined Lesions for Studying TLS *384*	
16.4	Gap-Lesion Plasmid Assay for Mammalian TLS *384*	
16.5	Some Lesions are Bypassed Most Effectively and Most Accurately by Specific Cognate TLS DNA Polymerases *387*	
16.6	Pivotal Role for Pol ζ in TLS Across a Wide Variety of DNA Lesions *388*	
16.7	Knocking-Down the Expression of TLS Polymerases using Small Interfering RNA Provides a useful Tool for the Analysis of TLS using the Gapped Plasmid Assay *388*	
16.8	Evidence that TLS Occurs by Two-Polymerase Mechanisms, in Combinations that Determine the Accuracy of the Process *391*	
16.9	Conclusions *393*	
	Acknowledgements *393*	
	References *394*	

17 **DNA Damage and Transcription Elongation: Consequences and RNA Integrity** *399*
Kristian Dreij, John A. Burns, Alexandra Dimitri, Lana Nirenstein, Taissia Noujnykh, and David A. Scicchitano

17.1	Introduction *399*	
17.2	DNA Repair *400*	
17.3	Transcription Elongation and DNA Damage *402*	
17.4	RNA Polymerases: A Brief Overview *402*	

17.5	RNA Polymerase Elongation Past DNA Damage	*407*
17.5.1	Abasic Sites, Single-Strand Nicks, and Gaps	*407*
17.5.2	Oxidative DNA Damage	*408*
17.5.3	Alkylated Bases in DNA	*412*
17.5.4	Intrastrand and Interstrand DNA Cross-links	*414*
17.5.5	"Bulky" DNA Adducts	*416*
17.6	Conclusions	*421*
	Acknowledgements	*428*
	References	*429*

Index *439*

Preface

The relationships between the chemical, structural, and biological aspects of DNA damage have long been parallel and overlapping research domains. More recently, interest has intensified in relating the structural characteristics of DNA damage with its ultimate manifestation – the development of human disease. New opportunities for moving the field forward and for gaining a better understanding of the molecular basis of diseases associated with DNA damage are emerging. Rapid advances in instrumentation and computing power are yielding new structural information on macromolecular biological systems and assemblies through high-resolution structural studies at the molecular level. The subject of this book, the chemical biology of DNA damage, offers the opportunity for considering DNA damage from the molecular perspective with a focus on both the chemical elements at the level of damage generation and the biological properties of the various types of damage from the structure–function points of view.

The topics covered in this book should be of interest to researchers who wish to gain an overview of the frontier areas of the field, as well as to students who wish to learn or deepen their knowledge in areas that touch on the molecular origins of disease via DNA damage. Another objective of this book is to foster communication between the chemical and biological communities of researchers by highlighting the molecular origins that unite these topics at a fundamental level. The time is ripe for promoting such a fundamental understanding since new information on the biological impact of chemically defined lesions is now becoming available at an increasing pace.

This book is divided into two parts. The focus of Part One is on the chemical aspects of DNA damage, while the emphasis of Part Two is on the structural and functional relationships of DNA lesions, and their processing by the cellular machineries of repair, replication, and transcription. Chapter 1 in Part One is intended as a brief overview of the vast field of DNA damage, and introduces the reader to the relationships between the chemical and structural aspects of DNA damage, and some of the known biological endpoints and correlations with human disease. Ample references are provided with an emphasis on authoritative, recently published reviews to guide the interested reader.

The two parts of this book, the chemical and the biological components, spring from the same tree and feed from the same roots – the chemical and structural

The Chemical Biology of DNA Damage. Edited by Nicholas E. Geacintov and Suse Broyde
© 2010 WILEY-VCH Verlag GmbH & Co. KGaA, Weinheim
ISBN: 978-3-527-32295-4

features of DNA lesions that after the normal structural features of the DNA molecule. If these lesions are not removed by cellular DNA repair mechanisms, DNA replication may be either inhibited entirely or occur in an error-prone manner with dire consequences for the cell. The molecular events underlying these phenomena are at the intersections of the chemical and biological disciplines at the frontiers of our current knowledge. It is our belief that a deeper understanding of the connections between these disciplines will lead to more effective strategies for preventing disease and to advanced therapeutic approaches for treating diseases such as human cancers.

New York, April 2010

Nicholas E. Geacintov
Suse Broyde

List of Contributors

Judy L. Bolton
University of Illinois at Chicago
College of Pharmacy
Department of Medicinal Chemistry
and Pharmacognosy (M/C 781)
833 S. Wood Street
Chicago, IL 60612-7231
USA

Suse Broyde
New York University
Department of Biology
100 Washington Square East
New York, NY 10003-6688
USA

John A. Burns
New York University
Department of Biology
100 Washington Square East
New York, NY 10003-6688
USA

Jean Cadet
CEA/Grenoble
Laboratoire "Lésions des Acides
Nucléiques"
SCIB-UMR-E No. 3 (CEA/UJF)
Institut Nanosciences et Cryogénie
17 avenue des Martyrs, 38054
Grenoble Cedex 9
France
Université de Sherbrooke
Faculté de médecine et des sciences
de la santé
Département de médecine nucléaire et
radiobiologie
3001, 12e Avenue Nord, Sherbrooke,
Québec
Canada J1H 5N4

Yuqin Cai
New York University
Department of Biology
100 Washington Square East
New York, NY 10003-6688
USA

Arthur J. Campbell
Stony Brook University
Department of Chemistry
Graduate Chemistry Building
Stony Brook, NY 11794-3400
USA

Sushil Chandani
Boston University
Biology Department
5 Cummington Street
Boston, MA 02215
USA

Bongsup Cho
University of Rhode Island
Biomedical and Pharmaceutical
Sciences
41 Lower College Road
Kingston, RI 02881
USA

Young-Jin Cho
Vanderbilt University
Center in Molecular Toxicology, and
the Vanderbilt Institute for Chemical
Biology
Department of Chemistry
1211 Medical Center Drive
Nashville, TN 37235
USA

Peter C. Dedon
Massachusetts Institute of Technology
Department of Biological Engineering
Center for Environmental Health
Sciences
77 Massachusetts Avenue
Cambridge, MA 02139
USA

Michael S. DeMott
Massachusetts Institute of Technology
Department of Biological Engineering
77 Massachusetts Avenue
Cambridge, MA 02139
USA

Alexandra Dimitri
New York University
Department of Biology
1009 Silver Center, 100 Washington
Square East
New York, NY 10003-6688
USA

Thierry Douki
CEA/Grenoble
Laboratoire "Lésions des Acides
Nucléiques"
SCIB-UMR-E No. 3 (CEA/UJF)
Institut Nanosciences et Cryogénie
17 avenue des Martyrs, 38054
Grenoble Cedex 9
France

Kristian Dreij
New York University
Department of Biology
1009 Silver Center, 100 Washington
Square East
New York, NY 10003-6688
USA

Martin Egli
Vanderbilt University School of
Medicine
Department of Biochemistry and
Molecular Toxicology Center
2200 Pierce Avenue
Nashville, TN 37232-0146
USA

Robert L. Eoff
Vanderbilt University School of
Medicine
Department of Biochemistry and
Molecular Toxicology Center
2200 Pierce Avenue
Nashville, TN 37232-0146
USA

Nicholas E. Geacintov
New York University
Chemistry Department
31 Washington Place
New York, NY 10003-5180
USA

F. Peter Guengerich
Vanderbilt University School of
Medicine
Department of Biochemistry and
Molecular Toxicology Center
2200 Pierce Avenue
Nashville, TN 37232-0146
USA

Thomas M. Harris
Vanderbilt University
Center in Molecular Toxicology, and
the Vanderbilt Institute for Chemical
Biology
Department of Chemistry
1211 Medical Center Drive
Nashville, TN 37235
USA

Hai Huang
Vanderbilt University
Center in Molecular Toxicology, and
the Vanderbilt Institute for Chemical
Biology
Department of Chemistry
1211 Medical Center Drive
Nashville, TN 37235
USA

Hye-Young Kim
Vanderbilt University
Center in Molecular Toxicology, and
the Vanderbilt Institute for Chemical
Biology
Department of Chemistry
1211 Medical Center Drive
Nashville, TN 37235
USA

Charles G. Knutson
Vanderbilt University School of
Medicine
A.B. Hancock Jr. Memorial Laboratory
for Cancer Research
Departments of Biochemistry,
Chemistry, and Pharmacology
Vanderbilt Institute of Chemical
Biology
Center in Molecular Toxicology
Vanderbilt-Ingram Cancer Center
2220 Pierce Avenue
Nashville, TN 37232-1046
USA

Alexander Kolbanovskiy
New York University
Department of Chemistry
31 Washington Place
New York, NY 10003-5180
USA

Marina Kolbanovskiy
New York University
Department of Chemistry
31 Washington Place
New York, NY 10003-5180
USA

Ivan D. Kozekov
Vanderbilt University
Center in Molecular Toxicology, and
the Vanderbilt Institute for Chemical
Biology
Department of Chemistry
1211 Medical Center Drive
Nashville, TN 37235
USA

Albena Kozekova
Vanderbilt University
Center in Molecular Toxicology, and
the Vanderbilt Institute for Chemical
Biology
Department of Chemistry
1211 Medical Center Drive
Nashville, TN 37235
USA

Konstantin Kropachev
New York University
Department of Chemistry
31 Washington Place
New York, NY 10003-5180
USA

Zvi Livneh
Weizmann Institute of Science
Department of Biological Chemistry
PO Box 26, Rehovot 76100
Israel

R. Stephen Lloyd
Oregon Health & Science University
Center for Research on Occupational
and Environmental Toxicology
3181 SW Sam Jackson Park Road
Portland, OR 97239-3098
USA

Edward L. Loechler
Boston University
Biology Department
5 Cummington Street
Boston, MA 02215
USA

Lawrence J. Marnett
Vanderbilt University School of
Medicine
A.B. Hancock Jr. Memorial Laboratory
for Cancer Research
Departments of Biochemistry,
Chemistry, and Pharmacology,
Vanderbilt Institute of Chemical
Biology
Center in Molecular Toxicology
Vanderbilt-Ingram Cancer Center
2220 Pierce Avenue
Nashville, TN 37232-1046
USA

Irina G. Minko
Oregon Health & Science University
Center for Research on Occupational
and Environmental Toxicology
3181 SW Sam Jackson Park Road
Portland, OR 97239-3098
USA

Lana Nirenstein
New York University
Department of Biology
1009 Silver Center, 100 Washington
Square East
New York, NY 10003-6688
USA

Taissia Noujnykh
New York University
Department of Biology
1009 Silver Center, 100 Washington
Square East
New York, NY 10003-6688
USA

Dinshaw J. Patel
Memorial Sloan-Kettering Cancer
Center
Structural Biology Program
1275 York Avenue
New York, NY 10065
USA

Trevor M. Penning
University of Pennsylvania
School of Medicine
Centers of Excellence in
Environmental Toxicology and Cancer
Pharmacology
Department of Pharmacology
3620 Hamilton Walk
Philadelphia, PA 19104-6084
USA

Jean-Luc Ravanat
CEA/Grenoble
Laboratoire "Lésions des Acides
Nucléiques"
SCIB-UMR-E No. 3 (CEA/UJF)
Institut Nanosciences et Cryogénie
17 avenue des Martyrs, 38054
Grenoble Cedex 9
France

Carmelo J. Rizzo
Vanderbilt University
Center in Molecular Toxicology, and
the Vanderbilt Institute for Chemical
Biology, Department of Chemistry
1211 Medical Center Drive
Nashville, TN 37235
USA

Orlando D. Schärer
Stony Brook University
Departments of Pharmacological
Sciences and Chemistry
Graduate Chemistry Building
Stony Brook, NY 11794-3400
USA

David A. Scicchitano
New York University
Department of Biology
100 Washington Square East
New York, NY 10003-6688
USA

Vladimir Shafirovich
New York University
Chemistry Department
31 Washington Place
New York, NY 10003-5180
USA

Michael P. Stone
Vanderbilt University
Center in Molecular Toxicology, and
the Vanderbilt Institute for Chemical
Biology, Department of Chemistry
1211 Medical Center Drive
Nashville, TN 37235
USA

Gregory R.J. Thatcher
University of Illinois at Chicago
College of Pharmacy
Department of Medicinal Chemistry
and Pharmacognosy (M/C 781)
833 S. Wood Street
Chicago, IL 60612-7231
USA

Robert J. Turesky
Wadsworth Center
Division of Environmental Health
Sciences
Empire State Plaza
Albany, NY 12201
USA

Hao Wang
Vanderbilt University
Center in Molecular Toxicology, and
the Vanderbilt Institute for Chemical
Biology, Department of Chemistry
1211 Medical Center Drive
Nashville, TN 37235
USA

Lihua Wang
New York University
Biology Department
100 Washington Square East
New York, NY 10003-6688
USA

Part One
Chemistry and Biology of DNA Lesions

1
Introduction and Perspectives on the Chemistry and Biology of DNA Damage
Nicholas E. Geacintov and Suse Broyde

1.1
Overview of the Field

The subject of this book, the chemical biology of DNA damage, is concerned with the chemistry that produces DNA damage, and the relationships between the structural features of the DNA lesions formed and their biological impact. The subjects and examples described illustrate the interdisciplinary approaches that can be used to bridge the gaps between the chemical aspects and biological endpoints of DNA damage, especially lesions generated by different endogenous and exogenous DNA-damaging agents. In Part One (Chapters 2–8), the focus is on the chemistry and biological impact of some representative and important DNA lesions. The topics of Part Two (Chapters 9–17) deal with recent and current research on the relationships between the chemical structure and physical properties of selected DNA lesions, and how the lesions are processed by the DNA repair, replication, and transcription machineries.

The chemistry of DNA damage is complex and the variety of DNA lesions is enormous. This book considers an important subset of DNA lesions that illustrate the relationships between the chemistry, structure, biochemistry, and biology of DNA damage. In this chapter, we provide a broad but brief overview of this vast field. Some of the established links between DNA damage and human diseases are highlighted. The objectives of this chapter are to situate the topics covered in this book within the overall field and to guide the interested reader to the original literature concerned with topics that either are or are not explicitly covered in the rest of the book.

We begin with an overview of the origins of DNA damage, followed by summaries of the relationships between DNA lesions and disease, and a brief overview of cellular DNA damage response (DDR) systems, and conclude with a brief description of the specific topics covered in this book and how they relate to the field overall.

The Chemical Biology of DNA Damage. Edited by Nicholas E. Geacintov and Suse Broyde
© 2010 WILEY-VCH Verlag GmbH & Co. KGaA, Weinheim
ISBN: 978-3-527-32295-4

1.2
DNA Damage – A Constant Threat

The human genome is under constant attack from endogenous and exogenous reactive chemical species. A variety of genotoxic agents can induce chemical transformation of the nucleotides or damage the phosphodiester backbone of DNA with deleterious consequences for the cell. The relationships between cellular DNA damage caused by endogenous and environmental genotoxic agents, the cellular response, and the development and prevention of human diseases and aging are areas of great current interest in the medical, biological, and chemical research communities [1].

It has been estimated that there are tens of thousands of DNA-damaging events per day suffered by the approximately 10^{13} cells within the human body [2] and that DNA damage associated with endogenous species is more extensive (greater than 75%) than damage caused by environmental factors [3]. Among the endogenous species that damage cellular DNA are reactive oxygen species (ROS) and reactive nitrogen species (RNS). These reactive intermediates are produced under conditions of oxidative stress, a consequence of normal metabolic activity, and the inflammatory response [3, 4]. Other forms of endogenous DNA damage are depurination (and to a lesser extent depyrimidination) that arise from the hydrolysis of the glycosidic bonds between the nucleobase and deoxyribose residues, thus leading to the formation of apurinic (or apyrimidinic) sites [5]. The hydrolytic deamination of cytosine can also occur spontaneously and give rise to uracil [6]. Both forms of DNA damage, if not repaired by the normally efficient cellular base excision repair (BER) mechanism, can result in the mutagenic insertion of an incorrect base during error-prone translesion synthesis when the DNA is replicated past the lesion.

Among the external causes of DNA damage are ionizing radiation and solar UV radiation. Sunlight has been called the most prominent and ubiquitous physical carcinogen in our natural environment [7]. There are ample epidemiological data and a wealth of supporting animal model experiments that indicate that solar UV radiation is a major cause of skin cancer among the white Caucasian populations in the Western world [8]. The UV portion of the solar spectrum in the 290- to 300-nm region is absorbed by DNA and forms cyclobutane pyrimidine dimers (CPDs) that have been linked to the etiology of skin cancer [9]. Ionizing radiation is routinely used in medical diagnostic and chemotherapeutic applications. There are different forms of radiation that generate a variety of DNA lesions that include double- and single-strand breaks, as well as oxidatively modified nucleobases and deoxyribose moieties. The human population is also continuously exposed to environmental pollutants that are present in air, water, and food [10]. Many of these chemicals are metabolized in human cells to highly reactive intermediates that react chemically with the nucleobases to form deleterious DNA strand breaks and a variety of DNA lesions or adducts that are readily detectable in human cells [11–13]. Fortunately, nature has devised a host of cellular defense or DNA repair mechanisms that have been described [14] and reviewed in a comprehensive monograph [15]. Some of the mechanisms that involve the removal of DNA lesions are discussed in Chapters 11 and 12. The effects of DNA lesions that escape repair can be bypassed during DNA replication by a damage tolerance mechanism that depends on the actions of a set of specialized polymerases [16, 17] or through homologous recombination mechanisms that leave the lesion intact on the damaged strand [18].

1.3
DNA Damage and Disease

1.3.1
The Inflammatory Response

Chronic inflammation in mammalian tissues can be caused by a variety of chemical, physical, and infectious factors that are not only cytotoxic, but can also increase the risk of malignant cell transformation and promote the development of various human cancers [19]. The inflammatory response includes the activation of macrophage and neutrophil cells that result in a complex spectrum of chemically reactive species that damage DNA and other biomolecules [4]. Activated macrophages overproduce nitric oxide (NO) and superoxide ($O_2^{\cdot-}$) that combine rapidly to form peroxynitrite ($ONOO^-$). The latter decomposes to reactive intermediates that can cause damage to DNA and other biomolecules (see Chapters 2–4). The activated neutrophils, on the other hand, contribute to the myeloperoxidase-mediated generation of hypochlorous acid (HOCl) – a potent oxidizing and halogenating agent [4]. While many of the DNA lesions formed are oxidized forms of DNA bases themselves [20, 21], more bulky DNA lesions can also arise from the endogenous peroxidation of lipids that generate highly reactive aldehyde derivatives that react with DNA [22] (see also Chapters 5 and 9). The generation of guanine radical intermediates also leads to the formation of cross-linking reactions with thymine [20, 23] as discussed in Chapters 3 and 4.

1.3.2
Reactive Oxygen and Nitrogen Species

DNA lesions caused by reactions with ROS and RNS that are byproducts of the inflammatory response have also been implicated in the etiology of neurological diseases such as Alzheimer's [24] and Parkinson's [25]. Furthermore, oxidatively generated DNA damage has been implicated in aging, based on the hypothesis that DNA damage accumulation contributes to this natural phenomenon [26–28]. The elevated concentrations of ROS and RNS alter the intracellular signaling pathways via inflammatory cytokines; this can result in an imbalance between oxidative damage of cellular DNA and DNA repair processes, causing the accumulation of DNA lesions in the genome. If not removed by cellular DNA repair mechanisms, the cytotoxic lesions may result in abnormal cell physiology, apoptosis, and cell death if DNA replication or transcription is inhibited, or may cause mutations and cancer if error-prone translesion bypass occurs.

1.3.3
Early Recognition of Environmentally Related Cancers: Polycyclic Aromatic Hydrocarbons

The connections between environmental chemicals and cancer have a long history [29], dating to the eighteenth century when the first correlation was made between exposure to soot and the high incidence of scrotal cancer among chimney sweeps in London. During the twentieth century, a combination of epidemiological and animal experiments has provided persuasive evidence that polycyclic aromatic hydrocarbons (PAHs), such as the well-known and representative compound benzo[a]pyrene (B[a]P), are key chemical carcinogens in soot and coal tar. While PAH compounds are chemically unreactive and are, at best, sparingly soluble in aqueous solutions, a seminal early observation documented that these and other bulky aromatic compounds are metabolically activated by microsomal P450 enzymes to oxygenated derivatives with higher water solubilities, thus facilitating their excretion [30]. Among these metabolites, however, are highly reactive electrophiles that can react chemically with nucleic acids to form covalent DNA adducts (Chapter 6). The link between DNA damage and cancer risk is difficult to establish in humans. However, decades of epidemiological evidence and studies of animal chemical carcinogenesis models point to DNA adducts as being of central importance in causing permanent genetic changes [10, 29, 31] that play an important role in the etiology of cancer. The overall hypothesis is that normal growth control is adversely affected when the mutations occur in critical codons of tumor suppressor genes or oncogenes [32–34]. Establishing causal relationships between human exposure to a suspected environmental carcinogen, the formation of DNA adducts, and cancer risk involves a complex series of steps. These steps include (i) the analysis of the tumorigenic activity of a suspected human carcinogen in animal model systems; (ii) the identification and development of specific chemical biomarkers (DNA and protein (e.g., albumin) adducts), urinary carcinogen metabolites, and nucleic acid adducts, (iii) establishing correlations between exposure, biomarkers, and the development of disease in animal models [10], and (iv) applying similar criteria, if feasible, to humans and connecting epidemiological evidence with biomarkers of disease.

1.3.4
Exposure to Environmental Cancer-Causing Substances

Many environmental chemical substances have been implicated in the etiology and promotion of human cancers and have been classified as such by the World Health Organization's International Agency for Research on Cancer (IARC). The IARC classifications are widely utilized to assess the degree of risk associated with human exposure to well characterized chemicals or mixtures (http://

monographs.iarc.fr/ENG/Classification/). For example, the well-known and widely studied PAH compound B[*a*]P has been classified by the IARC as a substance that is carcinogenic to humans. The PAH compounds are products of fossil fuel combustion and are therefore ubiquitous in our environment [35]. Other aromatic carcinogens associated with the human diet are the aromatic amines and heterocyclic amines (Chapters 7 and 10) that are produced by broiling meats at high temperatures. Such products are known to contribute to the etiology of gastrointestinal cancers as discussed in detail in Chapter 7 [10]. The well-established association between cigarette smoking and the high incidence of lung and other cancers is based on worldwide epidemiological data, and is supported by animal chemical carcinogenesis model studies [36]. There are over 50 carcinogens in cigarette smoke, including PAH compounds, 4-(methylnitrosamino)-1-(3-pyridyl)-1-butanone, 1,3-butadiene [36, 37], and aromatic amines and heterocyclic aromatic amines (Chapter 7). All of these compounds are metabolically activated to reactive intermediates that form premutagenic covalent adducts with DNA.

1.3.5
Aflatoxins

The aflatoxins are mycotoxins that are among the most toxic and cancer-causing substances in animals and humans that occur naturally [10, 38]. The members of the aflatoxin family are produced by fungal *Aspergillus* species that grow as contaminants in stored grains and other crops, particularly during storage in humid environments, and are thus dietary carcinogens. Extensive epidemiological studies have shown that chronic exposure to aflatoxins is associated with a high incidence of human hepatocellular carcinoma [38]. While there are more than a dozen types of naturally occurring aflatoxins, the B_1 type is the most toxic. Aflatoxin B_1 is metabolized in the liver to the highly reactive exo-8,9-epoxide that binds covalently to DNA [39]. There is a strong correlation between the DNA adducts formed, the $G \rightarrow T$ transversion mutation signature associated with this form of DNA damage, and the biological endpoint—carcinogenesis [38].

1.3.6
Aristolochic Acid

Another important example of a naturally occurring dietary carcinogen, aristolochic acid, has emerged recently [40]. A high incidence of human renal disease and urothelial carcinoma, termed Balkan endemic nephropathy (BEN), was noted in the Balkan areas of Europe. The occurrence of BEN was traced to the contamination of grains of wheat with seeds from the plant *Aristolochia clematitis*, native to these regions, that contain aristolochic acid. [40, 41]. It has also been found that users of certain herbal medicines that contain aristolochic acid can also develop nephropathy that is similar to BEN [42, 43]. It was shown recently that BEN is correlated with the binding of metabolically activated forms of aristolochic acid with DNA, that such adducts are present in renal tissues from patients suffering from BEN, and that these DNA lesions cause mutations in the *p53* gene [40]. Thus, there is considerable evidence that aristolochic acid is implicated in the etiology and perhaps progression of human cancers associated with BEN [40].

1.3.7
Estrogens

Endogenous human estrogens have been classified as human carcinogens by the IARC. The mechanism of action involves hormonal activity that is related to the binding of estrogens to the estrogen-responsive element and the stimulation of cell proliferation by such receptor-mediated processes. However, a genotoxic mechanism that involves the metabolic activation of human estrogens to *o*-quinone intermediates that bind to cellular DNA and promote mutations, if not repaired, has also been proposed [44–46] (Chapter 8). Equine estrogen *o*-quinone metabolites such as equilin and equilenin, which are commonly used in hormone replacement therapy applications, also form covalent adducts with DNA [47]. Furthermore, a mechanism involving ROS derived from the redox cycling between the catechol and *o*-quinone derivatives of endogenous human and equine estrogens also leads to oxidatively damaged DNA [47–49], and can contribute to the etiology and progression of cancers. Interestingly, certain PAH compounds such as B[*a*]P can be metabolically activated to similar *o*-quinone derivatives that can redox cycle by a similar mechanism and oxidatively damage nuclear DNA by an analogous ROS mechanism [50] (Chapter 6). The nature of the two different mechanisms associated with the etiology of human cancers – hormonal versus genotoxic – is a topic of substantial current interest [48] (Chapter 8).

1.4
DNA Damage and Chemotherapeutic Applications

Up till now we have focused on DNA damage caused by genotoxic exogenous and endogenous agents with the implication that such damage must be avoided for maintaining the integrity of the genome. In contrast, in cancer therapy applications, the opposite result is desired – to damage DNA in order to diminish the survival of tumor cells. Both ionizing radiation and chemotherapy are commonly utilized. In the case of ionizing radiation, double-strand breaks are the major but not unique forms of DNA damage, but other important intracellular and intercellular signaling pathways are induced that also play a critical role in destroying tumor tissue [51]. Platinum-based compounds are among the most extensively used agents in cancer chemotherapy applications [52], and cisplatin (*cis*-diamminedichloroplatinum[II]) is the original and the best-known representative of this group. In cells, cisplatin reacts with a variety of biomolecules, but its reaction with DNA plays a dominant role in killing tumor cells by generating double-strand breaks [53] and a variety of cross-linked adducts [54] that inhibit DNA synthesis. The active forms of cisplatin are the aquated forms $[Pt(NH_3)_2Cl(OH_2)]^+$ and $[Pt(NH_3)_2(OH_2)_2]^{2+}$ that react with purines in DNA at their N7 positions to form several intrastrand cross-linked products in a sequence-dependent manner. The relative abundance of intrastrand cross-linked adducts is d(G*pG*) $\gg$ d(A*pG*) $\gg$ d(G*pNpG*) [55], d(G*pC*) [56], and smaller fractions of interstrand cross-linked lesions [56, 57] are deemed important because they are not easily removed by DNA repair mechanisms [53]. While normal cells are also sensitive to DNA-damaging agents such as cisplatin, tumor cells proliferate more rapidly than normal cells and are thus more sensitive to attack by chemotherapeutic compounds. One of the important issues in chemotherapy is that cancer cells eventually become resistant to further treatment, which may be related to a decreased susceptibility to DNA repair mechanisms [1].

1.5
The Cellular DNA Damage Response (DDR)

Cellular DNA damage elicits a variety of complex, tightly regulated transient pathways of DDR that arrests cell cycle progression. This delay allows the cells to cope with the deleterious effects of DNA strand breaks or DNA base or sugar damage and preserves the integrity of the genome before cell cycle progression is resumed. During the cell cycle arrest, a DNA damage signal transduction pathway not only slows cell cycle progression by inhibiting DNA replication, but also promotes relevant DNA repair mechanisms that correct or remove the damage. If DNA repair is successful, the DDR response is deactivated and cell progression is resumed. On the other hand, prolonged, chronic DNA damage, or defects in the DDR system, may overwhelm the cellular defense systems, thus resulting in apoptosis (programmed cell death), senescence, or to error-prone DNA replication that can lead to enhanced mutation rates and ultimately to genomic instability and cancer [1, 58, 59].

In eukaryotic cells, the DNA is complexed with histone proteins and packaged into chromatin where it is less accessible to DNA repair, transcription, and replication proteins than naked DNA. One important question in the DDR pathway is how the DNA damage that occurs within chromatin is recognized and how the DDR response is initiated. In response to signaling mechanisms and histone-modifying enzymes, the chromatin structure is altered (remodeled), and access to DNA repair and other proteins becomes possible. These remodeling processes involve the modification of critical histones by reversible acetylation, methylation, or phosphorylation events at well-defined sites in the proteins [60]. The signal transduction pathways can be viewed in terms of distinct steps that involve protein sensors that sense the presence of DNA damage and are recruited to these sites. These events activate the transducing DDR protein kinases ATR (ATM and Rad3-related) and ATM (ataxia telangiectasia mutated) that, in turn, activate the checkpoint effector kinases Chk1 and Chk2, respectively. The latter occurs via the phosphorylation of SQ/TQ clusters in Chk1 and Chk2 at sites that critically affect protein–protein interactions [59, 60]. The Chk1 activation pathway via ATR is known to occur mainly in response to stalled replication forks arising from UV photodamage that involves the recognition of single-stranded DNA regions complexed with the single-stranded DNA-binding protein RPA (replication protein A) [61]. On the other hand, the activation of Chk2 by ATM involves the response to the processing of DNA double-strand breaks [62]. The question arises how the signal transduction mechanism can recognize the plethora of different DNA lesions. It has been suggested that the triggering of the effector kinase checkpoints involves common single-stranded DNA regions that are formed during the initial processing of a DNA lesion, which are independent of the physical or chemical nature of the lesion [58].

1.6
Repair Mechanisms that Remove DNA Lesions

In contrast to the cell signal transduction mechanism, there are various types of DNA repair mechanisms [14, 15, 63] that are specialized in removing different kinds of DNA lesions.

1.6.1
Repair of Single- and Double-Strand Breaks

Single-strand breaks are converted to double-strand breaks upon DNA replication. The double-strand breaks are repaired by one of two complex mechanisms: homologous recombination [64] and nonhomologous end-joining mechanisms [65]. In the nonhomologous end-joining mechanism, Ku proteins bind to the two ends of the DNA and recruit a number of end-processing proteins that effect the repair; errors can occur during this type of repair. On the other hand, in the homologous recombination system, the participation of a sister chromatid sequence favors more accurate repair.

1.6.2
Alkylating Agents

Small alkylating agents that readily form DNA lesions include the directly acting nitrogen mustard and mustard gas, and those that require enzymatic activation, such as alkylnitrosamines, methylnitrosourea, dimethylsulfate, vinyl chloride, butadiene, chloroacetaldehyde, formaldehyde, and so on. Among the well-known repair enzymes are O^6-alkylguanine methyltransferases (AGTs) that irreversibly transfer alkyl groups from O-alkylated nucleobases by a direct and error-free mechanism to an internal cysteine residue, thus deactivating the enzyme in the process ("suicide" enzyme) [66]. Another family of proteins, comprising AlkB from *Escherichia coli* and the human homologs ABH2 and ABH3, utilize a unique oxidative mechanism to remove alkyl groups from 1-methyladenines, 3-methylcytosines,

1-methylguanine, and 3-methylthymine. This mechanism is unprecedented because it involves an iron-oxo intermediate to oxidize the methyl substituents that leads to the regeneration of the normal nucleobases [67, 68].

1.6.3
Base Excision Repair

This mechanism repairs lesions such as uracil and relatively small oxidatively generated DNA lesions such as 8-oxoguanine [14]. It is an example of an excision repair mechanism that removes the damaged nucleobase and replaces it by the normal base (when the replacement proceeds in an error-free manner). BER (Chapter 11) is initiated by DNA glycosylases that catalyze the hydrolysis of the N-glycosidic bond, thus excising the damaged base from double-stranded DNA and leaving behind an apurinic (AP) site. A 5′-AP endonuclease cleaves the strand with the AP site leaving a 3′-OH terminal strand and a 5′-terminal deoxyribose phosphate group. The latter is excised by DNA deoxyribophosphodiesterase, the single nucleotide gap is filled in by the insertion of a nucleotide catalyzed by a polymerase, and the repair is completed by sealing the remaining nick with a DNA ligase.

1.6.4
Mismatch Excision Repair

Mismatched base pairs can arise from errors that occur during recombination repair. A variety of prokaryotic and eukaryotic mismatch repair proteins repair mismatched base pairs by a variety of mechanisms that involve the excision of one of the bases [69, 70].

1.6.5
Nucleotide Excision Repair

This is a complex multiprotein system in prokaryotes and eukaryotes that specializes in the repair of bulky lesions (Chapters 11 and 12). Both nucleotide excision repair (NER) systems recognize distortions and deviations from the normal DNA structure instead of the DNA lesions [63, 71]. Instead of excising the damaged lesion itself, entire lesion-containing DNA sequences, around 14 and 24–32 nucleotides in length in prokaryotes and eukaryotes, respectively, are excised. The resulting gap is filled in with nucleotides by a polymerase using the residual undamaged strand as a template and the nick is sealed by a ligase [63, 71]. There are two mechanisms of NER: (i) transcription-coupled (TC)-NER and (ii) global genomic (GG)-NER. Both mechanisms are similar except that TC-NER is triggered by RNA polymerases stalled at the sites of the DNA lesions, while GG-NER involves recognition of the DNA lesions by the eukaryotic NER factor XPC/HR23B (or currently named XPC-RAD23B, Chapter 11). The impact of DNA lesions on the elongation of transcripts catalyzed by RNA polymerases, as well as the differences between TC-NER and GG-NER, are discussed in detail in Chapter 17.

1.6.6
Translesion Bypass of Unrepaired Lesions by Specialized DNA Polymerases and RNA Polymerases

A specialized set of DNA polymerases cooperate to successfully replicate the strand containing the damaged nucleotide (Chapters 13–17). However, this mechanism of DNA damage tolerance is error-prone, and the fidelity of translesion bypass in human cells depends on the DNA lesion and the polymerase [17]. The progress of RNA polymerases depends generally on the size and shape of the DNA adduct, the local DNA sequence (Chapter 17), and the structure of the active site of the RNA polymerase [72, 73].

1.7
Relationships between the Chemical, Structural, and Biological Features of DNA Lesions

Advances in this area became feasible when technology became available for constructing oligodeoxyribonucleotides of defined base composition and sequence containing single, chemically defined DNA lesions [74]. The chapters in Part Two (Chapters 9–17) are devoted to different aspects of the structural and/or biological characteristics of such well-defined DNA lesions. The relationships between the structures of bulky aromatic DNA lesions and prokaryotic and eukaryotic NER susceptibilities are addressed in Chapters 10 and 12, respectively, while the general mechanisms of NER are reviewed in Chapter 11. The mutagenic characteristics of such site-specific DNA lesions have been studied extensively *in vitro* or in cellular environments [75–79]; specific examples are discussed in detail in Chapters 13–17.

Structural information on site-specific DNA lesions can be gained by high-resolution nuclear magnetic resonance (NMR) methods, X-ray crystallography, and molecular dynamic simulation methods. The latter, computational approach has the unique capability of providing insights into the dynamics of structural characteristics of DNA lesions by studying the evolution in time of dynamic ensembles (Chapter 14). The applications of NMR methods have, over the past two decades, yielded rich insights into the structural properties of bulky PAH-derived DNA adducts [80] (Chapter 12), adducts derived from aromatic amines (Chapter 10) and many other DNA lesions [81]. However, there are challenges in studying the structures of DNA lesions in solution by NMR methods, mainly because the lesions in solution can be highly mobile and heterogeneous in structure; this can make data interpretation difficult and defining structures even impossible in some cases. X-ray crystallography has provided some representative crystal structures of lesions in duplex DNA. Examples are intrastrand [82] and interstrand [83] cross-linked cisplatin adducts, a B[*a*]P 7,8-diol 9,10-epoxide-N^2-deoxyguanosine adduct [84], and CPD [85] in double-stranded DNA, and 6-4 pyrimidine–pyrimidone [86] and CPD [87] photodimers in DNA in complexes

1.7 Relationships between the Chemical, Structural, and Biological Features of DNA Lesions

with photolyases – enzymes that restore the photodimers to their undamaged condition. Considerable progress has been made recently in determining the crystallographic structures of lesions in DNA in complexes with polymerases and repair proteins [88]. These include a variety of oxidatively generated and other types of lesions in double-stranded DNA bound to BER proteins [89, 90]. Examples, include O^6-alkylguanine lesions in complexes with AGT [66] and in alkyltransferase-like proteins [66]. There are fewer examples of crystallographic structures of bulky lesions in complexes with NER-related proteins. The existing structures include a fluorescein-derived DNA lesion in a complex with the prokaryotic NER protein UvrB [91], CPD lesions in double-stranded DNA in a complex with Rad4, a yeast ortholog of the human DNA lesion-recognizing XPC/HR23B protein complex [92], and a 6-4 photodimer in double-stranded DNA in a complex with DDB1–DDB2 [93] – proteins that facilitate the identification and repair of UV photolesions in chromatin.

Considerable progress has been made in characterizing the structures of various DNA lesions in complexes with polymerases, including small [94, 95] and bulky adducts at or near the single/double-strand junctions positioned close to the active sites of polymerases [96–98]. Various aspects of these topics are addressed in Chapters 13–16. A number of reviews of this highly active field have been published [88, 95, 99–102]. While crystallography can provide outstanding resolutions of the structural properties of DNA lesions in proteins, challenges remain in growing crystals of sufficient high quality to yield high-resolution structures (resolutions below around 2 Å). This can be a serious obstacle in the case of DNA-containing lesions, where conformational flexibility and heterogeneity diminish the resolution. Of course, the same problem is encountered in solution NMR studies. Standard biochemical methods that provide detailed information on the error-prone or error-free rates of nucleotide incorporation opposite the lesion and neighboring DNA template sites provide valuable insights into the kinetic and potentially mutagenic bypass of DNA lesions catalyzed by polymerases [94, 95, 103]. These techniques, when complemented with NMR or crystallographic structural information, can provide valuable insights into the relationships between the structural features of DNA lesions and their impact on DNA replication.

Molecular dynamic simulation methods, in addition to being essential for interpreting NMR data at the atomic level, also augment experimental studies in a number of other ways [101]: modeling techniques can (i) yield structural information where experimental data has not yet been acquired, (ii) expand on experimental data through simulations that yield dynamic trajectories whose analysis provides unique information on lesion mobility, and (iii) provide thermodynamic insights by ensemble analysis using statistical mechanical methods. Furthermore, reaction mechanisms can now be determined with some confidence by combined quantum mechanical and molecular mechanical methods [104, 105].

Part One of the book, entitled *Chemistry and Biology of DNA Lesions*, will introduce the reader to the fundamentals of the chemical characterization, biochemical properties, and biological effects of some representative and important sets of DNA lesions, as well as their impact on human health when such information is available. The material covered provides insights into the approaches used in this field that can serve as a blueprint for analyzing other forms of DNA damage. In Chapter 2, M.S. DeMott and P.C. Dedon discuss the chemistry of RNS associated with the inflammatory response and the chemistry of deoxyribose oxidation. Chapter 3 by J. Cadet, T. Douki and J.-L. Ravanat describes the chemistry of ROS, and their reactions with the nucleobases of DNA *in vitro* and in cellular environments. The role of free radical reactions with DNA is addressed in Chapter 4 by V. Shafirovich and N.E. Geacintov. In Chapter 5, C.G. Knutsen and L.J. Marnett discuss the chemistry of lipid peroxidation and the types of endogenous DNA adducts formed under oxidative stress. The different metabolic pathways of PAHs and the DNA lesions formed are reviewed by T.M. Penning in Chapter 6. The chemistry of aromatic amines and heterocyclic aromatic amines present in food and tobacco smoke, as well as the variety of DNA adducts formed, is described by R.J. Turesky in Chapter 7. The last chapter in Part One, Chapter 8 by J.L. Bolton and G.R.J. Thatcher, deals with genotoxic estrogen pathways of reactions with DNA of endogenous estrogens and estrogen derivatives used in hormone replacement therapy.

Part Two, entitled *New Frontiers and Challenges: Understanding Structure–Function Relationships and Biological Activity*, is focused on topics that deal with recent and current research on the relationships between the chemical structure and physical properties of selected DNA lesions, and how they are processed by the DNA repair, replication, and transcription machineries. In Chapter 9, Stone *et al.* discuss the relationships between the structural features and functional properties of interstrand cross-linked lesions derived from the reactions of α,β-unsaturated aldehydes with DNA. Chapters 10–12 deal with different aspects of DNA repair, and Chapters 13–17 are focused on the impact of lesions on DNA replication and transcription. In Chapter 10, B. Cho describes the relationships between the structures and biochemical functions of aromatic amine-DNA adducts. O.D. Schärer and A.J. Campbell summarize the mechanisms of BER and NER (Chapter 11), while Y. Cai *et al.* address the molecular basis of the experimentally observed variable efficiencies of recognition and removal of PAH-derived bulky DNA adducts by the eukaryotic NER system (Chapter 12). The impact of structural features of different DNA lesions on translesion synthesis catalyzed by replicative and bypass DNA polymerases are discussed by R.L. Eoff, M. Egli, and F.P. Guengerich in Chapter 13, while insights into the relationships between the structures of DNA lesions in solution and their function in polymerases are discussed in Chapter 14 by S. Broyde *et al.* In Chapter 15, S. Chandani and E.L. Loechler provide an overview of the field and describe their work on mechanisms of tranlesion bypass utilizing model system *E. coli* polymerases. In Chapter 16, Z. Livneh summarizes the work of his laboratory on the molecular mechanisms of translesion DNA synthesis in human cells. Last, but not least, Chapter 17 by K. Dreij *et al.* represents a comprehensive description of the effects of structurally diverse DNA lesions on transcription elongation.

Acknowledgements

The writing of this article was supported by research grants from the National Cancer Institute, National Institutes of Health CA 099194 (N.E.G.), CA 112412 (N.E.G.), CA 28038 (S.B.), and CA 75449 (S.B.).

References

1. Jackson, S.P. and Bartek, J. (2009) The DNA-damage response in human biology and disease. *Nature*, **461**, 1071–1078.
2. Lindahl, T. and Barnes, D.E. (2000) Repair of endogenous DNA damage. *Cold Spring Harb. Symp. Quant. Biol.*, **65**, 127–133.
3. De Bont, R. and van Larebeke, N. (2004) Endogenous DNA damage in humans: a review of quantitative data. *Mutagenesis*, **19**, 169–185.
4. Dedon, P.C. and Tannenbaum, S.R. (2004) Reactive nitrogen species in the chemical biology of inflammation. *Arch. Biochem. Biophys.*, **423**, 12–22.
5. Lindahl, T. and Nyberg, B. (1972) Rate of depurination of native deoxyribonucleic acid. *Biochemistry*, **11**, 3610–3618.
6. Lindahl, T. and Nyberg, B. (1974) Heat-induced deamination of cytosine residues in deoxyribonucleic acid. *Biochemistry*, **13**, 3405–3410.
7. de Gruijl, F.R. (1999) Skin cancer and solar UV radiation. *Eur. J. Cancer*, **35**, 2003–2009.
8. Kojo, K., Jansen, C.T., Nybom, P., Huurto, L., Laihia, J., Ilus, T., and Auvinen, A. (2006) Population exposure to ultraviolet radiation in Finland 1920–1995: exposure trends and a time-series analysis of exposure and cutaneous melanoma incidence. *Environ. Res.*, **101**, 123–131.
9. de Gruijl, F.R. and Rebel, H. (2008) Early events in UV carcinogenesis – DNA damage, target cells and mutant p53 foci. *Photochem. Photobiol.*, **84**, 382–387.
10. Wogan, G.N., Hecht, S.S., Felton, J.S., Conney, A.H., and Loeb, L.A. (2004) Environmental and chemical carcinogenesis. *Semin. Cancer Biol.*, **14**, 473–486.
11. Poirier, M.C., Santella, R.M., and Weston, A. (2000) Carcinogen macromolecular adducts and their measurement. *Carcinogenesis*, **21**, 353–359.
12. Blair, I.A. (2008) DNA adducts with lipid peroxidation products. *J. Biol. Chem.*, **283**, 15545–15549.
13. Farmer, P.B. and Singh, R. (2008) Use of DNA adducts to identify human health risk from exposure to hazardous environmental pollutants: the increasing role of mass spectrometry in assessing biologically effective doses of genotoxic carcinogens. *Mutat. Res.*, **659**, 68–76.
14. Friedberg, E.C., Walker, G.C., Siede, W., Wood, R.D., Schultz, R.A., and Ellenberger, T. (2006) *DNA Repair and Mutagenesis*, 2nd edn, ASM Press, Washington, DC.
15. Siede, W., Kow, Y., and Doetsch, P. (eds) (2006) *DNA Damage Recognition*, Taylor & Francis, New York.
16. McCulloch, S.D. and Kunkel, T.A. (2008) The fidelity of DNA synthesis by eukaryotic replicative and translesion synthesis polymerases. *Cell Res.*, **18**, 148–161.
17. Loeb, L.A. and Monnat, R.J., Jr. (2008) DNA polymerases and human disease. *Nat. Rev. Genet.*, **9**, 594–604.
18. Adar, S., Izhar, L., Hendel, A., Geacintov, N., and Livneh, Z. (2009) Repair of gaps opposite lesions by homologous recombination in mammalian cells. *Nucleic Acids Res.*, **37**, 5737–5748.
19. Coussens, L.M. and Werb, Z. (2002) Inflammation and cancer. *Nature*, **420**, 860–867.
20. Cadet, J., Douki, T., and Ravanat, J.L. (2008) Oxidatively generated damage to

the guanine moiety of DNA: mechanistic aspects and formation in cells. *Acc. Chem. Res.*, **41**, 1075–1083.
21 Neeley, W.L. and Essigmann, J.M. (2006) Mechanisms of formation, genotoxicity, and mutation of guanine oxidation products. *Chem. Res. Toxicol.*, **19**, 491–505.
22 West, J.D. and Marnett, L.J. (2006) Endogenous reactive intermediates as modulators of cell signaling and cell death. *Chem. Res. Toxicol.*, **19**, 173–194.
23 Wang, Y. (2008) Bulky DNA lesions induced by reactive oxygen species. *Chem. Res. Toxicol.*, **21**, 276–281.
24 Markesbery, W.R. and Lovell, M.A. (2006) DNA oxidation in Alzheimer's disease. *Antioxid. Redox Signal.*, **8**, 2039–2045.
25 Tsang, A.H. and Chung, K.K. (2009) Oxidative and nitrosative stress in Parkinson's disease. *Biochim. Biophys. Acta*, **1792**, 643–650.
26 Campisi, J. and Vijg, J. (2009) Does damage to DNA and other macromolecules play a role in aging? If so, how? *J. Gerontol. A Biol. Sci. Med. Sci.*, **64**, 175–178.
27 Gruber, J., Schaffer, S., and Halliwell, B. (2008) The mitochondrial free radical theory of ageing – where do we stand? *Front. Biosci.*, **13**, 6554–6579.
28 Garinis, G.A., van der Horst, G.T., Vijg, J., and Hoeijmakers, J.H. (2008) DNA damage and ageing: new-age ideas for an age-old problem. *Nat. Cell Biol.*, **10**, 1241–1247.
29 Loeb, L.A. and Harris, C.C. (2008) Advances in chemical carcinogenesis: a historical review and prospective. *Cancer Res.*, **68**, 6863–6872.
30 Conney, A.H., Miller, E.C., and Miller, J.A. (1956) The metabolism of methylated aminoazo dyes. V. Evidence for induction of enzyme synthesis in the rat by 3-methylcholanthrene. *Cancer Res.*, **16**, 450–459.
31 Luch, A. (2005) Nature and nurture – lessons from chemical carcinogenesis. *Nat. Rev. Cancer*, **5**, 113–125.
32 Vogelstein, B. and Kinzler, K.W. (1992) Carcinogens leave fingerprints. *Nature*, **355**, 209–210.
33 Vogelstein, B. and Kinzler, K.W. (1993) The multistep nature of cancer. *Trends Genet.*, **9**, 138–141.
34 Vogelstein, B. and Kinzler, K.W. (2004) Cancer genes and the pathways they control. *Nat. Med.*, **10**, 789–799.
35 Bostrom, C.E., Gerde, P., Hanberg, A., Jernstrom, B., Johansson, C., Kyrklund, T., Rannug, A., Tornqvist, M., Victorin, K., and Westerholm, R. (2002) Cancer risk assessment, indicators, and guidelines for polycyclic aromatic hydrocarbons in the ambient air. *Environ. Health Perspect.*, **110** (Suppl. 3), 451–488.
36 Hecht, S.S. (1999) Tobacco smoke carcinogens and lung cancer. *J. Natl. Cancer Inst.*, **91**, 1194–1210.
37 Hecht, S.S. (2008) Progress and challenges in selected areas of tobacco carcinogenesis. *Chem. Res. Toxicol.*, **21**, 160–171.
38 Williams, J.H., Phillips, T.D., Jolly, P.E., Stiles, J.K., Jolly, C.M., and Aggarwal, D. (2004) Human aflatoxicosis in developing countries: a review of toxicology, exposure, potential health consequences, and interventions. *Am. J. Clin. Nutr.*, **80**, 1106–1122.
39 Croy, R.G., Essigmann, J.M., Reinhold, V.N., and Wogan, G.N. (1978) Identification of the principal aflatoxin B1-DNA adduct formed *in vivo* in rat liver. *Proc. Natl. Acad. Sci. USA*, **75**, 1745–1749.
40 Grollman, A.P., Shibutani, S., Moriya, M., Miller, F., Wu, L., Moll, U., Suzuki, N., Fernandes, A., Rosenquist, T., Medverec, Z., *et al.* (2007) Aristolochic acid and the etiology of endemic (Balkan) nephropathy. *Proc. Natl. Acad. Sci. USA*, **104**, 12129–12134.
41 Hranjec, T., Kovac, A., Kos, J., Mao, W., Chen, J.J., Grollman, A.P., and Jelakovic, B. (2005) Endemic nephropathy: the case for chronic poisoning by aristolochia. *Croat. Med. J.*, **46**, 116–125.
42 Debelle, F.D., Vanherweghem, J.L., and Nortier, J.L. (2008) Aristolochic acid nephropathy: a worldwide problem. *Kidney Int.*, **74**, 158–169.
43 Schmeiser, H.H., Stiborova, M., and Arlt, V.M. (2009) Chemical and molecular basis of the carcinogenicity

of *Aristolochia* plants. *Curr. Opin. Drug Discov. Devel.*, **12**, 141–148.
44 Yager, J.D. and Liehr, J.G. (1996) Molecular mechanisms of estrogen carcinogenesis. *Annu. Rev. Pharmacol. Toxicol.*, **36**, 203–232.
45 Santen, R., Cavalieri, E., Rogan, E., Russo, J., Guttenplan, J., Ingle, J., and Yue, W. (2009) Estrogen mediation of breast tumor formation involves estrogen receptor-dependent, as well as independent, genotoxic effects. *Ann. NY Acad. Sci.*, **1155**, 132–140.
46 Stack, D.E., Byun, J., Gross, M.L., Rogan, E.G., and Cavalieri, E.L. (1996) Molecular characteristics of catechol estrogen quinones in reactions with deoxyribonucleosides. *Chem. Res. Toxicol.*, **9**, 851–859.
47 Bolton, J.L., Pisha, E., Zhang, F., and Qiu, S. (1998) Role of quinoids in estrogen carcinogenesis. *Chem. Res. Toxicol.*, **11**, 1113–1127.
48 Bolton, J.L. and Thatcher, G.R. (2008) Potential mechanisms of estrogen quinone carcinogenesis. *Chem. Res. Toxicol.*, **21**, 93–101.
49 Bolton, J.L., Yu, L., and Thatcher, G.R. (2004) Quinoids formed from estrogens and antiestrogens. *Methods Enzymol.*, **378**, 110–123.
50 Park, J.H., Mangal, D., Frey, A.J., Harvey, R.G., Blair, I.A., and Penning, T.M. (2009) Aryl hydrocarbon receptor facilitates DNA strand breaks and 8-oxo-2′-deoxyguanosine formation by the aldo-keto reductase product benzo[*a*]pyrene-7,8-dione. *J. Biol. Chem.*, **284**, 29725–29734.
51 Prise, K.M., Schettino, G., Folkard, M., and Held, K.D. (2005) New insights on cell death from radiation exposure. *Lancet Oncol.*, **6**, 520–528.
52 Wang, D. and Lippard, S.J. (2005) Cellular processing of platinum anticancer drugs. *Nat. Rev. Drug Discov.*, **4**, 307–320.
53 Miyagawa, K. (2008) Clinical relevance of the homologous recombination machinery in cancer therapy. *Cancer Sci.*, **99**, 187–194.
54 Jung, Y. and Lippard, S.J. (2007) Direct cellular responses to platinum-induced DNA damage. *Chem. Rev.*, **107**, 1387–1407.
55 Eastman, A. (1986) Reevaluation of interaction of *cis*-dichloro(ethylenediamine)platinum(II) with DNA. *Biochemistry*, **25**, 3912–3915.
56 Zou, Y., Van Houten, B., and Farrell, N. (1994) Sequence specificity of DNA–DNA interstrand cross-link formation by cisplatin and dinuclear platinum complexes. *Biochemistry*, **33**, 5404–5410.
57 Eastman, A. (1985) Interstrand cross-links and sequence specificity in the reaction of *cis*-dichloro(ethylenediamine)platinum(II) with DNA. *Biochemistry*, **24**, 5027–5032.
58 Lazzaro, F., Giannattasio, M., Puddu, F., Granata, M., Pellicioli, A., Plevani, P., and Muzi-Falconi, M. (2009) Checkpoint mechanisms at the intersection between DNA damage and repair. *DNA Repair*, **8**, 1055–1067.
59 Stracker, T.H., Usui, T., and Petrini, J.H. (2009) Taking the time to make important decisions: the checkpoint effector kinases Chk1 and Chk2 and the DNA damage response. *DNA Repair*, **8**, 1047–1054.
60 Altaf, M., Saksouk, N., and Cote, J. (2007) Histone modifications in response to DNA damage. *Mutat. Res.*, **618**, 81–90.
61 Cimprich, K.A. and Cortez, D. (2008) ATR: an essential regulator of genome integrity. *Nat. Rev. Mol. Cell Biol.*, **9**, 616–627.
62 Oliver, A.W., Knapp, S., and Pearl, L.H. (2007) Activation segment exchange: a common mechanism of kinase autophosphorylation? *Trends Biochem. Sci.*, **32**, 351–356.
63 Hoeijmakers, J.H. (2001) Genome maintenance mechanisms for preventing cancer. *Nature*, **411**, 366–374.
64 San Filippo, J., Sung, P., and Klein, H. (2008) Mechanism of eukaryotic homologous recombination. *Annu. Rev. Biochem.*, **77**, 229–257.
65 Lieber, M.R. (2008) The mechanism of human nonhomologous DNA end joining. *J. Biol. Chem.*, **283**, 1–5.
66 Tubbs, J.L., Pegg, A.E., and Tainer, J.A. (2007) DNA binding, nucleotide

flipping, and the helix-turn-helix motif in base repair by O^6-alkylguanine-DNA alkyltransferase and its implications for cancer chemotherapy. *DNA Repair*, **6**, 1100–1115.

67 Aravind, L. and Koonin, E.V. (2001) The DNA-repair protein AlkB, EGL-9, and leprecan define new families of 2-oxoglutarate- and iron-dependent dioxygenases. *Genome Biol.*, **2**, RESEARCH007.

68 Sedgwick, B. (2004) Repairing DNA-methylation damage. *Nat. Rev. Mol. Cell Biol.*, **5**, 148–157.

69 Kunz, C., Saito, Y., and Schar, P. (2009) DNA Repair in mammalian cells: mismatched repair: variations on a theme. *Cell Mol. Life Sci.*, **66**, 1021–1038.

70 Acharya, S. and Fishel, R. (eds.) (2006) *Mechanism of DNA Mismatch Repair from Bacteria to Human*, Taylor & Francis, New York.

71 Truglio, J.J., Croteau, D.L., Van Houten, B., and Kisker, C. (2006) Prokaryotic nucleotide excision repair: the UvrABC system. *Chem. Rev.*, **106**, 233–252.

72 Dimitri, A., Burns, J.A., Broyde, S., and Scicchitano, D.A. (2008) Transcription elongation past O^6-methylguanine by human RNA polymerase II and bacteriophage T7 RNA polymerase. *Nucleic Acids Res.*, **36**, 6459–6471.

73 Dimitri, A., Goodenough, A.K., Guengerich, F.P., Broyde, S., and Scicchitano, D.A. (2008) Transcription processing at $1,N^2$-ethenoguanine by human RNA polymerase II and bacteriophage T7 RNA polymerase. *J. Mol. Biol.*, **375**, 353–366.

74 Basu, A.K. and Essigmann, J.M. (1988) Site-specifically modified oligodeoxynucleotides as probes for the structural and biological effects of DNA-damaging agents. *Chem. Res. Toxicol.*, **1**, 1–18.

75 Fernandes, A., Liu, T., Amin, S., Geacintov, N.E., Grollman, A. P., and Moriya, M. (1998) Mutagenic potential of stereoisomeric bay region (+)- and (−)-*cis-anti*-benzo[*a*]pyrenediolepoxide-N^2-2′-deoxyguanosine adducts in *Escherichia coli* and simian kidney cells. *Biochemistry*, **37**, 10164–10172.

76 Loechler, E.L. (1996) The role of adduct site-specific mutagenesis in understanding how carcinogen-DNA adducts cause mutations: perspective, prospects and problems. *Carcinogenesis*, **17**, 895–902.

77 Shachar, S., Ziv, O., Avkin, S., Adar, S., Wittschieben, J., Reissner, T., Chaney, S., Friedberg, E.C., Wang, Z., Carell, T., Geacintov, N., and Livneh, Z. (2009) Two-polymerase mechanisms dictate error-free and error-prone translesion DNA synthesis in mammals. *EMBO J.*, **28**, 383–393.

78 Wang, D., Kreutzer, D.A., and Essigmann, J.M. (1998) Mutagenicity and repair of oxidative DNA damage: insights from studies using defined lesions. *Mutat. Res.*, **400**, 99–115.

79 Shrivastav, N., Li, D., and Essigmann, J.M. (2010) Chemical biology of mutagenesis and DNA repair: cellular responses to DNA alkylation. *Carcinogenesis*, **31**, 59–70.

80 Geacintov, N.E., Cosman, M., Hingerty, B.E., Amin, S., Broyde, S., and Patel, D.J. (1997) NMR solution structures of stereoisomeric covalent polycyclic aromatic carcinogen-DNA adduct: principles, patterns, and diversity. *Chem. Res. Toxicol.*, **10**, 111–146.

81 Lukin, M. and de Los Santos, C. (2006) NMR structures of damaged DNA. *Chem. Rev.*, **106**, 607–686.

82 Takahara, P.M., Rosenzweig, A.C., Frederick, C.A., and Lippard, S.J. (1995) Crystal structure of double-stranded DNA containing the major adduct of the anticancer drug cisplatin. *Nature*, **377**, 649–652.

83 Coste, F., Malinge, J.M., Serre, L., Shepard, W., Roth, M., Leng, M., and Zelwer, C. (1999) Crystal structure of a double-stranded DNA containing a cisplatin interstrand cross-link at 1.63 A resolution: hydration at the platinated site. *Nucleic Acids Res.*, **27**, 1837–1846.

84 Karle, I., Yagi, H., Sayer, J., and DM, J. (2004) Crystal and molecular structure of a benzo[*a*]pyrene 7,8-diol 9,10-epoxide N^2-deoxyguanosine adduct: absolute configuration and conformation. *Proc. Natl. Acad. Sci. USA*, **101**, 1433–1438.

85 Park, H., Zhang, K., Ren, Y., Nadji, S., Sinha, N., Taylor, J.S., and Kang, C. (2002) Crystal structure of a DNA decamer containing a *cis-syn* thymine dimer. *Proc. Natl. Acad. Sci. USA*, **99**, 15965–15970.

86 Glas, A.F., Schneider, S., Maul, M.J., Hennecke, U., and Carell, T. (2009) Crystal structure of the T(6-4)C lesion in complex with a (6-4) DNA photolyase and repair of UV-induced (6-4) and Dewar photolesions. *Chemistry*, **15**, 10387–10396.

87 Mees, A., Klar, T., Gnau, P., Hennecke, U., Eker, A.P., Carell, T., and Essen, L.O. (2004) Crystal structure of a photolyase bound to a CPD-like DNA lesion after *in situ* repair. *Science*, **306**, 1789–1793.

88 Schneider, S., Schorr, S., and Carell, T. (2009) Crystal structure analysis of DNA lesion repair and tolerance mechanisms. *Curr. Opin. Struct. Biol.*, **19**, 87–95.

89 Fromme, J.C. and Verdine, G.L. (2004) Base excision repair. *Adv. Protein Chem.*, **69**, 1–41.

90 Hitomi, K., Iwai, S., and Tainer, J.A. (2007) The intricate structural chemistry of base excision repair machinery: implications for DNA damage recognition, removal, and repair. *DNA Repair*, **6**, 410–428.

91 Truglio, J.J., Karakas, E., Rhau, B., Wang, H., DellaVecchia, M.J., Van Houten, B., and Kisker, C. (2006) Structural basis for DNA recognition and processing by UvrB. *Nat. Struct. Mol. Biol.*, **13**, 360–364.

92 Min, J.H. and Pavletich, N.P. (2007) Recognition of DNA damage by the Rad4 nucleotide excision repair protein. *Nature*, **449**, 570–575.

93 Scrima, A., Konickova, R., Czyzewski, B.K., Kawasaki, Y., Jeffrey, P.D., Groisman, R., Nakatani, Y., Iwai, S., Pavletich, N.P., and Thoma, N.H. (2008) Structural basis of UV DNA-damage recognition by the DDB1–DDB2 complex. *Cell*, **135**, 1213–1223.

94 Pence, M.G., Blans, P., Zink, C.N., Hollis, T., Fishbein, J.C., and Perrino, F.W. (2009) Lesion bypass of N^2-ethyl-guanine by human DNA polymerase iota. *J. Biol. Chem.*, **284**, 1732–1740.

95 Guengerich, F.P. (2006) Interactions of carcinogen-bound DNA with individual DNA polymerases. *Chem. Rev.*, **106**, 420–452.

96 Batra, V.K., Shock, D.D., Prasad, R., Beard, W.A., Hou, E.W., Pedersen, L.C., Sayer, J.M., Yagi, H., Kumar, S., Jerina, D.M., *et al.* (2006) Structure of DNA polymerase β with a benzo[c]phenanthrene diol epoxide-adducted template exhibits mutagenic features. *Proc. Natl. Acad. Sci. USA*, **103**, 17231–17236.

97 Bauer, J., Xing, G., Yagi, H., Sayer, J.M., Jerina, D.M., and Ling, H. (2007) A structural gap in Dpo4 supports mutagenic bypass of a major benzo[a]pyrene dG adduct in DNA through template misalignment. *Proc. Natl. Acad. Sci. USA*, **104**, 14905–14910.

98 Ling, H., Sayer, J.M., Plosky, B.S., Yagi, H., Boudsocq, F., Woodgate, R., Jerina, D.M., and Yang, W. (2004) Crystal structure of a benzo[a]pyrene diol epoxide adduct in a ternary complex with a DNA polymerase. *Proc. Natl. Acad. Sci. USA*, **101**, 2265–2269.

99 Prakash, S., Johnson, R.E., and Prakash, L. (2005) Eukaryotic translesion synthesis DNA polymerases: specificity of structure and function. *Annu. Rev. Biochem.*, **74**, 317–353.

100 Friedberg, E.C., Lehmann, A.R., and Fuchs, R.P. (2005) Trading places: how do DNA polymerases switch during translesion DNA synthesis? *Mol. Cell*, **18**, 499–505.

101 Broyde, S., Wang, L., Rechkoblit, O., Geacintov, N.E., and Patel, D.J. (2008) Lesion processing: high-fidelity versus lesion-bypass DNA polymerases. *Trends Biochem. Sci.*, **33**, 209–219.

102 Broyde, S., Wang, L., Zhang, L., Rechkoblit, O., Geacintov, N.E., and Patel, D.J. (2008) DNA adduct structure–function relationships: comparing solution with polymerase structures. *Chem. Res. Toxicol.*, **21**, 45–52.

103 Lone, S., Townson, S.A., Uljon, S.N., Johnson, R.E., Brahma, A., Nair, D.T., Prakash, S., Prakash, L., and Aggarwal, A.K. (2007) Human DNA polymerase

kappa encircles DNA: implications for mismatch extension and lesion bypass. *Mol. Cell*, **25**, 601–614.

104 Wang, L., Yu, X., Hu, P., Broyde, S., and Zhang, Y. (2007) A water-mediated and substrate-assisted catalytic mechanism for *Sulfolobus solfataricus* DNA polymerase IV. *J. Am. Chem. Soc.*, **129**, 4731–4737.

105 Lin, P., Pedersen, L.C., Batra, V.K., Beard, W.A., Wilson, S.H., and Pedersen, L.G. (2006) Energy analysis of chemistry for correct insertion by DNA polymerase β. *Proc. Natl. Acad. Sci. USA*, **103**, 13294–13299.

2
Chemistry of Inflammation and DNA Damage: Biological Impact of Reactive Nitrogen Species

Michael S. DeMott and Peter C. Dedon

2.1
Introduction

Chronic inflammation is now viewed as a major cause of human disease, with DNA damage playing a potentially important role in the pathophysiology. As an acute response to injury and infection, inflammation is a critical feature of the immune system involving infiltration of lymphocytes, macrophages, and neutrophils into tissues at sites of infection and injury. Phagocytic macrophages and neutrophils become activated by a variety of cytokines, bacterial lipopolysaccharides, and cell debris that secrete a battery of chemically reactive oxygen species (ROS) and reactive nitrogen species (RNS) to eradicate microbial pathogens and process dead cells, as shown in Figure 2.1 [1–5]. However, there is mounting epidemiological evidence that inflammatory reactions sustained over decades play an important causal role in many human diseases [6, 7]. For example, inflammation in the vascular system leads to oxidative modification of lipids and lipoproteins that contribute to atherosclerosis [8–10]. Chemical irritation in the esophagus caused by gastric acid and bile in gastroesophageal reflux leads to Barrett's esophagus and esophageal cancer [11]. Moreover, there is a strong correlation between chronic inflammation and specific forms of cancer, including inflammatory bowel disease and colon cancer [12, 13], *Heliobacter pylori* infection and gastric cancer [14, 15], and *Schistosoma haematobium* infection and bladder cancer [16, 17].

One possible mechanism linking chronic inflammation with disease lies in the prolonged exposure of cells to the ROS and RNS secreted by macrophages and neutrophils. The chemical mediators of inflammation shown in Figure 2.1 span a wide range of reactions, including nitrosation, nitration, oxidation, and halogenation, all of which can cause cytotoxicity and mutation by reaction with cellular biomolecules. Activated macrophages generate large quantities of nitric oxide (NO) and superoxide (O_2^-) [5, 18, 19]. NO at low levels (nanomolar) is an important regulator of the cardiovascular, nervous, and immune systems [20–26], while high concentrations of NO (up to $1\,\mu M$) [27–29] produced by macrophages at sites of inflammation lead to pathophysiological reactions with oxygen and O_2^- to generate RNS. For example, as covered thoroughly in several recent reviews [2, 5, 30, 31],

The Chemical Biology of DNA Damage. Edited by Nicholas E. Geacintov and Suse Broyde
© 2010 WILEY-VCH Verlag GmbH & Co. KGaA, Weinheim
ISBN: 978-3-527-32295-4

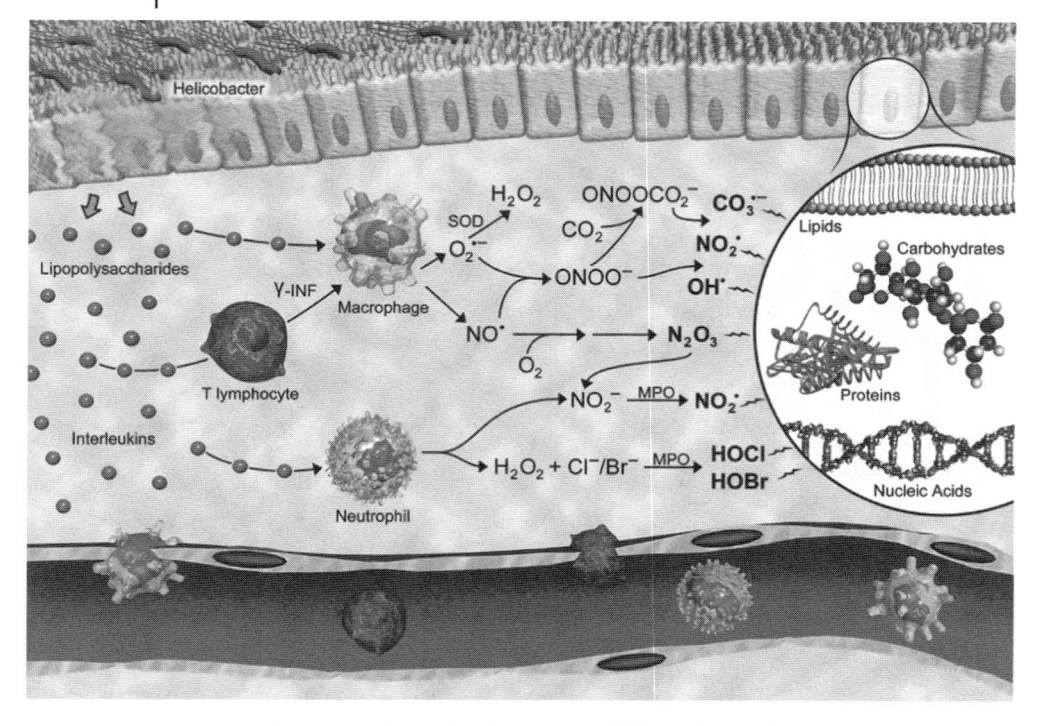

Figure 2.1 Chemical biology of the formation of RNS in inflammation.

auto-oxidation of NO generates nitrous anhydride (N_2O_3; Figure 2.1) – a potent nitrosating agent capable of deaminating proteins and DNA bases. The reaction of O_2^- and NO at diffusion-controlled rates leads to peroxynitrite ($ONOO^-$), which, in its protonated form, undergoes rapid ($t_{1/2}$ ~ 1 s) homolysis to yield hydroxyl radical ($^{\bullet}OH$) and the weak oxidant, nitrogen dioxide radical ($NO_2^{\bullet}$). Further reaction of $ONOO^-$ with carbon dioxide (CO_2) in tissues leads to formation of nitrosoperoxycarbonate ($ONOOCO_2^-$), which also undergoes homolytic scission ($t_{1/2}$ ~ 50 ms) to form carbonate radical anion ($CO_3^{\bullet-}$) and $NO_2^{\bullet}$. The neutrophil contribution to chemical mediators of inflammation arises from myeloperoxidase-mediated generation of hypochlorous acid (HOCl), a potent oxidizing and halogenating agent, and conversion of nitrite to $NO_2^{\bullet}$ via a nitryl chloride (NO_2Cl) intermediate [32–35].

These short-lived, highly reactive molecules cause damage to virtually all types of cellular biomolecules, including lipids, proteins, nucleic acids, carbohydrates, and small metabolites. In this chapter, we will survey the reactions of the chemical mediators of inflammation with DNA, with a focus on RNS. Due to space limitations, we will not be able to cover all of the complicated aspects of oxidatively damaged DNA related to hydroxyl radicals generated by Fenton chemistry, ionizing radiation, and singlet oxygen – topics that have been covered in several recent reviews [36–38].

2.2
DNA Oxidation and Nitration

Modeling studies of activated macrophages suggest that the intracellular concentration of ONOO$^-$ lies in the nanomolar range, which is around 10^6-fold higher than analogous estimates for N_2O_3 [39]. Possibly as a result of homolysis to the highly reactive ·OH radical, ONOO$^-$ causes mainly 2-deoxyribose oxidation in DNA (i.e., strand breaks and oxidized abasic sites) [5, 40], with a smaller amount of base damage [40, 41]. The CO_2 adduct of ONOO$^-$, ONOOCO$_2^-$, on the other hand, preferentially causes oxidation and nitration of G in DNA and probably RNA [5, 40, 42]. In one sense, CO_2 may be viewed as a catalyst for the reaction of ONOO$^-$ with G in DNA [43], by increasing the rate of G oxidation and regenerating CO_2 following $CO_3^{·-}$ one-electron oxidation of G. ONOOCO$_2^-$ is likely to be the major oxidizing species *in vivo* due to millimolar concentrations of CO_2 in tissues, and the large reaction rate constants of CO_2 and ONOO$^-$ [39]. Both ONOO$^-$ and ONOOCO$_2^-$ undergo homolysis to form the weakly oxidizing NO$_2^·$. With contributions from other sources, such as myeloperoxidase oxidation of nitrite (NO$_2^-$) [33, 44–46], the steady-state levels of NO$_2^·$ in activated macrophages have been estimated to be in the picomolar range [39].

The selectivity of RNS for oxidation of guanine (G) in DNA is due primarily to simple redox chemistry. The DNA bases dG, dA, dC, and dT have reduction potentials of 1.29, 1.42, 1.6, and 1.7 V, respectively (versus NHE [47]), which makes G the most easily oxidized base by both ·OH (2.3 V versus NHE [48]) and CO$_3^{·-}$ (1.7 V versus NHE [49]). However, the reduction potential of NO$_2^·$ (1.04 V versus NHE [48, 50]) means that it is incapable of oxidizing G, but it readily oxidizes 7,8-dihydro-8-oxoguanine (8-oxo-G; 0.74 V versus NHE [51]) – a major product of G oxidation (Figure 2.2).

2.2.1
Spectrum of Guanine Oxidation Products Caused by ONOO$^-$, ONOOCO$_2^-$, and NO$_2^·$

The major primary products arising from reactions of G with ONOO$^-$ and ONOOCO$_2^-$ are shown in Figure 2.2, and include 8-oxo-2′-deoxy-guanosine (8-oxo-dG) and 8-nitro-2′-deoxyguanosine (8-nitro-dG), which readily depurinates to release 8-nitro-G [40, 52], 5-guanidino-4-nitroimidazole (NitroIm), and 2,2-diamino-4-[(2-deoxy-β-D-*erythro*-pentofuranosyl)amino]-5(2H)-oxazolone (oxazolone (Ox)), which is the stable degradation product of 2-aminoimidazolone (Iz). Of some importance to understanding the determinants of the spectrum of DNA lesions at sites of inflammation is the fact that 8-oxo-G, a major primary product of G oxidation, is at least 1000-fold more reactive than the parent G [53] toward further oxidation ($E° = 0.74$ V versus NHE [54]) and its oxidation gives rise to a variety of more stable secondary products (Figure 2.2). Of these products, only the diastereomeric spiroiminodihydantoin (Sp) lesions have been detected in cells [55]. Interestingly, oxidation of G and 8-oxo-G by "pure" CO$_3^{·-}$ results in the predominant formation of the Sp lesions (40–60%) [56]. The spectrum of G oxidation products

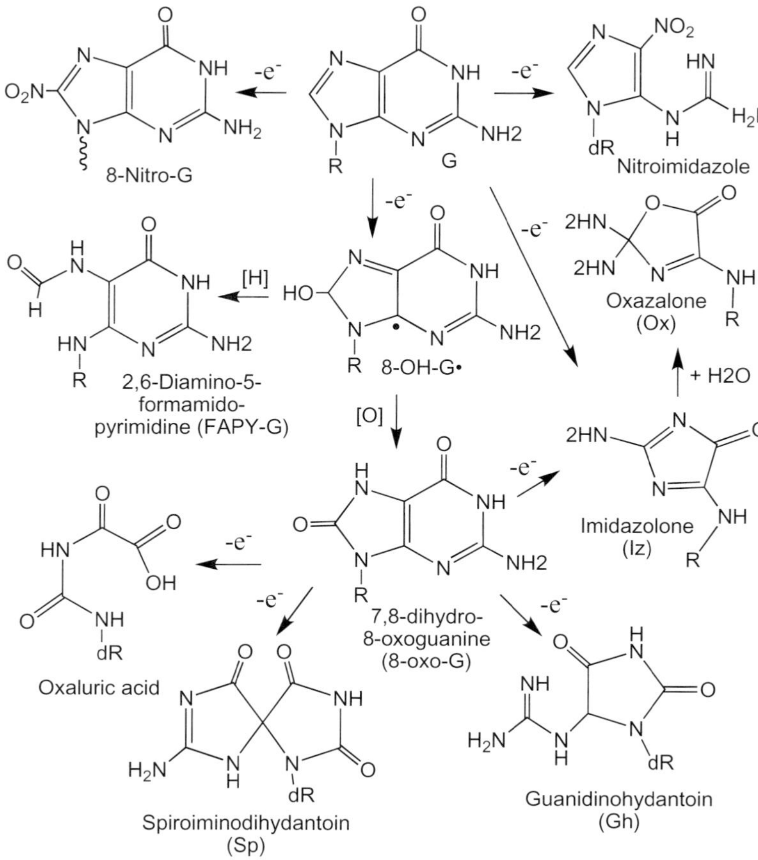

Figure 2.2 Products of G oxidation by RNS.

arising from reactions with ONOO⁻ and ONOOCO₂⁻ stands in contrast to the three major G oxidation products that arise from γ-irradiation of DNA: N-2,6-diamino-4-hydroxy-5-formamidopyrimidine (Fapy-G), 8-oxo-G, and Ox [57, 58]. The formamidopyrimidine lesions of G and A have not been observed among the major DNA oxidation products associated with RNS and inflammation, but further studies are needed [5].

An important feature of the spectrum of G oxidation products arising from reactions with RNS is the concentration dependence of the distributions of the different lesions. The products generated from the oxidation of G by ONOO⁻ depends on oxidant flux and concentration, since the formation of 8-oxo-G and Ox is dose-dependent [59]; the products dehydroguanidinohydantoin (DGh), 2,4,6-trioxo[1,3,5]triazinane-1-carboxamidine (CAC), and NO₂-DGh predominate at high ONOO⁻ fluxes [60, 61], while Sp and guanidinohydantoin (Gh) predominate at low fluxes and concentrations [59–61]. It is thus likely that the latter lesions, along with the direct G oxidation products, predominate in biological systems. These dose–response effects are likely due to the complexity of ONOO⁻ chemistry, with reactions of ONOO⁻ involving both oxidation and nucleophilic reactions in competition with water, and also the presence of NO$_2^•$ coupled with •OH oxidation reactions.

2.2.2
Base Oxidation Products as Biomarkers of Inflammation and Oxidative Stress

Recent discoveries about the chemistry of G oxidation and nitration by RNS and ROS justify a reconsideration of the use of 8-oxo-G as a biomarker of inflammation and oxidative stress. The relative ease of G oxidation to form 8-oxo-G and the relatively low reduction potential of 8-oxo-G pose problems for its use as a biomarker [41, 53]. The problem of 8-oxo-G formation as an artifact of DNA manipulation has been thoroughly addressed by the European Standards Committee on Oxidative DNA Damage comparison of identical samples analyzed in different laboratories using different methods [62, 63]. The results revealed mainly operator-dependent variations in 8-oxo-G levels over a 1000-fold range [63–67]. This problem is minimized by using precautions such as low temperatures, and the use of antioxidants and metal-chelating agents such as desferrioxiamine during the cell or tissue processing and DNA purification steps [64, 68–72].

The low stability of 8-oxo-G with respect to further oxidation is another complication. The relatively low redox potential of 8-oxo-G makes it significantly more susceptible to further oxidation than G by even weak oxidants such as NO$_2^•$ (1.04 V versus NHE [50]) and alkyl hydroperoxides (around 0.9 V versus NHE [73]). As discussed previously, the oxidation of 8-oxo-G leads to several new products (Figure 2.2) [5, 37, 74], most of which are more stable than 8-oxo-G itself, and thus potentially better candidates as biomarkers of inflammation and oxidative stress. However, even a 1000-fold difference in reactivity between G and 8-oxo-G means that G is still the major target when 8-oxo-G is present in DNA at the low steady-state levels of one lesion per 10^6 dG. Furthermore, the use of reducing agents (i.e., antioxidants) during the DNA isolation steps could lead to the reductive consumption of 8-oxo-G. The ease of both oxidation and reduction of 8-oxo-G poses a problem for developing analytical methods that yield the lowest possible background steady-state levels of 8-oxo-G. The question is how low is low enough? Further work is needed to define the quantities of 8-oxo-G and its secondary oxidation products in biological systems.

2.2.3
Charge Transfer as a Determinant of the Location of G Oxidation Products in DNA

The previous discussion of the selective oxidation of G in DNA by RNS was based on the significant differences in redox potentials of the canonical nucleobases. However, the secondary structure of DNA can significantly alter the redox potential of G oxidation in at least two ways: sequence context effects on the redox potential of G and charge transfer along the helix. The role of charge transfer in

DNA has been well characterized in model systems of one-electron oxidation involving photo-oxidation [75], and reactions mediated by pterins [76], anthraquinones [77], rhodium complexes [78], and riboflavin [79]. All of these agents oxidize G to form the G radical cation (G$^{•+}$), with the resulting electron hole migrating through the π-stack of B-DNA in competition with trapping to form stable products [77–79]. In all cases, it has been commonly observed that stable damage occurs at "hotspots" containing multiple adjacent Gs (e.g., GG, GGG). The basis for this selectivity has been proposed to involve sequence-specific variation in G ionization potentials, as calculated by Saito et al. [79, 80] and Senthilkumar et al. [81]. Saito et al. correlated the sequence-dependent ionization potential with the relative reactivity of G toward riboflavin-mediated photo-oxidation and observed an inverse correlation between ionization potential and reactivity, which is consistent with the observation of high levels of damage at runs of G – a sequence context that is characterized by the lowest ionization potential values of G [75, 79].

The role of charge transfer as a major determinant of G oxidation chemistry by RNS was called into question by recent studies [82]. Contrary to other one-electron oxidants, damage products arising from oxidation of G in DNA by $ONOOCO_2^-$ were found to occur most frequently in sequence contexts that conferred the highest G ionization potential (i.e., XGC), with low levels of damage at runs of G [82]. The spectrum of G oxidation products was also found to vary as a function of sequence context [82]. The basis for this phenomenon may lie with the generation of $NO_2^•$ along with the G-oxidizing $CO_3^{•-}$ in the decomposition of $ONOOCO_2^-$. Studies of "pure" $CO_3^{•-}$ generated by laser-induced photochemical reactions have revealed that the rate constant for the one-electron oxidation event is similar for contiguous and isolated Gs, but the subsequent charge transfer favors product formation at runs of G, and the sequence dependence of initial oxidation and final product formation are not correlated [83]. Further studies are needed to clarify the mechanisms involved in G oxidation.

2.3
DNA Deamination

Nitrosative deamination of nucleobases in DNA represents a second major form of DNA damage caused by RNS. The auto-oxidation of macrophage-generated NO leads to the formation of nitrous anhydride (N_2O_3) – a short-lived ($t_{1/2} \sim 0.2$ ms [84]) but potent nitrosating agent (reviewed in [5]). The reactivity of N_2O_3 makes it difficult to define intracellular concentrations, but recent calculations assuming a NO steady-state concentration of 1 µM, which lies at the upper end of estimates of biological levels of NO, are in the femtomolar range [39]. While the studies of Moeller et al. suggest a high rate of N_2O_3 formation in membranes due to the lipid solubility of NO and O_2 [85], an accounting of other reactions of NO with a more complete array of cellular constituents under biological conditions reveals that the membrane effects have little influence on the steady-state concentrations of N_2O_3 in cells [39]. To complicate matters further, the complexity of nucleobase deamina-

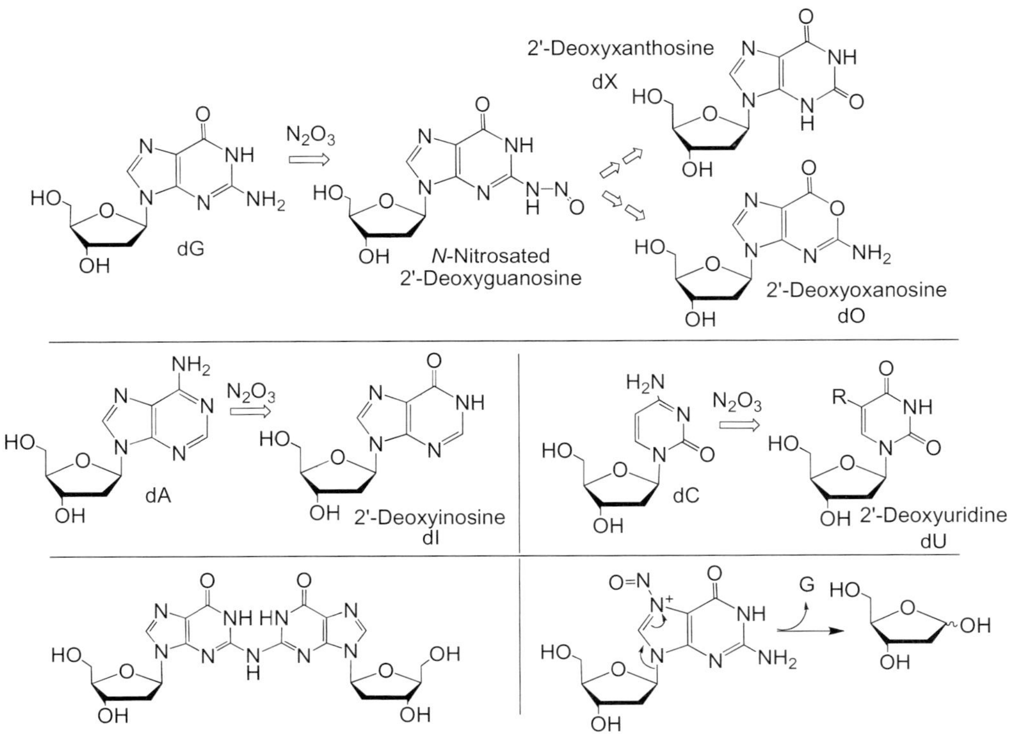

Figure 2.3 Nitrosative deamination of nucleobases in DNA.

tion derives from the many mechanisms affecting DNA, RNA, and the nucleotide pool, which include simple hydrolysis, enzymatically mediated deamination, and misincorporation of purine nucleotide biosynthetic intermediates. These are discussed in numerous review articles covering the various aspects of DNA and RNA deamination [5, 86–94].

The chemistry of deamination of nucleobases in DNA by N_2O_3 involves nitrosation of exocyclic amines of G, A, and C, with subsequent reactions leading to formation of hypoxanthine (2′-deoxyinosine/dI and inosine/rI as nucleosides) derived from A; uracil (2′-deoxyuridine/dU, uridine/rU) from C; xanthine (2′-deoxyxanthosine/dX, xanthosine/rX) and oxanine (2′-deoxyoxanosine/dO, oxanosine/rO) derived from G; and thymine from 5-methyl-C (Figure 2.3); as well as several novel derivatives including abasic sites, and inter- and intrastrand GG cross-links [5]. One factor contributing to the relatively slow progress in studies of nucleobase deamination has been the unsubstantiated claim that dX is too prone to rapid depurination to pose a mutagenic threat to cells [95, 96]. Several recent studies, however, revealed that, at neutral pH, dX is a relatively stable lesion capable of contributing to the mutagenic burden of cells [96–98], with a half-life of around 2 years under biological conditions [98].

A major problem with studies of nitrosative DNA damage is the means by which the gaseous NO is delivered into solution. For example, Nguyen et al. exposed DNA and TK6 cells to an approximately 20 mM total bolus (syringe) dose of NO, and found that dX and dI were formed to the extent of three and 10 lesions per 10^3 nucleotides, respectively, for isolated DNA, with 3-fold lower levels in cells [99]. Yields of dI and dX in DNA were found to be 15–100 times higher, respectively, than those observed with free dA and dG [99]. Similarly, Wink et al. found five dU per 10^3 nucleotides in DNA exposed to NO by bubbling the gas into solution until a 1 M concentration had been absorbed [100]. The use of NO-generating compounds, such as "NONOates," S-nitrosoglutathione, glyceryl trinitrate, and sodium nitroprusside, has advantages over these other methods in terms of a more controlled delivery of biologically relevant concentrations of NO, but they suffer from side-reactions involving generation of other radical species and some require bioactivation [101–103].

To avoid these problems, Deen et al. developed a Silastic tubing-based reactor to deliver controlled and biologically relevant steady-state concentrations of NO (submicromolar) and O_2 (100–200 μM) over prolonged periods [104]. Dong et al. used this reactor to define the spectrum of nitrosative DNA damage products arising under conditions approaching physiological: 0.7 μM NO and 180 μM O_2 [105]. While dX, dI, and dU were formed at nearly identical rates ($k = 1.2 \times 10^5 \, M^{-1} s^{-1}$), dO was not observed above the limit of detection of the liquid chromatography/mass spectrometry (LC/MS) analytical method (less than six lesions per 10^8 nucleotides) except when the DNA was exposed to nitrite at a pH < 4 [106]. Another important observation was the NO-induced production of abasic sites, which likely arise by nitrosation of the N7 positions of G and A with subsequent depurination. In conjunction with other studies of nitrosatively-induced GG cross-links [107], these results predicted the following spectrum of nitrosative DNA lesions in inflamed tissues: around 2% GG cross-links, 4–6% abasic sites, and 25–35% each of dX, dI, and dU. However, this prediction was not sustained in subsequent studies in cells and tissues.

Several studies have sought to measure DNA deamination products *in vivo*. For example, using human TK6 lymphoblastoid cells, a 12-h exposure to steady-state concentrations of NO (0.65 μM) and O_2 (190 μM) caused 1.7-, 1.8-, and 2.0-fold increases in dX, dI, and dU, respectively [108]. These results revealed that the cellular environment provides a roughly 4-fold protective effect against nitrosative deamination [106, 108], with significant elevations of X, I, and U only when the cells are exposed to toxic concentrations of NO and associated N_2O_3 [108]. Similar results have been obtained in animal models of inflammation [70, 109]. For example, studies with the SJL mouse model of NO overproduction revealed only modest increases in the nucleobase deamination products (10–30%) in inflamed tissues compared to control mice [70]. It should be noted that, in these studies, steady-state levels of these products are dependent both on the rate of formation and on the rate of repair or elimination. Active DNA repair pathways could explain a modest increase in the observable damage. Further studies will be needed to dissect these contributions. Interestingly, similar deamination products were seen from extracted RNA, although with a more pronounced effect, suggesting that RNA might be a more available target compared to DNA, which is sequestered and thus more protected in the nucleus (data not shown). Studies are also underway to explore any direct effects on the nucleotide pool as a further genetic target of RNS. Thus, the significance of N_2O_3 in the context of chronic inflammation remains unclear, but many of the technical approaches to sensitivity and specificity have undergone significant improvements in the last few years, allowing more rigorous identification and quantitation.

2.3.1
Problem of Oxanine

It was recognized a decade ago that nitrosative deamination of G by acidified nitrite led to the formation of two products: xanthine and oxanine (O) [110, 111]. Shuker *et al.* also observed the formation of O base in reactions of 2′-deoxyribonucleotides and calf thymus DNA with millimolar concentrations of the mutagenic nitrosating agent 1-nitrosoindole-3-acetonitrile under weakly buffered conditions (0.5 mM Tris) [112, 113]. These observations have led several groups to define the physicochemical and biological properties of O [96, 114–133].

The problem here is that O has never been detected in any biological system. Using highly sensitive high-performance liquid chromatography (HPLC)-coupled single and tandem quadrupole mass spectrometric methods (i.e., LC/MS and LC/MS-MS) with isotopically labeled internal standards, we were unable to detect O in DNA under biologically relevant conditions in purified DNA and cells exposed to high levels of NO and O_2 *in vitro* [106, 108] or in tissues from a mouse model of NO overproduction [70]. The fact that O has only been detected in DNA exposed to NO and nitrite at acidic pH suggests an acid-dependent partitioning of the G deamination chemistry. To explain the various observations, Glaser *et al.* have proposed a model in which, under conditions of neutral pH, the initially formed diazonium ion at N2 of G cannot undergo reactions leading to O due to the conformational restriction of double-stranded DNA and catalytic interference from the base-paired C [105]. The model accounts for the observed deamination products under different conditions and predicts that significant levels of O should form from G in nucleosides, nucleotides, and single-stranded DNA under conditions of nitrosative stress [105].

2.3.2
Analytical Methods and Artifacts

Among the variety of methods that have been developed to quantify nucleobase deamination products, including simple HPLC quantification [134], ^{32}P-postlabeling of 2′-deoxyribonucleosides released from DNA [135] and uptake and labeling of DNA with [^{3}H]uridine [136] or [^{3}H]2′-deoxyuridine [137], all lack the specificity and sensitivity required to detect background levels of dU in DNA. This problem has been overcome with the development of methods involving gas or liquid chromatography coupled to a mass spectrometer (i.e., GC/MS, LC/MS, LC/MS-MS). These approaches combine the resolving power of chromatography with the sensitivity and rigor of MS for identifying specific chemical species, and all possess sensitivities to detect at least five lesions per 10^8 nucleotides in 50 μg of DNA [106, 109, 138]. GC/MS approaches have been used to quantify U [139, 140], X [99, 141], and I [99, 141] as free bases, while LC/MS and LC/MS-MS methods have been developed for O base and dO [70, 106, 110, 112, 113, 142], I in RNA [143], and dX, dI, and dU in DNA [70, 106, 142].

A major problem associated with all of these methods involves adventitious deamination of nucleobases in DNA, and probably RNA, by deaminase enzymes present in all cells. This is illustrated by activation-induced cytidine deaminase (AID) that converts C to U in DNA as part of the immunoglobulin diversification process in B lymphocytes [144, 145], as well as the A deaminase involved in RNA editing [88, 146, 147] and purine nucleotide metabolism, and G deaminase involved in purine base salvage pathways [148, 149]. These enzymes have the potential to cause deamination of nucleobases in DNA during cell manipulations, DNA isolation, and DNA processing. We have experienced significant dA deaminase activity as a contaminant of commercial sources of alkaline phosphatase [106], as well as adventitious dC deaminase activity in human B lymphoblastoid TK6 cells [108], which may be related to AID protein in B cells [144, 145]. To overcome the problem of undesirable deaminase activity, several groups have employed deaminase inhibitors. Both coformycin, an inhibitor of the adenosine deaminase [150], and tetrahydrouridine, an inhibitor of cytidine deaminase, have been used to obtain essentially complete inhibition of deamination artifacts [70, 106, 108, 142]. Lim et al. have used *erythro*-9-(2-hydroxy-3-nonyl)adenine to inhibit adenosine deaminase [109, 138]. In addition to deaminases, deamination artifacts have also been observed as a result of the acid hydrolysis step frequently employed to obtain free bases for GC/MS analysis, and the high temperatures typically used for the derivatization step can accelerate hydrolytic deamination of G and A [109]. Reducing reaction temperatures for both the acid hydrolysis and derivatization reactions minimized artifact formation, but did not prevent it [109].

2.4
2′-Deoxyribose Oxidation

As noted earlier, ONOO$^-$ preferentially oxidizes 2-deoxyribose in DNA compared to nucleobases. While 2′-deoxyribose oxidation is often dismissed as "strand breaks," there is growing evidence that damage to the sugar moieties of DNA by endogenous processes plays an important role in the genetic toxicology of inflammation and oxidative stress. Oxidation of the five carbons in 2-deoxyribose leads to the formation of a variety of reactive electrophiles, both freely diffusible and covalently attached to DNA, as well as sugar fragments that require removal by DNA repair enzymes. The electrophilic 2′-deoxyribose oxidation products have been found to form protein–DNA cross-links, and protein and DNA adducts, while

2′-deoxyribose oxidation products, in general, contribute to complex DNA lesions caused by ionizing radiation and ·OH radicals, with closely opposed strand breaks and oxidized abasic sites [151, 152]. The chemical biology of 2′-deoxyribose oxidation has been reviewed recently [153], so this portion of the chapter on RNS and DNA damage will only survey the product spectrum.

Oxidation of each of the five positions in 2-deoxyribose in DNA occurs with ·OH and other strong oxidants with an initial hydrogen atom abstraction to form a carbon-centered radical that adds molecular oxygen to form a peroxyl radical [153]. The degradation of the peroxyl radical results in the formation of a variety of electrophilic and genotoxic products that differ for each position, with the more biologically relevant spectrum of products formed under aerobic conditions shown in Figure 2.4 [153]. The sole product arising from the 1′-oxidation of DNA by a variety of oxidants is the 2′-deoxyribonolactone abasic site (Figure 2.4) [154–159], which is more stable than a native abasic site (around 10- to 50-fold [160]). The ribonolactone undergoes a rate-limiting β-elimination reaction to form a butenolide species with a half-life of 20 h in single-stranded DNA and 32–54 h in duplex DNA, followed by a rapid δ-elimination to release electrophilic 5-methylene-2(5H)-furanone [160]. One potentially important consequence of the ribonolactone abasic site involves the observed formation of DNA–protein cross-links with DNA repair proteins, which involves a nucleophilic attack by a lysine on the carbonyl of the ribonolactone [161, 162].

Oxidation of the 2′-position is perhaps the least-studied 2′-deoxyribose oxidation chemistry and results in the formation of a D-erythrose abasic site (Figure 2.4) [163, 164]. With implications for the biological response to 2′-deoxyribose oxidation, this abasic site is substantially more stable to hydrolysis than the native and other oxidized abasic sites, with a half-life of 3 h in 0.1 M NaOH at 37 °C [165].

As shown in Figure 2.4, oxidation of the 3′-position has recently been demonstrated to partition along two pathways to form a strand break with 3′-phosphoglycolaldehyde, 5′-phosphate, and base propenoic acid residues; or a 3′-oxo-nucleoside residue that undergoes β/δ-eliminations to release 2-methylene-3(2H)-furanone [166–170]. An interesting feature of the 3′-chemistry involves "migration" of the radical to other sites in DNA. As proposed by Bryant-Friedrich *et al.*, the peroxyl radical formed initially at the 3′-position following 3′-oxidation causes intranucleotide abstraction of the 4′- or 5′-hydrogen atoms to produce the relevant 4′- and 5′-oxidized residues [169, 170]. Another interesting consequence of 3′-oxidation involves the conversion of the 3′-phosphoglycolaldehyde residue to glyoxal. Kasai *et al.*, originally demonstrated that glyoxal and its 1,N^2-imidazo adducts of dG (Figure 2.5) were generated in DNA subjected to oxidation by Fe(II)-EDTA [171]. We later observed that this glyoxal was derived from the 3′-phosphoglycolaldehyde residue (Figure 2.4) by a mechanism involving a radical-independent phosphate–phosphonate rearrangement that led to oxidation of the glycolaldehyde to glyoxal [172].

Recent studies have revealed novel facets of 4′-oxidation of 2-deoxyribose in DNA. In all cases, one pathway leads to a 2′-deoxypentos-4-ulose abasic site (Figure 2.4). In addition, the other two pathways both entail formation of a strand break

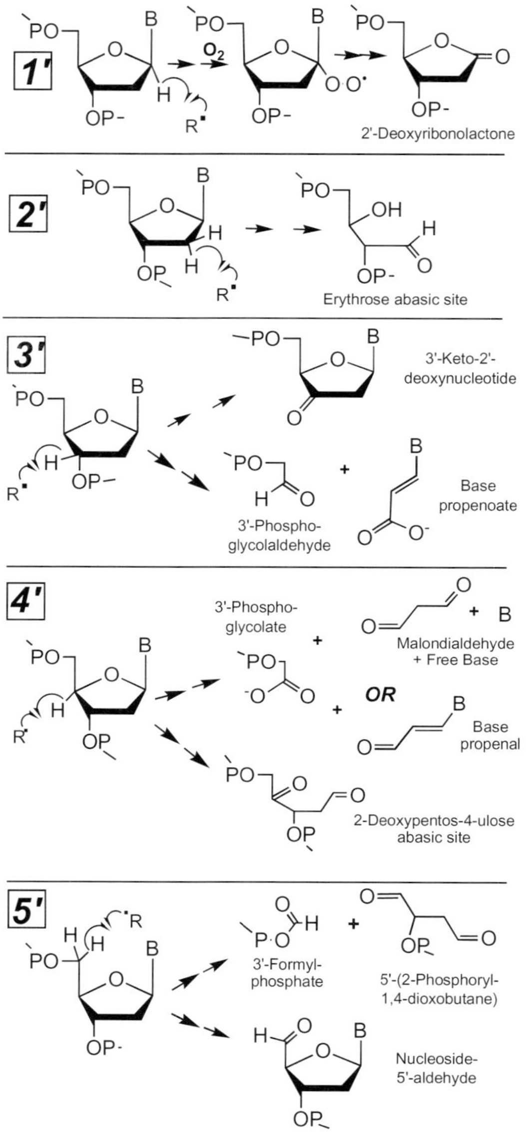

Figure 2.4 Chemistry of 2′-deoxyribose oxidation in DNA.

with a 3′-phosphoglycolate residue. However, the partner products to this residue vary depending on the oxidizing agent [5, 173, 174]. For peroxynitrite, bleomycin, and enediyne antibiotics, the variable product consists of a base propenal [5, 175], while γ-radiation and Fe^{+2}-EDTA were found to give rise to malondialdehyde and a free nucleobase [173, 174]. As discussed later in Section 2.5.1, base propenals

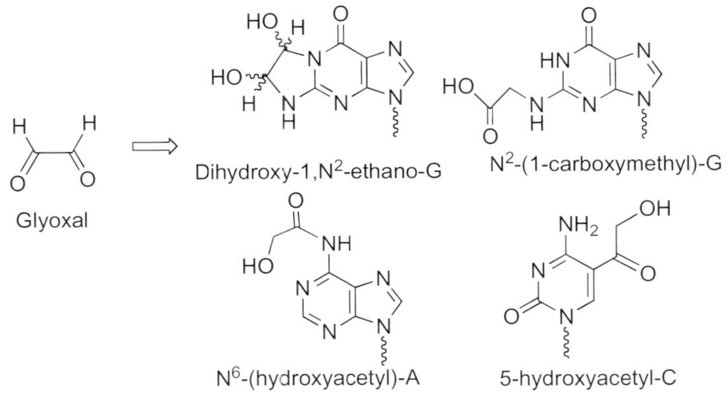

Figure 2.5 DNA adducts derived from glyoxal.

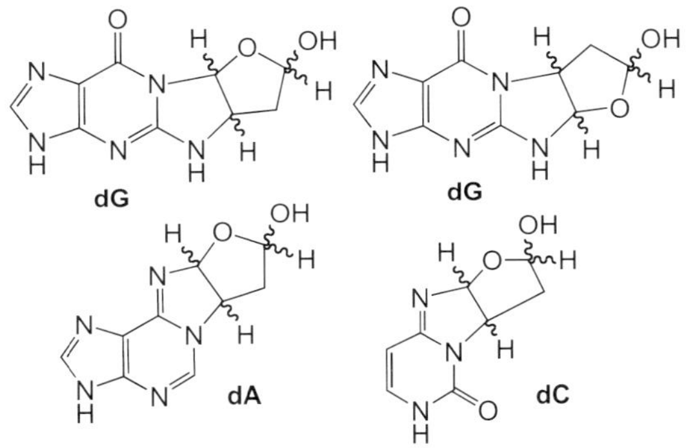

Figure 2.6 DNA adducts derived from the 5′-(2-phosphoryl-1,4-dioxobutane) residue of 5′-oxidation.

and the β-hydroxyacrolein tautomer of malondialdehyde are structural analogs that react with G to form the M_1G adduct [173, 176–178], with base propenals around 100-fold more reactive than malondialdehyde [176–178] and accounting for the majority of M_1G in cells [173].

5′-Oxidation also partitions to form two sets of products: a 3′-formylphosphate-ended fragment and a 2-phosphoryl-1,4-dioxo-2-butane residue that undergoes β-elimination to form a *trans*-butenedialdehyde species; or a nucleoside-5′-aldehyde residue that can undergo β/δ-eliminations to produce furfural (Figure 2.4) [175, 179]. The electrophilic species generated by 5′-oxidation have been found to form both DNA and protein adducts. In terms of DNA adducts, the 5′-(2-phosphoryl-1,4-dioxobutane) residue reacts with dC, dG, and dA to form bicyclic oxadiazabi-cyclo-(3.3.0)octaimine adducts illustrated in Figure 2.6 [180–184]. A second biological consequence of 5′-oxidation involves the reaction of the 3′-formylphosphate residue with ε-amino group of lysine in histone proteins (Figure 2.7) [185].

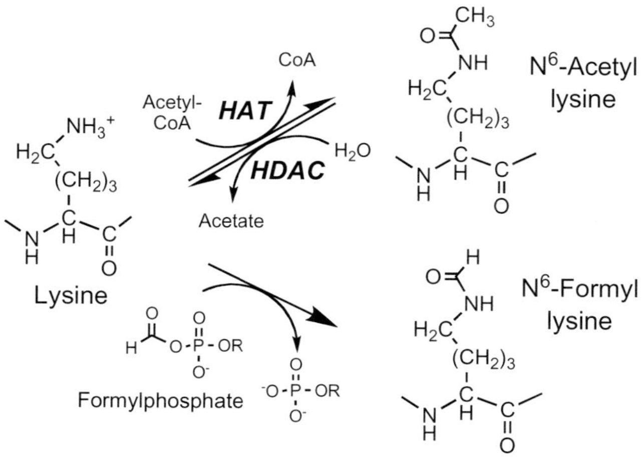

Figure 2.7 Lysine N-formylation by the 3′-formylphosphate residue of 5′-oxidation. HAT, histone acetyltransferase; HDAC, histone deacetylase.

Its chemical analogy to the physiologically important lysine N-acetylation suggests that lysine N-formylation may interfere with signaling mediated by histone modifications.

2.4.1
Variation of 2′-Deoxyribose Oxidation Chemistry as a Function of the Oxidant

Two sets of observations illustrate variations in 2′-deoxyribose oxidation chemistry as a function of the identity of the oxidant – a phenomenon that may complicate our understanding of the cellular responses to 2′-deoxyribose oxidation. The first observation involves variation in 4′-chemistry. As noted earlier, the presumably "pure" hydroxyl radicals produced by ionizing radiation and Fe^{+2}-EDTA cause 4′-oxidation consisting of the abasic site common to all oxidants, but the strand break pathway is comprised of 3′-phosphoglycolate, free base, and malondialdehyde. This contrasts with 4′-oxidants such as bleomycin and $ONOO^-$ that produce 3′-phosphoglycolate and a base propenal species (Figure 2.4). The mechanistic basis for this different partitioning of 4′-oxidation chemistry is unclear, but the consequences are apparent in the formation of the M_1G adduct, as discussed in Section 2.5.1. A second example involves recent studies suggesting that the identity of the oxidant also affects the overall proportions of oxidation of the five carbons of 2-deoxyribose. For example, the yield of 3′-phosphoglycolaldehyde residues varies for γ- and α-radiation [168]. While α-particle irradiation of DNA caused 0.13 residues/10^6 nucleotides/Gy, γ-radiation produced 1.5 residues/10^6 nucleotides/Gy [167, 168]. Correction for the different yields of total 2′-deoxyribose oxidation events for α- and γ-radiation revealed that the former is 7-fold more efficient at producing the 3′-phosphoglycolaldehyde than the latter, probably as a result of its higher linear energy transfer properties [167, 168]. Whether this difference reflects a shift between the several branches of 3′-chemistry or between the various sites in 2-deoxyribose remains to be determined.

2.5
Indirect Base Damage Caused by RNS

Beyond damage originating from direct attack on DNA by RNS, recent work has highlighted the importance of reactions of DNA with electrophilic products derived from oxidation of other cellular components, such as polyunsaturated fatty acids, proteins, and carbohydrates. For example, peroxidation of polyunsaturated fatty acids leads to a host of electrophiles capable of forming DNA adducts, with several recent reviews addressing the full spectrum of DNA adducts derived from the diversity of lipid peroxidation products, as well as the methods to quantify them in biological systems [186–188]. For example, linoleic acid, the most common polyunsaturated fatty acid in mammalian cell membranes, gives rise to several α,β-unsaturated aldehydes (Figure 2.8), such as *trans*-4-hydroxy-2-nonenal, acro-

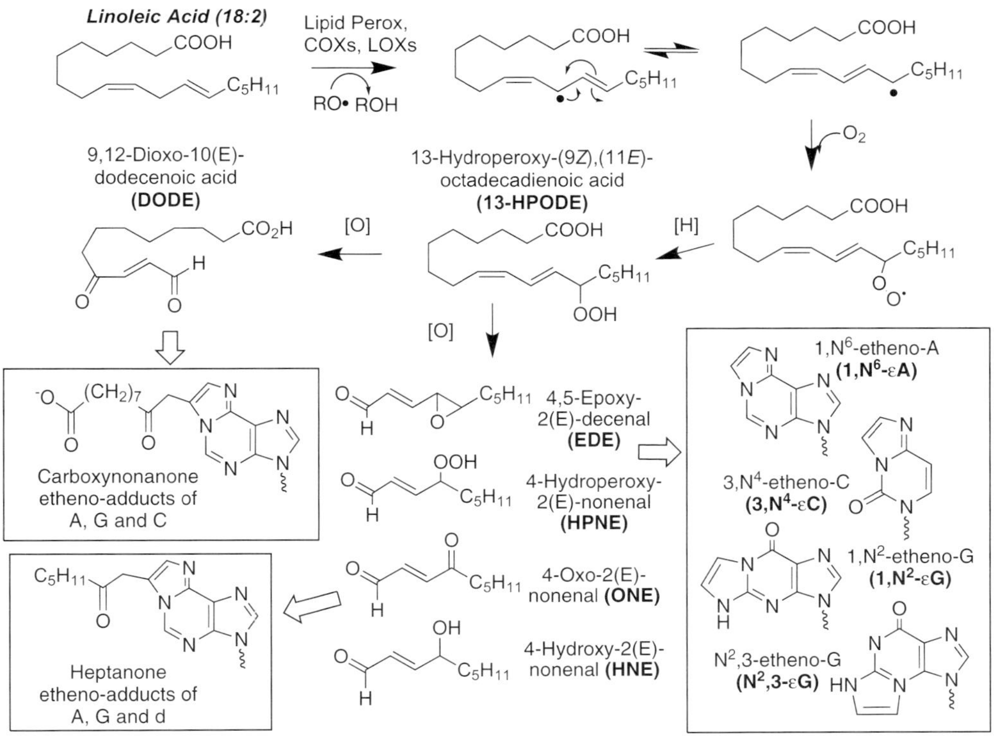

Figure 2.8 Lipid peroxidation-derived etheno adducts.

lein, and 4-oxo-2-nonenal [186, 189] that can react with A, G, and C to form substituted and unsubstituted etheno adducts. As shown in Figure 2.8, the unsubstituted etheno adducts are comprised of $1,N^6$-etheno-2′-deoxyadenosine, $3,N^4$-etheno-2′-deoxycytidine, and $1,N^2$-etheno-2′-deoxyguanosine [190–193]. Elevated levels of these lesions have been found under conditions of oxidative stress in human and mouse tissues [70, 186, 187, 194–196].

The more recently discovered substituted etheno adducts present a complicated situation due to the many possible structural permutations with the variety of polyunsaturated fatty acids and potential peroxidation products. The Blair group has identified two types of substituted etheno adducts *in vitro* and *in vivo*: heptanone and carboxynonanone adducts (Figure 2.8) [188, 191]. The reaction of 13-HPODE (Figure 2.8) or its degradation product, oxononenal, with G, C, and A leads to the formation of the respective heptanone adducts [188, 190, 197, 198], while 13-HPODE-derived DODE gives rise to the analogous carboxynonanone DNA adducts [199]. The lipid peroxidation products arising from reactions of cyclo-oxygenase and lipoxygenase with polyunsaturated fatty acids are stereospecific in nature, which suggests that the various stereoisomers of the lipid peroxidation products and their DNA adducts may be useful as biomarkers capable of distinguishing enzymatic versus chemically derived lipid damage products [188, 191].

This picture is further complicated by the reaction of α,β-unsaturated carbonyl-containing lipid peroxidation products with DNA to form more simple adducts resulting from Michael addition reactions such as the propano adducts shown in Figure 2.9. These DNA lesions are derived from hydroxynonenal, heptenal, pentenal, crotonaldehyde, and acrolein with strong regioselectivity. For example, HNE, crotonaldehyde, and other substituted enals (e.g., pentenal, heptenal) all react to form the 8-hydroxy-propano adducts with G in DNA (Figure 2.9), while acrolein has been observed to form both the 8- and 6-hydroxy-propano adducts, the latter by Michael addition starting at the heterocyclic nitrogen (e.g., N1 of G) [200–205]. Similar adducts form with A and C, though these are less well explored, while T forms the simple Michael adduct via its heterocyclic nitrogen (Figure 2.9) [201, 205–210]. With implications for the biological effects of the propano adduct

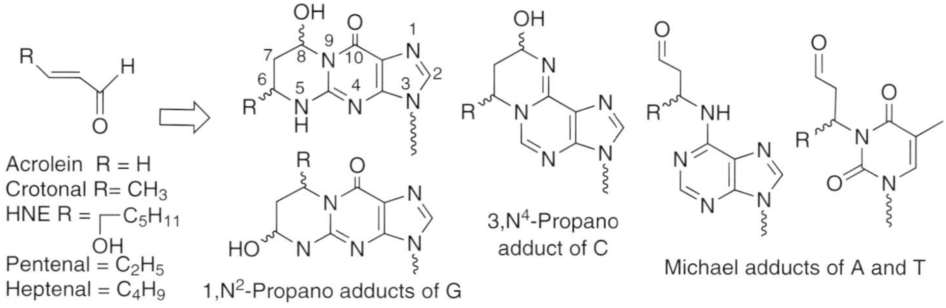

Acrolein R = H
Crotonal R = CH_3
HNE R = $-C_5H_{11}$ with OH
Pentenal = C_2H_5
Heptenal = C_4H_9

$1,N^2$-Propano adducts of G

$3,N^4$-Propano adduct of C

Michael adducts of A and T

Figure 2.9 Propano adducts.

structure in general, it has been observed that all types of propano adducts undergo ring-opening to aldehydic forms in DNA, with subsequent formation of cross-links with proteins and neighboring DNA bases [211–221]. With the advent of ultrasensitive analytical methods to quantify the propano adducts in cells and tissues [187], including LC/MS-MS [222] and ^{32}P-postlabeling [223, 224], the lesions have been detected in a variety of rodent and human tissues [187, 206, 222, 224–227].

2.5.1
Malondialdehyde and Related Adducts

The lipid peroxidation product, malondialdehyde, has been the subject of significant study for two decades since it was discovered that it reacts with DNA bases to form a variety of adducts, the most well characterized of which is the pyrimidopurinone adduct of G, M_1G (Figure 2.10). As such, the so-called malondialdehyde adducts have been discussed in numerous review articles [153, 187, 189, 228–233], as well as in Part One, Chapter 5 of this book. When base-paired in DNA or under basic conditions, M_1dG undergoes hydrolytic ring-opening to N^2-oxopropenyl-dG [234–239] – a phenomenon that has been exploited to develop an analytical method for quantifying M_1G [71, 240, 241]. Further, the exocyclic ring of M_1G is susceptible to direct nucleophilic attack with subsequent ring-opening [235], which suggests that the alkyl adduct has the potential to migrate to neighboring protein or nucleic acid nucleophiles.

While the M_1G adduct was originally described as a product of DNA reaction with malondialdehyde *in vitro*, recent observations suggest a more complicated chemical basis for its formation *in vivo* [173, 241]. As discussed in Section 2.4, base propenals arising from 4′-oxidation of DNA (Figure 2.4) are structural analogs of the DNA-reactive β-hydroxyacrolein enol tautomer of malondialdehyde (Figure 2.10) and have been shown to be significantly more reactive in forming M_1G in DNA as a result of the presence of the nucleobase as a better leaving group than hydroxide [176]. More recent studies in *Escherichia coli* revealed that base propenals, and not malondialdehyde, were the major source of M_1G in cells subjected to oxidative stress [173]. In light of the potential "mobility" of M_1G [234–239] and

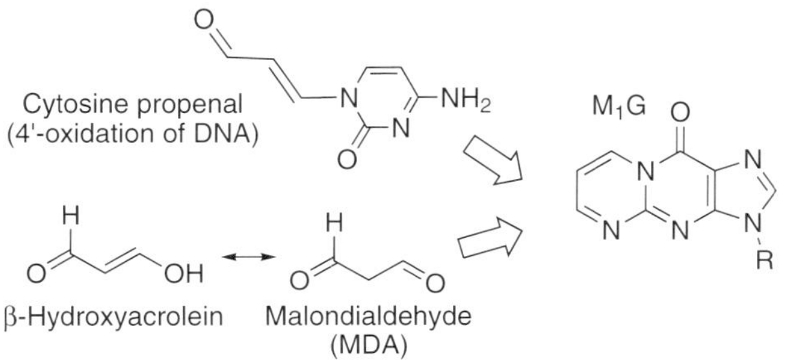

Figure 2.10 Formation of M_1dG.

the potential for transfer of the oxopropenyl group to and from DNA via N^ε-oxopropenyllysine adducts in histone proteins [178], it will be difficult to precisely define the source of M_1G adducts *in vivo*.

2.6
Conclusions

This chapter has addressed only a small portion of the emerging picture of the cellular damage produced by RNS. The spectrum of different DNA damage products, both direct and indirect, is expanding, as is the number of non-DNA targets for electrophilic products arising from attack of RNS on other cellular biomolecules, which has complicated models attempting to link inflammation and oxidative stress to diseases such as cancer. The formation of non-DNA adducts raises the issue of epigenetic influences of RNS on cell pathology. In this regard, the emergence of more sensitive and specific analytical technologies will provide insights into the interplay between the chemical biology of DNA damage and damage to other cellular components in the pathophysiology of RNS.

Acknowledgements

The authors extend grateful thanks to members of the Dedon research group for helpful discussions and to the National Institutes of Health for generous funding (grants CA026731, ES016450, CA103146, CA116318, ES002109).

References

1 Nathan, C. (2002) Points of control in inflammation. *Nature*, **420**, 846–852.
2 Sawa, T. and Ohshima, H. (2006) Nitrative DNA damage in inflammation and its possible role in carcinogenesis. *Nitric Oxide*, **14**, 91–100.
3 Coussens, L.M. and Werb, Z. (2002) Inflammation and cancer. *Nature*, **420**, 860–867.
4 Tan, T.T. and Coussens, L.M. (2007) Humoral immunity, inflammation and cancer. *Curr. Opin. Immunol.*, **19**, 209–216.
5 Dedon, P.C. and Tannenbaum, S.R. (2004) Reactive nitrogen species in the chemical biology of inflammation. *Arch. Biochem. Biophys.*, **423**, 12–22.
6 Schottenfeld, D. and Beebe-Dimmer, J. (2006) Chronic inflammation: a common and important factor in the pathogenesis of neoplasia. *CA Cancer J. Clin.*, **56**, 69–83.
7 Thun, M.J., Henley, S.J., and Gansler, T. (2004) Inflammation and cancer: an epidemiological perspective. *Novartis Found. Symp.*, **256**, 6–21; discussion 22–28, 49–52, 266–269.
8 Libby, P. (2002) Inflammation in atherosclerosis. *Nature*, **420**, 868–874.
9 Libby, P. (2006) Inflammation and cardiovascular disease mechansisms. *Am. J. Clin. Nutr.*, **83**, 456S–460S.
10 Shibata, N. and Glass, C.K. (2009) Regulation of macrophage function in inflammation and atherosclerosis. *J. Lipid. Res.*, **50**, S277–S281.
11 Olliver, J.R., Hardie, L.J., Gong, Y., Dexter, S., Chalmers, D., Harris, K.M.,

and Wild, C.P. (2005) Risk factors, DNA damage, and disease progression in Barrett's esophagus. *Cancer Epidemiol. Biomarkers Prev.*, **14**, 620–625.
12. Levin, B. (1992) Ulcerative colitis and colon cancer: biology and surveillance. *J. Cell Biochem. Suppl.*, **16G**, 47–50.
13. Farrell, R.J. and Peppercorn, M.A. (2002) Ulcerative colitis. *Lancet*, **359**, 331–340.
14. Asaka, M., Takeda, H., Sugiyama, T., and Kato, M. (1997) What role does helicobacter pylori play in gastric cancer? *Gastroenterology*, **113**, S56–S60.
15. Ebert, M.P., Yu, J., Sung, J.J., and Malfertheiner, P. (2000) Molecular alterations in gastric cancer: the role of *Helicobacter pylori*. *Eur. J. Gastroenterol. Hepatol.*, **12**, 795–798.
16. Mostafa, M.H., Sheweita, S.A., and O'Connor, P.J. (1999) Relationship between schistosomiasis and bladder cancer. *Clin. Microbiol. Rev.*, **12**, 97–111.
17. Badawi, A.F., Mostafa, M.H., Probert, A., and O'Connor, P.J. (1995) Role of schistosomiasis in human bladder cancer: evidence of association, aetiological factors, and basic mechanisms of carcinogenesis. *Eur. J. Cancer Prev.*, **4**, 45–59.
18. Ohshima, H. (2003) Genetic and epigenetic damage induced by reactive nitrogen species: implications in carcinogenesis. *Toxicol. Lett.*, **140–141**, 99–104.
19. Ohshima, H., Tatemichi, M., and Sawa, T. (2003) Chemical basis of inflammation-induced carcinogenesis. *Arch. Biochem. Biophys.*, **417**, 3–11.
20. Bredt, D. and Snyder, S. (1994) Nitric oxide: a physiologic messenger molecule. *Neuron*, **63**, 175–195.
21. Gross, S. and Wolin, M. (1995) Nitric oxide: pathophysiological mechanisms. *Annu. Rev. Physiol.*, **57**, 737–769.
22. Lancaster, J. (1992) Nitric oxide in cells. *Am. Sci.*, **80**, 248–259.
23. MacMicking, J., Xie, Q., and Nathan, C. (1997) Nitric oxide and macrophage function. *Annu. Rev. Immunol.*, **15**, 323–350.
24. Moncada, S., Palmer, R.M., and Higgs, E.A. (1991) Nitric oxide: physiology, pathophysiology, and pharmacology. *Pharmacol. Rev.*, **43**, 109–142.
25. Nathan, C. (1992) Nitric oxide as a secretory product of mammalian cells. *FASEB J.*, **6**, 3051–3064.
26. Thomas, D.D., Ridnour, L.A., Isenberg, J.S., Flores-Santana, W., Switzer, C.H., Donzelli, S., Hussain, P., Vecoli, C., Paolocci, N., Ambs, S., Colton, C.A., Harris, C.C., Roberts, D.D., and Wink, D.A. (2008) The chemical biology of nitric oxide: implications in cellular signaling. *Free Radic. Biol. Med.*, **45**, 18–31.
27. Miwa, M., Stuehr, D.J., Marletta, M.A., Wishnok, J.S., and Tannenbaum, S.R. (1987) Nitrosation of amines by stimulated macrophages. *Carcinogenesis*, **8**, 955–958.
28. Stuehr, D.J. and Marletta, M.A. (1987) Synthesis of nitrite and nitrate in murine macrophage cell lines. *Cancer Res.*, **47**, 5590–5594.
29. Lewis, R.S., Tamir, S., Tannenbaum, S.R., and Deen, W.M. (1995) Kinetic analysis of the fate of nitric oxide synthesized by macrophages *in vitro*. *J. Biol. Chem.*, **270**, 29350–29355.
30. Mancardi, D., Ridnour, L.A., Thomas, D.D., Katori, T., Tocchetti, C.G., Espey, M.G., Miranda, K.M., Paolocci, N., and Wink, D.A. (2004) The chemical dynamics of NO and reactive nitrogen oxides: a practical guide. *Curr. Mol. Med.*, **4**, 723–740.
31. Hughes, M.N. (2008) Chemistry of nitric oxide and related species. *Methods Enzymol.*, **436**, 3–19.
32. van der Vliet, A., Eiserich, J.P., Halliwell, B., and Cross, C.E. (1997) Formation of reactive nitrogen species during peroxidase-catalyzed oxidation of nitrite. A potential additional mechanism of nitric oxide-dependent toxicity. *J. Biol. Chem.*, **272**, 7617–7625.
33. Hazen, S.L., Zhang, R., Shen, Z., Wu, W., Podrez, E.A., MacPherson, J.C., Schmitt, D., Mitra, S.N., Mukhopadhyay, C., Chen, Y., Cohen, P.A., Hoff, H.F., and Abu-Soud, H.M. (1999) Formation of nitric oxide-derived oxidants by myeloperoxidase in monocytes: pathways for monocyte-mediated protein nitration and lipid

peroxidation *in vivo*. *Circ. Res.*, **85**, 950–958.

34 Wu, W., Chen, Y., and Hazen, S.L. (1999) Eosinophil peroxidase nitrates protein tyrosyl residues. Implications for oxidative damage by nitrating intermediates in eosinophilic inflammatory disorders. *J. Biol. Chem.*, **274**, 25933–25944.

35 Eiserich, J.P., Hristova, M., Cross, C.E., Jones, A.D., Freeman, B.A., Halliwell, B., and van der Vliet, A. (1998) Formation of nitric oxide-derived inflammatory oxidants by myeloperoxidase in neutrophils. *Nature*, **391**, 393–397.

36 Cadet, J., Douki, T., Pouget, J.P., and Ravanat, J.L. (2000) Singlet oxygen DNA damage products: formation and measurement. *Methods Enzymol.*, **319**, 143–153.

37 Cadet, J., Douki, T., and Ravanat, J.L. (2008) Oxidatively generated damage to the guanine moiety of DNA: mechanistic aspects and formation in cells. *Acc. Chem. Res.*, **41**, 1075–1083.

38 Pluskota-Karwatka, D. (2008) Modifications of nucleosides by endogenous mutagens–DNA adducts arising from cellular processes. *Bioorg. Chem.*, **36**, 198–213.

39 Lim, C.H., Dedon, P.C., and Deen, W.M. (2008) Kinetic analysis of intracellular concentrations of reactive nitrogen species. *Chem. Res. Toxicol.*, **21**, 2134–2147.

40 Tretyakova, N.Y., Burney, S., Pamir, B., Wishnok, J.S., Dedon, P.C., Wogan, G.N., and Tannenbaum, S.R. (2000) Peroxynitrite-induced DNA damage in the *supF* gene: correlation with the mutational spectrum. *Mutat. Res.*, **447**, 287–303.

41 Burney, S., Niles, J.C., Dedon, P.C., and Tannenbaum, S.R. (1999) DNA damage in deoxynucleosides and oligonucleotides treated with peroxynitrite. *Chem. Res. Toxicol.*, **12**, 513–520.

42 Yermilov, V., Yoshie, Y., Rubio, J., and Ohshima, H. (1996) Effects of carbon dioxide/bicarbonate on induction of DNA single-strand breaks and formation of 8-nitroguanine, 8-oxoguanine and base propenal mediated by peroxynitrite. *FEBS Lett.*, **399**, 67–70.

43 Shukla, P.K. and Mishra, P.C. (2008) Catalytic involvement of CO_2 in the mutagenesis caused by reactions of $ONOO^-$ with guanine. *J. Phys. Chem. B*, **112**, 4779–4789.

44 Armstrong, R. (2001) The physiological role and pharmacological potential of nitric oxide in neutrophil activation. *Int. Immunopharmacol.*, **1**, 1501–1512.

45 Gaut, J.P., Byun, J., Tran, H.D., Lauber, W.M., Carroll, J.A., Hotchkiss, R.S., Belaaouaj, A., and Heinecke, J.W. (2002) Myeloperoxidase produces nitrating oxidants *in vivo*. *J. Clin. Invest.*, **109**, 1311–1319.

46 Masuda, M., Suzuki, T., Friesen, M.D., Ravanat, J.L., Cadet, J., Pignatelli, B., Nishino, H., and Ohshima, H. (2001) Chlorination of guanosine and other nucleosides by hypochlorous acid and myeloperoxidase of activated human neutrophils. Catalysis by nicotine and trimethylamine. *J. Biol. Chem.*, **276**, 40486–40496.

47 Steenken, S. and Jovanovic, S.V. (1997) How easily oxidizable is DNA? One-electron reduction potentials of adenosine and guanosine radicals in aqueous solution. *J. Am. Chem. Soc.*, **119**, 617–618.

48 Buettner, G.R. (1993) The pecking order of free radicals and antioxidants: lipid peroxidation, α-tocopherol, and ascorbate. *Arch. Biochem. Biophys.*, **300**, 535–543.

49 Shafirovich, V., Dourandin, A., Huang, W., and Geacintov, N.E. (2001) The carbonate radical is a site-selective oxidizing agent of guanine in double-stranded oligonucleotides. *J. Biol. Chem.*, **276**, 24621–24626.

50 Stanbury, D.M. (1989) Reduction potentials involving inorganic free radicals in aqueous solution. *Adv. Inorg. Chem.*, **33**, 69–138.

51 Steenken, S., Jovanovic, S.V., Bietti, M., and Bernhard, K. (2000) The trap depth (in DNA) of 8-oxo-7,8-dihydro-2′-deoxyguanosine as derived from electron-transfer equilibria in aqueous solution. *J. Am. Chem. Soc.*, **122**, 2373–2374.

52 Shafirovich, V., Mock, S., Kolbanovskiy, A., and Geacintov, N.E. (2002) Photo-

chemically catalyzed generation of site-specific 8-nitroguanine adducts in DNA by the reaction of long-lived neutral guanine radicals with nitrogen dioxide. *Chem. Res. Toxicol.*, **15**, 591–597.

53 Uppu, R.M., Cueto, R., Squadrito, G.L., Salgo, M.G., and Pryor, W.A. (1996) Competitive reactions of peroxynitrite with 2′-deoxyguanosine and 7,8-dihydro-8-oxo-2′-deoxyguanosine (8-oxodG): relevance to the formation of 8-oxodG in DNA exposed to peroxynitrite. *Free Radic. Biol. Med.*, **21**, 407–411.

54 Yanagawa, H., Ogawa, Y., and Ueno, M. (1992) Redox ribonucleosides. Isolation and characterization of 5-hydroxyuridine, 8-hydroxyguanosine, and 8-hydroxyadenosine from torula yeast RNA. *J. Biol. Chem.*, **267**, 13320–13326.

55 Hailer, M.K., Slade, P.G., Martin, B.D., and Sugden, K.D. (2005) Nei deficient *Escherichia coli* are sensitive to chromate and accumulate the oxidized guanine lesion spiroiminodihydantoin. *Chem. Res. Toxicol.*, **18**, 1378–1383.

56 Joffe, A., Geacintov, N.E., and Shafirovich, V. (2003) DNA lesions derived from the site-selective oxidation of guanine by carbonate radical anions. *Chem. Res. Toxicol.*, **16**, 1528–1538.

57 Pouget, J.P., Frelon, S., Ravanat, J.L., Testard, I., Odin, F., and Cadet, J. (2002) Formation of modified DNA bases in cells exposed either to γ radiation or to high-LET particles. *Radiat. Res.*, **157**, 589–595.

58 Douki, T., Riviere, J., and Cadet, J. (2002) DNA tandem lesions containing 8-oxo-7,8-dihydroguanine and formamido residues arise from intramolecular addition of thymine peroxyl radical to guanine. *Chem. Res. Toxicol.*, **15**, 445–454.

59 Yu, H., Venkatarangan, L., Wishnok, J.S., and Tannenbaum, S.R. (2005) Quantitation of four guanine oxidation products from reaction of DNA with varying doses of peroxynitrite. *Chem. Res. Toxicol.*, **18**, 1849–1857.

60 Niles, J.C., Wishnok, J.S., and Tannenbaum, S.R. (2004) Spiroiminodihydantoin and guanidinohydantoin are the dominant products of 8-oxoguanosine oxidation at low fluxes of peroxynitrite: mechanistic studies with ^{18}O. *Chem. Res. Toxicol.*, **17**, 1510–1519.

61 Niles, J.C., Wishnok, J.S., and Tannenbaum, S.R. (2006) Peroxynitrite-induced oxidation and nitration products of guanine and 8-oxoguanine: structures and mechanisms of product formation. *Nitric Oxide*, **14**, 109–121.

62 EESCOOD (2002) Comparative analysis of baseline 8-oxo-7,8-dihydroguanine in mammalian cell DNA, by different methods in different laboratories: an approach to consensus. *Carcinogenesis*, **23**, 2129–2133.

63 Gedik, C.M. and Collins, A. (2005) Establishing the background level of base oxidation in human lymphocyte DNA: results of an interlaboratory validation study. *FASEB J.*, **19**, 82–84.

64 Cadet, J., D'Ham, C., Douki, T., Pouget, J.P., Ravanat, J.L., and Sauvaigo, S. (1998) Facts and artifacts in the measurement of oxidative base damage to DNA. *Free Radic. Res.*, **29**, 541–550.

65 Cadet, J., Douki, T., and Ravanat, J.L. (1997) Artifacts associated with the measurement of oxidized DNA bases. *Environ. Health Perspect.*, **105**, 1034–1039.

66 Collins, A., Cadet, J., Epe, B., and Gedik, C. (1997) Problems in the measurement of 8-oxoguanine in human DNA. Report of a workshop, DNA Oxidation, held in Aberdeen, UK, 19–21 January, 1997. *Carcinogenesis*, **18**, 1833–1836.

67 Collins, A.R., Cadet, J., Möller, L., Poulsen, H.E., and Viña, J. (2004) Are we sure we know how to measure 8-oxo-7,8-dihydroguanine in DNA from human cells? *Arch. Biochem. Biophys.*, **423**, 57–65.

68 Ravanat, J.L., Douki, T., Duez, P., Gremaud, E., Herbert, K., Hofer, T., Lasserre, L., Saint-Pierre, C., Favier, A., and Cadet, J. (2002) Cellular background level of 8-oxo-7,8-dihydro-2′-deoxyguanosine: an isotope based method to evaluate artefactual oxidation of DNA during its extraction and subsequent work-up. *Carcinogenesis*, **23**, 1911–1918.

69 Helbock, H.J., Beckman, K.B., Shigenaga, M.K., Walter, P.B., Woodall, A.A., Yeo, H.C., and Ames, B.N. (1998) DNA oxidation matters: the HPLC-electrochemical detection assay of 8-oxo-deoxyguanosine and 8-oxoguanine. *Proc. Natl. Acad. Sci. USA*, **95**, 288–293.

70 Pang, B., Zhou, X., Yu, H.-B., Dong, M., Taghizadeh, K., Wishnok, J.S., Tannenbaum, S.R., and Dedon, P.C. (2007) Lipid peroxidation dominates the chemistry of DNA adduct formation in a mouse model of inflammation. *Carcinogenesis*, **28**, 1807–1813.

71 Jeong, Y.C., Nakamura, J., Upton, P.B., and Swenberg, J.A. (2005) Pyrimido[1,2-a]-purin-10(3H)-one, M_1G, is less prone to artifact than base oxidation. *Nucleic Acids Res.*, **33**, 6426–6434.

72 Cadet, J., Douki, T., Gasparutto, D., and Ravanat, J.L. (2003) Oxidative damage to DNA: formation, measurement and biochemical features. *Mutat. Res.*, **531**, 5–23.

73 Das, T.N., Dhanasekaran, T., Alfassi, Z.B., and Neta, P. (1997) Reduction potential of the *tert*-butylperoxyl radical in aqueous solutions. 4th International Conference on Chemical Kinetics. National Institute of Standards and Technology, Gaithersburg, MD, USA.

74 Burrows, C.J., Muller, J.G., Kornyushyna, O., Luo, W., Duarte, V., Leipold, M.D., and David, S.S. (2002) Structure and potential mutagenicity of new hydantoin products from guanosine and 8-oxo-7,8-dihydroguanine oxidation by transition metals. *Environ. Health Perspect.*, **110** (Suppl. 5), 713–717.

75 Yang, X., Wang, X.B., Vorpagel, E.R., and Wang, L.S. (2004) Direct experimental observation of the low ionization potentials of guanine in free oligonucleotides by using photoelectron spectroscopy. *Proc. Natl. Acad. Sci. USA*, **101**, 17588–17592.

76 Ito, K. and Kawanishi, S. (1997) Photoinduced hydroxylation of deoxyguanosine in DNA by pterins: sequence specificity and mechanism. *Biochemistry*, **36**, 1774–1781.

77 Henderson, P.T., Jones, D., Hampikian, G., Kan, Y., and Schuster, G.B. (1999) Long-distance charge transport in duplex DNA: the phonon-assisted polaron-like hopping mechanism. *Proc. Natl. Acad. Sci. USA*, **96**, 8353–8358.

78 Hall, D.B., Holmlin, R.E., and Barton, J.K. (1996) Oxidative DNA damage through long-range electron transfer. *Nature*, **382**, 731–735.

79 Saito, I., Nakamura, T., Nakatani, K., Yoshioka, Y., Yamaguchi, K., and Sugiyama, H. (1998) Mapping of the hot spots for DNA damage by one-electron oxidation: efficacy of GG doublets and GGG triplets as a trap in long-range hole migration. *J. Am. Chem. Soc.*, **120**, 12686–12687.

80 Sugiyama, H. and Saito, I. (1996) Theoretical studies of GG-specific photocleavage of DNA via electron transfer: significant lowering of ionization potential and 5′-localization of homo of stacked GG bases in B-form DNA. *J. Am. Chem. Soc.*, **118**, 7063–7068.

81 Senthilkumar, K., Grozema, F.C., Guerra, C.F., Bickelhaupt, F.M., and Siebbeles, L.D.A. (2003) Mapping the sites of selective oxidation of guanines in DNA. *J. Am. Chem. Soc.*, **125**, 13658–13659.

82 Margolin, Y., Cloutier, J.F., Shafirovich, V., Geacintov, N.E., and Dedon, P.C. (2006) Paradoxical hotspots for guanine oxidation by a chemical mediator of inflammation. *Nat. Chem. Biol.*, **2**, 365–366.

83 Lee, Y.A., Yun, B.H., Kim, S.K., Margolin, Y., Dedon, P.C., Geacintov, N.E., and Shafirovich, V. (2007) Mechanisms of oxidation of guanine in DNA by carbonate radical anion, a decomposition product of nitrosoperoxycarbonate. *Chemistry*, **13**, 4571–4581.

84 Lewis, R.S., Tannenbaum, S.R., and Deen, W.M. (1995) Kinetics of N-nitrosation in oxygenated nitric oxide solutions at physiological pH: role of nitrous anhydride and effects of phosphate and chloride. *J. Am. Chem. Soc.*, **117**, 3933–3939.

85 Moller, M.N., Li, Q., Vitturi, D.A., Robinson, J.M., Lancaster, J.R., Jr., and Denicola, A. (2007) Membrane "lens" effect: focusing the formation of reactive nitrogen oxides from the *NO/

O$_2$ reaction. *Chem. Res. Toxicol.*, **20**, 709–714.
86 Kow, Y.W. (2002) Repair of deaminated bases in DNA. *Free Radic. Biol. Med.*, **33**, 886–893.
87 Visnes, T., Doseth, B., Pettersen, H.S., Hagen, L., Sousa, M.M., Akbari, M., Otterlei, M., Kavli, B., Slupphaug, G., and Krokan, H.E. (2009) Uracil in DNA and its processing by different DNA glycosylases. *Philos. Trans. R. Soc. Lond. B Biol. Sci.*, **364**, 563–568.
88 Anant, S. and Davidson, N.O. (2003) Hydrolytic nucleoside and nucleotide deamination, and genetic instability: a possible link between RNA-editing enzymes and cancer? *Trends Mol. Med.*, **9**, 147–152.
89 Barnes, D.E. and Lindahl, T. (2004) Repair and genetic consequences of endogenous DNA base damage in mammalian cells. *Annu. Rev. Genet.*, **38**, 445–476.
90 Pham, P., Bransteitter, R., and Goodman, M.F. (2005) Reward versus risk: DNA cytidine deaminases triggering immunity and disease. *Biochemistry*, **44**, 2703–2715.
91 Chelico, L., Pham, P., and Goodman, M.F. (2009) Stochastic properties of processive cytidine DNA deaminases AID and APOBEC3G. *Philos. Trans. R. Soc. Lond. B Biol. Sci.*, **364**, 583–593.
92 Goodman, J.E., Hofseth, L.J., Hussain, S.P., and Harris, C.C. (2004) Nitric oxide and p53 in cancer-prone chronic inflammation and oxyradical overload disease. *Environ. Mol. Mutagen.*, **44**, 3–9.
93 Cristalli, G., Costanzi, S., Lambertucci, C., Lupidi, G., Vittori, S., Volpini, R., and Camaioni, E. (2001) Adenosine deaminase: functional implications and different classes of inhibitors. *Med. Res. Rev.*, **21**, 105–128.
94 Yonekura, S.I., Nakamura, N., Yonei, S., and Zhang-Akiyama, Q.M. (2009) Generation, biological consequences and repair mechanisms of cytosine deamination in DNA. *J. Radiat. Res.*, **50**, 19–26.
95 Lindahl, T. (1993) Instability and decay of the primary structure of DNA. *Nature*, **362**, 709–714.
96 Suzuki, T., Matsumura, Y., Ide, H., Kanaori, K., Tajima, K., and Makino, K. (1997) Deglycosylation susceptibility and base-pairing stability of 2′-deoxyoxanosine in oligodeoxynucleotide. *Biochemistry*, **36**, 8013–8019.
97 Wuenschell, G.E., O'Connor, T.R., and Termini, J. (2003) Stability, miscoding potential, and repair of 2′-deoxyxanthosine in DNA: implications for nitric oxide-induced mutagenesis. *Biochemistry*, **42**, 3608–3616.
98 Vongchampa, V., Dong, M., Gingipalli, L., and Dedon, P. (2003) Stability of 2′-deoxyxanthosine in DNA. *Nucleic Acids Res.*, **31**, 1045–1051.
99 Nguyen, T., Brunson, D., Crespi, C.L., Penman, B.W., Wishnok, J.S., and Tannenbaum, S.R. (1992) DNA damage and mutation in human cells exposed to nitric oxide *in vitro*. *Proc. Natl. Acad. Sci. USA*, **89**, 3030–3034.
100 Wink, D.A., Kasprzak, K.S., Maragos, C.M., Elespuru, R.K., Misra, M., Dunams, T.M., Cebula, T.A., Koch, W.H., Andrews, A.W., Allen, J.S., and Keefer, L.K. (1991) DNA deaminating ability and genotoxicity of nitric oxide and its progenitors. *Science*, **254**, 1001–1003.
101 Thatcher, G.R. (2005) An introduction to NO-related therapeutic agents. *Curr. Top. Med. Chem.*, **5**, 597–601.
102 Feelisch, M. (1998) The use of nitric oxide donors in pharmacological studies. *Naunyn Schmiedebergs Arch. Pharmacol.*, **358**, 113–122.
103 Bennett, B.M., McDonald, B.J., Nigam, R., and Simon, W.C. (1994) Biotransformation of organic nitrates and vascular smooth muscle cell function. *Trends Pharmacol. Sci.*, **15**, 245–249.
104 Wang, C. and Deen, W.M. (2003) Nitric oxide delivery system for cell culture studies. *Ann. Biomed. Eng.*, **31**, 65–79.
105 Glaser, R., Wu, H., and Lewis, M. (2005) Cytosine catalysis of nitrosative guanine deamination and interstrand cross-link formation. *J. Am. Chem. Soc.*, **127**, 7346–7358.
106 Dong, M., Wang, C., Deen, W.M., and Dedon, P.C. (2003) Absence of 2′-deoxyxanosine and presence of abasic sites in DNA exposed to nitric

107 Caulfield, J.L., Wishnok, J.S., and Tannenbaum, S.R. (2003) Nitric oxide-induced interstrand cross-links in DNA. *Chem. Res. Toxicol.*, **16**, 571–574.

108 Dong, M. and Dedon, P.C. (2006) Relatively small increases in the steady-state levels of nucleobase deamination products in DNA from human TK6 cells exposed to toxic levels of nitric oxide. *Chem. Res. Toxicol.*, **19**, 50–57.

109 Lim, K.S., Huang, S.H., Jenner, A., Wang, H., Tang, S.Y., and Halliwell, B. (2006) Potential artifacts in the measurement of DNA deamination. *Free Radic. Biol. Med.*, **40**, 1939–1948.

110 Suzuki, T., Yamaoka, R., Nishi, M., Ide, H., and Makino, K. (1996) Isolation and characterization of a novel product, 2′-deoxyoxanosine, from 2′-deoxyguanosine, oligodeoxynucleotide, and calf thymus DNA treated with nitrous acid and nitric oxide. *J. Am. Chem. Soc.*, **118**, 2515–2516.

111 Suzuki, T., Kanaori, K., Tajima, K., and Makino, K. (1997) Mechanism and intermediate for formation of 2′-deoxyoxanosine. *Nucleic Acids Symp. Ser.*, (37), 313–314.

112 Lucas, L.T., Gatehouse, D., Jones, G.D.D., and Shuker, D.E.G. (2001) Characterization of DNA damage at purine residues in oligonucleotides and calf thymus DNA induced by the mutagen 1-nitrosoindole-3-acetonitrile. *Chem. Res. Toxicol.*, **14**, 158–164.

113 Lucas, L.T., Gatehouse, D., and Shuker, D.E. (1999) Efficient nitroso group transfer from *N*-nitrosoindoles to nucleotides and 2′-deoxyguanosine at physiological pH. A new pathway for *N*-nitroso compounds to exert genotoxicity. *J. Biol. Chem.*, **274**, 18319–18326.

114 Suzuki, T., Yamada, M., Furukawa, H., Kanaori, K., Tajima, K., and Makino, K. (1997) Detection of 2′-deoxyoxanosine by capillary electrophoresis. *Nucleic Acids Symp. Ser.*, (37), 239–240.

115 Suzuki, T., Yoshida, M., Yamada, M., Ide, H., Kobayashi, M., Kanaori, K., Tajima, K., and Makino, K. (1998) Misincorporation of 2′-deoxyoxanosine 5′-triphosphate by DNA polymerases and its implication for mutagenesis. *Biochemistry*, **37**, 11592–11598.

116 Suzuki, T., Yamada, M., Ishida, T., Morii, T., and Makino, K. (1999) Reactivity of 2′-deoxyoxanosine, a novel DNA lesion. *Nucleic Acids Symp. Ser.*, (42), 7–8.

117 Suzuki, T., Ide, H., Yamada, M., Endo, N., Kanaori, K., Tajima, K., Morii, T., and Makino, K. (2000) Formation of 2′-deoxyoxanosine from 2′-deoxyguanosine and nitrous acid: mechanism and intermediates. *Nucleic Acids Res.*, **28**, 544–551.

118 Suzuki, T., Yamada, M., Ide, H., Kanaori, K., Tajima, K., Morii, T., and Makino, K. (2000) Identification and characterization of a reaction product of 2′-deoxyoxanosine with glycine. *Chem. Res. Toxicol.*, **13**, 227–230.

119 Terato, H., Masaoka, A., Asagoshi, K., Honsho, A., Ohyama, Y., Suzuki, T., Yamada, M., Makino, K., Yamamoto, K., and Ide, H. (2002) Novel repair activities of AlkA (3-methyladenine DNA glycosylase II) and endonuclease VIII for xanthine and oxanine, guanine lesions induced by nitric oxide and nitrous acid. *Nucleic Acids Res.*, **30**, 4975–4984.

120 Nakano, T., Asagoshi, K., Terato, H., Suzuki, T., and Ide, H. (2005) Assessment of the genotoxic potential of nitric oxide-induced guanine lesions by *in vitro* reactions with *Escherichia coli* DNA polymerase I. *Mutagenesis*, **20**, 209–216.

121 Nakano, T., Katafuchi, A., Shimizu, R., Terato, H., Suzuki, T., Tauchi, H., Makino, K., Skorvaga, M., Van Houten, B., and Ide, H. (2005) Repair activity of base and nucleotide excision repair enzymes for guanine lesions induced by nitrosative stress. *Nucleic Acids Res.*, **33**, 2181–2191.

122 Nakano, T., Terato, H., Asagoshi, K., Ohyama, Y., Suzuki, T., Yamada, M., Makino, K., and Ide, H. (2001) Adduct formation between oxanine and amine derivatives. *Nucleic Acids Res. Suppl.*, **1**, 47–48.

123 Nakano, T., Terato, H., Asagoshi, K., Masaoka, A., Mukuta, M., Ohyama, Y., Suzuki, T., Makino, K., and Ide, H. (2003) DNA–protein cross-link formation mediated by oxanine. A novel genotoxic mechanism of nitric oxide-induced DNA damage. *J. Biol. Chem.*, **278**, 25264–25272.

124 Suzuki, T., Yamada, M., Ide, H., Kanaori, K., Tajima, K., Morii, T., and Makino, K. (2000) Influence of ring opening–closure equilibrium of oxanine, a novel damaged nucleobase, on migration behavior in capillary electrophoresis. *J. Chromatogr. A*, **877**, 225–232.

125 Hitchcock, T.M., Dong, L., Connor, E.E., Meira, L.B., Samson, L.D., Wyatt, M.D., and Cao, W. (2004) Oxanine DNA glycosylase activity from mammalian alkyladenine glycosylase. *J. Biol. Chem.*, **279**, 38177–38183.

126 Hitchcock, T.M., Gao, H., and Cao, W. (2004) Cleavage of deoxyoxanosine-containing oligodeoxyribonucleotides by bacterial endonuclease V. *Nucleic Acids Res.*, **32**, 4071–4080.

127 Dong, L., Mi, R., Glass, R.A., Barry, J.N., and Cao, W. (2008) Repair of deaminated base damage by *Schizosaccharomyces pombe* thymine DNA glycosylase. *DNA Repair*, **7**, 1962–1972.

128 Dong, L., Meira, L.B., Hazra, T.K., Samson, L.D., and Cao, W. (2008) Oxanine DNA glycosylase activities in mammalian systems. *DNA Repair*, **7**, 128–134.

129 Chen, H.J., Chiu, W.L., Lin, W.P., and Yang, S.S. (2008) Investigation of DNA–protein cross-link formation between lysozyme and oxanine by mass spectrometry. *ChemBioChem*, **9**, 1074–1081.

130 Pack, S.P., Doi, A., Nonogawa, M., Kamisetty, N.K., Devarayapalli, K.C., Kodaki, T., and Makino, K. (2007) Biophysical stability and enzymatic recognition of oxanine in DNA. *Nucleosides Nucleotides Nucleic Acids*, **26**, 1589–1593.

131 Chen, H.J., Hsieh, C.J., Shen, L.C., and Chang, C.M. (2007) Characterization of DNA–protein cross-links induced by oxanine: cellular damage derived from nitric oxide and nitrous acid. *Biochemistry*, **46**, 3952–3965.

132 Pack, S.P., Nonogawa, M., Kodaki, T., and Makino, K. (2005) Chemical synthesis and thermodynamic characterization of oxanine-containing oligodeoxynucleotides. *Nucleic Acids Res.*, **33**, 5771–5780.

133 Katafuchi, A., Nakano, T., Masaoka, A., Terato, H., Iwai, S., Hanaoka, F., and Ide, H. (2004) Differential specificity of human and *Escherichia coli* endonuclease III and VIII homologues for oxidative base lesions. *J. Biol. Chem.*, **279**, 14464–14471.

134 Kirsh, M.E., Cutler, R.G., and Hartman, P.E. (1986) Absence of deoxyuridine and 5-hydroxymethyldeoxyuridine in the DNA from three tissues of mice of various ages. *Mech. Ageing Dev.*, **35**, 71–77.

135 Green, D.A. and Deutsch, W.A. (1984) Direct determination of uracil in [^{32}P,uracil-^{3}H]poly(dA · dT) and bisulfite-treated phage PM2 DNA. *Anal. Biochem.*, **142**, 497–503.

136 Wickramasinghe, S.N. and Fida, S. (1993) Misincorporation of uracil into the DNA of folate- and B_{12}-deficient HL60 cells. *Eur. J. Haematol.*, **50**, 127–132.

137 Goulian, M., Bleile, B., and Tseng, B.Y. (1980) Methotrexate-induced misincorporation of uracil into DNA. *Proc. Natl. Acad. Sci. USA*, **77**, 1956–1960.

138 Lim, K.S., Jenner, A., and Halliwell, B. (2006) Quantitative gas chromatography mass spectrometric analysis of 2′-deoxyinosine in tissue DNA. *Nat. Protoc.*, **1**, 1995–2002.

139 Mashiyama, S.T., Courtemanche, C., Elson-Schwab, I., Crott, J., Lee, B.L., Ong, C.N., Fenech, M., and Ames, B.N. (2004) Uracil in DNA, determined by an improved assay, is increased when deoxynucleosides are added to folate-deficient cultured human lymphocytes. *Anal. Biochem.*, **330**, 58–69.

140 Caulfield, J.L., Wishnok, J.S., and Tannenbaum, S.R. (1998) Nitric oxide-induced deamination of cytosine and guanine in deoxynucleosides and oligonucleotides. *J. Biol. Chem.*, **273**, 12689–12695.

141 Spencer, J.P., Whiteman, M., Jenner, A., and Halliwell, B. (2000) Nitrite-induced deamination and hypochlorite-induced oxidation of DNA in intact human respiratory tract epithelial cells. *Free Radic. Biol. Med.*, **28**, 1039–1050.

142 Taghizadeh, K., McFaline, J.L., Pang, B., Sullivan, M., Dong, M., Plummer, E., and Dedon, P.C. (2008) Quantification of DNA damage products resulting from deamination, oxidation and reaction with products of lipid peroxidation by liquid chromatography isotope dilution tandem mass spectrometry. *Nat. Protoc.*, **3**, 1287–1298.

143 Polson, A.G., Crain, P.F., Pomerantz, S.C., McCloskey, J.A., and Bass, B.L. (1991) The mechanism of adenosine to inosine conversion by the double-stranded RNA unwinding/modifying activity: a high-performance liquid chromatography-mass spectrometry analysis. *Biochemistry*, **30**, 11507–11514.

144 Reynaud, C.A., Aoufouchi, S., Faili, A., and Weill, J.C. (2003) What role for aid: mutator, or assembler of the immunoglobulin mutasome? *Nat. Immunol.*, **4**, 631–638.

145 Krokan, H.E., Drablos, F., and Slupphaug, G. (2002) Uracil in DNA – occurrence, consequences and repair. *Oncogene*, **21**, 8935–8948.

146 Bass, B.L. (2002) RNA editing by adenosine deaminases that act on RNA. *Annu. Rev. Biochem.*, **71**, 817–846.

147 Bass, B.L. (1997) RNA editing and hypermutation by adenosine deamination. *Trends Biochem. Sci.*, **22**, 157–162.

148 Maynes, J.T., Yuan, R.G., and Snyder, F.F. (2000) Identification, expression, and characterization of *Escherichia coli* guanine deaminase. *J. Bacteriol.*, **182**, 4658–4660.

149 Snyder, F.F., Yuan, R.G., Bin, J.C., Carter, K.L., and McKay, D.J. (2000) Human guanine deaminase: cloning, expression and characterisation. *Adv. Exp. Med. Biol.*, **486**, 111–114.

150 Hong, M.Y. and Hosmane, R.S. (1997) Irreversible, tight-binding inhibition of adenosine deaminase by coformycins: inhibitor structural features that contribute to the mode of enzyme inhibition. *Nucleosides Nucleotides*, **16**, 1053–1057.

151 Povirk, L.F. and Goldberg, I.H. (1985) Endonuclease-resistant apyrimidinic sites formed by neocarzinostatin at cytosine residues in DNA: evidence for a possible role in mutagenesis. *Proc. Natl. Acad. Sci. USA*, **82**, 3182–3186.

152 Weinfeld, M., Rasouli-Nia, A., Chaudhry, M.A., and Britten, R.A. (2001) Response of base excision repair enzymes to complex DNA lesions. *Radiat. Res.*, **156**, 584–589.

153 Dedon, P.C. (2008) The chemical toxicology of 2-deoxyribose oxidation in DNA. *Chem. Res. Toxicol.*, **21**, 206–219.

154 Kappen, L.S. and Goldberg, I.H. (1989) Identification of 2-deoxyribonolactone at the site of neocarzinostatin-induced cytosine release in the sequence d(AGC). *Biochemistry*, **28**, 1027–1032.

155 Urata, H., Yamamoto, K., Akagi, M., Hiroaki, H., and Uesugi, S. (1989) A 2-deoxyribonolactone-containing nucleotide: isolation and characterization of the alkali-sensitive photoproduct of the trideoxyribonucleotide d(ApCpA). *Biochemistry*, **28**, 9566–9569.

156 Hwang, J.T., Tallman, K.A., and Greenberg, M.M. (1999) The reactivity of the 2-deoxyribonolactone lesion in single-stranded DNA and its implication in reaction mechanisms of DNA damage and repair. *Nucleic Acids Res.*, **27**, 3805–3810.

157 Jourdan, M., Garcia, J., Defrancq, E., Kotera, M., and Lhomme, J. (1999) 2′-Deoxyribonolactone lesion in DNA: refined solution structure determined by nuclear magnetic resonance and molecular modeling. *Biochemistry*, **38**, 3985–3995.

158 Kotera, M., Roupioz, Y., Defrancq, E., Bourdat, A.G., Garcia, J., Coulombeau, C., and Lhomme, J. (2000) The 7-nitroindole nucleoside as a photochemical precursor of 2′-deoxyribonolactone: access to DNA fragments containing this oxidative abasic lesion. *Chemistry*, **6**, 4163–4169.

159 Lenox, H.J., McCoy, C.P., and Sheppard, T.L. (2001) Site-specific generation of deoxyribonolactone

lesions in DNA oligonucleotides. *Org. Lett.*, **3**, 2415–2418.

160 Zheng, Y. and Sheppard, T.L. (2004) Half-life and DNA strand scission products of 2-deoxyribonolactone oxidative DNA damage lesions. *Chem. Res. Toxicol.*, **17**, 197–207.

161 Hashimoto, M., Greenberg, M.M., Kow, Y.W., Hwang, J.T., and Cunningham, R.P. (2001) The 2-deoxyribonolactone lesion produced in DNA by neocarzinostatin and other damaging agents forms cross-links with the base-excision repair enzyme endonuclease III. *J. Am. Chem. Soc.*, **123**, 3161–3162.

162 DeMott, M.S., Beyret, E., Wong, D., Bales, B.C., Hwang, J.T., Greenberg, M.M., and Demple, B. (2002) Covalent trapping of human DNA polymerase β by the oxidative DNA lesion 2-deoxyribonolactone. *J. Biol. Chem.*, **277**, 7637–7640.

163 Dizdaroglu, M., von Schulte-Frohlinde, D., and Sonntag C. (1977) Gamma-radiolyses of DNA in oxygenated aqueous solution. Structure of an alkali-labile site. *Z. Naturforsch.*, **32C**, 1021–1022.

164 Sugiyama, H., Tsutsumi, Y., Fujimoto, K., and Saito, I. (1993) Photoinduced deoxyribose C2′ oxidation in DNA: alkali-dependent cleavage of erythrose-containing sites via a retroaldol reaction. *J. Am. Chem. Soc.*, **115**, 4443–4448.

165 Kim, J., Weledji, Y.N., and Greenberg, M.M. (2004) Independent generation and characterization of a C2′-oxidized abasic site in chemically synthesized oligonucleotides. *J. Org. Chem.*, **69**, 6100–6104.

166 Sitlani, A., Long, E.C., Pyle, A.M., and Barton, J.K. (1992) DNA photocleavage by phenanthrenequinone diimine complexes with rhodium (III): shape-selective recognition and reaction. *J. Am. Chem. Soc.*, **114**, 2303–2312.

167 Collins, C., Awada, M.M., Zhou, X., and Dedon, P.C. (2003) Analysis of 3′-phosphoglycolaldehyde residues in oxidized DNA by gas chromatography/negative chemical ionization/mass spectrometry. *Chem. Res. Toxicol.*, **16**, 1560–1566.

168 Collins, C., Zhou, X., Wang, R., Barth, M.C., Jiang, T., Coderre, J.A., and Dedon, P.C. (2005) Differential oxidation of deoxyribose in DNA by γ- and α-particle radiation. *Radiat. Res.*, **163**, 654–662.

169 Bryant-Friedrich, A.C. (2004) Generation of a C-3′-thymidinyl radical in single-stranded oligonucleotides under anaerobic conditions. *Org. Lett.*, **6**, 2329–2332.

170 Lahoud, G.A., Hitt, A.L., and Bryant-Friedrich, A.C. (2006) Aerobic fate of the C-3′-thymidinyl radical in single-stranded DNA. *Chem. Res. Toxicol.*, **19**, 1630–1636.

171 Murata-Kamiya, N., Kamiya, H., Iwamoto, N., and Kasai, H. (1995) Formation of a mutagen, glyoxal, from DNA treated with oxygen free radicals. *Carcinogenesis*, **16**, 2251–2253.

172 Awada, M. and Dedon, P.C. (2001) Formation of the 1,N^2-glyoxal adduct of deoxyguanosine by phosphoglycolaldehyde, a product of 3′-deoxyribose oxidation in DNA. *Chem. Res. Toxicol.*, **14**, 1247–1253.

173 Zhou, X., Taghizadeh, K., and Dedon, P.C. (2005) Chemical and biological evidence for base propenals as the major source of the endogenous M_1dG adduct in cellular DNA. *J. Biol. Chem.*, **280**, 25377–25382.

174 Rashid, R., Langfinger, D., Wagner, R., von Schuchmann, H.P., and Sonntag C. (1999) Bleomycin versus OH-radical-induced malonaldehydic-product formation in DNA. *Int. J. Radiat. Biol.*, **75**, 101–109.

175 Dedon, P.C. and Goldberg, I.H. (1992) Free-radical mechanisms involved in the formation of sequence-dependent bistranded DNA lesions by the antitumor antibiotics bleomycin, neocarzinostatin, and calicheamicin. *Chem. Res. Toxicol.*, **5**, 311–332.

176 Dedon, P.C., Plastaras, J.P., Rouzer, C.A., and Marnett, L.J. (1998) Indirect mutagenesis by oxidative DNA damage: formation of the pyrimidopurinone adduct of deoxyguanosine by base propenal. *Proc. Natl. Acad. Sci. USA*, **95**, 11113–11116.

177 Plastaras, J.P., Dedon, P.C., and Marnett, L.J. (2002) Effects of DNA structure on oxopropenylation by the endogenous mutagens malondialdehyde and base propenal. *Biochemistry,* **41**, 5033–5042.

178 Plastaras, J.P., Riggins, J.N., Otteneder, M., and Marnett, L.J. (2000) Reactivity and mutagenicity of endogenous DNA oxopropenylating agents: base propenals, malondialdehyde, and N^ε-oxopropenyllysine. *Chem. Res. Toxicol.,* **13**, 1235–1242.

179 Chen, B., Bohnert, T., Zhou, X., and Dedon, P.C. (2004) 5′-(2-Phosphoryl-1,4-dioxobutane) as a product of 5′-oxidation of deoxyribose in DNA: elimination as *trans*-1,4-dioxo-2-butene and approaches to analysis. *Chem. Res. Toxicol.,* **17**, 1406–1413.

180 Chen, B., Vu, C.C., Byrns, M.C., Dedon, P.C., and Peterson, L.A. (2006) Formation of 1,4-dioxo-2-butene-derived adducts of 2′-deoxyadenosine and 2′-deoxycytidine in oxidized DNA. *Chem. Res. Toxicol.,* **19**, 982–985.

181 Byrns, M.C., Vu, C.C., Neidigh, J.W., Abad, J.-L., Jones, R.A., and Peterson, L.A. (2006) Detection of DNA adducts derived from the reactive metabolite of furan, *cis*-2-butene-1,4-dial. *Chem. Res. Toxicol.,* **19**, 414–420.

182 Byrns, M.C., Predecki, D.P., and Peterson, L.A. (2002) Characterization of nucleoside adducts of *cis*-2-butene-1,4-dial, a reactive metabolite of furan. *Chem. Res. Toxicol.,* **15**, 373–379.

183 Gingipalli, L. and Dedon, P.C. (2001) Reaction of *cis*- and *trans*-2-butene-1,4-dial with 2′-deoxycytidine to form stable oxadiazabicyclooctaimine adducts. *J. Am. Chem. Soc.,* **123**, 2664–2665.

184 Bohnert, T., Gingipalli, L., and Dedon, P.C. (2004) Reaction of 2′-deoxyribonucleosides with *cis*- and *trans*-1,4-dioxo-2-butene. *Biochem. Biophys. Res. Commun.,* **323**, 838–844.

185 Jiang, T., Zhou, X., Taghizadeh, T., Dong, M., and Dedon, P.C. (2006) *N*-Formylation of lysines in histone proteins as a secondary modification arising from oxidative DNA damage. *Proc. Natl. Acad. Sci. USA,* **104**, 60–65.

186 Bartsch, H. and Nair, J. (2004) Oxidative stress and lipid peroxidation-derived DNA-lesions in inflammation driven carcinogenesis. *Cancer Detect. Prev.,* **28**, 385–391.

187 Nair, U., Bartsch, H., and Nair, J. (2007) Lipid peroxidation-induced DNA damage in cancer-prone inflammatory diseases: a review of published adduct types and levels in humans. *Free Radic. Biol. Med.,* **43**, 1109–1120.

188 Blair, I.A. (2008) DNA adducts with lipid peroxidation products. *J. Biol. Chem.,* **283**, 15545–15549.

189 Marnett, L.J. (2002) Oxy radicals, lipid peroxidation and DNA damage. *Toxicology,* **181–182**, 219–222.

190 Lee, S.H., Arora, J.A., Oe, T., and Blair, I.A. (2005) 4-Hydroperoxy-2-nonenal-induced formation of 1,N^2-etheno-2′-deoxyguanosine adducts. *Chem. Res. Toxicol.,* **18**, 780–786.

191 Williams, M.V., Lee, S.H., Pollack, M., and Blair, I.A. (2006) Endogenous lipid hydroperoxide-mediated DNA-adduct formation in min mice. *J. Biol. Chem.,* **281**, 10127–10133.

192 Nair, J., Barbin, A., Velic, I., and Bartsch, H. (1999) Etheno DNA-base adducts from endogenous reactive species. *Mutat. Res.,* **424**, 59–69.

193 Lee, S.H., Oe, T., Arora, J.S., and Blair, I.A. (2005) Analysis of FeII-mediated decomposition of a linoleic acid-derived lipid hydroperoxide by liquid chromatography/mass spectrometry. *J. Mass Spectrom.,* **40**, 661–668.

194 Nair, J., De Flora, S., Izzotti, A., and Bartsch, H. (2007) Lipid peroxidation-derived etheno-DNA adducts in human atherosclerotic lesions. *Mutat. Res.,* **621**, 95–105.

195 Godschalk, R.W., Albrecht, C., Curfs, D.M., Schins, R.P., van Bartsch, H., Schooten F.J., and Nair, J. (2007) Decreased levels of lipid peroxidation-induced DNA damage in the onset of atherogenesis in apolipoprotein E deficient mice. *Mutat. Res.,* **621**, 87–94.

196 Meerang, M., Nair, J., Sirankapracha, P., Thephinlap, C., Srichairatanakool, S., Fucharoen, S., and Bartsch, H. (2008) Increased urinary 1,N^6-ethenodeoxyadenosine and 3,N^4-ethenodeoxycytidine excretion in thalassemia patients: markers for lipid peroxidation-induced

DNA damage. *Free Radic. Biol. Med.*, **44**, 1863–1868.

197 Rindgen, D., Lee, S.H., Nakajima, M., and Blair, I.A. (2000) Formation of a substituted 1,N^6-etheno-2′-deoxyadenosine adduct by lipid hydroperoxide-mediated generation of 4-oxo-2-nonenal. *Chem. Res. Toxicol.*, **13**, 846–852.

198 Pollack, M., Oe, T., Lee, S.H., Silva Elipe, M.V., Arison, B.H., and Blair, I.A. (2003) Characterization of 2′-deoxycytidine adducts derived from 4-oxo-2-nonenal, a novel lipid peroxidation product. *Chem. Res. Toxicol.*, **16**, 893–900.

199 Lee, S.H., Silva Elipe, M.V., Arora, J.S., and Blair, I.A. (2005) Dioxododecenoic acid: a lipid hydroperoxide-derived bifunctional electrophile responsible for etheno DNA adduct formation. *Chem. Res. Toxicol.*, **18**, 566–578.

200 Winter, C.K., Segall, H.J., and Haddon, W.F. (1986) Formation of cyclic adducts of deoxyguanosine with the aldehydes *trans*-4-hydroxy-2-hexenal and *trans*-4-hydroxy-2-nonenal *in vitro*. *Cancer Res.*, **46**, 5682–5686.

201 Chung, F.L., Young, R., and Hecht, S.S. (1984) Formation of cyclic 1,N^2-propanodeoxyguanosine adducts in DNA upon reaction with acrolein or crotonaldehyde. *Cancer Res.*, **44**, 990–995.

202 Chung, F.L. and Hecht., S.S. (1983) Formation of cyclic 1,N^2-adducts by reaction of deoxyguanosine with α-acetoxy-*N*-nitrosopyrrolidine, 4-(carbethoxynitrosamino)butanal, or crotonaldehyde. *Cancer Res.*, **43**, 1230–1235.

203 Douki, T., Odin, F., Caillat, S., Favier, A., and Cadet., J. (2004) Predominance of the 1,N^2-propano 2′-deoxyguanosine adduct among 4-hydroxy-2-nonenal-induced DNA lesions. *Free Radic. Biol. Med.*, **37**, 62–70.

204 Douki, T. and Ames, B.N. (1994) An HPLC-EC assay for 1,N^2-propano adducts of 2′-deoxyguanosine with 4-hydroxynonenal and other α,β-unsaturated aldehydes. *Chem. Res. Toxicol.*, **7**, 511–518.

205 Pan, J. and Chung., F.L. (2002) Formation of cyclic deoxyguanosine adducts from ω-3 and ω-6 polyunsaturated fatty acids under oxidative conditions. *Chem. Res. Toxicol.*, **15**, 367–372.

206 Zhang, S., Villalta, P.W., Wang, M., and Hecht, S.S. (2007) Detection and quantitation of acrolein-derived 1,N^2-propanodeoxyguanosine adducts in human lung by liquid chromatography-electrospray ionization-tandem mass spectrometry. *Chem. Res. Toxicol.*, **20**, 565–571.

207 Pawlowicz, A.J., Munter, T., Klika, K.D., and Kronberg, L. (2006) Reaction of acrolein with 2′-deoxyadenosine and 9-ethyladenine – formation of cyclic adducts. *Bioorg. Chem.*, **34**, 39–48.

208 Pawlowicz, A.J., Munter, T., Zhao, Y., and Kronberg, L. (2006) Formation of acrolein adducts with 2′-deoxyadenosine in calf thymus DNA. *Chem. Res. Toxicol.*, **19**, 571–576.

209 Pawlowicz, A.J., Klika, K.D., and Kronberg, L. (2007) The structural identification and conformational analysis of the products from the reaction of acrolein with 2′-deoxycytidine, 1-methylcytosine and calf thymus DNA. *Eur. J. Org. Chem.*, 1429–1437.

210 Pawlowicz, A.J. and Kronberg, L. (2008) Characterization of adducts formed in reactions of acrolein with thymidine and calf thymus DNA. *Chem. Biodivers.*, **5**, 177–188.

211 Cho, Y.J., Kim, H.Y., Huang, H., Slutsky, A., Minko, I.G., Wang, H., Nechev, L.V., Kozekov, I.D., Kozekova, A., Tamura, P., Jacob, J., Voehler, M., Harris, T.M., Lloyd, R.S., Rizzo, C.J., and Stone, M.P. (2005) Spectroscopic characterization of interstrand carbinolamine cross-links formed in the 5′-CpG-3′ sequence by the acrolein-derived γ-OH-1,N^2-propano-2′-deoxyguanosine DNA adduct. *J. Am. Chem. Soc.*, **127**, 17686–17696.

212 Cho, Y.J., Kozekov, I.D., Harris, T.M., Rizzo, C.J., and Stone, M.P. (2007) Stereochemistry modulates the stability of reduced interstrand cross-links arising from R- and S-α-CH$_3$-γ-OH-1,N^2-propano-2′-deoxyguanosine in the 5′-CpG-3′ DNA sequence. *Biochemistry*, **46**, 2608–2621.

213 Cho, Y.J., Wang, H., Kozekov, I.D., Kozekova, A., Kurtz, A.J., Jacob, J.,

Voehler, M., Smith, J., Harris, T.M., Rizzo, C.J., Lloyd, R.S., and Stone, M.P. (2006) Orientation of the crotonaldehyde-derived N^2-[3-oxo-1(S)-methylpropyl]-dG DNA adduct hinders interstrand cross-link formation in the 5′-CpG-3′ sequence. *Chem. Res. Toxicol.*, **19**, 1019–1029.

214 Cho, Y.J., Wang, H., Kozekov, I.D., Kurtz, A.J., Jacob, J., Voehler, M., Smith, J., Harris, T.M., Lloyd, R.S., Rizzo, C.J., and Stone, M.P. (2006) Stereospecific formation of interstrand carbinolamine DNA cross-links by crotonaldehyde- and acetaldehyde-derived α-CH$_3$-γ-OH-1,N^2-propano-2′-deoxyguanosine adducts in the 5′-CpG-3′ sequence. *Chem. Res. Toxicol.*, **19**, 195–208.

215 Huang, H., Wang, H., Lloyd, R.S., Rizzo, C.J., and Stone, M.P. (2009) Conformational interconversion of the *trans*-4-hydroxynonenal-derived (6S,8R,11S) 1,N^2-deoxyguanosine adduct when mismatched with deoxyadenosine in DNA. *Chem. Res. Toxicol.*, **22**, 187–200.

216 Huang, H., Wang, H., Qi, N., Kozekova, A., Rizzo, C.J., and Stone, M.P. (2008) Rearrangement of the (6S,8R,11S) and (6S,8R,11R) exocyclic 1,N^2-deoxyguanosine adducts of *trans*-4-hydroxynonenal to N^2-deoxyguanosine cyclic hemiacetal adducts when placed complementary to cytosine in duplex DNA. *J. Am. Chem. Soc.*, **130**, 10898–10906.

217 Huang, H., Wang, H., Qi, N., Lloyd, R.S., Rizzo, C.J., and Stone, M.P. (2008) The stereochemistry of *trans*-4-hydroxynonenal-derived exocyclic 1,N^2-2′-deoxyguanosine adducts modulates formation of interstrand cross-links in the 5′-CpG-3′ sequence. *Biochemistry*, **47**, 11457–11472.

218 Kozekov, I.D., Nechev, L.V., Moseley, M.S., Harris, C.M., Rizzo, C.J., Stone, M.P., and Harris, T.M. (2003) DNA interchain cross-links formed by acrolein and crotonaldehyde. *J. Am. Chem. Soc.*, **125**, 50–61.

219 Stone, M.P., Cho, Y.J., Huang, H., Kim, H.Y., Kozekov, I.D., Kozekova, A., Wang, H., Minko, I.G., Lloyd, R.S., Harris, T.M., and Rizzo, C.J. (2008) Interstrand DNA cross-links induced by α,β-unsaturated aldehydes derived from lipid peroxidation and environmental sources. *Acc. Chem. Res.*, **41**, 793–804.

220 Minko, I.G., Kozekov, I.D., Kozekova, A., Harris, T.M., Rizzo, C.J., and Lloyd, R.S. (2008) Mutagenic potential of DNA–peptide crosslinks mediated by acrolein-derived DNA adducts. *Mutat. Res.*, **637**, 161–172.

221 Sanchez, A.M., Kozekov, I.D., Harris, T.M., and Lloyd, R.S. (2005) Formation of inter- and intrastrand imine type DNA–DNA cross-links through secondary reactions of aldehydic adducts. *Chem. Res. Toxicol.*, **18**, 1683–1690.

222 Zhang, S., Villalta, P.W., Wang, M., and Hecht, S.S. (2006) Analysis of crotonaldehyde- and acetaldehyde-derived 1,N^2-propanodeoxyguanosine adducts in DNA from human tissues using liquid chromatography electrospray ionization tandem mass spectrometry. *Chem. Res. Toxicol.*, **19**, 1386–1392.

223 Pan, J., Davis, W., Trushin, N., Amin, S., Nath, R.G., Salem, N., Jr., and Chung, F.L. (2006) A solid-phase extraction/high-performance liquid chromatography-based ^{32}P-postlabeling method for detection of cyclic 1,N^2-propanodeoxyguanosine adducts derived from enals. *Anal. Biochem.*, **348**, 15–23.

224 Budiawan and Eder, E. (2000) Detection of 1,N^2-propanodeoxyguanosine adducts in DNA of Fischer 344 rats by an adapted ^{32}P-post-labeling technique after per os application of crotonaldehyde. *Carcinogenesis*, **21**, 1191–1196.

225 Chung, F.L., Nath, R.G., Nagao, M., Nishikawa, A., Zhou, G.D., and Randerath, K. (1999) Endogenous formation and significance of 1,N^2-propanodeoxyguanosine adducts. *Mutat. Res.*, **424**, 71–81.

226 Chung, F.L., Pan, J., Choudhury, S., Roy, R., Hu, W., and Tang, M.S. (2003) Formation of *trans*-4-hydroxy-2-nonenal- and other enal-derived cyclic DNA adducts from ω-3 and ω-6 polyunsaturated fatty acids and their roles in DNA repair and human p53 gene mutation. *Mutat. Res.*, **531**, 25–36.

227 Chung, F.L., Zhang, L., Ocando, J.E., and Nath, R.G. (1999) Role of 1,N^2-propanodeoxyguanosine adducts as endogenous DNA lesions in rodents and humans. *IARC Sci. Publ.*, **150**, 45–54.

228 Blair, I.A. (2001) Lipid hydroperoxide-mediated DNA damage. *Exp. Gerontol.*, **36**, 1473–1481.

229 Lee, S.H. and Blair, I.A. (2001) Oxidative DNA damage and cardiovascular disease. *Trends Cardiovasc. Med.*, **11**, 148–155.

230 Poirier, M.C. (1997) DNA adducts as exposure biomarkers and indicators of cancer risk. *Environ. Health Perspect.*, **105** (Suppl. 4), 907–912.

231 Marnett, L.J. (1994) DNA adducts of α,β-unsaturated aldehydes and dicarbonyl compounds. *IARC Sci. Publ.*, **125**, 151–163.

232 Marnett, L.J. (2000) Oxyradicals and DNA damage. *Carcinogenesis*, **21**, 361–370.

233 Marnett, L.J. and Burcham, P.C. (1993) Endogenous DNA adducts: potential and paradox. *Chem. Res. Toxicol.*, **6**, 771–785.

234 Mao, H., Schnetz-Boutaud, N.C., Weisenseel, J.P., Marnett, L.J., and Stone, M.P. (1999) Duplex DNA catalyzes the chemical rearrangement of a malondialdehyde deoxyguanosine adduct. *Proc. Natl. Acad. Sci. USA*, **96**, 6615–6620.

235 Schnetz-Boutaud, N., Daniels, J.S., Hashim, M.F., Scholl, P., Burrus, T., and Marnett, L.J. (2000) Pyrimido[1,2-α]purin-10(3H)-one: a reactive electrophile in the genome. *Chem. Res. Toxicol.*, **13**, 967–970.

236 Schnetz-Boutaud, N.C., Saleh, S., Marnett, L.J., and Stone, M.P. (2001) Structure of the malondialdehyde deoxyguanosine adduct m1g when placed opposite a two-base deletion in the (CpG)3 frameshift hotspot of the *Salmonella typhimurium hisd3052* gene. *Adv. Exp. Med. Biol.*, **500**, 513–516.

237 Wang, Y., Schnetz-Boutaud, N.C., Saleh, S., Marnett, L.J., and Stone, M.P. (2007) Bulge migration of the malondialdehyde OPdG DNA adduct when placed opposite a two-base deletion in the (CpG)3 frameshift hotspot of the *Salmonella typhimurium hisD3052* gene. *Chem. Res. Toxicol.*, **20**, 1200–1210.

238 Riggins, J.N., Pratt, D.A., Voehler, M., Daniels, J.S., and Marnett, L.J. (2004) Kinetics and mechanism of the general-acid-catalyzed ring-closure of the malondialdehyde-DNA adduct, N^2-(3-oxo-1-propenyl)deoxyguanosine (N^2OPdG-), to 3-(2′-deoxy-β-D-*erythro*-pentofuranosyl)pyrimido[1,2-α]purin-10(3H)-one (M$_1$dG). *J. Am. Chem. Soc.*, **126**, 10571–10581.

239 Riggins, J.N., Daniels, J.S., Rouzer, C.A., and Marnett, L.J. (2004) Kinetic and thermodynamic analysis of the hydrolytic ring-opening of the malondialdehyde-deoxyguanosine adduct, 3-(2′-deoxy-β-D-*erythro*-pentofuranosyl)-pyrimido[1,2-α]purin-10(3H)-one. *J. Am. Chem. Soc.*, **126**, 8237–8243.

240 Jeong, Y.C., Sangaiah, R., Nakamura, J., Pachkowski, B.F., Ranasinghe, A., Gold, A., Ball, L.M., and Swenberg, J.A. (2005) Analysis of M$_1$G-dR in DNA by aldehyde reactive probe labeling and liquid chromatography tandem mass spectrometry. *Chem. Res. Toxicol.*, **18**, 51–60.

241 Jeong, Y.C. and Swenberg, J.A. (2005) Formation of M$_1$G-dR from endogenous and exogenous ROS-inducing chemicals. *Free Radic. Biol. Med.*, **39**, 1021–1029.

3
Oxidatively Generated Damage to Isolated and Cellular DNA

Jean Cadet, Thierry Douki, and Jean-Luc Ravanat

3.1
Introduction

3.1.1
Overview and Summary

Emphasis is placed in this chapter on mechanistic aspects of the formation of single and clustered oxidatively generated damage in cellular DNA as the result of the reactions of the bases and the 2-deoxyribose moieties with singlet oxygen ($^1O_2$), hydroxyl radicals ($^{\bullet}OH$), two-photon photoionizing UV laser pulses, and hypochlorous acid (HOCl). The measurement of oxidatively generated damage was performed, once extracted DNA was either enzymatically or chemically hydrolyzed, by suitable chromatographic methods that mostly involve the combination of high-performance liquid chromatography with the electrospray ionization tandem mass spectrometry (HPLC-ESI/MS-MS) detection technique operating in the highly accurate multiple reactive monitoring (MRM) mode. The modifications thus identified were found to be generated in small amounts, typically at a frequency that in most cases is lower than one lesion per 10^6 normal bases. The formation of the 19 modified bases and nucleosides that have been identified so far in cellular DNA is accounted for by currently available mechanisms that were inferred from model studies involving free nucleosides, oligonucleotides, or isolated DNA. As one of the main conclusions, it is pointed out that the aqueous solutions that were used in the latter mechanistic studies appear to be a relevant medium for mimicking the oxidative degradation pathways of DNA that occur in cells.

3.1.2
Overview of Oxidatively Generated DNA Damage

Major progress has been achieved during the last two decades in achieving a better understanding of the degradation pathways of the bases and 2-deoxyribose moieties of isolated DNA and related model compounds upon exposure to $^1O_2$, $^{\bullet}OH$, one-electron oxidants, and HOCl. Extensive model studies have led to the isolation

The Chemical Biology of DNA Damage. Edited by Nicholas E. Geacintov and Suse Broyde
© 2010 WILEY-VCH Verlag GmbH & Co. KGaA, Weinheim
ISBN: 978-3-527-32295-4

and characterization of more than 80 modified nucleosides, including relatively unstable thymidine hydroperoxides and diastereomers of 5,6-dihydropyrimidines that result from oxidation reactions of DNA base moieties (for comprehensive reviews, see [1–9]). In addition, most of the degradation products arising from ˙OH-mediated hydrogen abstraction reactions from the methinic and methylenic groups of the 2-deoxyribose ring of nucleosides and DNA have been identified [10–12]. Relevant structural, kinetic, and reactivity information has been gained on radical precursors of most of the oxidation products of DNA components from numerous investigations that have involved electron spin resonance, laser flash, and pulse radiolysis measurements (for detailed surveys, see [13, 14]). The experimental data thus accumulated have led to the proposal of comprehensive oxidative degradation pathways for the 2-deoxyribose moiety and the four main nucleobases of isolated DNA and related model compounds, mostly in terms of the formation of single lesions. The conclusions based on these data have recently received further support from extensive theoretical studies based mostly on computational density functional theory calculations [15–19]. In contrast, there is still a large deficit of accurate information concerning the formation of oxidatively generated base and sugar damage in cellular DNA [9]. This can be mostly explained by the difficulties of measuring low amounts of oxidatively formed DNA damage in cells, typically within the range of a few lesions per 10^7 unmodified bases. Despite much efforts, one should note that the measurement of oxidized bases has been hampered for more than two decades by the occurrence of major artifacts such as either self-radiolysis processes [20] in the earlier determination of thymine glycol levels [21] or the spurious oxidation of normal bases during the derivatization step [22] of a questionable gas chromatography-based GC/MS assay [23]. In both cases this has led to the overestimation of the level of oxidized bases by factors of 2–4 orders of magnitude, making the data obtained by these methods questionable, and the related conclusions regarding the biological endpoints inaccurate. Another source of adventitious oxidation that is, however, of much lower impact has been identified more recently [22]. This involves the spurious oxidation of DNA by Fenton-type reactions during the extraction step and the subsequent work-up of the DNA samples. These issues are relevant to all of the methods that require DNA isolation by chromatographic assays. Optimized methods of DNA extraction that include the use of chelating agents for trapping transition metals are now available. As a result, a significant reduction of the contribution of adventitious oxidation processes has been reported [24, 25]. This issue was addressed by the European Standard Committee on Oxidative DNA Damage as part of a more general objective to compare and optimize existing methods for quantitatively measuring 8-oxo-7,8-dihydro-2′-deoxyguanosine (4) – a ubiquitous form of DNA oxidation damage [26, 27]. Thus, it was shown that HPLC coupled with either the electrochemical detection (ECD) technique that was introduced almost 25 years ago [28, 29] or the most recently available ESI/MS-MS [30] are both suitable for measuring 8-oxo-7,8-dihydro-2′-deoxyguanosine (4) accurately. However, a limiting factor is that at least 30 µg of DNA is required in order to minimize interference from spurious oxidation of the DNA that occurs during the extraction step [31]. It should be noted that the HPLC-ECD method is restricted to a few electroactive modified bases including 4, 8-oxo-7,8-dihydro-2′-deoxyadenosine (**34**), 5-hydroxypyrimidine bases, and formamidopyrimidine derivatives **22** and **34** arising from the opening of the imidazole ring of guanine and adenine, respectively (the numbers in bold refer to the structures shown in the figures below). In contrast, HPLC-ESI/MS-MS is more versatile and sensitive than HPLC-ECD, being capable of detecting at least a few femtomoles of **4** using the more recent version of an ESI-mass spectrometer operating in the MS^2 or MS^3 mode. Therefore, in this chapter, only the data that were obtained using optimized conditions of extraction with any of the two above accurate HPLC analytical tools are reported and critically reviewed.

3.2
Single Base Damage

Appropriate physical and chemical sources of oxidizing and halogenating species were used in order to induce a significant increase in the level of base and sugar damage with respect to the steady-state background of DNA modifications arising from oxidative metabolism. It may be added that this was achieved under conditions of acute exposure that do not allow for efficient DNA repair to occur due to the short time required for the oxidation reactions. $^1O_2$ was produced by either the thermal decomposition of a naphthalene endoperoxide or a type II photosensitization mechanism. Ionizing radiation also causes extensive and various kinds of DNA damage. The interactions of ionizing radiation with DNA in aqueous environment can occur either by *direct* or *indirect* mechanisms. In the direct effect, the ionizing radiation directly disrupts the chemical bonds in the DNA molecules. In the indirect effect, the ionizing radiation first disrupts the bonds in the water molecules, thus yielding the hydroxyl radicals, ˙OH, that then can interact with the DNA base or 2-deoxyribose residues. The indirect effect of ionizing radiation is predominant and was used by us to generate the highly reactive ˙OH, whereas efficient ionization of the purine and pyrimidine bases was achieved using high-intensity UV-C (below 290 nm wavelength) nanosecond laser pulses. The chloration of the three amino-substituted bases of DNA and RNA was performed using chemically generated HOCl.

3.2.1
Singlet Oxygen Oxidation of Guanine

Singlet oxygen, $^1O_2$, in the $^1\Delta_g$ state ($E = 22.4$ kcal/mol), exhibits dienophilic features, explaining its specific reactivity with guanine among all of the DNA bases. In contrast, molecular oxygen in its triplet state, $^3O_2$, may be considered as a biradical that therefore reacts efficiently mostly with carbon-centered radicals. $^1O_2$ can be generated in cells either by the photodynamic action of type II photosensitizers or as a side-product of enzymatic reactions mediated by myeloperoxidase [7, 32]. However, for mechanistic studies, the use of a clean chemical source of $^1O_2$

appears to be a better alternative in order to avoid the occurrence of side reactions that may be associated with the occurrence of one-electron oxidation processes and/or the generation of other reactive oxygen species (ROS) such as ˙OH. For this purpose, a suitably protected and thermolabile naphthalene endoperoxide that is able to penetrate cells and release $^1O_2$ at 37 °C in the vicinity of the nucleus was used [33]. It was found that $^1O_2$ oxidation of cellular DNA, as observed for isolated DNA under conditions where only a few guanine bases per 10^6 bases are modified, gives rise exclusively to 8-oxo-7,8-dihydro-2′-deoxyguanosine (4). This was inferred from HPLC-ESI/MS-MS analysis after enzymatic digestion of DNA into 2′-deoxyribonucleosides [34]. It was also shown using ^{18}O-labeled $^1O_2$ that the increase in the level of 4 was essentially due to singlet oxygen oxidation and not due to oxidative stress associated with the cellular incubation of the endoperoxide. The formation of 4 is accounted for by a Diels–Alder [4 + 2] cycloaddition of $^1O_2$ across the imidazole ring of the purine moiety of 1 giving rise to a diastereomeric pair of 4,8-endoperoxides 2 (Figure 3.1). Indirect support for such a mechanism was provided by a detailed nuclear magnetic resonance (NMR) analysis and characterization of a related endoperoxide resulting from the oxidation of 2′,3′,5′-O-tert-butyldimethylsilyl derivative of 8-methylguanosine by $^1O_2$ in CD_2Cl_2 at low temperature [35]. It is likely that the guanine endoperoxide 2 rearranges into the linear 8-hydroperoxy-2′-deoxyguanosine (3) [35, 36] as observed by a low-temperature NMR analysis of an organic solution of a 2′,3′,5-O-tert-butyldimethylsilyl derivative of 8-[^{13}C]guanosine derived from type II photosensitizer-mediated oxidation reactions [37]. A similar situation is expected to apply to the $^1O_2$ oxidation of 1 in both isolated and cellular DNA [38] in which 4 has been shown to be the predominant end-product of the reduction of the hydroperoxide precursor 3. The generation of 4R*- and 4S*-diastereomers of spiroiminodihydantoin 2′-deoxyribonucleosides 8 that are the main $^1O_2$ oxidation products of free 1 [39] through a complex sequence of reactions involving the transient formation of the quinonoid intermediate 6 [40] from 3 prior to its conversion into 5-hydroxy-8-oxo-7,8-dihydro-2′-deoxyguanosine (7) is, at best, a minor process in DNA. In addition, the forma-

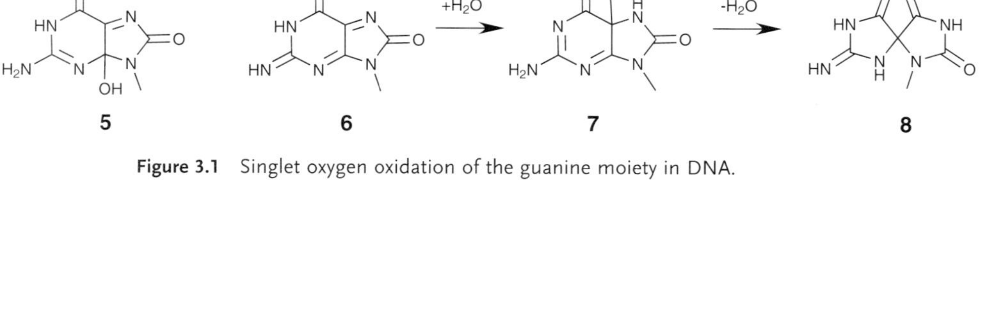

Figure 3.1 Singlet oxygen oxidation of the guanine moiety in DNA.

tion of 4-hydroxy-8-oxo-4,8-dihydro-2′-deoxyguanosine (**5**) that has been shown to occur with low efficiency by the oxidation of free **1** by $^1O_2$ [9] remains to be verified in double-stranded DNA. Evidence was recently provided [38] that in contrast to a previous claim [41], $^1O_2$ oxidation of guanine does not lead to the formation of 2,6-diamino-4-hydroxy-5-formamidopyrimidine (**22**), one of the main final degradation products arising from the hydration reaction of the guanine radical cation (Section 3.2.3); this rules out the possibility of charge transfer reaction mediated by singlet oxygen. Another clarification of a controversial issue concerned the putative DNA nicking activity of $^1O_2$ that was inferred from COMET assay analysis of cells upon incubation with naphthalene endoperoxide – this showed an almost total lack of generation of DNA strands breaks and/or alkali-labile sites [42].

It is now well documented that UV-A irradiation of several cell lines (for a review, see [43]) and human skin [44] leads to an increase in the steady-state level of **4** as the result of photodynamic activity [43]. Detailed mechanistic studies that involved the measurement of DNA strand breaks, together with oxidized pyrimidine bases and modified purine bases as DNA repair glycosylase-sensitive sites using a modified version of the altaline elution assay, have provided insights into the mechanisms of UV-A-sensitized oxidation of DNA. More specifically, the comparison in human monocytes of the UV-A-induced DNA oxidation product pattern with that obtained upon exposure to γ-radiation used as a reference for •OH reactions, has been of major diagnostic value [45]. Thus, the relative yield of formation of both DNA strand breaks and oxidized pyrimidine bases was much lower than the yield of purine base damage in the DNA of UV-A-irradiated cells, in comparison to analogous effects induced via the indirect effects of ionizing radiation (Table 3.1). This was accounted for by the predominant implication of $^1O_2$ oxidation that was estimated to be about 80% as the result of a type II

Table 3.1 Lesions and classes of damage to DNA of human THP-1 monocytes upon exposure to UV-A and ionizing radiation.*

Lesions	UV-A radiation (per kJ/m^2)	γ-Rays (per Gy)
8-Oxo-7,8-dihydro-2′-deoxyguanosine (**4**)[a]	0.98	11
2,6-Diamino-4-hydroxy-5-formamidopyrimidine (**22**)[b]	not determined	27
Strand breaks[c]	0.9	130
Endonuclease III-sensitive sites[d]	0.3	53
Formamidopyrimidine DNA glycosylase-sensitive sites[d]	1.9	48

The level of the lesions is expressed as number of modifications per 10^9 bases and per either kJ/m^2 (UV-C laser pulse dose range 0–150 kJ/m^2) or per Gy of ionizing radiation (dose range 0–40 Gy) [45].
a) HPLC-ECD.
b) HPLC-GC/MS.
c) COMET assay (single-strand breaks, double-strand breaks, and alkali-labile sites).
d) Modified COMET assay.
*From Ref. [45].

photosensitization mechanism [45]. The generation of the other 20% of **4** is explained in terms of ˙OH-mediated degradation pathways as the result of Fenton-type reactions that involve, in the initial step, the production of superoxide radical ($O_2^{\bullet-}$) and its subsequent spontaneous or enzymatic dismutation into H_2O_2 [45].

3.2.2
Hydroxyl Radical Reactions

As a ubiquitous ROS, ˙OH is known to react efficiently with the main purine and pyrimidine bases by its addition to unsaturated bonds. There is, however, a major exception – the hydrogen abstraction reaction from the methyl group of thymine **9**.

3.2.2.1 Thymine
Six stable oxidation products of thymine **9** have been detected and accurately measured mostly in the DNA of THP-1 human monocytes following exposure to γ-rays by HPLC-ESI/MS-MS methods [45–48]. This was achieved using the accurate isotopic dilution technique after suitable enzymatic digestion of extracted DNA [49]. The six oxidized nucleosides were generated in a linear manner as a function of the applied dose of low linear energy transfer (LET) γ-rays (within the dose range 90–450 Gy). Four *cis*- and *trans*-diastereomers of 5,6-dihydroxy-5,6-dihydrothymidine (**16**) and two methyl oxidation products including 5-(hydroxymethyl)-2′-deoxyuridine (**19**) and 5-formyl-2′-deoxyuridine (**20**) were obtained in this manner. It should be stressed that the radiation-induced formation of **16**, **19**, and **21** is rather low with values within the range of 29–97 lesions/10^9 bases/Gy (Table 3.2). These are much lower by factors varying from two to three orders of magnitude than the levels reported earlier using either the GC/MS assay or, more recently, HPLC/MS analytical tools operating in the relatively insensitive selective ion monitoring (SIM) mode. It was also found that exposure of the monocytes to

Table 3.2 Percent yields[a] of degradation products of thymine, guanine, and adenine in the DNA of THP-1 malignant human monocytes upon exposure to γ-rays [47].

Lesions	γ-Rays
Cis- and trans-5,6-dihydroxy-5,6-dihydrothymidine (**16**)	97
5-(Hydroxymethyl)-2′-deoxyuridine (**19**)	29
5-Formyl-2′-deoxyuridine (**20**)	22
8-Oxo-7,8-dihydro-2′-deoxyguanosine (**4**)	20
2,6-Diamino-4-hydroxy-5-formamidopyrimidine (**22**)	39
8-Oxo-7,8-dihydro-2′-deoxyadenosine (**33**)	3
4,6-Diamino-5-formamidopyrimidine (**34**)	5

a) Expressed in lesions per 10^9 nucleobases.

3.2 Single Base Damage

high-LET $^{12}C^{6+}$ particles led to similar trends with, however, a reduced efficiency in the formation of the oxidized nucleosides **16**, **19**, and **20**. This was even more pronounced when cells were irradiated with $^{36}Ag^{18+}$ ions that are associated with a higher LET value than $^{12}C^{6+}$ heavy particles. This observation may be rationalized in terms of the lower yield of ˙OH formation under exposure to high-LET radiation. Therefore, the indirect effect plays a major role in the radiation-induced degradation of cellular DNA.

Reasonable mechanisms that involve the predominant implication of ˙OH as the oxidizing species were proposed to explain the formation of **16**, **19**, and **20** by analogy with the comprehensive oxidative degradation pathways that have been proposed earlier for thymidine (**9**) in aerated aqueous solutions. The generation of the four cis- and trans-diastereomers of **16** in cellular DNA is rationalized in terms of the initial addition of ˙OH to the C5 and to a lesser extent the C6 carbon atoms of the thymine moiety as suggested by considering the results of redox titration pulse radiolysis experiments performed on the isolated base [50–52]. Addition of molecular oxygen to the reducing 6-yl (**10**) and to the oxidizing 5-yl (**11**) radicals thus generated (Figure 3.2) is a fast reaction that in aqueous solution is controlled by diffusion [53]. The resulting peroxyl radicals can be reduced, most likely by $O_2^{˙-}$ as shown in model studies [54], thus giving rise after protonation to 6-hydroperoxides **13** and 5-hydroperoxides **14**, respectively, whose two sets of four diastereomers have been fully characterized and their absolute configuration established [55, 56]. Reduction of the peroxidic bond of **13** and **14** that has been shown to occur in a stereospecific way [57] leads to the formation of corresponding thymidine glycols with retention of configuration at C5 and C6, respectively [58, 59]. It may be noted that N^1-(2-deoxy-β-D-*erythro*-pentofuranosyl)-5-hydroxy-5-methylhydantoin (**17**) and N-(2-deoxy-β-D-*erythro*-pentofuranosyl)formamide (**18**) that arise from the opening of the pyrimidine ring at the C5–C6 bond, followed either by ring rearrangement or hydrolysis, are also major decomposition products of **13** and **14**. However, these products have not yet been found to be generated in cellular DNA.

The radiation-induced formation of **19** and **20** in cellular DNA presents mechanistic similarities to those previously established for free thymidine and isolated DNA [55, 56]. Hydrogen abstraction from the methyl group of **9** by ˙OH generates the 5-(2′-deoxyuridilyl) methyl radical (**12**), which is efficiently converted into 5-(hydroperoxymethyl)-2′-deoxyuridine (**15**) after the transient formation of corresponding peroxyl radicals that are reduced prior to undergoing protonation. The hydroperoxide **15** has been shown to lead to the formation of **20** by the loss of a water molecule from the peroxidic functional group, whereas the generation of **19** is likely to result from the competitive reduction of the O–OH bond (Figure 3.2). It is striking and notable that the importance of the hydrogen atom abstraction pathway from the methyl group, as compared to the ˙OH addition pathway across the 5,6 double bond, is much higher in cellular DNA than in isolated **9**. This may be explained by the greater accessibility to ˙OH of the exocyclic methyl group that protrudes into the major groove in the double-stranded DNA structure, relative to the accessibility of the 5,6 ethylenic bond.

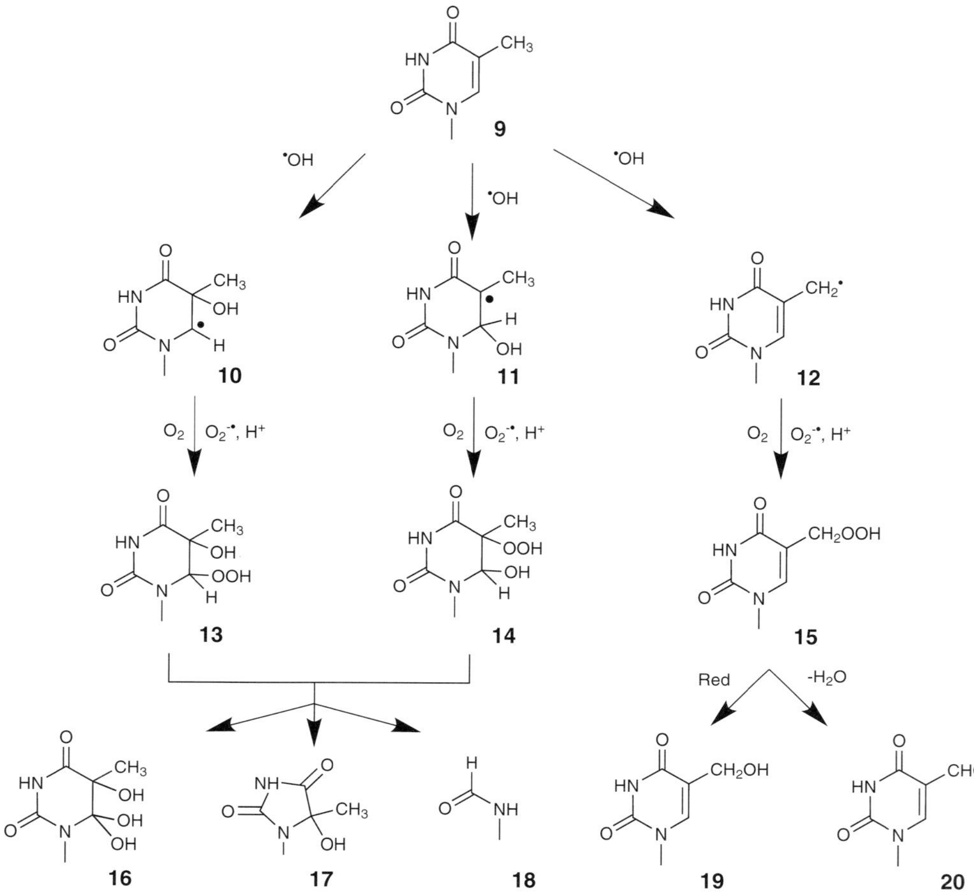

Figure 3.2 Hydroxyl radical-mediated oxidation of the thymine moiety in DNA.

3.2.2.2 Guanine

8-Oxo-7,8-dihydro-2′-deoxyguanosine (**4**) has been shown to be efficiently generated by $^1O_2$, ˙OH, one-electron oxidants, and peroxynitrite [1, 6, 7, 9] or as the result of intrastrand addition with thymine 5(6)-hydroxy-6(5)-hydroperoxides [60, 61]; this lesion is one of the main radiation-induced degradation products formed in cellular DNA following exposure to γ-rays and high-LET heavy ions [47, 48]. The accurate detection of **4** was performed by either HPLC-ECD or HPLC-ESI/MS-MS measurements after the latter oxidized guanine damage was quantitatively released enzymatically as a 2′-deoxyribonucleoside from extracted DNA. In addition, 2,6-diamino-4-hydroxy-5-formamidopyrimidine (**22**), the related opened imidazole ring compound, is generated in the DNA of γ-irradiated cells in about 2-fold higher yield than **4** (Table 3.2). A reliable assay that takes into consideration the known lability of the N-glycosidic bond of the formamidopyrimidine derivatives of purine

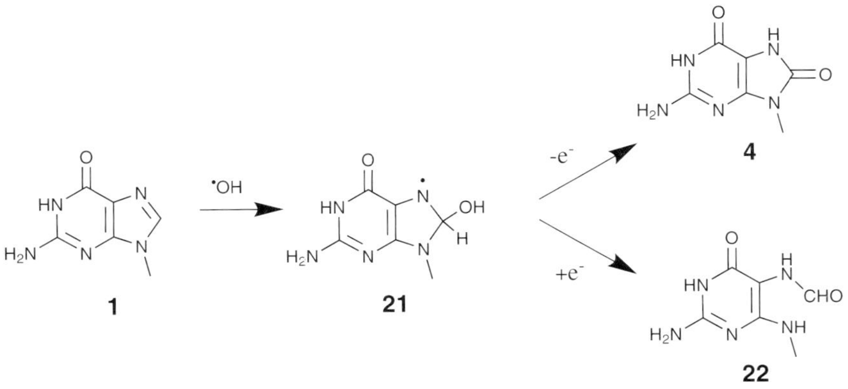

Figure 3.3 Oxidative degradation pathways of the guanine moiety of DNA upon ˙OH addition at C8.

2′-deoxyribonucleosides was used [49]. This allowed for the quantitative release of **22** whose formation was assessed by HPLC-ESI/MS-MS methods. In agreement with previous observations on the radiation-induced formation of thymidine degradation products, the higher LET values of the applied heavy particles with respect to γ photons led to a decrease in the yield of both **4** and **22** (Table 3.1). Again, this is strongly supportive of the major role played by ˙OH in the overall effects of ionizing radiation on the transformation of the guanine base into oxidized products in cellular DNA. Therefore, it can be proposed, in close agreement with previous mechanistic data inferred from model studies, that the radiation-induced formation of **4** and **22** in cellular DNA is accounted for by the degradation pathways depicted in Figure 3.3. Thus addition of ˙OH to the purine ring at C8 leads to the formation of reducing 8-hydroxy-7,8-dihydro-7-yl radical (**21**) [13] – the precursor of both **4** and **22**. One-electron oxidation of **21**, most likely by O_2 present in the nucleus, leads to the generation of **4** (Figure 3.3). On the other hand, the competitive one-electron reduction of **21** which appears to be a major process in cellular DNA, gives rise quantitatively to **22**. This occurs through the cleavage of the C8–N9 imidazole bond [61] with a high unimolecular rate ($k = 2 \times 10^5 \, s^{-1}$), at least in model compounds as inferred from pulse radiolysis measurements [62]. A second major competitive oxidative degradation pathway of **1** involves the transient formation of the highly oxidizing oxyl type radical **24** that has been proposed to arise from the dehydration of the predominant initially generated ˙OH adduct at C4, **23**. It has been shown in model studies involving nucleosides and oligonucleotides that $O_2^{\cdot-}$ [63], and not O_2 as proposed initially [64], is capable of adding to the C5 carbon of the related tautomeric radical **25**, which after protonation leads to the corresponding hydroperoxide **26**. This is followed by the nucleophilic addition of a water molecule across the 7,8 double bond of **26** giving rise to **27**. Opening of the 5,6 pyrimidine bond via a keto-hydroperoxide cleavage mechanism, followed by decarboxylation are the likely steps implicated in the generation of **28** [64, 65]. The

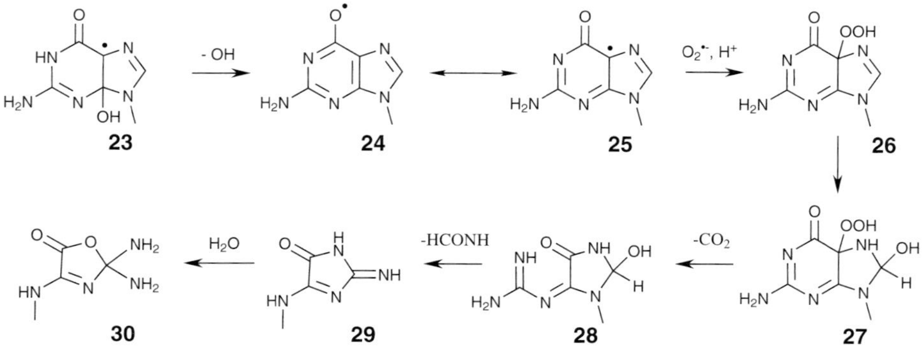

Figure 3.4 Oxidative degradation pathways of the guanine moiety of DNA upon ˙OH addition at C4.

release of a formamide molecule [66] that may be accounted for by ring-chain tautomerism of the carbinolamine function of **28**, is accompanied by an intramolecular cyclization reaction involving the guanidine residue, the overall rearrangement giving rise to 2-amino-5-[2-deoxy-β-D-*erythro*-pentofuranosyl)amino]-4*H*-imidazol-4-one (**29**). The latter compound is unstable in aqueous solution with a half-life of 10 h at neutral pH and 20 °C, and is gradually converted into 2,2-diamino-4-[(2-deoxy-β-D-*erythro*-pentofuranosyl)amino]-5(2*H*)-oxazolone (**30**) (Figure 3.4). Interestingly, **30** has been detected by HPLC/MS-MS in hepatic DNA of diabetic rats with a yield of formation that is, however, 10-fold lower than that of **4** [67]. This constitutes another example of the formation of such products in nuclear DNA that arise from the ˙OH-mediated degradation pathways previously established with free nucleosides and oligonucleotides in aerated aqueous solutions.

3.2.2.3 Adenine

Two degradation products of the adenine moiety **31** have been characterized that are structurally analogous to the products **5** and **22**, the two main ˙OH oxidatively generated guanine lesions; they have been detected by HPLC-ESI/MS-MS analysis in the DNA of human monocytes that had been exposed to γ-rays and accelerated heavy particles [47, 48]. These adenine products are 8-oxo-7,8-dihydro-2′-deoxyadenosine (**33**) and 4,6-diamino-5-formamidopyrimidine (**34**); however, these products are formed with an efficiency that is about 10-fold lower than that of related guanine decomposition products **4** and **22**. By analogy with the pathways of formation of radiation-induced damage to thymine and guanine in cellular DNA, it appears that the generation of **32** and **33** may be rationalized mostly in terms of initial ˙OH radical reactions with **31**. Therefore, a reasonable decomposition pathway of **31**, supported by earlier mechanistic studies, involves the transient formation of the 8-hydroxy-7,8-dihydroadenyl radical **32** as the result of ˙OH addition to C8 of the purine ring (Figure 3.5). The one-electron oxidation of precursor **32** leads to **33** [68], whereas its competitive one-electron reduction [13] gives rise to the formamidopyrimidine derivative **34** [69] through the scission of the 7,8

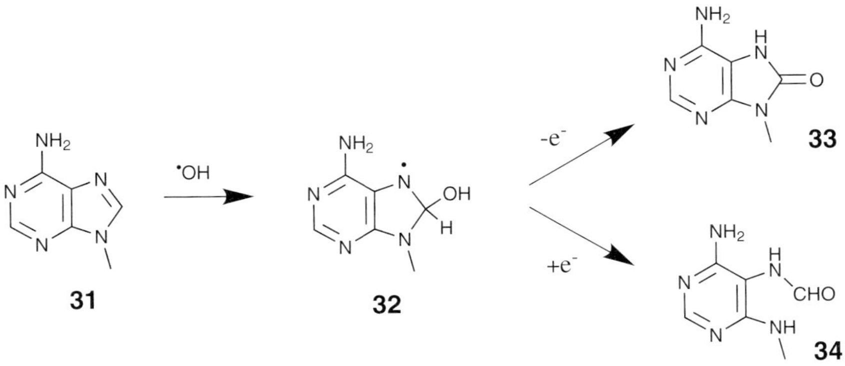

Figure 3.5 Oxidative degradation pathways of the adenine moiety of DNA upon ˙OH addition at C8.

bond of the imidazole ring. Recent attempts have been made using the sensitive and accurate HPLC-ESI/MS-MS method to search for the formation of 2-hydroxy-2′-deoxyadenosine, an ˙OH-mediated adenine decomposition product whose existence has been proposed for a long time, in the DNA of γ-irradiated human monocytes [70]. However, no detectable amounts of 2-hydroxy-2′-deoxyadenosine were found to be present in the nuclear DNA even at a dose of 200 Gy. This contrasts with the relatively high yield of the latter oxidized nucleoside, within the range of $0.1/10^6$ nucleosides/Gy that was found by CG/MS analysis of the DNA of γ-irradiated cells and mice [71, 72]. This represents another example of the overestimation of the levels of oxidatively generated damage to cellular DNA when the questionable CG/MS assay is applied.

3.2.3
One-Electron Oxidation of Nucleobases

The one-electron oxidation of nucleobases may result from the direct effect of ionizing radiation [73, 74], electron abstraction by type I photosensitizers [32], high-intensity UV laser pulses [75, 76], or reactions with oxidants such as $KBrO_3$ [77] and CO_3^- [78] – a decomposition product of nitrosoperoxycarbonate [79]. Abundant information is available in the literature on the chemical reactions of the pyrimidine and purine radical cations [9, 32] thus generated that may be accounted for by the initial deprotonation and/or hydration processes. The use of high-intensity UV-C nanosecond laser pulses has been shown to be an efficient way of producing ionized purine and pyrimidine bases and related radical cations in both isolated [76] and cellular DNA [48] by two-photon-induced ionization mechanisms. Even if bipyrimidine photoproducts such as cyclobutane pyrimidine dimers and pyrimidine–pyrimidone 6-4 photoproducts are also produced by monophotonic excitation [43], the two-quantum mechanism constitutes a better alternative for generating one-electron base oxidation products than ionizing radiation that predominantly induces •OH-mediated degradation products. Several oxidized nucleosides have been detected by HPLC-ESI/MS-MS analysis in the DNA of THP-1 neoplastic human monocytes following exposure to nanosecond UV-C pulses [48]. The predominant modified nucleoside has been identified as 8-oxo-7,8-dihydro-2′-deoxyguanosine (**4**). In addition, 5,6-dihydroxy-5,6-dihydrothymidine (**16**), 5-(hydroxymethyl)-2′-deoxyuridine (**19**), and 5-formyl-2′-deoxyuridine (**20**) are produced in much lower yields since, altogether, these minor products represent about 10% of the yield of **4**. The biphotonic-mediated formation of **4** and **16** may be accounted for by hydration reactions of the initially formed guanine and thymine radical cations **35** and **36**, leading to the transient generation of the reducing 8-hydroxy-7,8-dihydroguanyl (**21**) [80] and the oxidizing 6-hydroxy-5,6-dihydrothym-5-yl (**11**) [81] radicals, respectively (Figures 3.6 and 3.7). In a subsequent step, molecular oxygen is expected to oxidize the 8-yl radical, a critical step for the generation of **6**, whereas addition of O_2 to **11** represents the initial step of a sequence of reactions giving rise to **16** through the likely intermediacy of **14**. The competitive deprotonation of the thymine radical cation followed by fast addi-

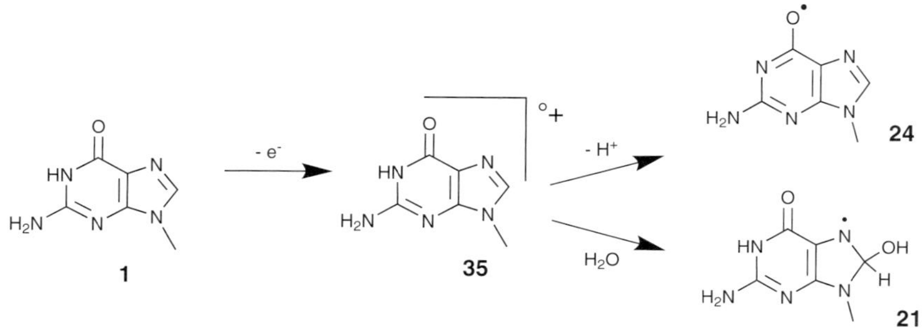

Figure 3.6 One-electron oxidation reactions of the guanine moiety of DNA.

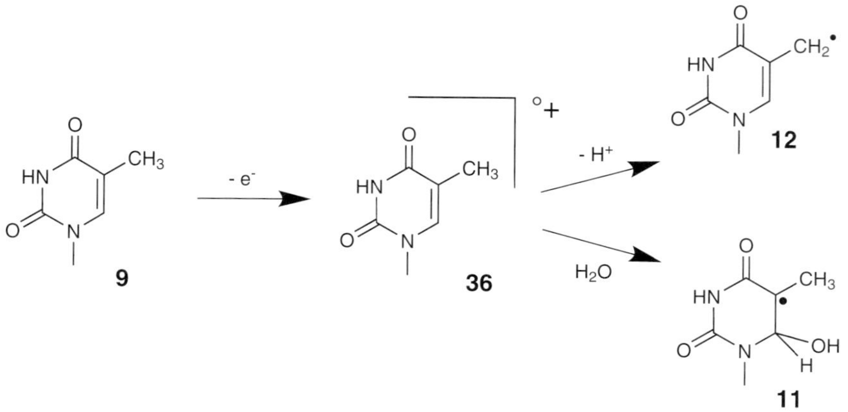

Figure 3.7 One-electron oxidation reactions of the adenine moiety of DNA.

tion of O_2 to the resulting 5-methyl-centered thymine radical **12** constitute the initial events of oxidative degradation pathways giving rise to **19** and **20** through the transient formation of 5-(hydroperoxymethyl)uracil (**15**), as depicted in Figure 3.2 for ˙OH-mediated reactions (Section 3.2.1).

The preferential formation of **4** is most likely explained in terms of charge transfer of the initially generated purine and pyrimidine radical cations to guanine sites that act as sinks of positive holes as previously shown in model studies involving double-stranded DNA. This appears to be the first evidence for the occurrence of positive hole migration within cellular DNA that may operate through several possible mechanisms including phonon-assisted polaron-like hopping, coherent super-exchange hopping, and multistep hopping [82–85].

3.2.4
HOCl Acid-Mediated Halogenation of Pyrimidine and Purine Bases

HOCl is enzymatically produced by myeloperoxidase during the inflammation processes [86]. It has been shown to exhibit both halogenating and one-electron oxidation features [87]. Evidence has been provided for the generation of the nucleoside derivatives of 5-chlorocytosine (**37**), 8-chloroguanine (**38**), and 8-chloroadenine (**39**) (Figure 3.8) in the DNA and RNA of SKM-1 cells upon exposure to HOCl [88], as previously observed in free nucleosides [89, 90]. This was inferred from accurate and quantitative HPLC-ESI/MS-MS measurements that allow for the detection of each of the halogenated ribo- and 2′-deoxyribonucleosides in the subfemtomole range [88]. 5-Chloro-2′-deoxycytidine (**37**) was found to be formed in larger yields than 8-chloro-2′-deoxyguanosine (**38**) and 8-chloro-2′-deoxyadenosine (**39**) in cellular DNA (Table 3.3) [88]. Another interesting feature deals with the higher susceptibility of RNA to HOCl-mediated halogenation of amino-substituted nucleobases with respect to nuclear DNA. One may also note that the levels of 5-chlorocytidine and 8-chloroguanosine were higher than those of 8-chloroadenosine (Table 3.3). Support for the use of **37** as an indicator of inflammation was provided by the observation of higher levels of the latter halogenated pyrimidine nucleoside in the DNA of diabetic patients with respect to those in healthy volunteers [91]. A final example of chloration reaction in cellular DNA is illustrated by the detection of 5-chlorocytosine in bacteria upon exposure to the myeloperoxidase–H_2O_2–Cl^- system of phagocytes [86].

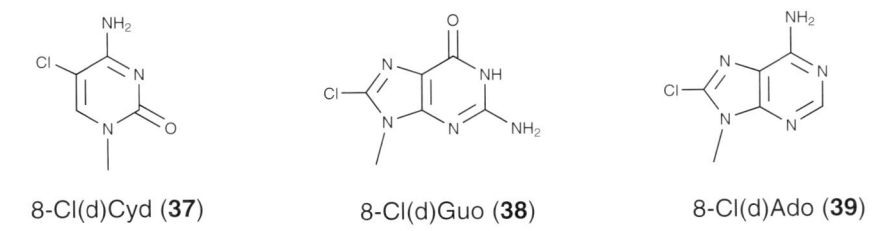

8-Cl(d)Cyd (**37**)   8-Cl(d)Guo (**38**)   8-Cl(d)Ado (**39**)

Figure 3.8 Chlorinated nucleosides formed in DNA and RNA.

Table 3.3 Yields[a] of chlorinated aminobases in the DNA and RNA of SKM-1 cells upon incubation with 300 µM HOCl for 10 min [88].

Chlorinated aminobases	DNA	RNA
5-Chlorocytosine (37)	9.8 ± 2.3	15.8 ± 0.5
8-Chloroguanine (38)	2.0 ± 0.4	16.2 ± 1.8
8-Chloroadenine (39)	1.5 ± 0.4	0.5 ± 0.4

a) Expressed in number of lesions per 10^6 nucleobases.

3.3
Tandem Base Lesions

Ionizing radiation and high-LET heavy ions are known to generate clustered lesions in cellular DNA through multiple radical and excitation events whose complexity increases with the LET value. However, the structures of these complex types of DNA damage remain mostly unidentified. This also applies, in part, to the heterogeneous class of double-strand breaks that can be detected by the neutral COMET assay [92], the pulsed-field gel electrophoresis method [93], or the sensitive immunodetection of γ-H2AX foci [94]. The occurrence of intrastrand tandem base modifications that arise from a single radical event initiated by either ·OH or one-electron oxidants has been initially discovered through the pioneering contributions of Box *et al.* [95, 96]. The participation of pyrimidine peroxyl radicals in addition reactions with a vicinal guanine base [97] has been demonstrated on the basis of a detailed mechanistic study [60] and further extended to vicinal pyrimidine bases [98, 99]. However, no evidence for the formation of these tandem modifications has been provided so far in cellular DNA. Another type of tandem base damage has been shown to involve the addition of either 5-(uracilyl)methyl [100–104] or the 5-(cytosilyl)methyl radical [104–107] to the C8 of vicinal purine bases on the basis of experimental evidence, while theoretical calculations gave further support to the mechanisms involved [19, 108]. Evidence was also provided for the ·OH-mediated formation of an intrastrand cross-link between the 6-hydroxy-5,6-dihydrocytosyl radical and a vicinal guanine leading, after dehydration to the generation of the G[8–5]C adduct **41** [109] (Figure 3.9). As a general trend, the addition reaction of the pyrimidine radicals is favored in the absence of oxygen and when the purine base is located on its 5'-side. Thus, it was shown using HPLC-ESI/MS-MS measurements that the G[8–5m]T **40** adduct is more efficiently generated in isolated DNA exposed to ·OH in oxygen-free aqueous solutions than the other T[5m–8]G (**41**) position isomer [101]. It may be added that the related A[8–5m]T and T[5m–8]A intrastrand cross-links are produced in lower yields [101]. Evidence has been recently provided for the formation of G[8–5m]T **40** and G[8–5]C **41** tandem base lesions as minor degradation products, however, in naked DNA exposed to the Cu(II)/H_2O_2/ascorbate or Fe(II)/H_2O_2/ascorbate oxidizing systems in aerated aqueous solutions [105, 107]. Both lesions have been accurately detected

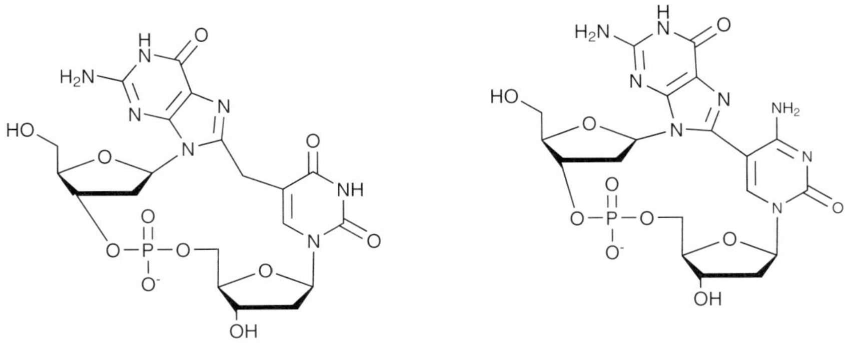

G[8-5m]T 40   **G[8-5]C 41**

Figure 3.9 Vicinal base lesions formed in DNA.

in the DNA of HeLa cells exposed to γ-rays after suitable enzymatic digestions followed by HPLC analysis with tandem MS and MS^3 detection [110, 111]. The yields of formation of G[8–5m]T [110] and G[8–5]C [111] cross-links that were shown to be more than two orders of magnitude lower than that of 5-formyl-2′-deoxyuridine (**20**) were 0.050 and 0.037 lesions/10^9 nucleosides/Gy, respectively.

3.4
Hydroxyl Radical-Mediated 2-Deoxyribose Oxidation Reactions

The main oxidation reactions of the 2-deoxyribose of DNA are mediated by ˙OH that are able to abstract hydrogen atoms from most of the positions with the exception of the methylene group at C2, which is a poorly reactive site [12, 14, 112]. An abundant literature is available on the degradation pathways that are derived from the reactions of osidic carbon-centered radicals and that lead in most cases to the formation of strand breaks [12, 14, 112]. However, there is one major exception that concerns the chemical reactions of the C1′ radical that is the precursor of 2-deoxyribonolactone [113]. Here, emphasis is placed on reactions of C4′ and C5′ radicals that may lead to the formation of base modifications either as tandem lesions or clustered damage.

3.4.1
Hydrogen Abstraction at C4′: Formation of Cytosine Adducts

Radiomimetic drugs of the enedyine family are known to induce DNA damage through hydrogen abstraction of the 2-deoxyribose moiety at mostly C4′ and C5′ positions [114]. The toxicity of such drugs is attributed to the possibility of generating two closely located single nicks on opposite strands, leading to highly genotoxic double-strand breaks. Among them is bleomycin – a minor groove binding drug used in human cancer therapy that is able to abstract a hydrogen atom in double-

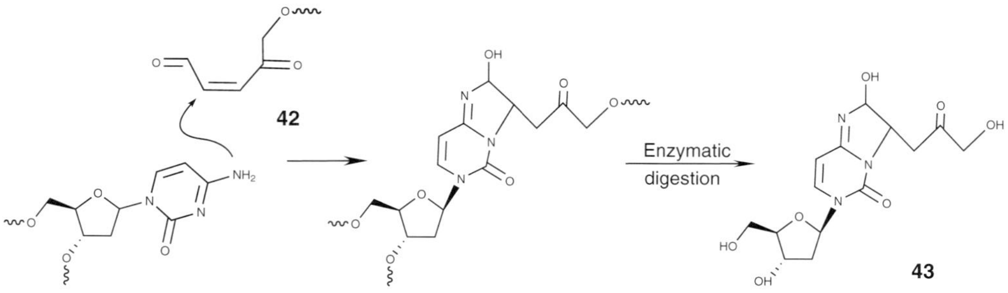

Figure 3.10 Formation of cytosine adducts (**43**) by reaction of reactive aldehyde arising from C4′ hydrogen abstraction in DNA.

stranded DNA mostly at C4′. Interestingly, the cascade of reactions occurring subsequently to hydrogen atom abstraction [10] may give rise to reactive aldehydic intermediates. For instance, hydrogen abstraction at C4′ generates a conjugated keto-aldehyde (**42**) that may further react with nucleophilic DNA bases (Figure 3.10). The cycloadducts that are formed with cytosine were characterized as the two pairs of diastereomers of 6-(2-deoxy-β-D-*erythro*-pentofuranosyl)-2-hydroxy-3(3-hydroxy-2-oxopropyl)-2,6-dihydroimidazo[1,2-c]pyrimidin-5(3*H*)-one (**43**) in both isolated and cellular DNA [115, 116]. The latter damage could be considered as clustered lesions whose formation is initiated by a single oxidation event (i.e., C4′ hydrogen abstraction produced either by bleomycin or ·OH). Indeed, the formation of cytosine adducts involves first the formation of strand breaks on the 3′-end relative to the hydrogen abstraction site as a result of a β-elimination reaction of the C4′ oxidized abasic site [116]. Then, the generated aldehyde (**42**) is able to react with a cytosine base, most probably located on the opposite DNA strand, thus creating a cross-link between the two strands. Work is currently in progress to better ascertain the structure of the interstrand cross-link. The yield of formation of such complex DNA damage in γ-irradiated cells has been determined to be similar to that of the deleterious double-strand breaks [116]. It may also be pointed out that the repair of the cytosine adducts (**43**) is rather low with a half-life of removal that was estimated to be 10 h. It has been recently shown that the reactive conjugated aldehyde (**42**) is also able to react with adenine to produce interstrand cross-links as demonstrated using photolabile precursors of the C4′ oxidized abasic site that has been site-specifically inserted into defined sequence DNA duplexes [117].

3.4.2
Hydrogen Atom Abstraction at C5′: Formation of Purine 5′,8-Cyclonucleosides

Evidence for the ·OH induced formation of purine 5′,8-cyclonucleosides has been provided in early model studies involving nucleosides [118], nucleotides [119, 120], and isolated DNA [121]. This may be explained in terms of ·OH-mediated hydrogen atom abstraction from the C5′-hydroxymethyl group followed by intramolecular cyclization that occurs with a rate constant $k = 1.6 \times 10^5 \, s^{-1}$ in the case of

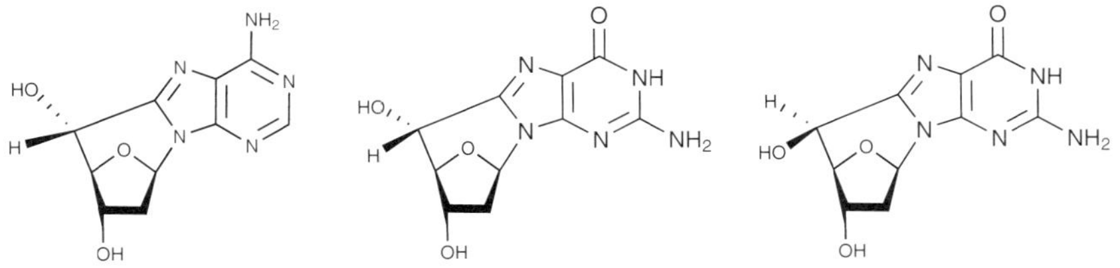

(5′R) 5′,8-cyclodAdo **44a**   (5′R) 5′,8-cyclodGuo **45a**   (5′S) 5′,8-cyclodGuo **45b**

Figure 3.11 Purine 5′,8-cyclonucleosides formed in DNA upon •OH-mediated hydrogen atom abstraction at C5′.

2′-deoxyadenosine [122], prior to a final oxidation step that is likely to involve molecular oxygen [123, 124]. Thus, 5′R- and 5′S-diastereomers of both 5′,8-cyclo-2′-deoxyadenosine (**44**) and 5′,8-cyclo-2′-deoxyguanosine (**45**) (Figure 3.11) have been found to be generated in DNA upon exposure to the indirect effects of ionizing radiation in aqueous solutions [125, 126]. The presence of molecular oxygen that efficiently reacts with the carbon-centered C5′ radical is expected to reduce the formation of the purine 5′,8-cyclonucleosides [127–129]. The radiation-induced formation of the 5′S-diastereomer of **44** and both 5′R- and 5′S-diastereomers of **45** [130, 131] has been tentatively detected in cellular DNA using GC/MS [126] and HPLC/MS [125] assays, respectively, that were both operating in the SIM mode using the isotopic dilution technique for the quantification. Thus, it was found that the levels of the 5′S form of **44** and both diastereomers of **45** were significantly increased in the DNA of normal human keratinocytes upon exposure to a dose of X-rays as moderate as 5 Gy. The yields of radiation-induced cyclonucleosides per 10^9 normal nucleosides/Gy were found to be 14, 100, and 30 lesions for (5′S)-5′,8-cyclo-2′-deoxyadenosine (**44a**) and the 5′S- and 5′R-diastereomers of 5′,8-cyclo-2′-deoxyguanosine (**45**), respectively [130]. These values appear to be rather high if compared with the yields of radiation-induced 8-oxo-7,8-dihydro-2′-deoxyguanosine (**4**) and 8-oxo-7,8-dihydro-2′-deoxyadenosine (**33**), as measured by HPLC-ESI/MS-MS in the DNA of human monocytes, which are 20 and 4 lesions/10^9 normal nucleosides/Gy, respectively [47]. This is not consistent with the fact that in isolated DNA exposed to γ-rays in partially deaerated aqueous solutions, the yield of (5′S)-cyclodAdo (**44b**) is about 200-fold lower than that of **4** (Belmadoui et al., in preparation). In that respect, one may question the accuracy of either HPLC associated with only a single quadrupole or the CG/MS assay when high sensitivity is required. Another questionable measurement method concerns the tentative detection of purine 5′,8-cyclonucleoside containing dinucleotides in cellular DNA by applying a ^{32}P-postlabeling assay [131]. In fact accurate detection of nucleosides in the subfemtomole range would necessitate at least HPLC-ESI/MS-MS and even in some cases the more reliable HPLC/MS^3. Recent examples of the latter application dealt with the measurement of G[8–5m]T and G[8–5]C tandem lesions in γ-irradiated DNA that are produced with very low frequencies, as expected also for the purine 5′,8-cyclonucleosides. It may be added that preliminary HPLC-ESI/

MS-MS measurements performed recently in our laboratory have shown that the yield of radiation-induced formation of (5′*R*)-cyclodAdo (**44a**) in cellular DNA is at best around 0.2 lesions/10^9 normal nucleotides/Gy. This is in agreement with the fact that the shielding effect exerted by cellular constituents on the extent of radiation-induced damage to DNA bases and sugar is about three orders of magnitude with respect to naked DNA [12].

3.5 Secondary Oxidation Reactions of Bases

There is an increasing interest in the possible occurrence in cellular DNA of secondary one-electron oxidation reaction products of several oxidized bases including 8-oxo-7,8-dihydroguanine (**4**), 5-hydroxyuracil, and 5-hydroxycytosine because the oxidation potentials are much lower than those of the canonical DNA bases, particularly those of guanine (**1**). There is a large body of information from model studies showing that **4**, whose oxidation potential is about 0.5 eV lower than that of 2′-deoxyguanosine (**1**) [132, 133], is a preferential target for numerous one-electron oxidizing agents and radicals [134–138] that can generate the corresponding radical cation **46** (Figure 3.12). Interestingly, the *R*- and *S*-diastereomers of the spiroiminodihydantoin nucleosides **8** have been shown to be the predominant one-electron oxidation products of both 2′-deoxyribonucleoside and ribonucleoside derivatives of **4** at neutral pH [134–140] as the result of an acyl shift of the transiently produced 5-hydroxy-8-oxo-7,8-dihydro-2′-deoxyguanosine (**7**) [139]; it was initially proposed that **7** is a relatively stable oxidation product of **4**. In addition, it has been shown that at acidic pH, the decomposition of **7** gives rise mostly to the guanidinohydantoin derivatives **47**. The accumulation of two diastereomers of **8** has been reported in the DNA of Nei-deficient *Escherichia coli* cells upon exposure to chromate ion [141]. The level of **8** that was assessed by HPLC-MS operating in the SIM mode was six lesions per 10^3 guanines upon treatment of TK3D11 bacterial cells with

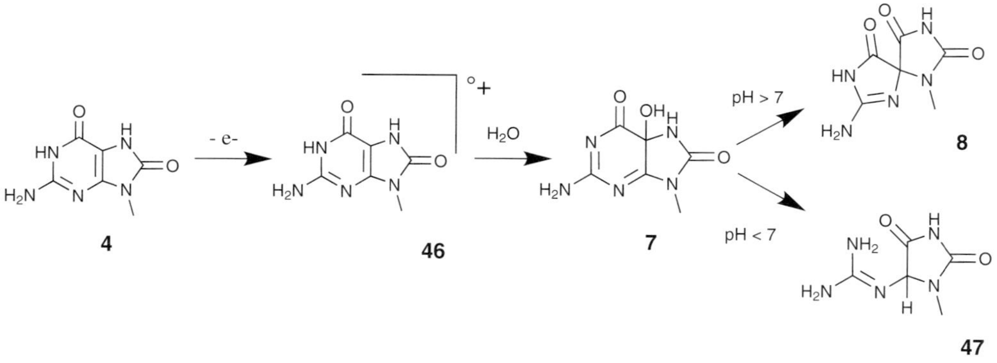

Figure 3.12 One-electron oxidation reactions of 8-oxo-7,8-dihydro-2′-deoxyguanosine (**4**).

500 μM Cr(VI) – a frequency which is about 20-fold higher than in wild-type cells. It may be stressed that in both bacterial cells, the level of **4** that is the likely precursor of **8** is much lower. However, the secondary oxidation of **4** appears to be an unlikely event that could explain the preferential formation of the latter secondary oxidation product **8** at the expense of **4**, particularly in wild-type cells [80]. It would be of interest to search for the formation of **8** using, for example, the HPLC-ESI/MS-MS technique which in the MRM mode is much more accurate than the HPLC-MS/SIM method that has been shown to lead to overestimated values of radiation-induced **4** and **33** in the DNA of human cells [142–144]. This should shed some light on the postulated sacrificial role played by **4** in protecting normal nucleobases within DNA against the damaging effects of one-electron oxidants [145, 146] that has been recently questioned [147]. One may wonder how **4**, with endogenously generated steady-state levels not exceeding at best a few residues per 10^6, could be preferentially oxidized in the presence of overwhelming amounts of nucleobases by one-electron oxidants. Even if one considers that hole transfer process within DNA helix can occur at a distance of up to 20 bp, this does not seem very likely.

3.6
Conclusions and Perspectives

Evidence is provided in this chapter that oxidatively generated base and sugar damage occurs within cellular DNA upon acute exposure to physical and chemical oxidizing agents. The measurement of the 16 modified nucleosides and nucleobases so far identified was achieved using mostly accurate and specific HPLC-ESI/MS-MS assays that have required optimization of DNA extraction protocols in order to minimize the occurrence of artifactual oxidation. It appears that the induced level of oxidized bases is rather low, typically a few modifications per 10^8 normal nucleosides/Gy of ionizing radiation with respect to a background level which, at its highest is within the range of a few lesions per 10^6 normal nucleosides, as has been shown for 8-oxo-7,8-dihydroguanine (**4**). The mechanisms of degradation of nucleobases by ·OH radicals, one-electron oxidants, $^1O_2$, and HOCl, which have been adopted from model studies performed in aqueous solutions, were found to be applicable in cells as well. Efforts should be made to search for still unidentified lesions in cells, including cytosine oxidation products, DNA–protein cross-links, and tandem modifications such as purine 5′,8-cyclonucleosides. Such efforts should benefit, at least partially, from the recent addition of more specific and sensitive HPLC-ESI/MS-MS instruments for this purpose.

Acknowledgements

Partial support was provided by the Marie Curie Research Training Network under contract MRTN-CT-2003-50586 (CLUSTOXDNA) and COST Action CM0603 on "Free Radicals in Chemical Biology" (CHEMBIORADICAL).

References

1 Cadet, J., Berger, M., Douki, T., and Ravanat, J.-L. (1997) Oxidative damage to DNA: formation, measurement and biological significance. *Rev. Physiol. Biochem. Pharmacol.*, **131**, 1–87.

2 Cadet, J., Douki, T., Gasparutto, D., and Ravanat, J.-L. (2003) Oxidative damage to DNA: formation, measurement and biochemical features. *Mutat. Res.*, **531**, 5–23.

3 Douki, T., Ravanat, J.-L., Angelov, D., Wagner, J.R., and Cadet, J. (2004) Effects of duplex stability on charge transfer efficiency within DNA. *Top. Curr. Chem.*, **236**, 1–25.

4 Gimisis, T. and Cismas, C. (2006) Isolation, characterization, and independent synthesis of guanine oxidation products. *Eur. J. Org. Chem.*, 1351–1378.

5 Pratviel, G. and Meunier, B. (2006) Guanine oxidation: one- and two-electron reactions. *Chem. Eur. J.*, **12**, 6018–6030.

6 Neeley, W.L. and Essigmann, J.M. (2006) Mechanisms of formation, genotoxicity, and mutation of guanine oxidation products. *Chem. Res. Toxicol.*, **19**, 491–505.

7 Cadet, J., Ravanat, J.-L., Martinez, G.R., Medeiros, M.H.G., and Di Mascio, P. (2006) Singlet oxygen oxidation of isolated and cellular DNA: product formation and mechanistic insights. *Photochem. Photobiol.*, **82**, 1219–1225.

8 Cadet, J. and Di Mascio, P. (2006) Peroxides in biological systems, in *Functional Groups in Organic Chemistry – The Chemistry of Peroxides*, vol. 2 (ed. Z. Rappoport), John Wiley & Sons, Inc., Hoboken, NJ, pp. 915–1000.

9 Cadet, J., Douki, T., and Ravanat, J.-L. (2008) Oxidatively generated damage to the guanine moiety of DNA: mechanistic aspects and formation in cells. *Acc. Chem. Res.*, **41**, 1075–1083.

10 Pogozelski, W.K. and Tullius, T.D. (1998) Oxidative strand scission of nucleic acids: routes initiated by hydrogen abstraction from the sugar moiety. *Chem. Rev.*, **98**, 1089–1107.

11 Aydogan, B., Marshall, D.T., Swarts, S.G., Turner, J.E., Boone, A.J., Richards, N.G., and Bolch, W.E. (2002) Site-specific OH attack to the sugar moiety of DNA: a comparison of experimental data with computational simulation. *Radiat. Res.*, **157**, 38–44.

12 Dedon, P.C. (2008) The chemical toxicology of 2-deoxyribose oxidation in DNA. *Chem. Res. Toxicol.*, **21**, 206–219.

13 Steenken, S. (1989) Purine bases, nucleosides, and nucleotides: aqueous solution redox chemistry and transformation reactions of their radical cations and e⁻ and OH adducts. *Chem. Rev.*, **89**, 503–520.

14 von Sonntag C. (ed.) (2006) *Free-Radical Induced DNA Damage and its Repair: A Chemical Perspective*, Springer, Heidelberg.

15 Grand, A., Morell, C., Labet, V., Cadet, J., and Eriksson, L.A. (2007) ˙H atom and ˙OH reactions with 5-metylcytosine. *J. Phys. Chem. A*, **111**, 8968–8972.

16 Munk, B.H., Burrows, C.J., and Schlegel, H.B. (2007) Exploration of mechanisms for the transformation of 8-hydroxyguanine radical to FAPyG by density functional theory. *Chem. Res. Toxicol.*, **20**, 432–444.

17 Munk, B.H., Burrows, C.J., and Schlegel, H.B. (2008) An exploration of mechanisms for the transformation of 8-oxoguanine to guanidinohydantoin and spiroiminodihydantoin by density functional theory. *J. Am. Chem. Soc.*, **130**, 5245–5256.

18 Labet, V., Grand, A., Cadet, J., and Eriksson, L.A. (2008) Deamination of the radical cation of base moiety of 2′-deoxycytidine: a theoretical study. *ChemPhysChem*, **9**, 1195–1203.

19 Labet, V., Morell, C., Grand, A., Cadet, J., Cimino, P., and Barone, V. (2008) Formation of cross-linked adducts between guanine and thymine mediated by hydroxyl radical and one-electron oxidation: a theoretical study. *Org. Biomol. Chem.*, **6**, 3300–3305.

20 Cadet, J. and Berger, M. (1985) Radiation-induced decomposition of the

purine bases within DNA and related model compounds. *Int. J. Radiat. Biol.*, **47**, 127–143.

21 Hariharan, P.V. and Cerutti, P.A. (1972) Formation and repair of γ-ray induced thymine damage in *Micrococcus radiodurans*. *J. Mol. Biol.*, **66**, 65–81.

22 Cadet, J., Douki, T., and Ravanat, J.-L. (1997) Artifacts associated with the measurement of oxidized DNA bases. *Environ. Health Perspect.*, **105**, 1034–1039.

23 Halliwell, B. and Dizdaroglu, M. (1992) The measurement of oxidative damage to DNA by HPLC and GC-MS techniques. *Free Radic. Res. Commun.*, **16**, 75–87.

24 Ravanat, J.-L., Douki, T., Duez, P., Gremaud, E., Herbert, K., Hofer, T., Lassere, L., Saint-Pierre, C., Favier, A., and Cadet, J. (2002) Cellular background of 8-oxo-7,8-dihydro-2′-deoxyguanosine: an isotope based method to evaluate artefactual oxidation of DNA during its extraction and subsequent work-up. *Carcinogenesis*, **23**, 1911–1918.

25 Chao, M.R., Yen, C.C., and Hu, C.W. (2008) Prevention of artifactual oxidation in determination of cellular 8-oxo-7,8-dihydro-2′-deoxyguanosine by isotope-dilution LC-MS/MS with automated solid-phase extraction. *Free Radic. Biol. Med.*, **44**, 464–473.

26 Collins, A.R., Cadet, J., Möller, L., Poulsen, H., and Vina, J. (2004) Are we sure we know how to measure 8-oxo-7,8-dihydroguanine in DNA from human cells? *Arch. Biochem. Biophys.*, **423**, 57–65.

27 ESCODD (2005) Establishing the background level of base oxidation in human lymphocyte DNA: results of an inter-laboratory validation study. *FASEB J.*, **19**, 82–84.

28 Floyd, R.A., Watson, J.J., Wong, P.K., Altmiller, D.H., and Rickard, R.C. (1986) Hydroxyl free radical adduct of deoxyguanosine: sensitive detection and mechanism of formation. *Free Radic. Res. Commun.*, **1**, 163–172.

29 Kasai, H. (1997) Analysis of a form of oxidative DNA damage, 8-hydroxy-2′-deoxyguanosine, as a marker of cellular oxidative stress during carcinogenesis. *Mutat. Res.*, **387**, 147–163.

30 Ravanat, J.L., Duretz, B., Guiller, A., Douki, T., and Cadet, J. (1998) Isotope dilution high-performance liquid chromatography-electrospray tandem mass spectrometry assay for the measurement of 8-oxo-7,8-dihydro-2′-deoxyguanosine in biological samples. *J. Chromatogr. B*, **715**, 349–356.

31 Badouard, C., Ménézo, Y., Panteix, G., Ravanat, J.-L., Douki, T., and Favier, A. (2008) Determination of new types of DNA lesions in human sperm. *Zygote*, **16**, 9–13.

32 Cadet, J. and Vigny, P. (1990) The photochemistry of nucleic acids, in *Bioorganic Photochemistry* vol. **1** (ed. H. Morrison), John Wiley & Sons, Inc., New York, pp. 1–272.

33 Martinez, G.R., Ravanat, J.-L., Medeiros, M.H.G., Cadet, J., and Di Mascio, P. (2000) Synthesis of a naphthalene endoperoxide as a source of ^{18}O-labeled singlet oxygen for mechanistic studies. *J. Am. Chem. Soc.*, **122**, 10212–10213.

34 Ravanat, J.L., Di Mascio, P., Martinez, G.R., Medeiros, M.H., and Cadet, J. (2000) Singlet oxygen induces oxidation of cellular DNA. *J. Biol. Chem.*, **275**, 40601–40604.

35 Sheu, C. and Foote, C.S. (1993) Endoperoxide formation in a guanosine derivative. *J. Am. Chem. Soc.*, **115**, 10446–10447.

36 Sheu, C., Kang, P., Khan, S., and Foote, C.S. (2002) Low-temperature photosensitized oxidation of a guanosine derivative and formation of an imidazole ring-opened product. *J. Am. Chem. Soc.*, **124**, 3905–3913.

37 Kang, P. and Foote, C.S. (2002) Formation of transient intermediates in low-temperature photosensitized oxidation of an 8-^{13}C-guanosine. *J. Am. Chem. Soc.*, **124**, 4865–4873.

38 Ravanat, J.-L., Saint-Pierre, C., Di Mascio, P., Martinez, G.R., Medeiros, M.H.G., and Cadet, J. (2001) Damage to isolated DNA mediated by singlet oxygen. *Helv. Chim. Acta*, **84**, 3702–3709.

39 Niles, J.C., Wishnok, J.S., and Tannenbaum, S.R. (2001) Spiroiminodi-

hydantoin is the major product of 8-oxo-7,8-dihydroguanosine with peroxynitrite in the presence of thiols and guanosine oxidation by methylene blue. *Org. Lett.*, **3**, 963–936.

40 Ye, Y., Muller, J.G., Luo, W., Mayne, C.L., Shallop, A.J., Jones, R.A., and Burrows, C.J. (2003) Formation of ^{13}C-, ^{15}N-, and ^{18}O-labeled guanidinohydantoin from guanosine oxidation with singlet oxygen. Implications for structure and mechanism. *J. Am. Chem. Soc.*, **125**, 13926–13927.

41 Boiteux, S., Gajewski, E., Laval, J., and Dizdaroglu, M. (1992) Substrate specificity of the *Escherichia coli* Fpg protein (formamidopyrimidine-DNA glycosylase): excision of purine lesions in DNA produced by ionizing radiation or photosensitization. *Biochemistry*, **31**, 106–110.

42 Ravanat, J.-L., Sauvaigo, S., Caillat, S., Martinez, G.R., Medeiros, M.H.G., Di Mascio, P., Favier, A., and Cadet, J. (2004) Singlet-oxygen-mediated damage to cellular DNA determined by the comet assay associated with DNA repair enzymes. *Biol. Chem.*, **385**, 7–20.

43 Cadet, J., Douki, T., and Sage, E. (2005) Ultraviolet radiation-mediated damage to cellular DNA. *Mutat. Res.*, **571**, 3–17.

44 Mouret, S., Baudouin, C., Charveron, M., Favier, A., Cadet, J., and Douki, T. (2006) Cyclobutane pyrimidine dimers are predominant DNA lesions in whole human skin exposed to UVA radiation. *Proc. Natl. Acad. Sci. USA*, **103**, 13765–13770.

45 Pouget, J.-P., Douki, T., Richard, M.-J., and Cadet, J. (2000) DNA damage induced in cells by γ and UVA radiation as measured by HPLC/GC-MS and HPLC-EC and comet assay. *Chem. Res. Toxicol.*, **13**, 541–549.

46 Pouget, J.-P., Ravanat, J.-L., Douki, T., Richard, M.-J., and Cadet, J. (1999) Measurement of DNA base damage in cells exposed to low doses of γ radiation: comparison between the HPLC/ECD and the comet assays. *Int. J. Radiat. Biol.*, **75**, 51–58.

47 Pouget, J.P., Frelon, S., Ravanat, J.-L., Testard, I., Odin, F., and Cadet, J. (2002) Formation of modified DNA to DNA in cells exposed to either γ radiation or high-LET particles. *Radiat. Res.*, **157**, 589–595.

48 Douki, T., Ravanat, J.L., Pouget, J.-P., Testard, I., and Cadet, J. (2006) Minor contribution of direct ionization to DNA base damage induced by heavy ions. *Int. J. Radiat. Biol.*, **82**, 119–127.

49 Frelon, S., Douki, T., Ravanat, J.-L., Pouget, J.-P., Tornabene, C., and Cadet, J. (2000) High-performance liquid chromatography–tandem mass spectrometry measurement of radiation-induced base damage to isolated and cellular DNA. *Chem. Res. Toxicol.*, **13**, 1002–1010.

50 von Sonntag, C. (ed.) (1987) *The Chemical Basis of Radiation Biology*, Taylor & Francis, London.

51 Fujita, S. and Steenken, S. (1981) Pattern of OH radical addition to uracil and methyl- and carboxyl-substituted uracils. Electron transfer of OH adducts with N,N,N',N'-tetramethyl-*p*-phenylenediamine and tetranitromethane. *J. Am. Chem. Soc.*, **103**, 2540–2545.

52 Jovanovic, S.V. and Simic, M.G. (1986) Mechanism of OH radical reactions with thymine and uracil derivatives. *J. Am. Chem. Soc.*, **108**, 5968–5972.

53 Isildar, M., Schuchmann, M.N., Schulte-Frohlinde, D., and von Sonntag, C. (1982) Oxygen uptake in the radiolysis of aqueous solutions of nucleic acids and their constituents. *Int. J. Radiat. Biol.*, **41**, 525–533.

54 Wagner, J.R., van Lier, J.E., and Johnston, L.J. (1990) Quinone sensitized electron transfer photooxidation of nucleic acids: chemistry of thymine and thymidine radical cations in aqueous solution. *Photochem. Photobiol.*, **52**, 333–343.

55 Cadet, J. and Téoule, R. (1975) Radiolyse γ de la thymidine en solution aqueuse aérée. 1.–Identification des hydroperoxydes. *Bull. Soc. Chim. Fr.*, 879–884.

56 Wagner, J.R., van Lier, J.E., Berger, M., and Cadet, J. (1994) Thymidine hydroperoxides. Structural assignment, conformational features, and thermal decomposition in water. *J. Am. Chem. Soc.*, **116**, 2235–2242.

57 Davies, A.G. (1961) O–O heterolysis: intermolecular nucleophilic substitution at oxygen, in *Organic Peroxides*, Butterworth, London, pp. 128–142.
58 Lutsig, M.J., Cadet, J., Boorstein, R.J., and Teebor, G.W. (1992) Synthesis of the diastereomers of thymidine glycol, determination of concentrations and rates of interconversion of their *cis–trans* epimers at equilibrium and demonstration of differential alkali lability within DNA. *Nucleic Acids Res.*, **20**, 4839–4845.
59 Douki, T., Delatour, T., Paganon, F., and Cadet, J. (1996) Measurement of oxidative damage at pyrimidine bases in γ-irradiated DNA. *Chem. Res. Toxicol.*, **9**, 1145–1151.
60 Douki, T., Rivière, J., and Cadet, J. (2002) DNA tandem lesions containing 8-oxo-7,8-dihydroguanine and formamide residues arise from intramolecular addition of thymine peroxyl radical to guanine. *Chem. Res. Toxicol.*, **15**, 445–454.
61 Douki, T., Martini, R., Ravanat, J.L., Turesky, R.J., and Cadet, J. (1997) Measurement of 2,6-diamino-4-hydroxy-5-formamidopyrimidine and 8-oxo-7,8-dihydroguanine in isolated DNA exposed to γ radiation in aqueous solution. *Carcinogenesis*, **18**, 2385–2891.
62 Candeias, L.P. and Steenken, S. (2000) Reaction of HO• with guanine derivatives in aqueous solution: formation of two different redox-active OH-adduct radical and their unimolecular transformation reactions. Properties of G(-H)•. *Chem. Eur. J.*, **6**, 475–484.
63 Misiaszek, R., Crean, C., Joffe, A., Geacintov, N.E., and Shafirovich, V. (2004) Oxidative DNA damage associated with combination of guanine and superoxide radicals and repair mechanisms via radical trapping. *J. Biol. Chem.*, **279**, 32106–32115.
64 Cadet, J., Berger, M., Buchko, G.W., Joshi, P., Raoul, S., and Ravanat, J.-L. (1994) 2,2-Diamino-4-[(3,5-di-*O*-acetyl-2′-deoxy-β-D-*erythro*-pentofuranosyl)amino]-5-(2*H*)-oxazolone. *J. Am. Chem. Soc.*, **116**, 7403–7404.
65 Raoul, S., Berger, M., Buchko, G.W., Joshi, P.C., Morin, B., Weinfeld, M., and Cadet, J. (1996) ^{1}H, ^{13}C and ^{15}N NMR analysis and chemical features of the two main radical oxidation products of 2′-deoxyguanosine: oxazolone and imidazolone nucleosides. *J. Chem. Soc. Perkin Trans. 2*, 371–381.
66 Vialas, C., Pratviel, G., Claporols, C., and Meunier, B. (1998) Efficient oxidation of 2′-deoxyguanosine by Mn-TMPyP/KHSO$_5$ to imidazolone dIz without formation of 8-Oxo-dG. *J. Am. Chem. Soc.*, **120**, 11548–11553.
67 Matter, B., Malejka-Giganti, D., Csallany, A.S., and Tretyakova, N. (2006) Quantitative analysis of the oxidative DNA lesion, 2,2-diamino-4-[(-2-deoxy-β-D-*erythro*-pentofuranosyl)amino]-5(2*H*)-oxazolone (oxazolone) *in vitro* and *in vivo* by isotope dilution-capillary HPLC-ESI-MS/MS. *Nucleic Acids Res.*, **34**, 5449–5460.
68 Berger, M., de Hazen, M., Nejjari, A., Fournier, J., Guignard, J., Pezerat, H., and Cadet, J. (1993) Radical oxidation reactions of the purine moiety of 2′-deoxyribonucleosides and DNA by iron-containing minerals. *Carcinogenesis*, **14**, 41–46.
69 Raoul, S., Bardet, M., and Cadet, J. (1995) Gamma irradiation of 2′-deoxyadenosine in oxygen-free aqueous solutions: identification and conformational features of formamidopyrimidine nucleoside derivatives. *Chem. Res. Toxicol.*, **8**, 924–933.
70 Frelon, S., Douki, T., and Cadet, J. (2002) Radical oxidation of the adenine moiety of nucleoside and DNA: 2-hydroxy-2′-deoxyadenosine is a minor decomposition product. *Free Radic. Res.*, **36**, 499–508.
71 Mori, T. and Dizdaroglu, M. (1994) Ionizing radiation causes greater DNA base damage in radiation-sensitive mutant M10 cells than in parent mouse lymphoma L5178Y cells. *Radiat. Res.*, **140**, 65–90.
72 Mori, T., Hori, Y. and Dizdaroglu, M. (1993) DNA base damage generated *in vivo* in hepatic chromatin of mice upon whole body γ-irradiation. *Int. J. Radiat. Biol.*, **64**, 645–650.
73 Becker, D. and Sevilla, M. (1993) The chemical consequences of radiation

damage to DNA. *Adv. Radiat. Biol.*, **17**, 121–180.

74 Symons, M.C.R. (1995) Electron-spin-resonance studies of radiation-damage to DNA and proteins. *Radiat. Phys. Chem.*, **45**, 837–845.

75 Angelov, D., Spassky, A., Berger, M., and Cadet, J. (1997) High-intensity UV laser photolysis of DNA and purine 2′-deoxyribonucleosides: formation of 8-oxopurine damage and oligonucleotide strand cleavage as revealed by HPLC and gel electrophoresis studies. *J. Am. Chem. Soc.*, **119**, 11373–11360.

76 Douki, T., Angelov, D., and Cadet, J. (2001) UV laser photolysis of DNA: effect of duplex stability on charge transfer efficiency. *J. Am. Chem. Soc.*, **123**, 11360–11366.

77 Ballmaier, D. and Epe, B. (2006) DNA damage by bromate: mechanism and consequences. *Toxicology*, **221**, 166–171.

78 Joffe, A., Geacintov, N.E., and Shafirovich, V. (2003) DNA lesions derived from the site selective oxidation of guanine by carbonate radical anions. *Chem. Res. Toxicol.*, **16**, 1528–1538.

79 Medinas, D.B., Cerchiaro, G., Trindale, D.F., and Augusto, O. (2007) The carbonate radical and related oxidants derived from bicarbonate buffer. *IUBMB Life*, **59**, 255–262.

80 Kasai, H., Yamaizumi, Z., Berger, M., and Cadet, J. (1992) Photosensitized formation of 7,8-dihydro-8-oxo-2′-deoxyguanosine (8-hydroxy-2′-deoxyguanosine) in DNA by riboflavin: a non singlet oxygen mediated reaction. *J. Am. Chem. Soc.*, **114**, 9692–9694.

81 Decarroz, C., Wagner, J.R., van Lier, J.E., Krishna, C.M., Riesz, P., and Cadet, J. (1986) Sensitized photo-oxidation of thymidine by 2-methyl-1,4-naphthoquinone. Characterization of the stable photoproducts. *Int. J. Radiat. Biol.*, **50**, 491–505.

82 Giese, B. and Spichty, M. (2000) Long-distance charge transport through DNA: quantification and extension of the hopping model. *ChemPhysChem*, **1**, 195–198.

83 Schuster, G.B. (2000) Long-range charge transfer in DNA: transient structural distortions control the distance dependence. *Acc. Chem. Res.*, **33**, 2253–2260.

84 Boon, E.M. and Barton, J.K. (2002) Charge transfer in DNA. *Curr. Opin. Struct. Biol.*, **12**, 320–329.

85 Osakada, Y., Kawai, K., Fujitsuka, M., and Majima, T. (2006) Charge transfer through DNA nanoscaled assembly programmable with DNA building blocks. *Proc. Natl. Acad. Sci. USA*, **103**, 18072–18076.

86 Henderson, J.P., Byun, J., and Heinecke, J.W. (1999) Molecular chlorine generated by the myeloperoxidase hydrogen-chloride system of phagocytes produces 5-chlorocytosine in bacterial DNA. *J. Biol. Chem.*, **274**, 33440–33448.

87 Suzuki, T., Friesen, M.D., and Ohshima, H. (2003) Identification of products formed by reaction of 3′,5′-di-O-actyl-2′-deoxyguanosine with hypochlorous acid or a myeloperoxidase–H_2O_2–Cl^- system. *Chem. Res. Toxicol.*, **16**, 382–389.

88 Badouard, C., Masuda, M., Nishino, H., Cadet, J., Favier, A., and Ravanat, J.-L. (2005) Detection of chlorinated DNA and RNA nucleosides by HPLC coupled to mass spectrometry as potential biomarkers of inflammation. *J. Chromatogr. B*, **827**, 26–31.

89 Masuda, M., Suzuki, T., Friesen, M.D., Ravanat, J.-L., Cadet, J., Pignatelli, B., Nishino, H., and Ohshima, H. (2001) Chlorination of guanosine and other nucleosides by hypochlorous acid and myeloperoxidase of activated human neutrophils. *J. Biol. Chem.*, **276**, 40486–40496.

90 Buyn, J., Henderson, J.P., and Heinecke, J.W. (2003) Identification and quantification of mutagenic halogenated cytosines by gas chromatography, fast atom bombardment, and electrospray ionization tandem mass spectrometry. *Anal. Biochem.*, **317**, 201–209.

91 Badouard, C., Douki, T., Faure, P., Halimi, S., Cadet, J., Favier, A., and Ravanat, J.-L. (2005) DNA lesions as biomarkers of inflammation and oxidative stress: a preliminary evaluation, in *Free Radicals and Diseases, Gene Expression, Cellular Metabolism and*

92 Olive, P.L. (2009) Impact of the comet assay in radiobiology. *Mutat. Res.*, **681**, 13–23.

93 Gradzka, I. (2005) A non-radioactive, PGFE-based assay for low levels of DNA double-strand breaks in mammalian cells. *DNA Repair*, **4**, 1129–1139.

94 Muslimovic, A., Ismail, I.H., Gao, Y., and Hammarsten, O. (2008) An optimized method for measurement of γ-H2AX in blood mononuclear and cultured cells. *Nat. Protoc.*, **3**, 1187–1193.

95 Box, H.C., Budzinski, E.E., Freund, H.G., Evans, M.S., Patrzyc, H.B., Wallace, J.C., and MacCubbin, A.E. (1993) Vicinal lesions in X-irradiated DNA? *Int. J. Radiat. Biol.*, **64**, 261–263.

96 Box, H.C., Freund, H.G., Budzinski, E.E., Wallace, J.C., and MacCubbin, A.E. (1995) Free radical-induced double base lesions. *Radiat. Res.*, **141**, 91–94.

97 Bourdat, A.G., Douki, T., Frelon, S., Gasparutto, D., and Cadet, J. (2000) Tandem base lesions are generated by hydroxyl radical within isolated DNA in aerated aqueous solution. *J. Am. Chem. Soc.*, **122**, 4549–4556.

98 Hong, I.S., Carter, K.N., Sato, K., and Greenberg, M.M. (2007) Characterization and mechanism of formation of tandem lesions in DNA by a nucleobase peroxyl radical. *J. Am. Chem. Soc.*, **129**, 4089–4098.

99 Ghosh, A., Joy, A., Schuster, G.B., Douki, T., and Cadet, J. (2008) Selective one-electron oxidation of duplex DNA oligomers: reactions at thymines. *Org. Biomol. Chem.*, **6**, 916–928.

100 Romieu, A., Bellon, S., Gasparutto, D., and Cadet, J. (2000) Synthesis and UV photolysis of oligodeoxynucleotides that contain 5-(phenyl-thiomethyl)-2′-deoxyuridine: a specific photolabile precursor of 5-(2′-deoxyuridilyl)methyl radical. *Org. Lett.*, **2**, 1085–1088.

101 Bellon, S., Ravanat, J.-L., Gasparutto, D., and Cadet, J. (2002) Cross-linked thymine-purine base tandem lesions: synthesis, characterization, and measurement in γ-irradiated DNA. *Chem. Res. Toxicol.*, **15**, 598–606.

102 Bellon, S., Gasparutto, D., Saint-Pierre, C., and Cadet, J. (2006) Guanine–thymine intrastrand cross-linked lesion containing oligonucleotides: from chemical synthesis to *in vitro* enzymatic replication. *Org. Biomol. Chem.*, **4**, 3831–3837.

103 Hong, H., Cao, H., and Wang, Y. (2006) Identification and quantification of a guanine–thymine instrastrand cross-link lesion induced by Cu(II)/H_2O_2/ascorbate. *Chem. Res. Toxicol.*, **19**, 614–621.

104 Wang, Y. (2008) Bulky DNA lesions induced by reactive oxygen species. *Chem. Res. Toxicol.*, **21**, 276–281.

105 Zhang, Q. and Wang, Y. (2003) Independent generation of 5-(2′-deoxycytidinyl)methyl radical and the formation of a novel crosslink lesion between 5-methylcytosine and guanine. *J. Am. Chem. Soc.*, **125**, 12795–12802.

106 Zhang, Q. and Wang, Y. (2005) Generation of 5-(2′-deoxycytidyl)methyl radical and the formation of intrastrand cross-link lesions in oligodeoxyribonucleotides. *Nucleic Acids Res.*, **33**, 1593–1603.

107 Cao, H. and Wang, Y. (2007) Quantification of oxidative single-base and intrastrand cross-link lesions in unmethylated and CpG-methylated DNA induced by Fenton-type reagents. *Nucleic Acids Res.*, **35**, 4833–4844.

108 Xerri, B., Morell, C., Grand, A., Cadet, J., Cimino, P., and Barone, V. (2006) Radiation-induced formation of DNA intrastrand crosslinks between thymine and adenine bases. A theoretical approach. *Org. Biomol. Chem.*, **4**, 3986–3992.

109 Gu, C. and Wang, Y. (2004) LC-MS/MS identification and yeast polymerase η bypass of a novel γ-irradiation-induced intrastrand cross-link lesion G[8–5]C. *Biochemistry*, **43**, 6745–6750.

110 Jiang, Y., Hong, H., Cao, H., and Wang, Y. (2007) *In vivo* formation and *in vitro* replication of a guanine–thymine intrastrand cross-link lesion. *Biochemistry*, **46**, 12757–12763.

111 Hong, H., Cao, H., and Wang, Y. (2007) Formation and genotoxicity of a guanine–cytosine intrastrand cross-link

lesion in vivo. *Nucleic Acids Res.*, **35**, 7118–7127.

112 Chen, B., Zhou, X., Taghizadeh, K., Chen, J., Stubbe, J., and Dedon, P.C. (2007) GC/MS methods to quantify the 2-deoxypentos-4-ulose and 3′-phosphoglycolate pathways of 4′ oxidation of 2-deoxyribose in DNA: application to DNA damage produced by γ radiation and bleomycin. *Chem. Res. Toxicol.*, **20**, 1701–1708.

113 Chatgilialoglu, C. and Gimisis, T. (1998) Fate of the C-14 peroxyl radical in the 2′-deoxyurdine system. *Chem. Commun.*, 1249–1259.

114 Dedon, P.C. and Goldberg, I.H. (1992) Free-radical mechanism involved in the formation of sequence-dependent bistranded DNA lesions by the antitumor antibiotics bleomycin, neocarzinostatin, and calicheamicin. *Chem. Res. Toxicol.*, **5**, 311–322.

115 Regulus, P., Spessotto, S., Gateau, M., Cadet, J., Favier, A., and Ravanat, J.-L. (2004) Detection of new radiation-induced DNA degradation lesions by liquid chromatography coupled to tandem mass spectrometry. *Rapid Commun. Mass Spectrom.*, **18**, 2223–2228.

116 Regulus, P., Duroux, B., Bayle, P.-A., Favier, A., Cadet, J., and Ravanat, J.-L. (2007) Oxidation of the sugar moiety of DNA by ionizing radiation or bleomycin could induce the formation of a cluster DNA lesion. *Proc. Natl. Acad. Sci. USA*, **104**, 14032–14037.

117 Sczepanski, J.T., Jacobs, A.C., and Greenberg, M.M. (2008) Self promoted interstrand cross-link formation by an abasic site. *J. Am. Chem. Soc.*, **130**, 9646–9647.

118 Mariaggi, N., Cadet, J., and Téoule, R. (1976) Cyclisation radicalaire de la désoxy-2′-adénosine en solution aqueuse, sous l'effet du rayonnement γ. *Tetrahedron*, **32**, 2385–2387.

119 Keck, K. (1968) Bildung von cyclonucleotiden bei bestrahlung wassiriger losungen von purinnucleotiden. *Z. Naturforsch.*, **23b**, 1034–1043.

120 Raleigh, J.A., Kremers, W., and Whitehouse, R. (1976) Radiation chemistry of nucleotides: 8,5′-cyclonucleotide formation and phosphate release initiated by hydroxyl radical attack on adenosine monophosphates. *Radiat. Res.*, **65**, 414–422.

121 Dizdaroglu, M. (1986) Free-radical-induced formation of an 8,5′-cyclo-2′-deoxyguanosine moiety in deoxyribonucleic acid. *Biochem. J.*, **238**, 247–254.

122 Chatgilialoglu, C., Guerra, M., and Mulazzani, Q.G. (2003) Model studies of DNA C5′ radicals. Selective generation and reactivity of 2′-deoxyadenos-5′-yl radical. *J. Am. Chem. Soc.*, **125**, 3839–1848.

123 Navacchia, M.L., Chatgilialoglu, C., and Montevecchi, P.C. (2006) C5′-Adenosyl radical cyclization. A stereochemical investigation. *J. Org. Chem.*, **71**, 4445–4452.

124 Chatgilialoglu, C., Bazzanini, R., Jimenez, L.B., and Miranda, M. (2007) (5′S)- and (5′R)-5′,8-cyclo-2′-deoxyguanosine: mechanistic insights on the 2′-deoxyguanosin-5′-yl radical cyclization. *Chem. Res. Toxicol.*, **20**, 1820–1824.

125 Dizdaroglu, M., Jaruga, P., and Rodriguez, H. (2001) Identification and quantification of 8,5′-cyclo-2′-deoxyadenosine in DNA by liquid chromatography mass spectrometry. *Free Radic. Biol. Med.*, **30**, 774–784.

126 Jaruga, P., Birincioglu, M., Rodriguez, H., and Dizdaroglu, M. (2002) Mass spectrometric assays for the tandem lesion 8,5′-cyclo-2′deoxyguanosine in mammalian DNA. *Biochemistry*, **41**, 3703–3711.

127 Jaruga, P., Theruvathu, J., Dizdaroglu, M., and Brooks, P.J. (2004) Complete release of (5′S)-8,5′-cyclo-2′-deoxyadenosine from dinucleotides, oligonucleotides and DNA, and direct comparison of its levels in cellular DNA with other oxidatively induced DNA lesions. *Nucleic Acids Res.*, **32**, e87.

128 Brooks, P.J. (2007) The case for 8,5′-cyclopurine-2′-deoxynucleosides as endogenous lesions that cause neurodegeneration in xeroderma pigmentosum. *Neuroscience*, **145**, 1407–1417.

129 Boussicault, F., Kaladis, P., Caminal, C., Mulazzani, Q.G., and Chatgilialoglu, C.

(2008) The fate of the C5' radicals of purine nucleosides under oxidative conditions. *J. Am. Chem. Soc.*, **130**, 8377–8385.

130 D'Errico, M., Parlanti, E., Teson, M., Bernades de Jesus, B.M., Degan, P., Calcagnile, A., Jaruga, P., Bjørås, M., Crescenzi, M., Pedrini, A.M., Egly, J.-M., Zambruno, G., Stefanini, M., Dizdaroglu, M., and Dogliotti, E. (2006) New functions of XPC in the protection of human skin cells from oxidative damage. *EMBO J.*, **25**, 4305–4315.

131 D'Errico, M., Parlanti, E., Teson, M., Degan, P., Lemma, T., Calcagnile, A., Iavarone, I., Jaruga, P., Ropolo, M., Pedrini, A.M., Orioli, D., Frosina, G., Zambruno, G., and Stefanini, M.E. (2007) The role of CSA in the response to oxidative DNA damage in human cells. *Oncogene*, **26**, 4336–4343.

132 Steenken, S. and Jovanovic, S.V. (1997) How easily oxidizable is DNA? One-electron reduction potentials of adenosine and guanosine radicals in aqueous solutions. *J. Am. Chem. Soc.*, **119**, 617–618.

133 Bernstein, R., Prat, F., and Foote, C.S. (1999) On the mechanism of DNA cleavage by fullerenes investigated in model systems. Electron transfer from guanosine and 8-oxoguanosine to C_{60}. *J. Am. Chem. Soc.*, **121**, 464–465.

134 Luo, W., Muller, J.G., Rachlin, E.M., and Burrows, C.J. (2000) Characterization of spiroiminodihydantoin as a product of one-electron oxidation of 8-oxo-7,8-dihydroguanosine. *Org. Lett.*, **2**, 613–616.

135 Luo, W., Muller, J.G., Rachlin, E.M., and Burrows, C.J. (2001) Characterization of hydantoin products from one-electron oxidation of 8-oxo-7,8-dihydroguanosine in a nucleoside model. *Chem. Res. Toxicol.*, **14**, 927–938.

136 Luo, W., Muller, J.G., and Burrows, C.J. (2001) The pH-dependent role of superoxide in riboflavin-catalyzed photooxidation of 8-oxo-7,8-dihydroguanosine. *Org. Lett.*, **3**, 2801–2804.

137 Ravanat, J.-L., Saint-Pierre, C., and Cadet, J. (2003) One-electron oxidation of the guanine moiety of 2'-deoxyguanosine: influence of 8-oxo-7,8-dihydro-2'-deoxyguanosine. *J. Am. Chem. Soc.*, **125**, 2030–2031.

138 Crean, C., Geacintov, N.E., and Shafirovich, V. (2005) Oxidation of 8-oxo-7,8-dihydroguanine by carbonate radical anions; insight from oxygen-18 labeling experiments. *Angew. Chem. Int. Ed.*, **44**, 5057–5060.

139 McCallum, J.E.B., Kuniyoshi, C.Y., and Foote, C.S. (2004) Characterization of 5-hydroxy-8-oxo-7,8-dihydroguanosine in the photosensitized oxidation of 8-oxo-7,8-dihydroguanosine and its rearrangement to spiroiminodihydantoin. *J. Am. Chem. Soc.*, **126**, 16777–16782.

140 Slade, P.G., Hailer, M.K., Martin, B.D., and Sugden, K.D. (2005) Guanine-specific oxidation of double-stranded DNA by Cr(VI) and ascorbic acid forms spiroiminodihydantoin and 8-oxo-2'-deoxyguanosine. *Chem. Res. Toxicol.*, **18**, 1140–1149.

141 Hailer, M.K., Slade, P.G., Martin, B.D., and Sugden, K.D. (2005) Nei deficient *Escherichia coli* are sensitive to chromate and accumulate the oxidized guanine lesion spiroiminodihydantoin. *Chem. Res. Toxicol.*, **18**, 1378–1383.

142 Tuo, J., Muftuoglu, M., Chen, C., Jaruga, P., Selzer, R.R., Brosh, R.M., Jr., Rodriguez, H., Dizdaroglu, M., and Bohr, V.A. (2001) The Cockayne syndrome group B gene product is involved in general genome base excision repair of 8-hydroxyguanine in DNA. *J. Biol. Chem.*, **276**, 45772–45779.

143 Tuo, J., Jaruga, P., Rodriguez, H., Dizdaroglu, M., and Bohr, V.A. (2002) The Cockayne syndrome group B gene product is involved in cellular repair of 8-hydroxyadenine in DNA. *J. Biol. Chem.*, **277**, 30832–30837.

144 Tuo, J., Jaruga, P., Rodriguez, H., Bohr, V.A., and Dizdaroglu, M. (2003) Primary fibroblasts of Cockayne syndrome patients are defective in cellular repair of 8-hydroxyguanine and 8-hydroxyadenine from oxidative stress. *FASEB J.*, **17**, 668–674.

145 Friedman, K.A. and Keller, A. (2001) On the non-uniform distribution of

guanine in introns of human genes: possible protection of exons against oxidation by proximal introns poly-G sequences. *J. Phys. Chem. B*, **105**, 11859–11865.

146 Friedman, K.A. and Keller, A. (2004) Guanosine distribution and oxidation resistance in eight eukaryotic genomes. *J. Am. Chem. Soc.*, **126**, 2368–2371.

147 Kanvah, S. and Schuster, G.B. (2005) The sacrificial role of easily oxidizable sites in the protection of DNA from damage. *Nucleic Acids Res.*, **33**, 5133–5138.

4
Role of Free Radical Reactions in the Formation of DNA Damage

Vladimir Shafirovich and Nicholas E. Geacintov

4.1
Introduction

In human cells, the major source of endogenous reactive oxygen/nitrogen species (ROS/RNS) is aerobic metabolism, which converts the inhaled oxygen to highly reactive intermediates [1]. In healthy tissues these ROS/RNS are present in extremely low concentrations and act as intracellular signaling species in diverse biological processes, such as the activation of transcription factors, apoptosis, and regulation of cell growth [2]. In contrast, increased concentrations of ROS/RNS are typically associated with various disorders. Overproduction of ROS/RNS is linked with inflammation – the primary reaction of tissues to eliminate pathogenic insult and injured tissue components in order to restore normal physiological cell functions [3, 4]. During the inflammation process, macrophages and neutrophils secrete various reactive species that include superoxide radical anion ($O_2^{\bullet-}$), hydrogen peroxide (H_2O_2), hydroxyl radical ($^{\bullet}OH$), nitric oxide ($^{\bullet}NO$), peroxynitrite ($ONOO^-$), and hydrochlorous acid (HOCl) [1, 3–5]. These reactive intermediates play a pivotal role in eliminating pathogens or biological insults and thus represent an effective defense mechanisms against cellular injury by a variety of mechanisms. Reactive molecules that have even pairs of electrons are H_2O_2, HOCl, and the $ONOO^-$ anion, while free radicals such as $O_2^{\bullet-}$, $^{\bullet}OH$, and $^{\bullet}NO$ have one unpaired electron, and therefore an odd number of electrons. Due to the presence of unpaired electrons, free radicals are typically more reactive than molecules with even numbers of electrons. These radicals can react chemically with diverse biomolecules (proteins, lipids, and DNA), thus inducing modifications and damage [1]. In healthy tissues, the harmful effects of ROS/RNS are minimized by enzymatic pathways (catalase, superoxide dismutase (SOD)) and nonenzymatic mechanisms involving antioxidants [1]. An excess of ROS/RNS results in the development of oxidative stress that leads to a cascade of signaling

The Chemical Biology of DNA Damage. Edited by Nicholas E. Geacintov and Suse Broyde
© 2010 WILEY-VCH Verlag GmbH & Co. KGaA, Weinheim
ISBN: 978-3-527-32295-4

events that trigger the production of proinflammatory cytokines and chemokines. In turn, chemical reactions of ROS/RNS with cellular DNA induce genomic alterations (point mutations, deletions, and rearrangements), which in normal cells are excised by the cellular DNA repair machinery. Chronic inflammation provides a microenvironment rich in (i) inflammatory cells, (ii) ROS/RNS, (iii) recurring DNA damage, (iv) cell-proliferating growth factors, and (v) other growth-supporting stimuli that increase the frequency of mutations. This further facilitates the progressive transformation of cells to the malignant state and the development of cancers [6–8].

4.2
Importance of Free Radical Reactions with DNA

4.2.1
Free Radical Mechanisms: General Considerations

The oxidation of biomolecules can be considered as a free radical chain reaction as is observed in the case of lipids, RH, and typically includes the following key steps [9]:

- Initiation:

$$\text{Production of } R^\bullet \tag{4.1}$$

- Propagation:

$$R^\bullet + O_2 \to ROO^\bullet \tag{4.2}$$

$$ROO^\bullet + RH \to ROOH + R^\bullet \tag{4.3}$$

- Termination:

$$ROO^\bullet + ROO^\bullet \to \text{nonradical products} \tag{4.4}$$

In this scheme, $R^\bullet$ is a carbon-centered radical produced by hydrogen atom abstraction from RH. The first step in the chain reaction is initiation that can be triggered by oxyl radicals produced by normal aerobic metabolism (which are overproduced under inflammatory conditions) or one-electron reduction of hydroperoxides, ROOH, by transition metal ions (Fe^{2+}, Cu^+) [10, 11]. The carbon-centered radicals, $R^\bullet$, rapidly react with oxygen with the rate constant k_2 (generally of the order of $10^9 M^{-1} s^{-1}$) to form peroxyl radicals, $ROO^\bullet$. These peroxyl radicals subsequently abstract hydrogen atoms from RH to form the hydroperoxide ROOH and another $R^\bullet$ radical at a significantly lower rate [12]. The propagation reactions (4.2) and (4.3) are terminated via the bimolecular combination of the radicals (4.4) to form molecular products. This scheme is particularly relevant to lipid molecules, where lipid hydroperoxides are the primary nonradical products of lipid peroxidation [9]. The free radical-mediated oxidation of nucleic acids differs from this mechanism and is discussed in the following sections.

4.2.2
Types of Free Radicals and their Reactions with Nucleic Acids

Among the oxyl radicals that are believed to be generated at sites of inflammation, hydroxyl radicals [13] and carbonate radical anions [14] can directly react with DNA to produce irreversible damage of nucleic acid bases.

The major source of $^\bullet$OH radicals *in vivo* is the reduction of hydrogen peroxide by transition metal ions such as Fe^{2+} and Cu^+ [15], known as the Fenton reaction:

$$H_2O_2 + Fe^{2+} \rightarrow {}^\bullet OH + OH^- + Fe^{3+} \tag{4.5}$$

Another potential source of $^\bullet$OH radicals *in vivo* is homolysis of peroxynitrite derived from extremely fast combination [16] of nitric oxide and superoxide radical anions ($k_6 = 6.7 \times 10^9 \, M^{-1} s^{-1}$) [17] that is overproduced in inflammatory cells:

$$^\bullet NO + O_2^{\bullet -} \rightarrow ONOO^- \tag{4.6}$$

In neutral solutions, peroxynitrite exists in the form of the unstable peroxynitrous acid ($pK_a = 6.5$–6.8) [18, 19] that decays with a rate constant [20] of $1.3 \, s^{-1}$ to form hydroxyl and nitrogen dioxide radicals with a yield $x \sim 0.30$ [21]:

$$ONOOH \rightarrow x({}^\bullet NO_2 + {}^\bullet OH) + (1-x)(NO_3^- + H^+) \tag{4.7}$$

The reduction potential [22] of the hydroxyl radical is $E^\circ = 1.9 \, V$ versus NHE; it is therefore an extremely strong electrophile that rapidly and unselectively reacts with DNA mostly via the addition to double bonds of the nucleobases, and by the abstraction of hydrogen atoms from either the 2-deoxyribose moieties or the methyl group of thymine.

In contrast, $CO_3^{\bullet -}$ radicals, with a reduction potential of $1.59 \, V$ versus NHE, selectively oxidize biomolecules by one-electron abstraction mechanisms [23]. It has been suggested that *in vivo*, $CO_3^{\bullet -}$ radicals are formed via homolysis of nitrosoperoxycarbonate derived from the combination ($k_8 = 2.9 \times 10^4 \, M^{-1} s^{-1}$) [24] of carbon dioxide and peroxynitrite:

$$ONOO^- + CO_2 \rightarrow ONOOCO_2^- \tag{4.8}$$

The nitrosoperoxycarbonate anion is highly unstable and decomposes to form nitrogen dioxide and carbonate radical anions with the yield $y \sim 0.33$ [21]:

$$ONOOCO_2^- \rightarrow y({}^\bullet NO_2 + CO_3^{\bullet -}) + (1-y)(NO_3^- + CO_2) \tag{4.9}$$

In vivo, the concentrations of CO_2 are relatively high due to the high levels of bicarbonate in intracellular (12 mM) and interstitial (30 mM) fluids [25]. Thus, the reaction of $ONOO^-$ with CO_2 is expected to be the major pathway of decay of peroxynitrite in biological systems [4, 5].

4.2.3
Methods for Studying Free Radical Reactions: Laser Flash Photolysis

Laser flash photolysis is one of the most efficient methods for the direct spectroscopic observation of free radicals and for monitoring the kinetics of formation and decay in real-time. This method is an extension of conventional flash photolysis method [26] that was invented by Norrish and Porter in 1949, and who were awarded by the Nobel Prize in 1967. We have used this approach to investigate the generation and reactions of free radicals with DNA. In this technique, a laser light pulse is used to produce short-lived intermediates in solution contained in an optical cuvette, and the kinetics of their formation and decay are monitored by transient absorption spectroscopy. The apparatus we used is shown in Figure 4.1.

The free radicals are first generated by a brief and intense laser pulse. The changes in the intensities of the analyzing (probe) light signal associated with the absorbance of light by the transient intermediates are monitored by the photomultiplier tube with its output signal fed to a digital oscilloscope. The laser provides a single-wavelength pulsed excitation (e.g., 308 nm from an XeCl excimer laser, or 532, 355, or 266 nm from a Nd:Yag laser with second, third, and fourth harmonic generation, respectively). The transient absorbance of the free radicals can be measured with a nanosecond time resolution (5–10 ns). The high reproducibility

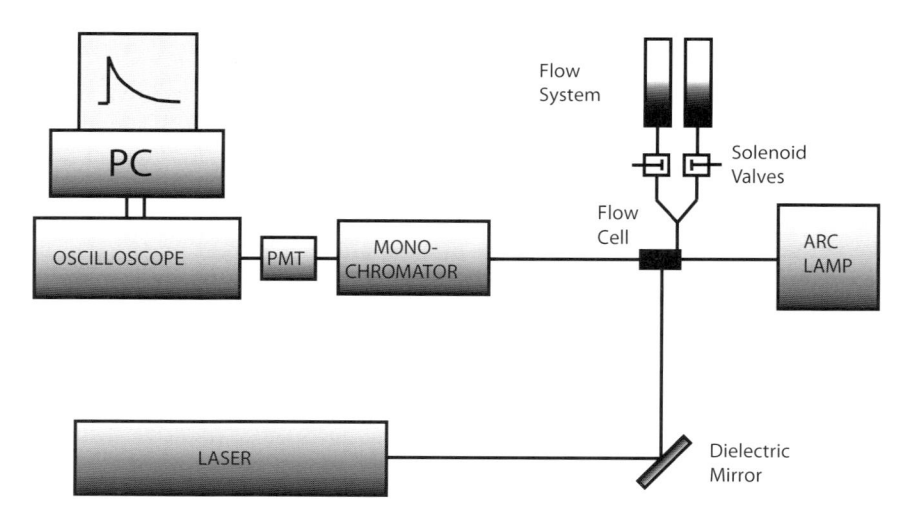

Figure 4.1 Laser flash photolysis setup with a flow system that injects either of two solutions into the sample cuvette that can be emptied after every laser flash and refilled with a fresh sample solution. PMT, photomultiplier tube. The steady-state arc lamp is usually a high-pressure xenon lamp. The light is incident on the sample and measures the transient absorbance after each laser pulse at a wavelength selected by the monochromator. All systems are controlled by a suitably programmed personal computer.

of the transient absorption signal allows for the monitoring of transient absorption spectra in the desired spectral wavelength range and time range of around 50 ns to seconds. In some studies of radical reactions, the actinic laser pulse induces irreversible changes in the sample that requires the replacement of the sample between successive laser pulses using a flow cell system (Figure 4.1).

4.2.4
Types of Radical Reactions and Kinetics

Free radicals are highly reactive species that can react with other radicals or molecules, or undergo spontaneous decomposition (fragmentation) with the formation of secondary radicals. Radical-to-radical reactions typically involve combination reactions of radical $R_1^\bullet$ with another radical $R_1^\bullet$ (radical recombination):

$$R_1^\bullet + R_1^\bullet \rightarrow R_1-R_1 \tag{4.10}$$

or with a radical $R_2^\bullet$ of a different structure:

$$R_1^\bullet + R_2^\bullet \rightarrow R_1-R_2 \tag{4.11}$$

Since each radical has one unpaired electron, the products have even number of electrons and typically are molecules or ions. An example of radical recombination is the combination of two hydroxyl radicals to form hydrogen peroxide:

$$^\bullet OH + ^\bullet OH \rightarrow H_2O_2 \tag{4.12}$$

An example of the combination of two different radicals is the reaction of nitric oxide and superoxide radicals that results in the formation of peroxynitrite (4.6). Radical recombination reactions follow second-order kinetics:

$$-d[R_1^\bullet]/dt = 2k_{10}[R_1^\bullet]^2 \tag{4.13}$$

with the following time dependence of the radical concentration:

$$[R_1^\bullet]^{-1} = [R_1^\bullet]_0^{-1} + 2k_{10}t \tag{4.14}$$

The kinetics of (4.11) are more complex because the initial concentrations of $R_1^\bullet$ and $R_2^\bullet$ radicals can be different, and the decay of radicals can also occur by other, competitive recombination reactions.

Reactions of free radicals with molecules (or ions) can occur via an addition reaction, for instance the addition of a C-centered radical to oxygen (4.2), hydrogen atom abstraction (4.3), or electron transfer mechanism (e.g., oxidation of CO_3^{2-} by $SO_4^{\bullet-}$). Since the total number of electrons is odd, one of the products is a new radical (e.g., (4.2) and (4.3)), except for the reactions with transition metal ions, such as the oxidation of superoxide radicals by Fe^{3+} ions:

$$O_2^{\bullet-} + Fe^{3+} \rightarrow O_2 + Fe^{2+} \tag{4.15}$$

In these reactions, the radical concentrations are typically much lower than the concentrations of the substrate (S) molecules (or ions), and the reaction kinetics follow pseudo-first order-reaction kinetics:

$$-d[R^\bullet]/dt = k[S] \tag{4.16}$$

with an exponential decrease in the radical concentration:

$$[R^\bullet] = [S]_0 \exp(-kt) \tag{4.17}$$

Radicals with reactive groups can undergo intra-radical reactions such as the unimolecular β-scission of *tert*-butyl radicals to form an acetone molecule and a secondary methyl radical:

$$(CH_3)_3CO^\bullet \rightarrow (CH_3)_2C{=}O + {}^\bullet CH_3 \tag{4.18}$$

These fragmentation reactions typically follow first-order kinetics.

Analysis of the kinetic data obtained by laser flash photolysis typically includes the following steps:

i) Determining the type of reaction kinetics (e.g., first or second order). This can be done by varying the initial concentrations of the radicals (e.g., using different energies of the actinic laser pulse) or substrate molecules.

ii) The decay profiles of the reactive species are measured, and the rate constants of the radical reactions are determined by fitting (4.14) and (4.17) to the experimentally determined transient absorption profiles (absorbance as a function of time).

4.2.5
Examples of DNA Radical Reactions

In our laser pulse-induced transient absorbance experiments, we have used three different photolysis methods for generating free radicals at a given time point and in a controlled fashion: (i) photoionization [27–31], (ii) photodissociation [14, 32–36], and (iii) charge transfer in metal complexes [37]. The energy of 193-nm photons from an ArF excimer laser, $E = 6.42\,eV$ is sufficient to induce single-photon ionization of nucleic acid bases [38–41]. However, this process is unselective because 193-nm light can ionize all four DNA bases. To improve the selectivity of photoionization, we have used a consecutive two-photon-induced ionization mechanism that provides a highly selective generation of radicals using photon energies that are much lower than the ionization potentials of the radical precursor molecules [27–31].

An example of this process is the two-photon ionization of 2-aminopurine (2AP) – a nucleic acid base analog that can site-specifically inserted into any DNA sequence [27–31]. We have shown that photoexcitation of 2AP, which has a broad absorption maximum at 305 nm, by 308-nm intensive nanosecond XeCl excimer laser pulses, induced efficient and selective photoionization of 2AP residues. Absorption of the first photon results in the formation of the 2AP singlet excited state (^{1}2AP) and absorption of the second photon causes photoionization according to:

$$2AP + h\nu \rightarrow {}^1 2AP \tag{4.19}$$

$$^1 2AP + h\nu \rightarrow 2AP^{\bullet +} + e^- \tag{4.20}$$

The energy delivered by a consecutive two-photon excitation of 2AP is $E = 7.77$ eV (the sum of the energy of ^{1}2AP, $\Delta E_{00} = 3.74$ eV, and the energy of the 308-nm photon, $E = 4.03$ eV), thus providing sufficient energy for the efficient photoionization of 2AP [27]. The lifetime of free ^{1}2AP in solution (around 10 ns) is sufficiently long for absorption of the second photon to occur within the same excimer laser pulse (halfwidth around 12 ns). In DNA, the ^{1}2AP lifetime is severely reduced due to quenching [42] by neighboring nucleobases, which decreases the probability of absorption of the second photon. Therefore, in double-stranded DNA, the yield of photoionization of ^{1}2AP is significantly lower than that of free 2AP molecules in the same solution. Nevertheless, the efficiency of the two-photon ionization of 2AP in DNA is sufficient for exploring charge transfer processes in double-stranded DNA by laser photolysis methods [27–31].

Other examples of generating free radicals utilized in our laboratory are the consecutive two-photon ionization of the aromatic pyrene residue in the benzo[a]pyrene derivative 7,8,9,10-tetrahydroxytetrahydrobenzo[a]pyrene (B[a]PT) or the covalent adducts derived from the reactions of racemic anti-r7,t8-dihydroxy-t9,10-epoxy-7,8,9,10-tetrahydrobenzo[a]pyrene (B[a]PDE) with the N^2-exocyclic amino groups of guanine or adenine bases in DNA [43–47]. The selective excitation of the pyrenyl residue of B[a]PT by intense 355-nm nanosecond laser pulses yields free radical products by a two-photon mechanism:

$$B[a]PT + h\nu \rightarrow {}^1B[a]PT \tag{4.21}$$

$$^1B[a]PT + h\nu \rightarrow B[a]PT^{\bullet +} + e^- \tag{4.22}$$

A similar mechanism applies to covalently bound B[a]PDE-DNA adducts and it has been shown that the 355-nm laser photoexcitation of the pyrenyl residues can be used to inject holes into double-stranded DNA at the site of modified guanine residues [47].

The 2AP and B[a]PT absorption bands are positioned above 300 nm, and are thus beyond the threshold of the absorption spectrum of DNA. Thus, the canonical DNA bases are not photoionized by either 308- or 355-nm nanosecond laser pulses used in these experiments.

The photoinduced dissociation of persulfate anions [23] using light above 300 nm in wavelength is yet another method for generating reactive radical intermediates that can oxidize DNA [14, 32–35]:

$$S_2O_8^{2-} + h\nu \rightarrow 2SO_4^{\bullet -} \tag{4.23}$$

The photolysis of carbonatotetrammine Co(III) complexes [37], $[Co(NH_3)_4CO_3]^+$ can be used for the generation of carbonate radical anions. In this method, the UV irradiation of the $[Co(NH_3)_4CO_3]^+$ complexes in aqueous solutions induces charge transfer from the coordinated CO_3^{2-} anions to Co(III), followed by a rapid hydrolysis of the Co(II) complex and the formation of free $CO_3^{\bullet -}$ radicals [48–50]:

$$[Co^{III}(NH_3)_4CO_3]^+ + h\nu \rightarrow Co^{II}_{aq} + 4NH_3 + CO_3^{\bullet -} \tag{4.24}$$

Using this method, $CO_3^{\bullet-}$ radicals can be generated within a wide pH range (3.0–13.0) [48–51].

Both (4.23) and (4.24) require only a single-photon absorption, and can thus be implemented by either by steady-state [32, 35–37] irradiation from a standard arc lamp or by a pulsed UV light source (e.g., 308-nm XeCl excimer laser pulses [14, 33, 34, 52, 53]).

4.2.6
Lifetimes of Free Radicals and Environmental Considerations

Free radicals are highly reactive species, and their lifetimes depend on the presence of other radicals and target molecules or ions. For instance, in neutral aqueous solutions, $SO_4^{\bullet-}$ radicals derived from the photodissociation of $S_2O_8^{2-}$ anions (4.23) decay mostly via recombination ($k_{25} = 1.1 \times 10^9 \, M^{-1} s^{-1}$) that results in the regeneration of the parent $S_2O_8^{2-}$ ion [33]:

$$SO_4^{\bullet-} + SO_4^{\bullet-} \to 2SO_4^{2-} \tag{4.25}$$

In our laser flash photolysis experiments, the concentrations of $SO_4^{\bullet-}$ radicals after the actinic laser flash (around 60 mJ/pulse/cm²) are typically approximately 10 μM and the lifetimes of the $SO_4^{\bullet-}$ radicals are $\tau_{25} = (2k_{25}[SO_4^{\bullet-}]_0)^{-1} \sim 50 \, \mu s$ [54]. Decreasing the laser pulse energy to around 10 mJ/pulse/cm² results in the reduction of the initial sulfate radical concentration to $[SO_4^{\bullet-}]_0 \sim 1.7 \, \mu M$ and an increase in the lifetimes of $SO_4^{\bullet-}$ radicals, τ_{25}, to approximately 300 μs.

In contrast, the decay of radicals in bimolecular reactions with other molecules (or ions), which are typically present at much higher concentrations than free radicals, does not depend on radical concentrations (or laser flash energy) and follows pseudo-first-order kinetics (4.16). An example of such a bimolecular reaction, is the one-electron oxidation ($k_{26} = 4.6 \times 10^6 \, M^{-1} s^{-1}$) of bicarbonate anions by sulfate radicals [33]:

$$SO_4^{\bullet-} + HCO_3^- \to SO_4^{2-} + CO_3^{\bullet-} \tag{4.26}$$

For instance, at $[HCO_3^-] = 0.1 \, M$, the decay of $SO_4^{\bullet-}$ radicals is mostly determined by (4.26) and the lifetime of the $SO_4^{\bullet-}$ radical can be estimated as $\tau_{26} = (k_{26}[HCO_3^-]_0)^{-1} = 2.2 \, \mu s$ [33].

When the oxidizing species, an electron acceptor, and the electron donor are both embedded within a biological macromolecule (e.g., in a protein or DNA molecules), the reaction kinetics are entirely different from those in solution in which both species can diffuse freely and encounter one another in order to undergo chemical reaction. An example of such intramolecular processes is the one-electron oxidation of guanine (G) by a 2AP neutral radical, both site-specifically positioned within a DNA duplex [28]. Here, both reaction partners are fixed within a DNA helix and the bimolecular reaction model is not suitable for describing the reaction kinetics (4.16). Instead, the kinetics of oxidation of G by 2AP(-H)$^{\bullet}$ radicals in double-stranded DNA follow first-order kinetics with the magnitudes

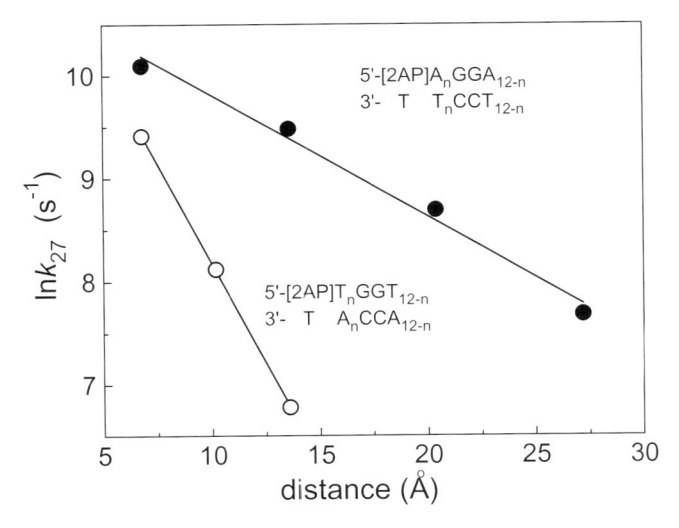

Figure 4.2 Rate constants of one-electron oxidation of G bases by 2AP neutral radicals in DNA duplexes as a function of distance between the G and 2AP bases. (Adapted from [28] and [55].)

of the rate constant depending on the number of bridging nucleobases between the electron donor and the acceptor [28, 55, 56]. The dependence of the rate constant of guanine oxidation by the 2AP(-H)˙ radicals on the number of intervening bases are described by:

$$k_{27} = k^\circ \exp(-\beta r) \quad (4.27)$$

where β is an attenuation parameter, and r is the distance between the G and the 2AP(-H)˙ residues. Plots of the linearized form of (4.27) are shown in Figure 4.2.

The slopes of these plots are a measure of the parameter β, which is a function of the intervening bases between the G and 2AP(-H)˙ residues. We found that in the case of intervening thymines, the parameter $\beta = 0.4\,\text{Å}^{-1}$ while in the case of intervening adenines $\beta = 0.12\,\text{Å}^{-1}$ in double-stranded DNA [28, 56].

4.2.7
Reactions of Free Radicals

The formation and reactions of guanine radicals in DNA and their reactions in the presence of carbon-centered radicals derived from lipid molecules provide instructive examples of free radical reactions in solutions involving these biologically relevant species. The reactivities of free radicals derived from biomolecules depend on their structures. The carbon-centered radicals produced by either by hydrogen atom abstraction or the addition of oxyl radicals to double bonds of polyunsaturated fatty acids (PUFAs) are primary intermediates of lipid peroxidation

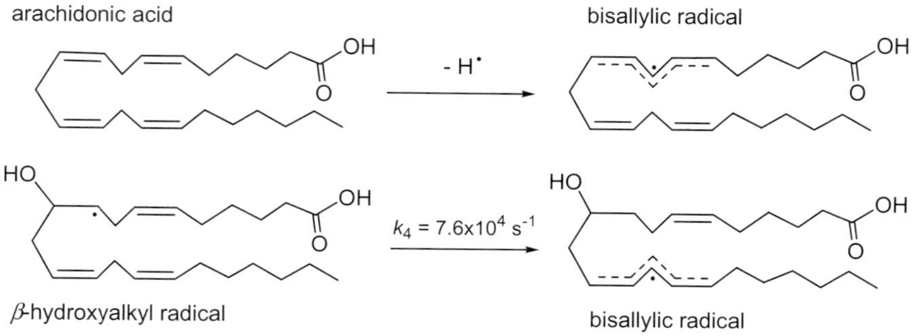

Figure 4.3 Formation of bisallylic radicals of arachidonic acid via inter- and intramolecular hydrogen atom abstraction mechanisms.

– a major contributor to the damage of cellular membranes and the generation of diverse reactive genotoxic intermediates [57–59] as discussed in detail by Knutsen and Marnett in Chapter 5. In PUFA molecules, the hydrogen atom abstractions by peroxyl radicals (chain propagation (4.3)) preferentially occur at the bisallylic positions with the weakest C–H bonds and result in the formation of bisallylic radicals, in which the unpaired electron is delocalized over two adjacent *cis* double bonds (Figure 4.3) [9].

The nonconjugated β-hydroxyalkyl radicals shown in the lower part of Figure 4.3 are derived from the direct addition of oxyl radicals (e.g., •OH) to the double bonds. These β-hydroxyalkyl radicals can give rise to bisallylic radicals by a hydrogen atom abstraction from a nearby bisallylic position. This hydrogen atom abstraction is feasible because the difference in the bond dissociation energies [60] of the C–H bonds in the -CH(OH)-CH$_2$- motif (94 ± 2 kcal/mol) and in the bisallylic methylene bridge (80 ± 1 kcal/mol) provides a sufficient driving force for this reaction [54]. Indeed, our own laser flash photolysis experiments have shown that β-hydroxyalkyl radicals derived from the oxidation of PUFAs such as arachidonic and linoleic acids rapidly transform to the bisallylic radicals (Figure 4.3) [54]. This *intramolecular* hydrogen atom abstraction can be monitored directly via the appearance of the characteristic narrow absorption band at 280 nm due to bisallylic radicals that can occur with the first-order rate constant of $7.5 \pm 0.7 \times 10^4 \, \text{s}^{-1}$ in the case of arachidonic acid. The C-centered lipid radicals rapidly react with molecular oxygen to form peroxyl radicals (chain propagation (4.2)); the rate constants of these fast reactions vary from $2.2 \pm 0.2 \times 10^8$ (bisallylic radicals) to $3.8 \pm 0.4 \times 10^9 \, \text{s}^{-1}$ (β-hydroxyalkyl radicals) [54].

The C-centered radicals of nucleic acid bases generated via hydrogen atom abstraction and addition of hydroxyl or other oxyl radicals to the double bonds also decay rapidly via bimolecular reactions with molecular oxygen. For instance, the rate constant for the reaction of O_2 with the C-centered 5-(2′-deoxyuridinyl) methyl radical (U–•CH$_2$) derived by hydrogen atom abstraction from the methyl group of thymidine was estimated as $2.2 \times 10^9 \, \text{M}^{-1} \text{s}^{-1}$ [61]. The radicals produced by the addition of •OH radicals to the C8 position of dG (8-HO-dG•) also rapidly react with O₂ with the rate constant of $4 \times 10^9 \, M^{-1} s^{-1}$ [62]. We note here that the reaction of ˙OH radicals with guanine is very complex and results in the formation of a series of radicals. These radicals exhibit different reactivities towards molecular oxygen and the yields of 8-HO-dG˙ radicals are typically around 17% [62]. The reactions of the C-centered radicals with molecular oxygen are very fast and their lifetimes in air-equilibrated aqueous solutions ($[O_2]$ = 0.27 mM) are typically 1–10 μs. In contrast, lifetimes of radicals generated by electron abstractions from guanine bases can vary from tens of nanoseconds to seconds (see below). Guanine is the most easily oxidizable canonical DNA base and reactions with various oxidizing species by one-electron transfer mechanisms generate the guanine radical cation, $G^{˙+}$ [63]. The free nucleoside radical cation, $dG^{˙+}$, is a weak acid [64] (pK_a = 3.9) that, according to the results of pulse radiolysis studies, deprotonates at pH 7.0 within around 50 ns with the rate constant of $1.8 \times 10^7 \, s^{-1}$ [65]. In double-stranded DNA, depending on the position within a DNA strand, the rate constant of deprotonation ($3 \times 10^6 \, s^{-1}$ or greater) is smaller than that of the free $dG^{˙+}$ cations [65], but is sufficiently high for a complete deprotonation of $G^{˙+}$ in the millisecond time range when G in DNA is oxidized by oxyl radicals (e.g., $CO_3^{˙-}$ radicals). The neutral guanine radical, G(-H)˙, remains strongly oxidizing with the reduction potential at pH 7, E_7 = 1.29 V versus NHE [63]. The free base dG(-H)˙ radical does not react with observable rates with molecular oxygen ($k \leq 10^2 \, M^{-1} s^{-1}$) [66]. Our own laser flash photolysis experiments have shown that the G(-H)˙ radicals derived from the site-selective oxidation of G bases with $CO_3^{˙-}$ radicals have lifetimes approaching seconds in the absence of reactive intermediates. However, in the presence of other radicals and nucleophiles that can react with the G(-H)˙ radicals, their lifetimes are significantly shorter [14, 33].

4.3
Mechanisms of Product Formation

4.3.1
Reactions of G(-H)˙ Radicals with Nucleophiles

Such reactions are illustrated by the oxidation of 5-GCT sequences by photochemically generated $CO_3^{˙-}$ radicals, which selectively oxidize G bases to form G(-H)˙ radicals [14, 33, 35]. To reduce the contributions of bimolecular $CO_3^{˙-}$ radical–radical reactions, the $CO_3^{˙-}$ radical were generated at very low concentrations. This is easily accomplished by the steady-state, continuous illumination with an arc lamp rather than photoexcitation with a laser pulse. Under these conditions, the oxidation of guanine in the single-stranded oligonucleotide, 5′-d(CCATCGCTACC) by $CO_3^{˙-}$ radicals generates a novel intrastrand cross-linked product as one of the major oxidation product [36, 37]. The cross-linked bases were excised from the oligonucleotide adduct by enzymatic digestion with nuclease P1 and alkaline phosphatase, and identified liquid chromatography tandem mass spectrometry

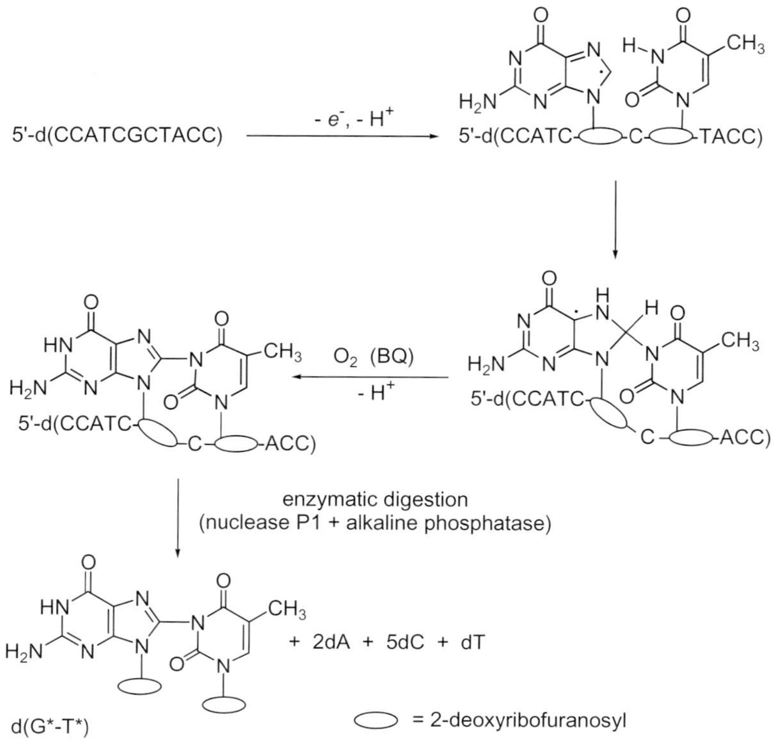

Figure 4.4 Intrastrand cross-links generated by the one-electron oxidation of 5'-GCT sequences.

(LC/MS-MS) analysis as d(G*–T*), in which guanine and thymine bases are linked by a covalent bond (Figure 4.4).

The structure of the d(G*–T*) excision fragment is unusual because G and T are not adjacent to one another in the parent oligonucleotide. In order to verify the structure of this cross-linked product in the oligonucleotide 5'-d(CCATCGCTACC), the same photo-oxidation method was used to irradiate the trinucleotide 5'-d(GpCpT) to generate the 5'-d(G*pCpT*) cross-linked product. The structure of this trinucleotide cross-linked product was confirmed by LC-MS/MS, and one- and two-dimensional nuclear magnetic resonance methods [36], and its properties were the same as those of the cross-linked trinucleotide product obtained by the digestion of the photoirradiated 11mer oligonucleotide.

The efficiency of G*–T* cross-link formation is strongly dependent on the base sequence context [36]. The highest efficiency of cross-linking was found in the 5'-d(GpCpT) sequence. Decreasing or increasing the number of C bases between the G and T reaction partners resulted in a reduction of the efficiency of cross-link formation (e.g., the yield of cross-linked products was lower in the 5'-d(GpT) than in the 5'-d(GCCT) sequence); in the case of the 5'-d(GCCCT) sequence, cross-linked products were not detected. The intrastrand 5-d(T*pG*) cross-links were not detected in the dinucleotide 5'-d(TpG) either, although there is a substantial yield in the isomeric 5'-d(GpT). This latter result suggests that cross-link formation is more efficient when the T residue is on the 3'-side rather than on the 5'-side of the guanine. In double-stranded DNA, intrastrand cross-links were found between adjacent G and T bases [36]. Interestingly, enzymatic digestion with nuclease P1 and alkaline phosphatase generates the 5'-d(G*pT*) fragment (i.e., this treatment does not excise the sugar residue between the G and the T bases). The $CO_3^{•-}$ radicals are not unique in their ability to generate cross-linked products.

We found that other one-electron oxidants with reduction potentials sufficient for the removal of an electron from guanine (e.g., $SO_4^{•-}$ radicals and photoexcited riboflavin molecules) also generate the 5'-G*CT* cross-linked products [36]. Formation of G(-H)• radicals is followed by a nucleophilic addition of the N3 atom of T to the C8 position of the G(-H)• radical (Figure 4.4). The radical adduct thus formed is further oxidized by O_2 (or another oxidant, such as 1,4-benzoquinone) to yield the G*–T* cross-linked products [36, 67]. This mechanism is supported by a significant increase of the cross-linked product yields when the pH is increased to pH ≥ 10. The deprotonation of T-N3(H) with pK_a = 9.7 [68] occurring in basic solutions, greatly enhances the nucleophilicity of T-N3. At pH 10, the yield of G*–T* cross-linked products is greater by a factor of around 5 relative to the yield measured at pH 7.5. Similar mechanisms have been proposed by Perrier *et al.* for the formation of cross-linked products between guanine and lysine mediated by the riboflavin photosensitization of d(TpGpT) in aerated solutions containing the KKK (lysine) tripeptide [69]. These authors suggested that the cross-linking reaction occurs via a nucleophilic addition of the ε-amino group of the lysine residue to the C8 atom of the $G^{•+}$ or G(-H)• radicals. The radical adduct thus formed is subsequently oxidized thus generating an $N^ε$-(guanin-8-yl)-lysine cross-link.

4.3.2
Combinations of G(-H)• and Oxyl Radicals

Laser pulse excitation can be utilized to generate high transient concentrations of free radicals that favor radical–radical combination reactions over nucleophilic addition reactions. An example of such bimolecular reactions is the combination of G(-H)• and •NO_2 radicals [31]. *In vivo*, nitrogen dioxide radicals are overproduced at sites of inflammation (reactions (4.7) and (4.9)). The reduction potential of nitrogen dioxide radicals [22], $E°$ = 1.04 V versus NHE is lower than the reduction potential of guanine neutral radicals [63], E_7 = 1.29 V versus NHE, and thus the oxidation of guanine by •NO_2 radicals is thermodynamically unfavorable. Indeed, our laser flash photolysis experiments have shown that •NO_2 radicals do not react with guanine with observable rates, but readily combine with G(-H)• radicals to form guanine nitro products [31].

The combination of G(-H)• and •NO_2 radicals was directly monitored by the appearance of the characteristic absorption band of nitroguanine products near 380 nm that correlates with the decay of the transient absorbance of G(-H)• radicals

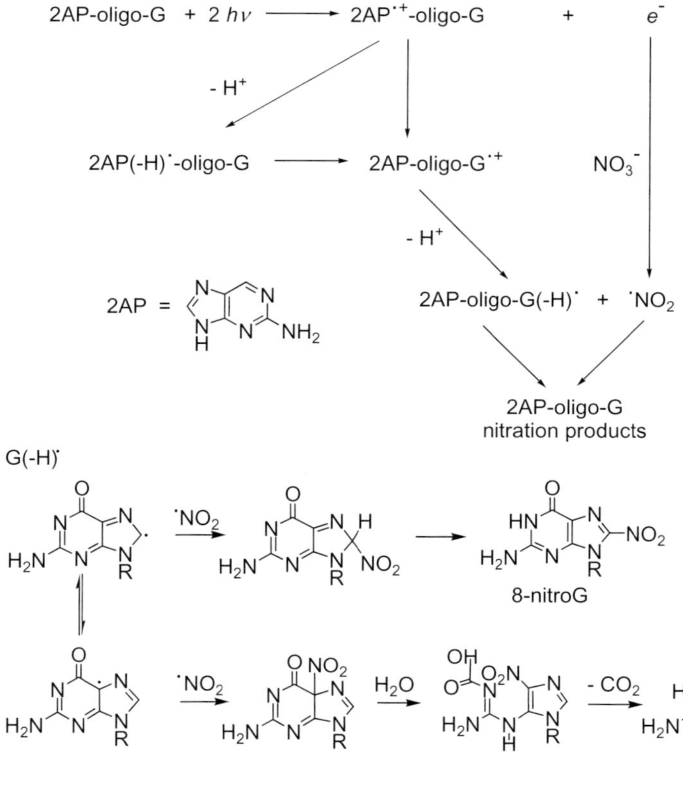

Figure 4.5 Generation of G(-H)˙ and ˙NO$_2$ radicals triggered by the two-photon ionization of 2AP radicals in DNA. These two radicals then combine to generate the 8-nitro-G and NIm products.

at 315 nm [31]. In these experiments, the ˙NO$_2$ and G(-H)˙ radicals in aqueous solutions were generated simultaneously by laser pulse excitation methods. The G(-H)˙ radicals positioned in the oligonucleotides were generated by the two-photon photoiozation of 2AP positioned near a G residue in the same oligonucleotide by an intense laser pulse (Figure 4.5). The photoionization of 2AP is accompanied by the ejection of an electron into the aqueous solution that are immediately hydrated to form "hydrated electrons." The ˙NO$_2$ radicals are then generated by the one-electron reduction of NO$_3^-$ anions by the hydrated electrons (Figure 4.5) [31].

The 2AP radical cations, 2AP˙$^+$, formed together with hydrated electrons by the photoionization process, rapidly deprotonate to the neutral radicals, 2AP(-H)˙ [31]. The latter radicals selectively oxidize nearby G bases to form G(-H)˙ radicals – a process that is completed within 100 μs after the actinic laser flash. Combination of the G(-H)˙ and ˙NO$_2$ radicals occurs on millisecond time scales with similar rate constants (around $4.4 \times 10^8 \, M^{-1} s^{-1}$) in both single- and double-stranded DNA [31]. The major nitro products isolated by high-performance liquid chromatography (HPLC) and identified by MS methods are 8-nitroguanine (8-nitro-G) and 5-guanidino-4-nitroimidazole (NIm) [31, 32, 34]. The formation of these products occurs via the addition of $^\bullet NO_2$ radicals to either the C8 or the C5 positions of G(-H)$^\bullet$ radicals (Figure 4.5).

Another interesting example of radical–radical reactions is the combination of G(-H)$^\bullet$ and $O_2^{\bullet-}$ radicals [29]. In healthy tissues, the concentrations of $O_2^{\bullet-}$ radicals is small and is regulated by SODs [70, 71]. These enzymes rapidly deactivate $O_2^{\bullet-}$ radicals by catalytic dismutation to oxygen and hydrogen peroxide:

$$2\, O_2^{\bullet-} + 2\, H^+ \rightarrow O_2 + H_2O_2 \tag{4.28}$$

which is further catalytically converted to O_2 and H_2O by catalases. However, $O_2^{\bullet-}$ radicals are overproduced under inflammatory conditions and in mitochondria [72], leading to the development of oxidative stress. In our laser flash photolysis experiments [29], in air-saturated solutions, the hydrated electrons derived from the two-photon ionization of the 2AP residues are rapidly and quantitatively scavenged by molecular oxygen to form $O_2^{\bullet-}$ radicals (Figure 4.6) [73].

The rate of combination of the $O_2^{\bullet-}$ and G(-H)$^\bullet$ radicals positioned in single- and double-stranded oligonucleotides can be determined from the decay kinetics of the characteristic narrow absorption band of G(-H)$^\bullet$ radicals at 315 nm [29]. The rate constant of this bimolecular combination reaction was found to be around $4.7 \times 10^8 \, M^{-1} s^{-1}$. The Cu,Zn-SOD reacts with $O_2^{\bullet-}$ radicals with nearly diffusion-controlled rates [74, 75] and thus dramatically enhances the lifetimes of the DNA-bound G(-H)$^\bullet$ radicals from 4–7 ms to 0.2–0.6 s in the presence of micromolar concentrations of Cu,Zn-SOD (around 5 µM) [29]. Thus, radical–radical combination reactions can play important roles in shortening the lifetimes of guanine radicals in DNA, as mentioned earlier.

In the absence of superoxide scavengers, the products of combination of G(-H)$^\bullet$ with $O_2^{\bullet-}$ radicals (generated by the capture of hydrated electrons by molecular oxygen present in air-saturated or O_2-containing solutions), isolated by HPLC and identified by MS methods, are mostly 2,5-diamino-4H-imidazolone (Iz) lesions and minor amounts of 8-oxoguanine (8-oxo-G) lesions (Figure 4.6) [29]. The formation of these products occurs via the addition of $O_2^{\bullet-}$ radicals to either the C5 or the C8 positions of G(-H)$^\bullet$ radicals (Figure 4.6). In agreement with this mechanism, $^{18}O_2$-labeling experiments showed that the O atom in the Iz lesion originates from $^{18}O_2$ in aqueous solutions [76]. Moreover, we found that in the presence of 1,4-benzoquinone added in small concentrations (around 100 µM), the yields of the Iz lesions become negligible because the addition of BQ can eliminate the $O_2^{\bullet-}$ and G(-H)$^\bullet$ radicals [29]. In contrast, ^{18}O-labeling experiments reveal that there are two pathways of 8-oxo-G formation, including either the addition of $O_2^{\bullet-}$ to the C8 position of G(-H)$^\bullet$ radicals (Figure 4.6) or the hydration of the G$^{\bullet+}$/G(-H)$^\bullet$ radicals [29, 77]. The two-electron reduction of the 8-HOO-G(-H) hydroperoxide formed leads to 8-oxo-G in which the added oxygen atom originates from $^{16}O_2$. Molecular oxygen is reduced to $O_2^{\bullet-}$ by hydrated electrons. Alternatively,

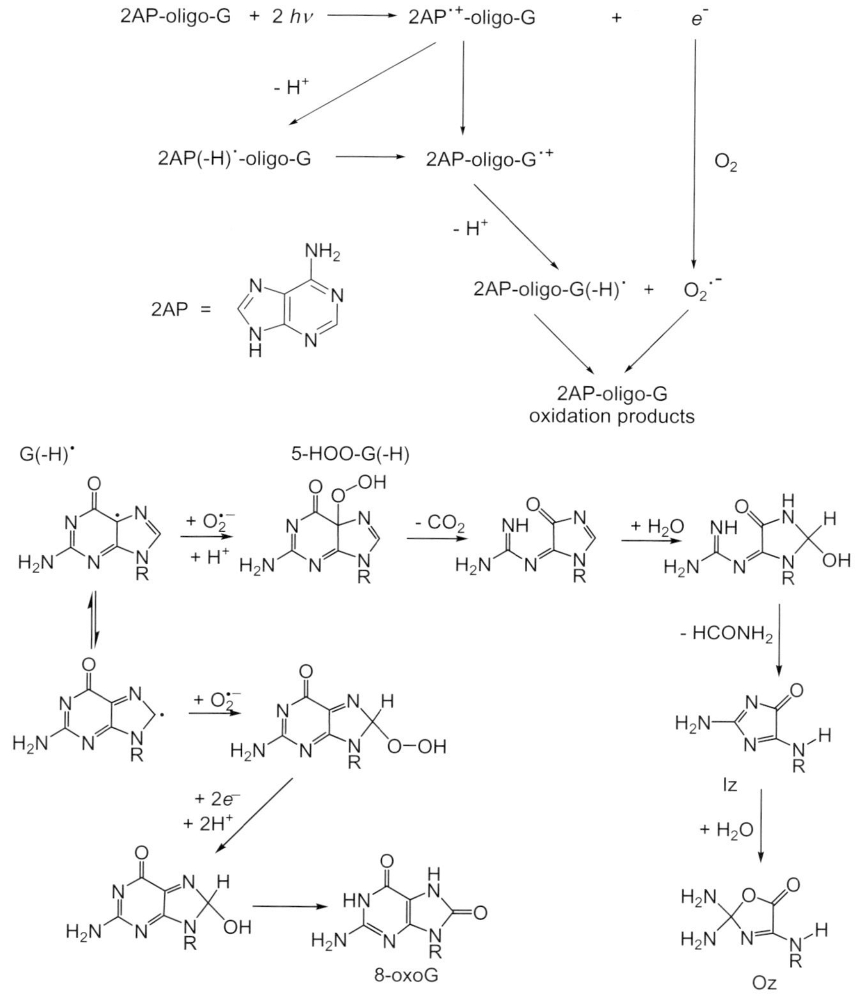

Figure 4.6 Mechanisms of generation of the oxidatively formed guanine products Iz and 8-oxo-G. The laser pulse-induced two-photon ionization of 2AP radicals in DNA generates 2AP$^{\bullet+}$/2AP(-H)$^{\bullet}$ radicals and hydrated electrons. The 2AP$^{\bullet+}$/2AP(-H)$^{\bullet}$ radicals oxidize a nearby guanine base within the same oligonucleotide molecule by a one-electron oxidation mechanism to form the G(-H)$^{\bullet}$ radicals. The latter combines with O$_2^{\bullet-}$ radicals to form the products 8-oxo-G, or Iz and Oz.

the O$_2^{\bullet-}$ radicals can be generated by the oxidation of hydrogen peroxide by Fe^{3+} ions. Indeed, the appearance of ^{18}O atom in 8-oxo-G has been detected during the course of oxidation of human A549 lung epithelial cells by H$_2^{18}$O$_2$ [78]. In laser pulse excitation experiments, a decrease in the ^{16}O$_2^{\bullet-}$ concentrations (by removal of ^{16}O$_2$ by bubbling inert gases through the solution) favors the hydration of guanine radicals by $H_2^{18}O$ that results in the formation of 8-HO-G$^•$ radicals [29]. Pulse radiolysis experiments have shown that the 8-HO-dG$^•$ radicals derived from the addition of $^•$OH radicals to the C8 position of dG are easily oxidized by weak oxidants (e.g., methylviologen, $Fe(CN)_6^{3-}$, oxygen [62], and benzoquinone [79]) to form 8-oxo-dG. In these mechanisms, the oxygen atom in 8-oxo-dG originates from the addition of $H_2^{18}O$ to the guanine radicals by the hydration mechanism. Cadet et al. showed that 8-oxo-dG formed in calf thymus DNA by a photoexcited riboflavin oxidation pathway also contains oxygen atoms derived from $H_2^{18}O$, thus pointing to a hydration mechanism [77]. Since in the latter experiments the photoexcitation was carried out by steady-state illumination, the concentrations of radical intermediates is low and thus the hydration of guanine radicals competes effectively with radical–radical combination mechanisms. In the case of photoexcitation experiments with high-intensity laser pulses, the concentrations of radical intermediates such as $^{16}O_2^{•-}$ is high and guanine radical–radical combination reactions dominate over the hydration of guanine radicals. These considerations explain the differences in the incorporation of ^{18}O or ^{16}O atoms into 8-oxo-G in either $^{18}O_2$ or $H_2^{18}O$ experiments because the relative hydration and radical–radical reaction pathways involving $G^{•+}/G(-H)^•$ radicals depend strongly on reaction conditions.

We also explored the oxidative modification of guanine by alkyl peroxyl radicals (ROO$^•$) formed in the course of lipid peroxidation (reactions (4.2)–(4.4)) [57]. The reduction potential of these radicals [80, 81], $E_7 = 1.0–1.1$ V versus NHE is lower than the reduction potential of G(-H)$^•$ radicals [63], $E_7 = 1.29$ V versus NHE, and the oxidation of guanine by ROO$^•$ radicals as well by $^•NO_2$ radicals is thermodynamically unfavorable. Our experiments have shown that the major role of ROO$^•$ radicals in the oxidation of guanine is their combination with G(-H)$^•$ radicals [54]. In these experiments, the peroxyl radicals were generated by the addition of O_2 to C-centered radicals of polyunsaturated fatty acids, such as arachidonic acid. As in the case of $O_2^{•-}$ radicals, one of the major products derived from the reactions of G(-H)$^•$ radicals and arachidonic acid peroxyl radicals are the imidazolone lesions formed together with the products of oxidation of 8-oxo-G intermediates, such as dehydroguanidinohydantoin (Gh$_{ox}$) and the diastereomeric spiroiminodihydantoin (Sp) lesions. The mechanisms of Sp and Gh$_{ox}$ formation from 8-oxo-G are considered in the next section.

4.3.3
Oxidation of 8-oxoG

There is abundant evidence that 8-oxo-G is ubiquitous in cellular DNA as discussed in detail by Cadet et al. in Chapter 3 [82, 83]. The major pathway of formation of this important marker of oxidative DNA damage involves the one-electron oxidation of 8-HO-G$^•$ radicals derived from either the hydration of the $G^{•+}/G(-H)^•$ radicals or the addition of $^•$OH radicals to the C8 position of G [62, 84]. In DNA, the redox potential of 8-oxo-G at pH 7, $E_7 = 0.74$ V versus NHE, indicates that it is more easily oxidized than any of the four natural nucleobases [85]. The primary

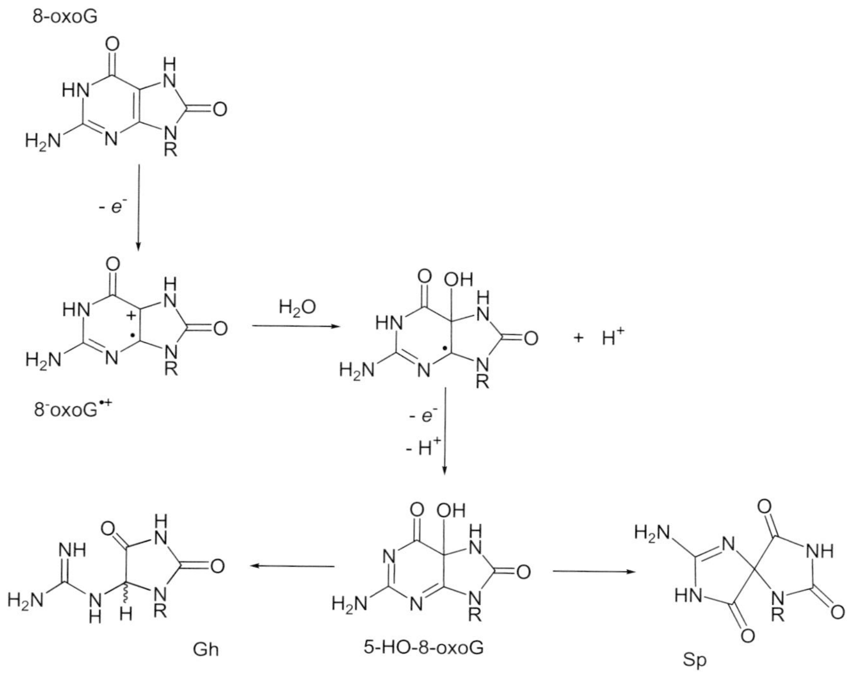

Figure 4.7 Consecutive two-electron oxidation of 8-oxo-G that result in the formation of diastereomeric Sp and Gh lesions.

product of the one-electron oxidation of 8-oxo-G is the radical cation 8-oxo-G$^{\bullet+}$, which can be detected by direct spectroscopic methods by the appearance of the characteristic absorption band at 325 nm [85, 86]. In neutral aqueous solutions, the radical cation (pK_a = 6.6) is in equilibrium with its neutral form, 8-oxo-G(-H)$^{\bullet}$, that has very similar spectroscopic characteristics. Our own laser flash photolysis experiments have shown that 8-oxo-G$^{\bullet+}$/8-oxo-G(-H)$^{\bullet}$ radicals do not react with observable rates with molecular oxygen, O_2, and their lifetimes are controlled by other radicals and nucleophiles [30, 31, 86] that may be present, as discussed earlier in this chapter.

The hydration of the 8-oxo-G$^{\bullet+}$/8-oxo-G(-H)$^{\bullet}$ radicals, followed by one-electron oxidation mechanisms, generates the 5-HO-8-oxo-G intermediate (Figure 4.7) – a precursor of the Sp and Gh products [87, 88].

The subsequent transformation of this 5-HO-8-oxo-G intermediate to form either Sp or Gh lesions depends on the pH of the solution. Tannenbaum *et al.* proposed that the partitioning of 5-HO-8-oxo-G(-H) into either Sp or Gh is determined by the different reactivities of the deprotonated and protonated forms of this adduct (pK_a ~ 5.8) [89]. Decreasing the pH below 5.8 favors pyrimidine ring-opening, followed by the formation of the Gh lesions; in contrast, the acyl shift leading to the Sp lesions dominates at pH > 5.8. This pH dependence qualitatively

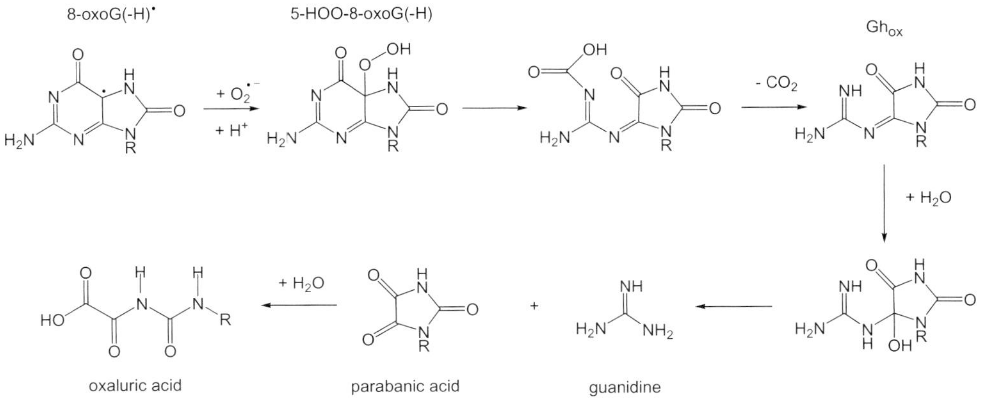

Figure 4.8 Formation of oxaluric acid lesions via the combination reaction of 8-oxo-G(-H)· with $O_2^{\bullet-}$ radicals.

agrees with the enhancement of the yield of Gh products at low pH that has been reported for the oxidation of 8-oxo-G by either peroxynitrite [89], photoexcited riboflavin, $IrCl_6^{2-}$ [90, 91], or Cr(VI) complexes [92].

The combination of $O_2^{\bullet-}$ with 8-oxo-G(-H)· radicals in double- and single-stranded DNA occurs with the rate constant of $(1.0–1.3) \times 10^8 \, M^{-1} \, s^{-1}$ [30]. The major end products of this reaction are the dehydroguanidinohydantoin lesions (Gh_{ox}) derived from the addition of $O_2^{\bullet-}$ to the C5 position of 8-oxo-G(-H)·, followed by the decomposition of the 5-HOO-8-oxo-G(-H) hydroperoxide (Figure 4.8). The Gh_{ox} lesions are unstable and slowly hydrolyze to oxaluric acid.

4.4
Biological Implications

The final oxidative modifications that arise from the initial one-electron abstraction from guanine in DNA are determined by further reactions of guanine radicals with free radicals or nucleophiles. Of the guanine lesions considered in this work, only 8-oxo-G and the 8-nitro-G lesions have been detected in humans [93]. Detailed studies by several groups have shown that the levels of 8-oxo-G in normal human cells are in the range of 0.3–4 per 10^6 molecules of intact G bases [94, 95]. The accurate measurement of the levels of 8-oxo-G in tissues under inflammatory conditions is a difficult problem due to the efficient removal of these lesions by base excision repair enzymes. The enhanced levels of 8-nitro-G bases that easily depurinate in DNA have been detected in tissues and biological fluids of patients with various diseases including chronic hepatitis C [96, 97], intrahepatic cholangiocarcinoma [98, 99], nasopharyngeal carcinoma [100], malignant fibrous histiocytoma [101, 102], *Helicobacter pylori* infection [103, 104], and inflammatory bowel disease [105, 106]. Further research should be focused on the detection of 5-

guanidino-4-nitroimidazole lesions which, *in vitro* experiments, are formed with nearly the same proportions as 8-nitro-G lesions, but do not undergo depurination. At a formal level, both 8-oxo-G and 8-nitro-G are products of a two-electron oxidation mechanism of guanine. Indeed, the formation of 8-nitro-G occurs in two steps – electron abstraction from guanine ("first hit"), followed by the combination of ˙NO_2 radicals ("second hit"); in other words, the interaction of guanine with two free radicals ("two hits") is required for the formation of nitration products. In contrast, the formation of 8-oxo-G requires only one-electron abstraction from guanine (a "single hit") and molecular oxygen molecules that are ubiquitous in cells can abstract the second electron after the hydration of the guanine radicals. The formation of the intrastrand G*–T* cross-links involves a similar sequence of events, a one-electron oxidation of guanine ("single hit"), the nucleophilic addition of T-N3 to G-C8, followed by the oxidation of the radical adduct by O_2. This mechanism of formation of G*–T* cross-links may be operative *in vivo* and further research should be directed towards determining whether these intrastrand cross-linked lesions are also formed in cellular DNA.

The Iz (or Oz, the Iz hydrolysis product) and Sp lesions are products of the four-electron oxidation of guanine. However, the formation of the Iz lesions requires only "two hits" of guanine by free radicals: the one-electron oxidation of guanine, followed by the combination reaction of guanine radicals with the superoxide radical anion – a three-electron oxidant. Recently, Oz lesions were detected at a level of two to six molecules of Oz per 10^7 molecules of intact G in liver DNA of diabetic and control rats maintained on a diet high in animal fat [107]. The formation of Sp lesions *in vitro* can occur via the intermediate formation of 8-oxo-G. Since the levels of 8-oxo-G in cells are quite low [94, 95], the further oxidation of 8-oxo-G to Sp by oxidants that can abstract electrons from both G and 8-oxo-G seems to be unlikely because guanine in cellular DNA is present in concentrations by a factor of around 10^6 greater than 8-oxo-G. The selective oxidation of 8-oxo-G can occur only by oxidants that are unable to oxidize guanine, but have a sufficient oxidation potential to oxidize 8-oxo-G. Under inflammatory conditions, the best candidates for such reaction are ˙NO_2 and ROO˙ radicals that do not react with G bases, but can potentially oxidize 8-oxo-G. Recently, the Sp lesions have been detected in *Escherichia coli* treated with Cr(VI) [108]. The search for Sp lesions *in vivo*, especially in tissues under inflammatory conditions, is a critically important task that might further clarify the pathways of oxidatively generated damage in DNA generated by reactive intermediates that are products of the inflammatory response.

Acknowledgements

This work was supported by the NIEHS (5 R01 ES 011589-07). The content is solely the responsibility of the authors and does not necessarily represent the official views of the NIEHS or NIH.

References

1 Valko, M., Leibfritz, D., Moncol, J., Cronin, M.T., Mazur, M., and Telser, J. (2007) *Int. J. Biochem. Cell Biol.*, **39**, 44–84.
2 Hancock, J.T., Desikan, R., and Neill, S.J. (2001) *Biochem. Soc. Trans.*, **29**, 345–350.
3 Grisham, M.B., Jourd'heuil, D., and Wink, D.A. (2000) *Aliment. Pharmacol. Ther.*, **14** (Suppl. 1), 3–9.
4 Dedon, P.C. and Tannenbaum, S.R. (2004) *Arch. Biochem. Biophys.*, **423**, 12–22.
5 Pacher, P., Beckman, J.S., and Liaudet, L. (2007) *Physiol. Rev.*, **87**, 315–424.
6 Kuper, H., Adami, H.O., and Trichopoulos, D. (2000) *J. Intern. Med.*, **248**, 171–183.
7 Balkwill, F. and Mantovani, A. (2001) *Lancet*, **357**, 539–545.
8 Coussens, L.M. and Werb, Z. (2002) *Nature*, **420**, 860–867.
9 Yin, H. and Porter, N.A. (2005) *Antioxid. Redox Signal.*, **7**, 170–184.
10 Gardner, H.W. (1989) *Free Radic. Biol. Med.*, **7**, 65–86.
11 Dix, T.A. and Aikens, J. (1993) *Chem. Res. Toxicol.*, **6**, 2–18.
12 Neta, P., Grodkowski, J., and Ross, A.B. (1996) *J. Phys. Chem. Ref. Data*, **25**, 709–1050.
13 Breen, A.P. and Murphy, J.A. (1995) *Free Radic. Biol. Med.*, **18**, 1033–1077.
14 Shafirovich, V., Dourandin, A., Huang, W., and Geacintov, N.E. (2001) *J. Biol. Chem.*, **276**, 24621–24626.
15 Valko, M., Izakovic, M., Mazur, M., Rhodes, C.J., and Telser, J. (2004) *Mol. Cell. Biochem.*, **266**, 37–56.
16 Beckman, J.S., Beckman, T.W., Chen, J., Marshall, P.A., and Freeman, B.A. (1990) *Proc. Natl. Acad. Sci. USA*, **87**, 1620–1624.
17 Huie, R.E. and Padmaja, S. (1993) *Free Radic. Res. Commun.*, **18**, 195–199.
18 Løgager, T. and Sehested, K. (1993) *J. Phys. Chem.*, **97**, 6664–6669.
19 Pryor, W.A. and Squadrito, G.L. (1995) *Am. J. Physiol.*, **268**, L699–L722.
20 Koppenol, W.H., Moreno, J.J., Pryor, W.A., Ischiropoulos, H., and Beckman, J.S. (1992) *Chem. Res. Toxicol.*, **5**, 834–842.
21 Goldstein, S., Lind, J., and Merenyi, G. (2005) *Chem. Rev.*, **105**, 2457–2470.
22 Stanbury, D.M. (1989) *Adv. Inorg. Chem.*, **33**, 69–138.
23 Neta, P., Huie, R.E., and Ross, A.B. (1988) *J. Phys. Chem. Ref. Data*, **17**, 1027–1284.
24 Lymar, S.V. and Hurst, J.K. (1995) *J. Am. Chem. Soc.*, **117**, 8867–8868.
25 Carola, R., Harely, J.P., and Noback, C.R. (1990) *Human Anatomy and Physiology*, McGraw-Hill, New York.
26 Norrish, R.G.W. and Porter, G. (1949) *Nature*, **164**, 658.
27 Shafirovich, V., Dourandin, A., Huang, W., Luneva, N.P., and Geacintov, N.E. (1999) *J. Phys. Chem. B*, **103**, 10924–10933.
28 Shafirovich, V., Dourandin, A., Huang, W., Luneva, N.P., and Geacintov, N.E. (2000) *Phys. Chem. Chem. Phys.*, **2**, 4399–4408.
29 Misiaszek, R., Crean, C., Joffe, A., Geacintov, N.E., and Shafirovich, V. (2004) *J. Biol. Chem.*, **279**, 32106–32115.
30 Misiaszek, R., Uvaydov, Y., Crean, C., Geacintov, N.E., and Shafirovich, V. (2005) *J. Biol. Chem.*, **280**, 6293–6300.
31 Misiaszek, R., Crean, C., Geacintov, N.E., and Shafirovich, V. (2005) *J. Am. Chem. Soc.*, **127**, 2191–2200.
32 Shafirovich, V., Mock, S., Kolbanovskiy, A., and Geacintov, N.E. (2002) *Chem. Res. Toxicol.*, **15**, 591–597.
33 Joffe, A., Geacintov, N.E., and Shafirovich, V. (2003) *Chem. Res. Toxicol.*, **16**, 1528–1538.
34 Joffe, A., Mock, S., Yun, B.H., Kolbanovskiy, A., Geacintov, N.E., and Shafirovich, V. (2003) *Chem. Res. Toxicol.*, **16**, 966–973.
35 Crean, C., Geacintov, N.E., and Shafirovich, V. (2005) *Angew. Chem. Int. Ed.*, **44**, 5057–5060.
36 Crean, C., Uvaydov, Y., Geacintov, N.E., and Shafirovich, V. (2008) *Nucleic Acids Res.*, **36**, 742–755.

37 Crean, C., Lee, Y.A., Yun, B.H., Geacintov, N.E., and Shafirovich, V. (2008) *ChemBioChem*, **9**, 1985–1991.
38 Candeias, L.P. and Steenken, S. (1992) *J. Am. Chem. Soc.*, **114**, 699–704.
39 Melvin, T., Botchway, S.W., Parker, A.W., and O'Neill, P.J. (1995) *J. Chem. Soc. Chem. Commun.*, 653–654.
40 Melvin, T., Botchway, S.W., Parker, A.W., and O'Neill, P. (1996) *J. Am. Chem. Soc.*, **118**, 10031–10036.
41 Angelov, D., Spassky, A., Berger, M., and Cadet, J. (1997) *J. Am. Chem. Soc.*, **119**, 11373–11380.
42 Larsen, O.F.A., Van Stokkum, I.H.M., De Weerd, F.L., Vengris, M., Aravindakumar, C.T., Van Grondelle, R., Geacintov, N.E., and Van Amerongen, H. (2004) *Phys. Chem. Chem. Phys.*, **6**, 154–160.
43 Shafirovich, V.Y., Levin, P.P., Kuzmin, V.A., Thorgeirsson, T.E., Kliger, D.S., and Geacintov, N.E. (1994) *J. Am. Chem. Soc.*, **116**, 63–72.
44 O'Connor, D., Shafirovich, V.Y., and Geacintov, N.E. (1994) *J. Phys. Chem.*, **98**, 9831–9839.
45 Shafirovich, V.Y., Courtney, S.H., Ya, N., and Geacintov, N.E. (1995) *J. Am. Chem. Soc.*, **117**, 4920–4929.
46 Kuzmin, V.A., Dourandin, A., Shafirovich, V., and Geacintov, N.E. (2000) *Phys. Chem. Chem. Phys.*, **2**, 1531–1535.
47 Yun, B.H., Lee, Y.A., Kim, S.K., Kuzmin, V., Kolbanovskiy, A., Dedon, P.C., Geacintov, N.E., and Shafirovich, V. (2007) *J. Am. Chem. Soc.*, **129**, 9321–9332.
48 Cope, V.W. and Hoffman, M.Z. (1972) *J. Chem. Soc. Chem. Commun.*, 227–228.
49 Chen, S.-N., Cope, V.W., and Hoffman, M.Z. (1973) *J. Phys. Chem.*, **77**, 1111–1116.
50 Ferraudi, G. and Perkovic, M. (1993) *Inorg. Chem.*, **32**, 2587–2590.
51 Kuimova, M.K., Cowan, A.J., Matousek, P., Parker, A.W., Sun, X.Z., Towrie, M., and George, M.W. (2006) *Proc. Natl. Acad. Sci. USA*, **103**, 2150–2153.
52 Lee, Y.A., Yun, B.H., Kim, S.K., Margolin, Y., Dedon, P.C., Geacintov, N.E., and Shafirovich, V. (2007) *Chem. Eur. J.*, **13**, 4571–4581.
53 Lee, Y.A., Durandin, A., Dedon, P.C., Geacintov, N.E., and Shafirovich, V. (2008) *J. Phys. Chem. B*, **112**, 1834–1844.
54 Crean, C., Geacintov, N.E., and Shafirovich, V. (2008) *Chem. Res. Toxicol.*, **21**, 358–373.
55 Shafirovich, V. and Geacintov, N.E. (2004) *Top. Curr. Chem.*, **237**, 129–157.
56 Shafirovich, V. and Geacintov, N.E. (2005) in *Charge Transfer in DNA* (ed. H.-A. Wagenknecht), Wiley-VCH Verlag GmbH, Weinheim, pp. 175–196.
57 Marnett, L.J. (2002) *Toxicology*, **181–182**, 219–222.
58 Blair, I.A. (2008) *J. Biol. Chem.*, **283**, 15545–15549.
59 Schneider, C., Porter, N.A., and Brash, A.R. (2008) *J. Biol. Chem.*, **283**, 15539–15543.
60 Luo, Y.R. and Kerr, A. (2006) in *CRC Handbook of Chemistry and Physics*, 87th edn (ed. D.R. Lide), CRC Press, Boca Raton, FL, pp. 9–60.
61 Hong, I.S., Ding, H., and Greenberg, M.M. (2006) *J. Am. Chem. Soc.*, **128**, 485–491.
62 Candeias, L.P. and Steenken, S. (2000) *Chem. Eur. J.*, **6**, 475–484.
63 Steenken, S. and Jovanovic, S.V. (1997) *J. Am. Chem. Soc.*, **119**, 617–618.
64 Candeias, L.P. and Steenken, S. (1989) *J. Am. Chem. Soc.*, **111**, 1094–1099.
65 Kobayashi, K. and Tagawa, S. (2003) *J. Am. Chem. Soc.*, **125**, 10213–10218.
66 Al-Sheikhly, M. (1994) *Radiat. Phys. Chem.*, **44**, 297–301.
67 Crean, C., Geacintov, N.E., and Shafirovich, V. (2008) *Free Radic. Biol. Med.*, **45**, 1125–1134.
68 Knobloch, B., Linert, W., and Sigel, H. (2005) *Proc. Natl. Acad. Sci. USA*, **102**, 7459–7464.
69 Perrier, S., Hau, J., Gasparutto, D., Cadet, J., Favier, A., and Ravanat, J.L. (2006) *J. Am. Chem. Soc.*, **128**, 5703–5710.
70 Fridovich, I. (1989) *J. Biol. Chem.*, **264**, 7761–7764.
71 Fridovich, I. (1995) *Annu. Rev. Biochem.*, **64**, 97–112.
72 Van Houten, B., Woshner, V., and Santos, J.H. (2006) *DNA Repair*, **5**, 145–152.

73 Bielski, B.H.J., Cabelli, D.E., Arudi, R.L., and Ross, A.B. (1985) *J. Phys. Chem. Ref. Data*, **14**, 1041–1100.
74 Klug, D., Rabani, J., and Fridovich, I. (1972) *J. Biol. Chem.*, **247**, 4839–4842.
75 Rotilio, G., Bray, R.C., and Fielden, E.M. (1972) *Biochim. Biophys. Acta*, **268**, 605–609.
76 Cadet, J., Berger, M., Buchko, G.W., Joshi, P.C., Raoul, S., and Ravanat, J.-L. (1994) *J. Am. Chem. Soc.*, **116**, 7403–7404.
77 Kasai, H., Yamaizumi, Z., Berger, M., and Cadet, J. (1992) *J. Am. Chem. Soc.*, **114**, 9692–9694.
78 Hofer, T., Seo, A.Y., Prudencio, M., and Leeuwenburgh, C. (2006) *Biol. Chem.*, **387**, 103–111.
79 Simic, M.G. and Jovanovic, S.V. (1986) in *Mechanisms of DNA Damage and Repair: Implications for Carcinogenesis and Risk Assessment* (eds M.G. Simic, L. Grossman, and A.C. Upton), Plenum Press, New York, pp. 39–49.
80 Das, T.N., Dhanasekaran, T., Alfassi, Z.B., and Neta, P. (1998) *J. Phys. Chem. A*, **102**, 280–284.
81 Jovanovic, S.V., Jankovic, I., and Josimovic, L. (1992) *J. Am. Chem. Soc.*, **114**, 9018–9021.
82 Beckman, K.B. and Ames, B.N. (1997) *J. Biol. Chem.*, **272**, 19633–19636.
83 Cadet, J., Bellon, S., Berger, M., Bourdat, A.G., Douki, T., Duarte, V., Frelon, S., Gasparutto, D., Muller, E., Ravanat, J.L., *et al.* (2002) *Biol. Chem.*, **383**, 933–943.
84 Steenken, S. (1989) *Chem. Rev.*, **89**, 503–520.
85 Steenken, S., Jovanovic, S.V., Bietti, M., and Bernhard, K. (2000) *J. Am. Chem. Soc.*, **122**, 2373–2374.
86 Shafirovich, V., Cadet, J., Gasparutto, D., Dourandin, A., Huang, W., and Geacintov, N.E. (2001) *J. Phys. Chem. B*, **105**, 586–592.
87 Luo, W., Muller, J.G., Rachlin, E.M., and Burrows, C.J. (2000) *Org. Lett.*, **2**, 613–616.
88 McCallum, J.E., Kuniyoshi, C.Y., and Foote, C.S. (2004) *J. Am. Chem. Soc.*, **126**, 16777–16782.
89 Niles, J.C., Wishnok, J.S., and Tannenbaum, S.R. (2004) *Chem. Res. Toxicol.*, **17**, 1510–1519.
90 Leipold, M.D., Muller, J.G., Burrows, C.J., and David, S.S. (2000) *Biochemistry*, **39**, 14984–14992.
91 Kornyushyna, O., Berges, A.M., Muller, J.G., and Burrows, C.J. (2002) *Biochemistry*, **41**, 15304–15314.
92 Sugden, K.D., Campo, C.K., and Martin, B.D. (2001) *Chem. Res. Toxicol.*, **14**, 1315–1322.
93 Neeley, W.L. and Essigmann, J.M. (2006) *Chem. Res. Toxicol.*, **19**, 491–505.
94 Collins, A., Gedik, C., Vaughan, N., Wood, S., White, A., Dubois, J., Duez, P., Dehon, G., Rees, J.-F., Loft, S., *et al.* (2002) *Carcinogenesis*, **23**, 2129–2133.
95 Collins, A.R., Cadet, J., Moller, L., Poulsen, H.E., and Vina, J. (2004) *Arch. Biochem. Biophys.*, **423**, 57–65.
96 Horiike, S., Kawanishi, S., Kaito, M., Ma, N., Tanaka, H., Fujita, N., Iwasa, M., Kobayashi, Y., Hiraku, Y., Oikawa, S., *et al.* (2005) *J. Hepatol.*, **43**, 403–410.
97 Kaito, M., Horiike, S., Tanaka, H., Fujita, N., and Adachi, Y. (2006) *J. Gastroenterol.*, **41**, 713–714.
98 Pinlaor, S., Sripa, B., Ma, N., Hiraku, Y., Yongvanit, P., Wongkham, S., Pairojkul, C., Bhudhisawasdi, V., Oikawa, S., Murata, M., *et al.* (2005) *World J. Gastroenterol.*, **11**, 4644–4649.
99 Pinlaor, S., Hiraku, Y., Yongvanit, P., Tada-Oikawa, S., Ma, N., Pinlaor, P., Sithithaworn, P., Sripa, B., Murata, M., Oikawa, S., *et al.* (2006) *Int. J. Cancer*, **119**, 1067–1072.
100 Ma, N., Kawanishi, M., Hiraku, Y., Murata, M., Huang, G.W., Huang, Y., Luo, D.Z., Mo, W.G., Fukui, Y., and Kawanishi, S. (2008) *Int. J. Cancer*, **122**, 2517–2525.
101 Hoki, Y., Murata, M., Hiraku, Y., Ma, N., Matsumine, A., Uchida, A., and Kawanishi, S. (2007) *Oncol. Rep.*, **18**, 1165–1169.
102 Hoki, Y., Hiraku, Y., Ma, N., Murata, M., Matsumine, A., Nagahama, M., Shintani, K., Uchida, A., and Kawanishi, S. (2007) *Cancer Sci.*, **98**, 163–168.
103 Ma, N., Adachi, Y., Hiraku, Y., Horiki, N., Horiike, S., Imoto, I., Pinlaor, S., Murata, M., Semba, R., and Kawanishi, S. (2004) *Biochem. Biophys. Res. Commun.*, **319**, 506–510.

104 Kawanishi, S., Hiraku, Y., Pinlaor, S., and Ma, N. (2006) *Biol. Chem.*, **387**, 365–372.

105 Ding, X., Hiraku, Y., Ma, N., Kato, T., Saito, K., Nagahama, M., Semba, R., Kuribayashi, K., and Kawanishi, S. (2005) *Cancer Sci.*, **96**, 157–163.

106 Sawa, T. and Ohshima, H. (2006) *Nitric Oxide*, **14**, 91–100.

107 Matter, B., Malejka-Giganti, D., Csallany, A.S., and Tretyakova, N. (2006) *Nucleic Acids Res.*, **34**, 5449–5460.

108 Hailer, M.K., Slade, P.G., Martin, B.D., and Sugden, K.D. (2005) *Chem. Res. Toxicol.*, **18**, 1378–1383.

5
DNA Damage Caused by Endogenously Generated Products of Oxidative Stress

Charles G. Knutson and Lawrence J. Marnett

5.1
Lipid Peroxidation

Phospholipids in all cell membranes contain mainly saturated fatty acids at the *sn*-1 position and unsaturated fatty acids at the *sn*-2 position of the glycerol portion of the phospholipid head group. Many of the unsaturated fatty acyl groups are polyunsaturated (linoleic acid, linolenic acid, arachidonic acid, etc.). The presence of the polyunsaturated fatty acyl moieties renders them sensitive to autoxidation. The bisallylic methylene hydrogens between the double bonds of polyunsaturated fatty acids (PUFAs) are the targets of free radical abstraction by reactive oxygen and nitrogen species (hydroxyl radical, peroxynitrite, etc.) that arise during oxidative stress, inflammation, metabolism, and cell signaling. The initial removal of a bisallylic hydrogen produces a carbon-centered pentadienyl radical that is delocalized across adjacent double bonds. The pentadienyl radical reacts rapidly with molecular oxygen at a diffusion-controlled rate to form peroxyl radicals. Peroxyl radicals may propagate the reaction by abstracting a bisallylic hydrogen from a neighboring PUFA to yield a lipid hydroperoxide and a new carbon-centered radical. It is estimated that initiation of one PUFA peroxyl radical may propagate as many as 10–200 additional free radical reactions (depending on the degree of unsaturation in the lipid environment). Lipid peroxidation is terminated by reaction of two radicals, forming a nonradical species. See also Chapter 2 for additional discussion of topics related to the present chapter.

A competing reaction for the peroxyl radical is intramolecular cyclization. This occurs when the oxidized PUFA contains three or more double bonds with an inner peroxyl radical (a peroxyl radical positioned between double bonds). The inner peroxyl radical may react with an unconjugated double bond three atoms away, producing an endoperoxide. The resulting carbon-centered radical may react with oxygen or cyclize to a bicyclic endoperoxide (Figure 5.1). Products resulting from lipid peroxidation are numerous and complex with regard to regiochemistry and stereochemistry; the multiplicity of products is proportional to the degree of unsaturation [1].

The Chemical Biology of DNA Damage. Edited by Nicholas E. Geacintov and Suse Broyde
© 2010 WILEY-VCH Verlag GmbH & Co. KGaA, Weinheim
ISBN: 978-3-527-32295-4

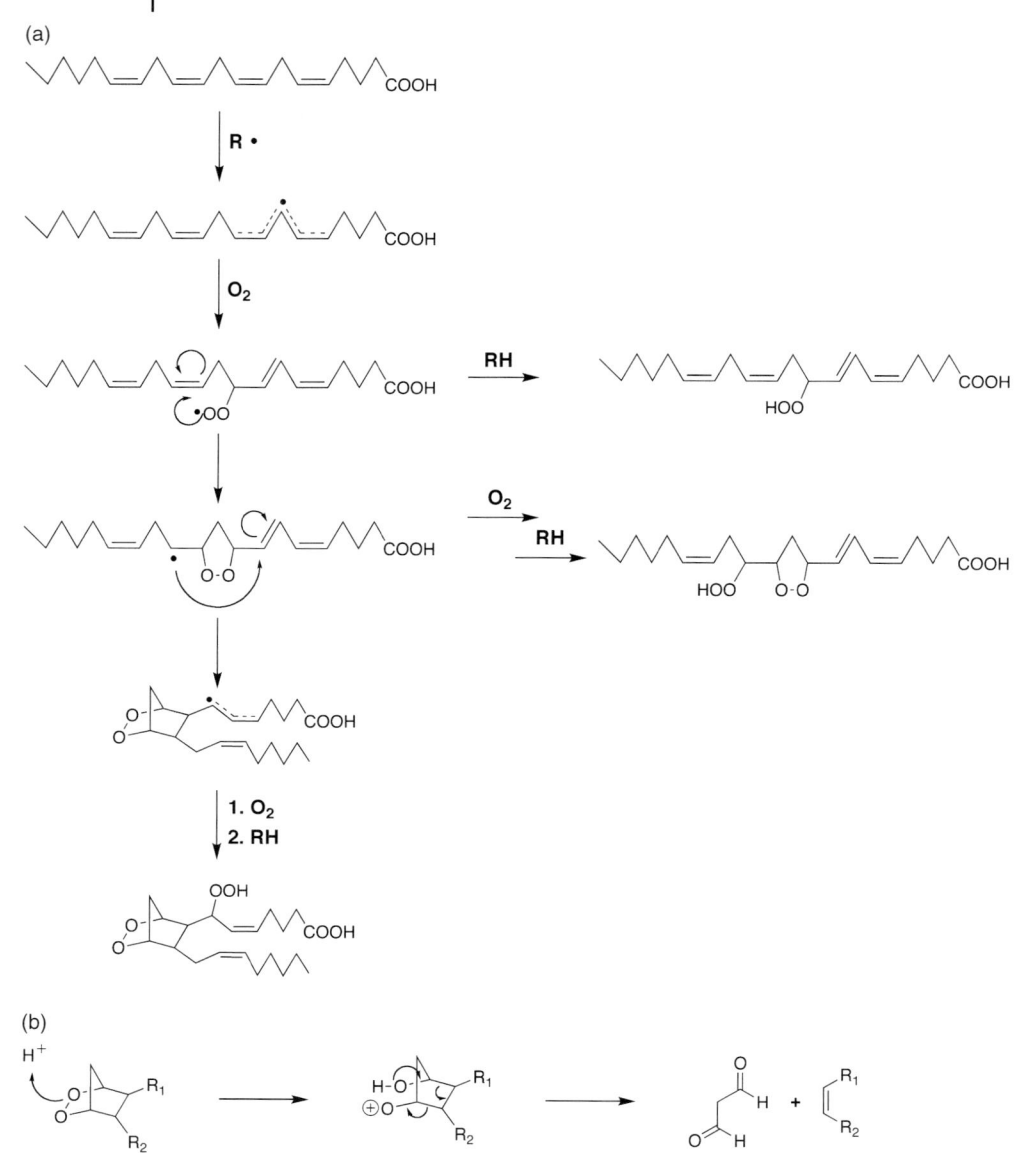

Figure 5.1 Autoxidation of arachidonic acid and β-scission of a bicyclic endoperoxide to generate MDA. (a) Following initial hydrogen abstraction from C7 on arachidonic acid, an inner peroxyl radical forms leading to the production of lipid hydroperoxides, endoperoxides, and bicyclic endoperoxides. (Figure adapted from [1].) (b) Bicyclic endoperoxides undergo acid-catalyzed ring-scission to produce MDA.

The initial products derived from lipid peroxidation are subject to decomposition reactions, which may be facilitated by the presence of metals and other one-electron-donating species. In particular, the lipid hydroperoxide and bicyclic endoperoxide products are susceptible to degradation. Bicyclic endoperoxides undergo acid-catalyzed ring-scission across the endoperoxide to yield the three-carbon dialdehyde, malondialdehyde (MDA) (Figure 5.1) [2]. MDA is also produced enzymatically from the endoperoxide-metabolizing enzyme thromboxane synthase [3, 4]. MDA is an abundant product of lipid peroxidation and reacts with DNA nucleophiles (see discussion below).

Lipid hydroperoxides are produced enzymatically by lipoxygenase and cyclooxygenase enzymes as well as by nonenzymatic lipid peroxidation [5–9]. The lipid hydroperoxides 9- and 13-hydroperoxyoctadecadienoic acid (HPODE, derived from linoleic acid) and 5- and 15-hydroperoxyeicosatetraenoic acid (HPETE, derived from arachidonic acid) are precursors of several reactive bifunctional electrophiles. Various mechanisms for the decomposition of lipid hydroperoxides have been implicated, such as homolytic cleavage, autoxidation, and Hock cleavage [10–14]. In the presence of vitamin C or transition metals (Fe^{2+}, Cu^{1+}), 13-HPODE and 15-HPETE decompose to 4-hydroperoxy-2-nonenal, 4-oxo-2-nonenal, 4-hydroxy-2-nonenal, and 4,5-epoxy-2-decenal, among other products (Figure 5.2) [10–12, 14–16]. The decomposition of 9-HPODE and 5-HPETE produces 4-hydroperoxy-2-nonenal, 4-oxo-2-nonenal, and 4-hydroxy-2-nonenal, but not 4,5-epoxy-2-decenal [14]. 4-Oxo-2-nonenal is the dehydration product of 4-hydroperoxy-2-nonenal [17] and 4-hydroxy-2-nonenal is mainly produced from the reduction of 4-hydroperoxy-2-nonenal [10, 12]. The decomposition of 13-HPODE also yields 9,12-dioxo-10-dodecenoic acid, which is a carboxylate-containing analog of 4-oxo-2-nonenal [18].

Two other important bifunctional electrophiles produced during lipid peroxidation are acrolein and crotonaldehyde, which are 3- and 4-carbon α,β-unsaturated aldehydes, respectively. Acrolein is produced during the autoxidation of PUFAs, presumably via β-cleavage of alkoxy radical intermediates derived from 3-hydroperoxy-1-alkene precursors [19, 20]. Carbohydrate, glycerol, and amino acid degradation also gives rise to acrolein [20]. Crotonaldehyde is principally produced from ω-3 PUFAs, although the mechanism of formation remains unclear [21, 22]. Acrolein and crotonaldehyde are also environmental contaminants, which arise during the combustion of organic material [23, 24].

5.2
2′-Deoxyribose Peroxidation

Oxidants such as hydroxyl radical or peroxynitrite also react with the 2′-deoxyribose scaffold of DNA. Each hydrogen on the 2′-deoxyribose ring is subject to free radical-mediated abstraction and subsequent diffusion-limited reaction with molecular oxygen. Each site of peroxidation leads to a separate and overlapping array of products. Following hydrogen abstraction on 2′-deoxyribose by reactive oxygen

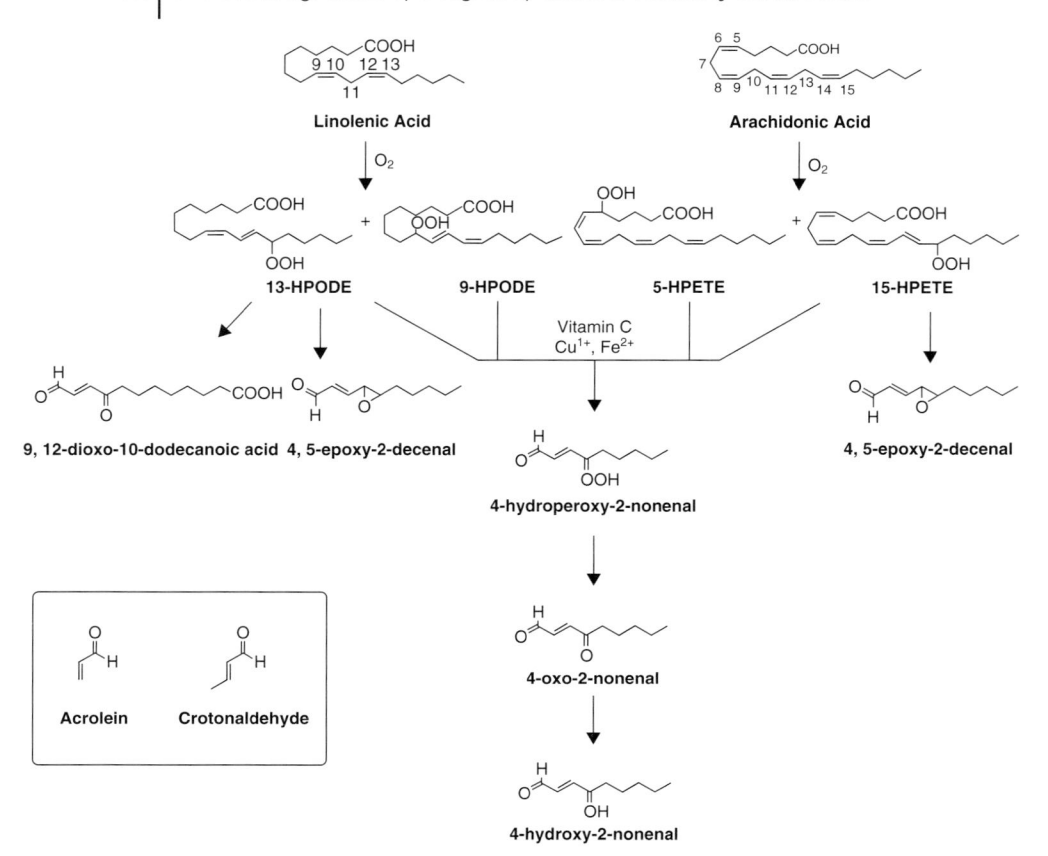

Figure 5.2 Non-enzymatic degradation of 9- and 13-HPODE and 5- and 15-HPETE. (a) Transition metals and vitamin C mediate the decomposition of lipid hydroperoxides and produce several reactive, bifunctional electrophiles: 4-hydroperoxy-2-nonenal, 4-oxo-2-nonenal, 4-hydroxy-2-nonenal, 4,5-epoxy-2-decenal, and 9,12-dioxo-10-dodecenoic acid. (Inset) α,β-Unsaturated aldehydes acrolein and crotonaldehyde are also products of lipid peroxidation.

species (ROS), a peroxyl radical is rapidly formed leading to additional intra- and intermolecular reactions, producing abasic sites and strand breaks. DNA strand breaks and abasic sites are thought to significantly contribute to the toxicity and mutagenicity induced by ROS. Several side-products derived from fragmentation of the 2′-deoxyribose ring are produced, such as MDA, base propenal, and base propenoate. These electrophiles, which are generated within DNA, react with nucleic acids.

Chemistry at the 4′-position of 2′-deoxyribose has been extensively studied due to the availability of reagents producing radical formation at this site (bleomycin, calicheamicin, peroxynitrite, and Fe^{2+}-EDTA) [25–28]. 4′-Free radical formation

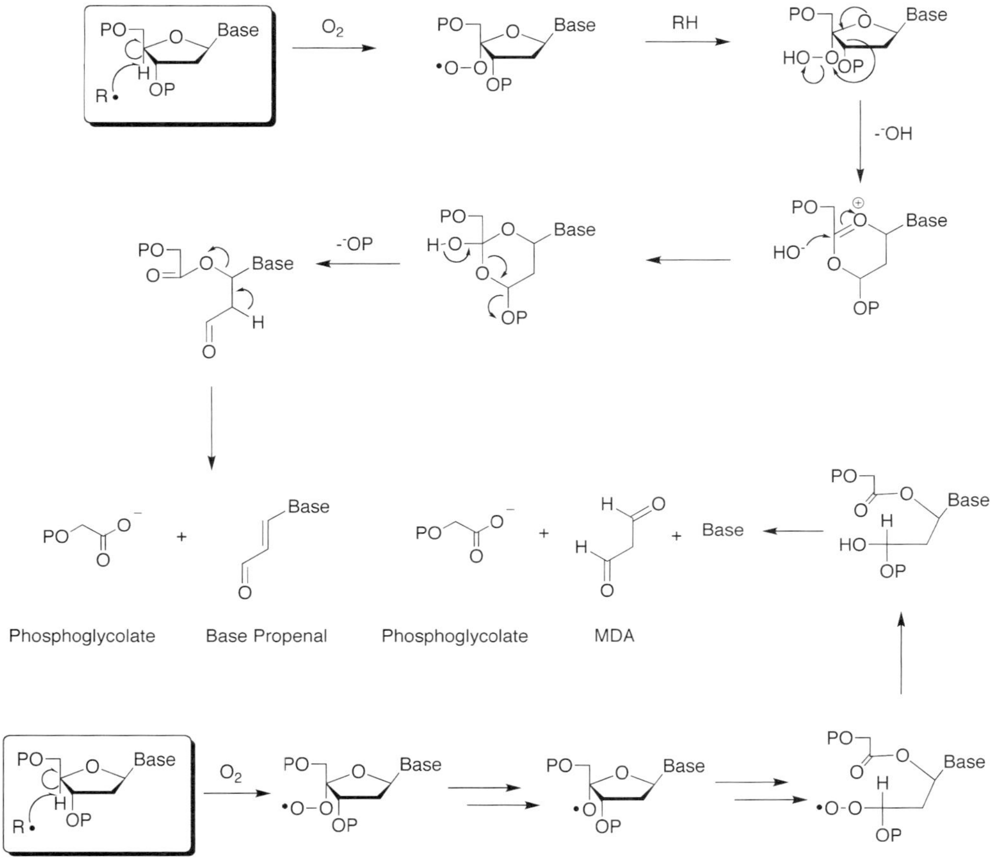

Figure 5.3 4′-Autoxidation of 2′-deoxyribose. 4′-Autoxidation of 2′-deoxyribose leads to the production of a free phosphoglycolate with either a base propenal or MDA and the base. (Figure adapted from [30].)

produces either an abasic site or a strand break with a free 3′-phosphoglycolate terminus. Fragmentation products derived from the 2′-deoxyribose moiety following a strand break depend on the identity of the free radical initiator [27]. These include either base propenal or MDA and a free base, which appear to be derived from separate mechanisms (Figure 5.3) [28–30].

5.3
Reactions of MDA and β-Substituted Acroleins with DNA Bases

The bifunctional nature of many lipid peroxidation products provides two reactive centers. Addition reactions to pyrimidines and purines often confer exocyclic

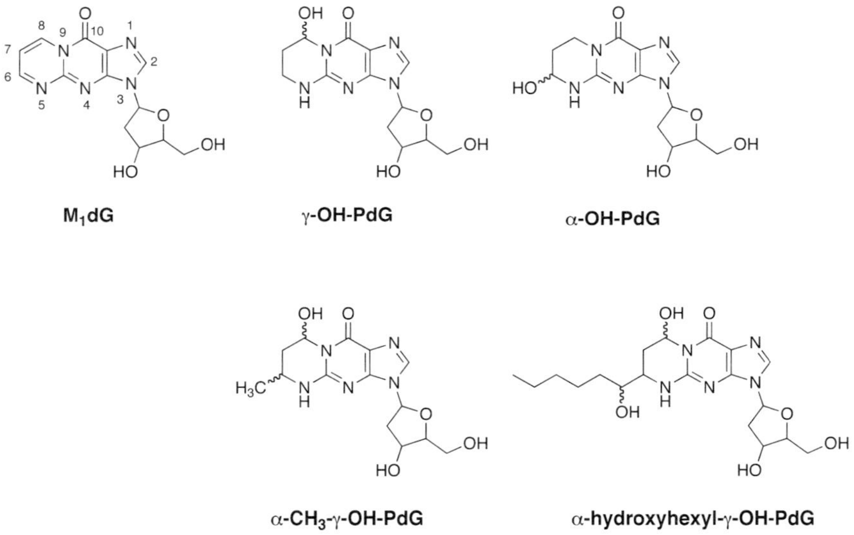

Figure 5.4 Pyrimidopurinone DNA adducts derived from MDA and related β-substituted acroleins.

structures on the nucleobases, which form across the Watson–Crick base-pairing interface. Exocyclic rings are either five- or six-membered and exhibit varying degrees of saturation. These lesions are highly mutagenic and repaired by multiple mechanisms.

MDA is an abundant product of lipid peroxidation and one of the first products identified to arise from the oxidation of lipids [31–33]. MDA is a bifunctional electrophile with pH-dependent reactivity, which exists as the β-hydroxyacrolein tautomer in polar solvents and forms an enolate at physiological pH (pK_a = 4.46) [34, 35]. The structurally related base propenals are β-substituted acroleins that are similar to β-hydroxyacrolein in their reactivity with nucleophiles, but they are not ionizable at physiological pH.

Moschel and Leonard discovered that 2′-deoxyguanosine (dG) reacts under acidic conditions with α-substituted MDA analogs [36]. The products of these reactions are pyrimidopurinones that are fluorescent. Guanosine and dG react with MDA under acidic conditions to yield two products: a pyrimidopurinone molecule (similar to that found by Moschel and Leonard), incorporating one equivalent of MDA, and an oxadiazabicyclononene derivative (a pair of enantiomers), incorporating two equivalents of MDA [37–39]. The MDA-dG adducts are named based on the number of MDA equivalents added to dG (M_1dG = 1 equiv. and M_2dG = 2 equiv. added to dG) (Figure 5.4). Base propenals also react with dG to add 1 equiv. MDA to form M_1dG [25].

At neutral pH, dG is the most reactive nucleoside toward MDA (dG > 2′-deoxyadenosine (dA) ≫ 2′-deoxycytidine (dC)). The MDA adducts of dA and dC are

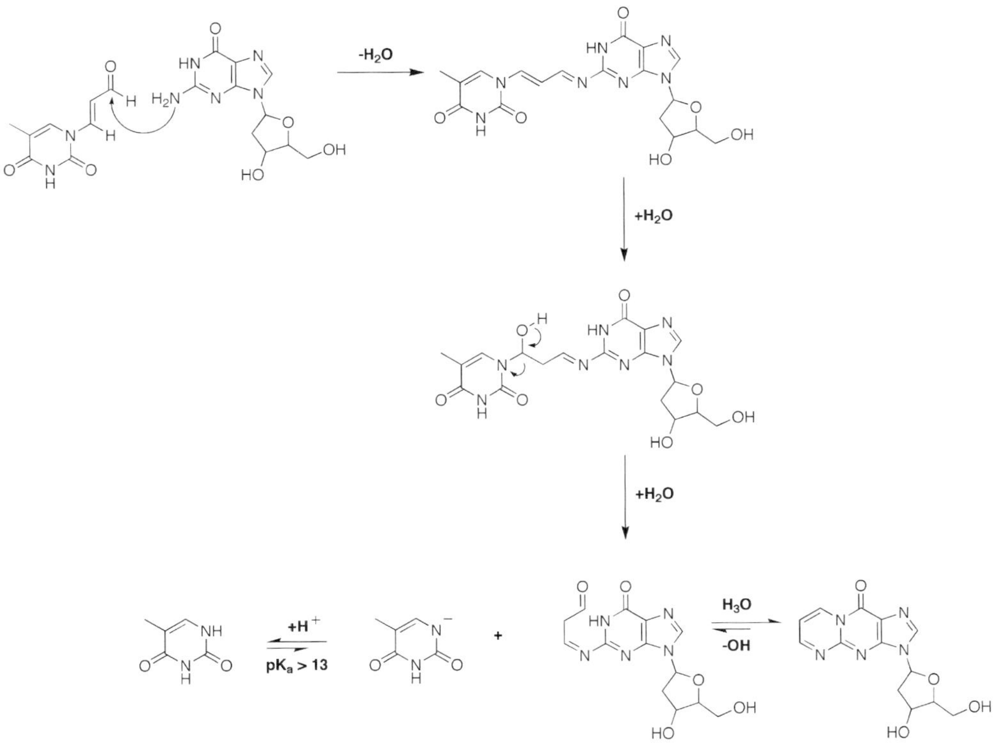

Figure 5.5 Reaction mechanism for the 1,2-addition of β-substituted acroleins (thymine propenal) with dG. (Figure adapted from [46].)

formed in low yields (below 1%), and MDA does not react with thymidine [40]. Single and multimeric additions of MDA are observed in reactions with dA and dC [41–44]. One equivalent of MDA adds to the exocyclic amino group yielding oxopropenyl derivatives: N^6-(3-oxo-1-propenyl)-dA (OPdA or M$_1$dA) and N^4-(3-oxo-1-propenyl)-dC (OPdC or M$_1$dC). Multimeric adducts arise from reactions of polymers of MDA *in situ* and are not the product of successive additions of MDA to the oxopropenyl derivatives [41–43]. The *in vivo* significance of the multimeric adducts is currently unknown.

The reaction mechanism of β-substituted acroleins with dG proceeds via 1,2-addition of the exocyclic N^2-amino group to the aldehyde, followed by cyclization onto the ring nitrogen (N1). Cyclization is preceded by hydrolysis of the β-substituent forming a 3-oxo-1-propenyl intermediate on the exocyclic N^2 (Figure 5.5) [45]. The extent of reactivity for β-substituted acroleins with dG is dependent on the pK_a of the conjugate acid for the leaving group. The order of reactivity observed in reactions of base propenals or β-hydroxyacrolein with dG is: adenine propenal (9.8) > cytosine propenal (12.2) > thymine propenal (>13) > β-hydroxyacrolein (15.7) (pK_a values for the leaving groups are given in parentheses) [46]. The use of the pK_a for β-hydroxyacrolein overstates the reactivity of its conjugate base because β-hydroxyacrolein exists principally as an enolate ion at neutral pH. In the case of the MDA-enolate ion, the oxide is an exceptionally poor leaving group. Initial approach of the MDA-enolate ion to DNA may also be affected by charge repulsion from the negatively charged phosphodiester backbone of DNA. Thus, M_1dG is more likely to arise from neutral β-hydroxyacrolein analogs such as base propenal. Recent studies from the Dedon and Swenberg Laboratories substantiate this [25, 47, 48].

In the context of double-stranded DNA, the positioning of the N^2 atom of dG impacts its reactivity towards β-substituted acroleins [49]. The N^2 position of dG is located in the minor groove and is involved in hydrogen bonding. Investigations into the DNA structural requirements leading to the formation of M_1dG demonstrate that access of electrophiles (MDA, base propenal) to the minor groove is critical for reaction. Single-stranded DNA is more reactive than double-stranded DNA towards MDA and base propenal. Additionally, factors that limit access to the minor groove (high ionic strength, minor groove binding agents) reduce M_1dG formation, while factors that enhance access to the minor grove (intercalating agents, major groove binders) increase M_1dG formation.

5.4
Stability of M_1dG: Hydrolytic Ring-Opening and Reaction with Nucleophiles

M_1dG is subject to hydrolytic ring-opening in basic solutions resulting in the formation of the 3-oxo-1-propenyl moiety on the exocyclic N^2 position of dG (OPdG) [50]. This structure is analogous to the OPdA and OPdC structures. M_1dG undergoes ring-opening by the addition of hydroxide ion to C8 of the exocyclic ring producing a transient carbinolamine anion, which rapidly ring-opens by breaking the C8–N9 bond [51] (Figure 5.6). The ring-opened OPdG has a pK_a of 6.9, so greater than 50% is deprotonated at physiological pH [52]. This negative charge on OPdG impacts further reactions with nucleophiles (see below). Ring-closure of OPdG is subject to a general acid catalysis. This reaction is biphasic with a fast protonation step (protonation of N1) preceding the cyclization to 8-hydroxy-propenodeoxyguanosine and followed by rate-limiting dehydration [52].

The reaction of M_1dG with nucleophiles is not limited to reactions with hydroxide. Amines, hydroxylamines, and hydrazines react with M_1dG to form imine, oxime, and hydrazone derivatives of OPdG [53–55]. These reactions principally occur by direct addition to C8, but hydroxylamines appear to react with the free aldehyde (C8) of OPdG [54, 55].

In single-stranded oligonucleotides, M_1dG is present in the ring-closed form [51], but in duplex DNA the conformation of M_1dG is sequence-dependent. When positioned opposite dC residues in double-stranded DNA, M_1dG ring-opens

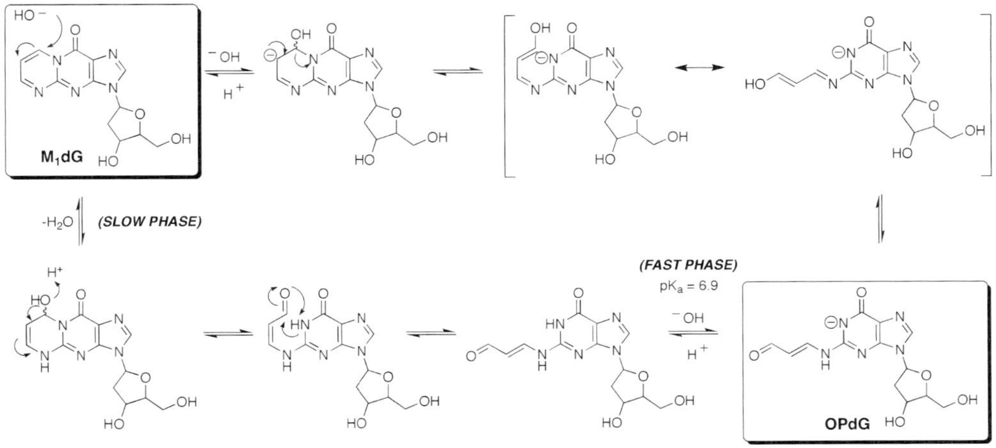

Figure 5.6 Hydrolytic ring-opening of M$_1$dG to OPdG and general acid-catalyzed ring-closure of OPdG to M$_1$dG.

rapidly to OPdG; however, when opposite dT residues, M$_1$dG remains as a pyrimidopurinone [56, 57]. It is hypothesized that duplex annealing catalyzes the ring-opening of M$_1$dG to OPdG, which suggests that the dC residue on the complimentary strand directs the addition of water to C8 of M$_1$dG prior to the stacking of incoming nucleotides [51]. Once positioned in DNA, OPdG extends into the minor groove. *Eco*RI cleavage of adduct-containing duplexes is retarded by the ring-closed form, M$_1$dG, and stable ring-closed analogs, but not by the ring-opened OPdG and other structural analogs incapable of ring-closure. However, the DNA-binding of *Eco*RI is unaffected by the presence of an alkyl substituent in the minor groove, which is consistent with the binding of the *Eco*RI dimer to its recognition sequence via interactions in the major groove [58].

The 3-oxo-1-propenyl moiety of OPdG, OPdA, and OPdC represents an electrophilic center that can react with primary amines. However, in the case of OPdG, which has a pK_a of 6.9, this reaction is not favorable. OPdG carries a net negative charge on the pyrimidine at N1 in the ring-opened state, which makes the aldehyde less reactive through resonance interactions. The methylated analog of OPdG, (N1-methyl)-OPdG, raises the pK_a to 8.2, which renders the molecule susceptible to nucleophilic addition at physiological pH. OPdA and OPdC have pK_a values of 10.5, and are neutral at physiological pH. OPdA, OPdC, and (N1-methyl)-OPdG readily react with the primary amine, *N*-α-acetyl-lysine, at physiological pH to form stable conjugates that can be isolated by high-performance liquid chromatography and studied spectroscopically. Thus, the acidity of the 3-oxo-1-propenyl nucleosides impacts their reactivity with nucleophiles, and the subsequent stability of cross-links [59].

5.5
Propano Adducts

Acrolein is the prototype α,β-unsaturated aldehyde, and reacts across N^2 and N1 of dG to add three carbons and generate the exocyclic tetrahydropyrimidopurinone ring (propano ring) (Figure 5.4). The reaction occurs via Michael addition of the olefin to either N1 or N^2 followed by 1,2-addition of the aldehyde to N^2 or N1, respectively. 1,N^2-8-Hydroxy-propanodeoxyguanosine (γ-OH-PdG) and 1,N^2-6-OH-PdG (α-OH-PdG) are the principal products with the former being the major isomer [60, 61].

Reactions of dG with longer chain enals result in ring-closure to give a single regioisomer due to the steric affects of the additional alkyl group. For longer chain enals such as crotonaldehyde and 4-hydroxy-2-nonenal, the N^2-amino of dG reacts exclusively at the β-carbon of the enal followed by ring-closure of the aldehyde at N1. The resulting propano ring contains a hydroxyl at C8 and the alkyl chain extending from C6. Crotonaldehyde reacts with dG to produce α-CH_3-γ-OH-PdG in a mixture of four diastereomers (Figure 5.4) [61, 62]. 4-Hydroxy-2-nonenal addition to dG is identical to that observed with crotonaldehyde, resulting in the formation of α-hydroxyhexyl-γ-OH-PdG (Figure 5.4) [63]. Reaction of 4-hydroxy-2-nonenal with dG produces four diastereomers at the chiral centers (C6 and C8) of the propano ring.

The propano ring is unsaturated and puckered in structure, which is distinctly different from the planar, aromatic character of the pyrimidopurinone ring of M_1dG. Like M_1dG, the propano rings derived from acrolein, crotonaldehyde, and 4-hydroxy-2-nonenal have been shown to ring-open in duplex DNA [61, 64–66]. In the ring-opened state, the extended oxopropyl side-chain may interact with neighboring nucleic acids and protein to form cross-links. Several intra- and interstrand cross-links have been identified for the propano adducts [67–72]. It is proposed that the stereochemistry of substituents on the propano ring impacts the biological reactivity and cross-linking of these adducts in the context of DNA [73]. The multiple conformations of the propano adducts likely contributes to the broad array of genetic mutations induced by these lesions [74–78].

5.6
Etheno Adducts

The addition of two carbons across the exocyclic and ring nitrogens of dA, dG, and dC produces an etheno bridge, which imparts fluorescent properties to the nucleosides. Originally, etheno adducts were the synthetic target of Leonard et al., who were interested in obtaining fluorescent nucleotide derivatives to probe for the biological activity and structure of nucleotide-containing enzymes and tRNA molecules [79–81]. Some of the reagents commonly used to synthesize etheno adducts were found to be mutagens, such as chloroacetaldehyde [82]. Chloroacetaldehyde was identified as a metabolite of the industrial contaminant, vinyl chloride

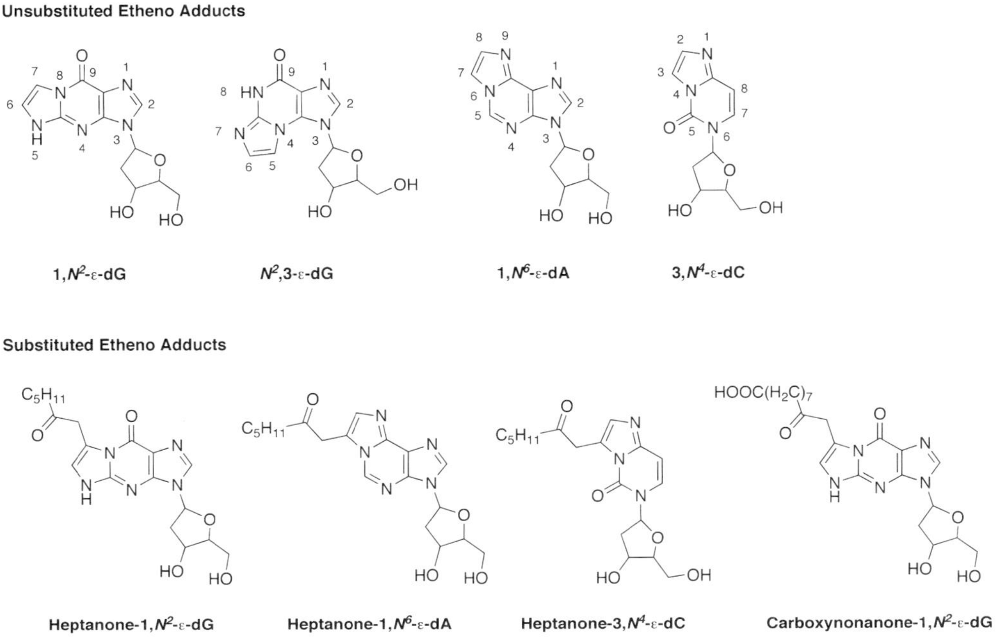

5.6 Etheno Adducts

Unsubstituted Etheno Adducts

1,N^2-ε-dG N^2,3-ε-dG   1,N^6-ε-dA   3,N^4-ε-dC

Substituted Etheno Adducts

Heptanone-1,N^2-ε-dG   Heptanone-1,N^6-ε-dA   Heptanone-3,N^4-ε-dC   Carboxynonanone-1,N^2-ε-dG

Figure 5.7 Unsubstituted and substituted etheno adducts.

[83]. Later 1-chlorooxirane (another metabolite of vinyl chloride) was found to be more reactive with DNA than chloroacetaldehyde [84]. Further study into the metabolism and reactivity of other alkyl halides revealed many of these molecules were either directly damaging to DNA or underwent metabolic transformation to reactive electrophiles capable of modifying DNA; etheno DNA adducts were an abundant and important product of their reactivity.

Four etheno adducts arise from exposure to alkyl halides: 1,N^6-ε-dA, 1,N^2-ε-dG, N^2,3-ε-dG, and 3,N^4-ε-dC (Figure 5.7). However, etheno adducts are found in genomic DNA of humans and animals in the absence of exposure to alkyl halides, and, therefore, must also have an endogenous origin [85–87]. Based on previous work demonstrating the reactivity of substituted 2,3-epoxy-aldehydes [88, 89], Sodum and Chung established the formation of 1,N^2-ε-dG upon reacting dG with an epoxide derived from 4-hydroxy-2-nonenal (2,3-epoxy-4-hydroxy-nonenal) [90]. El Ghissassi et al. also demonstrated that transition metal-mediated decomposition of microsomes supplemented with arachidonic acid in the presence of dA and dC led to the production of etheno adducts 1,N^6-ε-dA and 3,N^4-ε-dC, respectively [91]. Despite these observations, no clear electrophiles emerged as a precursor to the production of etheno adducts [92].

Lee and Blair provided a critical link from lipid peroxidation to the formation of etheno adducts by demonstrating the reactivity of 4-hydroperoxy-2-nonenal, 4-oxo-2-nonenal, and 4,5-epoxy-2-decenal with nucleosides. Both 4-hydroperoxy-

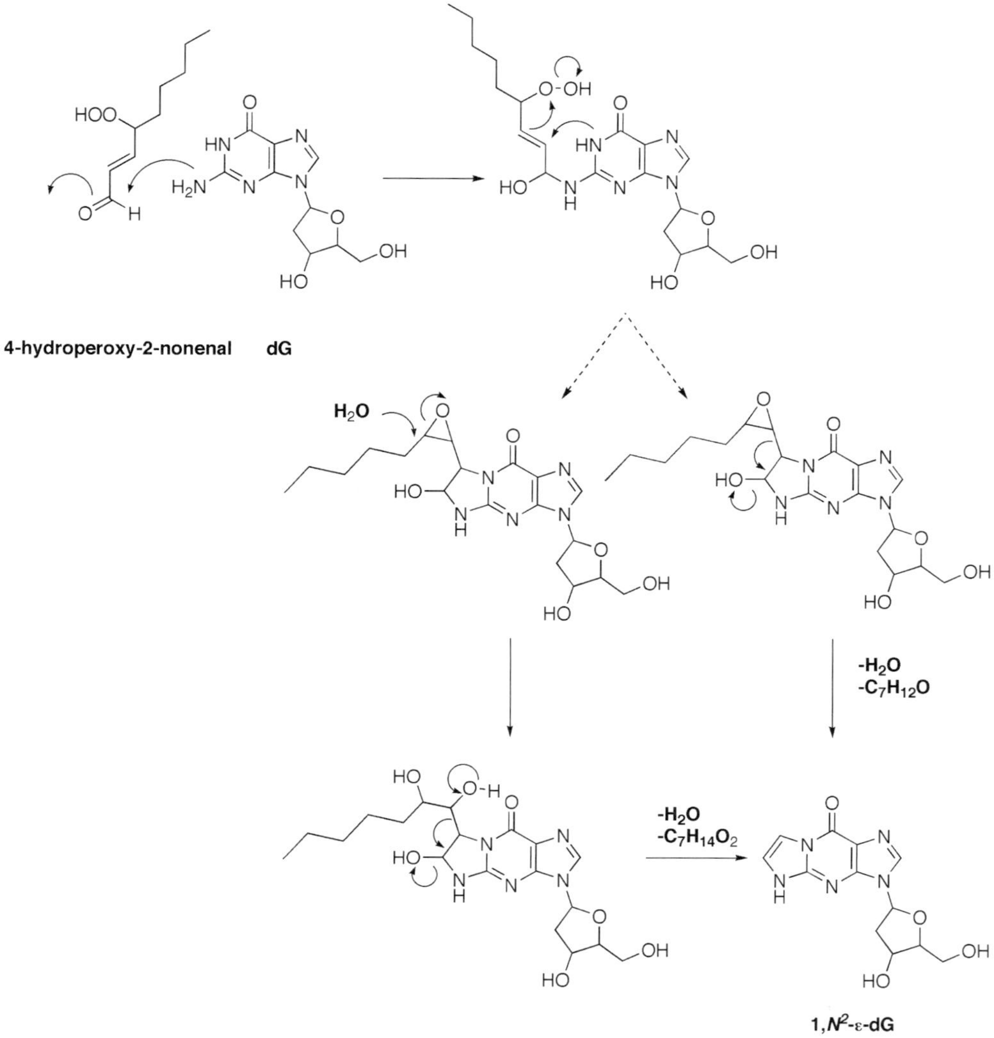

Figure 5.8 Proposed mechanism of formation for 1,N^2-ε-dG from reaction of 4-hydroperoxy-2-nonenal with dG. (Figure adapted from [93].)

2-nonenal and 4,5-epoxy-2-decenal react with dG and dA to form two unsubstituted etheno adducts, 1,N^6-ε-dA and 1,N^2-ε-dG [93, 94]. 4-Oxo-2-nonenal reacts with dG, dA, and dC to form substituted etheno adducts [95–97]. The proposed mechanism of addition by 4-hydroperoxy-2-nonenal proceeds by 1,2-addition of the nucleoside's exocyclic N-amino group to the aldehyde, followed by intramolecular cyclization to an epoxide intermediate, which either decomposes directly or sequentially dehydrates to the etheno product (Figure 5.8) [93]. 4,5-Epoxy-2-

decenal is proposed to react by initial 1,2-addition to the aldehyde, with the epoxide remaining intact; ring-closure promotes epoxide opening [98]. Tautomerization and solvolysis of the side-chain hydroxyl yields a substituted etheno adduct, with the hydroxyl proximal to the etheno ring. Retro-aldol cleavage then yields the etheno adduct and 2-octenal.

Reactions of 4-oxo-2-nonenal with dG, dA, and dC produce substituted etheno adducts containing a heptanone chain extended from the etheno carbon proximal to the ring nitrogen (Figure 5.7) [95–97]. 4-Oxo-2-nonenal appears to be most reactive with dC [97, 99]. The reaction of 4-oxo-2-nonenal with nucleosides proceeds by 1,2-addition of the exocyclic N-amino group to the aldehyde followed by intramolecular cyclization on the ring nitrogen. Other substituted etheno adducts have been identified, such as the carboxylate containing substituted etheno adduct derived from 9,12-dioxo-10-dodecenoate, carboxynonanone-1,N^2-ε-dG (Figure 5.7) [14, 18, 100]. Unlike the propano and pyrimidopurinone adducts, the etheno adducts are not subject to ring-opening, although etheno adducts may react under basic or acidic conditions to produce hydrated derivatives [96, 101, 102].

5.7
Mutagenicity of Peroxidation-Derived Adducts

MDA and β-substituted acroleins are mutagenic in several tester strains of *Salmonella typhimurium*. The strain sensitivity of mutation indicates that adducts formed from these bifunctional electrophiles induce base-pair substitutions, are repaired by nucleotide excision repair (NER), and require the action of translesion DNA polymerases (Pol II, IV, or V) to induce mutations. Interestingly, MDA induces frameshift mutations in *Salmonella*, whereas acrolein, crotonaldehyde, and 4-hydroxy-2-nonenal do not [103–106].

The spectrum of mutations induced by MDA was obtained by random mutagenesis experiments in which MDA-modified single-stranded phage DNA was transformed into *Escherichia coli*. Progeny phage were screened for the presence of mutations and mutant plaques sequenced. Base-pair substitutions (G → T, C → T, and A → G) and frameshift mutations (mainly single base additions occurring in runs of reiterated bases) were observed [107]. Although the specific types of adducts formed in these experiments cannot unequivocally be linked to their genetic consequences (i.e., it is unknown which adducts induced a particular mutation), the pattern of base-pair substitutions corresponds to the chemistry of MDA modification to DNA (i.e., a G → T transversion implies modification of dG by MDA; as described above, MDA forms adducts with dG, dA, and dC).

The major adducts produced from bifunctional electrophiles have all been evaluated by site-specific methodology in which a single adduct of known structure is built into a single-stranded vector that replicates in bacterial or mammalian cells [74–77, 108–117]. The vectors are introduced into recipient hosts and, following replication, the progeny are isolated and the outcome of replication at the site of the adduct is determined. Table 5.1 displays the types of mutations that are

Table 5.1 Mutations induced by peroxidation-derived adducts in site-specific mutagenesis experiments.

Adduct	Bacteria	Mammalian cells	References
	G → T; G → A frameshifts in (CpG)$_n$	G → T; G → A frameshifts in (CpG)$_n$	[108, 109]
	–	G → T; G → C	[74]
	G → A; G → T; G → C	G → T; G → C; G → A	[74, 75, 77, 78]
	–	G → T; G → C; G → A	[76]
	G → T; G → A	G → A	[111, 112]
	G → T; G → A	G → T; G → C; G → A	[111, 113]
	A → T; A → C; A → G	A → T; A → C; A → G	[115, 114]
	C → T; C → A	C → T; C → A	[116]
	C → G	C → T; C → A	[117]

observed in bacterial and mammalian cells. Inspection of the data and cited references in Table 5.1 enables several general conclusions to be reached.

i) All of the adducts are mutagenic and the percentage of mutations is typically between 1 and 10%. Higher mutation frequencies are observed in some cases (e.g., 1,N^6-ε-dA, heptanone-3,N^4-ε-dC in mammalian cells). It is important to note that a mutation frequency of even 1% is extremely high compared to the mutation frequency observed with unmodified DNA (typically of the order of one mutation per 10^8–10^{10} replication events).

ii) Different adducts induce different patterns of mutations.

iii) The types of mutations induced in bacterial and mammalian cells are similar for individual adducts.

iv) Error-prone, translesion polymerases appear responsible for inducing mutations for those adducts where host factors have been evaluated. For example, when site-specific vectors containing M_1dG or PdG are replicated in *E. coli* strains lacking *umuD*, the mutation frequency declines by over 90% [108, 118]. UmuD is a component of the *E. coli* translesion polymerase, Pol V.

v) The induction of frameshift mutations is sequence-dependent. M_1dG induces frameshift mutations in both *E. coli* and monkey kidney cells but only if it is placed in a reiterated sequence (e.g., (CpG)$_n$) [109]. The ability of M_1dG to induce frameshift mutations may explain the ability of MDA to induce this family of mutations in *Salmonella* as discussed above.

Most of the adducts in Table 5.1 also have been evaluated for their effects on DNA replication *in vitro* using a range of replicative and translesion polymerases. Consideration of the results for each adduct is beyond the scope of this chapter but a few trends are worth highlighting.

i) All of the adducts block replication to some extent.

ii) The outcome of replication to full-length does not correspond to the kinetics of single nucleotide incorporation opposite the adduct. The nucleotide that appears opposite the site of the adduct in the full-length duplex product is frequently not the nucleotide that is the most readily inserted opposite the adduct in single-nucleotide experiments. The key determinant of the identity of the full-length product is the efficiency with which the various template/primers containing a single nucleotide opposite the adduct are extended.

iii) The exocyclic adducts derived from peroxidation products are very poorly extended by replicative polymerases but are more readily extended by translesion, Y-family DNA polymerases.

iv) The products of *in vitro* DNA replication frequently do not correspond to those observed *in vivo*. Whether this reflects differences in either the DNA polymerases used *in vitro* compared to those operative *in vivo* or the absence of accessory factors in the *in vitro* experiments is uncertain.

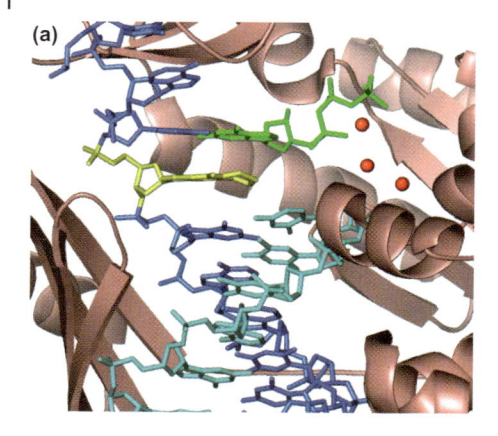

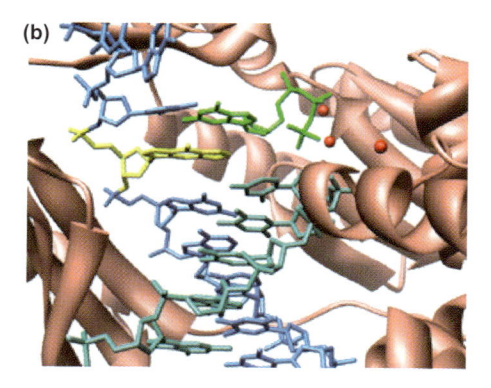

Figure 5.9 Structures of the ternary complexes of *Sulfolobus sulfataricus* Dpo4 with template/primers containing PdG (a) or 1,N^2-ε-dG (b) and an incoming dGP (courtesy of Martin Egli).

Structures of template/primers containing PdG or 1,N^2-ε-dG bound to the Y-family polymerase Dpo4 have been determined (Figure 5.9) [119, 120]. An incoming nucleoside triphosphate is present in both ternary complexes. Both complexes reveal a type II structure in which the incoming dNTP complexes with the base 5′ to the adduct in the template strand. Y-family polymerases have much larger, more open active sites than replicative polymerases and do not induce sharp bending of the template strand as it enters the active site [121]. This enables two bases to be present in the active site simultaneously in some cases, notably Dpo4. When the 3′-template base is a DNA adduct, the incoming dNTP can hydrogen bond to the 5′-base, which also is in the active site thereby avoiding a steric clash with the adduct. The type II structure explains the origin of one-base deletions, which are frequently observed as products of *in vitro* replication of adducted template/primers [119, 120].

5.8
Repair of DNA Damage

NER and base excision repair (BER) appear to play complementary roles in the repair of adducts derived from bifunctional aldehydes although a comprehensive analysis has not been conducted. The deletion of the *uvrA* gene of *E. coli* abolishes NER and increases the mutation frequency of M_1dG residues introduced on singly adducted vectors [108]. Similar deletions of genes associated with BER have no impact on M_1dG-induced mutations. *uvrA* deletion increases the mutagenicity of the M_1dG analog, PdG, to an extent similar to the increase observed with M_1dG (around 4-fold) [122]. Incubation of a 156mer oligonucleotide containing a single PdG residue with the purified bacterial NER complex, UvrABC, releases a 12mer oligonucleotide containing the PdG residue. Similar experiments with the mammalian NER complex isolated from CHO cell nuclei release 25–30mer oligonucleotides containing PdG [122]. Thus, *in vivo* and *in vitro* experiments indicate that M_1dG and PdG are repaired in bacterial and mammalian cells by NER.

The ability of human cell (HeLa) extracts to repair the exocyclic 4-hydroxy-2-nonenal-dG adducts has been investigated using plasmid DNAs randomly modified by 4-hydroxy-2-nonenal [123]. Incubation of modified plasmids with HeLa nuclear extracts in the presence of $[^{32}P]dCTP$ led to the incorporation of ^{32}P indicative of repair synthesis. ^{32}P-Postlabeling of the modified plasmid revealed that approximately 50–60% of the 4-hydroxy-2-nonenal-dG adducts initially present in the plasmid were removed on incubation with the extracts. This analysis also indicated that two of the four isomeric 4-hydroxy-2-nonenal-dG adducts were preferentially removed. The findings that M_1dG, PdG, and the 4-hydroxy-2-nonenal-dG adducts are all repaired by bacterial and mammalian NER suggests that other exocyclic dG adducts are repaired by this pathway.

In vivo and *in vitro* experiments suggest that mismatch repair may play a role in recognition and removal of M_1dG adducts under certain conditions. Electroporation of duplex vectors containing M_1dG or PdG residues into *E. coli* strains deficient in the mismatch repair gene, *mutS*, leads to a *decrease* in mutation frequency compared to wild-type *E. coli* [124]. Since MutS is the protein that binds to mismatches in DNA, one interpretation of this observation is that MutS binds to M_1dG or PdG and protects the adducts from recognition and repair by the NER complex (Figure 5.10). The ability of MutS protein to bind to M_1dG- or PdG-containing oligonucleotides was confirmed by gel-shift experiments, and the kinetics of association and dissociation were determined using surface plasmon resonance. The rate of association for MutS with a M_1dG T mismatch is comparable to that of binding to a G T mismatch (3.39 ± 0.03 versus $1.36 \pm 0.07 \times 10^5 \, M^{-1} s^{-1}$) but the dissociation rate is higher (7.82 ± 0.07 versus $2.44 \pm 0.04 \, s^{-1}$) [124]. Thus, the affinity for binding to either M_1dG or PdG mismatches is lower than that for binding to G T mismatches. *In vivo* experiments in which PdG residues were introduced on differentially methylated duplex vectors ("+" or "−" strand) suggest that the

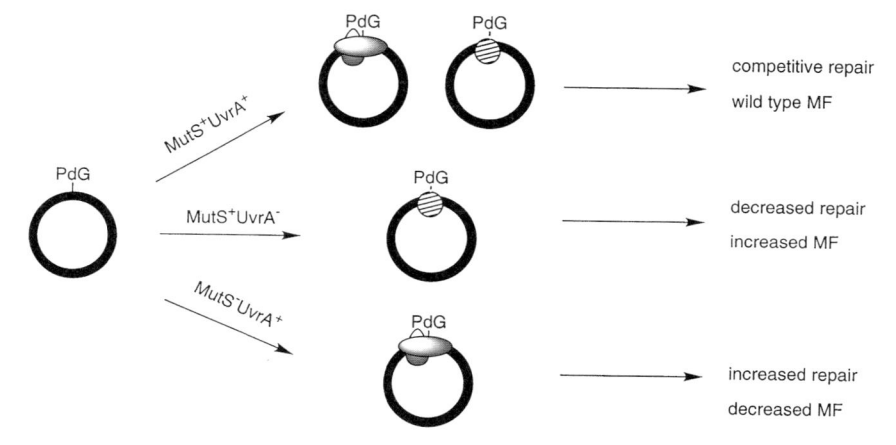

Figure 5.10 Competition between NER and MutS for binding to and repair of M_1dG or PdG. MF; mutation frequency. (Reproduced from [124] with permission.)

bacterial methyl-directed mismatch repair system catalyzes removal of PdG residues if the nonadducted strand is methylated. However, there is no repair if the adducted strand is methylated. The relative contribution of mismatch repair versus NER for removal of M_1dG and PdG residues on genomic DNA is uncertain in either prokaryotes or eukaryotes.

The repair mechanisms for the unsubstituted etheno adducts are incompletely characterized, but they appear different than those for M_1dG and PdG. Each unsubstituted etheno adduct is subject to BER by DNA glycosylases. $1,N^6$-ε-dA is repaired by the inducible 3-methyladenine glycosylase (AlkA) in bacteria and the analogous enzyme, alkylpurine-DNA N-glycosylase, in mammalian cells [125, 126]. $1,N^2$-ε-dG is a substrate for the latter enzyme [125, 127], $N^2,3$-ε-dG is a substrate for AlkA [128], and $3,N^4$-ε-dC is removed by the mismatch specific thymine-DNA glycosylase [129–131]. Glycosylases catalyze the hydrolytic removal of the adducted base from the helix and leaves an abasic site. The abasic site is nicked, removed, and the gap filled in by a polymerase. The remaining nick is sealed by DNA ligase [132]. Cleavage products formed by glycosylase action are, therefore, released from DNA as the base adducts. *In vitro* experiments with nuclear extracts from several different mammalian cell lines provide no evidence for the existence of glycosylases that cleave either M_1dG or PdG.

$1,N^6$-ε-dA and $3,N^4$-ε-dC are also substrates for repair in *E. coli* by AlkB. AlkB is an Fe^{2+}- and α-ketoglutarate-dependent oxygenase that hydroxylates methylated nucleotides to release the methyl group as formaldehyde [133]. It catalyzes direct DNA repair without cleavage of the glycosidic or phosphodiester bonds (e.g., $1\text{-}CH_3\text{-}dG \rightarrow dG$) (Figure 5.11). AlkB also catalyzes the oxygenation of the etheno double bond of $1,N^6$-ε-dA to form an epoxide (Figure 5.11). This epoxide hydrolyzes to a vicinal diol, which releases glyoxal and regenerates dA [133]. AlkB in *E. coli* only functions on single-stranded DNA, whereas the multiple analogous enzymes

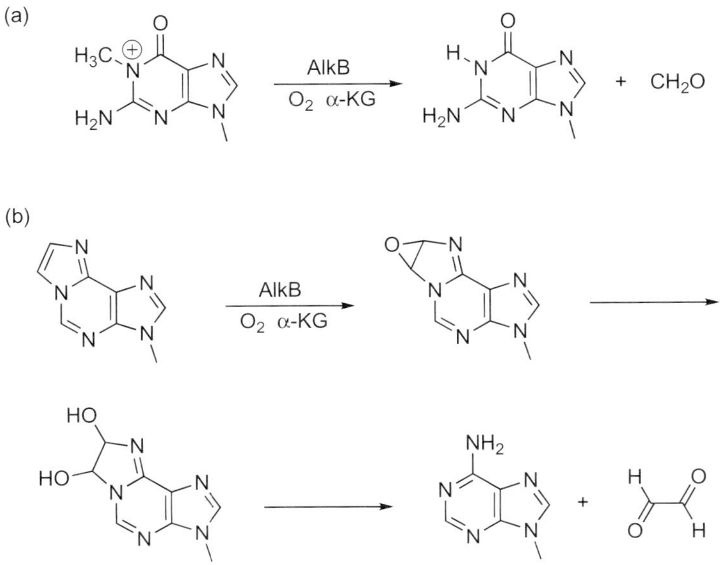

Figure 5.11 Repair of (a) 1-CH$_3$-dG and (b) 1,N^6-ε-dA by AlkB.

in mammalian cells do not work on single-stranded DNA. This explains a curious observation from early site-specific mutagenicity experiments conducted with 1,N^6-ε-dA and 3,N^4-ε-dC in *E. coli* and COS-7 cells. These adducts were essentially nonmutagenic in *E. coli*, but highly mutagenic in COS-7 cells [114, 116]. These observations were originally interpreted to indicate that the replication systems that bypass 1,N^6-ε-dA in bacteria and mammals are very different. However, the discovery of AlkB and its preference for single-stranded substrates reveals that the difference in mutagenicity may actually be due to a difference in DNA repair. Indeed, when AlkB is knocked out in *E. coli*, 1,N^6-ε-dA is highly mutagenic and displays a similar mutation spectrum to that observed in COS-7 cells [133]. The full range of substrates of AlkB and the roles of homologous enzymes in mammalian cell DNA repair has not been elucidated.

5.9
Assessment of DNA Damage

Implicit in the discussion of the chemistry of formation for DNA adducts *in vitro* is the generation of DNA adducts *in vivo*. The previously described pyrimidopurinone, propano, and etheno adducts are found in human tissue samples and cultured human cell lines. The estimated levels of these adducts range from one to 10 adducts/10^8 bp DNA. Owing to the relatively low abundance of these lesions amid a preponderance of impurities, highly sensitive and specific analytical methodologies are required to accurately quantify their levels. Immunoaffinity capture coupled to tandem mass spectrometry analysis is a commonly used approach for the purification and quantification of DNA adducts.

A challenging problem in the field of chemical carcinogenesis is the use of endogenous DNA adduct levels as a diagnostic measure of disease progression. Several attempts have been made to address this question in animal studies, but few attempts have been made in human populations [134, 135]. The steady-state levels of adducts found in DNA are a composite of two factors: the rate of adduct formation and the rate of adduct removal (repair). These rates cannot be determined from a single examination of genomic DNA. In humans, adduct measurements from DNA are typically made from tissues samples collected from biopsies. While direct assessments can be made about the tissue from which the sample is taken, these procedures are invasive and discomforting to the patient. Furthermore, repeated measurements from an individual (i.e., multiple biopsies) would be very challenging, and require a significant investment in time and expense. As a result, noninvasive strategies that utilize an easily accessible biological matrix, such as urine, represent an alternative means of assessing DNA adduct formation in human populations. This methodology relies on the purging of DNA adducts from cells and is often attributed to the repair-dependent removal of adducts. A caveat to this analysis is that it is an indirect measurement and the origin of adducts (site of formation) is unknown as well as the levels of adducts in specific tissue. Furthermore, it is unclear if other factors (such as the generation of adducts in the nucleotide pool, the elimination of adducts from apoptotic cells, or the intake of adducts from the diet) contribute to urinary levels. Ultimately, a noninvasive strategy for the routine analysis of DNA adducts offers greater applicability to population-based and time-dependent studies, while minimizing the cost and labor of sample collection.

As a prerequisite to monitoring DNA adducts from urine samples, our laboratory has conducted a series of experiments to examine the biological processing and elimination of endogenously occurring adducts $in\ vivo$. The pyrimidopurinone adduct, M_1dG, has been the focus of several investigations. M_1dG undergoes a single enzymatic oxidation at the 6-position of the pyrimido ring to 6-oxo-M_1dG $in\ vitro$ and $in\ vivo$ (Figure 5.12) [136]. Approximately 50% of M_1dG is converted to 6-oxo-M_1dG when administered intravenously to rats [137, 138]. The extent of metabolite formation is quantitative at dosing levels as low as 750 fmol [137]. In addition to excretion in urine, 6-oxo-M_1dG also is eliminated in bile, which suggests that transport proteins may be involved in the disposition of DNA adducts $in\ vivo$. Additionally, studies on the elimination of DNA adducts are also suggestive of biliary elimination [139, 140].

$In\ vitro$ observations on the oxidation of other structurally similar exocyclic adducts suggest additional lesions are likely to undergo biological processing $in\ vivo$. The base adduct M_1G is oxidized at the 6-position of the pyrimido ring to 6-oxo-M_1G, which is identical to what is observed for its 2'-deoxynucleoside, M_1dG. However, the absence of the 2'-deoxyribose moiety permits the ensuing oxidation of 6-oxo-M_1G at the 2-position to yield 2,6-dioxo-M_1G (Figure 5.12) [141]. $1,N^2$-ε-

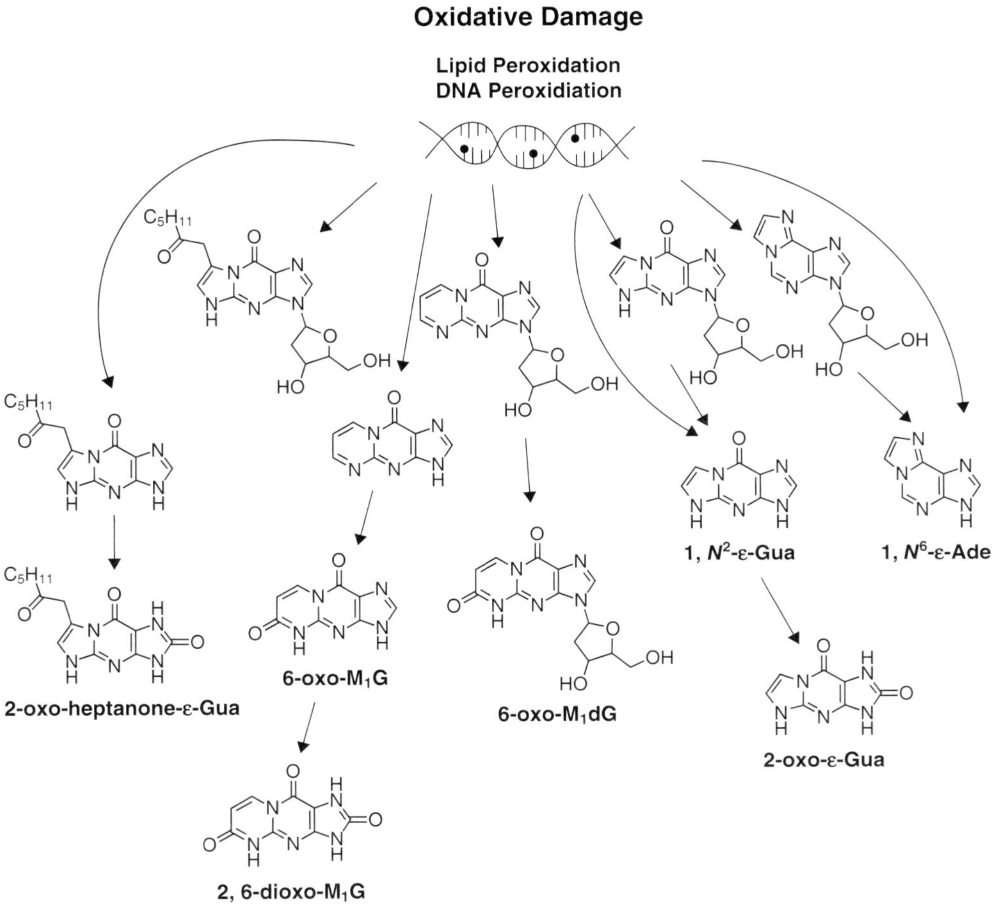

Oxidative Damage

Figure 5.12 Enzymatic oxidation and glycolytic cleavage of exocyclic DNA adducts.

Gua, which differs from M_1G by a single carbon in the exocyclic ring, undergoes oxidation, but not on the exocyclic ring. $1,N^2$-ε-dG is oxidized on the imidazole ring to produce 2-oxo-ε-Gua (Figure 5.12) [142]. The substituted etheno adduct heptanone-$1,N^2$-ε-Gua also undergoes oxidation on the imidazole ring to produce 2-oxo-heptanone-ε-Gua [142]. Comparing the rates of turnover for M_1dG, M_1G, 6-oxo-M_1G, and $1,N^2$-ε-dG reveals that the bases are better substrates than the 2'-deoxynucleoside (around 5-fold) [141]. Neither $1,N^2$-ε-dG nor heptanone-$1,N^2$-ε-dG is subject to oxidation, which suggests that the presence of 2'-deoxyribose prohibits oxidation on the imidazole ring.

Xanthine oxidoreductase and aldehyde oxidase are the principal enzymes involved in the oxidation of exocyclic DNA adducts. Their role in purine catabolism and in the oxidation of nitrogen-rich heterocycles is well documented [143–145].

In addition to oxidation, some adduct nucleosides are subject to glycolytic cleavage, which yields the free base. Both 1,N^2-ε-dG and 1,N^6-ε-dA are glycolytically cleaved in vitro in rat liver cytosol to 1,N^2-ε-Gua and 1,N^6-ε-Ade, respectively [142]. Purified purine nucleoside phosphorylase also catalyzes the glycolytic cleavage of 1,N^6-ε-dA. Of note, both 1,N^6-ε-dA and 1,N^6-ε-Ade have been detected in urine from humans and rats [146, 147]. Based on in vitro metabolism, 1,N^6-ε-Ade is likely to be present at higher concentrations in urine samples, and may be the best target for analytical investigation.

Understanding the in vitro and in vivo routes of biological processing should help guide the development of analytical methodologies for the detection of endogenous adducts in human populations. Several exocyclic adducts have been characterized as good substrates for oxidation. The end-products of biological processing may therefore be reliable targets for assessing the formation of endogenous adducts from biological matrixes in vivo.

5.10
Conclusions

Peroxidation of polyunsaturated fatty acyl moieties on phospholipids generates many electrophilic products including a series of bifunctional aldehydes capable of reacting with DNA to form adducts. Most of the adducts contain exocyclic rings that block Watson–Crick base-pairing. They are highly mutagenic if not repaired by NER, BER, or oxidation, and induce a range of base-pair substitutions and, in some cases, frameshift mutations. Error-prone translesion polymerases appear responsible for lesion bypass and mutation induction. Several of the adducts are present at significant levels in genomic DNA of healthy individuals testifying to the importance of peroxidation as a source of endogenous DNA damage. Development of facile biomarkers for use in population-based studies should help identify the factors responsible for the production of this class of adducts and hopefully identify individuals at increased risk for development of genetic diseases including cancer.

Acknowledgements

Research in the Marnett laboratory is supported by a research grant from the NIH (R37CA87819). We are grateful to Martin Egli for assistance with molecular graphics.

References

1 Porter, N.A., Caldwell, S.E., and Mills, K.A. (1995) Lipids, **30**, 277–290.

2 Pryor, W.A. and Stanley, J.P. (1975) J. Org. Chem., **40**, 3615–3617.

3 Diczfalusy, U., Falardeau, P., and Hammarstrom, S. (1977) *FEBS Lett.*, **84**, 271–274.
4 Hecker, M. and Ullrich, V. (1989) *J. Biol. Chem.*, **264**, 141–150.
5 Brash, A.R. (1999) *J. Biol. Chem.*, **274**, 23679–23682.
6 Blair, I.A. (2001) *Exp. Gerontol.*, **36**, 1473–1481.
7 Khan, N.A. (1961) *Pak. J. Sci.*, **13**, 178–183.
8 Bergström, S. (1945) *Arkiv Kemi. Mineral. Geol.*, **21A**, 11–18.
9 Bergström, S. (1945) *Nature*, **156**, 717–718.
10 Lee, S.H. and Blair, I.A. (2000) *Chem. Res. Toxicol.*, **13**, 698–702.
11 Lee, S.H., Oe, T. and Blair, I.A. (2001) *Science*, **292**, 2083–2086.
12 Schneider, C., Tallman, K.A., Porter, N.A., and Brash, A.R. (2001) *J. Biol. Chem.*, **276**, 20831–20838.
13 Pryor, W.A. and Porter, N.A. (1990) *Free Radic. Biol. Med.*, **8**, 541–543.
14 Jian, W., Lee, S.H., Arora, J.S., Silva Elipe, M.V., and Blair, I.A. (2005) *Chem. Res. Toxicol.*, **18**, 599–610.
15 Lee, S.H., Williams, M.V., Dubois, R.N., and Blair, I.A. (2005) *J. Biol. Chem.*, **280**, 28337–28346.
16 Williams, M.V., Lee, S.H., and Blair, I.A. (2005) *Rapid Commun. Mass Spectrom.*, **19**, 849–858.
17 Lee, S.H., Oe, T., Arora, J.S., and Blair, I.A. (2005) *J. Mass Spectrom.*, **40**, 661–668.
18 Lee, S.H., Silva Elipe, M.V., Arora, J.S., and Blair, I.A. (2005) *Chem. Res. Toxicol.*, **18**, 566–578.
19 Pan, X., Kaneko, H., Ushio, H., and Ohshima, T. (2005) *Eur. J. Lipid Sci. Technol.*, **107**, 228–238.
20 Stevens, J.F. and Maier, C.S. (2008) *Mol. Nutr. Food Res.*, **52**, 7–25.
21 Pan, J. and Chung, F.L. (2002) *Chem. Res. Toxicol.*, **15**, 367–372.
22 Chung, F.L., Pan, J., Choudhury, S., Roy, R., Hu, W., and Tang, M.S. (2003) *Mutat. Res.*, **531**, 25–36.
23 Chung, F.L., Chen, H.J., and Nath, R.G. (1996) *Carcinogenesis*, **17**, 2105–2011.
24 Uchida, K., Kanematsu, M., Morimitsu, Y., Osawa, T., Noguchi, N., and Niki, E. (1998) *J. Biol. Chem.*, **273**, 16058–16066.
25 Dedon, P.C., Plastaras, J.P., Rouzer, C.A., and Marnett, L.J. (1998) *Proc. Natl. Acad. Sci. USA*, **95**, 11113–11116.
26 Dedon, P.C. and Tannenbaum, S.R. (2004) *Arch. Biochem. Biophys.*, **423**, 12–22.
27 Dedon, P.C. (2007) *Chem. Res. Toxicol.*, **21**, 206–219.
28 Rashid, R., Langfinger, D., Wagner, R., Schuchmann, H.P., and von Sonntag, C. (1999) *Int. J. Radiat. Biol.*, **75**, 101–109.
29 Chen, B., Zhou, X., Taghizadeh, K., Chen, J., Stubbe, J., and Dedon, P.C. (2007) *Chem. Res. Toxicol.*, **20**, 1701–1708.
30 Breen, A.P. and Murphy, J.A. (1995) *Free Radic. Biol. Med.*, **18**, 1033–1077.
31 Kohn, H.I. and Liversedge, M. (1944) *J. Pharmacol.*, **82**, 292–300.
32 Wilbur, K.M., Bernheim, M.L.C., and Shapiro, O.W. (1949) *Arch. Biochem. Biophys.*, **24**, 305–313.
33 Sinnhuber, R.O., Yu, T.C., and Tu, T.C. (1958) *Food Res.*, **23**, 626–633.
34 Bothner-By, A.A. and Harris, R.K. (1965) *J. Org. Chem.*, **30**, 254–257.
35 Osman, M.M. (1972) *Helv. Chim. Acta*, **55**, 239–244.
36 Moschel, R.C. and Leonard, N.J. (1976) *J. Org. Chem.*, **41**, 294–300.
37 Seto, H., Akiyama, K., Okuda, T., Hashimoto, T., Takesue, T., and Ikemura, T. (1981) *Chem. Lett.*, **6**, 707–708.
38 Seto, H., Okuda, T., Takesue, T., and Ikemura, T. (1983) *Bull. Chem. Soc. Jpn.*, **56**, 1799–1802.
39 Marnett, L.J., Basu, A.K., Ohara, S.M., Weller, P.E., Rahman, A.F.M.M., and Oliver, J.P. (1986) *J. Am. Chem. Soc.*, **108**, 1348–1350.
40 Basu, A.K., O'Hara, S.M., Valladier, P., Stone, K., Mols, O., and Marnett, L.J. (1988) *Chem. Res. Toxicol.*, **1**, 53–59.
41 Nair, V., Turner, G.A., and Offerman, R.J. (1984) *J. Am. Chem. Soc.*, **106**, 3370–3371.
42 Stone, K., Ksebati, M.B., and Marnett, L.J. (1990) *Chem. Res. Toxicol.*, **3**, 33–38.
43 Stone, K., Uzieblo, A., and Marnett, L.J. (1990) *Chem. Res. Toxicol.*, **3**, 467–472.
44 Chaudhary, A.K., Reddy, G.R., Blair, I.A., and Marnett, L.J. (1996) *Carcinogenesis*, **17**, 1167–1170.

45 Reddy, G.R. and Marnett, L.J. (1996) *Chem. Res. Toxicol.*, **9**, 12–15.
46 Plastaras, J.P., Riggins, J.N., Otteneder, M., and Marnett, L.J. (2000) *Chem. Res. Toxicol.*, **13**, 1235–1242.
47 Zhou, X., Taghizadeh, K., and Dedon, P.C. (2005) *J. Biol. Chem.*, **280**, 25377–25382.
48 Jeong, Y.C. and Swenberg, J.A. (2005) *Free Radic. Biol. Med.*, **39**, 1021–1029.
49 Plastaras, J.P., Dedon, P.C., and Marnett, L.J. (2002) *Biochemistry*, **41**, 5033–5042.
50 Reddy, G.R. and Marnett, L.J. (1995) *J. Am. Chem. Soc.*, **117**, 5007–5008.
51 Riggins, J.N., Daniels, J.S., Rouzer, C.A., and Marnett, L.J. (2004) *J. Am. Chem. Soc.*, **126**, 8237–8243.
52 Riggins, J.N., Pratt, D.A., Voehler, M., Daniels, J.S., and Marnett, L.J. (2004) *J. Am. Chem. Soc.*, **126**, 10571–10581.
53 Niedernhofer, L.J., Riley, M., Schnetz-Boutaud, N., Sanduwaran, G., Chaudhary, A.K., Reddy, G.R., and Marnett, L.J. (1997) *Chem. Res. Toxicol.*, **10**, 556–561.
54 Schnetz-Boutaud, N., Daniels, J.S., Hashim, M.F., Scholl, P., Burrus, T., and Marnett, L.J. (2000) *Chem. Res. Toxicol.*, **13**, 967–970.
55 Otteneder, M., Plastaras, J.P., and Marnett, L.J. (2002) *Chem. Res. Toxicol.*, **15**, 312–318.
56 Mao, H., Reddy, G.R., Marnett, L.J., and Stone, M.P. (1999) *Biochemistry*, **38**, 13491–13501.
57 Mao, H., Schnetz-Boutaud, N.C., Weisenseel, J.P., Marnett, L.J., and Stone, M.P. (1999) *Proc. Natl. Acad. Sci. USA*, **96**, 6615–6620.
58 VanderVeen, L.A., Druckova, A., Riggins, J.N., Sorrells, J.L., Guengerich, F.P., and Marnett, L.J. (2005) *Biochemistry*, **44**, 5024–5033.
59 Szekely, J., Rizzo, C.J., and Marnett, L.J. (2008) *J. Am. Chem. Soc.*, **130**, 2195–2201.
60 Chung, F.L., Roy, K.R., and Hecht, S.S. (1988) *J. Org. Chem.*, **53**, 14–17.
61 Chung, F.L., Young, R., and Hecht, S.S. (1984) *Cancer Res.*, **44**, 990–995.
62 Chung, F.L. and Hecht, S.S. (1983) *Cancer Res.*, **43**, 1230–1235.
63 Winter, C.K., Segall, H.J., and Haddon, W.F. (1986) *Cancer Res.*, **46**, 5682–5686.
64 de los Santos, C., Zaliznyak, T., and Johnson, F. (2001) *J. Biol. Chem.*, **276**, 9077–9082.
65 Kozekov, I.D., Nechev, L.V., Moseley, M.S., Harris, C.M., Rizzo, C.J., Stone, M.P., and Harris, T.M. (2003) *J. Am. Chem. Soc.*, **125**, 50–61.
66 Wang, H., Marnett, L.J., Harris, T.M., and Rizzo, C.J. (2004) *Chem. Res. Toxicol.*, **17**, 144–149.
67 Kim, H.Y., Voehler, M., Harris, T.M., and Stone, M.P. (2002) *J. Am. Chem. Soc.*, **124**, 9324–9325.
68 Cho, Y.J., Wang, H., Kozekov, I.D., Kurtz, A.J., Jacob, J., Voehler, M., Smith, J., Harris, T.M., Lloyd, R.S., Rizzo, C.J., and Stone, M.P. (2006) *Chem. Res. Toxicol.*, **19**, 195–208.
69 Kozekov, I.D., Nechev, L.V., Sanchez, A., Harris, C.M., Lloyd, R.S., and Harris, T.M. (2001) *Chem. Res. Toxicol.*, **14**, 1482–1485.
70 Hecht, S.S., McIntee, E.J., and Wang, M. (2001) *Toxicology*, **166**, 31–36.
71 Wang, M., McIntee, E.J., Cheng, G., Shi, Y., Villalta, P.W., and Hecht, S.S. (2000) *Chem. Res. Toxicol.*, **13**, 1149–1157.
72 Wang, M., McIntee, E.J., Cheng, G., Shi, Y., Villalta, P.W., and Hecht, S.S. (2001) *Chem. Res. Toxicol.*, **14**, 423–430.
73 Wang, H., Kozekov, I.D., Harris, T.M., and Rizzo, C.J. (2003) *J. Am. Chem. Soc.*, **125**, 5687–5700.
74 Sanchez, A.M., Minko, I.G., Kurtz, A.J., Kanuri, M., Moriya, M., and Lloyd, R.S. (2003) *Chem. Res. Toxicol.*, **16**, 1019–1028.
75 VanderVeen, L.A., Hashim, M.F., Nechev, L.V., Harris, T.M., Harris, C.M., and Marnett, L.J. (2001) *J. Biol. Chem.*, **276**, 9066–9070.
76 Fernandes, P.H., Wang, H., Rizzo, C.J., and Lloyd, R.S. (2003) *Environ. Mol. Mutagen.*, **42**, 68–74.
77 Yang, I.Y., Hossain, M., Miller, H., Khullar, S., Johnson, F., Grollman, A., and Moriya, M. (2001) *J. Biol. Chem.*, **276**, 9071–9076.
78 Yang, I.Y., Johnson, F., Grollman, A.P., and Moriya, M. (2002) *Chem. Res. Toxicol.*, **15**, 160–164.
79 Barrio, J.R., Secrist, J.A., 3rd, and Leonard, N.J. (1972) *Biochem. Biophys. Res. Commun.*, **46**, 597–604.

80 Sattsangi, P.D., Leonard, N.J., and Frihart, C.R. (1977) *J. Org. Chem.*, **42**, 3292–3296.
81 Secrist, J.A., 3rd, Barrio, J.R., Leonard, N.J., and Weber, G. (1972) *Biochemistry*, **11**, 3499–3506.
82 McCann, J., Simmon, V., Streitwieser, D., and Ames, B.N. (1975) *Proc. Natl. Acad. Sci. USA*, **72**, 3190–3193.
83 Guengerich, F.P., Crawford, W.M., Jr., and Watanabe, P.G. (1979) *Biochemistry*, **18**, 5177–5182.
84 Guengerich, F.P. (1992) *Chem. Res. Toxicol.*, **5**, 2–5.
85 Bartsch, H., Barbin, A., Marion, M.J., Nair, J., and Guichard, Y. (1994) *Drug Metab. Rev.*, **26**, 349–371.
86 Nair, J., Barbin, A., Guichard, Y., and Bartsch, H. (1995) *Carcinogenesis*, **16**, 613–617.
87 Misra, R.R., Chiang, S.Y., and Swenberg, J.A. (1994) *Carcinogenesis*, **15**, 1647–1652.
88 Goldschmidt, B.M., Blazej, T.P., and Van Duuren, B.L. (1968) *Tetrahedron Lett.*, **13**, 1583–1585.
89 Nair, V. and Offerman, R.J. (1985) *J. Org. Chem.*, **50**, 5627–5631.
90 Sodum, R.S. and Chung, F.L. (1988) *Cancer Res.*, **48**, 320–323.
91 el Ghissassi, F., Barbin, A., Nair, J., and Bartsch, H. (1995) *Chem. Res. Toxicol.*, **8**, 278–283.
92 Douki, T., Odin, F., Caillat, S., Favier, A., and Cadet, J. (2004) *Free Radic. Biol. Med.*, **37**, 62–70.
93 Lee, S.H., Arora, J.A., Oe, T., and Blair, I.A. (2005) *Chem. Res. Toxicol.*, **18**, 780–786.
94 Lee, S.H., Oe, T., and Blair, I.A. (2002) *Chem. Res. Toxicol.*, **15**, 300–304.
95 Lee, S.H., Rindgen, D., Bible, R.H., Jr., Hajdu, E., and Blair, I.A. (2000) *Chem. Res. Toxicol.*, **13**, 565–574.
96 Rindgen, D., Nakajima, M., Wehrli, S., Xu, K., and Blair, I.A. (1999) *Chem. Res. Toxicol.*, **12**, 1195–1204.
97 Pollack, M., Oe, T., Lee, S.H., Silva Elipe, M.V., Arison, B.H., and Blair, I.A. (2003) *Chem. Res. Toxicol.*, **16**, 893–900.
98 Petrova, K.V., Jalluri, R.S., Kozekov, I.D., and Rizzo, C.J. (2007) *Chem. Res. Toxicol.*, **20**, 1685–1692.
99 Kawai, Y., Uchida, K., and Osawa, T. (2004) *Free Radic. Biol. Med.*, **36**, 529–541.
100 Sodum, R.S. and Chung, F.L. (1991) *Cancer Res.*, **51**, 137–143.
101 Guengerich, F.P., Persmark, M., and Humphreys, W.G. (1993) *Chem. Res. Toxicol.*, **6**, 635–648.
102 Rindgen, D., Lee, S.H., Nakajima, M., and Blair, I.A. (2000) *Chem. Res. Toxicol.*, **13**, 846–852.
103 Mukai, F.H. and Goldstein, B.D. (1976) *Science*, **191**, 868–869.
104 Basu, A.K. and Marnett, L.J. (1983) *Carcinogenesis*, **4**, 331–333.
105 Basu, A.K. and Marnett, L.J. (1984) *Cancer Res.*, **44**, 2848–2854.
106 Basu, A.K., Marnett, L.J., and Romano, L.J. (1984) *Mutat. Res.*, **129**, 39–46.
107 Benamira, M., Johnson, K., Chaudhary, A., Bruner, K., Tibbetts, C., and Marnett, L.J. (1995) *Carcinogenesis*, **16**, 93–99.
108 Fink, S.P., Reddy, G.R., and Marnett, L.J. (1997) *Proc. Natl. Acad. Sci. USA*, **94**, 8652–8657.
109 VanderVeen, L.A., Hashim, M.F., Shyr, Y., and Marnett, L.J. (2003) *Proc. Natl. Acad. Sci. USA*, **100**, 14247–14252.
110 Yang, I.Y., Chan, G., Miller, H., Huang, Y., Torres, M.C., Johnson, F., and Moriya, M. (2002) *Biochemistry*, **41**, 13826–13832.
111 Langouet, S., Mican, A.N., Muller, M., Fink, S.P., Marnett, L.J., Muhle, S.A., and Guengerich, F.P. (1998) *Biochemistry*, **37**, 5184–5193.
112 Akasaka, S. and Guengerich, F.P. (1999) *Chem. Res. Toxicol.*, **12**, 501–507.
113 Fernandes, P.H., Kanuri, M., Nechev, L.V., Harris, T.M., and Lloyd, R.S. (2005) *Environ. Mol. Mutagen.*, **45**, 455–459.
114 Pandya, G.A. and Moriya, M. (1996) *Biochemistry*, **35**, 11487–11492.
115 Basu, A.K., Wood, M.L., Niedernhofer, L.J., Ramos, L.A., and Essigmann, J.M. (1993) *Biochemistry*, **32**, 12793–12801.
116 Moriya, M., Zhang, W., Johnson, F., and Grollman, A.P. (1994) *Proc. Natl. Acad. Sci. USA*, **91**, 11899–11903.
117 Pollack, M., Yang, I.Y., Kim, H.Y., Blair, I.A., and Moriya, M. (2006) *Chem. Res. Toxicol.*, **19**, 1074–1079.
118 Fink, S.P., Reddy, G.R., and Marnett, L.J. (1996) *Chem. Res. Toxicol.*, **9**, 277–283.

119 Wang, Y., Musser, S.K., Saleh, S., Marnett, L.J., Egli, M., and Stone, M.P. (2008) *Biochemistry*, **47**, 7322–7334.

120 Zang, H., Goodenough, A.K., Choi, J.Y., Irimia, A., Loukachevitch, L.V., Kozekov, I.D., Angel, K.C., Rizzo, C.J., Egli, M., and Guengerich, F.P. (2005) *J. Biol. Chem.*, **280**, 29750–29764.

121 Yang, W. (2005) *FEBS Lett.*, **579**, 868–872.

122 Johnson, K.A., Fink, S.P., and Marnett, L.J. (1997) *J. Biol. Chem.*, **272**, 11434–11438.

123 Choudhury, S., Pan, J., Amin, S., Chung, F.L., and Roy, R. (2004) *Biochemistry*, **43**, 7514–7521.

124 Johnson, K.A., Mierzwa, M.L., Fink, S.P., and Marnett, L.J. (1999) *J. Biol. Chem.*, **274**, 27112–27118.

125 Singer, B., Antoccia, A., Basu, A.K., Dosanjh, M.K., Fraenkel-Conrat, H., Gallagher, P.E., Kusmierek, J.T., Qiu, Z.H., and Rydberg, B. (1992) *Proc. Natl. Acad. Sci. USA*, **89**, 9386–9390.

126 Saparbaev, M., Kleibl, K., and Laval, J. (1995) *Nucleic Acids Res.*, **23**, 3750–3755.

127 Saparbaev, M., Langouet, S., Privezentzev, C.V., Guengerich, F.P., Cai, H., Elder, R.H., and Laval, J. (2002) *J. Biol. Chem.*, **277**, 26987–26993.

128 Dosanjh, M.K., Chenna, A., Kim, E., Fraenkelconrat, H., Samson, L., and Singer, B. (1994) *Proc. Natl. Acad. Sci. USA*, **91**, 1024–1028.

129 Saparbaev, M. and Laval, J. (1998) *Proc. Natl. Acad. Sci. USA*, **95**, 8508–8513.

130 Hang, B., Chenna, A., Rao, S., and Singer, B. (1996) *Carcinogenesis*, **17**, 155–157.

131 Hang, B., Medina, M., Fraenkel-Conrat, H., and Singer, B. (1998) *Proc. Natl. Acad. Sci. USA*, **95**, 13561–13566.

132 Huffman, J.L., Sundheim, O., and Tainer, J.A. (2005) *Mutat. Res.*, **577**, 55–76.

133 Delaney, J.C., Smeester, L., Wong, C., Frick, L.E., Taghizadeh, K., Wishnok, J.S., Drennan, C.L., Samson, L.D., and Essigmann, J.M. (2005) *Nat. Struct. Mol. Biol.*, **12**, 855–860.

134 Meira, L.B., Bugni, J.M., Green, S.L., Lee, C.W., Pang, B., Borenshtein, D., Rickman, B.H., Rogers, A.B., Moroski-Erkul, C.A., McFaline, J.L., Schauer, D.B., Dedon, P.C., Fox, J.G., and Samson, L.D. (2008) *J. Clin. Invest.*, **118**, 2516–2525.

135 Pang, B., Zhou, X., Yu, H., Dong, M., Taghizadeh, K., Wishnok, J.S., Tannenbaum, S.R., and Dedon, P.C. (2007) *Carcinogenesis*, **28**, 1807–1813.

136 Otteneder, M.B., Knutson, C.G., Daniels, J.S., Hashim, M., Crews, B.C., Remmel, R.P., Wang, H., Rizzo, C., and Marnett, L.J. (2006) *Proc. Natl. Acad. Sci. USA*, **103**, 6665–6669.

137 Knutson, C.G., Skipper, P.L., Liberman, R.G., Tannenbaum, S.R., and Marnett, L.J. (2008) *Chem. Res. Toxicol.*, **21**, 1290–1294.

138 Knutson, C.G., Wang, H., Rizzo, C.J., and Marnett, L.J. (2007) *J. Biol. Chem.*, **282**, 36257–36264.

139 Wang, M. and Hecht, S.S. (1997) *Chem. Res. Toxicol.*, **10**, 772–778.

140 Dutta, S.P., Mittelman, A., and Chheda, G.B. (1980) *Biochem. Med.*, **23**, 179–184.

141 Knutson, C.G., Akingbade, D., Crews, B.C., Voehler, M., Stec, D.F., and Marnett, L.J. (2007) *Chem. Res. Toxicol.*, **20**, 550–557.

142 Knutson, C.G., Rubinson, E.H., Akingbade, D., Anderson, C.S., Stec, D.F., Petrova, K.V., Kozekov, I.D., Guengerich, F.P., Rizzo, C.J., and Marnett, L.J. (2009) *Biochemistry*, **48**, 800–809.

143 Krenitsky, T.A., Neil, S.M., Elion, G.B., and Hitchings, G.H. (1972) *Arch. Biochem. Biophys.*, **150**, 585–599.

144 Beedham, C. (2002) Molybdenum hydroxylases, in *Enzyme Systems that Metabolise Drugs and Other Xenobiotics* (ed. C. Ioannides), John Wiley & Sons, Ltd, Chichester, pp. 147–187.

145 Kitamura, S., Sugihara, K., and Ohta, S. (2006) *Drug Metab. Pharmacokinet.*, **21**, 83–98.

146 Yen, T.Y., Holt, S., Sangaiah, R., Gold, A., and Swenberg, J.A. (1998) *Chem. Res. Toxicol.*, **11**, 810–815.

147 Chen, H.J. and Chang, C.M. (2004) *Chem. Res. Toxicol.*, **17**, 963–971.

6
Polycyclic Aromatic Hydrocarbons: Multiple Metabolic Pathways and the DNA Lesions Formed

Trevor M. Penning

6.1
Introduction

Polycyclic aromatic hydrocarbons (PAHs) represent a class of compounds that contain two or more fused benzene rings. They are environmental pollutants and the most ubiquitous, benzo[a]pyrene (B[a]P), has been upgraded by the International Agency for Research on Cancer to a Group 1 or known human carcinogen [1]. PAHs are products of fossil fuel combustion; they are a component of fine particulate matter (size 2.5 µm); and as a consequence contaminate the air we breathe, the soil and water supply, and enter the food chain [2, 3]. They are also introduced artificially into smoked, cured, and barbecued food [4, 5]. Finally, they are present as a complex mixture in tobacco smoke and second-hand smoke, and are suspect causative agents in human lung cancer [6].

Not all PAHs are carcinogens in animal models, and structural features required to cause tumors generally require the presence of four or more fused benzene rings and the presence of a bay region [7, 8]. Distortion in planarity of the PAHs becomes more pronounced when the bay region is either methylated or closed by an additional benzo ring to create a "fjord" region. This increase in distortion is associated with an increase in carcinogenicity (Scheme 6.1) [9, 10].

The PAHs that are tumorigenic are often referred to as procarcinogens since they require metabolic activation to biological reactive intermediates that can cause DNA adducts that will lead to mutation. Three different pathways of PAH activation have been proposed (Scheme 6.2). Each of these pathways gives rise to their own distinctive DNA adducts. These pathways include: (i) the radical cation pathway (mediated by P450 peroxidases and other peroxidases, e.g., horseradish peroxidase, prostaglandin H synthase (PHS), myeloperoxidase) to yield PAH-depurinating adducts [11]; (ii) the diol epoxide pathway (mediated by P450 mono-oxygenases, e.g., P450 1A1/1A2 and 1B1) to yield stable bulky diol epoxide-DNA adducts [12–16]; and (iii) the PAH o-quinone pathway (mediated by aldo-keto reductases AKR1A1 and AKR1C1–1C4), which form bulky stable DNA adducts, depurinating DNA adducts, and oxidatively modified DNA bases [17–20].

The Chemical Biology of DNA Damage. Edited by Nicholas E. Geacintov and Suse Broyde
© 2010 WILEY-VCH Verlag GmbH & Co. KGaA, Weinheim
ISBN: 978-3-527-32295-4

Bay-Region PAH

Chrysene

Benzo[a]pyrene

Benz[a]anthracene

Methylated Bay-Region PAH

5-Methylchrysene

7,12-Dimethylbenz[a]anthracene

Fjord-Region PAH

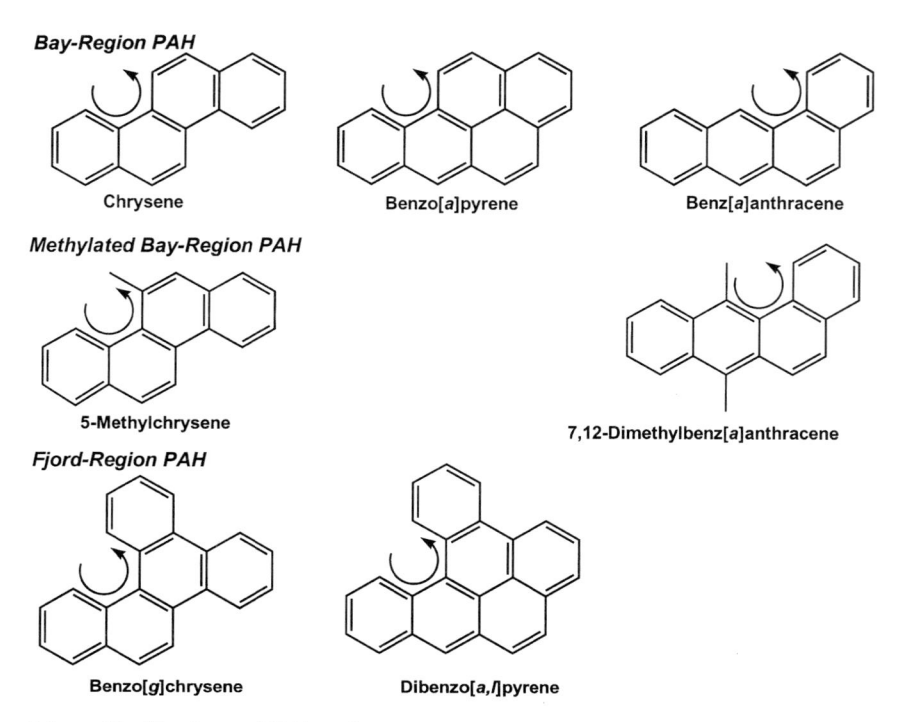

Benzo[g]chrysene

Dibenzo[a,l]pyrene

Scheme 6.1 Structures of PAH carcinogens.

If these PAH-derived DNA adducts are unrepaired they can give rise to mutation in critical growth control genes, such as K-*ras* leading to activation of the proto-oncogene or *p53* leading to inactivation of this tumor suppressor gene. These genes are of the most interest since K-*ras* and *p53* show signature mutations in lung cancer patients. In K-*ras*, G → T transversions predominate at the 12th and 13th codon, and A → T transversions predominate at the 61st codon [21–23]. These mutations render K-Ras GTPase-deficient so that it is constitutively activated. In p53, a series of "hotspot" mutations are observed in the DNA-binding domain of this transcription factor that render it transcriptionally incompetent. The mutation pattern observed in these "hotspots" is predominately G → T transversions [24–26]. Each of the three pathways of PAH activation must account for the structure–activity relationship of the bay region and its substitution, and provide routes to G → T or A → T transversions in proto-oncogenes and tumor suppressor genes.

Knowing which pathways of PAH activation prevail provides a rationale for the detection and quantitation of PAH-derived DNA adducts in population-based studies. Measurement of these adducts provides a method for biomonitoring PAH exposure and provides a risk assessment for susceptibility to diseases such as lung cancer. This could be particularly useful if it is assumed that there is a correlation between DNA adducts, mutational load, and, ultimately, tumor formation.

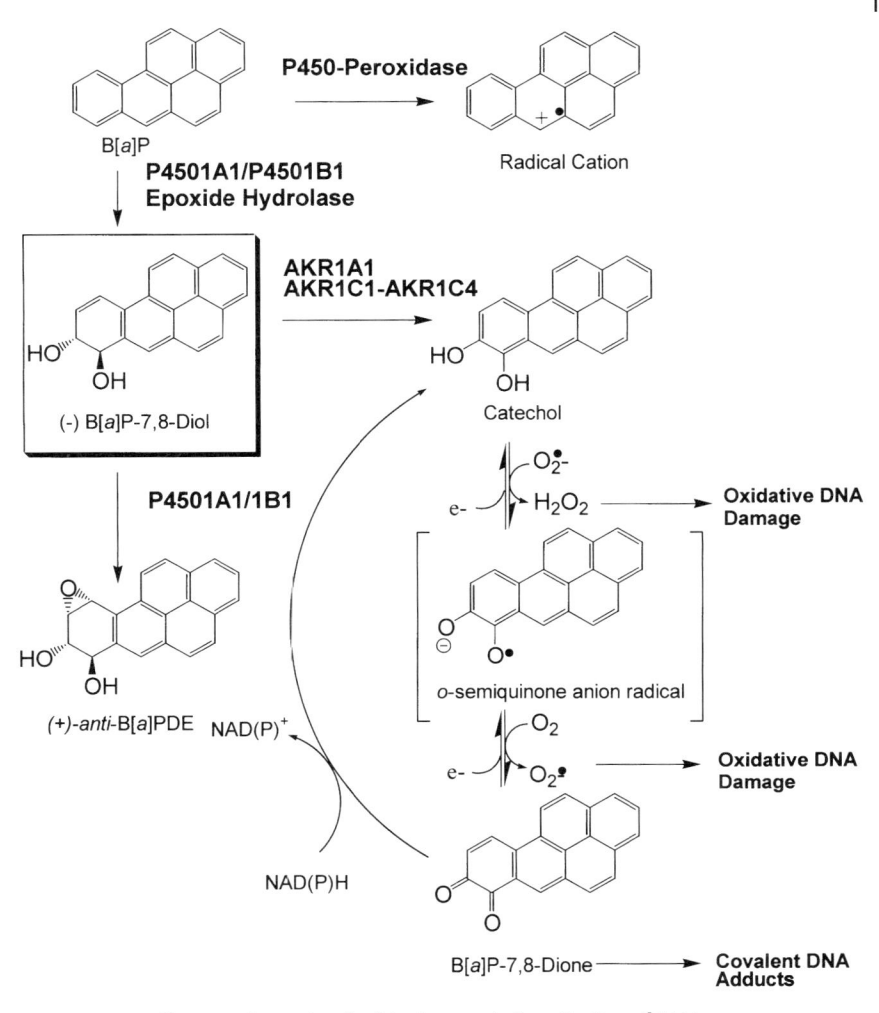

Scheme 6.2 Different pathways involved in the metabolic activation of PAHs.

Detection and quantitation of PAH-derived DNA adducts requires authentic synthetic standards to act as internal standards and therefore a knowledge of adduct structure. Adduct structure can often be used to provide information on the mechanism of PAH activation that must have occurred. Unfortunately, this is quite challenging since PAH-DNA adducts can be difficult to detect and quantitate since they may be present in only $1:10^8$ nucleotides. State-of-the art methods requiring stable isotope dilution liquid chromatography/multiple reaction monitoring/mass spectrometry (LC/MRM/MS) are necessary to provide sufficient sensitivity to detect these adducts reliably. Additionally, PAH-DNA adducts have different rates of repair and yield different rates of mutation. Thus, each adduct must be assessed so that it can be ranked according to its half-life and miscoding potential.

The focus of this chapter is on the three different routes of PAH activation that have been reported and how each of these pathways gives rise to its own distinct spectrum of PAH-derived DNA adducts. The detection of these adducts in target tissues, their repair mechanisms, and whether these adducts can produce the expected G → T or A → T transversions is considered. Finally, the reliability of existing analytical methods to detect these adducts with precision and accuracy is discussed.

6.2
Radical Cation Pathway

6.2.1
Metabolic Activation of PAHs

PAHs can be metabolically activated by the peroxidase cycle of P450 enzymes and other heme-containing peroxidases (e.g., PHS). In the peroxidase catalytic mechanism ROOH is cleaved to yield ROH and water, and requires $2H^+$ plus $2e^-$. These electrons come from Fe(III) and create the equivalent of Fe=O^+ (Fe(V)) (Compound I), which then needs a coreductant to regenerate Fe(IV) and then Fe(III) in two one-electron reduction steps. PAHs with electrophilic centers can provide these electrons by acting as coreductants. In the case of B[a]P, the electrons are preferentially removed from C6 to generate a radical cation (Scheme 6.3). B[a]P radical cations give rise to B[a]P-1,6-dione, -3,6-dione, and -6,12-dione, which are well-characterized B[a]P metabolites [27]. B[a]P radical cations can also be intercepted by DNA to yield a series of B[a]P-depurinating adducts. Thus, if B[a]P is incubated with 3-methylcholanthrene-induced rat liver microsomes in the presence of a peroxide substrate and calf thymus DNA, B[a]P radical cation depurinating DNA adducts can be detected. These adducts correspond to B[a]P-6-C8-Gua, B[a]P-N7-Gua, and B[a]P-6-N7-Ade (I–III; Scheme 6.4) [28]. Identical adducts were

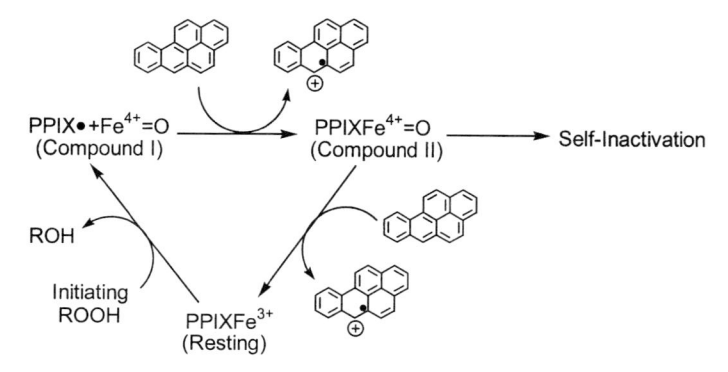

Scheme 6.3 Metabolic activation of PAHs to radical cations.

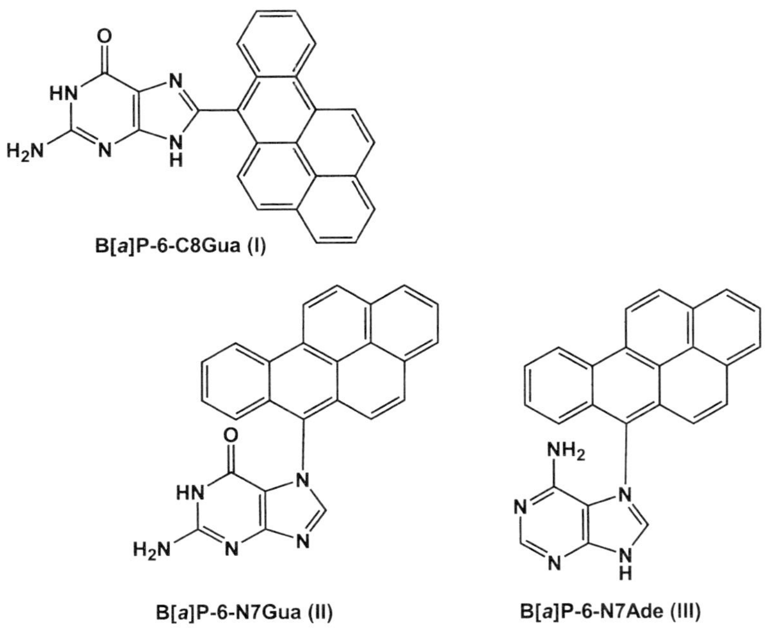

Scheme 6.4 Depurinating DNA adducts formed from radical cations.

subsequently detected when [³H]B[*a*]P was incubated with rat nuclei isolated from 3-methylcholanthrene-induced Wistar rats [29].

6.2.2
Radical Cation DNA Adducts

PAH radical cation depurinating DNA adducts have since been detected *in vitro* with some of the most potent PAH procarcinogens, 7,12-dimethylbenz[*a*]anthracene (DMBA) and dibenzo[*a*,*l*]pyrene (DB[*a*,*l*]P) [30, 31]. These depurinating adducts have also been detected in two end organs susceptible to PAH carcinogenesis – SENCAR mouse skin and Sprague-Dawley rat mammary gland following PAH exposure [32, 33].

Depurinating adducts leave behind an abasic site on the DNA. Replicative DNA polymerase will insert an A opposite an abasic site so that depurination provides a straightforward route to either G → T or A → T transversions when the daughter strand is replicated [34]. As radical cation depurinating adducts result in *N*-glycosidic bond cleavage, repair of the abasic site becomes important. Apurinic/apyrimidinic endonuclease creates a strand break at the site of the lesion, and the gap is then repaired by phosphodiesterase, DNA polymerase, and DNA ligase [35], as discussed in Chapter 11.

In mouse skin studies, a correlation between radical cation depurinating adducts and mutations in the H-*ras* gene was noted after treatment with B[*a*]P, DMBA, and DB[*a*,*l*]P. The mutations observed were G → T transversions at codon 13 or A → T transversions at codon 61, which could thus be explained by the formation of radical cation depurinating DNA adducts that were detected [36]. In a small cohort of seven women who were exposed to either household coal-smoke or smoked cigarettes, three women had modest amounts of the B[*a*]P-6-N7-Gua adduct that exceeded the B[*a*]P-6-N7-Ade adduct by 200- to 300-fold [37]. The levels of other PAH-derived DNA lesions were not reported.

Attempts have been made by others to compare the levels of depurinating radical cation DNA adducts with diol epoxide-DNA adducts in cell culture models. Stable diol epoxide-DNA adducts and depurinating adducts were measured in MCF-7 cells (high expressors of P450) and HL-60 cells (high peroxidase expressors) in PAH-naïve cells and cells pretreated with PAHs for enzyme induction. Stable-bulky diol epoxide-DNA adducts were quantified by ^{32}P-postlabeling coupled to reversed-phase high-performance liquid chromatography (RP-HPLC) and depurinating adducts were estimated from abasic sites measured either as chemically induced strand breaks or by reaction with the aldehyde-reactive probe (ARP)-based enzyme-linked immunosorbent assay (ELISA) method [38, 39]. The ARP reagent reacts with the aldehyde of the ring-opened anomeric sugar that remains after depurination. The results overwhelmingly favored the formation of diol epoxide-DNA adducts in these cell-based systems [40–42].

6.2.3
Limitations of the Radical Cation Pathway

There are limitations to the radical cation pathway that need to be considered:

i) The production of the depurinating adducts *in vitro* has been performed solely with rat liver microsomes in the presence of bulk DNA as a trapping agent or in isolated rat liver nuclei. Thus, the discrete human recombinant P450 enzymes that can generate these reactive intermediates remain unidentified.

ii) Radical cation depurinating DNA adducts have not been detected in cell culture models and estimates exist that the rate of repair of abasic sites is rapid (i.e., 10 000 events/cell/day) [43]. This is because a high level of spontaneous depurination exists in the native DNA and the repair mechanisms have evolved to avoid mutation.

iii) The radical cations are highly reactive and short-lived, and it is difficult to imagine how they would survive transit from the endoplasmic reticulum to the nucleus and reach naked DNA without being scavenged. The problem also exists in the experiments performed in the isolated nuclei since radical cations would still have to traverse the chromatin and find uncoiled DNA.

iv) The methods used to detect radical cation depurinating DNA adducts do not provide unequivocal structural assignment. The method of detection has been predominately cochromatography with synthetic standards by HPLC or capillary electrophoresis using fluorescence line narrowing spectrometry (FLNS). However, these methods do not include the use of an internal standard and FLNS does not provide unambiguous structural identity.

v) Attempts to measure radical cation depurinating DNA adducts in reasonable quantities in humans exposed to PAH has met with limited success.

vi) While radical cation depurinating DNA adducts have been observed in bay region, methylated bay region, and fjord region PAH, the mechanism of cation formation does not mandate the presence of these structural features in PAH.

vii) The more stable radical cation metabolites are the extended diones, B[a]P-1,6-dione, -3-6-dione, and -6,12-dione, which are redox-active. This raises the possibility that radical cation-derived metabolites can also produce reactive oxygen species (ROS) and oxidatively damaged bases in DNA. It is conceivable that ROS-derived lesions may be more important than the depurinating adducts thus far measured.

6.3
Diol Epoxides

6.3.1
Metabolic Activation of PAHs

The most widely accepted pathway of PAH activation to yield DNA adducts involves the formation of *anti*- or *syn*-diol epoxides in the bay region [12, 13]. Their formation is mediated by the sequential reaction of cytochrome P450 and epoxide hydratase, and is best described using the example of B[a]P.

B[a]P induces its own metabolism by binding to the aryl-hydrocarbon receptor (AhR), which is translocated to the nucleus and binds with its heterodimeric partner (ARNT) to the xenobiotic response element on the promoter regions of the *CYP1A1/1B1* genes [44, 45]. This leads to increased expression of P450 1A1 and 1B1, where P450 1B1 may be the more relevant enzyme in extrahepatic tissues [15]. NADPH-dependent P450 monoxygenation in the non-K region leads to the formation of an arene oxide (B[a]P-7S,8S-oxide). The oxide is a substrate for epoxide hydratase which yields the corresponding *trans*-dihydrodiol ((−)-7R,8R-dihydroxy-dihydro-B[a]P). Further, monoxygenation of the *trans*-dihydrodiol yields the corresponding *anti*-diol epoxide ((+)-7β,8α-dihydroxy-7,8-dihydro-9α,10α-oxo-B[a]P or 7R,8S-dihydroxy-7,8-dihydro-9S,10R-oxo-B[a]P (*anti*-B[a]PDE)) [12–16]. The metabolic pathway in rat and human liver microsomes is stereoselective so that the 7R,8R-dihydrodiol represents about 80% of the total dihydrodiol formed. The 7R,8R-dihydrodiol is preferentially converted to the (+)-*anti*-B[*a*]PDE as opposed to the (+)-*syn*-B[*a*]PDE. The (+)-*anti*-B[*a*]PDE can be either hydrolyzed in water via *trans* or *cis* ring-opening to yield a series of four stereospecific tetraols or it can be conjugated with glutathione. Similar pathways of metabolic activation have been observed for PAHs that contain bay regions (phenanthrene, chrysene, and benz[*a*]anthracene) [46–48], methylated bay regions (5-methylchrysene and DMBA) [9, 49, 50], and fjord regions (benzo[*g*]chrysene (B[*g*]C) and DB[*a,l*]P) [10, 51].

6.3.2
Diol Epoxide-DNA Adducts

(+)-*anti*-B[*a*]PDE can undergo *trans* or *cis* ring-opening via attack by the N^2-exocyclic amino group on deoxyguanosine (dGuo) to yield four possible diastereomeric adducts or it can undergo *trans* or *cis* ring-opening via attack by the N^6-exocyclic amino group on deoxyadenosine (dAdo) to yield four different diastereomeric adducts, making a total of eight adducts (**IV–XI**) An additional eight adducts are possible from the minor (+)-*syn*-B[*a*]PDE isomer, making 16 possible diastereomers in all. Of these, the major one formed with B[*a*]P is (+)-*trans-anti*-B[*a*]PDE-N^2-dGuo adduct (**IV**). The stereochemistry of this major adduct is 7R,8S,9R,10S and is a common feature of many of the PAH diol epoxide-DNA adducts (Scheme 6.5) [52]. In double-stranded DNA (+)-*trans-anti*-B[*a*]PDE-N^2-dGuo adducts occupy the minor groove 5′ to the adducted G [53].

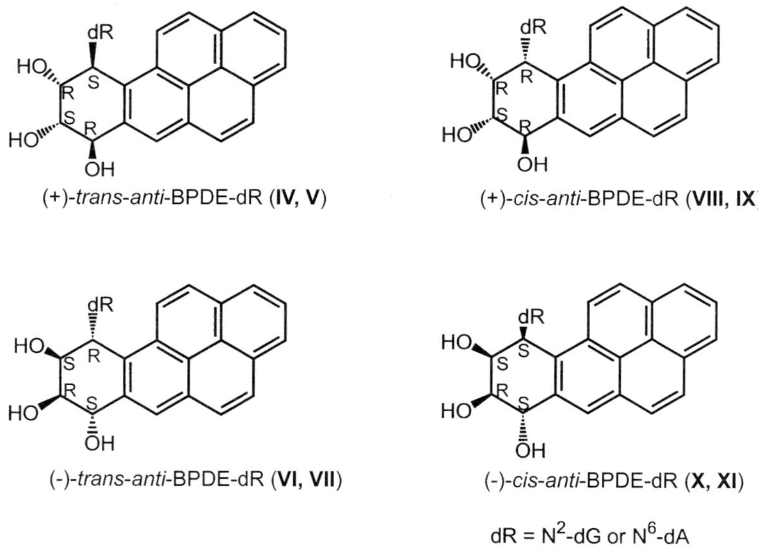

(+)-*trans-anti*-BPDE-dR (**IV, V**)

(+)-*cis-anti*-BPDE-dR (**VIII, IX**)

(-)-*trans-anti*-BPDE-dR (**VI, VII**)

(-)-*cis-anti*-BPDE-dR (**X, XI**)

dR = N^2-dG or N^6-dA

Scheme 6.5 Diol epoxide-DNA adducts – different stereochemistries.

(+)-*anti*-B[*a*]PDE-N^2-dGuo adducts have been detected with isolated nucleosides and in bulk DNA upon reaction with (+)-*anti*-B[*a*]PDE [54]. They have also been detected in cell culture models (hamster embryo fibroblasts, MCF-7 cells, and human lung cells) [55–59]; in tumor sites (e.g., SENCAR mouse skin [60] and the A/J mouse lung model of B[*a*]P carcinogenesis [61]), and the repaired adducts have been detected in the plasma of humans [62]. Perhaps, the most convincing early evidence for the formation of (+)-*anti*-B[*a*]PDE-N^2-dGuo were studies in which it was isolated in sufficient quantity from SENCAR mouse skin to validate its structure by ^{1}H-nuclear magnetic resonance (NMR) [60]. Adduct detection methods in humans have relied on ^{32}P-postlabeling [63] and ELISA-based methods [62, 64], HPLC fluorescence [65], and gas chromatography/MS detection of phenanthrene tetraols as a biomarker for PAH activation [66]. More recently, stable isotope dilution LC/MRM/MS methods have been developed that have sufficient sensitivity to supersede the ^{32}P-postlabeling methods for the detection of (+)-*trans-anti*-B[*a*]PDE-N^2-dGuo adducts in cell culture models and in human lungs of PAH-exposed individuals [58, 67]. The application of these methods to detect *anti*-B[*a*]PDE-DNA adducts in the lungs of never, former, and current smokers showed that the adducts could only be detected in one of 26 samples, and could only account for a small fraction of the total covalent PAH-DNA adducts detected in the same sample by ^{32}P-postlabeling [67].

The attention given the (+)-*trans-anti*-B[*a*]PDE-N^2-dGuo adduct and related adducts from other PAHs is warranted since there is compelling evidence that (+)-*anti*-B[*a*]PDE is an ultimate carcinogen. It is the B[*a*]P metabolite that is consistently the most mutagenic in the Ames test [68] and it is most tumorigenic in the newborn mouse lung model of B[*a*]P carcinogenesis [69]. Treatment of proto-oncogenic K-*ras* followed by transfection will transform NIH 3T3 cells and the isolated K-*ras* contains a transforming mutation at the 12th and 61st codons [70]. Ligation-mediated polymerase chain reaction shows that (±)-*anti*-B[*a*]PDE will preferentially form DNA adducts in p53 at "hotspots" that are mutated in patients in lung cancer [71]. This last observation has led to the proposition that B[*a*]P may contribute to the causation of tobacco-related lung cancer.

(+)-*anti*-B[*a*]PDE and related diol epoxides clearly form bulky stable covalent adducts with DNA, raising the issue as to how they may cause mutations in the first place. In order to cause mutations, the lesions must avoid repair prior to replication. Excision of the adduct attached to an 24–32mer oligonucleotide is accomplished by nucleotide excision repair (NER). There are two types of NER: global genome (GG)-NER and transcription-coupled (TC)-NER. The GG-NER pathway is operative throughout the entire genome, while TC-NER is triggered when the RNA polymerase stalls at the site of the lesion during transcription (see [72, 73] and Chapter 17). The NER apparatus includes a helicase that unwinds the DNA once the lesion is recognized and two endonucleases, one operating on the 5′-side and the other on the 3′-side of the lesion, which incise the damaged strand (see Chapter 11 for further details). TC-NER leads to the persistence of the lesion on the nontranscribed strand, which results in a marked strand bias in mutations. The excised 24–32mer containing the adduct is digested by nucleases to the deoxyribonucleoside leading to the detection of (+)-*trans-anti*-B[*a*]PDE-N^2-dGuo in the plasma and urine. Yet another repair mechanism operative in cells is the bypass of lesions catalyzed by specialized translesion bypass polymerases [74, 75]. These polymerases take over during replication when a high-fidelity replicative polymerase is stalled by the bulky lesion. However, the bypass DNA polymerases often have low fidelity and processivity is low; this can lead to the misincorporation of a nucleobase on the growing strand, leading to mutations. In elegant studies by Zhao *et al.*, it was found that the recruitment of two different bypass polymerases (Pol η and Pol ζ plus Rev1) were required to produce a G → T transversion from a *anti*-B[*a*]PDE-DNA adduct in yeast cells [76], indicating that the route to these signature mutations is less than straightforward. The requirement for Pol ζ in combination with a second bypass polymerase for transiting this lesion has been recently demonstrated in mammalian cells [77].

6.3.3
Limitations of the Diol Epoxide Pathway

There are some limitations to the diol epoxide pathway:

i) Whereas it is clear that diol epoxide-DNA adducts do form, it is less clear as to which are the most relevant P450 isoforms responsible for the formation of diol epoxides in tumor sites. P450 1A1 and 1B1 knockout mice have been shown to produce more diol epoxide-DNA adducts than intact wild-type mice [78–80]. These data suggest that the inducible P450 isoforms play a protective role (i.e., they produce more *anti*-B[*a*]PDE that can be rapidly hydrolyzed to tetraols or conjugated by glutathione *S*-transferases for elimination). This would imply that low levels of *anti*-B[*a*]PDE escape detection by phase II enzymes since they are present at levels much lower than K_m. These conclusions are supported by recent studies in the human bronchoalveolar (H358) cell line that showed, using stable-isotope dilution LC/MS methods, that less (+)-*anti*-B[*a*]PDE-N^2-dGuo adducts were formed in cells in which P450 1A1/1B1 was induced by 2,3,7,8-tetrachlorodibenzo-*p*-dioxin than in its absence [59].

ii) The mechanism by which (+)-*anti*-B[*a*]PDE-N^2-dGuo adducts causes G → T transversions in either K-*ras* or *p53* is less than straightforward, as noted above.

iii) (+)-*anti*-B[*a*]PDE adducts do not account for all the stable PAH-DNA adducts at PAH-exposed sites.

iv) (+)-*anti*-B[*a*]PDE does not provide an explanation for the presence of oxidative DNA lesions in PAH-exposed sites [81, 82].

The diol epoxide pathway does, however, account for the structure–activity relationships in the bay region, methylated bay region, and fjord region PAH, since diol epoxides would not form without this structural feature.

6.4 PAH o-Quinones

6.4.1 Metabolic Activation of PAH *trans*-Dihydrodiols by AKRs

A third pathway of PAH activation is the NAD(P)$^+$-dependent oxidation of non-K-region (−)-R,R-*trans*-dihydrodiols to yield electrophilic and redox-active PAH o-quinones catalyzed by AKRs [17]. Using B[*a*]P as an example (−)-7R,8R-dihydroxy-dihydroB[*a*]P is oxidized by human recombinant AKR1A1 to the corresponding B[*a*]P-7,8-dione [19]. By contrast, AKR1B1 and AKR1B10 will oxidize the (+)-7S,8S-dihydroxy-dihydroB[*a*]P to B[*a*]P-7,8-dione [83], whereas AKR1C1–AKR1C4 will oxidize both stereoisomers of (±)-B[*a*]P-7,8-dihydrodiol to yield the dione [20]. In this reaction, the AKRs catalyze a formal dihydrodiol dehydrogenation to yield a ketol that spontaneously rearranges to form a catechol, 7,8-dihydroxy-B[*a*]P. This catechol is unstable and undergoes two one-electron oxidations (auto-oxidation). The first one-electron oxidation in air yields the o-semiquinone anion radical and hydrogen peroxide, and the second one-electron oxidation in air yields the fully oxidized o-quinone and superoxide anion radical [84, 85].

The mechanism of auto-oxidation was elaborated using measurements of ROS coupled with spin-trapping agents and is as follows. The superoxide anion acts as an initiating radical and acts as a base to remove a proton from the catechol producing the catecholate anion and a hydroperoxy radical. The hydroperoxy radical abstracts an electron from the catecholate anion to yield an o-semiquinone radical and hydrogen peroxide. The o-semiquinone radical then reacts with molecular oxygen to produce the o-quinone and superoxide anion, which can then act as a propagating radical [85]. A consequence of this auto-oxidation is the production of ROS.

B[*a*]P-7,8-dione is a Michael acceptor that undergoes 1,4- and 1,6-Michael addition reactions with cellular nucleophiles including DNA and RNA. Alternatively, in the presence of NADPH it can undergo both enzymatic and nonenzymatic reduction back to the corresponding catechol, which can then undergo a further round of auto-oxidation. This establishes a futile redox cycle in which the PAH catechol and o-quinone can be interconverted multiple times leading to the amplification of ROS. This will continue to occur until cellular reducing equivalent is depleted, creating oxidative stress and a pro-oxidant state [86–88]. The reactive o-quinone produced by this pathway can be trapped with 2-mercaptoethanol and characterized by NMR and electrospray ionization/LC/MS [19, 20, 84].

Human recombinant AKRs have been shown to catalyze identical reactions for bay region *trans*-dihydrodiols (phenanthrene, chrysene, B[*a*]P, benz[*a*]anthracene), methylated bay region *trans*-dihydrodiols (5-methylchrysene and 7,12-DMBA), and fjord region *trans*-dihydrodiols (B[*g*]C) [18–20]. Thus far, no human AKR has been show to be an efficient catalyst of the oxidation of the potent proximate carcinogen 11R,12R-dihydroxy-dihydro-DB[*a,l*]P. Interestingly, this metabolic pathway of

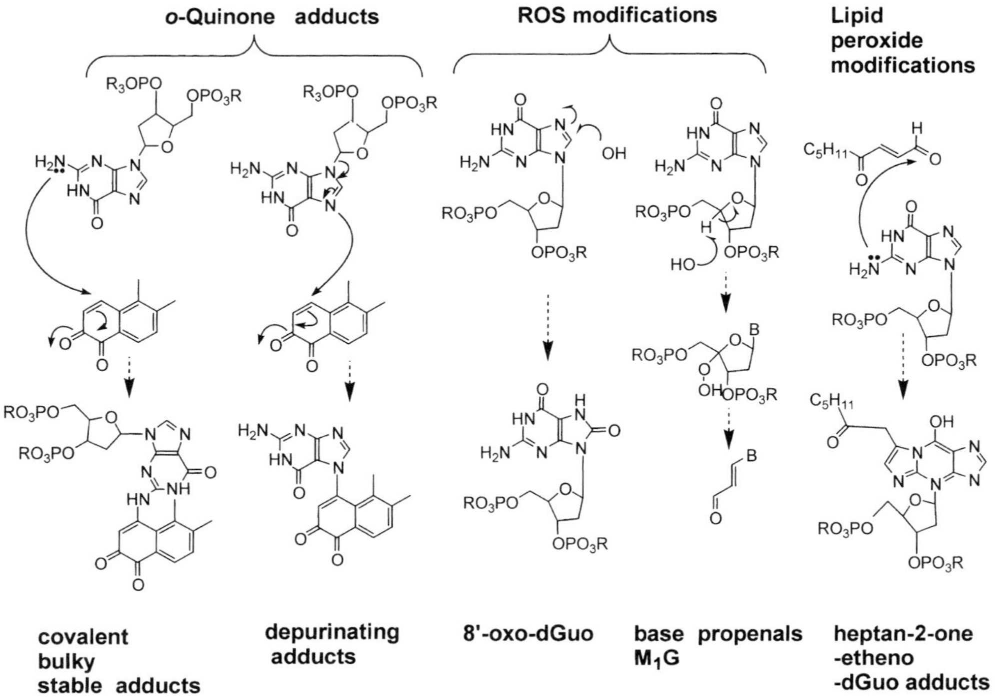

Scheme 6.6 Classes of PAH o-quinone-derived DNA adducts.

PAH activation is reminiscent of that proposed for the catechol estrogens derived from the 4-hydroxylation of 17β-estradiol and the equine estrogens used in hormone replacement therapeutics proposed by the Bolton [89–91] (Chapter 8) and Cavalieri groups [92, 93].

6.4.2
PAH o-Quinone-Derived DNA Adducts

Since the AKR pathway produces electrophilic and redox-active PAH o-quinones, two major types of DNA lesions are possible. These are the covalent adducts in which the PAH o-quinones are bound to bases in DNA and the oxidative lesions that can arise due to the production of ROS (Scheme 6.6).

6.4.2.1 Covalent PAH o-Quinone-DNA Adducts
Two types of covalent PAH o-quinone-DNA adducts are possible. The first type involves the formation of stable covalent adducts that may result from either 1,4- or 1,6-Michael addition of the N^2-exocyclic amino group of dGuo or the N^6-exocyclic amino group of dAde. Evidence exists for these adducts in reactions of B[a]P-7,8-dione with either deoxyribonucleosides or bulk DNA [94, 95]. In the former case,

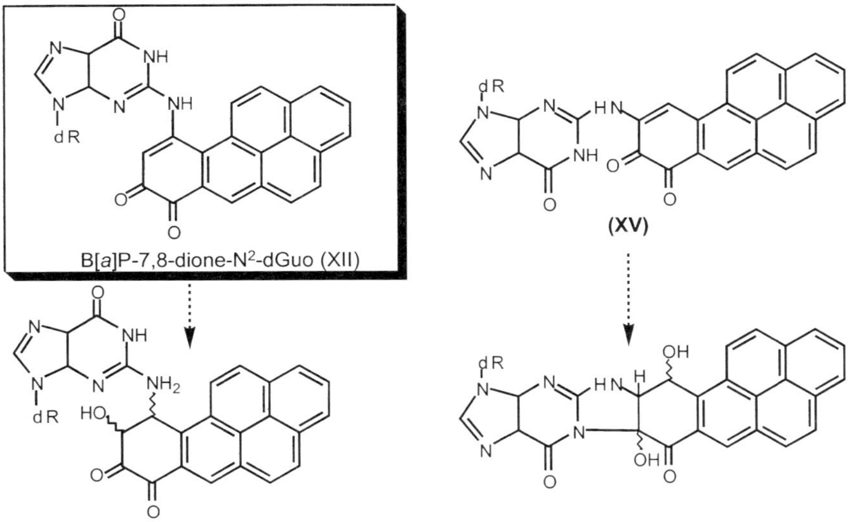

Scheme 6.7 Unusual stable PAH o-quinone dGuo adducts.

reaction of [^{3}H]B[a]P-7,8-dione with oligo-p(dG)$_{10}$ followed by digestion to the deoxyribonucleosides gave a single adduct which coeluted on RP-HPLC with the adduct obtained by digesting calf thymus DNA pretreated with the radioactive dione. These data suggested that a stable B[a]P-7,8-dione-N^2-dGuo (XII) adduct had formed (Scheme 6.7). At higher temperatures (55 °C) in the presence of 50% dimethylformamide, four isomeric adducts resulted from reaction with the N^2-exocyclic amino group of dGuo and corresponded to either 1,4-Michael addition adducts that were hydrated (XIII–XIV) or cyclized hydrated adducts that resulted from a 1,6-Michael addition (XVI–XVII) [95]. Two isomeric adducts were also observed from reaction of dAdo with B[a]P-7,8-dione and corresponded to cyclized hydrated adducts that resulted from 1,6-Michael addition of the N1 ring nitrogen to the dione (XX–XXI) (Scheme 6.8) [95]. These adducts were thoroughly characterized by a combination of NMR and MS. These adducts were also observed in calf thymus DNA when milligram quantities of B[a]P-7,8-dione were used and ^{32}P-postlabeling was used as the detection method [96].

The second type of PAH o-quinone covalent adduct involves formation of depurinating adducts that result from 1,4- or 1,6-Michael addition of the N7 ring nitrogen of guanine or adenine. Evidence for these adducts were observed *in vitro* by reaction of PAH o-quinones with deoxyribonucleosides under acidic conditions or by reaction of PAH o-quinones with calf thymus DNA [97]. These adducts were fully characterized by NMR and LC/MS. Recently, interest in the covalent PAH o-quinone adducts has diminished based on the accumulated evidence that PAH

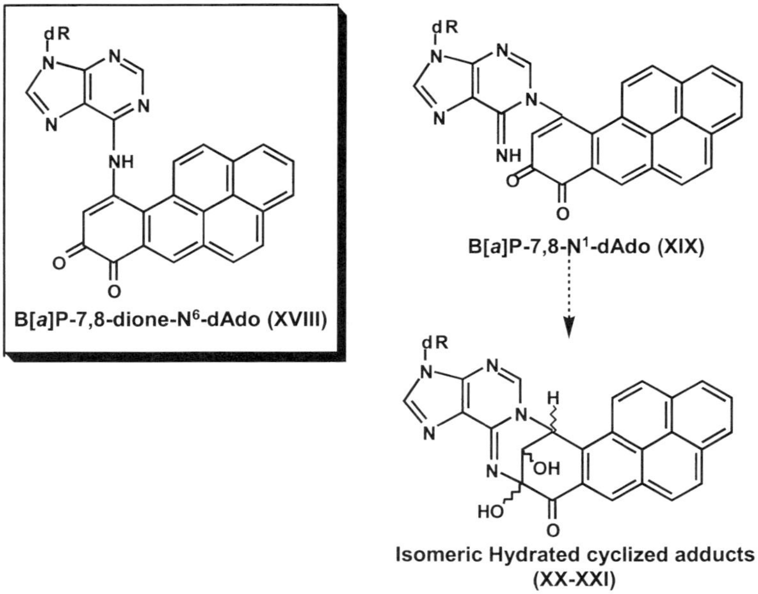

Scheme 6.8 Unusual stable PAH o-quinone dAdo adducts.

o-quinones are highly mutagenic when they redox cycle in the presence of NADPH and $CuCl_2$, suggesting that oxidative DNA lesions from this pathway are the most important [98–100]. It is worth emphasizing that the formation of covalent PAH o-quinone-DNA adducts via Michael addition proceeds through ketol and catechol intermediates that could redox cycle bound to DNA. This provides a potential mechanism where ROS amplification could occur on the DNA itself and in this instance the covalent quinone adducts would act as "Trojan horses" for oxidative DNA damage.

6.4.2.2 Oxidative DNA Lesions from PAH o-Quinones

The first type of oxidative lesion that can occur is attack of the nucleobase by ROS [101]. The base most susceptible to attack by ROS is guanine and the product can be 8-oxo-dGuo or the ring-opened formamdiopyrimdine adducts. Under complete redox cycling conditions, nanomolar concentrations of PAH o-quinones can produce large quantities of 8-oxo-dGuo in salmon testis DNA using RP-HPLC and an electrochemical (EC) detection method [102]. At higher concentrations, DNA strand scission is observed [103]. To quantitate these lesions versus others that may form, the ARP was used to detect aldehydic sites in DNA under different reaction conditions. In the absence of redox cycling the ARP assay will detect abasic sites that form as a result of PAH o-quinone depurinating adduct production. Under redox cycling conditions this assay would measure aldehydic sites that may result from N-glycosidic bond cleavage and if the assay is coupled with specific base excision repair (BER) enzymes, aldehydic sites to specific lesions can be detected. Enzymes used in the coupled assay were human oxoguanine glycosylase (OGG1, which specifically removes 8-oxo-Gua) and endonuclease III (which removes oxidized pyrimidines). Using this assay to rank PAH o-quinone-derived DNA lesions it was found that the order of adduct levels was 8-oxo-dGuo >> oxidized pyridmines = abasic sites >> depurinated adducts. The ARP assay coupled with hOOG1 was validated by independently measuring 8-oxo-dGuo formation by RP-HPLC-EC methods [104].

To determine whether AKRs can mediate PAH-dependent 8-oxo-dGuo formation in a relevant cell model, A549 human lung adenocarcinoma cells that express high levels of AKR1C1–AKR1C3 were selected. In addition, a sensitive immunoaffinity capture stable isotope dilution LC/MRM/MS assay was developed to measure 8-oxo-dGuo. In this cell-based assay, the background adduct level was 2.2 8-oxo-dGuo adducts per 10^7 dGuo which is an order of magnitude lower than generally reported [105, 106]. This assay used strict controls to eliminate the adventitious oxidation of guanine that can occur in the DNA isolation and work-up. This was achieved by using Chelex-treated buffers and desferoxamine to prevent Fenton chemistry from trace transition metals. Treatment of A549 cells with B[a]P-trans-7,8-dihydrodiol (AKR substrate) and B[a]P-7,8-dione (AKR product) led to increased ROS formation that could not be detected with either (+)-anti-B[a]PDE or B[a]P-trans-4,5-dihydrodiol. In addition, treatments with B[a]P-trans-7,8-dihydrodiol and B[a]P-7,8-dione led to the elevated formation of 8-oxo-dGuo. The level of 8-oxo-dGuo detected was increased further in the presence of a catechol-O-methyl transferase (COMT) inhibitor, suggesting that 8-oxo-dGuo formation was dependent on the redox cycling of the catechol (7,8-dihydroxy-B[a]P) with the quinone (B[a]P-7,8-dione) [105]. These data suggest for the first time that in addition to measuring stable anti-diol epoxide-DNA adducts, the formation of 8-oxo-dGuo and other oxidative lesions should be taken into account when considering the total DNA adduct burden associated with PAH exposures.

Mutagenic studies show that 8-oxo-dGuo is a highly mutagenic lesion and when unrepaired it will form Hoogstein base pairs with adenine. This mismatch will yield a G → T transversion when the daughter strand is replicated and offers a straightforward route to this mutation [74, 107, 108]. Recently, an in vitro p53 mutagenicity assay was established using a yeast reporter gene. In this assay, PAH o-quinones were only mutagenic when they redox cycled and under redox cycling conditions they were 20- to 80-fold more mutagenic than (+)-anti-B[a]PDE. Direct linear correlations were noted in these assays between p53 mutagenic frequency and adduct number measured as 8-oxo-dGuo (detected by RP-HPLC-EC) or (+)-trans-anti-B[a]PDE-N^2-dGuo (detected by LC/MS). The slopes and correlation coefficients were almost identical for both adducts, suggesting they were equally mutagenic. However, considerably less B[a]P-7,8-dione was required to make these lesions than (±)-anti-B[a]PDE. When the mutations in p53 were analyzed, B[a]P-7,8-dione preferentially caused G → T transversions consistent with the formation of 8-oxo-dGuo and its processing by replicative DNA polymerase [98, 100]. By contrast (+)-anti-B[a]PDE preferentially mutated G residues consistent with the formation of (+)-*trans-anti*-B[a]PDE-N^2-dGuo adducts; however, G → T transversions were the least frequent and may be explained by the complement of bypass DNA polymerases present in the yeast strain used. In this yeast reporter gene assay the spectrum of mutations (mutations by codon) was essentially random. However, when there was biological selection for dominance, mutations at hotspot "G" bases was revealed [100]. These data suggest that adduct chemistry preferentially targets the guanine base and that biological selection may play a role in defining the mutational spectrum observed in lung cancer.

Since 8-oxo-dGuo is a nonbulky DNA lesion, it is repaired by BER rather than by NER [109]. Both monofunctional and bifunctional glycosylases exist to correct the damage. If a replicative DNA polymerase misincorporates an A opposite the 8-oxo-dGuo this base mismatch is recognized by MutY, a monofunctional glycosylase that cleaves the *N*-glycosidic bond to release Ade and forms an abasic site [110]. This abasic site is repaired by apurinic/apyrimidinic endonuclease as described earlier. If a C is incorporated opposite the 8-oxo-dGuo, a bifunctional glycosylase (MutM or OGG1) removes 8-oxo-Gua and cleaves the sugar to form an aldehyde and an overt strand break [111]. The strand break is repaired by endonuclease III, DNA polymerase, and DNA ligase. Loss of heterozygosity and polymorphism in human OGG1 is associated with the development of lung cancer [112, 113]. This finding suggests that during lung cancer development, bronchial epithelial cells may have an increased mutational load because of their inability to repair 8-oxo-dGuo.

The second type of oxidative damage that can occur in DNA results from the attack of the deoxyribose moiety by ROS [101]. A common lesion that occurs as a result of the removal of a hydrogen atom from C4′ of the deoxyribose is the formation of a base propenal. Base propenals give rise to malondialdehyde (MDA) which reacts with guanine bases to yield MDA-dG adducts (M_1G adducts). Although, MDA can be derived from lipid peroxidation, the general consensus is that the M_1G adduct comes primarily from the hydrolysis of base propenals [114]. (See Chapters 2 and 5 for details concerning these lesions.) Thus far no attempt has been made to measure these adducts as a result of the AKR pathway of PAH activation.

The third type of oxidative damage that can occur in DNA would result from the addition of bifunctional electrophiles (4-hydroxy-2-nonenal and 4-oxo-2-nonenal) that result from lipid peroxidation to the nucleobases. The etheno adducts formed with 4-hydroxy-2-nonenal can also be derived from vinyl chloride; however, the heptano-etheno adducts derived from 4-oxo-2-nonenal can only be derived from lipid peroxidation [115]. Thus far no attempt has been made to measure these adducts as a result of the AKR pathway of PAH activation.

6.4.3
Limitations of the PAH *o*-Quinone Pathway

A number of limitations exist with the AKR pathway of PAH activation:

i) PAH o-quinones have not been shown to be tumorigenic in models of B[a]P carcinogenesis.

ii) 8-oxo-dGuo adducts are a common oxidative DNA lesion and detection in sites of B[a]P induced tumors does not necessarily mean that they come from B[a]P-7,8-dione. They could also be produced from the extended B[a]P-1,6-dione, -3,6-dione, or -6,12-dione formed as a result of the radical cation pathway. However, if a COMT inhibitor enhances 8-oxo-dGuo formation from B[a]P this would implicate a redox-cycle between 7,8-dihydroxy-B[a]P and B[a]P-7,8-dione.

iii) If oxidative DNA lesions from PAH are relevant, 8-oxo-dGuo may not be the ultimate lesion on this nucleobase. In models of oxidative stress, 8-oxo-dGuo is often a reaction intermediate, and gives rise to spiroiminohydantoin and guanidohydantoin adducts that have higher mutagenic potential [108, 116, 117].

iv) No attempt has been made to measure either M_1G adducts or the heptano-etheno adducts that arise due to lipid peroxidation in PAH-exposed sites.

v) Until an AKR that oxidizes 11R,12R-dihydroxy-dihydro-DB[a,l]P is found, the AKR pathway cannot account for the tumorigenicity of this potent proximate carcinogen. The o-quinone pathway does account for the structure–activity relationships in the bay region, methylated bay region and fjord regions of PAH, since only non-K region *trans*-dihydrodiols are substrates for AKRs [118].

6.5
Future Directions

The question that remains is what is the most important PAH-DNA adduct in PAH-mediated carcinogenesis? There may be no single answer to this question since it will depend upon the pathways of PAH activation that exist in the target organ, the expression of NER and BER enzymes, and lesion bypass DNA polymerases. These may also be influenced by genetic variants that exist in these enzymes. Perhaps a more focused question is what is the most important PAH-DNA adduct in human lung bronchial epithelial cells, if it is assumed that PAHs are causative agents in lung cancer in both smokers and never smokers alike. At the minimum this would require the development of stable isotope dilution LC/MS assays for each of the major lesions of relevance and their quantitation within the same DNA sample. This challenge is made more difficult when it is realized that for B[a]P alone, each pathway of activation (radical cation, diol epoxide, or o-quinone) yields multiple distinct DNA adducts. The approach could be simplified if one representative DNA adduct from each pathway were selected, but this assumes they are equally abundant and that large differences in rates of repair do not exist between adducts. This is made more difficult for the AKR pathway, since the tendency would be to now focus on 8-oxo-dGuo lesions; however, other oxidative DNA lesions could be important, including those that would come from lipid peroxidation. Clearly, much still needs to be done to clarify this open question.

Acknowledgements

During the writing of this chapter T.M.P. was supported by NIH grants P30-ES013508, R01-ES015857, and R01-CA39504.

References

1 Straff, K., Baan, R., Grosse, Y., Secretan, B., El Ghissassi, F., and Cogliano, V. (2005) Carcinogenicity of polycyclic aromatic hydrocarbons. Policy Watch. *Lancet Oncol.*, **6**, 931–932.
2 Beak, S.O., Field, R.A., Goldstein, M.E., Kirk, P.W., Lester, J.N., and Perry, R. (1991) A review of atmospheric polycyclic aromatic hydrocarbons: sources, fate and behavior. *Water Air Soil Pollut.*, **60**, 279–300.
3 Marr, L.C., Kirchstetter, T.W., and Harley, R.A. (1999) Characterization of polycyclic aromatic hydrocarbons in motor vehicle fuels and exhaust emissions. *Environ. Sci. Technol.*, **33**, 3091–3099.
4 Lijinsky, W. and Ross, A.E. (1967) Production of carcinogenic polynuclear hydrocarbons in the cooking of food. *Food Cosmet. Toxicol.*, **5**, 343–347.
5 Doremire, M.E., Harmon, G.E., and Pratt, D.E. (1979) 3,4-Benzopyrene in charcoal grilled meats. *J. Food Sci.*, **44**, 622–623.
6 Hecht, S.S. (1999) Tobacco smoke carcinogens and lung cancer. *J. Natl. Cancer Inst.*, **91**, 1194–1210.
7 Dipple, A. (1984) Polycyclic aromatic hydrocarbon carcinogenesis: an introduction, in *Polycyclic Hydrocarbons and Carcinogenesis* (ed. R.G. Harvey), American Chemical Society, Washington, DC, pp. 1–18.
8 Lehr, R.E., Kumar, S., Levin, W., Wood, A.W., Chang, R.L., Conney, A.H., Yagi, H., Sayer, J.M., and Jerina, D.M. (1985) The bay-region theory of polycyclic aromatic hydrocarbon carcinogenesis, in *Polycyclic Hydrocarbons and Carcinogenesis* (ed. R.G. Harvey), American Chemical Society, Washington, DC, pp. 63–84.
9 Hecht, S.S., Amin, S., Melkikian, A.A., LaVoie, E.J., and Hoffmann, D. (1985) Effects of methyl and fluorine substitution on the metabolic activation and tumorigenicity of polycyclic aromatic hydrocarbons, in *Polycyclic Hydrocarbons and Carcinogenesis* (ed. R.G. Harvey), American Chemical Society, Washington, DC, pp. 85–105.
10 Cavalieri, E.L., Higginbotham, S., RamaKrishna, N.V.S., Devanesan, P.D., Todorovic, R., Rogan, E.G., and Salmasi, S. (1991) Comparative dose–response tumorigenicity studies of dibenzo[a,l]pyrene versus 7,12-dimethylbenz[a]anthracene, benzo[a]pyrene and two dibenzo[a,l] pyrene dihydrodiols in mouse skin and rat mammary gland. *Carcinogenesis*, **12**, 1939–1944.
11 Cavalieri, E.L. and Rogan, E.G. (1995) Central role of radical cations in the metabolic activation of polycyclic aromatic hydrocarbons. *Xenobiotica*, **25**, 677–688.
12 Gelboin, H.V. (1980) Benzo[a]pyrene metabolism, activation and carcinogenesis: role and regulation of mixed function oxidases and related enzymes. *Physiol. Rev.*, **60**, 1107–1166.
13 Conney, A.H. (1982) Induction of microsomal enzymes by foreign chemicals and carcinogenesis by

polycyclic aromatic hydrocarbons: G. H. A. Clowes Memorial Lecture. *Cancer Res.*, **42**, 4875–4917.

14. Shimada, T., Martin, M.V., Pruess-Schwartz, D., Marnett, L.J., and Guengerich, F.P. (1989) Roles of individual human cytochrome P-450 enzymes in the bioactivation of benzo[a]pyrene, 7,8-dihydroxy-7,8-dihydrobenzo[a]pyrene and other dihydrodiol derivatives of polycyclic aromatic hydrocarbons. *Cancer Res.*, **49**, 6304–6312.

15. Shimada, T., Hayes, C.L., Yamazaki, H., Amin, S., Hecht, S.S., Guengerich, F.P., and Sutter, T.R. (1996) Activation of chemically diverse procarcinogens by human cytochrome P450 1B1. *Cancer Res.*, **56**, 2979–2984.

16. Shimada, T., Yamazaki, H., Mimura, M., Wakamiya, N., Ueng, Y.-F., Guengerich, F.P., and Inui, I. (1996) Characterization of microsomal cytochrome P450 enzymes involved in the oxidation of xenobiotic chemicals in human fetal livers and adult lungs. *Drug Metab. Dispos.*, **24**, 515–522.

17. Penning, T.M., Burczynski, M.E., Hung, C.-F., McCoull, K.D., Palackal, N.T., and Tsuruda, L.S. (1999) Dihydrodiol dehydrogenases and polycyclic aromatic hydrocarbon activation: generation of reactive and redox-active o-quinones. *Chem. Res. Toxicol.*, **12**, 1–18.

18. Burczynski, M.E., Harvey, R.G., and Penning, T.M. (1998) Expression and characterization of four recombinant human dihydrodiol dehydrogenase isoforms: oxidation of *trans*-7,8-dihydroxy-7,8-dihydrobenzo[a]pyrene to the activated o-quinone metabolite benzo[a]pyrene-7,8-dione. *Biochemistry*, **37**, 6781–6790.

19. Palackal, N.T., Burczynski, M.E., Harvey, R.G., and Penning, T.M. (2001) The ubiquitous aldehyde reductase (AKR1A1) oxidizes proximate carcinogen *trans*-dihydrodiols to o-quinones: potential role in polycyclic aromatic hydrocarbon activation. *Biochemistry*, **40**, 10901–10910.

20. Palackal, N.T., Lee, S.H., Harvey, R.G., Blair, I.A., and Penning, T.M. (2002) Activation of polycyclic aromatic hydrocarbon *trans*-dihydrodiol proximate carcinogens by human aldo-keto reductase (AKR1C) enzymes and their functional overexpression in human lung adenocarcinoma (A549) cells. *J. Biol. Chem.*, **277**, 24799–24808.

21. Yamamoto, F. and Perucho, M. (1984) Activation of human c-K-*ras* oncogene. *Nucleic Acid Res.*, **12**, 8873–8885.

22. Rodenhuis, S. and Siebos, R.J. (1990) The *ras* oncogenes in human lung cancer. *Am. Rev. Respir. Dis.*, **142**, S27–S30.

23. Capella, G., Cronauer-Mitra, S., Pienado, M.A., and Perucho, M. (1991) Frequency and spectrum of mutations at codons 12 and 13 of the c-K-*ras* gene in human tumors. *Environ. Health Perspect.*, **93**, 125–131.

24. Top, B., Mooj, W.J., Klaver, S.G., Berrigter, L., Wisman, P., Elbers, H.R., Viser, S., and Rodenhuis, S. (1995) Comparative analysis of *p53* gene mutations and protein accumulation in human non-small-cell lung cancer. *Int. J. Cancer*, **64**, 83–91.

25. Gealy, R., Zhang, L., Siegfried, J.M., Luketich, J.D., and Keohavong, P. (1999) Comparison of mutations in the *p53* and K-*ras* genes in lung carcinomas from smoking and non-smoking women. *Cancer Epidemiol. Biomarkers Prev.*, **8**, 297–302.

26. Vahakanga, K.H., Bennett, W.P., Castren, K., Welsh, J.A., Khan, M.A., Bjomek, B., Alavanja, M.C., and Harris, C.C. (2001) *p53* and K-*ras* mutations in lung cancers from former and never smoking women. *Cancer Res.*, **61**, 4350–4356.

27. Lorentzen, R.J., Caspary, W.J., Lesko, S.A., and Ts'o, P.O.P. (1975) The autoxidation of 6-hydroxybenzo[a]pyrene and 6-oxobenzo[a]pyrene radical, reactive metabolites of benzo[a]pyrene. *Biochemistry*, **14**, 3970–3977.

28. Devanesan, P.D., Rama-Krishna, N.V.S., Todorovic, R., Rogan, E.G., Cavalieri, E.L., Jeong, H., Jankowiak, R., and Small, G.J. (1992) Identification and quantitation of benzo[a]pyrene-DNA adducts formed by rat liver microsomes *in vitro*. *Chem. Res. Toxicol.*, **5**, 302–309.

29 Devanesan, P.D., Higginbotham, S., Ariese, F., Jankowiak, R., Suh, M., Small, G.J., Cavalieri, E., and Rogan, E. (1996) Depurinating and stable benzo[a]pyrene-DNA adducts formed in isolated rat liver nuclei. *Chem. Res. Toxicol.*, **9**, 1113–1116.

30 RamaKrishna, N.V.S., Devanesan, P.D., Rogan, E.G., Cavalieri, E.L., Jeong, H., Jankowiak, R., and Small, G.J. (1992) Mechanism of metabolic activation of the potent carcinogen 7,12-dimethylbenz[a]anthracene. *Chem. Res. Toxicol.*, **5**, 220–226.

31 Li, K.-M., Todorovic, R., Rogan, E.G., Cavalieri, E.L., Ariese, F., Suh, M., Jankowiak, R., and Small, G.J. (1995) Identification and quantitation of dibenz[a,l]pyrene-DNA adducts formed by rat liver microsomes *in vitro*: preponderance of depurinating DNA adducts. *Biochemistry*, **34**, 8043–8049.

32 Chen, L., Devanesan, P.D., Higginbotham, S., Ariese, F., Jankowiak, R., Small, G.J., Rogan, E.G., and Cavalieri, E. (1996) Expanded analysis of benzo[a]pyrene-DNA adducts formed *in vitro* and in mouse skin: their significance in tumor initiation. *Chem. Res. Toxicol.*, **9**, 897–903.

33 Cavalieri, E.L., Rogan, E.L., Li, K.-M., Todorovic, R., Ariese, F., Jankowiak, R., Grubor, N., and Small, G.J. (2005) Identification and quantification of the depurinating DNA adducts formed in mouse skin treated with dibenzo[a,l]pyrene (DB[a,l]P) or its metabolites and in rat mammary gland treated with DB[a,l]P. *Chem. Res. Toxicol.*, **18**, 976–983.

34 Sagher, D. and Strauss, B. (1983) Insertion of nucleotides opposite apurinic/apyrimidinic sites in deoxyribonucleic acid during *in vitro* synthesis. Uniqueness of adenine nucleotides. *Biochemistry*, **22**, 4518–4526.

35 Wilson, D.M. and Barsky, D. (2001) The major human abasic endonuclease: formation, consequences and repair of abasic lesions in DNA. *Mutat. Res.*, **485**, 283–307.

36 Chakravarti, D., Pelling, J.C., Cavalieri, E.L., and Rogan, E.G. (1995) Relating aromatic hydrocarbon-induced DNA adducts and c-H-*ras*-mutations in mouse skin papillomas: the role of apurinic sites. *Proc. Natl. Acad. Sci. USA*, **92**, 10422–10426.

37 Casale, G.P., Singhal, M., Bhattacharya, S., RamaNathan, R., Roberts, K.P., Barbaccia, D.C., Zhao, J., Jankowiak, R., Gross, M.L., Cavalieri, E.L., Smith, G.J., Reannard, S.I., Mumford, J.L., and Shen, M. (2001) Detection and quantitation of depurinated benzo[a]pyrene-adducted DNA bases in the urine of cigarette smokers and women exposed to household coal smoke. *Chem. Res. Toxicol.*, **14**, 192–201.

38 Nakamura, J., Walker, V.E., Upton, P.B., Chiang, S.Y., Kow, Y.W., and Swenberg, J.A. (1998) Highly sensitive apurinic/apyrimidinic site assay can detect spontaneous and chemically induced depurination under physiological conditions. *Cancer Res.*, **58**, 222–225.

39 Nakamura, J. and Swenberg, J.A. (1999) Endogenous apurinic/apyrimidinic sites in genomic DNA of mammalian tissues. *Cancer Res.*, **59**, 2522–2526.

40 MelendezColon, V.J., Smith, C.A., Seidel, A., Luch, A., Platt, K.L., and Baird, W.M. (1997) Formation of stable adducts and absence of depurinating DNA adducts in cells and DNA treated with the potent carcinogen dibenzo[a,l]pyrene or its diol-epoxides. *Proc. Natl. Acad. Sci. USA*, **94**, 13542–13547.

41 Melendez-Colon, V.J., Luch, A., Siedel, A., and Baird, W.M. (1999) Comparison of cytochrome P450- and peroxidase-dependent metabolic activation of the potent carcinogen dibenzo[a,l]pyrene in human cell lines: formation of stable DNA adducts and absence of a detectable increase in apurinic sites. *Cancer Res.*, **59**, 1412–1426.

42 Melendez-Colon, V.J., Luch, A., Seidel, A., and Baird, W.M. (2000) Formation of stable DNA adducts and apurinic sites upon metabolic activation of bay and fjord region polycyclic aromatic hydrocarbons in human cell cultures. *Chem. Res. Toxicol.*, **13**, 10–17.

43 De Bont, R. and van Larebeke, N. (2004) Endogenous DNA damage in humans: a review of quantitative data. *Mutagenesis*, **19**, 169–185.

44 Denison, M.S., Fisher, J.M., and Whitlock, J.P. Jr. (1988) The DNA recognition site for the dioxin–Ah receptor complex. Nucleotide sequence and functional analysis. *J. Biol. Chem.*, **263**, 17221–17224.

45 Sutter, T.R., Tang, Y.M., Hayes, C.L., Wo, Y.-Y.P., Jabs, W., Li, X., Yin, H., Cody, C.W., and Greenlee, W.F. (1994) Complete cDNA sequence of a human dioxin-inducible mRNA identifies a new gene subfamily of cytochrome P450 that maps to chromosome 2. *J. Biol. Chem.*, **269**, 13092–13099.

46 Nordqvist, M., Thakker, D.R., Vyas, K.P., Yagi, H., Levin, W., Ryan, D.E., Thomas, P.E., Conney, A.H., and Jerina, D.M. (1981) Metabolism of chrysene and phenanthrene to bay region diol-epoxides by rat liver enzymes. *Mol. Pharmacol.*, **19**, 168–178.

47 Wood, A.W., Levin, W., Ryan, D., Thomas, P.E., Yagi, H., Mah, H.D., Thakker, D.R., Jerina, D.M., and Conney, A.H. (1977) High mutagenicity of metabolically activated chrysene-1,2-dihydrodiol. Evidence for bay region activation of chrysene. *Biophys. Res. Commun.*, **78**, 847–854.

48 Levin, W., Thakker, D.R., Wood, A.W., Chang, R.L., Lehr, R.E., Jerina, D.M., and Conney, A.H. (1978) Evidence that benz[a]anthracene-3,4-diol-1,2-epoxide is an ultimate carcinogen on mouse skin. *Cancer Res.*, **38**, 1705–1710.

49 Melikian, A.A., LaVoie, E.J., Hecht, S.S., and Hoffmann, D. (1982) Influence of a bay-region methyl group on formation of 5-methyl-chrysene dihydrodiol epoxide: DNA adducts in mouse skin. *Cancer Res.*, **42**, 1239–1242.

50 Slaga, T.J., Gleason, G.L., DiGiovanni, J., Sukumaran, K.B., and Harvey, R.G. (1979) Potent tumor initiating activity of the 3,4-dihydrodiol of 7,12-dimethylbenz[a]anthracene in mouse skin. *Cancer Res.*, **39**, 1934–1936.

51 Agarwal, R., Coffing, S.L., Baird, W.M., Kiselyov, A.S., Harvey, R.G., and Dipple, A. (1997) Metabolic activation of benzo[g]chrysene in the human mammary carcinoma cell line MCF-7. *Cancer Res.*, **57**, 415–419.

52 Shukla, R., Jelinsky, S., Liu, T., Geactinov, N.E., and Loechler, E.L. (1997) How stereochemistry affects mutagenesis by N^2-deoxyguanosine adducts of 7,8-dihydroxy-9,10-epoxy-7,8,9,10-tetrahydrobenzo[a]pyrene: configuration of the adduct bond is more important than those of hydroxyl groups. *Biochemistry*, **36**, 13263–132639.

53 Zhang, N., Lin, C., Huang, X., Kolbanovskiy, A., Hingerty, B.E., Amin, S., Broyde, S., Geactinov, N.E., and Patel, D.J. (2005) Methylation of cytosine at C5 in a CpG sequence context causes a conformational switch of a benzo[a]pyrene diol-epoxide-N^2-guanine adduct in DNA from a minor grove alignment to intercalation with base displacement. *J. Mol. Biol.*, **346**, 951–965.

54 Jennette, K.W., Jeffery, A.M., Blobstein, S.H., Beland, F.A., Harvey, R.G., and Weinstein, I.B. (1977) Nucleoside adducts from the *in vitro* reaction of benzo[a]pyrene-7,8-dihydrodiol-9,10-oxide or benzo[a]pyrene-4,5-oxide with nucleic acids. *Biochemistry*, **16**, 932–938.

55 Smolarek, T.A. and Baird, W.M. (1984) Benzo[a]pyrene-induced alterations in the binding of benzo[a]pyrene to DNA in hamster embryo cell cultures. *Carcinogenesis*, **5**, 1065–1069.

56 Baird, W.M., Lau, H.H., Schmerold, I., Cofing, S.L., Brozich, S.L., Lee, H., and Harvey, R.G. (1993) Analysis of polycyclic aromatic hydrocarbon-DNA adducts by post-labeling with the β-emittters ^{35}S-phosphorothioate and ^{32}P-phosphate, immobilized boronate chromatography and high performance liquid chromatography. *IARC Sci. Publ.*, **124**, 217–226.

57 Kleiner, H.E., Reed, M.J., and DiGiovanni, J. (2003) Naturally occurring coumarins inhibit human cytochrome P450 and block benzo[a]pyrene and 7,12-dimethylbenz[a]anthracene DNA adduct formation in MCF-7 cells. *Chem. Res. Toxicol.*, **16**, 415–422.

58 Ruan, Q., Kim, H.-Y.H., Jiang, H., Penning, T.M., Harvey, R.G., and Blair, I.A. (2006) Quantification of benzo[a]pyrene diol epoxide DNA-adducts by

stable isotope dilution liquid chromatography/tandem mass spectrometry. *Rapid Commun. Mass Spectrom.*, **20**, 1369–1380.
59 Ruan, Q., Gelhaus, S.L., Penning, T.M., Harvey, R.G., and Blair, I.A. (2007) Aldo-keto reductase- and cytochrome P450-dependent formation of benzo[a]pyrene-derived DNA adducts in human bronchoalveolar cells. *Chem. Res. Toxicol*, **20**, 424–431.
60 Koreeda, M., Moore, P.D., Wislocki, P.G., Levin, W., Conney, A.H., Yagi, H., and Jerina, D.M. (1978) Binding of benzo[a]pyrene-7,8-diol-9,10-epoxides to DNA, RNA and protein of mouse skin occurs with high stereoselectivity. *Science*, **199**, 778–781.
61 Nesnow, S., Ross, J.A., Mass, M.J., and Stoner, G.D. (1998) Mechanistic relationships between DNA adducts, oncogene mutations and lung tumorigenesis in strain A mice. *Exp. Lung Res.*, **24**, 395–495.
62 Motoykiewicz, G., Malusecka, E., Grzybowska, E., Chorazy, M., Zhang, Y.J., Perera, F.P., and Santella, R.M. (1995) Immunohistochemical quantitation of polycyclic aromatic hydrocarbon-DNA adducts in human lymphocytes. *Cancer Res.*, **55**, 1417–1422.
63 Reddy, M., Gupta, R.C., Randerath, E., and Randerath, K. (1984) ^{32}P-Post-labeling test for covalent DNA binding of chemicals *in vivo*: application to a variety of aromatic carcinogens and methylating agents. *Carcinogenesis*, **5**, 231–243.
64 Santella, R.M., Yang, X.Y., Hsieh, L.L., and Young, T.L. (1990) Immunologic methods for the detection of carcinogen DNA adducts in humans. *Prog. Clin. Res.*, **340C**, 247–257.
65 Pavanello, S., Favretto, D., Brugnone, F., Mastrangelo, G., Pra, G.D., and Clonfero, E. (1999) HPLC/fluorescence determination of *anti*-BPDE-DNA adducts in mononuclear white blood cells from PAH-exposed humans. *Carcinogenesis*, **20**, 431–435.
66 Hecht, S., Chen, M., Yagi, H., Jerina, D.M., and Carmella, S.G. (2003) r-1,t-2,3,c-4-tetrahydroxy-1,2,3,4-tetrahydrophenanthrene in human urine: a potential biomarker for assessing polycyclic aromatic hydrocarbon metabolic activation. *Cancer Epidemiol. Biomarkers Prev. Res.*, **12**, 1501–1508.
67 Beland, F.A., Churchwell, M.L., Von Tungeln, L.S., Che, S., Fu, P.P., Culp, S.J., Schoket, B., Gyorffy, E., Mianarovits, J., Poirier, M.C., Bowman, E.D., Weston, A., and Doergee, D.R. (2005) High-performance liquid chromatography electrospray ionization tandem mass spectrometry for the detection and quantitation of benzo[a]pyrene-DNA adducts. *Chem. Res. Toxicol.*, **18**, 1306–1315.
68 Malaveille, C., Kuroki, T., Sims, P., Grover, P.L., and Bartsch, H. (1977) Mutagenicity of isomeric diol-epoxides of benzo[a]pyrene and benz[a]anthracene in *S. typhimurium* TA98 and TA100 and in V79 Chinese hamster cells. *Mutat. Res.*, **44**, 313–326.
69 Kapitulnik, J., Wislocki, P.G., Levin, W., Yagi, H., Jerina, D.M., and Conney, A.H. (1978) Tumorigenicity studies with diol-epoxides of benzo[a]pyrene which indicate that (+)-*trans*-7β,8α-dihydroxy-9α,10α-epoxy-7,8,9,10-tetrahydrobenzo[a]pyrene is an ultimate carcinogen in newborn mice. *Cancer Res.*, **38**, 354–358.
70 Marshall, C.J., Vousden, K.H., and Phillips, D.H. (1984) Activation of c-Ha-*ras*-1 proto-oncogene by *in vitro* chemical modification with a chemical carcinogen, benzo[a]pyrene diol-epoxide. *Nature*, **310**, 585–589.
71 Denissenko, M.F., Pao, A., Tang, M.-S., and Pfeifer, G.P. (1996) Preferential formation of benzo[a]pyrene adducts at lung cancer mutational hotspots in *p53*. *Science*, **274**, 430–432.
72 Hanawalt, P.C., Ford, J.M., and Lloyd, D.R. (2003) Functional characterization of global genomic DNA repair and its implications for cancer. *Mutat. Res.*, **54**, 107–114.
73 Fousterri, M. and Mullenders, L.H. (2008) Transcription-coupled nucleotide excision repair in mammalian cells: molecular mechanisms and biological effects. *Cell Res.*, **18**, 73–84.
74 McCulloch, S.D. and Kunkel, T.A. (2008) The fidelity of DNA synthesis by

74 eukaryotic replicative and translesion synthesis polymerases. *Cell Res.*, **18**, 148–161.
75 Yang, W. and Woodgate, R. (2007) What a difference a decade makes: insights into translesion DNA synthesis. *Proc. Natl. Acad. Sci. USA*, **104**, 15591–15598.
76 Zhao, B., Wang, J., Geactinov, N.E., and Wang, Z. (2006) Pol η, pol ζ and Rev1 together are required for G to T transversion mutations induced by the (+)- and (–)-*trans-anti*-BPDE-N^2-dG DNA adducts in yeast cells. *Nucleic Acid Res.*, **34**, 417–425.
77 Schcahar, S., Ziv, O., Avkin, S., Adar, S., Wittschieben, J., Reibner, T., Chaney, S., Friedberg, E.C., Wang, Z., Carell, T., Geacintov, N., and Lineh, Z. (2009) Two-polymerase mechanisms dictate error-free and error-prone translesion DNA synthesis in mammals. *EMBO J.*, **28**, 383–393.
78 Uno, S., Dalton, T.P., Shertzer, H.G., Genter, M.B., Warshawsky, D., Talaska, G., and Nebert, D.W. (2001) Benzo[a]pyrene-induced toxicity: paradoxical protection in $Cyp1a1^{-/-}$ knockout mice having increased hepatic B[a]P-DNA adduct levels. *Biochem. Biophys. Res. Commun.*, **289**, 1049–1056.
79 Uno, S., Dalton, T.P., Derkenne, S., Curran, C.P., Miller, M.L., Shertzer, H.G., and Nebert, D.W. (2004) Oral exposure to benzo[a]pyrene in the mouse: detoxication by inducible cytochrome P450 is more important than metabolic activation. *Mol. Pharmacol.*, **65**, 1225–1237.
80 Uno, S., Dalton, T.P., Dragin, N., Curran, C.P., Derkenne, S., Miller, M.L., Shertzer, H.G., Gonzalez, F.J., and Nebert, D.W. (2006) Oral benzo[a]pyrene in *Cyp1* knockout mouse lines: CYP1A1 important in detoxication, CYP1B1 metabolism required for immune damage independent of total body burden and clearance rate. *Mol. Pharmacol.*, **69**, 1103–1114.
81 Frenkel, K. (1992) Carcinogen-mediated oxidant formation and oxidative DNA damage. *Pharmacol. Ther.*, **53**, 127–166.
82 Frenkel, K., Wei, L., and Wei, H. (1995) 7,12-Dimethylbenz[a]anthracene induces oxidative DNA modification *in vivo*. *Free Radic. Biol. Med.*, **19**, 373–380.
83 Quinn, A., Harvey, R.G., and Penning, T.M. (2008) Oxidation of PAH *trans*-dihydrodiols by human aldo-keto reductase AKR1B10. *Chem. Res. Toxicol.*, **21**, 2207–2215.
84 Smithgall, T.E., Harvey, R.G., and Penning, T.M. (1988) Spectroscopic identification of *ortho*-quinones as the products of polycyclic aromatic *trans*-dihydrodiol oxidation catalyzed by dihydrodiol dehydrogenase. A potential route of proximate carcinogen metabolism. *J. Biol. Chem.*, **263**, 1814–1820.
85 Penning, T.M., Ohnishi, S.T., Ohnishi, T., and Harvey, R.G. (1996) Generation of reactive oxygen species during the enzymatic oxidation of polycyclic aromatic hydrocarbon *trans*-dihydrodiols catalyzed by dihydrodiol dehydrogenase. *Chem. Res. Toxicol.*, **9**, 84–92.
86 Flowers-Geary, L., Harvey, R.G., and Penning, T.M. (1992) Examination of polycyclic aromatic hydrocarbon o-quinones produced by dihydrodiol dehydrogenase as substrates for redox-cycling in rat liver. *Biochem. (Life. Sci. Adv.)*, **11**, 49–58.
87 Flowers-Geary, L., Harvey, R.G., and Penning, T.M. (1993) Cytotoxicity of polycyclic aromatic hydrocarbon o-quinones in rat and human hepatoma cells. *Chem. Res. Toxicol.*, **6**, 252–260.
88 Flowers-Geary, L., Bleczinski, W., Harvey, R.G., and Penning, T.M. (1996) Cytotoxicity and mutagenicity of polycyclic aromatic hydrocarbon o-quinones produced by dihydrodiol dehydrogenase. *Chem. Biol. Inter.*, **99**, 55–72.
89 Bolton, J.L., Psiha, E., Zhang, F., and Qiu, S. (1998) Role of quinoids in estrogen carcinogenesis. *Chem. Res. Toxicol.*, **11**, 13–27.
90 Zhang, F., Chen, Y., Pisha, E., Shen, L., Xiong, Y., van Breeman, R.B., and Bolton, J.L. (1999) The major metabolite of equilin, 4-hydroxyequilin, autoxidizes to an o-quinone which isomerizes to the potent cytotoxin 4-hydroxyequilenin o-quinone. *Chem. Res. Toxicol.*, **12**, 204–213.

91 Zhang, F., Swanson, S.M., van Breemen, R.B., Liu, X., Yang, Y., Gu, C., and Bolton, J.L. (2001) Equine estrogen metabolite 4-hydroxyequilenin induces DNA damage in rat mammary tissues: formation of single-strand breaks, apurinic sites and stable adducts and oxidized bases. *Chem. Res. Toxicol.*, **14**, 1654–1659.

92 Cavalieri, E.L. and Rogan, E.G. (2004) A unifying mechanism in the initiation of cancer and other diseases by catechol quinones. *Ann. NY Acad. Sci.*, **1028**, 247–257.

93 Yue, W., Santen, R.J., Wang, J.P., Li, Y., Verdearme, M.F., Bocchinfuso, W.P., Korach, K.S., Devanesan, P., Todorovic, R., Rogan, E.G., and Cavalieri, E.L. (2003) Genotoxic metabolites of estradiol in breast: potential mechanism of estradiol induced carcinogenesis. *J. Steroid Biochem. Mol. Biol.*, **86**, 477–486.

94 Shou, M., Harvey, R.G., and Penning, T.M. (1993) Reactivity of benzo[a]pyrene-7,8-dione with DNA. Evidence for the formation of deoxyguanosine adducts. *Carcinogenesis*, **14**, 475–482.

95 Balu, N., Padgett, W.T., Lambert, G.R., Swank, A.E., Richard, A.M., and Nesnow, S. (2004) Identification and characterization of novel stable deoxyguanosine and deoxyadenosine adducts of benzo[a]pyrene-7,8-quinone from reactions at physiological pH. *Chem. Res. Toxicol.*, **17**, 827–838.

96 Balu, N., Padgett, W.T., Nelson, G.B., Lambert, G.R., Ross, J.A., and Nesnow, S. (2006) Benzo[a]pyrene-7,8-quinone-3′-mononucleotide adduct standards for ^{32}P post-labeling analyses: detection of benzo[a]pyrene-7,8-quinone-calf thymus DNA adducts. *Anal. Biochem.*, **355**, 213–223.

97 McCoull, K.D., Rindgen, D., Blair, I.A., and Penning, T.M. (1999) Synthesis and characterization of polycyclic aromatic hydrocarbon o-quinone depurinating N7-guanine adducts. *Chem. Res. Toxicol.*, **12**, 237–246.

98 Yu, D., Berlin, J.A., Penning, T.M., and Field, J.M. (2002) Reactive oxygen species generated by PAH o-quinones cause change-in-function mutations in p53. *Chem. Res. Toxicol.*, **15**, 832–842.

99 Shen, Y., Troxel, A.B., Vedantam, S., Penning, T.M., and Field, J.M. (2006) Comparison of p53 mutations induced by PAH o-quinones with those caused by anti-benzo[a]pyrene-diol-epoxide *in vitro*: role of reactive oxygen and biological selection. *Chem. Res. Toxicol.*, **19**, 1441–1450.

100 Park, J.-H., Gelhaus, S., Vedantam, S., Olivia, A., Batra, A., Blair, I.A., Field, J., and Penning, T.M. (2008) The pattern of p53 mutations caused by PAH o-quinones is driven by 8-oxo-dGuo formation while the spectrum of mutations is determined by biological selection for dominance. *Chem. Res. Toxicol.*, **21**, 1039–1049.

101 Breen, A.P. and Murphy, J.P. (1995) Reactions of oxyl radicals with DNA. *Free Radic. Biol. Med.*, **18**, 1033–1077.

102 Park, J.-H., Gopishetty, S., Szewczuk, L.M., Troxel, A.B., Harvey, R.G., and Penning, T.M. (2005) Formation of 8-oxo-7,8-dihydro-2′-deoxyguanosine (8-oxo-dGuo) by PAH o-quinones: involvement of reactive oxygen species and copper(II)/copper(I) redox cycling. *Chem. Res. Toxicol.*, **18**, 1026–1037.

103 Flowers, L., Ohnishi, S.T., and Penning, T.M. (1997) DNA strand scission by polycyclic aromatic hydrocarbon o-quinones: role of reactive oxygen species Cu(II)/Cu(I) redox cycling, and o-semiquinone anion radicals. *Biochemistry*, **36**, 8640–8648.

104 Park, J.-H., Troxel, A.B., Harvey, R.G., and Penning, T.M. (2006) Polycyclic aromatic hydrocarbon (PAH) o-quinones produced by the aldo-keto reductases (AKRs) generate abasic sites, oxidized pyrimidines, and 8-oxo-dGuo via reactive oxygen species. *Chem. Res. Toxicol.*, **19**, 719–728.

105 Park, J.H., Mangal, D., Tacka, K.A., Quinn, A.M., Harvey, R.G., Blair, I.A., and Penning, T.M. (2008) Evidence for the aldo-keto reductase pathway of polycyclic aromatic *trans*-dihydrodiol activation in human lung A549 cells. *Proc. Natl. Acad. Sci. USA*, **105**, 6846–6851.

106 Mangal, D., Vudathala, D., Park, J.-H., Lee, S.H., Penning, T.M., and Blair, I.A. (2009) Analysis of 7,8-dihydro-8-oxo-2′-deoxyguanosine in cellular DNA during

oxidative stress. *Chem. Res. Toxicol.*, 788–797.
107 Shibutani, S., Takeshita, M., and Grollman, A.P. (1991) Insertion of specific bases during DNA synthesis past the oxidation-damaged base 8-oxo-dG. *Nature*, **349**, 431–434.
108 Henderson, P.T., Delaney, J.C., Gu, F., Tannenbuam, S.R., and Essigmann, J.M. (2002) Oxidation of 7,8-dihydro-8-oxoguanine affords lesions that are potent sources of replication errors *in vivo*. *Biochemistry*, **41**, 914–921.
109 Zharkov, D.O. (2008) Base excision DNA repair. *Cell. Mol. Life Sci.*, **65**, 1544–1565.
110 Fromme, J., Banerjee, A., Huang, S.J., and Verdine, G.L. (2004) Structural basis for removal of adenine mispaired with 8-oxoguanine by MutY adenine DNA glycosylase. *Nature*, **427**, 652–656.
111 Lu, R., Nash, H.M., and Verdine, G.L. (1997) A mammalian DNA repair enzyme that excises oxidatively damaged guanines maps to a locus frequently lost in lung cancer. *Curr. Biol.*, **7**, 397–407.
112 Chevillard, S., Radicella, J.P., Levalois, C., Lebeau, J., Poupon, M.F., Oudard, S., Dutrillaux, B., and Boiteux, S. (1998) Mutations in *OGG1*, a gene involved in the repair of oxidative DNA damage, are found in human lung and kidney tumors. *Oncogene*, **16**, 3083–3086.
113 Wikman, H., Risch, A., Klimek, F., Schmezer, P., Spiegelhalder, B., Dienemann, H., Kayser, K., Schulz, V., Drings, P., and Bartsch, H. (2000) hOOG1 polymorphism and loss of heterozygosity (LOH): significance for lung cancer susceptibility in a Caucasian population. *Int. J. Cancer*, **88**, 932–937.
114 Zhou, X., Taghizadeh, K., and Dedon, P.C. (2005) Chemical and biological evidence for base propenals as the major source of the endogenous M_1dG adduct in cellular DNA. *J. Biol. Chem.*, **280**, 25377–25382.
115 Blair, I.A. (2008) DNA adducts with lipid peroxidation products. *J. Biol. Chem.*, **283**, 15545–15549.
116 Luo, W., Muller, J.G., Rachin, E.M., and Burrows, C.J. (2000) Characterization of spiroiminodihydantoin as a product of one-electron oxidation of 8-oxo-7,8-dihydroguanosine. *Org. Lett.*, **2**, 613–616.
117 Ye, Y., Muller, J.G., Luo, W., Mayne, C.L., Shallop, A.J., Jones, R.A., and Burrows, C.J. (2003) Formation of ^{13}C, ^{15}N, and ^{18}O-labeled guandinohydantoin from guanosine oxidation with singlet oxygen. Implications for structure and mechanism. *J. Am. Chem. Soc.*, **125**, 13926–13927.
118 Smithgall, T.E., Harvey, R.G., and Penning, T.M. (1986) Regio- and stereospecificity of homogeneous 3α-hydroxysteroid/dihydrodiol dehydrogenase for *trans*-dihydrodiol metabolites of polycyclic aromatic hydrocarbons. *J. Biol. Chem.*, **261**, 6184–6191.

7
Aromatic Amines and Heterocyclic Aromatic Amines: From Tobacco Smoke to Food Mutagens
Robert J. Turesky

7.1
Introduction

Aromatic amines (AAs) and heterocyclic aromatic amines (HAAs) are ubiquitous environmental and dietary contaminants, many of which are carcinogens. Compounds of both classes contain an exocyclic amino group, which is a prerequisite for their genotoxicity (Figure 7.1). In this chapter, the sources of exposure, mechanisms of metabolic activation, formation and detection of DNA adducts, and biological effects of AA-DNA and HAA-DNA adducts are highlighted.

7.2
Exposure and Cancer Epidemiology

The dye, chemical, and rubber-manufacturing industries were major sources of occupational exposure to AAs, until the first half of the twentieth century [1]. AAs, such as aniline, were key intermediates in the developing textile dye industry in Europe during the 1800s. The occurrence of urinary bladder tumors among workers in dyestuff factories was first reported by Rehn in 1895 [2], who attributed these cancers to the patients' occupation, from which evolved the term aniline cancer [1]. Aniline, however, was not carcinogenic in experimental animals, but 4-aminobiphenyl (ABP), 2-naphthylamine (2-NA), and benzidine (Bz), contaminants in aniline dyes, were shown to be carcinogenic [1]. The liver, intestine, urinary bladder, and female mammary gland are target organs of cancer development in rodents or dogs exposed to these compounds. Numerous epidemiological studies have demonstrated that occupational exposure to AAs, including ABP, 2-NA, and Bz, is a cause of bladder cancer in humans [3–5]. 2-Aminofluorene (AF) and N^2-acetylaminofluorene (AAF) are perhaps the most well-studied among the AAs; these compounds were originally developed as pesticides but never used as intended because they were discovered to be animal carcinogens [3]. The biochemistry and biological effects of AF and AAF, and the genotoxic properties of their DNA adducts, have been extensively reviewed [6–8].

The Chemical Biology of DNA Damage. Edited by Nicholas E. Geacintov and Suse Broyde
© 2010 WILEY-VCH Verlag GmbH & Co. KGaA, Weinheim
ISBN: 978-3-527-32295-4

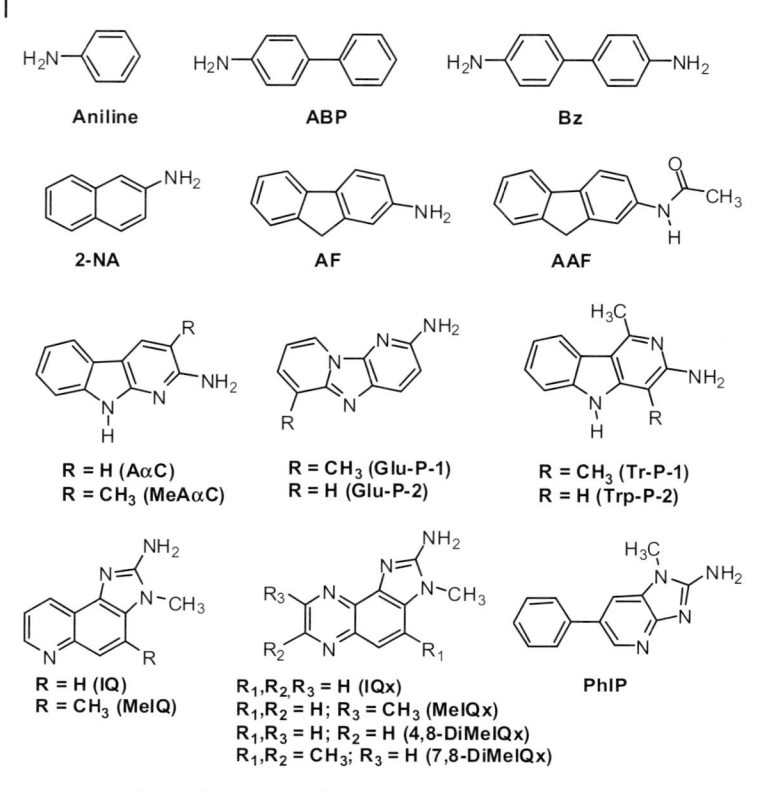

Figure 7.1 Chemical structures of common AAs and HAAs.

Although the amounts of carcinogenic AAs produced in dyestuff, chemical, and rubber factories have been greatly reduced, exposure to AAs and their oxidized nitroaromatic derivatives still occurs. Some of these compounds can be found in various color additives [9, 10], paints [11], food colors [12], leather and textile dyes [13, 14], fumes from heated cooking oils [15] and fuels [16], and tobacco smoke [17]. Several AAs have also recently been identified at trace levels in commercial hair dyes [18, 19].

The discovery of HAAs is more recent and occurred 30 years ago when researchers in Japan discovered these substances (reviewed in [20]), and showed that the charred parts and smoke generated from broiled fish and beef contained substances that exhibited potent activities in *Salmonella typhimurium*-based mutagenicity assays. More than 20 HAAs are formed in meats, fish, and poultry that result from common household cooking practices; HAA levels can range from below 1 to greater than 500 ppb [21, 22]. There are two major classes of HAAs. The "pyrolytic HAAs" arise during the high-temperature pyrolysis (above 250 °C) of some individual amino acids, including glutamic acid and tryptophan, or during the pyrolysis of proteins [21]. HAAs of the second class, aminoimidazoarenes (AIAs), occur in meats cooked at temperatures commonly used in household kitchens (150–250 °C). The Maillard reaction plays an important role in the formation of AIAs [23]; creatine, free amino acids, and hexoses present in uncooked meats are the precursors of AIAs. Both classes of HAAs have been identified in tobacco smoke condensate [24].

Many of the HAAs studied induce tumors at multiple sites in rodents during long-term feeding studies. The target organs include the liver, stomach, colon, and pancreas, and prostate in males, and mammary gland in females [21, 25]. A number of epidemiological studies have reported positive correlations between the frequency of eating well-done cooked meats and human cancers in these target organs, suggesting that HAAs are multisite carcinogens in humans as well [22].

7.3
Enzymes of Metabolic Activation and Genetic Polymorphisms

AAs and HAAs are mutagenic to bacteria and to mammalian cells. The use of various cell lines transfected with recombinant phase I and phase II enzymes has helped to elucidate critical enzymes involved in bioactivation [26]. Metabolic activation occurs by cytochrome P450-mediated N-oxidation of the exocyclic amino group, to form the N-hydroxy-AAs and -HAAs [27, 28]. This reaction is catalyzed principally by hepatic P450 1A2, and by P450 1A1 and 1B1 in extrahepatic tissues ([28, 29] and references therein). Peroxidases can also contribute to bioactivation; these enzymes are thought to be of particular importance in breast tissue [30]. The N-hydroxy-AA and -HAA metabolites can react directly with DNA, or else can undergo further biotransformation by sulfotransferases (SULTs) or N-acetyltransferases (NATs) to produce unstable esters. These esters are the penultimate metabolites; they undergo heterolytic cleavage to produce nitrenium ions that form adducts with DNA (Figure 7.2) [31, 32].

Genetic polymorphisms in the enzymes catalyzing the activation and/or detoxification of AAs and HAAs may account for interindividual differences in susceptibility to these carcinogens [33]. Humans show a large interindividual variation in the expression of P450 1A2. Environmental and dietary factors [34, 35], varying levels of CpG methylation [36], and genetic polymorphisms of the upstream 5′-regulatory region of the P450 1A2 gene [37, 38] alter the levels of P450 1A2 mRNA expression, and the level of expression of the P450 1A2 protein can vary by a greater than 60-fold range [39, 40]. The catalytic efficiency of human P450 1A2 is vastly superior to the efficacy of the rat ortholog, in effecting the N-oxidation of 2-amino-3,4-dimethylimidazo[4,5-f]quinoxaline (MeIQx) and 2-amino-1-methyl-6-phenylimidazo[4,5-b]pyridine (PhIP), to produce genotoxicants [40]. Moreover, human P450 1A2 poorly catalyzes the detoxication of these HAAs by ring-oxidation, while rat P450 1A2 carries out ring-oxidation reactions as the major pathways of metabolism [41]. The interspecies differences in enzyme expression and catalysis must be considered, when the human health risks of these procarcinogens are assessed.

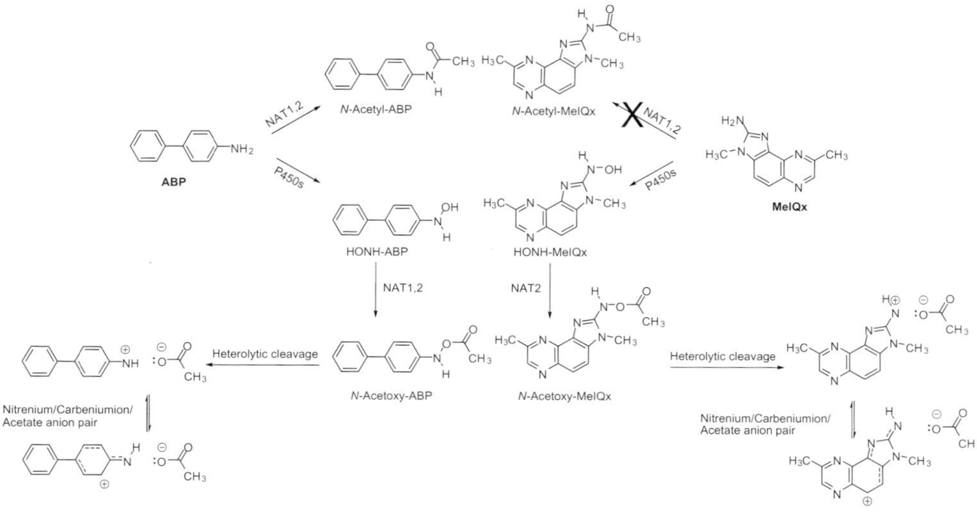

Figure 7.2 Metabolism of ABP and MeIQx as prototypes of AAs and HAAs. NAT enzymes effectively detoxicate AAs, but not HAAs, by N-acetylation, which is viewed as an important detoxication pathway. NATs also catalyze the formation of reactive N-acetoxy esters of N-hydroxy-AAs and -HAAs, which are formed by P450s. The N-acetoxy metabolites undergo heterolytic cleavage to produce the nitrenium/carbenium ion/acetate anion pair, which binds to DNA. The dG-N^2 adducts have been reported to involve a covalent bond at the C3 atom of 4-ABP, while the dG-N^2 and dA-N^6 adducts involve the C5 atom of MeIQx and IQ, indicating charge delocalization of the nitrenium over the AA and HAA moieties.

There are two distinct NAT isoenzymes (designated NAT1 and NAT2). NAT2 is expressed primarily in the liver, whereas NAT1 appears to be more prominently expressed in extrahepatic tissues HAAs [42, 43]. More than 20 genetic polymorphisms have been identified for both *NAT* genes that can affect the catalytic activity of NATs toward AAs and HAAs [42, 44]. NAT enzymes have a dual role in the metabolism of AAs and HAAs. The N-acetylation of aromatic monoamines is catalyzed by both NAT isoforms and is considered to be a detoxification reaction; the resulting acetamides are generally poor substrates for P450-mediated N-oxidation (Figure 7.2) [45]. Bz, an aromatic diamine, is an exception. N-Acetylation of one of the amine groups of Bz facilitates P450-mediated N-oxidation of the nonacetylated amine group, to form the reactive N′-acetyl-N-hydroxyamino-benzidine (N-acetyl-HONH-Bz) metabolite. The N-hydroxy-AAs are also substrates for NATs and undergo O-acetylation to form the reactive N-acetoxy intermediates, which bind to DNA [46]. Many HAAs do not undergo N-acetylation by either NAT isoform, but the HONH-HAA metabolites undergo O-acetylation, primarily by NAT2, to form the N-acetoxy species, which binds to DNA (Figure 7.2) [32].

The relationship between NAT2 polymorphisms and bladder cancer risk of AAs is well documented – the elevated cancer risk has been attributed to the lower capacity of slow N-acetylator individuals to detoxicate AAs [47]. However, the epidemiological data on the relationship between NAT2 genetic polymorphism in susceptibility to various cancers related to HAA exposures are inconsistent [42, 48]. Since both phase I and phase II enzymes are required to bioactivate HAAs, the risk may be particularly elevated in individuals who are both rapid N-oxidizers and rapid O-acetylators [49, 50].

The SULTs belong to a superfamily of genes that is divided into two subfamilies: the phenol SULTs (SULT1), and the hydroxysteroid SULTs (SULT2). A third, less well-characterized subfamily of brain-specific SULT has also been documented [51]. SULT1A1 and SULT1A2 preferentially catalyze the sulfation of small planar phenols, such as 4-nitrophenol, and the estrogens, while SULT1A3 preferentially catalyzes the sulfate conjugation of monoamine derivatives. Human SULT1A1 and SULT1A2 catalyze the binding of the N-hydroxy metabolites of ABP, PhIP, and 2-amino-9H-pyrido[2,3-b]indole (AαC) to DNA; however, the N-hydroxy metabolites of MeIQx and 2-amino-3-methylimidazo[4,5-f]quinoline (IQ) are not substrates for either SULT [26]. SULT1E1, which is under hormonal regulation, catalyzes the binding of N-hydroxy-2-amino-1-methyl-6-phenylmidazo[4,5-b]pyridine (HONH-PhIP) to DNA in cultured human mammary cells; SULT1E1 may play a role in carcinogen bioactivation in breast tissue [52].

Epidemiological studies on genetic polymorphisms of SULTs and the cancer risk posed by AAs or HAAs have provided inconsistent results. A G → A transition at codon 213 (CGC/Arg to CAC/His) of the *SULT1A1* gene is a common genetic polymorphism [53]. The gene product of this allele has substantially lower enzyme activity than does the gene product of the wild-type allele. Women who harbor the *Arg/Arg* genotype (*SULT1A1*1*) and who frequently ate well-done cooked meats were found to be at an elevated risk for breast cancer (odds ratio 3.6); no such association was evident for women with the *His/His* genotype in this subpopulation of women who ate meat [54]. Thus, the *His/His* genotype, characterized by poor bioactivation of HONH-HAAs and -AAs [55], was protective against breast cancer in this exposed population. However, polymorphisms in *SULT1A1* and *SULT1A2* have not been demonstrated to influence colorectal risk [56], and the variant *SULT1A1*2* allele did not show an association with prostate cancer risk [57].

7.4
Reactivity of *N*-Hydroxy-AAs and *N*-Hydroxy-HAAs with DNA

The chemical reactivity of DNA differs between N-hydroxy-AAs and -HAAs. The covalent DNA binding of N-hydroxy-AAs with DNA is acid-catalyzed – adduct formation increases approximately 5- to 10-fold for each unit of pH decrease between pH 7 and 5. This enhanced reactivity at acidic pH has been ascribed to the formation of the nitrenium ion [31]. In contrast to arylhydroxylamines, the AIA-type N-hydroxy-HAAs show a reactivity with DNA that is only modestly enhanced under acidic conditions [58]. However, the addition of acetic anhydride to the reaction increases the level of adduct formation of some N-hydroxy-AIA-type of HAAs with DNA by more than 10-fold, suggesting that the reactive N-acetoxy intermediates are formed *in situ*. The N-acetoxy derivatives of IQ and MeIQx [58] have lifetimes of seconds or less, but N-acetoxy-PhIP has been isolated and characterized by mass spectrometry [59]. The imidazo moiety of AIAs can facilitate the formation of the oxime tautomer, which can influence the chemical reactivity of N-hydroxy-AIAs with DNA. (Oximes are R_1R_2(C=N–OH) compounds with R_1 and R_2 bound to the oxime carbon atom, where R_1 is an organic side-chain, while R_2 can be either a hydrogen atom or an organic side-chain.) The oxime structure favors O-acetylation of the N-hydroxy-AIAs by acetic anhydride, to produce the N-acetoxy intermediates, instead of N-acetylation to form the acetamides. Indeed, phase II glucuronide conjugates of the oximes [60] and conjugates of the imines of AIAs have been reported [61, 62].

7.5
Syntheses of AA-DNA and HAA-DNA Adducts

The first syntheses of AA- and HAA-DNA adducts were carried out by biomimetic reactions of the N-hydroxy-AA or -HAA intermediates with deoxynucleosides or DNA, in the presence of ketene gas, or under acidic pH conditions, or acetic anhydride, so as to facilitate the formation of the nitrenium ion. The yields of adducts were generally only a few percent or lower. The principal adducts are formed at the C8 atom of dG and the exocyclic amino groups of AAs and HAAs (Figure 7.3). The C8 position of dG is only weakly nucleophilic: dG-C8 adducts (and presumably dA-C8 adducts) are proposed to be rearrangement products that are preceded by electrophilic substitution at the nucleophilic N7 atom of dG (Figure 7.4) [63, 64]. Charge delocalization of the nitrenium ion over the aromatic and heterocyclic aromatic ring moieties of these carcinogens also occurs: the resultant carbenium ion resonance forms undergo nucleophilic attack by the 2-NH_2 group of dG and the 6-NH_2 group of adenine, to form ring-substituted adducts (Figure 7.3) [31, 65, 66].

Nonbiomimetic approaches have recently been used to synthesize AA- and HAA-DNA adducts in high yields. The key reaction step is the Buchwald–Hartwig reaction of cross-coupling of primary and secondary amines with aryl halides, which has been incorporated into the high-yield synthesis of dG-C8 and dG-N^2 adducts of IQ, N-(deoxyguanosin-8-yl)-PhIP (dG-C8-PhIP) [67–69], and dG and dA adducts of AAs [70]. The phosphoramidites of these adducts have been site-specifically incorporated into oligonucleotides, which have been used to explore the effect of adducts in perturbations of DNA structure and the fidelity of polymerases during translesion synthesis of these adducted oligonucleotides.

7.6
Biological Effects of AA-DNA and HAA-DNA Adducts

The conformational changes in DNA induced by aromatic amine–purine base modifications are important determinants of the adduct's biological activity and

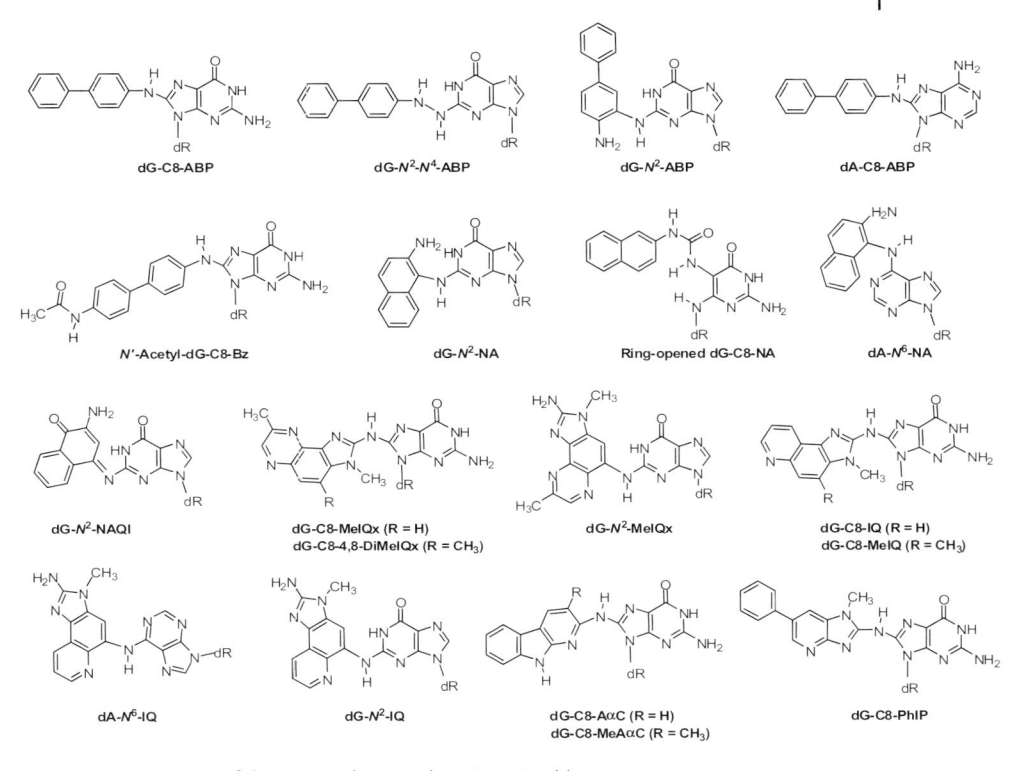

Figure 7.3 Structures of the principal AA- and HAA-DNA adducts.

Figure 7.4 Mechanisms of dG-C8-AA and HAA adduct formation via an N^7-dG intermediate. Two mechanisms of rearrangement of the putative N^7-dG adducts have been proposed [63, 64].

propensity to provoke frameshift mutations and base-pair substitutions during translesional synthesis [7, 71, 72]. The potential for an adduct to induce mutations or to block polymerase activity is a function not only of the adduct structure, but also of the sequence context of the oligonucleotide. The nuclear magnetic resonance (NMR) solution structures of DNA duplexes covalently adducted with AAs [73–75] and HAAs [67, 76, 77], and the biological effects and miscoding properties of these adducts during translesion syntheses with polymerases [6, 7, 71, 72, 78–80] have been extensively reviewed; these topics are also discussed in Chapter 10.

The structural conformation of the carcinogen adduct also influences the adduct's persistence and rate of removal *in vivo*. On the basis of NMR solution structural studies, the dG-C8-AA and dG-HAA adducts, in some oligonucleotide sequence contexts (discussed in greater detail in Chapter 10), adopt the glycosidic torsion angle in the *syn* conformation; that structural adaptation can result in greater perturbation of the DNA duplex at the site of carcinogen adduction than that seen for the glycosidic linkage of the dG-N^2 adducts, which often adopt the normally occurring *anti* conformation [67, 73–77]. The extent of structural perturbations induced by these adducts in duplex DNA is believed to result in variable levels of recognition and enzymatic removal of the adduct, by the nucleotide excision repair complex, as discussed by Cho in Chapter 10. The dG-C8 adducts of AAF and IQ, which exist in the *syn* conformation, are rapidly removed from tissues in rodents; however, 3-(deoxyguanosin-N^2-yl)-AAF (dG-N^2-AAF) and 5-(deoxyguanosin-N^2-yl)-IQ (dG-N^2-IQ), which occur in the *anti* conformation, persist [81, 82]. *N*-(Deoxyguanosin-8-yl)-IQ (dG-C8-IQ) was the major adduct formed in all tissues of nonhuman primates 24 h after a single dose of IQ. However, in nonhuman primates chronically treated with IQ for up to 3.6 years, the level of dG-C8-IQ only increased by several-fold in tissues composed of slowly dividing cells, while the level of dG-N^2-IQ increased by up to 90-fold so that this adduct became the major lesion [82]. Owing to their persistence, the dG-N^2 adducts may play a significant role in the tumorigenic properties of AAs and HAAs [82, 83].

7.7
Bacterial Mutagenesis

AAs are frameshift mutagens in Ames *S. typhimurium* tester strain TA1538 [31]. When revertants were expressed as a function of DNA binding, *N*-acetyl-HONH-Bz was the most mutagenic AA, followed by *N*-hydroxy-2-aminofluorene (HONH-AF), and then by *N*-hydroxy-2-naphthylamine (HONH-2-NA) and *N*-hydroxy-4-aminobiphenyl (HONH-ABP) (the latter two showed approximately the same number of revertants per adduct). In a similar study conducted with *S. typhimurium* strain TA 1535, which detects base-substitution revertants, only HONH-2-NA induced revertants [31]. ABP in the TA100 strain, also base substitution-specific and containing plasmid pKM101 (which enhances the mutagenic potential of some genotoxicants through error-prone repair DNA polymerases), induced revertants at levels 5- to 10-fold higher than the levels in the frameshift-specific TA98 strain [84].

2-Amino-3,4-dimethylimidazo[4,5-f]quinoline (MeIQ), IQ, and MeIQx rank among the most potent mutagens ever tested in the Ames bacterial reversion assay, while PhIP and AαC are, respectively, around 200- and 1000-fold weaker in potency [21]. Many HAA-DNA lesions are repaired, since their mutagenic potencies are 100-fold lower in the $uvrB^+$ proficient S. typhimurium strain [85]. HAAs preferentially induce frameshift mutations in S. typhimurium, but point mutations also occur. The high propensity of some HAAs to induce frameshift revertant mutations in S. typhimurium TA98 and TA1538 tester strains is attributed to a preference by these compounds to react at a site about 9 bp upstream of the original CG deletion in the $hisD^+$ gene, within a run of GC repeats [86]. This nature of this "hotspot" is consistent with the formation of dG-HAA adducts, which may lead to CG deletions during translesional DNA synthesis. The mutagenicity of HAAs in other bacterial genes such as the lacZ, lacZα, and lacI of Escherichia coli also reveals that mutations occur primarily at GC pairs. The types of mutations are dependent upon the DNA sequence context and assay system used ([87] and references within).

The mutagenic potencies of even some of the weaker HAAs can be increased by up to 250-fold in S. typhimurium TA1538/1,8-DNP-derived strains that have been engineered to express NAT or SULT proteins, or mammalian cells [26, 88], thereby demonstrating the importance of xenobiotic metabolism enzymes in the biological potencies of these genotoxicants. The range of genotoxic potencies of these HAAs is generally much narrower in mammalian cell assays than in the bacterial assays [26, 89]. The marked difference in genotoxic potencies is attributed to the differing metabolic activation systems, differing gene loci endpoints, and differing base sequence contexts and neighboring base effects on the lesions, all of which can affect the repair of these lesions (see Chapter 10) and mutation frequencies.

7.8
Mammalian Mutagenesis

In mammalian cells, base-pair substitutions at guanines are prominent mutations of AAs and HAAs; however, frameshift mutations at guanines also occur. These mutational events are consistent with the notion that guanine is a principal target for AA- and HAA-DNA adduct formation. PhIP-induced mutations at the hypoxanthine phosphoribosyltransferase (hprt) locus in human lymphoblastoid cells occur predominantly through GC $\rightarrow$ TA transversions [90], while CG $\rightarrow$ AT and GC $\rightarrow$ TA transversions have been reported at PhIP-induced mutants in the dihydrofolate reductase (dhfr) and adenine phosphoribosyltransferase (aprt) genes of CHO cells [91, 92]. PhIP also predominantly induced GC $\rightarrow$ TA transversions at the hprt locus in Chinese hamster V79 cells; however, 13% of the mutants displayed a −1 frameshift mutation in the 5′-GGGA-3′ sequence [93] – a sequence that is identical to a sequence mutated in the *Apc* tumor suppressor gene of rat colon tumors induced by PhIP [94]. IQ was also strongly mutagenic at the *hprt* locus in human lymphoblastoid cells [95]. The majority of the mutations occurred at GC base pairs, suggesting that either the dG-C8-IQ or dG-N^2-IQ is the premutagenic lesion. ABP and HONH-ABP also strongly induced mutations at the *hprt* locus in human uroepithelial cells; the nature of the mutants was not characterized [96].

7.9
Mutagenesis in Transgenic Rodents

The transgenic Muta Mouse and Big Blue mouse and rat models have been used to assess the mutagenicity of IQ [97, 98], MeIQ [99], MeIQx [100], PhIP [101], AαC [102], and ABP [103] in various organs, with either *lacZ* or *lacI* as the target genes. In studies on MeIQ, IQ, and PhIP, the mutations in the *lacI* gene were consistent with known mutations, in the Ha-*ras* and *Apc* genes in HAA-induced tumors of rodents [104, 105]. The induction of mutations in the *cII* transgene of liver and colon of IQ-treated rats [98], and in the small and large intestines of mice treated with MeIQ, PhIP, and AαC has also been reported [102, 106]: a high-fold induction of both GC → AT transitions and GC → TA transversions in this gene occurred in the colon of male rats and mice [107, 108], and was consistent with the near-exclusive reaction of each of these HAAs with dG [109]. Additionally, –G frameshifts in homopolymeric runs of guanine bases around the nucleotide 179 –(G:C) of the *cII* gene have been seen in the colon of rodents following PhIP treatment [107, 108]. The –G frameshift, or –1 deletion mutations, arise when the polymerase fails to insert any nucleotide opposite the lesion and continues synthesis using the next unmodified base as a template. The –G frameshift mutation in the GGGA sequence is a frequent signature mutation of PhIP in the *Apc* tumor suppressor [94] and *hprt* genes [93, 110], and in the *lacI* gene of the colon and mammary gland of Big Blue mice [104]; however, this sequence occurs only once in the *cII* gene and no mutations were observed at this position in the transgene.

Treatment with ABP significantly increased the frequency of mutations in the liver *cII* transgene in both genders of neonatal Big Blue B6C3F$_1$ transgenic mice, but not in the adult mice, following treatment with a regimen of ABP known to induce tumors in neonatal mice [111]. Sequence analysis of *cII* mutant DNA revealed that ABP induced a unique spectrum of mutations in the neonatal mice, characterized by a high frequency of GC → TA transversions, while the mutational spectrum in the ABP-treated adults was similar to that in the untreated mice. The authors suggested that neonates are more sensitive than adults to ABP, because the relatively high levels of cell division in the developing animal facilitate the conversion of DNA damage into mutations [111]. The mutational spectrum data are consistent with the data seen for the tetracycline resistance gene of the plasmid pBR322, where ABP modification resulted in GC → TA and GC → CG transversions; notably, the –G frameshift mutations occurred with a lower frequency [112].

Reporter transgenes can represent characteristic mutations that can also occur in cancer-related genes; however, relationships among DNA adduct formation levels, mutation frequencies, and cancer incidences of AAs and HAAs have not shown any quantitative correlations [99]. Moreover, mutations in transgenes also occur in organs that do not develop tumors. Some of the inconsistencies can be attributed to differences in cell proliferation rates among different organs that affect mutation frequencies. Additionally, the numbers of mutations and types of genetic alterations required for cancer development likely vary among different organs [99].

7.10
Genetic Alterations in Oncogenes and Tumor Suppressor Genes

Mutations have been observed in oncogenes and tumor suppressor genes, in various organs of experimental animals during long-term feeding studies with AAs or HAAs. These data have been summarized in references [105, 113, 114] and are only briefly reviewed here. In studies on ABP in the neonatal wild-type B6C3F$_1$ mouse, ABP preferentially induced CG → AT mutations (reflecting GC → TA transversions in the noncoding strand) in H-*ras* codon 61, followed in frequency by GC → CG mutations [111]. However, in CD-1 mice, ABP primarily induced AT → TA transversions in H-*ras* codon 61; this molecular feature is consistent with the formation of the *N*-(deoxyadenosin-8-yl)-ABP (dA-C8-ABP) adduct. In the C57B1/10J mouse strain, mutations in codon 61 of the H-*ras* gene were not significantly implicated in chemical induction of liver tumors by ABP [115], suggesting that the incidence of *ras* mutations in chemically induced mouse liver tumors is strain-dependent.

About 50% of human bladder cancers contain a mutation in the tumor suppressor *p53* gene [116]. The spectrum of mutations in the *p53* gene in smokers and nonsmokers with bladder cancer are base substitutions and occur at GC and AT base pairs. Codon 285 of the *p53* gene, a mutational hotspot at a non-CpG site in bladder cancer, was found to be the preferential binding site for HONH-ABP *in vitro* [117]. Moreover, C5 cytosine methylation greatly enhanced HONH-ABP binding at CpG sites, whereas two other mutational hotspots at CpG sites, codons 175 and 248, became preferential binding sites for HONH-ABP only after being methylated. The distribution of ABP–DNA adducts was mapped in the *p53* gene at the nucleotide-sequence level in human bladder cells (HTB-1) treated with HONH-ABP, and mutational hotspots in bladder cancer at codons 175, 248, 280, and 285 were found to be the preferential sites for ABP adduct formation [118]. The authors suggested that ABP contributes to the mutational spectrum in the *p53* gene of human bladder cancer; their data provided some molecular evidence that links ABP to bladder cancer. However, the roles of methylation status and transcriptional activity in the mutational spectrum in the *p53* gene induced by ABP have yet to be determined.

Genetic alterations in rat colon adenocarcinomas induced by IQ, PhIP, or the glutamic acid pyrolysate mutagen 2-amino-6-methyldiprido[1,2-*a*:3′,2′-*d*]imidazole were examined for *ras* family gene mutations, in rats. The K-*ras* mutations were rare and no mutations were detected in either the N-*ras* or Ha-*ras* genes for any of these tumors. Similarly, *p53* gene mutations were not detected in any rat colon tumors induced by these HAAs even though 60–70% of human colon cancers have mutations in the *p53* gene [105, 113]. Therefore, HAAs may represent suitable model compounds for investigations of sporadic colon carcinogenesis, which does not involve mutations in the *p53* gene. Sporadic mutations were found in either Ha-*ras* or Ki-*ras* and the *p53* genes in rat Zymbal gland tumors induced by several HAAs (see [105, 113]).

Mutations in the *APC* gene plays a major role in human colon carcinogenesis and mutation of this gene is considered as an initial or very early event in human colon carcinogenesis. Alterations of the *Apc* gene were more prominent in PhIP-induced than in IQ-induced carcinogenesis in the rat colon [94, 105, 113] and they featured a guanine deletion from 5′-GGGA-3′ sequences. This specific GC base-pair deletion in 5′-GTGGGA-3′ at codon 635 of the *Apc* gene was detected as an early mutation in the colon of male rats exposed to PhIP for only 1 week, when probed by the mismatch amplification mutation assay [119]. This 5′-GTGGGAT-3′ sequence around codon 635 in the rat is conserved in the human *Apc* gene and mutations at this locus may represent a signature mutation of PhIP in humans [94].

7.11
AA-DNA and HAA-DNA Adduct Formation in Experimental Animals and Methods of Detection

The first studies were conducted on the measurements of AA-DNA adducts in experimental animals and employed tritium-labeled carcinogens. Adduct identification was achieved by high-performance liquid chromatography with radiometric detection [83]. More recent methods for adduct detection and quantification include: ^{32}P-postlabeling [120]; immunohistochemistry (IHC) [121]; gas chromatography with negative ion chemical ionization-mass spectrometry (GC-NICI-MS) of alkaline-treated DNA – a technique that cleaves the bond between the guanyl C8 atom and the amino group of AAs or HAAs [122]; accelerator mass spectrometry (AMS) for the detection of tritiated or ^{14}C-labeled adducts [123]; and liquid chromatography/electrospray ionization/tandem MS (LC/ESI/MS-MS) methods [109, 124].

Five DNA adducts of 4-ABP are formed when HONH-ABP is reacted with calf thymus DNA at pH 5.0 [31] (Figure 7.3). The *N*-(deoxyguanosin-8-yl)-ABP (dG-C8-ABP) adduct is the principal one and accounts for 80% of the total adducts formed, followed by dA-C8-ABP (15% of total adducts) and then *N*-(deoxyguanosin-N^2-yl)-

ABP (dG-N^2-N^4-ABP) (around 5% of the total adducts) (Figure 7.3) The dG-N^2-N^4-ABP is unusual in that it contains a hydrazine linkage. Two other minor dG adducts have been identified: 3-(deoxyguanosin-N^2-yl)-ABP (dG-N^2-ABP) and N-(deoxyguanosin-N^2-yl)-4-azobiphenyl [125, 126].

DNA adducts of ^{3}H-labeled ABP were measured in urothelial cells of male Beagle dogs [83]: dG-C8-ABP accounted for 76% of the total binding, followed by dG-N^2-N^4-ABP (15%) and then by dA-C8-ABP (9%). These adducts were formed in target (urothelium) and nontarget (liver) tissues of dogs 2 days after the oral administration of [^{3}H]ABP [83]. DNA adducts of ABP were quantified, by ^{32}P-post-labeling and IHC, in liver and bladder of male and female BALB/c mice, following treatment with ABP at a range of concentrations (up to 220 ppm) in the drinking water for 28 days [127]. The principal adduct in both tissues, for both genders, was dG-C8-ABP. A comparison between DNA adduct formation and tumorigenesis indicated a linear correlation between adduct levels and incidence of liver tumors in female mice. The relationship between adducts and tumorigenesis was distinctly nonlinear in the bladders of male mice and tumor incidence rose rapidly at doses above the 50-ppm dose of ABP. Toxicity and cell proliferation may have increased the tumor incidence in the bladder.

Three DNA adducts are formed by the reaction of HONH-2-NA with DNA *in vitro*, at pH 5.0 (Figure 7.3) [31]. The major adduct was identified as an imidazole ring-opened derivative of N-(deoxyguanosin-8-yl)-2-NA (dG-C8-NA, 50% of the total adducts); there were lower levels of 1-(deoxyguanosin-N^2-yl)-2-NA (dG-N^2-NA, 30% of total adducts), and 1-(deoxyadenosin-N^6-yl)-2-NA (dA-N^6-NA, 15% of total adducts). These same three DNA adducts were formed in target (urothelium) and nontarget (liver) tissues of dogs 2 days after oral administration of [^{3}H]2-NA [83]. A 4-fold higher binding level of 2-NA was found in the urothelial DNA than in liver DNA. The major adduct in both tissues was the ring-opened dG-C8-NA; there were lower levels of dA-N^6-NA and dG-N^2-NA. The dG-N^2 adduct persisted in the liver, and both this adduct and the ring-opened dG-C8-NA adduct persisted in the bladder. The differential loss of adducts indicates that active repair processes are present in both tissues and the relative persistence of the ring-opened dG-C8-NA adduct in the target, but not the nontarget tissue, suggests that this adduct is a critical lesion for the initiation of urinary bladder tumors.

Peroxidative enzymes, such as prostaglandin H synthase (PHS), an arachidonic acid-dependent peroxidase, catalyze both the N-oxidation and ring-oxidation of 2-NA; a major ring-oxidation product is 2-amino-1-naphthol [128]. When PHS was used to catalyze the binding of 2-NA to DNA, the same three adducts arising from N-hydroxy-2-NA were detected. In addition, three other adducts were formed from 2-imino-1-naphthoquinone, the oxidative product of 2-amino-1-naphthol. The major product was characterized as N^4-(deoxyguanosin-N^2-yl)-2-amino-1,4-naphthoquinoneimine (dG-N^2-NAQI) (Figure 7.3) [83, 128]. This peroxidatively derived DNA adduct, along with two other minor adducts, accounted for approximately 60% of the total DNA binding that was obtained by incubation of 2-NA with PHS *in vitro*. In the dog, the DNA adducts derived from 2-imino-1-naphthoquinone accounted for approximately 20% of the 2-NA bound to urothelial DNA, but they were not detected in liver DNA [128]. The remaining adduction products in urothelium were derived from HONH-2-NA. Thus, PHS expressed in the bladder could play a significant role in bioactivation of 2-NA directly in the bladder, and could contribute to carcinogenesis of 2-NA and other AAs that serve as substrates of PHS.

The reaction of calf thymus DNA with N'-acetyl-HONH-Bz at pH 5 gives rise to N-(deoxyguanosin-8-yl)-N'-acetylbenzidine (N'-acetyl-dG-C8-Bz; Figure 7.3) [46, 129]. The structural isomer, N-(deoxyguanosin-8-yl)-N-acetylbenzidine, was shown not to occur in rodents [46]. The nonacetylated derivative, N-(deoxyguanosin-8-yl)-benzidine, was also shown not to form in rat and mouse liver DNA [46]. However, benzidine diimine, a reactive intermediate formed during the enzymatic peroxidation of Bz, can undergo deprotonation of the cationic diimine to form its nitrenium ion, which reacts with dG to form N-(deoxyguanosin-8-yl)-benzidine. This adduct is formed *in vivo* in the dog urothelium [130].

The doses of HAAs employed in many experimental animal studies exceeded daily human exposures by more than a million-fold [32, 109]. ^{32}P-Postlabeling, followed by thin-layer chromatography, was often used to discern these DNA adducts [32]. In early studies, a myriad of lesions were detected; however, many of these adduction products were subsequently shown to be incompletely digested oligomers of the dG-C8 adducts [82, 131, 132]. The dG-C8 adducts of IQ, MeIQ, MeIQx, 2-amino-3,4,8-trimethylimidazo[4,5-f]quinoxaline (4,8-DiMeIQx), PhIP, AαC, 2-amino-3-methyl-9H-pyrido[2,3-b]indole (MeAαC), and the glutamic acid and tryptophan pyrolysate mutagens were prominent lesions (see [32, 109] and references therein) (Figure 7.3). The isomeric dG-N^2 adducts of IQ and MeIQx also have been identified in rodents [82, 109]. Sensitive MS techniques were recently employed to detect a hydrazine-linked N^7-dG adduct with IQ, and a dA adduct of IQ *in vitro* [66], and a minor dA adduct of MeIQx formed *in vivo* in liver of rats [133]. Bond formation within these dA adducts is postulated to occur between the N^6 atom of adenine and the C5 atom of the IQ or MeIQx heteronucleus to form 5-(deoxyadenosin-N^6-yl)-IQ (dA-N^6-IQ) or 5-(deoxyadenosin-N^6-yl)-MeIQx (dA-N^6-MeIQx).

For many HAAs, DNA adduct formation is greatest in the liver, a result perhaps attributable to the high levels of P450 1A2 expression [32]. However, adducts are formed in all tissues investigated. DNA adduct formation of MeIQx [134] and IQ [82] in liver was studied over a wide range of doses, and adduct formation was found to occur in a near linear dose–response relationship. It was found that adducts were formed at dose levels approaching human exposures. In contrast to many other HAAs, PhIP shows levels of adduct formation that are lower in liver than in extrahepatic tissues; adduct levels are particularly elevated in colon and pancreas [135], and in prostate of male rodents [25] and mammary glands of female rodents [136]. Both glutathione S-transferases (GST) and UDP-glucuronosyltransferases, which are expressed at high levels in the liver, mediate the detoxication of reactive PhIP metabolites [137, 138], thus accounting for the relatively lower level of PhIP–DNA adduct formation in the liver.

7.12
AA-DNA and HAA-DNA Adduct Formation in Humans

dG-C8-ABP was first detected in human urinary bladder tissue biopsy samples by ^{32}P-postlabeling [139]. Subsequently, it was detected in lung and urinary bladder mucosa by GC-NICI-MS methods; it was found at levels ranging from below 0.32 to 49.5 adducts/10^8 nucleotides in the lung and from below 0.32 to 3.94 adducts/10^8 nucleotides in bladder samples [122]. Subsequently, dG-C8-ABP was detected, by IHC, ^{32}P-postlabeling, or GC-NICI-MS methods in bladder and lung tissues from smokers and ex-smokers [120], and by IHC in the liver of Taiwanese subjects with hepatocellular carcinoma [140]. In these studies, the ABP adduct levels did not correlate with the numbers of cigarettes smoked per day or the length of smoking history; environmental exposure to 4-nitrobiphenyl may have contributed to ABP-DNA adduct formation. However, the DNA present in the induced sputum of smokers, representing DNA of the lower respiratory tract, was shown to possess significantly higher levels of ABP-DNA adducts than were found in the sputum of nonsmokers, when assessed by IHC [141]. ABP-DNA adducts were also detected in female breast tissue biopsy samples, when visualized by IHC [121]; the women's smoking status was correlated with the levels of ABP-DNA in normal tissues adjacent to the tumors, but not in tumorigenic tissue. ABP-DNA adducts were also detected in laryngeal biopsies by IHC [142] and adduct levels were significantly higher in smokers than were the levels measured in tissue from nonsmokers.

Another environmental source of exposure to ABP is hair dyes [18, 19]. The relationship between ABP-DNA adduct levels and hair-dye usage has only been examined in one study, which determined the levels of ABP-DNA adduct, by ^{32}P-postlabeling, in exfoliated breast epithelial cells from lactating mothers [143]. The adduct levels were associated with the use of hair coloring products (odds ratio 11.2), but not with tobacco usage, in a statistically significant manner.

Recent studies have employed LC/ESI/MS-MS methods to quantitate dG-C8-ABP in human tissues. dG-C8-ABP was detected in urinary bladder epithelium in 12 out of 27 subjects in the DNA extracted from tumor tissue or nontumor surrounding tissue. The levels of adducts ranged from five to 80 adducts/10^9 bases, but a correlation was not observed between tobacco smoking and adduct levels [144]. In another study, dG-C8-ABP was identified in six of 12 human pancreas samples [145]. The levels ranged anywhere from one to 60 adducts/10^8 nucleotides; again, there was no observable correlation between the level of adducts and smoking preference, age, or gender. The prediction of the relationship between ABP exposure from tobacco smoke and adduct levels is not straightforward, being confounded by environmental exposure to 4-nitrobiphenyl and a variable persistence of dG-C8-ABP in the tissues. Since the activation or detoxification processes of ABP metabolism, as well as DNA repair mechanisms, can be tissue-specific, a correlation between tobacco usage and DNA adducts in different tissues may not exist.

The N'-acetyl-dG-C8-Bz adduct was identified in exfoliated urothelial cells of workers in factories manufacturing Bz in India [5]. This finding supports the hypothesis that acetylation represents an activation step for at least one Bz-related adduct in humans, analogous to the pathway proposed for Bz activation in rodents [46].

dG-C8-MeIQx was detected in the colon and kidney of some individuals at levels of around several adducts per 10^9 DNA bases, when assayed by ^{32}P-postlabeling [146]. A GC/MS assay, based upon alkaline hydrolysis of putative dG-C8-HAA adducts, was employed to measure the levels of PhIP adducts in colorectal mucosa [147] and in lymphocytes from colorectal cancer subjects [148]; the levels were found to be in the range of several adducts per 10^8 DNA bases. In the latter study, the adduct was found in about 30% of the subjects, and the adduct levels varied by a factor of 10 between the lowest and the highest level. The level of PhIP adducts was not significantly higher in smokers or high meat consumers than individuals who ate meat less frequently. The investigators observed that a subset of younger individuals carrying two mutated *GSTA1* alleles had higher adduct levels than homozygous wild-type and heterozygous subjects. This observation is consistent with the reported activity of the GSTA1 protein in the detoxication of *N*-acetoxy-PhIP [137, 149].

DNA adduct(s) of PhIP were frequently detected in human breast tissue by IHC [150], at adduct levels above one adduct/10^7 bases. An interactive effect of well-done meat consumption and NAT2 genotype on the level of PhIP-DNA adducts was observed. The dG-C8-PhIP adduct was frequently detected, by the ^{32}P-postlabeling method, in exfoliated breast epithelial cells of lactating mothers, at levels above one adduct/10^7 bases [151]. The frequent presence and high levels of PhIP adducts were also detected by IHC in prostate tissue [152]. PhIP-DNA adducts were also found in breast tissue (using AMS methods) of female cancer patients who had received biologically relevant doses of [^{14}C]PhIP (20 μg PhIP/70 kg body weight) via oral administration prior to surgery [153]. However, the estimates of PhIP-DNA adducts ranged from 26 to 480 adducts/10^{12} nucleotides [153], or nearly 10 000-fold lower than the levels of adducts reported by IHC or ^{32}P-postlabeling cited above. Two major limitations of the IHC and ^{32}P-postlabeling methods are the poor specificity of the assay and the lack of spectroscopic information that the techniques provide on the structure of the DNA adducts. Hence, the true estimate of the level of the adduct and it's identity remain equivocal. The large discrepancy in estimates of PhIP adducts measured by IHC and ^{32}P-postlabeling as opposed to the precise AMS method suggests that these biochemical assays are detecting a variety of lesions in addition to or other than dG-C8-PhIP. Current LC/ESI/MS-MS-based methods are able to accurately measure AA- and HAA-DNA adducts at levels of below one adduct/10^8 DNA bases and can also provide spectral data for the identification of the DNA adducts being assayed [109, 124, 145]. There is an urgent need to screen and quantitate HAA-DNA adducts by LC/ESI/MS-MS-based methods, to accurately determine adduct levels in human tissues.

7.13
Future Directions

The role of AAs in human bladder cancer has been clearly demonstrated through epidemiological studies of factory workers exposed to very high levels of AAs. However, the causal role of HAAs in human cancer remains controversial, because the amounts of HAAs present in the diet are generally low. The upper limit of daily human exposure to total HAAs in the United States was estimated to be 9 ng/kg body weight/day [154]; this exposure level is about 100 000- to 1 million-fold lower than the doses required to induce cancer in experimental laboratory animals [21]. Epidemiological studies entailing defined human exposures to HAAs, and employing DNA adducts and other biomarkers, including urinary metabolites, to demonstrate genetic damage, in conjunction with genetic polymorphisms in xenobiotic metabolism enzymes and DNA repair enzymes implicated in HAA genotoxicity [155], will help to clarify the role of HAAs in diet-linked human cancers. The numbers of site-specific mutagenesis studies and the known signature mutations of HAA-DNA adducts are few, and little research has been directed toward the elucidation of the mechanisms of HAA adduct recognition and enzyme repair processes (the current knowledge is reviewed in Chapter 10). The cellular response to specific AA-DNA and HAA-DNA adducts under low levels of adduct modification in cell systems, other than for ABP adducts [156], also has not been investigated. Studies that fill these gaps should shed light on cellular responses and differences in biological potency of specific AA-DNA and HAA-DNA lesions, and their roles in human cancers.

Acknowledgements

A portion of the work presented in this article that was conducted in the author's laboratory was supported by RO1CA122320 from the National Cancer Institute.

References

1 Weisburger, J.H. and Weisburger, E.K. (1966) Chemicals as causes of cancer. *Chem. Eng. News*, **44**, 124–142.

2 Rehn, L. (1895) Blasengeschwulste bei Fuchsinarbeitern. *Arch. Klin. Chir.*, **50**, 588–600.

3 Clayson, D.B. (1981) Specific aromatic amines as occupational bladder carcinogens. *Natl. Cancer Inst. Monogr.*, 15–19.

4 Vineis, P. and Pirastu, R. (1997) Aromatic amines and cancer. *Cancer Causes Control*, **8**, 346–355.

5 Rothman, N., Bhatnagar, V.K., Hayes, R.B., Zenser, T.V., Kashyap, S.K., Butler, M.A., Bell, D.A., Lakshmi, V., Jaeger, M., Kashyap, R., Hirvonen, A., Schulte, P.A., Dosemeci, M., Hsu, F., Parikh, D.J., Davis, B.B., and Talaska, G. (1996) The impact of interindividual variation in NAT2 activity on benzidine urinary metabolites and urothelial DNA adducts in exposed workers. *Proc. Natl. Acad. Sci. USA*, **93**, 5084–5089.

6 Heflich, R.H. and Neft, R.E. (1994) Genetic toxicity of 2-acetylaminoflu-

orene, 2-aminofluorene and some of their metabolites and model metabolites. *Mutat. Res.*, **318**, 73–114.

7 Hoffmann, G.R. and Fuchs, R.P. (1997) Mechanisms of frameshift mutations: insight from aromatic amines. *Chem. Res. Toxicol.*, **10**, 347–359.

8 Neumann, H.G. (2007) Aromatic amines in experimental cancer research: tissue-specific effects, an old problem and new solutions. *Crit. Rev. Toxicol.*, **37**, 211–236.

9 Stavric, B., Klassen, R., and Miles, W. (1979) Gas-liquid chromatographic-mass spectrometric determination of alpha- and beta-naphthylamines in FD&C Red No. 2 (amaranth). *J. Assoc. Off. Anal. Chem.*, **62**, 1020–1026.

10 Davis, V.M. and Bailey, J.E., Jr. (1993) Chemical reduction of FD&C yellow No. 5 to determine combined benzidine. *J. Chromatogr.*, **635**, 160–164.

11 Garrigos, M.C., Reche, F., Marin, M.L., and Jimenez, A. (2002) Determination of aromatic amines formed from azo colorants in toy products. *J. Chromatogr. A*, **976**, 309–317.

12 Lancaster, F.E. and Lawrence, J.F. (1999) Determination of benzidine in the food colours tartrazine and sunset yellow FCF, by reduction and derivatization followed by high-performance liquid chromatography. *Food Addit. Contam.*, **16**, 381–390.

13 Oh, S.W., Kang, M.N., Cho, C.W., and Lee, M.W. (1997) Detection of carcinogenic amines from dyestuffs or dyed substrates. *Dyes Pigments*, **33**, 119–135.

14 Cioni, F., Bartolucci, G., Pieraccini, G., Meloni, S., and Moneti, G. (1999) Development of a solid phase microextraction method for detection of the use of banned azo dyes in coloured textiles and leather. *Rapid Commun. Mass Spectrom.*, **13**, 1833–1837.

15 Chiang, T.A., Pei-Fen, W., Ying, L.S., Wang, L.F., and Ko, Y.C. (1999) Mutagenicity and aromatic amine content of fumes from heated cooking oils produced in Taiwan. *Food Chem. Toxicol.*, **37**, 125–134.

16 Tokiwa, H., Nakagawa, R., and Horikawa, K. (1985) Mutagenic/carcinogenic agents in indoor pollutants; the dinitropyrenes generated by kerosene heaters and fuel gas and liquefied petroleum gas burners. *Mutat. Res.*, **157**, 39–47.

17 Patrianakos, C. and Hoffmann, D. (1979) Chemical studies on tobacco smoke LXIV. On the analysis of aromatic amines in cigarette smoke. *J. Assoc. Off. Anal. Chem.*, **3**, 150–154.

18 Turesky, R.J., Freeman, J.P., Holland, R.D., Nestorick, D.M., Miller, D.W., Ratnasinghe, D.L., and Kadlubar, F.F. (2003) Identification of aminobiphenyl derivatives in commercial hair dyes. *Chem. Res. Toxicol.*, **16**, 1162–1173.

19 Akyuz, M. and Ata, S. (2008) Determination of aromatic amines in hair dye and henna samples by ion-pair extraction and gas chromatography-mass spectrometry. *J. Pharm. Biomed. Anal.*, **47**, 68–80.

20 Sugimura, T., Nagao, N., Kawachi, T., Honda, M., Yahagi, T., Seino, Y., Stao, S., Matsukura, N., Matsushima, T., Shirai, A., Sawamura, M., and Matsumoto, H. (1977) Mutagen-carcinogens in food, with special reference to highly mutagenic pyrolytic products in broiled foods, in *Origins of Human Cancer, Book C* (eds H.H. Hiatt, J.D. Watson, and J.A. Winstein), Cold Spring Harbor Laboratory Press, Cold Spring Harbor, NY, pp. 1561–1577.

21 Sugimura, T., Wakabayashi, K., Nakagama, H., and Nagao, M. (2004) Heterocyclic amines: mutagens/carcinogens produced during cooking of meat and fish. *Cancer Sci.*, **95**, 290–299.

22 Knize, M.G. and Felton, J.S. (2005) Formation and human risk of carcinogenic heterocyclic amines formed from natural precursors in meat. *Nutr. Rev.*, **63**, 158–165.

23 Jagerstad, M., Skog, K., Grivas, S., and Olsson, K. (1991) Formation of heterocyclic amines using model systems. *Mutat. Res.*, **259**, 219–233.

24 Manabe, S., Tohyama, K., Wada, O., and Aramaki, T. (1991) Detection of a carcinogen, 2-amino-1-methyl-6-phenylimidazo[4,5-b]pyridine, in cigarette smoke condensate. *Carcinogenesis*, **12**, 1945–1947.

25 Shirai, T., Sano, M., Tamano, S., Takahashi, S., Hirose, M., Futakuchi, M., Hasegawa, R., Imaida, K., Matsumoto, K., Wakabayashi, K., Sugimura, T., and Ito, N. (1997) The prostate: a target for carcinogenicity of 2-amino-1-methyl-6-phenylimidazo[4,5-b]pyridine (PhIP) derived from cooked foods. *Cancer Res.*, **57**, 195–198.

26 Glatt, H. (2006) Metabolic factors affecting the mutagenicity of heterocyclic amines, in *Acrylamide and Other Hazardous Compounds in Heat-Treated Foods* (eds K. Skog and J. Alexander), Woodhead, Cambridge, pp. 358–404.

27 Butler, M.A., Iwasaki, M., Guengerich, F.P., and Kadlubar, F.F. (1989) Human cytochrome P-450$_{PA}$ (P450IA2), the phenacetin *O*-deethylase, is primarily responsible for the hepatic 3-demethylation of caffeine and *N*-oxidation of carcinogenic arylamines. *Proc. Natl. Acad. Sci. USA*, **86**, 7696–7700.

28 Turesky, R.J. (2005) Interspecies metabolism of heterocyclic aromatic amines and the uncertainties in extrapolation of animal toxicity data for human risk assessment. *Mol. Nutr. Food Res.*, **49**, 101–117.

29 Shimada, T. and Guengerich, F.P. (1991) Activation of amino-α-carboline, 2-amino-1-methyl-6-phenylimidazo[4,5-b]pyridine, and a copper phthalocyanine cellulose extract of cigarette smoke condensate by cytochrome P-450 enzymes in rat and human liver microsomes. *Cancer Res.*, **51**, 5284–5291.

30 Josephy, P.D. (1996) The role of peroxidase-catalyzed activation of aromatic amines in breast cancer. *Mutagenesis*, **11**, 3–7.

31 Beland, F.A., Beranek, D.T., Dooley, K.L., Heflich, R.H., and Kadlubar, F.F. (1983) Arylamine-DNA adducts *in vitro* and *in vivo*: their role in bacterial mutagenesis and urinary bladder carcinogenesis. *Environ. Health Perspect.*, **49**, 125–134.

32 Schut, H.A. and Snyderwine, E.G. (1999) DNA adducts of heterocyclic amine food mutagens: implications for mutagenesis and carcinogenesis. *Carcinogenesis*, **20**, 353–368.

33 Turesky, R.J. (2004) The role of genetic polymorphisms in metabolism of carcinogenic heterocyclic aromatic amines. *Curr. Drug Metab.*, **5**, 169–180.

34 Conney, A.H. (1982) Induction of microsomal enzymes by foreign chemicals and carcinogenesis by polycyclic aromatic hydrocarbons: G. H. A. Clowes Memorial Lecture. *Cancer Res.*, **42**, 4875–4917.

35 Sinha, R., Rothman, N., Brown, E.D., Mark, S.D., Hoover, R.N., Caporaso, N.E., Levander, O.A., Knize, M.G., Lang, N.P., and Kadlubar, F.F. (1994) Pan-fried meat containing high levels of heterocyclic aromatic amines but low levels of polycyclic aromatic hydrocarbons induces cytochrome P4501A2 activity in humans. *Cancer Res.*, **54**, 6154–6159.

36 Hammons, G.J., Yan-Sanders, Y., Jin, B., Blann, E., Kadlubar, F.F., and Lyn-Cook, B.D. (2001) Specific site methylation in the 5′-flanking region of CYP1A2 interindividual differences in human livers. *Life Sci.*, **69**, 839–845.

37 Nakajima, M., Yokoi, T., Mizutani, M., Kinoshita, M., Funayama, M., and Kamataki, T. (1999) Genetic polymorphism in the 5′-flanking region of human CYP1A2 gene: effect on the CYP1A2 inducibility in humans. *J. Biochem.*, **125**, 803–808.

38 Sachse, C., Bhambra, U., Smith, G., Lightfoot, T.J., Barrett, J.H., Scollay, J., Garner, R.C., Boobis, A.R., Wolf, C.R., and Gooderham, N.J. (2003) Polymorphisms in the cytochrome P450 CYP1A2 gene (*CYP1A2*) in colorectal cancer patients and controls: allele frequencies, linkage disequilibrium and influence on caffeine metabolism. *Br. J. Clin. Pharmacol.*, **55**, 68–76.

39 Belloc, C., Baird, S., Cosme, J., Lecoeur, S., Gautier, J.-C., Challine, D., de Waziers, I., Flinois J.-P., and Beaune, P.H. (1996) Human cytochrome P450 expressed in *Escherichia coli*: production of specific antibodies. *Toxicology*, **106**, 207–219.

40 Turesky, R.J., Constable, A., Richoz, J., Varga, N., Markovic, J., Martin, M.V., and Guengerich, F.P. (1998) Activation of heterocyclic aromatic amines by rat

and human liver microsomes and by purified rat and human cytochrome P450 1A2. *Chem. Res. Toxicol.*, **11**, 925–936.

41 Turesky, R.J., Parisod, V., Huynh-Ba, T., Langouet, S., and Guengerich, F.P. (2001) Regioselective differences in C_8- and N-oxidation of 2-amino-3,8-dimethylimidazo[4,5-*f*]quinoxaline by human and rat liver microsomes and cytochromes P450 1A2. *Chem. Res. Toxicol.*, **14**, 901–911.

42 Hein, D.W., Doll, M.A., Fretland, A.J., Leff, M.A., Webb, S.J., Xiao, G.H., Devanaboyina, U.S., Nangju, N.A., and Feng, Y. (2000) Molecular genetics and epidemiology of the NAT1 and NAT2 acetylation polymorphisms. *Cancer Epidemiol. Biomarkers Prev.*, **9**, 29–42.

43 Hein, D.W. (2002) Molecular genetics and function of NAT1 and NAT2: role in aromatic amine metabolism and carcinogenesis. *Mutat. Res.*, **506–507**, 65–77.

44 Hein, D.W. (2006) N-Acetyltransferase 2 genetic polymorphism: effects of carcinogen and haplotype on urinary bladder cancer risk. *Oncogene*, **25**, 1649–1658.

45 Cohen, S.M., Boobis, A.R., Meek, M.E., Preston, R.J., and McGregor, D.B. (2006) 4-Aminobiphenyl and DNA reactivity: case study within the context of the 2006 IPCS human relevance framework for analysis of a cancer mode of action for humans. *Crit. Rev. Toxicol.*, **36**, 803–819.

46 Kennelly, J.C., Beland, F.A., Kadlubar, F.F., and Martin, C.N. (1984) Binding of N-acetylbenzidine and N,N'-diacetylbenzidine to hepatic DNA of rat and hamster *in vivo* and *in vitro*. *Carcinogenesis*, **5**, 407–412.

47 Garcia-Closas, M., Malats, N., Silverman, D., Dosemeci, M., Kogevinas, M., Hein, D.W., Tardon, A., Serra, C., Carrato, A., Garcia-Closas, R., Lloreta, J., Castano-Vinyals, G., Yeager, M., Welch, R., Chanock, S., Chatterjee, N., Wacholder, S., Samanic, C., Tora, M., Fernandez, F., Real, F.X., and Rothman, N. (2005) NAT2 slow acetylation, GSTM1 null genotype, and risk of bladder cancer: results from the Spanish Bladder Cancer Study and meta-analyses. *Lancet*, **366**, 649–659.

48 Hein, D.W. (2000) N-Acetyltransferase genetics and their role in predisposition to aromatic and heterocyclic amine-induced carcinogenesis. *Toxicol. Lett.*, **112–113**, 349–356.

49 Lang, N.P., Butler, M.A., Massengill, J.P., Lawson, M., Stotts, R.C., Hauer-Jensen, M., and Kadlubar, F.F. (1994) Rapid metabolic phenotypes for acetyltransferase and cytochrome P4501A2 and putative exposure to food-borne heterocyclic amines increase the risk for colorectal cancer or polyps. *Cancer Epidemiol. Biomarkers Prev.*, **3**, 675–682.

50 Le Marchand, L., Hankin, J.H., Wilkens, L.R., Pierce, L.M., Franke, A., Kolonel, L.N., Seifried, A., Custer, L.J., Chang, W., Lum-Jones, A., and Donlon, T. (2001) Combined effects of well-done red meat, smoking, and rapid N-acetyltransferase 2 and CYP1A2 phenotypes in increasing colorectal cancer risk. *Cancer Epidemiol. Biomarkers Prev.*, **10**, 1259–1266.

51 Nowell, S. and Falany, C.N. (2006) Pharmacogenetics of human cytosolic sulfotransferases. *Oncogene*, **25**, 1673–1678.

52 Lewis, A.J., Walle, U.K., King, R.S., Kadlubar, F.F., Falany, C.N., and Walle, T. (1998) Bioactivation of the cooked food mutagen N-hydroxy-2-amino-1-methyl-6-phenylimidazo[4,5-*b*]pyridine by estrogen sulfotransferase in cultured human mammary epithelial cells. *Carcinogenesis*, **19**, 2049–2053.

53 Raftogianis, R.B., Wood, T.C., and Weinshilboum, R.M. (1999) Human phenol sulfotransferases SULT1A2 and SULT1A1: genetic polymorphisms, allozyme properties, and human liver genotype–phenotype correlations. *Biochem. Pharmacol.*, **58**, 605–616.

54 Zheng, W., Xie, D., Cerhan, J.R., Sellers, T.A., Wen, W., and Folsom, A.R. (2001) Sulfotransferase 1A1 polymorphism, endogenous estrogen exposure, well-done meat intake, and breast cancer risk. *Cancer Epidemiol. Biomarkers Prev.*, **10**, 89–94.

55 Nowell, S., Ambrosone, C.B., Ozawa, S., MacLeod, S.L., Mrackova, G., Williams, S., Plaxco, J., Kadlubar, F.F., and Lang, N.P. (2000) Relationship of phenol sulfotransferase activity (SULT1A1) genotype to sulfotransferase phenotype in platelet cytosol. *Pharmacogenetics*, **10**, 789–797.

56 Moreno, V., Glatt, H., Guino, E., Fisher, E., Meinl, W., Navarro, M., Badosa, J.M., and Boeing, H. (2005) Polymorphisms in sulfotransferases SULT1A1 and SULT1A2 are not related to colorectal cancer. *Int. J. Cancer*, **113**, 683–686.

57 Steiner, M., Bastian, M., Schulz, W.A., Pulte, T., Franke, K.H., Rohring, A., Wolff, J.M., Seiter, H., and Schuff-Werner, P. (2000) Phenol sulphotransferase SULT1A1 polymorphism in prostate cancer: lack of association. *Arch. Toxicol.*, **74**, 222–225.

58 Turesky, R.J., Lang, N.P., Butler, M.A., Teitel, C.H., and Kadlubar, F.F. (1991) Metabolic activation of carcinogenic heterocyclic aromatic amines by human liver and colon. *Carcinogenesis*, **12**, 1839–1845.

59 Frandsen, H., Grivas, S., Andersson, R., Dragsted, L., and Larsen, J.C. (1992) Reaction of the N^2-acetoxy derivative of 2-amino-1-methyl-6-phenylimidazo[4,5-b]pyridine (PhIP) with 2′-deoxyguanosine and DNA. Synthesis and identification of N^2-(2′-deoxyguanosin-8-yl)-PhIP. *Carcinogenesis*, **13**, 629–635.

60 Kaderlik, K.R., Mulder, G.J., Turesky, R.J., Lang, N.P., Teitel, C.H., Chiarelli, M.P., and Kadlubar, F.F. (1994) Glucuronidation of N-hydroxy heterocyclic amines by human and rat liver microsomes. *Carcinogenesis*, **15**, 1701.

61 Styczynski, P.B., Blackmon, R.C., Groopman, J.D., and Kensler, T.W. (1993) The direct glucuronidation of 2-amino-1-methyl-6-phenylimidazo[4,5-b]pyridine (PhIP) by human and rabbit liver microsomes. *Chem. Res. Toxicol.*, **6**, 846–851.

62 Snyderwine, E.G., Turesky, R.J., Turteltaub, K.W., Davis, C.D., Sadrieh, N., Schut, H.A., Nagao, M., Sugimura, T., Thorgeirsson, U.P., Adamson, R.H., and Thorgeirsson, S.S. (1997) Metabolism of food-derived heterocyclic amines in nonhuman primates. *Mutat. Res.*, **376**, 203–210.

63 Humphreys, W.G., Kadlubar, F.F., and Guengerich, F.P. (1992) Mechanism of C8 alkylation of guanine residues by activated arylamines: evidence for initial adduct formation at the N^7 position. *Proc. Natl. Acad. Sci. USA*, **89**, 8278–8282.

64 Kennedy, S.A., Novak, M., and Kolb, B.A. (1997) Reactions of ester derivatives of carcinogenic N-(4-biphenylyl) hydroxylamine and the corresponding hydroxamic acid with purine nucleosides. *J. Am. Chem. Soc.*, **119**, 7654–7664.

65 Turesky, R.J., Rossi, S.C., Welti, D.H., O'Lay, J., Jr., and Kadlubar, F.F. (1992) Characterization of DNA adducts formed *in vitro* by reaction of N-hydroxy-2-amino-3-methylimidazo[4,5-f]quinoline and N-hydroxy-2-amino-3,8-dimethylimidazo[4,5-f]quinoxaline at the C-8 and N^2 atoms of guanine. *Chem. Res. Toxicol.*, **5**, 479–490.

66 Jamin, E.L., Arquier, D., Canlet, C., Rathahao, E., Tulliez, J., and Debrauwer, L. (2007) New insights in the formation of deoxynucleoside adducts with the heterocyclic aromatic amines PhIP and IQ by means of ion trap MS^n and accurate mass measurement of fragment ions. *J. Am. Soc. Mass Spectrom.*, **18**, 2107–2118.

67 Elmquist, C.E., Stover, J.S., Wang, Z., and Rizzo, C.J. (2004) Site-specific synthesis and properties of oligonucleotides containing C8-deoxyguanosine adducts of the dietary mutagen IQ. *J. Am. Chem. Soc.*, **126**, 11189–11201.

68 Stover, J.S. and Rizzo, C.J. (2007) Synthesis of oligonucleotides containing the N^2-deoxyguanosine adduct of the dietary carcinogen 2-amino-3-methylimidazo[4,5-f]quinoline. *Chem. Res. Toxicol.*, **20**, 1972–1979.

69 Bonala, R., Torres, M.C., Iden, C.R., and Johnson, F. (2006) Synthesis of the PhIP adduct of 2′-deoxyguanosine and its incorporation into oligomeric DNA. *Chem. Res. Toxicol.*, **19**, 734–738.

70 Lakshman, M.K. (2005) Synthesis of biologically important nucleoside

analogs by palladium-catalyzed C–N bond formation. *Cur. Org. Synth.*, **2**, 83–112.

71 Broyde, S., Wang, L., Zhang, L., Rechkoblit, O., Geacintov, N.E., and Patel, D.J. (2008) DNA adduct structure–function relationships: comparing solution with polymerase structures. *Chem. Res. Toxicol.*, **21**, 45–52.

72 Delaney, J.C. and Essigmann, J.M. (2008) Biological properties of single chemical-DNA adducts: a twenty year perspective. *Chem. Res. Toxicol.*, **21**, 232–252.

73 Patel, D.J., Mao, B., Gu, Z., Hingerty, B.E., Gorin, A., Basu, A.K., and Broyde, S. (1998) Nuclear magnetic resonance solution structures of covalent aromatic amine-DNA adducts and their mutagenic relevance. *Chem. Res. Toxicol.*, **11**, 391–407.

74 Lukin, M. and de Los, S.C. (2006) NMR structures of damaged DNA. *Chem. Rev.*, **106**, 607–686.

75 Zaliznyak, T., Bonala, R., Johnson, F., and de Los, S.C. (2006) Structure and stability of duplex DNA containing the 3-(deoxyguanosin-N^2-yl)-2-acetylaminofluorene (dG(N^2)-AAF) lesion: a bulky adduct that persists in cellular DNA. *Chem. Res. Toxicol.*, **19**, 745–752.

76 Brown, K., Hingerty, B.E., Guenther, E.A., Krishnan, V.V., Broyde, S., Turteltaub, K.W., and Cosman, M. (2001) Solution structure of the 2-amino-1-methyl-6-phenylimidazo[4,5-b]pyridine C8-deoxyguanosine adduct in duplex DNA. *Proc. Natl. Acad. Sci. USA*, **98**, 8507–8512.

77 Wang, F., DeMuro, N.E., Elmquist, C.E., Stover, J.S., Rizzo, C.J., and Stone, M.P. (2006) Base-displaced intercalated structure of the food mutagen 2-amino-3-methylimidazo[4,5-f]quinoline in the recognition sequence of the NarI restriction enzyme, a hotspot for –2 bp deletions. *J. Am. Chem. Soc.*, **128**, 10085–10095.

78 Choi, J.Y., Stover, J.S., Angel, K.C., Chowdhury, G., Rizzo, C.J., and Guengerich, F.P. (2006) Biochemical basis of genotoxicity of heterocyclic arylamine food mutagens: human DNA polymerase η selectively produces a two-base deletion in copying the N^2-guanyl adduct of 2-amino-3-methylimidazo[4,5-f]quinoline but not the C8 adduct at the NarI G3 site. *J. Biol. Chem.*, **281**, 25297–25306.

79 Stover, J.S., Chowdhury, G., Zang, H., Guengerich, F.P., and Rizzo, C.J. (2006) Translesion synthesis past the C8- and N^2-deoxyguanosine adducts of the dietary mutagen 2-Amino-3-methylimidazo[4,5-f]quinoline in the NarI recognition sequence by prokaryotic DNA polymerases. *Chem. Res. Toxicol.*, **19**, 1506–1517.

80 Shibutani, S., Fernandes, A., Suzuki, N., Zhou, L., Johnson, F., and Grollman, A.P. (1999) Mutagenesis of the N-(deoxyguanosin-8-yl)-2-amino-1-methyl-6-phenylimidazo[4,5-b]pyridine DNA adduct in mammalian cells. Sequence context effects. *J. Biol. Chem.*, **274**, 27433–27438.

81 Culp, S.J., Poirier, M.C., and Beland, F.A. (1993) Biphasic removal of DNA adducts in a repetitive DNA sequence after dietary administration of 2-acetylaminofluorene. *Environ. Health Perspect.*, **99**, 273–275.

82 Turesky, R.J., Box, R.M., Markovic, J., Gremaud, E., and Snyderwine, E.G. (1997) Formation and persistence of DNA adducts of 2-amino-3-methylimidazo[4,5-f]quinoline in the rat and nonhuman primates. *Mutat. Res.*, **376**, 235–241.

83 Beland, F.A. and Kadlubar, F.F. (1985) Formation and persistence of arylamine DNA adducts *in vivo*. *Environ. Health Perspect.*, **62**, 19–30.

84 Chung, K.T., Chen, S.C., Wong, T.Y., Li, Y.S., Wei, C.I., and Chou, M.W. (2000) Mutagenicity studies of benzidine and its analogs: structure–activity relationships. *Toxicol. Sci.*, **56**, 351–356.

85 Felton, J.S., Knize, M.G., Dolbeare, F.A., and Wu, R. (1994) Mutagenic activity of heterocyclic amines in cooked foods. *Environ. Health Perspect.*, **102** (Suppl. 6), 201–204.

86 Fuscoe, J.C., Wu, R., Shen, N.H., Healy, S.K., and Felton, J.S. (1988) Base-change analysis of revertants of the *hisD3052* allele in *Salmonella*

typhimurium. *Mutat. Res.*, **201**, 241–251.

87 Josephy, P.D. (2000) The *Escherichia coli lacZ* reversion mutagenicity assay. *Mutat. Res.*, **455**, 71–80.

88 Metry, K.J., Zhao, S., Neale, J.R., Doll, M.A., States, J.C., McGregor, W.G., Pierce, W.M., Jr., and Hein, D.W. (2007) 2-Amino-1-methyl-6-phenylimidazo [4,5-*b*] pyridine-induced DNA adducts and genotoxicity in Chinese hamster ovary (CHO) cells expressing human CYP1A2 and rapid or slow acetylator *N*-acetyltransferase 2. *Mol. Carcinog.*, **46**, 553–563.

89 Wu, R.W., Tucker, J.D., Sorensen, K.J., Thompson, L.H., and Felton, J.S. (1997) Differential effect of acetyltransferase expression on the genotoxicity of heterocyclic amines in CHO cells. *Mutat. Res.*, **390**, 93–103.

90 Morgenthaler, P.M. and Holzhauser, D. (1995) Analysis of mutations induced by 2-amino-1-methyl-6-phenylimidazo[4,5-*b*]pyridine (PhIP) in human lymphoblastoid cells. *Carcinogenesis*, **16**, 713–718.

91 Carothers, A.M., Yuan, W., Hingerty, B.E., Broyde, S., Grunberger, D., and Snyderwine, E.G. (1994) Mutation and repair induced by the carcinogen 2-(hydroxyamino)-1-methyl-6-phenylimidazo[4,5-*b*]pyridine (*N*-OH-PhIP) in the dihydrofolate reductase gene of Chinese hamster ovary cells and conformational modeling of the dG-C8-PhIP adduct in DNA. *Chem. Res. Toxicol.*, **7**, 209–218.

92 Wu, R.W., Wu, E.M., Thompson, L.H., and Felton, J.S. (1995) Identification of *aprt* gene mutations induced in repair-deficient and P450-expressing CHO cells by the food-related mutagen/carcinogen, PhIP. *Carcinogenesis*, **16**, 1207–1213.

93 Yadollahi-Farsani, M., Gooderham, N.J., Davies, D.S., and Boobis, A.R. (1996) Mutational spectra of the dietary carcinogen 2-amino-1-methyl-6-phenylimidazo[4,5-*b*]pyridine (PhIP) at the Chinese hamsters *hprt* locus. *Carcinogenesis*, **17**, 617–624.

94 Kakiuchi, H., Watanabe, M., Ushijima, T., Toyota, M., Imai, K., Weisburger, J.H., Sugimura, T., and Nagao, N. (1995) Specific 5′-GGGA-3′ → 5′-GGA-3′ mutation of the *Apc* gene in rat colon tumors induced by 2-amino-1-methyl-6-phenylimidazo[4,5-*b*]pyridine. *Proc. Natl. Acad. Sci. USA*, **92**, 910–914.

95 Leong-Morgenthaler, P.M., Op, H.V., Jaccaud, E., and Turesky, R.J. (1998) Mutagenicity of 2-amino-3-methylimidazo[4,5-*f*]quinoline in human lymphoblastoid cells. *Carcinogenesis*, **19**, 1749–1754.

96 Bookland, E.A., Reznikoff, C.A., Lindstrom, M., and Swaminathan, S. (1992) Induction of thioguanine-resistant mutations in human uroepithelial cells by 4-aminobiphenyl and its *N*-hydroxy derivatives. *Cancer Res.*, **52**, 1615–1621.

97 Bol, S.A., Horlbeck, J., Markovic, J., de Boer, J.G., Turesky, R.J., and Constable, A. (2000) Mutational analysis of the liver, colon and kidney of Big Blue rats treated with 2-amino-3-methylimidazo[4,5-*f*]quinoline. *Carcinogenesis*, **21**, 1–6.

98 Moller, P., Wallin, H., Vogel, U., Autrup, H., Risom, L., Hald, M.T., Daneshvar, B., Dragsted, L.O., Poulsen, H.E., and Loft, S. (2002) Mutagenicity of 2-amino-3-methylimidazo[4,5-*f*] quinoline in colon and liver of Big Blue rats: role of DNA adducts, strand breaks, DNA repair and oxidative stress. *Carcinogenesis*, **23**, 1379–1385.

99 Nagao, M., Ochiai, M., Okochi, E., Ushijima, T., and Sugimura, T. (2001) *LacI* transgenic animal study: relationships among DNA-adduct levels, mutant frequencies and cancer incidences. *Mutat. Res.*, **477**, 119–124.

100 Itoh1, T., Suzuki, T., Nishikawa, A., Furukawa, F., Takahashi, M., Xue, W., Sofuni, T., and Hayashi, M. (2000) *In vivo* genotoxicity of 2-amino-3,8-dimethylimidazo[4, 5-*f*]quinoxaline in *lacI* transgenic (Big Blue) mice. *Mutat. Res.*, **468**, 19–25.

101 Lynch, A.M., Gooderham, N.J., and Boobis, A.R. (1996) Organ distinctive mutagenicity in MutaMouse after short-term exposure to PhIP. *Mutagenesis*, **11**, 505–509.

102 Zhang, X.B., Felton, J.S., Tucker, J.D., Urlando, C., and Heddle, J.A. (1996) Intestinal mutagenicity of two carcinogenic food mutagens in

transgenic mice: 2-amino-1-methyl-6-phenylimidazo[4,5-*b*]pyridine and amino(α)carboline. *Carcinogenesis*, **17**, 2259–2265.
103 Fletcher, K., Tinwell, H., and Ashby, J. (1998) Mutagenicity of the human bladder carcinogen 4-aminobiphenyl to the bladder of MutaMouse transgenic mice. *Mutat. Res.*, **400**, 245–250.
104 Okonogi, H., Ushijima, T., Zhang, X.B., Heddle, J.A., Suzuki, T., Sofuni, T., Felton, J.S., Tucker, J.D., Sugimura, T., and Nagao, M. (1997) Agreement of mutational characteristics of heterocyclic amines in *lacI* of the Big Blue mouse with those in tumor related genes in rodents. *Carcinogenesis*, **18**, 745–748.
105 Nagao, M., Ushijima, T., Toyota, M., Inoue, R., and Sugimura, T. (1997) Genetic changes induced by heterocyclic amines. *Mutat. Res.*, **376**, 161–167.
106 Itoh, T., Kuwahara, T., Suzuki, T., Hayashi, M., and Ohnishi, Y. (2003) Regional mutagenicity of heterocyclic amines in the intestine: mutation analysis of the *cII* gene in lambda/*lacZ* transgenic mice. *Mutat. Res.*, **539**, 99–108.
107 Stuart, G.R., Thorleifson, E., Okochi, E., de Boer, J.G., Ushijima, T., Nagao, M., and Glickman, B.W. (2000) Interpretation of mutational spectra from different genes: analyses of PhIP-induced mutational specificity in the *lacI* and *cII* transgenes from colon of Big Blue rats. *Mutat. Res.*, **452**, 101–121.
108 Smith-Roe, S.L., Hegan, D.C., Glazer, P.M., and Buermeyer, A.B. (2005) Mlh1-dependent suppression of specific mutations induced *in vivo* by the food-borne carcinogen 2-amino-1-methyl-6-phenylimidazo [4,5-*b*]pyridine (PhIP). *Mutat. Res.*, **594**, 101–112.
109 Turesky, R.J. and Vouros, P. (2004) Formation and analysis of heterocyclic aromatic amine-DNA adducts *in vitro* and *in vivo*. *J. Chromatogr. B Analyt. Technol. Biomed. Life Sci.*, **802**, 155–166.
110 Glaab, W.E., Kort, K.L., and Skopek, T.R. (2000) Specificity of mutations induced by the food-associated heterocyclic amine 2-amino-1-methyl-6-phenylimidazo-[4,5-*b*]-pyridine in colon cancer cell lines defective in mismatch repair. *Cancer Res.*, **60**, 4921–4925.
111 Chen, T., Mittelstaedt, R.A., Beland, F.A., Heflich, R.H., Moore, M.M., and Parsons, B.L. (2005) 4-Aminobiphenyl induces liver DNA adducts in both neonatal and adult mice but induces liver mutations only in neonatal mice. *Int. J. Cancer*, **117**, 182–187.
112 Melchior, W.B., Jr., Marques, M.M., and Beland, F.A. (1994) Mutations induced by aromatic amine DNA adducts in pBR322. *Carcinogenesis*, **15**, 889–899.
113 Nagao, M. (2000) Mutagenicity, in *Food Borne Carcinogens Heterocyclic Amines* (eds M. Nagao and T. Sugimura), John Wiley & Sons, Ltd, Chichester, pp. 163–195.
114 Turesky, R.J. (2002) Heterocyclic aromatic amine metabolism, DNA adduct formation, mutagenesis, and carcinogenesis. *Drug Metab. Rev.*, **34**, 625–650.
115 Lord, P.G., Hardaker, K.J., Loughlin, J.M., Marsden, A.M., and Orton, T.C. (1992) Point mutation analysis of *ras* genes in spontaneous and chemically induced C57Bl/10J mouse liver tumours. *Carcinogenesis*, **13**, 1383–1387.
116 Olivier, M., Eeles, R., Hollstein, M., Khan, M.A., Harris, C.C., and Hainaut, P. (2002) The IARC TP53 database: new online mutation analysis and recommendations to users. *Hum. Mutat.*, **19**, 607–614.
117 Feng, Z., Hu, W., Rom, W.N., Beland, F.A., and Tang, M.S. (2002) 4-Aminobiphenyl is a major etiological agent of human bladder cancer: evidence from its DNA binding spectrum in human p53 gene. *Carcinogenesis*, **23**, 1721–1727.
118 Feng, Z., Hu, W., Rom, W.N., Beland, F.A., and Tang, M.S. (2002) N-hydroxy-4-aminobiphenyl-DNA binding in human p53 gene: sequence preference and the effect of C5 cytosine methylation. *Biochemistry*, **41**, 6414–6421.
119 Burnouf, D., Miturski, R., Nagao, M., Nakagama, H., Nothisen, M., Wagner, J., and Fuchs, R.P. (2001) Early detection of 2-amino-1-methyl-6-phenylimidazo[4,5-*b*]pyridine (PhIP)-

induced mutations within the *Apc* gene of rat colon. *Carcinogenesis*, **22**, 329–335.

120 Culp, S.J., Roberts, D.W., Talaska, G., Lang, N.P., Fu, P.P., Lay, J.O., Jr., Teitel, C.H., Snawder, J.E., Von Tungeln, L.S., and Kadlubar, F.F. (1997) Immunochemical, ^{32}P-postlabeling, and GC/MS detection of 4-aminobiphenyl-DNA adducts in human peripheral lung in relation to metabolic activation pathways involving pulmonary N-oxidation, conjugation, and peroxidation. *Mutat. Res.*, **378**, 97–112.

121 Faraglia, B., Chen, S.Y., Gammon, M.D., Zhang, Y., Teitelbaum, S.L., Neugut, A.I., Ahsan, H., Garbowski, G.C., Hibshoosh, H., Lin, D., Kadlubar, F.F., and Santella, R.M. (2003) Evaluation of 4-aminobiphenyl-DNA adducts in human breast cancer: the influence of tobacco smoke. *Carcinogenesis*, **24**, 719–725.

122 Lin, D., Lay, J.O., Jr., Bryant, M.S., Malaveille, C., Friesen, M., Bartsch, H., Lang, N.P., and Kadlubar, F.F. (1994) Analysis of 4-aminobiphenyl-DNA adducts in human urinary bladder and lung by alkaline hydrolysis and negative ion gas chromatography-mass spectrometry. *Environ. Health Perspect.*, **102** (Suppl. 6), 11–16.

123 Dingley, K.H., Roberts, M.L., Velsko, C.A., and Turteltaub, K.W. (1998) Attomole detection of ^{3}H in biological samples using accelerator mass spectrometry: application in low-dose, dual-isotope tracer studies in conjunction with ^{14}C accelerator mass spectrometry. *Chem. Res. Toxicol.*, **11**, 1217–1222.

124 Doerge, D.R., Churchwell, M.I., Marques, M.M., and Beland, F.A. (1999) Quantitative analysis of 4-aminobiphenyl-C8-deoxyguanosyl DNA adducts produced *in vitro* and *in vivo* using HPLC-ES-MS. *Carcinogenesis*, **20**, 1055–1061.

125 Hatcher, J.F. and Swaminathan, S. (2002) Identification of N-(deoxyguanosin-8-yl)-4-azobiphenyl by ^{32}P-postlabeling analyses of DNA in human uroepithelial cells exposed to proximate metabolites of the environmental carcinogen 4-aminobiphenyl. *Environ. Mol. Mutagen.*, **39**, 314–322.

126 Swaminathan, S. and Hatcher, J.F. (2002) Identification of new DNA adducts in human bladder epithelia exposed to the proximate metabolite of 4-aminobiphenyl using ^{32}P-postlabeling method. *Chem. Biol. Interact.*, **139**, 199–213.

127 Poirier, M.C., Fullerton, N.F., Smith, B.A., and Beland, F.A. (1995) DNA adduct formation and tumorigenesis in mice during the chronic administration of 4-aminobiphenyl at multiple dose levels. *Carcinogenesis*, **16**, 2917–2921.

128 Yamazoe, Y., Miller, D.W., Weis, C.C., Dooley, K.L., Zenser, T.V., Beland, F.A., and Kadlubar, F.F. (1985) DNA adducts formed by ring-oxidation of the carcinogen 2-naphthylamine with prostaglandin H synthase *in vitro* and in the dog urothelium *in vivo*. *Carcinogenesis*, **6**, 1379–1387.

129 Martin, C.N., Beland, F.A., Roth, R.W., and Kadlubar, F.F. (1982) Covalent binding of benzidine and N-acetylbenzidine to DNA at the C-8 atom of deoxyguanosine *in vivo* and *in vitro*. *Cancer Res.*, **42**, 2678–2686.

130 Yamazoe, Y., Roth, R.W., and Kadlubar, F.F. (1986) Reactivity of benzidine diimine with DNA to form N-(deoxyguanosin-8-yl)-benzidine. *Carcinogenesis*, **7**, 179–182.

131 Pfau, W., Brockstedt, U., Sohren, K.D., and Marquardt, H. (1994) ^{32}P-Postlabelling analysis of DNA adducts formed by food-derived heterocyclic amines: evidence for incomplete hydrolysis and a procedure for adduct pattern simplification. *Carcinogenesis*, **15**, 877–882.

132 Ochiai, M., Nakagama, H., Turesky, R.J., Sugimura, T., and Nagao, M. (1999) A new modification of the ^{32}P-post-labeling method to recover IQ-DNA adducts as mononucleotides. *Mutagenesis*, **14**, 239–242.

133 Bessette, E.E., Goodenough, A.K., Langouet, S., Yasa, I., Kozekov, I.D., Spivack, S.D., and Turesky, R.J. (2009) Screening for DNA adducts by data-dependent constant neutral loss-triple stage mass spectrometry with

a linear quadrupole ion trap mass spectrometer. *Anal. Chem.*, **81**, 809–819.

134 Turteltaub, K.W., Felton, J.S., Gledhill, B.L., Vogel, J.S., Southon, J.R., Caffee, M.W., Finkel, R.C., Nelson, D.E., Proctor, I.D., and Davis, J.C. (1990) Accelerator mass spectrometry in biomedical dosimetry: relationship between low-level exposure and covalent binding of heterocyclic amine carcinogens to DNA. *Proc. Natl. Acad. Sci. USA*, **87**, 5288–5292.

135 Lin, D., Kaderlik, K.R., Turesky, R.J., Miller, D.W., Lay, J.O., Jr., and Kadlubar, F.F. (1992) Identification of N-(deoxyguanosin-8-yl)-2-amino-1-methyl-6-phenylimidazo [4,5-*b*]pyridine as the major adduct formed by the food-borne carcinogen, 2-amino-1-methyl-6-phenylimidazo[4,5-*b*]pyridine, with DNA. *Chem. Res. Toxicol.*, **5**, 691–697.

136 Ghoshal, A., Davis, C.D., Schut, H.A.J., and Snyderwine, E.G. (1995) Possible mechanisms for PhIP-DNA adduct formation in the mammary gland of female Sprague-Dawley rats. *Carcinogenesis*, **16**, 2725–2731.

137 Lin, D.-X., Meyer, D.J., Ketterer, B., Lang, N.P., and Kadlubar, F.F. (1994) Effects of human and rat glutathione-S-transferase on the covalent binding of the N-acetoxy derivatives of heterocyclic amine carcinogens *in vitro:* a possible mechanism of organ specificity in their carcinogenesis. *Cancer Res.*, **54**, 4920–4926.

138 Malfatti, M.A. and Felton, J.S. (2004) Human UDP-glucuronosyltransferase 1A1 is the primary enzyme responsible for the N-glucuronidation of N-hydroxy-PhIP *in vitro*. *Chem. Res. Toxicol.*, **17**, 1137–1144.

139 Talaska, G., al Juburi, A.Z., and Kadlubar, F.F. (1991) Smoking related carcinogen-DNA adducts in biopsy samples of human urinary bladder: identification of N-(deoxyguanosin-8-yl)-4-aminobiphenyl as a major adduct. *Proc. Natl. Acad. Sci. USA*, **88**, 5350–5354.

140 Wang, L.Y., Chen, C.J., Zhang, Y.J., Tsai, W.Y., Lee, P.H., Feitelson, M.A., Lee, C.S., and Santella, R.M. (1998) 4-Aminobiphenyl DNA damage in liver tissue of hepatocellular carcinoma patients and controls. *Am. J. Epidemiol.*, **147**, 315–323.

141 Besaratinia, A., Van Straaten, H.W., Kleinjans, J.C., and Van Schooten, F.J. (2000) Immunoperoxidase detection of 4-aminobiphenyl- and polycyclic aromatic hydrocarbons-DNA adducts in induced sputum of smokers and non-smokers. *Mutat. Res.*, **468**, 125–135.

142 Flamini, G., Romano, G., Curigliano, G., Chiominto, A., Capelli, G., Boninsegna, A., Signorelli, C., Ventura, L., Santella, R.M., Sgambato, A., and Cittadini, A. (1998) 4-Aminobiphenyl-DNA adducts in laryngeal tissue and smoking habits: an immunohistochemical study. *Carcinogenesis*, **19**, 353–357.

143 Ambrosone, C.B., Abrams, S.M., Gorlewska-Roberts, K., and Kadlubar, F.F. (2007) Hair dye use, meat intake, and tobacco exposure and presence of carcinogen-DNA adducts in exfoliated breast ductal epithelial cells. *Arch. Biochem. Biophys.*, **464**, 169–175.

144 Zayas, B., Stillwell, S.W., Wishnok, J.S., Trudel, L.J., Skipper, P., Yu, M.C., Tannenbaum, S.R., and Wogan, G.N. (2007) Detection and quantification of 4-ABP adducts in DNA from bladder cancer patients. *Carcinogenesis*, **28**, 342–349.

145 Ricicki, E.M., Soglia, J.R., Teitel, C., Kane, R., Kadlubar, F., and Vouros, P. (2005) Detection and quantification of N-(deoxyguanosin-8-yl)-4-aminobiphenyl adducts in human pancreas tissue using capillary liquid chromatography-microelectrospray mass spectrometry. *Chem. Res. Toxicol.*, **18**, 692–699.

146 Totsuka, Y., Fukutome, K., Takahashi, M., Takashi, S., Tada, A., Sugimura, T., and Wakabayashi, K. (1996) Presence of N^2-(deoxyguanosin-8-yl)-2-amino-3,8-dimethylimidazo[4,5-*f*]quinoxaline (dG-C8-MeIQx) in human tissues. *Carcinogenesis*, **17**, 1029–1034.

147 Friesen, M.D., Kaderlik, K., Lin, D., Garren, L., Bartsch, H., Lang, N.P., and Kadlubar, F.F. (1994) Analysis of DNA adducts of 2-amino-1-methyl-6-

phenylimidazo[4,5-*b*]pyridine in rat and human tissues by alkaline hydrolysis and gas chromatography/electron capture mass spectrometry: validation by comparison with ^{32}P-postlabeling. *Chem. Res. Toxicol.*, **7**, 733–739.

148 Magagnotti, C., Pastorelli, R., Pozzi, S., Andreoni, B., Fanelli, R., and Airoldi, L. (2003) Genetic polymorphisms and modulation of 2-amino-1-methyl-6-phenylimidazo[4,5-*b*]pyridine (PhIP)-DNA adducts in human lymphocytes. *Int. J. Cancer*, **107**, 878–884.

149 Coles, B.F. and Kadlubar, F.F. (2003) Detoxification of electrophilic compounds by glutathione S-transferase catalysis: determinants of individual response to chemical carcinogens and chemotherapeutic drugs? *Biofactors*, **17**, 115–130.

150 Zhu, J., Chang, P., Bondy, M.L., Sahin, A.A., Singletary, S.E., Takahashi, S., Shirai, T., and Li, D. (2003) Detection of 2-amino-1-methyl-6-phenylimidazo[4,5-*b*]pyridine-DNA adducts in normal breast tissues and risk of breast cancer. *Cancer Epidemiol. Biomarkers Prev.*, **12**, 830–837.

151 Gorlewska-Roberts, K., Green, B., Fares, M., Ambrosone, C.B., and Kadlubar, F.F. (2002) Carcinogen-DNA adducts in human breast epithelial cells. *Environ. Mol. Mutagen.*, **39**, 184–192.

152 Tang, D., Liu, J.J., Rundle, A., Neslund-Dudas, C., Savera, A.T., Bock, C.H., Nock, N.L., Yang, J.J., and Rybicki, B.A. (2007) Grilled meat consumption and PhIP-DNA adducts in prostate carcinogenesis. *Cancer Epidemiol. Biomarkers Prev.*, **16**, 803–808.

153 Lightfoot, T.J., Coxhead, J.M., Cupid, B.C., Nicholson, S., and Garner, R.C. (2000) Analysis of DNA adducts by accelerator mass spectrometry in human breast tissue after administration of 2-amino-1-methyl-6-phenylimidazo[4,5-*b*]pyridine and benzo[*a*]pyrene. *Mutat. Res.*, **472**, 119–127.

154 Bogen, K.T. and Keating, G.A. (2001) US dietary exposures to heterocyclic amines. *J. Expo. Anal. Environ. Epidemiol.*, **11**, 155–168.

155 Sinha, R. (2002) An epidemiologic approach to studying heterocyclic amines. *Mutat. Res.*, **506–507**, 197–204.

156 Luo, W., Fan, W., Xie, H., Jing, L., Ricicki, E., Vouros, P., Zhao, L.P., and Zarbl, H. (2005) Phenotypic anchoring of global gene expression profiles induced by *N*-hydroxy-4-acetylaminobiphenyl and benzo[*a*]pyrene diol epoxide reveals correlations between expression profiles and mechanism of toxicity. *Chem. Res. Toxicol.*, **18**, 619–629.

8
Genotoxic Estrogen Pathway: Endogenous and Equine Estrogen Hormone Replacement Therapy

Judy L. Bolton and Gregory R.J. Thatcher

8.1
Risks of Estrogen Exposure

Experimental and epidemiological data strongly associate excessive estrogen exposure to hormone-dependent cancers, particularly breast and endometrial cancer [1]. The longer women are exposed to estrogens, either through early menarche and late menopause and/or through hormone replacement therapy (HRT), the higher is the risk of developing these cancers. In the past it was thought that the purported benefits of HRT, which included the relief of menopausal symptoms, and decrease in coronary heart disease, osteoporosis, stroke, and Alzheimer's disease, justified the use of long-term HRT. However, the release of the initial results from the Women's Health Initiative Study in July 2002 cast serious doubt on this paradigm for the treatment of postmenopausal women [2]. The estrogen-plus-progestin arm was halted 3 years early due to significant increases in breast cancer, coronary heart disease, stroke, and pulmonary embolism, with more recent data suggesting an increase in vascular dementia in women aged over 65 years on HRT [3]. In 2004, the estrogen-only arm was halted because of increased incidence of stroke [4]. A recent analysis of data from the National Cancer Institute's Surveillance, Epidemiology, and End Results registries showed that age-adjusted incidence rate of breast cancer fell sharply (6.7%) in 2003 compared to 2002, which seemed to be related to the drop in the use of HRT [5]. Similar trends have since been reported in other industrialized countries [6]. Finally, a reanalysis of nine prospective studies has shown that exposure to estrogens is associated with an increase in breast cancer risk with evidence of a dose–response relationship [7]. These troubling findings highlight the urgent need for a full understanding of all the deleterious effects of estrogens, including their potential to initiate and/or promote the carcinogenic process.

The Chemical Biology of DNA Damage. Edited by Nicholas E. Geacintov and Suse Broyde
© 2010 WILEY-VCH Verlag GmbH & Co. KGaA, Weinheim
ISBN: 978-3-527-32295-4

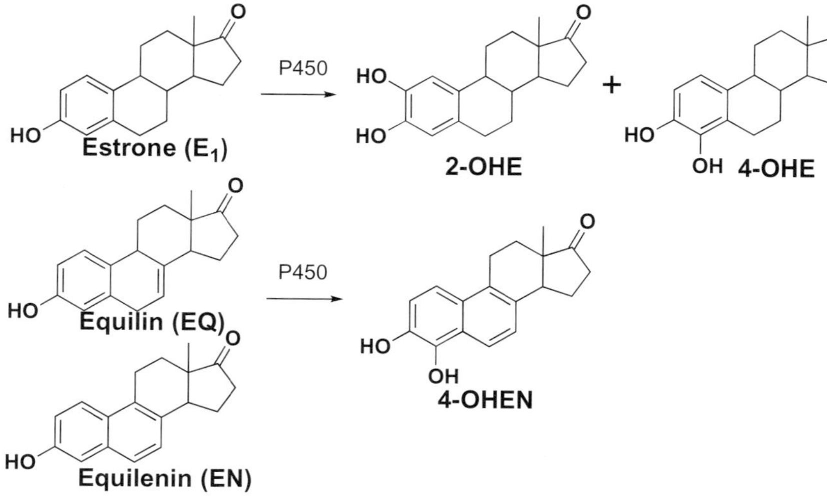

Scheme 8.1 Major estrogens present in Premarin and formation of catechol estrogen metabolites. The 17-ketone analogs are shown. *In vivo*, they are in equilibrium with the 17-hydroxyl derivatives.

Most of the epidemiological studies on HRT and cancer risk, including the Women's Health Initiative Study trial discussed above, have been conducted with Premarin® (Wyeth-Ayerest), which remains the estrogen replacement treatment of choice and one of the most widely prescribed drugs in North America [8]. Premarin was approved by the Food and Drug Administration in the 1940s, yet very little is known about the metabolism and potential toxic metabolites that could be produced from the various equine estrogens, which make up approximately 50% of the estrogens in Premarin [9–13] (Scheme 8.1). It is known that treating hamsters for 9 months with either estrone, equilin plus equilenin, or sulfatase-treated Premarin resulted in 100% kidney tumor incidences and abundant tumor foci [10]. Furthermore, in a small clinical trial of 596 postmenopausal women, a significant increase in endometrial hyperplasia was found in those women receiving a daily dose of 0.625 mg of Premarin [14]. Nevertheless, HRT is still the most effective remedy for relief of symptoms of menopause such as sleeplessness, hot flashes, and mood swings, provides protection against early menopausal bone loss, and lowers the risk of colon cancer [2, 15]. For these reasons, women continue to use HRT [16] in spite of the well-recognized risks [17]. Although the sales of standard-dose Premarin prescriptions (0.625 mg/day) have decreased by 33% since July 2002 when the National Heart, Blood, and Lung Institute terminated the clinical trial on the long-term risks and benefits of estrogen-plus-progestin therapy, more recently the sales of low-dose Premarin preparations (0.45 mg/day) have been rising [16]. This highlights the need to understand the deleterious effects of equine estrogens and the influence of dosage.

8.2 Mechanisms of Estrogen Carcinogenesis

8.2.1 Hormonal Mechanism

The mechanisms of estrogen carcinogenesis are not well understood [18–22]. One major pathway considered to be important is the extensively studied hormonal pathway, by which estrogen stimulates cell proliferation through nuclear estrogen receptor (ER)-mediated signaling pathways, thus resulting in an increased risk of genomic mutations during DNA replication (Scheme 8.2, using equilenin as an example) [23–26]. A similar "nongenomic pathway," potentially involving newly discovered membrane-associated ERs, also appears to regulate extranuclear estrogen signaling pathways and is able to increase gene transcription [27, 28]. Recent studies have also shown the presence of ERα and ERβ in the mitochondria of various cells and tissues that may be involved in deregulation of mitochondrial bioenergetics, contributing to estrogen-related cancers [29]. Cross-talk between these genomic and second-messenger pathways probably have important roles in estrogenic control of cell proliferation, inhibition of apoptosis, and induction of DNA damage. The hormonal activity of estrogen metabolites as ER agonists has less frequently been studied as a potential contributor to carcinogenesis; however,

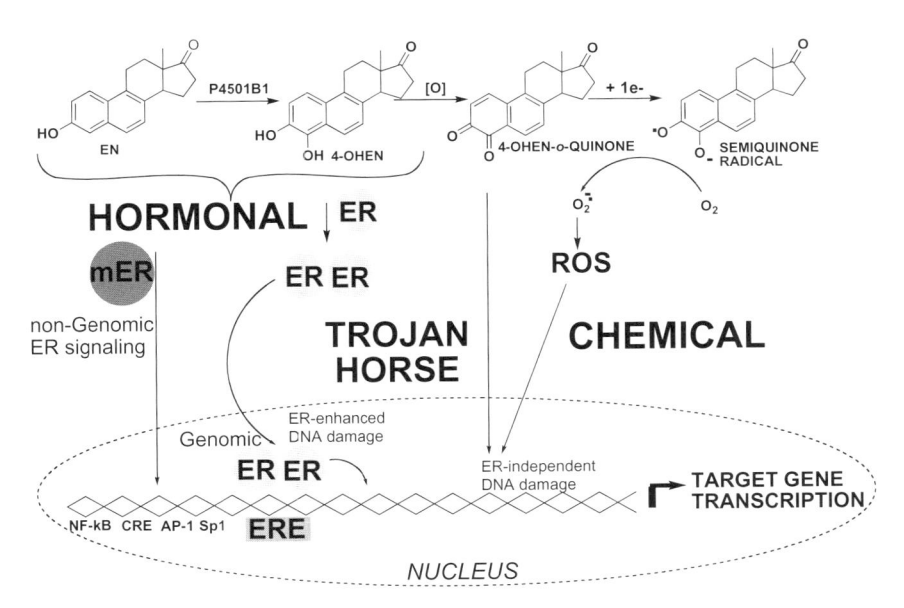

Scheme 8.2 Summary of potential carcinogenic mechanisms for estrogens using equilenin as an example. mER, membrane-associated estrogen receptor; ERE, estrogen response element; NF-κB, nuclear factor κB; CRE, cyclic AMP-responsive element; AP-1, activator protein-1; Sp1, steroidogenic protein-1.

recent work with methoxy ether metabolites of catechol estrogens suggests that instead of catechol-*O*-methyl transferase (COMT)-catalyzed detoxification of catechol estrogens, these ether metabolites may have hormonal properties similar to the parent estrogens that may contribute to the hormonal carcinogenesis mechanism [30].

8.2.2
Chemical Mechanism

8.2.2.1 Oxidative DNA Damage

Chemical carcinogenesis represents an alternative mechanism, describing the capacity of reactive intermediates derived from estrogen metabolism to cause DNA damage by electrophilic and oxidative reactions leading to genotoxicity [31]. For example, endogenous estrogens (Scheme 8.3) and equine estrogens (Scheme 8.2) are oxidized to *o*-quinones, which are electrophiles as well as potent redox-active compounds [32]. They can undergo redox cycling with the semiquinone radical generating superoxide radicals mediated through cytochrome P450/P450 reductase. The conversion of superoxide anion radicals to hydrogen peroxide, formed by the enzymatic or spontaneous dismutation of superoxide anion radicals, in the presence of trace amounts of iron or other transition metals gives rise to hydroxyl radicals. The hydroxyl radicals are powerful oxidizing agents that may be responsible for damage to essential macromolecules. In support of this mechanism,

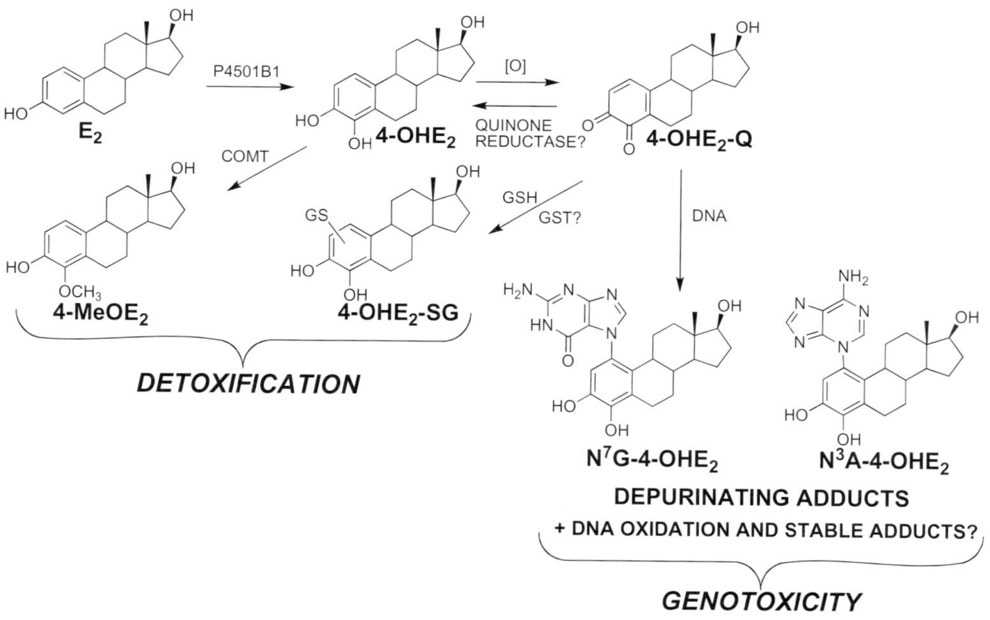

Scheme 8.3 Metabolism of estradiol to benign and mutagenic metabolites forming apurinic DNA adducts.

various free radical toxicities have been reported in hamsters treated with 17β-estradiol, including DNA single-strand breaks [33, 34], 8-oxo-dG formation [35–37], and chromosomal abnormalities [22, 38, 39]. Recently, it has also been shown that 4-hydroxyestradiol also induces oxidative stress and apoptosis in human mammary epithelial cells (MCF-10A), although the concentrations used in this study (25 μM) have questionable physiological relevance [40].

The equilenin catechol, 4-OHEN, which is the major phase 1 metabolite of both equilin and equilenin (Scheme 8.1), is also capable of causing DNA single-strand breaks and oxidative damage to DNA bases both *in vitro* and *in vivo* [41–44]. Injection of 4-OHEN into the mammary fat pads of Sprague-Dawley rats resulted in a dose-dependent increase in single-strand breaks and oxidized bases as analyzed by the COMET assay [43]. In addition, extraction of mammary tissue DNA, hydrolysis to deoxynucleosides, and analysis by liquid chromatography/tandem mass spectrometry (liquid chromatographyLC/mass spectrometryMS-MS) showed the formation of 8-oxo-dG as well as 8-oxo-dA. In mice treated with equilenin, the levels of 8-oxo-dG were increased 1.5-fold in the uterus [44]. In women, a recent study evaluated the potential of HRT to induce DNA damage in peripheral blood leukocytes of postmenopausal women using the COMET assay [45]. Significant increases in DNA damage were observed between women receiving 0.625 mg/day conjugated equine estrogens or conjugated equine estrogens plus medroxyprogesterone acetate as compared to the control group that had never received HRT. Finally, the excessive production of reactive oxygen species (ROS in breast cancer tissue has been linked to metastasis of tumors in women with breast cancer [46–49]. These and other data provide evidence for a mechanism of estrogen-induced tumor initiation/promotion by redox cycling of estrogen metabolites generating ROS, which damage DNA.

8.2.2.2 DNA Adducts

Estrogen quinoids can directly damage cellular DNA leading to genotoxic effects [21, 50–58]. Cavalieri's group has reported that the major DNA adducts produced from 4-hydroxyestradiol-*o*-quinone are depurinating N^7-guanine and N^3-adenine adducts resulting from 1,4-Michael addition both *in vitro* and *in vivo* (Scheme 8.3) [21, 37, 54, 57, 59–61]. Interestingly, they have recently concluded that only the N^3-adenine adduct is likely to induce mutations since this adduct depurinates instantaneously whereas the N^7-guanine adduct takes hours to hydrolyze [61, 62]. In contrast, the considerably more rapid isomerization of the 2-hydroxyestradiol-*o*-quinone to the corresponding quinone methides results in 1,6-Michael addition products with the exocyclic amino groups of adenine and guanine (Scheme 8.4) [60, 63]. Unlike the N^3- and N^7-purine DNA adducts, these adducts are stable which may alter their rate of repair and relative mutagenicity *in vivo*. A depurinating N^3-adenine adduct of 2-hydroxyestradiol quinone methide has recently been reported in reactions with adenine and DNA (Scheme 8.4) [61]. The levels of this adduct were considerably lower than corresponding depurinating adducts observed with similar experiments with 4-hydroxyestradiol-*o*-quinone, which may explain why 2-hydroxylation is considered a benign metabolic pathway whereas 4-hydroxylation

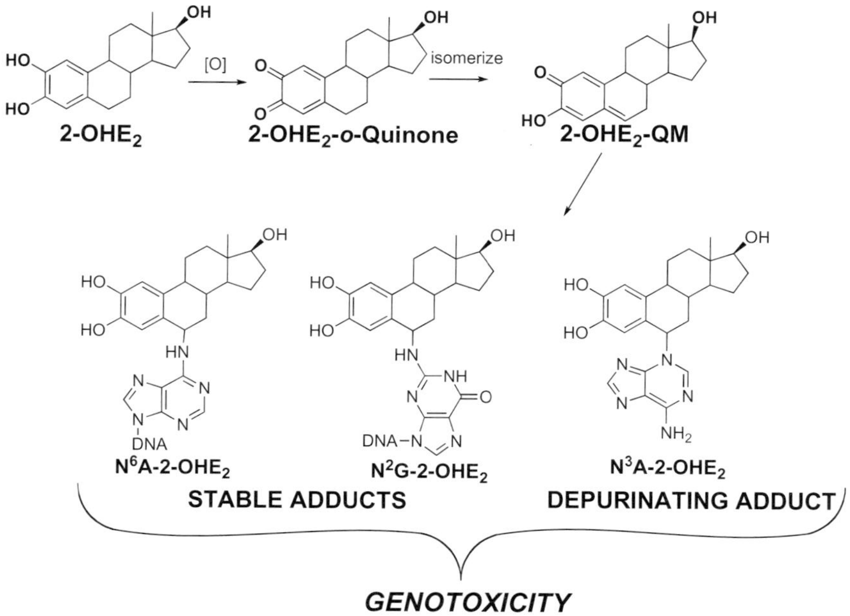

Scheme 8.4 Metabolism of equilenin benign and mutagenic metabolites forming stable DNA adducts.

results in carcinogenesis. Finally, this same study [61] suggested that depurinating DNA adducts of estrogen quinoids were formed in much greater abundance compared to stable bulky adducts, implying a causal role for these adducts in estrogen carcinogenesis; however, the depurinating adducts were analyzed by different methods (high-performance liquid chromatography with electrochemical detection) as compared to the stable adducts (^{32}P-postlabeling/thin-layer chromatography) making direct quantitative comparisons problematic. The mutagenic properties of 2-hydroxyestrogen quinone methide-derived stable DNA adducts have been evaluated using oligonucleotides containing site-specific adducts transfected into simian kidney (COS-7) cells where G→T and A→T mutations were observed [64]. It is important to mention that stable DNA adducts have been detected by ^{32}P-postlabeling in Syrian hamster embryo cells treated with estradiol and its catechol metabolites [65]. The rank order of DNA adduct formation which correlated with cellular transformation was 4-hydroxyestradiol > 2-hydroxyestradiol > estradiol. Finally, stable bulky adducts of 4-hydroxyestrone and 4-hydroxyestradiol corresponding to alkylation of guanine have been detected in human breast tumor tissue [66]. These data suggest that the relative importance of depurinating adducts versus stable DNA adducts in catechol estrogen carcinogenesis remains unclear.

Recently, there have been efforts to correlate depurinating estrogen DNA adducts with breast cancer risk. Ratios of depurinating DNA adducts to their respective estrogen metabolites were significantly higher in high-risk women (12

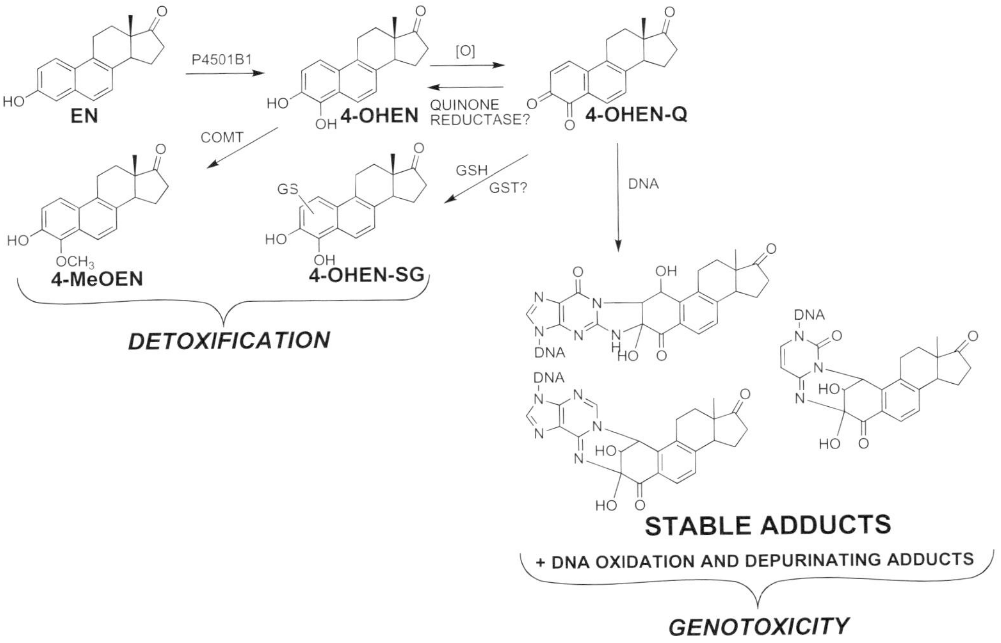

Scheme 8.5 Stable quinone methide DNA adducts formed from 2-hydroxyestradiol.

subjects) and women with breast cancer (17 subjects) compared to healthy women (46 subjects) [57]. However, another much smaller study (six subjects total) did not have the precision to conclude if the levels of depurinating estrogen DNA adducts were elevated in breast tissue from cancer patients [67]. More importantly, the levels of depurinating DNA adducts were close to the detection limits of the instrument (20–70 fmol/g tissue) – two orders of magnitude less than reported in an earlier Cavalieri study [68]. It is difficult to compare the results from the more recent Cavalieri study [57] with the Gross results [67], since only ratios of adducts were reported instead of fmol/g tissue. As a result, it is still not clear if depurinating estrogen DNA adducts can be used as biomarkers for breast cancer risk.

For the major equine estrogens (equilin and equilenin and 17β-ol derivatives) the data strongly suggests that the majority of DNA damage results from reactions of 4-OHEN-o-quinone through a combination of oxidative damage (i.e., single-strand cleavage and oxidation of DNA bases) and through generation of apurinic sites, as well as formation of stable bulky cyclic adducts (Scheme 8.2 and 8.5) [56]. For example, a depurinating guanine adduct was detected in *in vivo* experiments with rats treated with 4-OHEN, following LC/MS-MS analysis of extracted mammary tissue [43]. However, isolation of mammary tissue DNA, hydrolysis to deoxynucleosides, and analysis by LC/MS-MS also showed the formation of stable cyclic deoxyguanosine and deoxyadenosine adducts as well as the above-mentioned oxidized bases and single-strand breaks. Interestingly, the ratio of the diastereomeric adducts detected *in vivo* differs from *in vitro* experiments, suggesting that there are differences in the response of these stereoisomeric lesions to DNA replication and repair enzymes [69–72]. Finally, in a recent report, highly sensitive nano LC/MS-MS techniques were used to analyze the DNA in five human breast tumor and five adjacent tissue samples, including samples from donors with a known history of Premarin-based HRT [66]. While the sample size is small and the history of the patients is not fully known, cyclic 4-OHEN-dC, -dG, and -dA stable adducts were detected for the first time in four out of the 10 samples. These results suggest that 4-OHEN has the potential to be carcinogenic through the formation of a variety of DNA lesions *in vivo*.

8.2.2.3 Protection against DNA Damage

If catechol estrogen-induced DNA damage is a major mechanism contributing to estrogen carcinogenesis, it should be possible to lower the level of DNA damage, which may lead to a reduction in breast cancer risk. A number of protection mechanisms have been proposed including preventing the formation of estrone/estradiol with aromatase inhibitors (Scheme 8.6i) [73]. However, this strategy is not practical for healthy women since it places women into chemical menopause and removes all benefits of estrogens, including protection from osteoporosis. Another obvious strategy would therefore be the inhibition of CYP1B1 [74], which catalyzes 4-hydroxyestrogen formation (Scheme 8.6ii). Studies with CYP1B1 knockout mice demonstrated that animals lacking this gene developed normally and showed no noticeable deficiencies. Furthermore, CYP1B1 knockout mice showed strong resistance to 7,12-dimethylbenz[*a*]anthracene (DMBA)-induced tumor formation [75]. These studies provide evidence for the potential efficacy and safety of a chemopreventative agent for estrogen carcinogenesis that blocks CYP1B1 expression or activity. However, chemoprevention strategies based on inhibition of P450s are probably not practical due to the lack of isoform selectivity manifested by inhibitors. Alternatively, agents that control regulation of CYP1B1 maybe a more persuasive approach to chemopreventive therapy.

If it is not practical to prevent formation of the catechols/*o*-quinones, it may be possible to enhance their rate of detoxification. This could be achieved by COMT-catalyzed methylation of catechol estrogens (Scheme 8.6iii), reduction of estrogen quinones by quinone reductase (QR; Scheme 8.6iv), scavenging of estrogen semi-quinone radicals by antioxidants (Scheme 8.6v), or conjugation of estrogen quinones with thiols such as glutathione (GST; Scheme 8.6vi) [76]. It has been shown

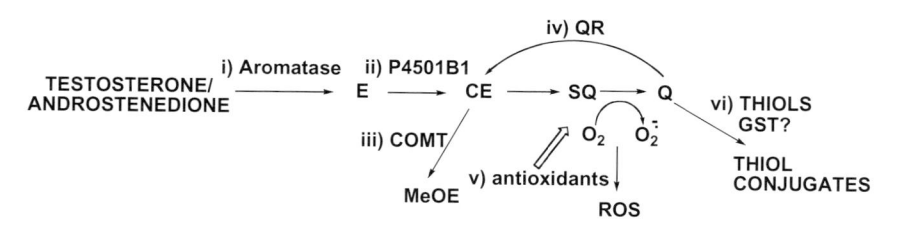

Scheme 8.6 Strategies to protect against estrogen-induced DNA damage.

that treatment of MCF-10F nontumorigenic breast epithelial cells with 4-hydroxyestradiol and the COMT inhibitor Ro41-0960 resulted in 3- to 4-fold increases in the levels of depurinating N^3-adenine and N^7-guanine adducts [77]. Similarly, knockdown of COMT expression increased neoplastic transformation of immortalized human endometrial glandular cells treated with 4-hydroxyestradiol [78]. As far as a link between genetic polymorphisms in COMT and risk of breast cancer are concerned, the data are equivocal [79].

It has been reported that induction of NQO1 activity protects against estrogen-induced oxidative DNA damage *in vitro* and *in vivo* [80]. These correlative findings were supported by findings that NQO1 downregulation led to increased levels of estrogen quinone metabolites and enhanced estrogen-induced transformation in MCF-10A nontumorigenic breast epithelial cells. Since epidemiological evidence indicates that genetically deficient NQO1 is a risk factor for the development of cancer [81], it is quite reasonable to hypothesize that NQO1 deficiency plays an important role in estrogen-dependent cancer etiology. A recent report showed that 4-hydroxyestrone *o*-quinone was observed to be a substrate for NQO1; however, the acceleration of NADPH-dependent reduction by NQO1 over the nonenzymatic reaction was less than 10-fold and at more relevant nanomolar concentrations of substrate was less than 2-fold [82]. These results indicate that a key role for NQO1 in direct detoxification of 4-hydroxyestrogen quinones is problematic.

Antioxidants such as resveratrol and melatonin have also been shown to reduce the levels of depurinating catechol estrogen DNA adducts [76, 83], presumably by reducing the catechol estrogen semiquinone radical to the catechol (Scheme 8.6v). Similar results were observed with *N*-acetylcysteine although this protective mechanism was probably a combination of antioxidant effects and nucleophilic scavenging of the estrogen quinones (Scheme 8.6v and vi). *N*-Acetylcysteine also reduced 4-hydroxyestradiol-induced transformation of normal mouse epithelial cells [84]. These preliminary studies suggest that it may be possible to reduce estrogen-dependent cancer risk by modulation of estrogen metabolism and detoxification of reactive intermediates.

8.3
Estrogen Receptor as a Trojan Horse (Combined Hormonal/Chemical Mechanism)

Estrogens that are potent ER agonists and are oxidized to electrophilic and redox-cycling metabolites have the potential to contribute to the initiation, promotion, and progression of hormone-sensitive cancers as dual-mechanism carcinogens (combined hormonal/chemical mechanisms discussed above). If catechol estrogens represent good estrogenic ligands, the ER would be capable of translocation of these genotoxins to the nucleus where oxidative DNA damage would be amplified, even at lower concentrations. The ER would act as a Trojan horse and ER-positive cells would be highly sensitive to DNA damage (Scheme 8.2). We have preliminary data that this mechanism may play a role in catechol estrogen-induced DNA damage [41, 42]. We have examined the effect of ER status on the relative ability of 4-OHEN and 4,17β-OHEN to induce DNA damage in ER-negative cells (MDA-MB-231), ERα-positive cells (S30), and ERβ-positive cells (β41). The data showed that both 4-OHEN and 4,17β-OHEN induced concentration-dependent DNA single-strand cleavage in all three cell lines. However, cells containing ERs had significantly higher DNA damage. The endogenous catechol estrogen metabolite 4-hydroxyestrone was considerably less effective at inducing DNA damage in breast cancer cell lines as compared to 4-OHEN [41]. Recently, we have shown that the rate of 4-OHEN-induced DNA damage was significantly enhanced in ERα-positive cells, whereas ER status had no effect on the rate of the nonestrogenic quinone menadione-induced damage [85]. Imaging of ROS induced by 4-OHEN showed that accumulation occurs selectively in the nucleus of ERα-positive cells within 5 min, whereas in ER-negative cells or menadione-treated cells, no selectivity was observed. Our data suggest that the genotoxic effects of 4-OHEN could be related to its ability to induce DNA damage in hormone-sensitive cells *in vivo* and that these effects may be potentiated by the ER. The Trojan horse model would also hold for alternative nuclear receptors and transcription factors and other ligands, such as environmental estrogen genotoxins and polyaromatic hydrocarbon metabolites in the case of the arylhydrocarbon receptor [86].

8.4
Conclusions and Future Directions

Receptor-mediated responses to hormones are a plausible and probably necessary mechanism for hormonal carcinogenesis. The results of research over the past few years add considerable support for a direct genotoxic effect of hormones or their associated byproducts such as ROS. Current knowledge does not allow a conclusion as to whether either of these mechanisms is the major determinant of hormonally induced cancer. It is entirely possible that both mechanisms contribute to and are necessary for carcinogenesis. Given the direct link between excessive exposure to estrogens, metabolism of estrogens, and increased risk of breast cancer, it is crucial that factors that affect the formation, reactivity, and cellular targets of estrogen quinoids be thoroughly explored.

Acknowledgements

This work is supported by NIH grants CA102590, CA79870, and CA73638.

References

1 Chen, W.Y. (2008) Exogenous and endogenous hormones and breast cancer. *Best Pract. Res. Clin. Endocrinol. Metab.*, **22**, 573–585.

2 Rossouw, J.E., Anderson, G.L., Prentice, R.L., LaCroix, A.Z., Kooperberg, C., Stefanick, M.L., Jackson, R.D., Beresford, S.A., Howard, B.V., Johnson, K.C.,

Kotchen, J.M., and Ockene, J. (2002) Risks and benefits of estrogen plus progestin in healthy postmenopausal women: principal results from the Women's Health Initiative randomized controlled trial. *J. Am. Med. Ass.*, **288**, 321–333.

3 Shumaker, S.A., Legault, C., Rapp, S.R., Thal, L., Wallace, R.B., Ockene, J.K., Hendrix, S.L., Jones, B.N., Assaf, A.R., Jackson, R.D., Kotchen, J.M., Wassertheil-Smoller, S., and Wactawski-Wende, J. (2003) Estrogen plus progestin and the incidence of dementia and mild cognitive impairment in postmenopausal women. The Women's Health Initiative memory study: a randomized controlled trial. *J. Am. Med. Ass.*, **289**, 2651–2662.

4 Brass, L.M. (2004) Hormone replacement therapy and stroke: clinical trials review. *Stroke*, **35**, 2644–2647.

5 Ravdin, P.M., Cronin, K.A., Howlader, N., Berg, C.D., Chlebowski, R.T., Feuer, E.J., Edwards, B.K., and Berry, D.A. (2007) The decrease in breast-cancer incidence in 2003 in the United States. *N. Engl. J. Med.*, **356**, 1670–1674.

6 Kumle, M. (2008) Declining breast cancer incidence and decreased HRT use. *Lancet*, **372**, 608–610.

7 Key, T., Appleby, P., Barnes, I., and Reeves, G. (2002) Endogenous sex hormones and breast cancer in postmenopausal women: reanalysis of nine prospective studies. *J. Natl. Cancer Inst.*, **94**, 606–616.

8 Wysowski, D.K. and Governale, L.A. (2005) Use of menopausal hormones in the United States, 1992 through June, 2003. *Pharmacoepidemiol. Drug Saf.*, **14**, 171–176.

9 Purdy, R.H., Moore, P.H., Williams, M.C., Goldzheher, H.W., and Paul, S.M. (1982) Relative rates of 2- and 4-hydroxyestrogen synthesis are dependent on both substrate and tissue. *FEBS Lett.*, **138**, 40–44.

10 Li, J.J., Li, S.A., Oberley, T.D., and Parsons, J.A. (1995) Carcinogenic activities of various steroidal and nonsteroidal estrogens in the hamster kidney: relation to hormonal activity and cell proliferation. *Cancer Res.*, **55**, 4347–4351.

11 Sarabia, S.F., Zhu, B.T., Kurosawa, T., Tohma, M., and Liehr, J.G. (1997) Mechanism of cytochrome P450-catalyzed aromatic hydroxylation of estrogens. *Chem. Res. Toxicol.*, **10**, 767–771.

12 Zhang, F., Chen, Y., Pisha, E., Shen, L., Xiong, Y., van Breemen, R.B., and Bolton, J.L. (1999) The major metabolite of equilin, 4-hydroxyequilin, autoxidizes to an o-quinone which isomerizes to the potent cytotoxin 4-hydroxyequilenin-o-quinone. *Chem. Res. Toxicol.*, **12**, 204–213.

13 Bhavnani, B.R. (1998) Pharmacokinetics and pharmacodynamics of conjugated equine estrogens: chemistry and metabolism. *Proc. Soc. Exp. Biol. Med.*, **217**, 6–16.

14 Judd, H.L., Mebane-Sims, I., Legault, C., Wasilauskas, C., Johnson, S., Merino, M., Barrett-Connor, B., and Trabal, J. (1996) Effects of hormone replacement therapy on endometrial histology in postmenopausal women. *J. Am. Med. Ass.*, **275**, 370–375.

15 Hays, J., Ockene, J.K., Brunner, R.L., Kotchen, J.M., Manson, J.E., Patterson, R.E., Aragaki, A.K., Shumaker, S.A., Brzyski, R.G., LaCroix, A.Z., Granek, I.A., and Valanis, B.G. (2003) Effects of estrogen plus progestin on health-related quality of life. *N. Engl. J. Med.*, **348**, 1839–1854.

16 Hersh, A.L., Stefanick, M.L., and Stafford, R.S. (2004) National use of postmenopausal hormone therapy: annual trends and response to recent evidence. *J. Am. Med. Ass.*, **291**, 47–53.

17 Zumoff, B. (1998) Does postmenopausal estrogen administration increase the risk of breast cancer? Contributions of animal, biochemical, and clinical investigative studies to a resolution of the controversy. *Proc. Soc. Exp. Biol. Med.*, **217**, 30–37.

18 Yager, J.D. and Davidson, N.E. (2006) Estrogen carcinogenesis in breast cancer. *N. Engl. J. Med.*, **354**, 270–282.

19 Russo, J., Hu, Y.F., Yang, X., and Russo, I.H. (2000) Developmental, cellular, and molecular basis of human breast cancer. *J. Natl. Cancer Inst. Monogr.*, **27**, 17–37.

20 Jefcoate, C.R., Liehr, J.G., Santen, R.J., Sutter, T.R., Yager, J.D., Yue, W., Santner, S.J., Tekmal, R., Demers, L., Pauley, R., Naftolin, F., Mor, G., and Berstein, L. (2000) Tissue-specific synthesis and oxidative metabolism of

estrogens. *J. Natl. Cancer Inst. Monogr.*, **27**, 95–112.

21 Cavalieri, E., Chakravarti, D., Guttenplan, J., Hart, E., Ingle, J., Jankowiak, R., Muti, P., Rogan, E., Russo, J., Santen, R., and Sutter, T. (2006) Catechol estrogen quinones as initiators of breast and other human cancers: implications for biomarkers of susceptibility and cancer prevention. *Biochim. Biophys. Acta*, **1766**, 63–78.

22 Russo, J. and Russo, I.H. (2006) The role of estrogen in the initiation of breast cancer. *J. Steroid Biochem. Mol. Biol.*, **102**, 89–96.

23 Feigelson, H.S. and Henderson, B.E. (1996) Estrogens and breast cancer. *Carcinogenesis*, **17**, 2279–2284.

24 Henderson, B.E. and Feigelson, H.S. (2000) Hormonal carcinogenesis. *Carcinogenesis*, **21**, 427–433.

25 Nandi, S., Guzman, R.C., and Yang, J. (1995) Hormones and mammary carcinogenesis in mice, rats, and humans: a unifying hypothesis. *Proc. Natl. Acad. Sci. USA*, **92**, 3650–3657.

26 Flototto, T., Djahansouzi, S., Glaser, M., Hanstein, B., Niederacher, D., Brumm, C., and Beckmann, M.W. (2001) Hormones and hormone antagonists: mechanisms of action in carcinogenesis of endometrial and breast cancer. *Horm. Metab. Res.*, **33**, 451–457.

27 Revankar, C.M., Cimino, D.F., Sklar, L.A., Arterburn, J.B., and Prossnitz, E.R. (2005) A transmembrane intracellular estrogen receptor mediates rapid cell signaling. *Science*, **307**, 1625–1630.

28 Song, R.X., Fan, P., Yue, W., Chen, Y., and Santen, R.J. (2006) Role in$\alpha$of receptor complexes in the extranuclear actions of estrogen receptor breast cancer. *Endocr. Relat. Cancer*, **13** (Suppl. 1), S3–S13.

29 Chen, J.Q., Brown, T.R., and Yager, J.D. (2008) Mechanisms of hormone carcinogenesis: evolution of views, role of mitochondria. *Adv. Exp. Med. Biol.*, **630**, 1–18.

30 Chang, M., Peng, K.W., Kastrati, I., Overk, C.R., Qin, Z.H., Yao, P., Bolton, J.L., and Thatcher, G.R. (2007) Activation of estrogen receptor-mediated gene transcription by the equine estrogen metabolite, 4-methoxyequilenin, in human breast cancer cells. *Endocrinology*, **148**, 4793–4802.

31 Liehr, J.G. (2001) Genotoxicity of the steroidal oestrogens oestrone and oestradiol: possible mechanism of uterine and mammary cancer development. *Hum. Reprod. Update*, **7**, 273–281.

32 Bolton, J.L., Trush, M.A., Penning, T.M., Dryhurst, G., and Monks, T.J. (2000) Role of quinones in toxicology. *Chem. Res. Toxicol.*, **13**, 135–160.

33 Roy, D. and Liehr, J.G. (1999) Estrogen, DNA damage and mutations. *Mutat. Res.*, **424**, 107–115.

34 Nutter, L.M., Ngo, E.O., and Abul-Hajj, Y.J. (1991) Characterization of DNA damage induced by 3,4-estrone-o-quinone in human cells. *J. Biol. Chem.*, **266**, 16380–16386.

35 Lavigne, J.A., Goodman, J.E., Fonong, T., Odwin, S., He, P., Roberts, D.W., and Yager, J.D. (2001) The effects of catechol-O-methyltransferase inhibition on estrogen metabolite and oxidative DNA damage levels in estradiol-treated MCF-7 cells. *Cancer Res.*, **61**, 7488–7494.

36 Rajapakse, N., Butterworth, M., and Kortenkamp, A. (2005) Detection of DNA strand breaks and oxidized DNA bases at the single-cell level resulting from exposure to estradiol and hydroxylated metabolites. *Environ. Mol. Mutagen.*, **45**, 397–404.

37 Cavalieri, E., Frenkel, K., Liehr, J.G., Rogan, E., and Roy, D. (2000) Estrogens as endogenous genotoxic agents – DNA adducts and mutations. *J. Natl. Cancer Inst. Monogr.*, **27**, 75–93.

38 Li, J.J., Gonzalez, A., Banerjee, S., Banerjee, S.K., and Li, S.A. (1993) Estrogen carcinogenesis in the hamster kidney: role of cytotoxicity and cell proliferation. *Environ. Health Perspect.*, **5**, 259–264.

39 Banerjee, S.K., Banerjee, S., Li, S.A., and Li, J.J. (1994) Induction of chromosome aberrations in Syrian hamster renal cortical cells by various estrogens. *Mutat. Res.*, **311**, 191–197.

40 Chen, Z.H., Na, H.K., Hurh, Y.J., and Surh, Y.J. (2005) 4-Hydroxyestradiol induces oxidative stress and apoptosis in human mammary epithelial cells:

possible protection by NF-κB and ERK/MAPK. *Toxicol. Appl. Pharmacol.*, **208**, 46–56.

41 Chen, Y., Liu, X., Pisha, E., Constantinou, A.I., Hua, Y., Shen, L., van Breemen, R.B., Elguindi, E.C., Blond, S.Y., Zhang, F., and Bolton, J.L. (2000) A metabolite of equine estrogens, 4-hydroxyequilenin, induces DNA damage and apoptosis in breast cancer cell lines. *Chem. Res. Toxicol.*, **13**, 342–350.

42 Liu, X., Yao, J., Pisha, E., Yang, Y., Hua, Y., van Breemen, R.B., and Bolton, J.L. (2002) Oxidative DNA damage induced by equine estrogen metabolites: role of estrogen receptor alpha. *Chem. Res. Toxicol.*, **15**, 512–519.

43 Zhang, F., Swanson, S.M., van Breemen, R.B., Liu, X., Yang, Y., Gu, C., and Bolton, J.L. (2001) Equine estrogen metabolite 4-hydroxyequilenin induces DNA damage in the rat mammary tissues: formation of single-strand breaks, apurinic sites, stable adducts, and oxidized bases. *Chem. Res. Toxicol.*, **14**, 1654–1659.

44 Okamoto, Y., Chou, P.H., Kim, S.Y., Suzuki, N., Laxmi, Y.R., Okamoto, K., Liu, X., Matsuda, T., and Shibutani, S. (2008) Oxidative DNA damage in XPC-knockout and its wild mice treated with equine estrogen. *Chem. Res. Toxicol.*, **21**, 1120–1124.

45 Ozcagli, E., Sardas, S., and Biri, A. (2005) Assessment of DNA damage in postmenopausal women under hormone replacement therapy. *Maturitas*, **51**, 280–285.

46 Malins, D.C., Polissar, N.L., and Gunselman, S.J. (1996) Progession of human breast cancers to the metastatic state is linked to hydroxyl radical-induced DNA damage. *Proc. Natl. Acad. Sci. USA*, **93**, 2557–2563.

47 Malins, D.C., Anderson, K.M., Jaruga, P., Ramsey, C.R., Gilman, N.K., Green, V.M., Rostad, S.W., Emerman, J.T., and Dizdaroglu, M. (2006) Oxidative changes in the DNA of stroma and epithelium from the female breast: potential implications for breast cancer. *Cell Cycle*, **5**, 1629–1632.

48 Karihtala, P. and Soini, Y. (2007) Reactive oxygen species and antioxidant mechanisms in human tissues and their relation to malignancies. *APMIS*, **115**, 81–103.

49 Benz, C.C. and Yau, C. (2008) Ageing, oxidative stress and cancer: paradigms in parallax. *Nat. Rev. Cancer*, **8**, 875–879.

50 Bolton, J.L., Yu, L., and Thatcher, G.R. (2004) Quinoids formed from estrogens and antiestrogens. *Methods Enzymol.*, **378**, 110–123.

51 Prokai-Tatrai, K. and Prokai, L. (2005) Impact of metabolism on the safety of estrogen therapy. *Ann. NY Acad. Sci.*, **1052**, 243–257.

52 Liehr, J.G. (2000) Role of DNA adducts in hormonal carcinogenesis. *Regul. Toxicol. Pharmacol.*, **32**, 276–282.

53 Russo, J. and Russo, I.H. (2004) Genotoxicity of steroidal estrogens. *Trends Endocrinol. Metab.*, **15**, 211–214.

54 Li, K.M., Todorovic, R., Devanesan, P., Higginbotham, S., Kofeler, H., Ramanathan, R., Gross, M.L., Rogan, E.G., and Cavalieri, E.L. (2004) Metabolism and DNA binding studies of 4-hydroxyestradiol and estradiol-3,4-quinone *in vitro* and in female ACI rat mammary gland *in vivo*. *Carcinogenesis*, **25**, 289–297.

55 Chakravarti, D., Mailander, P.C., Li, K.M., Higginbotham, S., Zhang, H.L., Gross, M.L., Meza, J.L., Cavalieri, E.L., and Rogan, E.G. (2001) Evidence that a burst of DNA depurination in SENCAR mouse skin induces error-prone repair and forms mutations in the H-*ras* gene. *Oncogene*, **20**, 7945–7953.

56 Bolton, J.L. and Thatcher, G.R. (2008) Potential mechanisms of estrogen quinone carcinogenesis. *Chem. Res. Toxicol.*, **21**, 93–101.

57 Gaikwad, N.W., Yang, L., Muti, P., Meza, J.L., Pruthi, S., Ingle, J.N., Rogan, E.G., and Cavalieri, E.L. (2008) The molecular etiology of breast cancer: evidence from biomarkers of risk. *Int. J. Cancer*, **122**, 1949–1957.

58 Zhang, Q. and Gross, M.L. (2008) Efficient synthesis, liquid chromatography purification, and tandem mass spectrometric characterization of estrogen-modified DNA Bases. *Chem. Res. Toxicol.*, **21**, 1244–1252.

59 Saeed, M., Rogan, E., Fernandez, S.V., Sheriff, F., Russo, J., and Cavalieri, E.

(2007) Formation of depurinating N^3 adenine and N^7 guanine adducts by MCF-10F cells cultured in the presence of 4-hydroxyestradiol. *Int. J. Cancer*, **120**, 1821–1824.

60 Stack, D.E., Byun, J., Gross, M.L., Rogan, E.G., and Cavalieri, E.L. (1996) Molecular characteristics of catechol estrogen quinones in reactions with deoxyribonucleosides. *Chem. Res. Toxicol.*, **9**, 851–859.

61 Zahid, M., Kohli, E., Saeed, M., Rogan, E., and Cavalieri, E. (2006) The greater reactivity of estradiol-3,4-quinone vs estradiol-2,3-quinone with DNA in the formation of depurinating adducts: implications for tumor-initiating activity. *Chem. Res. Toxicol.*, **19**, 164–172.

62 Saeed, M., Zahid, M., Gunselman, S.J., Rogan, E., and Cavalieri, E. (2005) Slow loss of deoxyribose from the N^7deoxyguanosine adducts of estradiol-3,4-quinone and hexestrol-3',4'-quinone. Implications for mutagenic activity. *Steroids*, **70**, 29–35.

63 Debrauwer, L., Rathahao, E., Jouanin, I., Paris, A., Clodic, G., Molines, H., Convert, O., Fournier, F., and Tabet, J.C. (2003) Investigation of the regio- and stereo-selectivity of deoxyguanosine linkage to deuterated 2-hydroxyestradiol by using liquid chromatography/ESI-ion trap mass spectrometry. *J. Am. Soc. Mass Spectrom.*, **14**, 364–372.

64 Terashima, I., Suzuki, N., and Shibutani, S. (2001) Mutagenic properties of estrogen quinone-derived DNA adducts in simian kidney cells. *Biochemistry*, **40**, 166–172.

65 Hayashi, N., Hasegawa, K., Barrett, J.C., and Tsutsui, T. (1996) Estrogen-induced cell transformation and DNA adduct formation in cultured Syrian hamster embryo cells. *Mol. Carcinogenesis*, **16**, 149–156.

66 Embrechts, J., Lemiere, F., Dongen, W.V., Esmans, E.L., Buytaert, P., van, Marck, E., Kockx, M., and Makar, A. (2003) Detection of estrogen DNA-adducts in human breast tumor tissue and healthy tissue by combined nano LC-nano ES tandem mass spectrometry. *J. Am. Soc. Mass Spectrom.*, **14**, 482–491.

67 Zhang, Q., Aft, R.L., and Gross, M.L. (2008) Estrogen carcinogenesis: specific identification of estrogen-modified nucleobase in breast tissue from women. *Chem. Res. Toxicol.*, **21**, 1509–1513.

68 Markushin, Y., Zhong, W., Cavalieri, E.L., Rogan, E.G., Small, G.J., Yeung, E.S., and Jankowiak, R. (2003) Spectral characterization of catechol estrogen quinone (CEQ)-derived DNA adducts and their identification in human breast tissue extract. *Chem. Res. Toxicol.*, **16**, 1107–1117.

69 Ding, S., Shapiro, R., Geacintov, N.E., and Broyde, S. (2003) Conformations of stereoisomeric base adducts to 4-hydroxyequilenin. *Chem. Res. Toxicol.*, **16**, 695–707.

70 Yasui, M., Laxmi, Y.R., Ananthoju, S.R., Suzuki, N., Kim, S.Y., and Shibutani, S. (2006) Translesion synthesis past equine estrogen-derived 2'-deoxyadenosine DNA adducts by human DNA polymerases eta and kappa. *Biochemistry*, **45**, 6187–6194.

71 Ding, S., Shapiro, R., Geacintov, N.E., and Broyde, S. (2007) 4-Hydroxyequilenin-adenine lesions in DNA duplexes: stereochemistry, damage site, and structure. *Biochemistry*, **46**, 182–191.

72 Kolbanovskiy, A., Kuzmin, V., Shastry, A., Kolbanovskaya, M., Chen, D., Chang, M., Bolton, J.L., and Geacintov, N.E. (2005) Base selectivity and effects of sequence and DNA secondary structure on the formation of covalent adducts derived from the equine estrogen metabolite 4-hydroxyequilenin. *Chem. Res. Toxicol.*, **18**, 1737–1747.

73 Castrellon, A.B. and Gluck, S. (2008) Chemoprevention of breast cancer. *Expert Rev. Anticancer Ther.*, **8**, 443–452.

74 Bruno, R.D. and Njar, V.C. (2007) Targeting cytochrome P450 enzymes: a new approach in anti-cancer drug development. *Bioorg. Med. Chem.*, **15**, 5047–5060.

75 Gonzalez, F.J. (2002) Transgenic models in xenobiotic metabolism and toxicology. *Toxicology*, **181–182**, 237–239.

76 Zahid, M., Gaikwad, N.W., Ali, M.F., Lu, F., Saeed, M., Yang, L., Rogan, E.G., and Cavalieri, E.L. (2008) Prevention of estrogen-DNA adduct formation in MCF-10F cells by resveratrol. *Free Radic. Biol. Med.*, **45**, 136–145.

77 Zahid, M., Saeed, M., Lu, F., Gaikwad, N., Rogan, E., and Cavalieri, E. (2007) Inhibition of catechol-*O*-methyltransferase increases estrogen-DNA adduct formation. *Free Radic. Biol. Med.*, **43**, 1534–1540.

78 Salama, S.A., Kamel, M., Awad, M., Nasser, A.H., Al-Hendy, A., Botting, S., and Arrastia, C. (2008) Catecholestrogens induce oxidative stress and malignant transformation in human endometrial glandular cells: protective effect of catechol-*O*-methyltransferase. *Int. J. Cancer*, **123**, 1246–1254.

79 Bugano, D.D., Conforti-Froes, N., Yamaguchi, N.H., and Baracat, E.C. (2008) Genetic polymorphisms, the metabolism of estrogens and breast cancer: a review. *Eur. J. Gynaecol. Oncol.*, **29**, 313–320.

80 Montano, M.M., Chaplin, L.J., Deng, H., Mesia-Vela, S., Gaikwad, N., Zahid, M., and Rogan, E. (2007) Protective roles of quinone reductase and tamoxifen against estrogen-induced mammary tumorigenesis. *Oncogene*, **26**, 3587–3590.

81 Cornblatt, B.S., Ye, L., Dinkova-Kostova, A.T., Erb, M., Fahey, J.W., Singh, N.K., Chen, M.S., Stierer, T., Garrett-Mayer, E., Argani, P., Davidson, N.E., Talalay, P., Kensler, T.W., and Visvanathan, K. (2007) Preclinical and clinical evaluation of sulforaphane for chemoprevention in the breast. *Carcinogenesis*, **28**, 1485–1490.

82 Chandrasena, R.E., Edirisinghe, P.D., Bolton, J.L., and Thatcher, G.R. (2008) Problematic detoxification of estrogen quinones by NAD(P)H-dependent quinone oxidoreductase and glutathione-S-transferase. *Chem. Res. Toxicol.*, **21**, 1324–1329.

83 Zahid, M., Gaikwad, N.W., Rogan, E.G., and Cavalieri, E.L. (2007) Inhibition of depurinating estrogen-DNA adduct formation by natural compounds. *Chem. Res. Toxicol.*, **20**, 1947–1953.

84 Venugopal, D., Zahid, M., Mailander, P.C., Meza, J.L., Rogan, E.G., Cavalieri, E.L., and Chakravarti, D. (2008) Reduction of estrogen-induced transformation of mouse mammary epithelial cells by N-acetylcysteine. *J. Steroid Biochem. Mol. Biol.*, **109**, 22–30.

85 Wang, Z., Wijewickrama, G.T., Peng, K.W., Dietz, B., Yuan, L., van Breemen, R.B., Thatcher, G.R.J., and Bolton, J.L. (2009) Estrogen receptor α enhances the rate of oxidative DNA damage by targeting an equine estrogen catechol metabolite to the nucleus. *J. Biol. Chem.*, **284**, 8633–8642.

86 Burczynski, M.E. and Penning, T.M. (2000) Genotoxic polycyclic aromatic hydrocarbon *ortho*-quinones generated by aldo-keto reductases induce CYP1A1 via nuclear translocation of the aryl hydrocarbon receptor. *Cancer Res.*, **60**, 908–915.

Part Two
New Frontiers and Challenges: Understanding Structure–Function Relationships and Biological Activity

The Chemical Biology of DNA Damage. Edited by Nicholas E. Geacintov and Suse Broyde
© 2010 WILEY-VCH Verlag GmbH & Co. KGaA, Weinheim
ISBN: 978-3-527-32295-4

9
Interstrand DNA Cross-Linking 1,N^2-Deoxyguanosine Adducts Derived from α,β-Unsaturated Aldehydes: Structure–Function Relationships

Michael P. Stone, Hai Huang, Young-Jin Cho, Hye-Young Kim, Ivan D. Kozekov, Albena Kozekova, Hao Wang, Irina G. Minko, R. Stephen Lloyd, Thomas M. Harris, and Carmelo J. Rizzo

9.1
Introduction

The α,β-unsaturated aldehydes (enals) acrolein, crotonaldehyde, and 4-hydroxynonenal (4-HNE) (Scheme 9.1) are endogenous byproducts of lipid peroxidation that result from cellular oxidative stress [1–4]. Human exposure to acrolein and crotonaldehyde also occurs from exogenous sources, such as tobacco smoke [5] and automobile exhaust [6]. These enals react as bis-electrophiles with the exocyclic amino groups and ring nitrogen atoms of DNA nucleobases to give exocyclic adducts [7]. Acrolein reacts principally with dG resulting in the exocyclic adduct 3-(2-deoxy-β-D-*erythro*-pentofuranosyl)-5,6,7,8-tetrahydro-8-hydroxypyrimido[1,2-α]purin-10(*3H*)-one (γ-OH-PdG, **9**) [8, 9]. The regioisomeric 3-(2-deoxy-β-D-*erythro*-pentofuranosyl)-5,6,7,8-tetrahydro-6-hydroxypyrimido[1,2-α]purin-10(*3H*)-one (α-OH-PdG, **10**) has also been observed [9, 10]. The γ-OH-PdG adduct (**9**) exists as a mixture of C8-OH epimers. With crotonaldehyde, addition at N^2-dG creates a new stereocenter at C6. Of the four possible products, the two with the *trans* relative configurations at C6 and C8 (**11** and **12**) predominate [9, 11]. Adducts **11** and **12** also form through the reaction of dG with 2 equiv. of acetaldehyde [5, 12, 13]. The corresponding 4-HNE-derived 1,N^2-dG adducts possess an additional stereocenter on the C6 side-chain, resulting in four observable diastereomers (**13–16**).

The 1,N^2-dG exocyclic adducts from acrolein (**9** and **10**), crotonaldehyde (**11** and **12**), and 4-HNE (**13–16**) exist in human and rodent DNA [2–4, 10, 14], and are implicated in the etiologies of human cancers. For example, the binding pattern of acrolein-DNA adducts is similar to the p53 mutational pattern in human lung cancer, implicating acrolein as a tobacco-related lung carcinogen [15]. Indeed, acrolein is mutagenic in bacterial and mammalian cells [16, 17], including human cells [18, 19], and it is carcinogenic in rats [20]. Crotonaldehyde is genotoxic and mutagenic in human lymphoblasts [21], and it induces liver tumors in rodents [22]. While 4-HNE induces a DNA damage response in *Salmonella typhimurium* [23] it is inactive in bacterial mutagenesis assays [16]. However, it causes mutations in

The Chemical Biology of DNA Damage. Edited by Nicholas E. Geacintov and Suse Broyde
© 2010 WILEY-VCH Verlag GmbH & Co. KGaA, Weinheim
ISBN: 978-3-527-32295-4

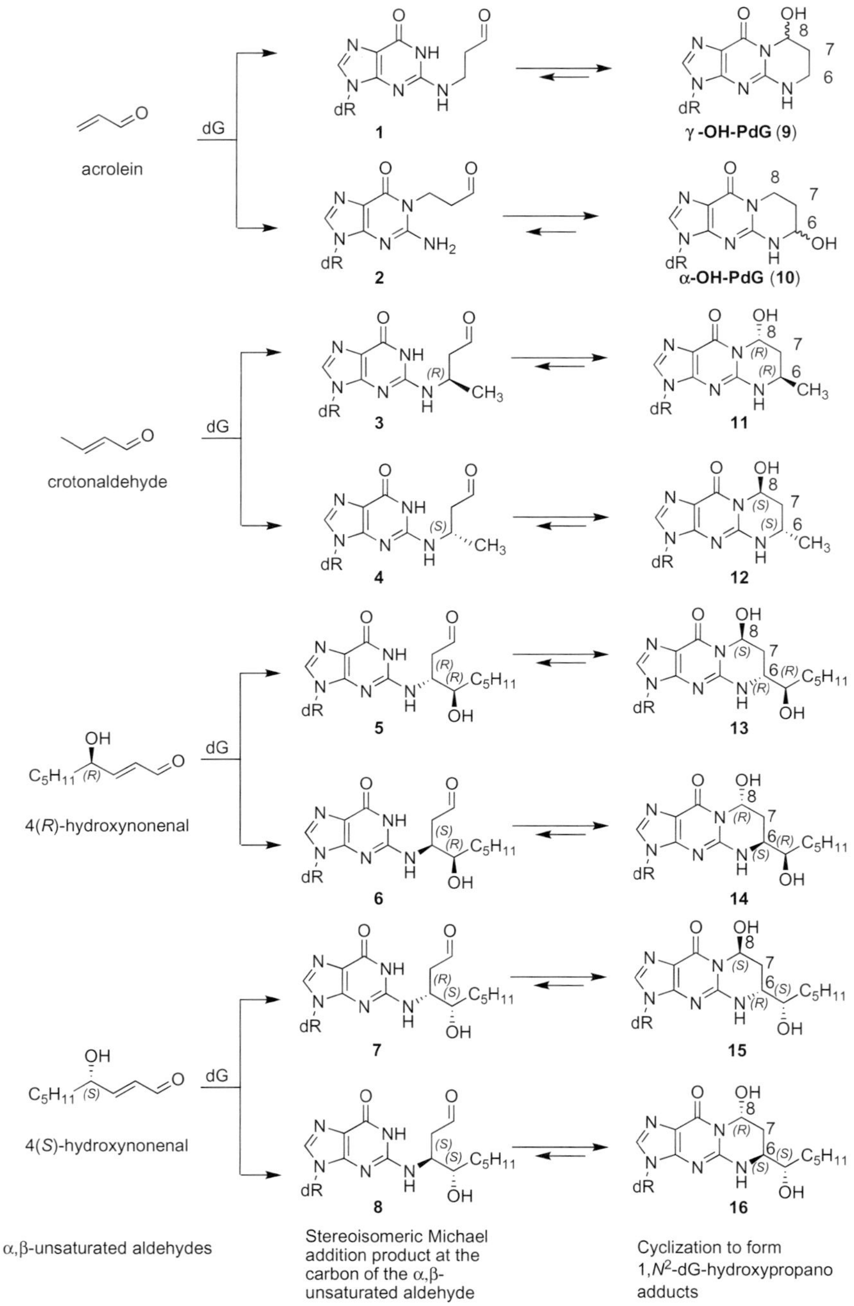

Scheme 9.1 1,N^2-dG cyclic adducts arising from Michael addition of acrolein, crotonaldehyde, and 4-HNE to dG.

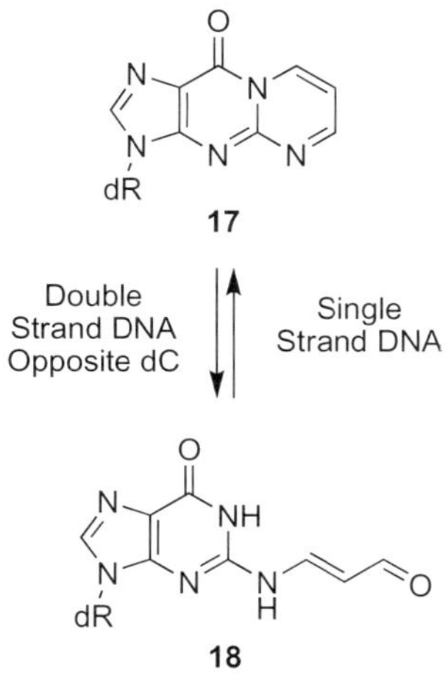

Scheme 9.2 Chemistry of the malondialdehyde-derived M₁dG adduct.

V79 CHO cells [24]. Individuals suffering from Wilson's disease and hemochromatosis have been reported to contain DNA mutations in liver cells, which are attributed to 4-HNE-dG adducts [25]. Site-specific mutagenesis in COS-7 cells reveals that these 1,N^2-dG adducts induce predominantly G → T transversions [26–28].

The malondialdehyde-derived adduct **17** opens to aldehyde **18** when placed opposite dC in DNA (Scheme 9.2) [29–31]. The enal adducts of interest herein represent lower oxidation state homologs of **17**, which suggested that they might undergo similar chemistry. This hypothesis was confirmed by de los Santos *et al.* [32] who showed that the γ-OH-PdG adduct (**9**) existed as the ring-opened aldehyde N^2-(3-oxopropyl)-dG (**1**) when placed opposite dC. We further hypothesized that aldehyde (**1**) could subsequently react with dG in the complementary DNA strand, forming interstrand cross-links. The likelihood of such cross-links was suggested from previous studies of acrolein-treated DNA [19]. The chemical nature of the cross-link could be carbinolamine (**19**), imine (**20**), pyrimidopurinone (**21**), or an equilibrium mixture of the three (Scheme 9.3) [33, 34]. This chapter summarizes emerging structure–function relationships associated with interstrand cross-linking chemistry arising from the N^2-(3-oxopropyl)-dG lesions derived from 1,N^2-dG adducts of acrolein, crotonaldehyde, and 4-HNE.

9.2
Interstrand Cross-Linking Chemistry of the γ-OH-PdG Adduct (9)

When the γ-OH-PdG adduct **9** was site-specifically synthesized in the 5′-CpG-3′ sequence and annealed to its complement, interstrand cross-linking reached a level of 50% yield after 7 days at 25 °C [33, 35]. Mass spectrometry (MS) analysis suggested that the cross-link involved is a carbinolamine linkage (**19**), in

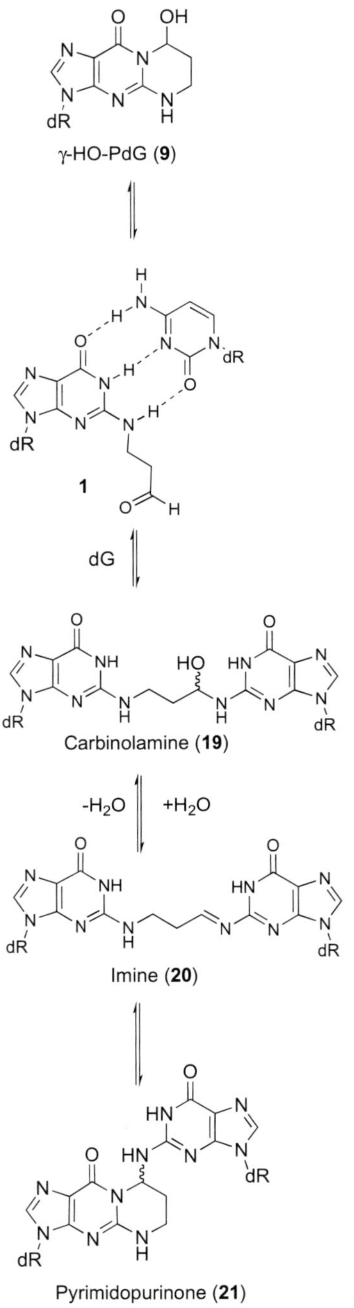

Scheme 9.3 DNA interstrand cross-link formation mediated by the acrolein-derived γ-OH-PdG adduct.

equilibrium with either or both the imine (**20**) or pyrimidopurinone (**21**) forms [33]. Enzymatic digestion of the cross-linked duplex yielded diastereomeric pyrimidopurinone bis-nucleoside cross-links **21**, which are structurally related to those arising from acetaldehyde-treated DNA [12]. Reduction of **21** afforded N^2-dG:N^2-dG bis-nucleosides tethered by a trimethylene chain [33]. If the cross-linked duplex was reduced with NaB(CN)H$_3$ prior to its digestion, N^2-(3-hydroxypropyl)-dG resulting from the reduction of γ-OH-PdG (**9**) and the reduced cross-link were observed.

Despite the fact that the cross-link could be reductively trapped, Nuclear magnetic resonance (NMR) experiments utilizing γ-^{13}C-γ-OH-PdG failed to detect the imine linkage, suggesting that in duplex DNA at equilibrium, it must be present only at low levels [36]. The identification of the major chemical cross-link species as the carbinolamine (**19**) and not the pyrimidopurinone (**21**) [33] was accomplished by isotope-edited NMR experiments [36, 37]. An oligodeoxynucleotide containing γ-OH-$^{15}N^2$-PdG in the 5′-CpG-3′ sequence was annealed with its complementary strand. A ^{15}N-heteronuclear single quantum coherence-filtered spectrum revealed the nuclear Overhauser effect between $^{15}N^2$H and the imino proton of the same nucleotide, precluding the pyrimidopurinone structure (**21**). When an oligodeoxynucleotide containing γ-^{13}C-PdG was annealed with the complementary oligodeoxynucleotide in which the targeted dG was ^{15}N-labeled at the exocyclic amino group, a triple resonance ^{1}H^{13}C^{15}N experiment revealed a correlation between the γ-^{13}C carbinol and the ^{15}N amine [37].

9.3
Interstrand Cross-Linking by the α-CH$_3$-γ-OH-PdG Adducts Derived from Crotonaldehyde

As noted, Michael addition of N^2-dG to crotonaldehyde creates a stereocenter at C6 [9, 11] and the abilities of these diastereomeric adducts to form interstrand cross-links in the 5′-CpG-3′ sequence depended upon stereochemistry at the C6 carbon. After 20 days, 40% cross-link formation occurred for the 6R-diastereomer **11**, whereas less than 5% cross-link was observed for the 6S-diastereomer **12** [35]. Digestion of the cross-links yielded the bis-nucleoside pyrimidopurinones analogous to **21** [35], identical to those isolated from acetaldehyde-treated DNA [12]. The presence of some amount of the imine linkage analogous to **20** was inferred since the cross-link was reductively trapped [13, 35]. However, the imine linkage could not be detected by NMR in duplex DNA. Using isotope-edited NMR experiments [38], the carbinolamine form of the 6R cross-link analogous to **19** was the only spectroscopically detectable cross-link species present.

9.4
Interstrand Cross-Linking by 4-HNE

Only the 4-HNE adduct with the (6S,8R,11S) configuration (**16**) formed an interstrand cross-link in the 5′-CpG-3′ sequence [39]. Note that this configuration of the 4-HNE adduct possesses the same relative stereochemistry at C6 as the 6R configuration of the crotonaldehyde adduct (**11**). Digestion yielded the pyrimidopurinone bis-nucleoside cross-link analogous to **21** [39]. Interstrand cross-linking proceeded slowly, but after 2 months, the yield was 85% [39]. When placed complementary to dC in the 5′-CpG-3′ sequence, (6S,8R,11S) adduct **16** and (6R,8S,11R) adduct **13** opened to the corresponding N^2-dG aldehydic rearrangement products, **8** and **5**, respectively. Thus, the formation of the interstrand cross-link by the adduct of (6S,8R,11S) stereochemistry **16** and the lack of cross-link formation by the adduct of (6R,8S,11R) stereochemistry **13** was not attributable to inability to undergo ring-opening to the aldehydes **5** and **8** in duplex DNA. Instead, the aldehydic adducts **5** and **8** existed in equilibrium with diastereomeric cyclic hemiacetals, **22** and **23** and **24** and **25**, respectively (Scheme 9.4) [40]. The cyclic hemiacetals were the predominant species present at equilibrium. In both

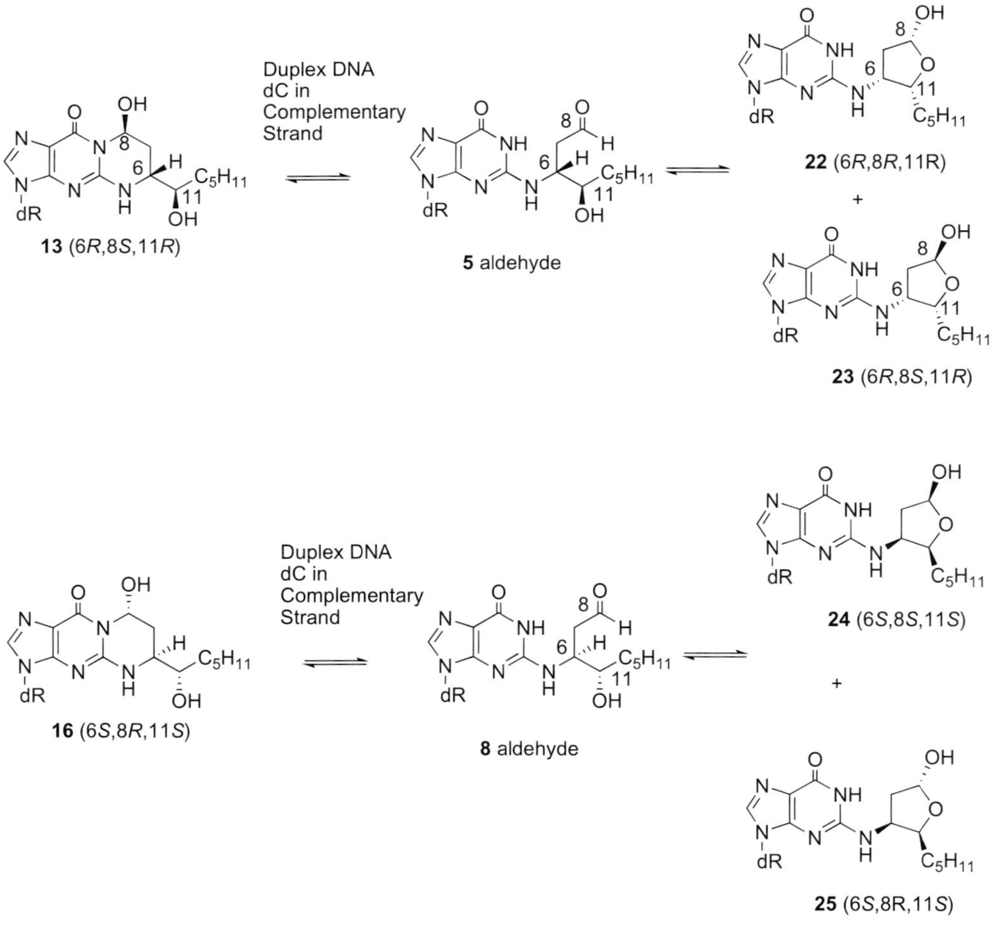

Scheme 9.4 Formation of cyclic hemiacetals by the stereoisomeric 4-HNE adducts.

instances, the *trans* configuration of the HNE H6 and H8 protons (i.e., cyclic hemiacetals **23** and **25**) was preferred. The presence of cyclic hemiacetals **22** and **23** and **24** and **25** in duplex DNA was significant as they masked the aldehyde species **5** and **8** necessary for interstrand cross-link formation.

9.5
Carbinolamine Cross-Links Maintain Watson–Crick Base-Pairing

The explanation as to why the acrolein-induced cross-link preferred the carbinolamine (**19**) structure [36, 37], and not the pyrimidopurinone structure (**21**) [33], was provided by isotope-edited ^{15}N-NMR data. These experiments also revealed the preferred stereochemistry of the carbinolamine linkage **19** [41]. The *R* configuration of the carbinolamine linkage was the major species, constituting more than 80% of the cross-linked species. The *R* carbinolamine linkage, located in the minor groove, maintained the tandem cross-linked base pairs with minimal structural perturbations (Figure 9.1). The two cross-linked guanine N^2 atoms were in the gauche conformation with respect to the linkage. The *anti* conformation of the hydroxyl group with respect to C^α of the tether minimized steric interaction. It also predicted the formation of a hydrogen bond between the carbinol and cytosine O^2 located in the 5′-neighbor GC base pair. This might, in part, explain the thermal stability of carbinolamine cross-link **19** and the stereochemical preference for the

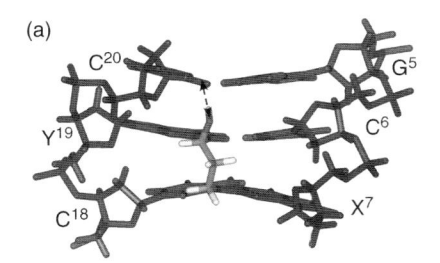

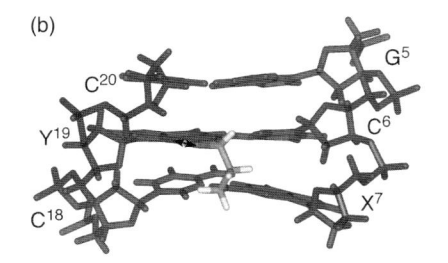

Figure 9.1 Interstrand cross-link arising from γ-OH-PdG adduct **9** observed from the minor groove. (a) Structure of the *R* carbinolamine cross-link **19**. (b) Predicted structure of the *S* carbinolamine cross-link **19**. The predicted hydrogen bonds of the carbinol group are illustrated by dashed arrows.

R configuration. The imino resonances of the cross-linked base pairs were observed at temperatures as high as 65 °C, suggesting that they were shielded from exchange with solvent by the presence of cross-link **19**. In addition, at the 5′-neighbor GC base pair, the imino resonance remained sharp at 55 °C, but broadened at 65 °C. In contrast, at the 3′-neighbor AT base pair, the thymine imino resonance was broadened at 55 °C. These results corroborated modeling studies that had predicted that the carbinolamine linkage **19** maintained Watson–Crick hydrogen-bonding at both of the cross-linked CG pairs [37]. The results also corroborated work in which a N^2-dG : N^2-dG trimethylene linkage, a surrogate for the carbinolamine cross-link **19**, was constructed in a self-complementary duplex 5′-d(AGGCXCCT)$_2$; X represents the linked guanines [42]. In contrast, the dehydration of carbinolamine **19** to imine **20**, or cyclization of the latter to pyrimidopurinone **21**, would have disrupted Watson–Crick pairing at one or both of the cross-linked base pairs. As for the γ-OH-PdG adduct **9**, modeling revealed that the carbinolamine linkage of the crotonaldehyde mediated cross-links analogous to **19** also maintained Watson–Crick bonding at the cross-linked base pairs.

9.6
Role of DNA Sequence

Interstrand cross-linking by 1,N^2-dG adducts of acrolein, crotonaldehyde, and 4-HNE is specific to the 5′-CpG-3′ sequence. When γ-OH-PdG **9** was engineered into 5′-d(CGTACXCATGC)-3′, containing both the 5′-CpG-3′ and 5′-GpC-3′ sequences, and hybridized to a complement with a specific $^{15}N^2$-dG label, only the 5′-CpG-3′ cross-link was identified after digestion and analysis by MS [35]. When the N^2-dG : N^2-dG trimethylene linkage was constructed in the self-complementary duplex d(TCCXCGGA)$_2$, its structure was distorted and its T_m was reduced [43]. Additionally, the prediction that the formation of a hydrogen bond between the carbinol and cytosine O^2 located in the 5′-neighbor GC base pair suggested that the identity of the 5′-neighbor base pair might also modulate formation of the cross-link **19** in the 5′-CpG-3′ sequence.

9.7
Role of Stereochemistry in Modulating Cross-Linking

Structural studies utilizing saturated analogs of crotonaldehyde-derived 6R and 6S cross-links **11** and **12** indicated that both retained Watson–Crick hydrogen bonds at the cross-linked base pairs (Figure 9.2) [44]. However, the 6S-diastereomer **12** showed lower stability. Whereas the CH$_3$ group was positioned in the center of the minor groove for the 6R-diastereomer **11**, for the 6S-diastereomer **12** the CH$_3$ was positioned in the 3′-direction and it sterically interfered with the DNA duplex [44]. These results were consistent with modeling studies [38]. Lao and Hecht [13] also concluded that the pyrimidopurinone cross-link arising from the 6R stereo-

9.7 Role of Stereochemistry in Modulating Cross-Linking

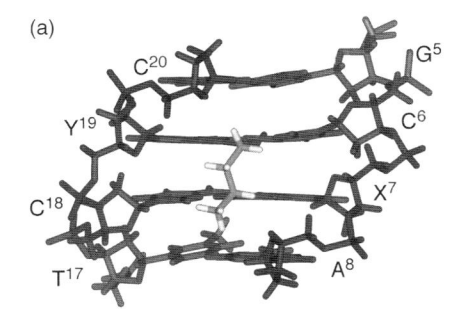

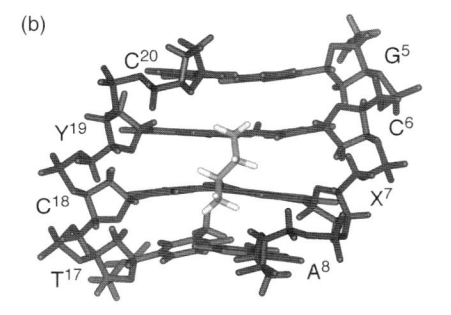

Figure 9.2 Structures of reduced cross-links arising from crotonaldehyde. (a) The 6R cross-link oriented in the center of the minor groove. (b) The 6S cross-link interfered sterically with the DNA and exhibited lower stability. X^7 = 6R- or 6S-crotonaldehyde-adducted dG in the 5′-CpG-3′ sequence; Y^{19} = cross-linked dG in the complementary strand.

chemistry analogous to **21** exhibited a more favorable orientation of the C6 CH$_3$ group.

Additional studies of the crotonaldehyde-derived 6S-diastereomer **12** were performed at pH 9.3. This pH favors hydrolytic ring-opening to the ring-opened aldehyde adduct. The aldehyde group of the ring-opened 6S adduct **12** was oriented in the 3′-direction within the minor groove (Figure 9.3) [45]. Consequently, the aldehyde was distant to the exocyclic amine of the guanine involved in cross-linking, explaining why this diastereomer generated cross-links less efficiently than the 6R-diastereomer **11** [45]. These observations corroborated modeling studies [38]. A similar observation was made for the HNE-derived diastereomeric (6S,8R,11S) **24** and **25** and (6R,8S,11R) **22** and **23** cyclic hemiacetals [46]. The (6S,8R,11S) cyclic hemiacetals **24** and **25** oriented in the 5′-direction, while the (6R,8S,11R) cyclic hemiacetals **22**–**23** oriented in the 3′-direction. Potential energy minimizations of the duplexes containing the two diastereomeric aldehydes predicted that the (6S,8R,11S) aldehyde **8** oriented in the 5′-direction, while the (6R,8S,11R) aldehyde **5** oriented in the 3′-direction. This stereochemical dependence of the orientation of the aldehydic group suggested a structural basis to explain, in part, why the (6S,8R,11S) stereoisomer **16** formed interchain

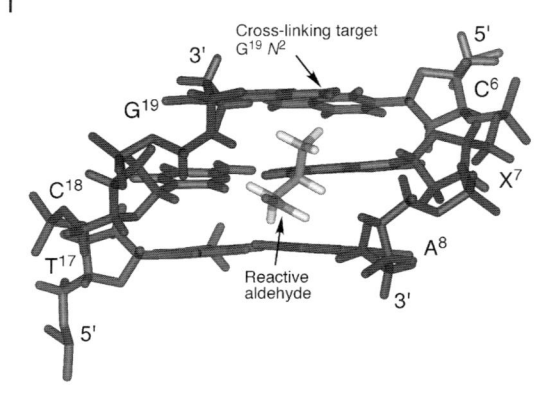

Figure 9.3 Base pairs C^6G^{19}, X^7C^{18}, and A^8T^{17} in the oligodeoxynucleotide containing N^2-(3-oxo-1S-methyl-propyl)-dG adduct **12**. The orientation of the aldehyde does not favor cross-linking to the target G^{19} N^2.

cross-links in the 5′-CpG-3′ sequence, whereas the (6R,8S,11R) stereoisomer **13** did not.

9.8
Biological Significance

A major goal of continuing research is to demonstrate that these acrolein, crotonaldehyde, and HNE-derived interstrand cross-links are present *in vivo*, utilizing MS-based analysis [10, 47–49]. Since the cross-links equilibrate with non-cross-linked species and require the presence of the 5′-CpG-3′ sequence, they may be present at very low levels in tissue samples. Nevertheless, it has been reported that acrolein preferentially binds at 5′-CpG-3′ sites – a consequence of cytosine methylation at these sequences [15].

It is anticipated that these cross-links, if present *in vivo*, will interfere with DNA replication and transcription. Further, in humans, interstrand cross-link repair requires the cooperation of multiple proteins belonging to different biological pathways, including, but not limited to, nucleotide excision repair, homologous recombination, translesion DNA synthesis, double-strand break repair, and the Fanconi anemia pathway [50–55]. Current models suggest that interstrand cross-link repair is initiated by dual incisions around the cross-link in one of the two affected strands. In particular, this "unhooking" can be performed by the endonucleolytic activity of the XPF/ERCC1 (xeroderma pigmentosum group F/excision repair cross-complementation group 1) complex. The result is a gap that may be filled by pairing of the 3′-end of the preincised strand with the homologous sequence, followed by DNA synthesis. Alternatively, the complementary strand with the cross-link attached may be used as a template for translesion DNA synthesis. Once the integrity of one DNA strand is restored, the second strand may be repaired by conventional nucleotide excision repair. When repair is concomitant with replication, a DNA double-strand break is formed; thus, additional biological processing would be required to tolerate interstrand cross-links [51, 52].

9.9
Conclusions

The $1,N^2$-dG adducts of acrolein, crotonaldehyde, and 4-HNE yield interstrand cross-links in the 5′-CpG-3′ sequence. These arise via opening of the 8-hydroxypropano ring to the corresponding aldehydes, which undergo attack by the N^2-amino group of the cross-strand dG in the 5′-CpG-3′ sequence. In the case of 4-HNE, the reactive aldehydes are present at low concentration due to their rearrangement into diastereomeric pairs of cyclic hemiacetals. The cross-links arising from acrolein and crotonaldehyde exist in duplex DNA as carbinolamine linkages, which enable the cross-linked CG base pairs to maintain Watson–Crick hydrogen-bonding with minimal distortion of the duplex. The cross-linking chemistry of crotonaldehyde and 4-HNE depends upon the stereochemistry of the C6 carbon, which orients the reactive aldehydes within the minor groove in the 5′-CpG-3′ sequence favoring the 6R configuration for crotonaldehyde and the stereochemically equivalent 6S configuration for 4-HNE.

Acknowledgements

This work was funded by NIH program project grant PO1 ES-05355 (M.P.S., I.D.K., T.M.H., R.S.L., and C.J.R.). The Vanderbilt University Center in Molecular Toxicology is funded by NIH grant P30 ES-00267. Vanderbilt University and NIH grant RR-05805 provided additional funding for NMR instrumentation.

References

1 Burcham, P.C. (1998) Genotoxic lipid peroxidation products: their DNA damaging properties and role in formation of endogenous DNA adducts. *Mutagenesis*, **13**, 287–305.

2 Chung, F.L., Nath, R.G., Nagao, M., Nishikawa, A., Zhou, G.D., and Randerath, K. (1999) Endogenous formation and significance of $1,N^2$-propanodeoxyguanosine adducts. *Mutat. Res.*, **424**, 71–81.

3 Chung, F.L., Zhang, L., Ocando, J.E., and Nath, R.G. (1999) Role of $1,N^2$-propanodeoxyguanosine adducts as endogenous DNA lesions in rodents and humans. *IARC Sci. Publ.*, **150**, 45–54.

4 Nair, U., Bartsch, H., and Nair, J. (2007) Lipid peroxidation-induced DNA damage in cancer-prone inflammatory diseases: a review of published adduct types and levels in humans. *Free Radic. Biol. Med.*, **43**, 1109–1120.

5 Hecht, S.S., Upadhyaya, P., and Wang, M. (1999) Reactions of α-acetoxy-N-nitrosopyrrolidine and crotonaldehyde with DNA. *IARC Sci. Publ.*, **150**, 147–154.

6 Nath, R.G., Ocando, J.E., Guttenplan, J.B., and Chung, F.L. (1998) $1,N^2$-propanodeoxyguanosine adducts: potential new biomarkers of smoking-induced DNA damage in human oral tissue. *Cancer Res.*, **58**, 581–584.

7 Esterbauer, H., Schaur, R., and Zollner, H. (1991) Chemistry and biochemistry of 4-hydroxynonenal, malonaldehyde and related aldehydes. *Free Radic. Biol. Med.*, **11**, 81–128.

8 Galliani, G. and Pantarotto, C. (1983) The reaction of guanosine and 2'-deoxyguanosine with acrolein. *Tetrahedron Lett.*, **24**, 4491–4492.

9 Chung, F., Young, R., and Hecht, S.S. (1984) Formation of 1,N^2-propanodeoxyguanosine adducts in DNA upon reaction of acrolein or crotonaldehyde. *Cancer Res.*, **44**, 990–995.

10 Zhang, S., Villalta, P., Wang, M., and Hecht, S. (2007) Detection and quantitation of acrolein-derived 1,N^2-propanodeoxyguanosine adducts in human lung by liquid chromatography-electrospray ionization-tandem mass spectrometry. *Chem. Res. Toxicol.*, **20**, 565–571.

11 Eder, E. and Hoffman, C. (1992) Identification and characterization of deoxyguanosine-crotonaldehyde adducts. Formation of 7,8 cyclic adducts and 1,N^2,7,8 bis-cyclic adducts. *Chem. Res. Toxicol.*, **5**, 802–808.

12 Wang, M., McIntee, E.J., Cheng, G., Shi, Y., Villalta, P.W., and Hecht, S.S. (2000) Identification of DNA adducts of acetaldehyde. *Chem. Res. Toxicol.*, **13**, 1149–1157.

13 Lao, Y. and Hecht, S.S. (2005) Synthesis and properties of an acetaldehyde-derived oligonucleotide interstrand cross-link. *Chem. Res. Toxicol.*, **18**, 711–721.

14 Chung, F.L., Nath, R.G., Ocando, J., Nishikawa, A., and Zhang, L. (2000) Deoxyguanosine adducts of t-4-hydroxy-2-nonenal are endogenous DNA lesions in rodents and humans: detection and potential sources. *Cancer Res.*, **60**, 1507–1511.

15 Feng, Z., Hu, W., Hu, Y., and Tang, M.S. (2006) Acrolein is a major cigarette-related lung cancer agent: preferential binding at p53 mutational hotspots and inhibition of DNA repair. *Proc. Natl. Acad. Sci. USA*, **103**, 15404–15409.

16 Marnett, L.J., Hurd, H.K., Hollstein, M.C., Levin, D.E., Esterbauer, H., and Ames, B.N. (1985) Naturally occurring carbonyl compounds are mutagens in *Salmonella* tester strain TA104. *Mutat. Res.*, **148**, 25–34.

17 Smith, R.A., Cohen, S.M., and Lawson, T.A. (1990) Acrolein mutagenicity in the V79 assay. *Carcinogenesis*, **11**, 497–498.

18 Curren, R.D., Yang, L.L., Conklin, P.M., Grafstrom, R.C., and Harris, C.C. (1988) Mutagenesis of xeroderma pigmentosum fibroblasts by acrolein. *Mutat. Res.*, **209**, 17–22.

19 Kawanishi, M., Matsuda, T., Nakayama, A., Takebe, H., Matsui, S., and Yagi, T. (1998) Molecular analysis of mutations induced by acrolein in human fibroblast cells using *supF* shuttle vector plasmids. *Mutat. Res.*, **417**, 65–73.

20 Cohen, S.M., Garland, E.M., St John, M., Okamura, T., and Smith, R.A. (1992) Acrolein initiates rat urinary bladder carcinogenesis. *Cancer Res.*, **52**, 3577–3581.

21 Czerny, C., Eder, E., and Runger, T.M. (1998) Genotoxicity and mutagenicity of the α,β-unsaturated carbonyl compound crotonaldehyde (butenal) on a plasmid shuttle vector. *Mutat. Res.*, **407**, 125–134.

22 Chung, F.L., Tanaka, T., and Hecht, S.S. (1986) Induction of liver tumors in F344 rats by crotonaldehyde. *Cancer Res.*, **46**, 1285–1289.

23 Benamira, M. and Marnett, L.J. (1992) The lipid peroxidation product 4-hydroxynonenal is a potent inducer of the SOS response. *Mutat. Res.*, **293**, 1–10.

24 Cajelli, E., Ferraris, A., and Brambilla, G. (1987) Mutagenicity of 4-hydroxynonenal in V79 Chinese hamster cells. *Mutat. Res.*, **190**, 169–171.

25 Hussain, S.P., Raja, K., Amstad, P.A., Sawyer, M., Trudel, L.J., Wogan, G.N., Hofseth, L.J., Shields, P.G., Billiar, T.R., Trautwein, C., Hohler, T., Galle, P.R., Phillips, D.H., Markin, R., Marrogi, A.J., and Harris, C.C. (2000) Increased p53 mutation load in nontumorous human liver of Wilson disease and hemochromatosis: oxyradical overload diseases. *Proc. Natl. Acad. Sci. USA*, **97**, 12770–12775.

26 Kanuri, M., Minko, I., Nechev, L., Harris, T., Harris, C., and Lloyd, R. (2002) Error prone translesion synthesis past γ-hydroxypropano deoxyguanosine, the primary acrolein-derived adduct in

27 Fernandes, P., Wang, H., Rizzo, C., and Lloyd, R. (2003) Site-specific mutagenicity of stereochemically defined 1,N^2-deoxyguanosine adducts of trans-4-hydroxynonenal in mammalian cells. *Environ. Mol. Mutagen.*, **42**, 68–74.

28 Fernandes, P., Kanuri, M., Nechev, L., Harris, T., and Lloyd, R. (2005) Mammalian cell mutagenesis of the DNA adducts of vinyl chloride and crotonaldehyde. *Environ. Mol. Mutagen.*, **45**, 455–459.

29 Marnett, L.J. (1999) Lipid peroxidation-DNA damage by malondialdehyde. *Mutat. Res.*, **424**, 83–95.

30 Mao, H., Schnetz-Boutaud, N.C., Weisenseel, J.P., Marnett, L.J., and Stone, M.P. (1999) Duplex DNA catalyzes the chemical rearrangement of a malondialdehyde deoxyguanosine adduct. *Proc. Natl. Acad. Sci. USA*, **96**, 6615–6620.

31 Riggins, J.N., Daniels, J.S., Rouzer, C.A., and Marnett, L.J. (2004) Kinetic and thermodynamic analysis of the hydrolytic ring-opening of the malondialdehyde-deoxyguanosine adduct, 3-(2′-deoxy-β-D-erythro-pentofuranosyl)-pyrimido[1,2-α]purin-10(3H)-one. *J. Am. Chem. Soc.*, **126**, 8237–8243.

32 de los Santos, C., Zaliznyak, T., and Johnson, F. (2001) NMR characterization of a DNA duplex containing the major acrolein-derived deoxyguanosine adduct γ-OH-1,-N^2-propano-2′-deoxyguanosine. *J. Biol. Chem.*, **276**, 9077–9082.

33 Kozekov, I.D., Nechev, L.V., Sanchez, A., Harris, C.M., Lloyd, R.S., and Harris, T.M. (2001) Interchain cross-linking of DNA mediated by the principal adduct of acrolein. *Chem. Res. Toxicol.*, **14**, 1482–1485.

34 Stone, M.P., Cho, Y.-J., Huang, H., Kim, H.-Y., Kozekov, I.D., Kozekova, A., Wang, H., Lloyd, R.S., Harris, T.M., and Rizzo, C.J. (2008) Interstrand DNA cross-links induced by α,β-unsaturated aldehydes derived from lipid peroxidation and environmental sources. *Acc. Chem. Res.*, **41**, 793–804.

35 Kozekov, I.D., Nechev, L.V., Moseley, M.S., Harris, C.M., Rizzo, C.J., Stone, M.P., and Harris, T.M. (2003) DNA interchain cross-links formed by acrolein and crotonaldehyde. *J. Am. Chem. Soc.*, **125**, 50–61.

36 Kim, H.Y., Voehler, M., Harris, T.M., and Stone, M.P. (2002) Detection of an interchain carbinolamine cross-link formed in a CpG sequence by the acrolein DNA adduct γ-OH-1,N^2-propano-2′-deoxyguanosine. *J. Am. Chem. Soc.*, **124**, 9324–9325.

37 Cho, Y.J., Kim, H.Y., Huang, H., Slutsky, A., Minko, I.G., Wang, H., Nechev, L.V., Kozekov, I.D., Kozekova, A., Tamura, P., Jacob, J., Voehler, M., Harris, T.M., Lloyd, R.S., Rizzo, C.J., and Stone, M.P. (2005) Spectroscopic characterization of interstrand carbinolamine cross-links formed in the 5′-CpG-3′ sequence by the acrolein-derived γ-OH-1,N^2-propano-2′-deoxyguanosine DNA adduct. *J. Am. Chem. Soc.*, **127**, 17686–17696.

38 Cho, Y.J., Wang, H., Kozekov, I.D., Kurtz, A.J., Jacob, J., Voehler, M., Smith, J., Harris, T.M., Lloyd, R.S., Rizzo, C.J., and Stone, M.P. (2006) Stereospecific formation of interstrand carbinolamine DNA cross-links by crotonaldehyde- and acetaldehyde-derived α-CH_3-γ-OH-1,N^2-propano-2′-deoxyguanosine adducts in the 5′-CpG-3′ sequence. *Chem. Res. Toxicol.*, **19**, 195–208.

39 Wang, H., Kozekov, I.D., Harris, T.M., and Rizzo, C.J. (2003) Site-specific synthesis and reactivity of oligonucleotides containing stereochemically defined 1,N^2-deoxyguanosine adducts of the lipid peroxidation product trans-4-hydroxynonenal. *J. Am. Chem. Soc.*, **125**, 5687–5700.

40 Huang, H., Wang, H., Qi, N., Kozekova, A., Rizzo, C.J., and Stone, M.P. (2008) Rearrangement of the (6S,8R,11S) and (6R,8S,11R) exocyclic 1,N^2-deoxyguanosine adducts of trans-4-hydroxynonenal to N^2-deoxyguanosine cyclic hemiacetal adducts when placed complementary to cytosine in duplex DNA. *J. Am. Chem. Soc.*, **130**, 10898–10906.

41 Huang, H., Kim, H.-Y., Kozekov, I.D., Cho, Y.-J., Wang, H., Kozekova, A., Harris, T.M., Rizzo, C.J., and Stone, M.P. (2009) Stereospecific formation of

the *R*-γ-hydroxytrimethylene interstrand N^2-dG:N^2-dG cross-link ansing from the γ-OH-1,N^2-propano-2'-deoxyguanosine adduct in the 5'-CpG-3' sequence. *J. Am. Chem. Soc.*, **31**, 8416–8424.

42 Dooley, P.A., Tsarouhtsis, D., Korbel, G.A., Nechev, L.V., Shearer, J., Zegar, I.S., Harris, C.M., Stone, M.P., and Harris, T.M. (2001) Structural studies of an oligodeoxynucleotide containing a trimethylene interstrand cross-link in a 5'-(CpG) motif: model of a malondialdehyde cross-link. *J. Am. Chem. Soc.*, **123**, 1730–1739.

43 Dooley, P.A., Zhang, M., Korbel, G.A., Nechev, L.V., Harris, C.M., Stone, M.P., and Harris, T.M. (2003) NMR determination of the conformation of a trimethylene interstrand cross-link in an oligodeoxynucleotide duplex containing a 5'-d(GpC) motif. *J. Am. Chem. Soc.*, **125**, 62–72.

44 Cho, Y.J., Kozekov, I.D., Harris, T.M., Rizzo, C.J., and Stone, M.P. (2007) Stereochemistry modulates the stability of reduced interstrand cross-links arising from *R*- and *S*-α-CH$_3$-γ-OH-1,N^2-propano-2'-deoxyguanosine in the 5'-CpG-3' DNA sequence. *Biochemistry*, **46**, 2608–2621.

45 Cho, Y.J., Wang, H., Kozekov, I.D., Kozekova, A., Kurtz, A.J., Jacob, J., Voehler, M., Smith, J., Harris, T.M., Rizzo, C.J., Lloyd, R.S., and Stone, M.P. (2006) Orientation of the crotonaldehyde-derived N^2-[3-oxo-1(*S*)-methyl-propyl]-dG DNA adduct hinders interstrand cross-link formation in the 5'-CpG-3' sequence. *Chem. Res. Toxicol.*, **19**, 1019–1029.

46 Huang, H., Wang, H., Qi, N., Lloyd, R.S., Rizzo, C.J., and Stone, M.P. (2008) The stereochemistry of *trans*-4-hydroxynonenal-derived exocyclic 1,N^2-deoxyguanosine adducts modulates formation of inter-strand cross-links in the 5'-CpG-3' sequence. *Biochemistry*, **47**, 11457–11472.

47 Goodenough, A.K., Schut, H.A., and Turesky, R.J. (2007) Novel LC-ESI/MS/MS^n method for the characterization and quantification of 2'-deoxyguanosine adducts of the dietary carcinogen 2-amino-1-methyl-6-phenylimidazo[4,5-β]pyridine by 2-D linear quadrupole ion trap mass spectrometry. *Chem. Res. Toxicol.*, **20**, 263–276.

48 Ricicki, E.M., Luo, W., Fan, W., Zhao, L.P., Zarbl, H., and Vouros, P. (2006) Quantification of *N*-(deoxyguanosin-8-yl)-4-aminobiphenyl adducts in human lymphoblastoid TK6 cells dosed with *N*-hydroxy-4-acetylaminobiphenyl and their relationship to mutation, toxicity, and gene expression profiling. *Anal. Chem.*, **78**, 6422–6432.

49 Stout, M.D., Jeong, Y.C., Boysen, G., Li, Y., Sangaiah, R., Ball, L.M., Gold, A., and Swenberg, J.A. (2006) LC/MS/MS method for the quantitation of *trans*-2-hexenal-derived exocyclic 1,N^2-propano-deoxyguanosine in DNA. *Chem. Res. Toxicol.*, **19**, 563–570.

50 Nojima, K., Hochegger, H., Saberi, A., Fukushima, T., Kikuchi, K., Yoshimura, M., Orelli, B.J., Bishop, D.K., Hirano, S., Ohzeki, M., Ishiai, M., Yamamoto, K., Takata, M., Arakawa, H., Buerstedde, J.M., Yamazoe, M., Kawamoto, T., Araki, K., Takahashi, J.A., Hashimoto, N., Takeda, S., and Sonoda, E. (2005) Multiple repair pathways mediate tolerance to chemotherapeutic cross-linking agents in vertebrate cells. *Cancer Res.*, **65**, 11704–11711.

51 Niedernhofer, L.J., Lalai, A.S., and Hoeijmakers, J.H. (2005) Fanconi anemia (cross)linked to DNA repair. *Cell*, **123**, 1191–1198.

52 Noll, D.M., Mason, T.M., and Miller, P.S. (2006) Formation and repair of interstrand cross-links in DNA. *Chem. Rev.*, **106**, 277–301.

53 Mirchandani, K.D. and D'Andrea, A.D. (2006) The Fanconi anemia/BRCA pathway: a coordinator of cross-link repair. *Exp. Cell Res.*, **312**, 2647–2653.

54 Patel, K.J. and Joenje, H. (2007) Fanconi anemia and DNA replication repair. *DNA Repair*, **6**, 885–890.

55 Kennedy, R.D. and D'Andrea, A.D. (2005) The Fanconi anemia/BRCA pathway: new faces in the crowd. *Genes Dev.*, **19**, 2925–2940.

10
Structure–Function Characteristics of Aromatic Amine-DNA Adducts
Bongsup Cho

10.1
Introduction

Aromatic amines are among the most notorious chemicals in the environment that have been implicated in human cancers [1, 2]. Other ubiquitous pollutants are the nitroaromatics [3] and food-borne heterocyclic amines [4, 5]. The metabolic profiles of these compounds closely resemble those of the aromatic amines and they have been classified as probable human carcinogens. Mutagenesis is a critical initial step in tumorigenesis [6, 7] and the formation of aromatic amine-DNA adducts is a hallmark of the initiation of cancer. The link between adduct formation and carcinogenesis has been inferred for various human tissues [8, 9]. Therefore, it is important to understand how the DNA lesions are replicated and repaired *in vivo*. The molecular players involved (polymerases, repair proteins) that are critical for processing the DNA damage must be identified, characterized, and understood [10–14] in order to devise rational strategies for chemoprevention [15], drug development [16], and risk assessment [17].

Upon *in vivo* activation, aromatic amines are converted into highly reactive *N*-acetoxy and/or sulfonated ester derivatives, which interact directly with cellular DNA to form adducts [18–21]. The major targets for such arylamination reactions are the C8 and the exocyclic amino positions of purine nucleosides, particularly guanine (dG) residues in DNA. For example, the prototype carcinogen 2-aminofluorene (AF) and its derivatives produce dG-C8-AF and dG-C8-2-acetylaminofluorene (AAF), along with smaller amounts of persistent dG-N^2-AAF adducts [18] (Figure 10.1a). This is in contrast to DNA lesions derived from polycyclic aromatic hydrocarbons (PAHs); these are mostly the N^2-dG and/or N^6-dA adducts, formed through substitution reactions with either chiral dihydrodiol epoxides and/or quinoids (see Chapter 6 and Figure 10.1c for the structure of a PAH-DNA adduct derived from benzo[*a*]pyrene) [21, 23]. Figure 10.1(a and b) lists all the major aromatic amine-DNA adducts whose nuclear magnetic resonance (NMR) structures in aqueous solutions have been solved. See also Chapter 7, where Turseky provides further insights into the structures of adducts derived from the chemical reactions of aromatic amines with DNA with a focus on food-derived mutagens.

The Chemical Biology of DNA Damage. Edited by Nicholas E. Geacintov and Suse Broyde
© 2010 WILEY-VCH Verlag GmbH & Co. KGaA, Weinheim
ISBN: 978-3-527-32295-4

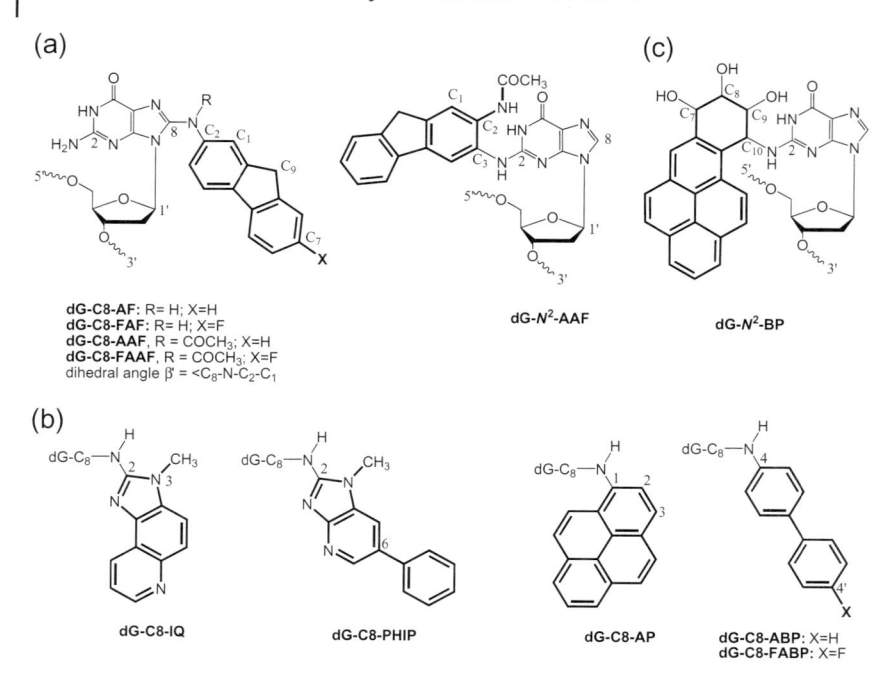

Figure 10.1 Chemical structures of dG-modified aromatic amine and B[a]P adducts whose solution DNA structures have been elucidated. (a) AF, 2-aminofluorene; FAF, 7-fluoro-2-aminofluorene; AAF, 2-acetylaminofluorene; FAAF, 7-fluoro-2-acetylaminofluorene. (b) IQ, 2-amino-3-methylimidazo[4,5-f]quinoline; PhIP, 2-amino-1-methyl-6-phenylimidazo[4,5-b]pyridine; AP, 1-aminopyrene; ABP, 4-aminobiphenyl; FABP, 4'-fluoro-4-aminobiphenyl. (c) B[a]P; carbon atoms 7, 8, 9, and 10 are stereogenic thus yielding a multiplicity of stereoisomeric DNA adducts (see Chapters 6, 12, and 14–16, and [2] and [22]).

Elucidating the mechanisms of aromatic amine mutagenesis has been a challenge from the viewpoint of structure–function relationships. A major effort has been directed towards obtaining detailed solution structures of short DNA duplexes containing a single specific lesion [22, 24–28]. The expectation was that such structures would provide structural insights into the mechanisms of mutagenesis. The progress in the last three decades has been substantial, embracing the enormous diversity of aromatic amine-DNA adduct structures. Earlier milestones on this topic have been reviewed, first by Patel [26], and more recently by Cho [27] and de los Santos [28]. Several crystal and model structures involving aromatic amine lesions and polymerases have recently appeared in the literature [29, 30]. Here, I summarize recent developments on the structure–function relationships of aromatic amine-DNA adducts. A particular emphasis is given to solution data on amine-DNA adduct structures, sequence-induced conformational heterogeneities, and their implication for mutation and repair.

10.2
Major Conformational Motifs

10.2.1
Fully Complementary DNA Duplexes

Aromatic amine-DNA adducts adopt three prototypical conformational motifs in the context of fully paired duplexes. They are classified broadly as the major groove B-type (B), base-displaced stacked (S), and minor groove "wedge" (W) conformations, depending on the location of the bulky carcinogens with respect to the B-DNA duplex. These conformers are illustrated in Figure 10.2, using dG-C8-AF (B and S) and dG-N^2-AAF (W) as model adducts. The modified dG in the B-conformer adopts an *anti* glycosidyl conformation, with Watson–Crick hydrogen bonds retained at the lesion site, and the carcinogenic aromatic amine moiety protrudes into the major groove (Figure 10.2a). This is in contrast to the *syn* glycosidic bond conformation observed for the S-conformer, in which the carcinogen is intercalated, with the damaged guanine and its partner cytosine extruded to make room for the inserted aromatic ring system of the carcinogen (Figure 10.2b). For the W-conformer, the carcinogen is in the minor groove (Figure 10.2c).

The population balance between these differing conformations is governed by the specific nature of the aromatic carcinogen as well as the neighboring base sequence contexts. A predominant (90–95%) B-conformer has been experimen-

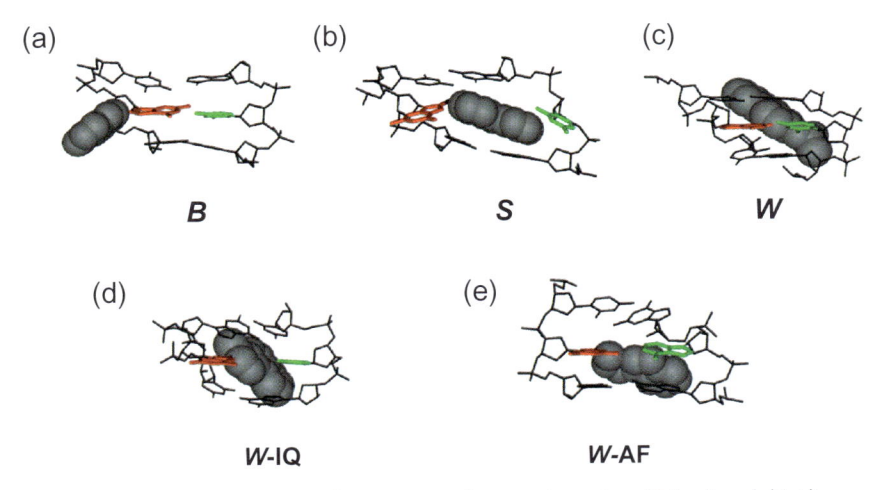

Figure 10.2 Major groove views of arylamine-induced conformational motifs in duplex DNA using AF as an example. (a) B (B-type), (b) S (stacked), and (c) W (wedge) conformers; and W-conformeric (d) IQ- and (e) AF-dG adducts. The modified dGs and the complementary dC (a–d) and dA (d) are indicated with red and green sticks, respectively and the carcinogenic aminofluorene moiety is highlighted in gray Corey–Pauling–Koltun (CPK; space-filling) representation.

tally elucidated for dG-C8-4-aminobiphenyl (ABP) (Figure 10.1b), the major adduct derived from the human bladder carcinogen ABP [31]. By contrast, the large planar dG-C8-1-aminopyrene (AP) adduct (Figure 10.1b), derived from the environmental mutagen 1-nitropyrene, exclusively adopts the intercalating S-conformer that lacks the Watson–Crick base pair at the lesion site [32]. Clearly, the coplanarity of the adduct is important for its stacking capability in intercalation, which seems to modulate the S/B conformational balance. The two phenyl rings of ABP can twist with respect to each other, whereas the fused AP ring system is electron-rich, large, and coplanar. The W-conformer of the dG-N^2-AAF adduct (Figure 10.1a) fits neatly into the minor groove of the B-DNA duplex, with the curvature of the aromatic ring system following the right-hand screw thread of the minor groove (Figure 10.2c) [33]. The nearly perfect Watson–Crick base-pairing at the lesion site sandwiches the hydrophobic and planar fluorene moiety tightly within the walls of the minor groove, minimizing its exposure to water and resulting in an exceptional thermal stabilization of an oligonucleotide duplex, relative to the analogous unmodified duplex (ΔT_m = 6.2 °C and favorable free energy change $\Delta\Delta G_{25°C}$ = −1.8 kcal/mol) stabilities [33]. The duplex melting point, T_m, is defined as the midtemperature point of the dissociation of the double strands to single strands. The ΔT_m is defined as T_m(modified duplex) − T_m(unmodified duplex) and $\Delta\Delta G_{25°C}$ is $\Delta G_{25°C}$(modified duplex) − $\Delta G_{25°C}$ (unmodified duplex).

Stone et al. have studied the structures of the dG-C8 adduct derived from the heterocyclic amine food mutagen 2-amino-3-methylimidazo[4,5-f]quinoline (IQ) (Figure 10.1b) in a 12mer duplex. The IQ adduct was positioned individually at the three different Gs in the fully base-paired NarI 12mer duplex of central sequence (5'- ... CTCG$_1$G$_2$CG$_3$CC ... -3') [34–36]. The NarI sequence is a mutational hotspot in Escherichia coli [37]. This study was motivated by the novel findings that, despite similar chemical reactivities, the G$_3$ adduct exhibited −2 frameshift mutations with unusually greater efficiency (around 100-fold) than adducts at the G$_1$ or G$_2$ positions [37, 38]. The G$_3$-IQ adduct was found to adopt a typical S-conformation, in which the complementary dC is completely out of the helix [34]. However, the G$_1$ and G$_2$ adducts exist as W-like conformers, in which the modified dG adopts a syn conformation and the isoquinoline portion of IQ is pushed into the minor groove (Figure 10.2d) [35, 36]. The ^{1}H-NMR data are consistent with the disruption of Watson–Crick hydrogen-bonding at the lesion site. Moreover, these IQ structures emphasize the importance of sequence effects – base displacement is favored when the flanking nucleotides are cytosine.

10.2.2
Other Sequence Contexts

The structures of aromatic amine adducts have also been studied in other sequence contexts [26, 27]. The dG-C8-AP adduct at a single/double-strand junction, which models the situation at a replication fork, shares the features of the S-conformation, regardless of whether there is a dC partner, a mismatched dA, or no base opposite the lesion. The modified dG is displaced into the major groove, while the carcinogen is inserted in its place. Similar conformational features were observed for dG-C8-AF at a single/double-strand junction, although there were subtle differences in the rotameric possibilities about the β'-linkage (C_8–N–C_2–C_1, Figure 10.1a), which governs rotation about the long axis of the aromatic ring system, and the position of the complementary base at the lesion site, possibly due to the smaller size of AF. The AF duplexes with −1 or −2 deletion sites adopt predominantly S-type conformations, in which the carcinogen is inserted into a bulge structure. These so-called slipped mutagenic intermediates (SMIs) have been implicated in frameshift mutations [38].

In 1989, the Patel group reported a W-type structure for an 11mer DNA duplex containing the dG-C8-AF:dA mispair at the lesion site [39]. This work was born out of earlier difficulties (e.g., conformational exchange) in characterizing the same AF in the fully complementary duplex (i.e., dG-C8-AF:dC). It was considered plausible that the G:A mismatch structure would be more conformationally stable because G:A mismatch mutations, producing G → T transversion mutations, are most frequently observed for dG-C8-AF and other bulky lesions ("A-rule") [40, 41]. The mismatch-induced W-structure (shown in Figure 10.2e) is different from that of the N^2-AAF adduct discussed above (Figure 10.2c) [33], as there are no Watson–Crick hydrogen bonds at the G:A mismatched lesion site and the glycosidic bond adopted the *syn* conformation. The exact position of the AF in the minor groove was pH-dependent [39]. For example, the protonated N1 of dA in the acidic pH ranges (5.12–6.75) was proposed to hydrogen bond with O6 of the modified dG, which displaces the carcinogen further away from the helix axis, relative to its position under neutral pH conditions at pH 6.90. Similar results were obtained for dG-C8-AF mismatched with other purines (dG, dI) [42].

10.3
Conformational Heterogeneity

In 1994, Cho *et al.* [43] and Eckel and Krugh [44, 45] independently reported for the first time an S/B equilibrium of AF-modified duplexes, in two different length and sequence contexts (15mer TG*A, around 55% B and 10mer AG*G, 60–70% B, respectively). The Patel group subsequently succeeded in elucidating the structure of the fully paired AF duplex in the CG*C sequence context (see above), which allowed them to characterize the S-conformer in greater detail (around 70% S) [46]. Three duplexes were also studied, in which AF is individually bonded at the three Gs in the *Nar*I hotspot sequence (CTCG$_1$G$_2$CG$_3$CCATC) [47]. The Patel group found that the AF at G$_3$ afforded an equal mixture of S- and B-conformers, whereas AF at G$_1$ and G$_2$ favored the B-conformer by around 70 and 90%, respectively. This is in contrast to the IQ adduct discussed above [34–36], which showed no such conformational exchange in the same *Nar*I sequence context. However, it should be noted that in both cases the lesion at the G$_3$ hotspot resulted in a consistently greater proportion of the base-displaced S-conformer. A PhIP-dG adduct, which is derived from the heterocyclic amine food-borne mutagen 2-amino-1-methyl-6-

phenylimidazo[4,5-*b*]pyridine (Figure 10.1b), afforded a roughly 85 : 15 mixture of S- and B-conformers [48].

The dG-C8-AAF adduct differs from dG-C8-AF in that the central nitrogen at the adduct linkage is acetylated, thus limiting the conformational flexibility at the lesion site (Figure 10.1a). Approximately 70% of an AAF-modified, fully paired 9mer duplex exhibited S-conformer characteristics [49]. In this conformer, the acetyl group protrudes into the major groove with its carbonyl oxygen located *trans* to the C8 of the modified dG. Similarly, the major portions (70–80%) of the dG-C8-AAF-modified duplexes, in which the undamaged strand contained one or two fewer residues than the damaged one (termed −1 and −2 deletion duplexes), exhibit the features of S-like structures, in which the modified dG adopts a *syn* conformation, and the carcinogen is inserted into the bulged double helix structure termed a SMI [50, 51]. These duplexes exhibited exceptional thermal stabilities (ΔT_m = 11–15 °C), presumably due to well-stacked SMI structures. The Cho group has provided dynamic ^{19}F-NMR evidence that dG-C8-AAF exists exclusively in the S-conformation, with a novel *cis/trans* equilibrium of the acetyl group [52]. The strong preference for dG-C8-AAF to adopt S-, while dG-C8-AF can adopt both S- and B-conformations, could explain why dG-C8 AAF adducts block polymerases more efficiently [53], since the S-conformation with its abnormal *syn* glycosidic bond conformation is not capable of forming a Watson–Crick pair [29].

It is clear from the discussion above that aromatic amine-induced conformational heterogeneity is modulated not just by differences in ring size, coplanarity, and topology of the carcinogen, but also by the adduct linkage position to the base (e.g., C8 or N^2-dG) and the sequence context. ABP is structurally similar to AF, differing only by the lack of one methylene (-CH$_2$-) bridge (Figure 10.1). Consequently, the two six-membered rings in ABP are not coplanar as they are in AF. PhIP and IQ (Figure 10.1b) are heterocyclic in nature and similar in size, but possess slightly different molecular topologies. The size of PhIP is between the sizes of AF and AP. However, PhIP is as flexible as ABP because the remote phenyl ring rotates with respect to the intercalated imidazopyridine ring. The phenyl ring rotation may explain the unusual temperature sensitivity observed for the PhIP-induced S/B conformer equilibrium (90% S-conformer at 25 °C, 85% at 20 °C, 68% at 1 °C) [48]. Not surprisingly, modeling studies showed that the single-aromatic ring aniline adducted to dG-C8 exclusively adopts a B-conformation and distorts the DNA structure to a lesser extent than the bulkier AF and AP adducts [54]. For the dG-C8-AF adduct, studies of flanking sequence base effects on biological outcomes have provided interesting insights and hypotheses, as detailed below.

10.3.1
Sequence Effects on the S/B Conformational Heterogeneity

Meneni *et al.* [55] recorded ^{19}F-NMR spectra of 16 fully paired 7-fluoro-2-a minofluorene (FAF)-modified 12mer duplexes (5′-CTTCT*N*G*NCCTC-3′), in which the bases (*N* = G, A, C, T) flanking the lesion (G*) were varied systematically. The purpose was to examine the effects of the flanking base sequences on

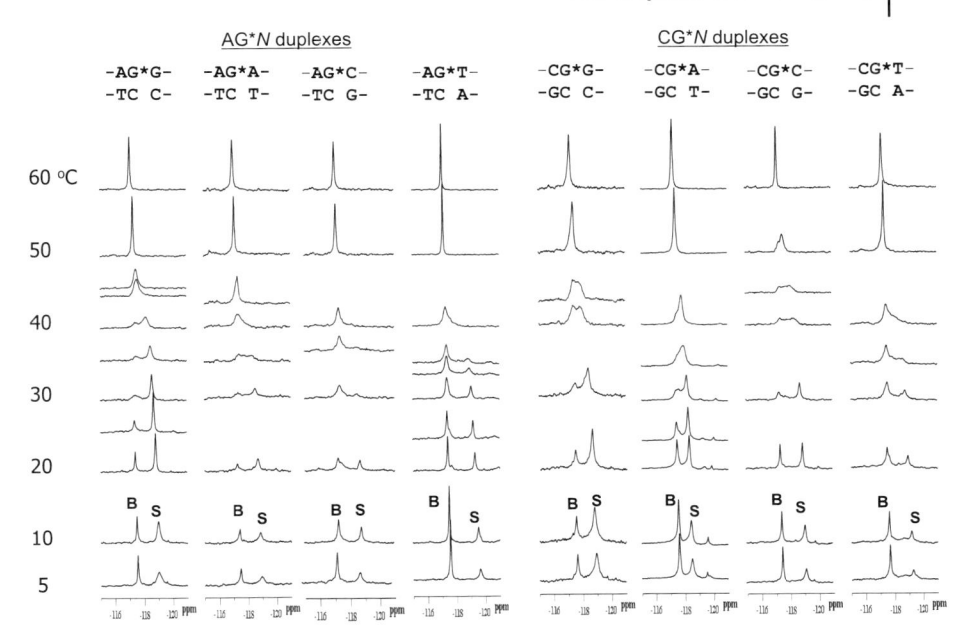

Figure 10.3 Dynamic ^{19}F-NMR spectra (−115 to −121 ppm) of FAF-modified 12mer AG*N and CG*N duplexes (5′-CTTCTNG*NCCTC-3′/5′-GAGGNCNAGAAG-3′; G* = FAF, N = G, A, C, T) [55]. Specific S/B signal assignments were made based on ring current and H/D isotope effects, as well as NOESY/exchange characteristics [55, 56]. The S-conformer was more susceptible to a ring current effect, which enabled the base-displaced FAF to be better shielded than the B-conformer. The downfield B-conformer signals exhibited consistently greater H/D shielding effects (0.13–0.22 ppm) than the upfield S-conformer signals (below 0.06 ppm) measured at 20 °C.

the AF-induced S/B heterogeneity. The studies used fluorine-tagged FAF (Figure 10.1a) as a conformational probe, which permitted the recording of dynamic ^{19}F-NMR spectra. Figure 10.3 shows the results of the AG*N and CG*N series duplexes. All the duplexes manifested a pair of resonances with varying intensities, corresponding to an S/B equilibrium (S, upfield or on the right side, and B, downfield or on the left side). Signal assignments were based on analysis of nuclear Overhauser effect spectroscopy (NOESY)/exchange spectra and D$_2$O isotope effects [57]. The results indicated that the S/B population ratios are particularly sensitive to the 3′-flanking bases (N); that is, a 3′-purine flanking base promotes the S-conformer. The orders of preference for S-conformer were 3′-G (64–68%) > A (56–61%) ≥ C (40–53%) > T (36–37%), all measured at 20 °C. Overall, the sequence effect on the S/B population ratio was substantial, ranging from 20% S (TG*C) (unpublished data) to 68% S (AG*G) (Figure 10.3), at 20 °C.

Fluorescence spectroscopy was used to gain insights into the origins of the 3′-flanking sequence effects. Jain et al. [58] studied fully paired AG*N duplexes (5′-CTTCT(2-AP)G*NCCTC-3′), in which the 5′A is replaced with the versatile fluorophore, 2-aminopurine (2-AP). Conformation-specific stacking differences of 2-AP with the lesion (S ≫ B) were evidenced by a sequence-dependent variation in the intensity of the position of emission maxima. Acrylamide fluorescence quenching experiments were also utilized to probe solvent exposure, where B-conformer adducts are expected to be more solvent-exposed than S and W adduct conformations. The fluorescence intensity and Stern–Volmer (K_{sv}) acrylamide quenching constants were decreased for the modified duplexes relative to the control. These results suggest that there is greater 2-AP stacking in the duplex upon adduct formation, and thus less solvent exposure and susceptibility to quenching by the fluorescence quencher acrylamide. Van der Waals energy and quantum mechanical calculations showed a similar sequence specific stacking interaction between the carcinogen and flanking base pairs [55]. Quantum mechanical calculations provided a more complete picture, which emphasized the importance of electrostatic interactions and solvation in sequence-dependence effects.

10.3.2
Conformational Dynamics of the S/B Heterogeneity

The kinetic and thermodynamic nature of the AF-induced S/B heterogeneity has been studied by analyzing UV melting [59] and ^{19}F-NMR data [55]. Van't Hoff thermodynamic results reflected the sequence-dependent S/B conformational heterogeneity due to differences in the polarity of the local environment and the adduct conformations at the lesion site. Line shape simulation of the dynamic ^{19}F-NMR data (Figure 10.3) indicated a small energy difference ($\Delta G° < 500$ cal/mol) between S- and B-conformers [55]. The rate constants at the coalescing temperature ($k_C > 1200\,\text{s}^{-1}$) are faster than the average rates of spontaneous base-pair opening in a typical B-form DNA, yet the hydrophobic AF moiety promotes intercalation with an exchange time of milliseconds ($\tau = 1/k = 2 - 5$ ms). This is potentially significant because the low energy barrier allowing the S/B equilibrium could be relevant physiologically when the DNA-adducted base comes into contact with repair proteins or polymerases. Eckel and Krugh [44] have coined the term "mutagenic switch" to describe such a scenario. Such an S/B equilibrium is reminiscent of the base flipping that is a component of the mechanism utilized by base excision repair proteins such as uracil DNA glycosylase [60], which excises uracils from DNA; rather than flipping a base, however, it is the carcinogen that is flipping between helix-inserted (S) and helix-external (B) positions (Figure 10.2a and b).

10.3.3
Base Sequence Context and Mutagenesis

The old paradigm of base substitution mutagenesis presumed that a lesion adopts a single structure, thus producing a single kind of mutation. This turned out to be an oversimplification because a DNA adduct produces a complex mutational spectrum, which varies with the sequence contexts and the biological systems used. As such, Seo et al. [61] have hypothesized that "a single adduct adopts multiple conformations, each of which is capable of producing different mutational outcomes." This is reasonable because adduct-induced mutations may be infrequent biological events and could be due to the replication of certain minor conformers that evade repair. As an example of sequence context effects in mutagenic outcome, AF-induced point mutations in GG*C sequences were observed to be 50-fold higher than in the TG*C context in mammalian COS-7 cells [40]. The presence of a G at the 5′-side of the lesion increased the mutational frequency substantially. It should be pointed out, however, that mutation results are also influenced greatly by the types of polymerases used (i.e., high-fidelity versus bypass polymerases). These results, taken together, illustrate the complexity of bulky carcinogen adduct-induced mutagenesis [24, 61].

10.3.4
Dependence of Nucleotide Excision Repair by *E. coli* UvrABC Proteins on Adduct Conformation

The incision rates of DNA duplexes containing single dG-C8-AF or dG-C8-AAF by the prokaryotic UvrABC apparatus were found to be significantly faster when the lesions are positioned at the G_3 of the *Nar*I mutational hotspot sequence (CG*C) than when they are positioned in a random TG*A sequence [62]. Zou *et al.* demonstrated that dG-C8-AF and dG-C8-AAF adducts in the TG*T sequence are incised more efficiently than in the CG*C context in *E. coli* [63]. A similar sequence dependence was observed with the (+)-*trans*-benzo[*a*]pyrene diol expoxide (B[*a*]PDE)-N^2-dG adduct (positioned in the minor groove of B-DNA duplexes, labeled dG-N^2-BP in Figure 10.1c) using UvrABC proteins from *E. coli* and in the thermophilic *Bacillus caldotenax* [64]; the overall structure of B[*a*]PDE-N^2-dG adducts is shown in Figure 10.1(c), while the stereochemical characteristics and properties of these benzo[*a*]pyreneB[*a*]P-derived adducts are discussed in Chapters 6,12, and 14–17). Kalam *et al.* [65, 66] have shown that the oxidative damage lesions Fapy (formamidopyrimidine)-dG and 8-oxo-dG in simian kidney (COS-7) cells were mutagenic, inducing primarily targeted G → T transversions with substantially greater mutational frequency in the TG*T relative to the CG*C sequence context.

The strong sequence dependence of the repair and the mutational outcomes of dG-C8-AF are likely the result of differences in the S/B equilibrium [40, 55, 61, 62, 67]. It can be hypothesized that aromatic amine lesions exist mostly in an equilibrium of S- and B-conformers, and that their relative population ratios determine the biological outcomes. The question then becomes which conformation, S or B, would be recognized as a defect in nucleotide excision repair (NER). To address this intriguing question, Meneni *et al.* [55] have carried out NER experiments for the FAF-modified AG*N and CG*N duplex series in the *E. coli* UvrABC system. Initially, they conducted control experiments using the 12mer sequence 5′-CTTCTAGGCCTC-3′, in which each of the two guanines was individually modified with either AF or FAF [67]. The repair profiles (Figure 10.4a) of the dG-C8-AF and FAF duplexes were found to be essentially identical, indicating their conformational characteristics and NER response are the same. These results ensured

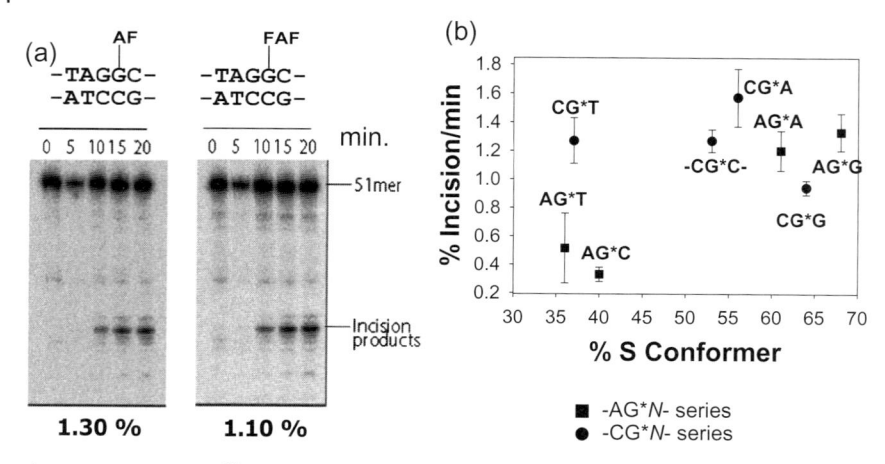

Figure 10.4 Incision efficiency (% DNA incision/min) of (a) AF versus FAF on the same 12mer (5'-CTTCTAGG*CCTC-3') sequence; and (b) the -AG*N- and -CG*N- sequences versus percentage S-conformer by *E. coli* UvrABC nuclease [55].

that the incision efficiencies observed with the FAF adducts could be extrapolated to those of the AF adducts (Figure 10.1a).

Figure 10.4(b) shows a plot of the *E. coli* incision efficiency for the FAF-modified AG*N and CG*N series against percentage S-conformer population, measured by ^{19}F-NMR [55, 57]. An excellent correlation (R^2 = 0.93) was obtained when only the AG*N series was considered. Consistently greater incision efficiencies were observed for duplexes that are predominantly S-conformers, such as the AG*G (68%) and AG*A (61%) sequences. No such clear-cut correlation was found for the CG*N series. Consequently, a relatively poor correlation was obtained when all eight data entries are considered together. Nevertheless, the trend persisted. For example, low NER was associated with a low percentage S-conformation (AG*C, AG*T, CG*T) and high NER was associated with a high percentage S-conformation (AG*G, AG*A, CG*G, CG*A, CG*C). The lesions with higher S/B ratios generally appeared to be more susceptible to NER repair than lesions with lower S/B ratios (except CG*T). Moreover, the CG*N series exhibited greater incision efficiencies than the AG*N counterparts [55].

NER is the major cellular pathway for removing DNA damage that would otherwise lead to mutagenesis [12–14, 68, 69]. Deficiencies in NER are genetically associated with human diseases such as xeroderma pigmentosum and Cockayne syndrome (Chapters 12 and 18). A hallmark of NER is its ability to repair a wide variety of bulky lesions. Although human NER proteins show no homology to their prokaryotic counterparts, the overall molecular strategy is analogous: recognition, helix unwinding, incision, and patch [13]. Several mechanisms have been proposed to explain how the NER machinery recognizes lesions situated within the expanse of intact genomic DNA. There is no single one-size-fits-all mechanism. The early concept of "thermal/thermodynamic probing" has gained support and evolved into more complex "bipartite" or "multipartite" mechanisms [70–73]. The latter mechanisms focus on the quality of the Watson–Crick base pairs at or near the lesion site and the adduct structures. Indirect readout mechanisms, which emphasize the local conformational distortion of either the modified [74] or the complementary strand [75] at the modification site, are among the suggested mechanistic possibilities that connect NER efficiencies with thermodynamic probing mechanisms and the multipartite concept. Chapters 11 and 12 provide more detailed insights into the NER process.

The molecular basis for the novel "conformation-specific *E. coli* NER" for AF lesions is not clear. The S-conformation is structurally similar to the conformations of repair-prone base-displaced *cis*-B[*a*]PDE-N^2-dG adducts. These *cis* adducts lack Watson–Crick base-pairing at the lesion site, and are characterized by the displacement of the modified guanine residue and the partner C into the major or minor grooves of the B-DNA duplex [22], resembling the dG-C8-AF S-conformation. It has been found, however, that the classically intercalated fjord region PAH-dA adducts with normal Watson–Crick base pairs lack NER susceptibility although the stereochemically similar bay region PAH-dA substrates (e.g., the B[*a*]P-dA adducts) are good substrates of NER [76]. However, adduct intercalation usually contributes to thermal and thermodynamic stabilization, which is a well-known negative indicator for NER [72]. In this regard, it is worth noting the unusually persistent, and probably NER-resistant dG-N^2-AAF adducts, whose W-conformation duplex is minimally perturbed and thermodynamically stabilized (Figure 10.2c) [33]. Also, as mentioned above, dG-C8-AF and dG-C8-AAF adducts in the TG*T sequence context are more susceptible to incision in *E. coli* than in the CG*C context [63]. It was argued that T:A flanking base pairs allow the local bends and flexibility and thus greater conformational mobility [63, 66]. This is consistent with the lower T_m and negative $\Delta G°$ values, as well as the lesion conformational flexibility detected for the AF- or FAF-modified TG*T duplexes [77]. By contrast, the corresponding CG*C duplex exhibited two well-resolved ^{19}F signals for S- and B-conformers (Figure 10.3). Interestingly, the CG*N duplexes exhibited collectively greater T_m and more negative $\Delta G°$ than their AG*N counterparts, yet they are found to be better substrates for repair.

Taken together, NER recognition of adducts involves thermal and thermodynamic stabilities and structural/dynamic properties that include the quality of Watson–Crick base-pairing, the lesion conformational flexibility, and adduct–DNA groove interactions [70–76, 78]. The results presented here illustrate the complexity of adducts with multiple conformations interacting with repair proteins; understanding these phenomena is crucial to elucidating the first step in damage recognition in NER.

10.3.5
Conformational Heterogeneity in Translesion Synthesis

Conformational heterogeneity at the replication fork can dictate the dynamics of protein–DNA interactions in the active site pocket of a polymerase, thus affecting

10 Structure–Function Characteristics of Aromatic Amine-DNA Adducts

(a)

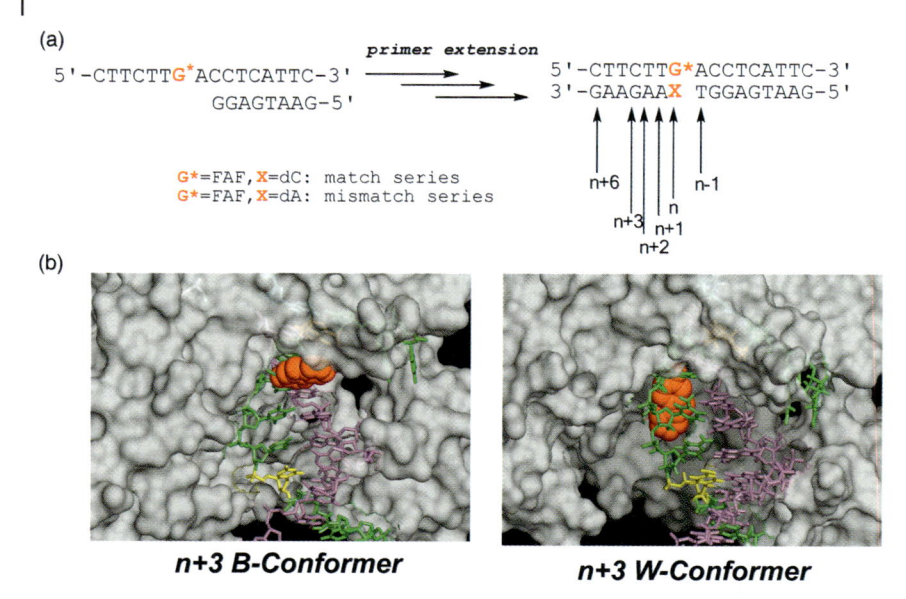

```
                    primer extension
5'-CTTCTTG*ACCTCATTC-3'    ———————→    5'-CTTCTTG*ACCTCATTC-3'
         GGAGTAAG-5'                    3'-GAAGAAX TGGAGTAAG-5'

    G*=FAF, X=dC: match series                    n+6       n-1
    G*=FAF, X=dA: mismatch series                          n
                                               n+3  n+1
                                                    n+2
```

(b)

n+3 B-Conformer   **n+3 W-Conformer**

Figure 10.5 (a) Template/primer sequence contexts used for dynamic ^{19}F-NMR and DSC experiments [79]. (b) Example $n + 3$ translesion synthesis models of BF complexed with AF-modified template/primer DNA: $n + 3$ dC-match (B-type conformer) and $n + 3$ dA-mismatch (W-conformer) [56]. The carcinogenic aminofluorene moiety is shown in red CPK representation. Template and primer sequences are colored green and magenta, respectively. For illustrative purpose, the $n + 1$ and $n + 6$ positions relative to the primer terminus n in the template strand are colored orange (under the surface) and yellow, respectively.

the choice of incoming dNTP substrates. Liang and Cho [79] have examined the adduct conformations at various downstream primer positions with respect to an FAF lesion paired with either dC or dA (match and mismatch, respectively) (Figure 10.5a). Dynamic ^{19}F-NMR/circular dichroism [56, 80] indicated that the matched template/primers adopted a B/S equilibrium, whereas the mismatched series preferred B/W-conformers (not shown). In both cases, the conformational preference increased as a function of the length of the primer. Modeling (Figure 10.5b) [56] indicated that the carcinogen moiety in the minor groove of the W-conformer would cause a significant steric clash with the tight-packed amino acid residues at the DNA binding site of the *Bacillus* fragment (BF), a well-characterized replicative DNA polymerase [29]. As for the B-conformer, the carcinogenic moiety resides in the solvent-exposed major groove throughout the replication/translocation process. These results imply that the potentially mutagenic W-conformer in the FAF:dA mismatch case is present in the active site of polymerases.

Differential scanning calorimetry (DSC) experiments were utilized to examine the thermodynamic contribution of AF heterogeneity on replication fidelity [79]. Unlike the UV-based data, DSC-derived thermodynamic data are model-

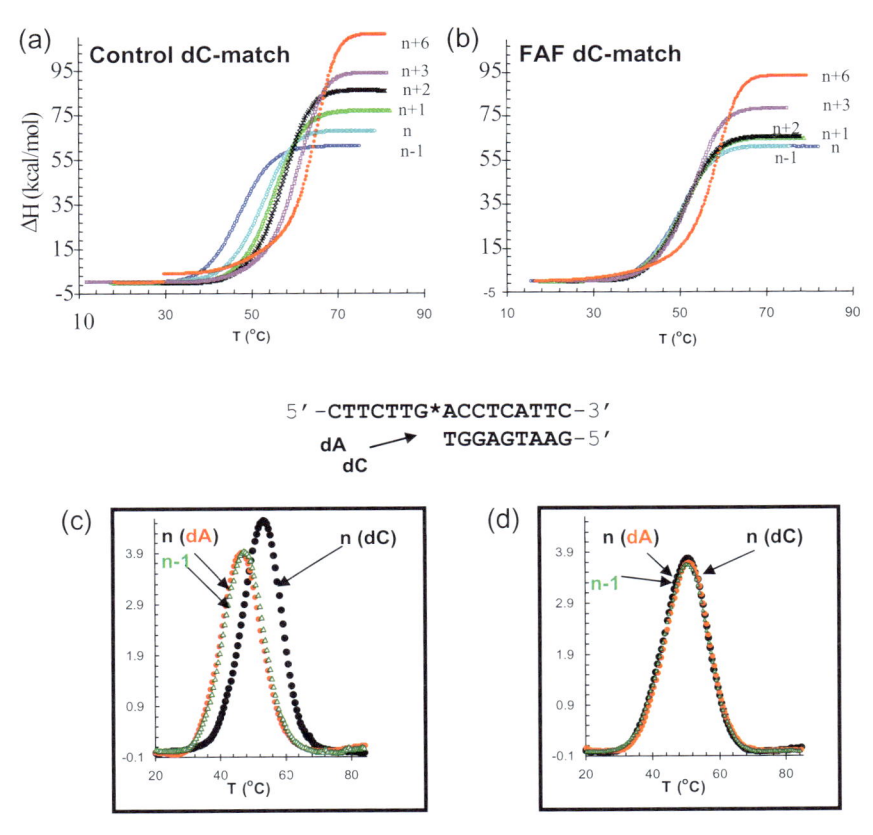

Figure 10.6 Enthalpy change (ΔH) versus temperature (T, °C) for (a) the control and (b) FAF-modified dC-match series (see Figure 10.6a for primer sequence definition). Also shown are DSC curves of an n − 1 (green) → n transition with insertion of dC (black) and dA (red) opposite (c) unmodified and (d) FAF-modified dG* at the primer-terminus junction (n). These melting curves were obtained in 20 mM phosphate buffer containing 0.1 M NaCl at pH 7.0.

independent, thus yielding accurate measurements of enthalpy and T_m. As shown in Figure 10.6(a), an insertion of the correct dC at the primer-terminus position n resulted in a substantial increase in ΔH (7.6 kcal/mol) and the subsequent primer extensions ($n + 1, \ldots, n + 6$) took place with an incremental increase in enthalpy. Similar profiles were obtained for T_m and the free energy of duplex dissociation, ΔG. In contrast, incorporation of dC opposite the FAF lesion showed little change in the duplex melting thermodynamics until the $n + 2$ position was reached and then the normal duplex trend resumed starting at $n + 3$ (Figure 10.6b). It is important to note that the thermodynamic destabilization induced by the lesion was not localized to the lesion site, but was extended further to the 5′-downstream sites. Entropy around the lesion site played a key role in the lesion-induced thermodynamic destabilization. Although no polymerases were involved, these results are consistent with previous primer extension kinetic data by Miller and Grollman [81], which showed significant reductions up to a million fold (10^2–10^6) in incorporation rates around the lesion and several 5′-downstream sites. These results also support a model in which adduct-induced heterogeneity at the replication fork affects polymerase function through longer-range DNA–protein interactions. However, kinetic effects undoubtedly play a key role in the processing of this bulky lesion, and the nature of the polymerase is likely to determine the balance between kinetic and thermodynamic effects [82]. In the same study [79], it was found that the addition of dC at unmodified primer-terminus position n lesion was favored over A ($\Delta\Delta G = 1.7$ kcal/mol, $\Delta\Delta H = 9.1$ kcal/mol) (Figure 10.6c), whereas no such thermodynamic advantage was observed at the same prime-terminus position n with lesion ($\Delta\Delta G \sim 0$ kcal//mol) (Figure 10.6d). These results provide valuable thermodynamic insights into the G → T transversion mutations that are frequently observed for bulky carcinogens, including aromatic amines [40].

10.3.6
Sequence Effects on the Conformational Stability of SMIs

The propensity for frameshift mutation depends on the characteristics of the base inserted opposite the lesion, including the 5′-sequence context surrounding the adduct, the rate of translesion DNA synthesis, and the nature of the carcinogens and the polymerases [83–85]. As noted earlier, an unusually higher mutation frequency of −2 deletions was observed when dG-C8-AAF was located at G_3, the hotspot in the *Nar*I sequence (-$G_1G_2CG_3CN$-) [37, 38]. Interestingly, the high mutational activity of this G_3 hotspot was found to depend considerably on the nature of the nucleotide, N (C ≫ T) [85]. The propensity for deletion mutation depends on the ability of AAF at G_3 to base-pair correctly with incoming dC during translesion synthesis and on the subsequent formation of SMIs.

Jain *et al.* [86] have conducted structural studies to probe the conformational basis of the effect of N on the propensity of the G_3 adduct to cause −2 mutation. Data indicated that the *Nar*I-dC/−2 deletion duplex (N = dC) (Figure 10.7) containing two bulged out unpartnered bases, namely CG_3 with modified G_3 adopted an intercalated conformer exclusively, whereas multiple conformers were detected for the *Nar*I-dT/−2 deletion duplex (N = dT). This dramatic sequence effect must have originated from the 3′-next flanking base (dC versus dT), which is the only difference between the two bulge structures. The 3′-G[FAF]CT-induced heterogeneity could be regarded as an equilibrium between the B- and S-SMI conformers (Figure 10.7b). It is likely that the carcinogen in S-SMI is well-accommodated in the bulge pocket with significant stacking with the 3′-flanking G:C base pair, whereas the carcinogen is fully exposed in B-SMI. In both cases, the immediate 3′-flanking base pair is G:C. These results support the hypothesis that the conformational stability of an SMI is a critical determinant of the efficacy of −2 deletions in *Nar*I.

It has been shown that the G_3-induced deletions are efficient in both *E. coli* Pol II exo$^-$ and exonuclease-free Klenow fragments [87]. These results imply a common AAF-induced SMI structure in the *Nar*I sequence. The suggestion is that such a

(a) 5'-C-T-C-G-G-C-G*-C-N-A-T-C-3'
     3'-C-A-G-C-C - - G-N-T-A-G-5'

G*=FAF;   N=C: NarI-dC/-2;   N=T: NarI-dT/-2

(b)

NarI-dC/-2 deletion duplex

```
        C
         \
          G
'5—G—G   C—C—3'
 |||  ||| ||| |||
 C—C—G—G-5'
```
S-conformeric SMI (S-SMI) ⬭ FAF adduct

NarI-dT/-2 deletion duplex

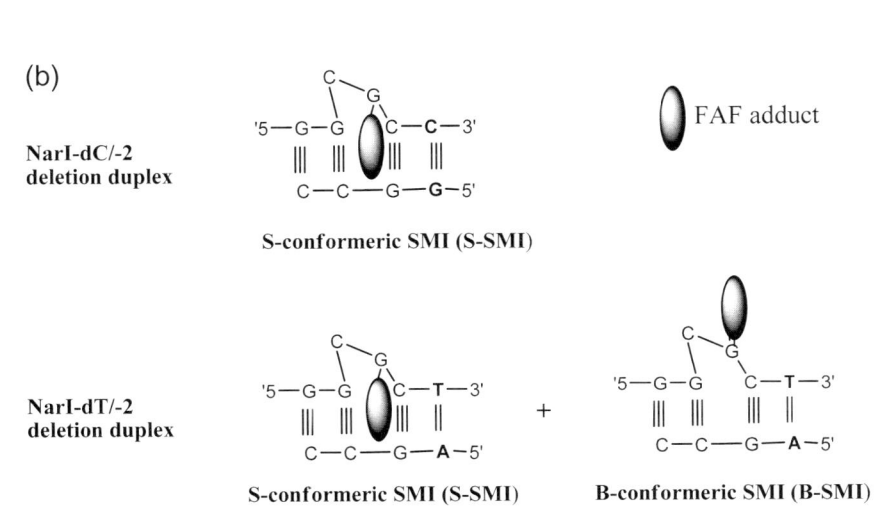

```
        C
         \
          G
'5—G—G   C—T—3'
 |||  ||| ||| ||
 C—C—G—A-5'
```
S-conformeric SMI (S-SMI)   +

```
        C
         \
          G
'5—G—G   C—T—3'
 |||  ||| ||| ||
 C—C—G—A-5'
```
B-conformeric SMI (B-SMI)

Figure 10.7 Proposed translesion synthesis for NarI-induced −2 deletion mutations. Incorporation of correct dC opposite dG-C8-FAF and subsequent nucleotide incorporation (e.g., dC) can lead to formation of stable S- and B-conformer SMIs (S-SMI and B-SMI, respectively). (a) Sequences for NarI-dC/–2 and NarI-dT/–2 deletion duplexes. (b) The NarI-dC/–2 deletion duplex resulted in an exclusive formation of S-SMI, whereas the NarI-dT/–2 deletion duplex produced a complex mixture of S- and B-SMI. The conformational stability of the SMI is strongly modulated by the nature (TA versus CG) of the next flanking base pairs at the 3' side of the lesion site (bold).

bulge structure would be quite different in the polymerase active site, compared to a structure produced by a similar modification in a non-*NarI* sequence. Replication of a *NarI*-based −2 deletion, generated by replication of the food mutagen dG-C8-IQ adduct, was strongly influenced by both the local sequence and the regioisomeric nature of the adduct in Dpo4 and Klenow exo⁻ [88].

10.4 Structures of DNA Lesion–DNA Polymerase Complexes

NMR solution structures provide the atomic-resolution conformational details of aromatic amine adducts in solution [26–28]. Obvious questions remain as to whether such structures are relevant to polymerase binding and, ultimately, to the cellular environment. Recent crystal structures of dG-C8-AF provide stunning views of how they are placed in the pre- and postinsertion site of the BF polymerase [29]. BF is analogous to the extensively used Klenow fragment, but lacks functional 5′–3′ exonuclease activity. In the preinsertion site, the modified dG adopts the *syn* conformation. The carcinogen ring in the preinsertion site has extensive van der Waals interactions with the polymerase. In transiting to the postinsertion site, the modified dG rotates to *anti*, leaving the Watson–Crick base-pairing intact, with the carcinogen in the major groove pocket (i.e., B-conformer). A structure of the dG-C8-AAF adduct was solved in T7 DNA polymerase [30]. Here the AAF was flipped out of the active site in the *syn* conformation (i.e., S-conformer) as observed in solution. The carcinogen moiety was inserted into a hydrophobic pocket behind the O helix, thus blocking the incoming nucleotide, which would cause polymerase blockage. Binding of the N-deacetylated AF to the polymerase was weak; consequently, electron densities were low around the lesion and no structure was resolved. The heterogeneity may stem from an S/B-type conformational heterogeneity in the crystal complex similar to the one observed in solution.

Crystal structures of aromatic amine adducts with lesion bypass polymerases are not yet available. These translesion bypass or Y-family polymerases possess more spacious active sites compared to the high-fidelity polymerases discussed above [89, 90]. Modeling studies with Dpo4, the prototype archeal Y-family bypass polymerase, suggested that translocation would be impeded much more on the major groove side of the modified template by C8-substituted AAF and PhIP dG-adducts, compared to the N^2-substituted (+)-*trans-anti*-B[a]P adduct (see Figure 10.1c) [90]. This may explain the more facile bypass observed for the N^2- compared to the C8-adducts in primer extension studies. Solution NMR structures of aromatic amine lesions complexed with either replicative or bypass polymerases are not yet available. Obtaining such information would be a major technical challenge. Comparisons of available data, however, seem to indicate that, in many cases, the favored adduct solution conformations are also manifested within crystalline complexes with polymerases [90].

10.5
Conclusions

Our studies have revealed that the structures of aromatic amine-DNA adducts are determined primarily by their size and coplanarity, as well as the nature of the adduct linkage (C8, N^2, etc). It has been shown that aromatic amine lesions exist primarily in three well-defined conformational categories (S, B, W) and their population balance is strongly influenced by the sequences surrounding the lesion site. It is believed that the S/B/W ratios of aromatic amine adducts, not the subtle structural differences at the lesion site (e.g., rotamers, C8 versus N^2 linkage, etc), determine the nature of the conformation-specific repair and mutational outcomes. As such, the available data points towards a new paradigm: lesion bypass (either error-free or error-prone) depends on various factors, including the thermodynamic and conformational characteristics of the lesion at the replication fork, and the types of polymerases involved. Therefore, it is crucial to elucidate systematically how these lesions are repaired and replicated at the atomic and molecular levels. The key molecular players (adduct structures, polymerases, repair proteins) that produce adverse outcomes must be identified, characterized, and understood, in order to devise appropriate chemoprevention approaches, and to develop drugs and risk assessment strategies [10–17, 68–76].

Mutation is the product of a balancing act between the efficiencies of DNA repair systems and the fidelity of the polymerases [91, 92]. The conformation-specific nature of repair and mutation observed for aromatic amine adducts (i.e., dG-C8-AF) is intriguing, and offers opportunities for both excitement and challenges. Hence, it is important to obtain conformation-specific structure–function relationships in simulated biological environments. Such knowledge would be crucial for pinpointing which adduct conformation is responsible for which mutation, the earliest step in the induction of cancer. Finally, it should be emphasized that the key to every biological problem must be sought in the cell. While the rigorous nature of the *in vitro* structural work (solution and crystalline solid) presented in this chapter is a first step, it is important to move forward by attaining basic structural knowledge in cellular environments. Such translational research will undoubtedly be technically challenging, but is the ultimate goal.

Acknowledgements

I gratefully acknowledge the former members of my laboratory for their major contribution on aromatic amine-DNA projects: Li Zhou, Srinivasa Meneni, Fengting Liang, and Nidhi Jain. Financial support for our research was provided by NIH grant R01 CA098296. This research was supported in part by RI-EPSCoR Proteomic (0554548) and RI-INBRE (P20 RR016457) core facilities.

References

1 Neumann, H.G. (2007) Aromatic amines in experimental cancer research: tissue-specific effects, an old problem and new solutions. *Crit. Rev. Toxicol.*, **37**, 211–236.

2 Luch, A. (2005) Nature and nurture – lessons from chemical carcinogenesis. *Nat. Rev. Cancer*, **5**, 113–125.

3 Purohit, V. and Basu, A.K. (2000) Mutagenicity of nitroaromatic compounds. *Chem. Res. Toxicol.*, **13**, 673–692.

4 Turesky, R.J. (2002) Heterocyclic aromatic amine metabolism, DNA adduct formation, mutagenesis, and carcinogenesis. *Drug Metab. Rev.*, **34**, 625–650.

5 Schut, H.A. and Snyderwine, E.G. (1999) DNA adducts of heterocyclic amine food mutagens: implications for mutagenesis and carcinogenesis. *Carcinogenesis*, **20**, 353–368.

6 Sarasin, A. (2003) An overview of the mechanisms of mutagenesis and carcinogenesis. *Mutat. Res.*, **544**, 99–106.

7 McGregor, W.G., Wei, D., Chen, R.H., Maher, V.M., and McCormick, J.J. (1997) Relationship between adduct formation, rates of excision repair and the cytotoxic

and mutagenic effects of structurally-related polycyclic aromatic carcinogens. *Mutat. Res.*, **376**, 143–152.

8 Takaska, G. (2003) Aromatic amines and human urinary bladder cancer: exposure sources and epidemiology. *J. Environ. Sci. Health C Environ. Carcinog. Ecotoxicol. Rev.*, **21**, 29–43.

9 Feng, Z., Hu, W., Rom, W.N., Beland, F.A., and Tang, M.S. (2002) 4-Aminobiphenyl is a major etiological agent of human bladder cancer: evidence from its DNA binding spectrum in human *p53* gene. *Carcinogenesis*, **23**, 1721–1727.

10 Singer, B. and Hang, B. (2000) Nucleic acid sequence and repair: role of adduct, neighbor bases and enzyme specificity. *Carcinogenesis*, **21**, 1071–1078.

11 Branzei, D. and Foiani, M. (2007) Interplay of replication checkpoints and repair proteins at stalled replication forks. *DNA Repair*, **6**, 994–1003.

12 Gillet, L.C.J. and Scharer, O.D. (2006) Molecular mechanisms of mammalian global genome nucleotide excision repair. *Chem. Rev.*, **106**, 253–276.

13 Sancar, A. and Reardon, J.T. (2004) Nucleotide excision repair in *E. coli* and man. *Adv. Protein Chem.*, **69**, 43–71.

14 Friedberg, E.C., Lehmann, A.R., and Fuchs, R.P. (2005) Trading places: how do DNA polymerases switch during translesion DNA synthesis? *Mol. Cell*, **18**, 11205–11215.

15 Izzotti, A., Bagnasco, M., Cartiglia, C., Longobardi, M., Balansky, R.M., Merello, A., Lubet, R.A., and De Flora, S. (2005) Chemoprevention of genome, transcriptome, and proteome alterations induced by cigarette smoke in rat lung. *Eur. J. Cancer*, **41**, 1864–1874.

16 Berdis, A.J. (2008) DNA polymerases as therapeutic targets. *Biochemistry*, **47**, 8253–8260.

17 Farmer, P.B. and Singh, R. (2008) Use of DNA adducts to identify human health risk from exposure to hazardous environmental pollutants: the increasing role of mass spectrometry in assessing biologically effective doses of genotoxic carcinogens. *Mutat. Res.*, **659**, 68–76.

18 Beland, F.A. and Kadlubar, F.F. (1990) in *Handbook of Experimental Pharmacology* (eds C.S. Cooper and P.L. Grover), Springer, Heidelberg, pp. 267–325.

19 Melchior, W.B., Jr., Marques, M.M., and Beland, F.A. (1994) Mutations induced by aromatic amine DNA adducts in pBR322. *Carcinogenesis*, **15**, 889–899.

20 Heflich, R.H. and Neft, R.E. (1994) Genetic toxicity of 2-acetylaminofluorene, 2-aminofluorene and some of their metabolites and model metabolites. *Mutat. Res.*, **318**, 73–114.

21 Dipple, A. (1995) DNA adducts of chemical carcinogens. *Carcinogenesis*, **16**, 437–441.

22 Geacintov, N.E., Cosman, M., Hingerty, B.E., Amin, S., Broyde, S., and Patel, D.J. (1997) NMR solution structures of stereoisometric covalent polycyclic aromatic carcinogen-DNA adduct: principles, patterns, and diversity. *Chem. Res. Toxicol.*, **10**, 111–146.

23 Szeliga, J. and Dipple, A. (1995) DNA adduct formation by polycyclic aromatic hydrocarbon dihydrodiol epoxides. *Chem. Res. Toxicol.*, **11**, 1–11.

24 Guengerich, F.P. (2006) Interactions of carcinogen-bound DNA with individual DNA polymerases. *Chem. Rev.*, **106**, 420–452.

25 Patel, D.J. (1992) Covalent carcinogenic guanine-modified DNA lesions solution structures of adducts and crosslinks. *Curr. Opin. Struct. Biol.*, **2**, 345–353.

26 Patel, D.J., Mao, B., Gu, Z., Hingerty, B.E., Gorin, A., Basu, A.K., and Broyde, S. (1998) Nuclear magnetic resonance solution structures of covalent aromatic amine-DNA adducts and their mutagenic relevance. *Chem. Res. Toxicol.*, **11**, 391–407.

27 Cho, B.P. (2004) Dynamic conformational heterogeneities of carcinogen-DNA adducts and their mutagenic relevance. *J. Environ. Sci. Health C Environ. Carcinog. Ecotoxicol. Rev.*, **22**, 57–90.

28 Lukin, M. and de los Santos, C. (2006) NMR structures of damaged DNA. *Chem. Rev.*, **106**, 607–686.

29 Hsu, G.W., Kiefer, J.R., Burnouf, D., Becherel, O.J., Fuchs, R.P., and Beese, L.S. (2004) Observing translesion synthesis of an aromatic amine DNA adduct by a high-fidelity DNA polymerase. *J. Biol. Chem.*, **279**, 50280–50285.

30 Dutta, S., Li, Y., Johnson, D., Dzantiev, J., Richardson, C.C., Romano, L.J., and Ellenberger, T. (2004) Crystal structures of 2-acetylaminofluorene and 2-aminofluorene in complex with T7 DNA polymerase reveal mechanisms of mutagenesis. *Proc. Natl. Acad. Sci. USA*, **101**, 16186–16191.

31 Cho, B.P., Beland, F.A., and Marques, M.M. (1992) NMR structural studies of a 15-mer DNA sequence from a *ras* protooncogene, modified at the first base of codon 61 with the carcinogen 4-aminobiphenyl. *Biochemistry*, **31**, 9587–9602.

32 Mao, B., Vyas, R.R., Hingerty, B.E., Broyde, S., Basu, A.K., and Patel, D.J. (1996) Solution conformation of the N-(deoxyguanosin-8-yl)-1-aminopyrene ([AP]dG) adduct opposite dC in a DNA duplex. *Biochemistry*, **35**, 12659–12670.

33 Zaliznyak, T., Bonala, R., Johnson, F., and de los Santos, C. (2006) Structure and stability of duplex DNA containing the 3-(deoxyguanosin-N^2-yl)-2-acetylaminofluorene (dG(N^2)-AAF) lesion: a bulky adduct that persists in cellular DNA. *Chem. Res. Toxicol.*, **19**, 745–752.

34 Wang, F., DeMuro, N.E., Elmquist, C.E., Stover, J.S., Rizzo, C.J., and Stone, M.P. (2006) Base-displaced intercalated structure of the food mutagen 2-amino-3-methylimidazo[4,5-*f*]-quinoline in the recognition sequence of the *Nar*I restriction enzyme, a hotspot for –2 bp deletions. *J. Am. Chem. Soc.*, **128**, 10085–10095.

35 Elmquist, C.E., Wang, F., Stover, J.S., Stone, M.P., and Rizzo, C.J. (2007) Conformational differences of the C8-deoxyguanosine adduct of 2-amino-3-methylimidazo[4,5-*f*]quinoline (IQ) within the *Nar*I recognition sequence. *Chem. Res. Toxicol.*, **20**, 445–454.

36 Wang, F., Elmquist, C.E., Stover, J.S., Rizzo, C.J., and Stone, M.P. (2007) DNA sequence modulates the conformation of the food mutagen 2-amino-3-methylimidazo[4,5-*f*]quinoline in the recognition sequence of the *Nar*I restriction enzyme. *Biochemistry*, **46**, 8498–8516.

37 Burnouf, D., Koehl, P., and Fuchs, R.P.P. (1989) Single adduct mutagenesis: strong effect of the position of a single acetylaminofluorene adduct within a mutation hot spot. *Proc. Natl. Acad. Sci. USA*, **86**, 4147–4151.

38 Fuchs, R.P.P. and Fujii, S. (2007) Translesion synthesis in *Escherichia coli*: lessons from the *Nar*I mutation hot spot. *DNA Repair (Amst.)*, **6**, 1032–1041.

39 Norman, D., Abuaf, P., Hingerty, B.E., Live, D., Grunberger, D., Broyde, S., and Patel, D.J. (1989) NMR and computational characterization of the N-(deoxyguanosin-8-yl)aminofluorene adduct [(AF)G] opposite adenosine in DNA: (AF)G[*syn*] · A[*anti*] pair formation and its pH dependence. *Biochemistry*, **28**, 7462–7476.

40 Shibutani, S., Suzuki, N., Tan, X., Johnson, F., and Grollman, A.P. (2001) Influence of flanking sequence context on the mutagenicity of acetylaminofluorene-derived DNA adducts in mammalian cells. *Biochemistry*, **27**, 3717–3722.

41 Devadoss, B., Lee, I., and Berdis, A.J. (2008) Enhancing the "A-rule" of translesion DNA synthesis: promutagenic DNA synthesis using modified nucleoside triphosphates. *Biochemistry*, **47**, 8253–8260.

42 Abuaf, P., Hingerty, B.E., Broyde, S., and Grunberger, D. (1995) Solution conformation of the N-(deoxyguanosin-8-yl)aminofluorene adduct opposite deoxyinosine and deoxyguanosine in DNA by NMR and computational characterization. *Chem. Res. Toxicol.*, **8**, 369–378.

43 Cho, B.P., Beland, F.A., and Marques, M.M. (1994) NMR structural studies of a 15-mer DNA duplex from a *ras* protooncogene modified with the carcinogen 2-aminofluorene: conformational heterogeneity. *Biochemistry*, **33**, 1373–1384.

44 Eckel, L.M. and Krugh, T.R. (1994) 2-Aminofluorene modified DNA duplex exists in two interchangeable conformations. *Nat. Struct. Biol.*, **1**, 89–94.

45 Eckel, L.M. and Krugh, T.R. (1994) Structural characterization of two interchangeable conformations of a 2-aminofluorene-modified DNA oligomer

46 Mao, B., Gu, Z., Hingerty, B.E., Broyde, S., and Patel, D.J. (1998) Solution structure of the aminofluorene [AF]-intercalated conformer of the *anti* [AF]-C^8-dG adduct opposite dC in a DNA duplex. *Biochemistry*, **37**, 81–94.

47 Mao, B., Gu, Z., Hingerty, B.E., Broyde, S., and Patel, D.J. (1998) Solution structure of the aminofluorene [AF]-external conformer of the *anti* [AF]-C^8-dG adduct opposite dC in a DNA duplex. *Biochemistry*, **37**, 95–106.

48 Brown, K., Hingerty, B.E., Guenther, E.A., Krishnan, V.V., Broyde, S., Turteltaub, K.W., and Cosman, M. (2001) Solution structure of the 2-amino-1-methyl-6-phenylimidazo[4,5-b]pyridine C8-deoxyguanosine adduct in duplex DNA. *Proc. Natl. Acad. Sci. USA*, **98**, 8507–8512.

49 O'Handley, S.F., Sanford, D.G., Xu, R., Lester, C.C., Hingerty, B.E., Broyde, S., and Krugh, T.R. (1993) Structural characterization of an N-acetyl-2-aminofluorene (AAF) modified DNA oligomer by NMR, energy minimization, and molecular dynamics. *Biochemistry*, **32**, 2481–2497.

50 Milhe, C., Dhalluin, C., Fuchs, R.P., and Lefevre, J.F. (1994) NMR evidence of the stabilisation by the carcinogen N-2-acetylaminofluorene of a frameshift mutagenesis intermediate. *Nucleic Acids Res.*, **22**, 4646–4652.

51 Milhe, C., Fuchs, R.P., and Lefevre, J.F. (1996) NMR data show that the carcinogen N-2-acetylaminofluorene stabilises an intermediate of –2 frameshift mutagenesis in a region of high mutation frequency. *Eur. J. Biochem.*, **235**, 120–127.

52 Cho, B.P. and Zhou, L. (1999) Probing the conformational heterogeneity of the acetylaminofluorene-modified 2′-Deoxyguanosine and DNA by ^{19}F NMR spectroscopy. *Biochemistry*, **38**, 7572–7583.

53 Lindsley, J.E. and Fuchs, R.P. (1994) Use of single-turnover kinetics to study bulky adduct bypass by T7 DNA polymerase. *Biochemistry*, **33**, 764–772.

54 Shapiro, R., Ellis, S., Hingerty, B.E., and Broyde, S. (1998) Effect of ring size on conformations of aromatic amine-DNA adducts: the aniline-C8 guanine adduct resides in the B-DNA major groove. *Chem. Res. Toxicol.*, **11**, 335–341.

55 Meneni, S.R., Shell, S.M., Gao, L., Jurecka, P., Lee, W., Sponer, J., Zou, Y., Chiarelli, M.P., and Cho, B.P. (2007) Spectroscopic and theoretical insights into sequence effects of aminofluorene-induced conformational heterogeneity and nucleotide excision repair. *Biochemistry*, **46**, 11263–11278.

56 Meneni, S.R., Liang, F., and Cho, B.P. (2007) Examination of the long-range effects of aminofluorene-induced conformational heterogeneity and its relevance to the mechanisms of translesion DNA synthesis. *J. Mol. Biol.*, **366**, 1387–1400.

57 Zhou, L., Rajabzadeh, G., Traficante, D.D., and Cho, B.P. (1997) Conformational heterogeneity of arylamine-modified DNA: ^{19}F NMR evidence. *J. Am. Chem. Soc.*, **119**, 5384–5389.

58 Jain, N., Reshetnyak, Y.K., Gao, L., Chiarelli, M.P., and Cho, B.P. (2008) Fluorescence probing of aminofluorene-induced conformational heterogeneity in DNA duplexes. *Chem. Res. Toxicol.*, **21**, 445–452.

59 Meneni, S.R., D'Mello, R., Norigian, G., Baker, G., Gao, L., Chiarelli, M.P., and Cho, B.P. (2006) Sequence effects of aminofluorene-modified DNA duplexes: thermodynamic and circular dichroism properties. *Nucleic Acids Res.*, **34**, 755–763.

60 Jiang, Y.L., McDowell, L.M., Poliks, B., Studelska, D.R., Cao, C., Potter, G.S., Schaefer, J., Song, F., and Stivers, J.T. (2004) Recognition of an unnatural difluorophenyl nucleotide by uracil DNA glycosylase. *Biochemistry*, **43**, 15429–15438.

61 Seo, K.Y., Jelinsky, S.A., and Loechler, E.L. (2000) Factors that influence the mutagenic patterns of DNA adducts from chemical carcinogens. *Mutat. Res.*, **463**, 215–246.

62 Mekhovich, O., Tang, M., and Romano, L.J. (1998) Rate of incision of N-acetyl-2-aminofluorene and N-2-aminofluorene adducts by UvrABC nuclease is

adduct- and sequence-specific: comparison of the rates of UvrABC nuclease incision and protein–DNA complex formation. *Biochemistry*, **37**, 571–579.
63 Zou, Y., Shell, S.M., Utzat, C.D., Luo, C., Yang, Z., Geacintov, N.E., and Basu, A.K. (2003) Effects of DNA adduct structure and sequence context on strand opening of repair intermediates and incision by UvrABC nuclease. *Biochemistry*, **42**, 12654–12661.
64 Ruan, Q., Liu, T., Kolbanovskiy, A., Liu, Y., Ren, J., Skovraga, M., Zou, Y., Lader, J., Malkani, B., Amin, S., Van Houten, B., and Geacintov, N.E. (2007) Sequence context- and temperature-dependent nucleotide excision repair of a benzo[a]pyrene diol epoxide-guanine DNA adduct catalyzed by thermophilic UvrABC proteins. *Biochemistry*, **46**, 7006–7015.
65 Kalam, M.A., Haraguchi, K., Chandani, S., Loechler, E.L., Moriya, M., Greenberg, M.M., and Basu, A.K. (2006) Genetic effects of oxidative DNA damages: comparative mutagenesis of the imidazole ring-opened formamidopyrimidines (Fapy lesions) and 8-oxo-purines in simian kidney cells. *Nucleic Acids Res.*, **34**, 2305–2315.
66 Kalam, M.A. and Basu, A.K. (2005) Mutagenesis of 8-oxoguanine adjacent to an abasic site in simian kidney cells: tandem mutations and enhancement of G–T transversions. *Chem. Res. Toxicol.*, **18**, 1187–1192.
67 Meneni, S., Shell, S.M., Zou, Y., and Cho, B.P. (2007) Conformation-specific recognition of carcinogen-DNA adduct in *E. coli*. nucleotide excision repair. *Chem. Res. Toxicol.*, **20**, 6–10.
68 Wood, R.D. (1999) DNA damage recognition during nucleotide excision repair in mammalian cells. *Biochemistry*, **81**, 39–44.
69 Riedl, T., Hanaoka, F., and Egly, J.M. (2003) The comings and goings of nucleotide excision repair factors on damaged DNA. *EMBO J.*, **22**, 5293–5303.
70 Min, J.H. and Pavletich, N.P. (2007) Recognition of DNA damage by the Rad4 nucleotide excision repair protein. *Nature*, **449**, 570–575.
71 Sharer, O.D. (2007) Achieving broad substrate specificity in damage recognition by binding accessible nondamaged DNA. *Mol. Cell*, **28**, 184–186.
72 Geacintov, N.E., Broyde, S., Buterin, T., Naegeli, H., Wu, M., Yan, S., and Patel, D.J. (2002) Thermodynamic and structural factors in the removal of bulky DNA adducts by the nucleotide excision repair machinery. *Biopolymers*, **65**, 202–210.
73 Maillard, O., Camenisch, U., Blagoev, K.B., and Naegeli, H. (2008) Versatile protection from mutagenic DNA lesions conferred by bipartite recognition in nucleotide excision repair. *Mutat. Res.*, **658**, 271–286.
74 Isaacs, R.J. and Spielmann, H.P. (2004) A model for initial DNA lesion recognition by NER and MMR based on local conformational flexibility. *DNA Repair*, **4**, 455–464.
75 Dip, R., Camenisch, U., and Naegeli, H. (2004) Mechanisms of DNA damage recognition and strand discrimination in human nucleotide excision repair. *DNA Repair*, **2**, 1409–1423.
76 Buterin, T., Hess, M.T., Luneva, N., Geacintov, N.E., Amin, S., Kroth, H., Seidel, A., and Naegeli, H. (2000) Unrepaired fjord region polycyclic aromatic hydrocarbon-DNA adducts in *ras* codon 61 mutational hot spots. *Cancer Res.*, **60**, 1849–1856.
77 Jain, N., Meneni, S., and Cho, B.P. (2009) Influence of flanking sequence context on the conformational flexibility of aminofluorene-modified dG adduct in dA mismatch DNA duplexes. *Nucleic Acids Res.*, **37**, 1628–1637.
78 Cai, Y., Patel, D.J., Geacintov, N.E., and Broyde, S. (2007) Dynamics of a benzo[a]pyrene-derived guanine DNA lesion in TGT and CGC sequence contexts: enhanced mobility in TGT explains conformational heterogeneity, flexible bending, and greater susceptibility to nucleotide excision repair. *J. Mol. Biol.*, **374**, 292–305.
79 Liang, F. and Cho, B.P. (2007) Probing the thermodynamics of aminofluorene-induced trans-lesion DNA synthesis by differential scanning calorimetry. *J. Am. Chem. Soc.*, **129**, 12108–12109.
80 Liang, F., Meneni, S.R., and Cho, B.P. (2006) Induced circular dichroism

characteristics as conformational probes for carcinogenic aminofluorene-DNA adducts. *Chem. Res. Toxicol.*, **19**, 1040–1043.

81 Miller, H. and Grollman, A.P. (1997) Kinetics of DNA polymerase I (Klenow fragment exo⁻) activity on damaged DNA templates: effect of proximal and distal template damage on DNA synthesis. *Biochemistry*, **36**, 15336–15342.

82 Minetti, A.A., Remeta, D., Miller, H., Gelfand, C.A., Plum, G.E., Grollman, A.P., and Breslauer, K.J. (2003) The thermodynamics of template-directed DNA synthesis: base insertion and extension enthalpies. *Proc. Natl. Acad. Sci. USA*, **100**, 14719–14724.

83 Shibutani, S., Suzuki, N., and Grollman, A.P. (1998) Mutagenic specificity of (acetylamino)-fluorene-derived DNA adducts in mammalian cells. *Biochemistry*, **37**, 12034–12041.

84 Tan, X., Suzuki, N., Grollman, A.P., and Shibutani, S. (2002) Mutagenic events in *Escherichia coli* and mammalian cells generated in response to acetylaminofluorene-derived DNA adducts positioned in the *Nar*I restriction enzyme site. *Biochemistry*, **41**, 14255–14262.

85 Broschard, T.H., Koffel-Schwartz, N., and Fuchs, R.P.P. (1999) Sequence-dependent modulation of frameshift mutagenesis at the *Nar*I-derived mutation hot spots. *J. Mol. Biol.*, **288**, 191–199.

86 Jain, N., Li, Y., Zhang, L., Meneni, S., and Cho, B.P. (2007) Probing the sequence effects on *Nar*I-induced −2 frameshift mutagenesis by dynamic ^{19}F NMR, UV and CD spectroscopy. *Biochemistry*, **46**, 13310–13321.

87 Gill, J.P. and Romano, L.J. (2005) Mechanism for N-acetyl-2-aminofluorene-induced frameshift mutagenesis by *Escherichia coli* DNA polymerase I (Klenow fragment). *Biochemistry*, **44**, 15387–15395.

88 Stover, J.S., Chowdhury, G., Zang, H., Guengerich, F.P., and Rizzo, C.J. (2006) Translesion synthesis past the C8- and N^2-deoxyguanosine adducts of the dietary mutagen 2-amino-3-methylimidazo[4,5-*f*]quinoline in the *Nar*I recognition sequence by prokaryotic DNA polymerases. *Chem. Res. Toxicol.*, **19**, 1506–1517.

89 Broyde, S., Wang, L., Rechkoblit, O., Geacintov, N.E., and Patel, D.J. (2008) Lesion processing: high-fidelity versus lesion-bypass DNA polymerases. *Trends Biochem. Sci.*, **33**, 209–219.

90 Broyde, S., Wang, L., Zhang, L., Rechkoblit, O., Geacintov, N.E., and Patel, D.J. (2008) DNA adduct structure–function relationships: comparing solution with polymerase structures. *Chem. Res. Toxicol.*, **21**, 45–52.

91 Bielas, J.H., Loeb, K.R., Rubin, B.P., True, L.D., and Loeb, L.A. (2006) Human cancers express a mutator phenotype. *Proc. Natl. Acad. Sci. USA*, **103**, 18238–18242.

92 Loeb, L.A., Bielas, J.H., and Beckman, R.A. (2008) Cancers exhibit a mutator phenotype: clinical implications. *Cancer Res.*, **68**, 3551–3557.

11
Mechanisms of Base Excision Repair and Nucleotide Excision Repair
Orlando D. Schärer and Arthur J. Campbell

11.1
General Features of Base Excision and Nucleotide Excision Repair

As discussed in previous chapters of this book, damage to DNA occurs frequently (10^5–10^6 lesions/cell/day), and can arise from endogenous and exogenous sources. Endogenous damage is an unavoidable consequence of the presence of water, oxygen, and alkylating agents in our cells. Endogenous lesions are often relatively small and lead to the addition or substitution of one or more atoms on the pyrimidine or purine bases (Figure 11.1a). Furthermore, a number of pre-eminent endogenous lesions exist, including, for example, uracil, 8-oxoguanine (8-oxo-G), or thymine glycol (TG), justifying the existence of repair systems that specifically reverse these forms of DNA damage. Consequently, repair systems that address endogenous lesions have evolved to deal with specific lesions or a small set of structurally related lesions [1]. The prototypical repair system for endogenous lesions is base excision repair (BER), which is initiated by DNA glycosylases that recognize a small set of DNA lesions and excise them from DNA by cleaving the glycosidic bond linking the base to the sugar moiety of the backbone. At least 12 different DNA glycosylases are known to exist in humans. In BER, the activities for recognizing and removing the damaged base resides in the enzyme initiating the repair process [2], leading in most cases to the formation of an abasic site intermediate. This intermediate is then processed by a common set of enzymes that ultimately enable the restoration of the original DNA sequence.

By contrast, exogenous lesions tend to be more diverse in structure and typically cause a more dramatic alteration to the DNA bases. Such lesions may result from exposure to solar UV irradiation, ingestion of mutagens present in cooked foods, the inhalation of industrial exhaust, or exposure to chemotherapeutic agents (Figure 11.1b) [3]. Such adducts, often referred to as bulky lesions, may occur with varying frequencies depending on the conditions and extent of exposure. Repair systems dealing with exogenous lesions, namely nucleotide excision repair (NER), have therefore evolved to deal with structurally diverse lesions and must have the ability to adapt to lesions that may suddenly arise in the environment. The recognition of all lesions in NER is achieved by the same set of core enzymes, and follows a general

The Chemical Biology of DNA Damage. Edited by Nicholas E. Geacintov and Suse Broyde
© 2010 WILEY-VCH Verlag GmbH & Co. KGaA, Weinheim
ISBN: 978-3-527-32295-4

(a) Typical Base Excision Repair Substrates

Uracil (U) | **Hypoxanthine (Hyp)** | **8-Oxoguanine (8-oxoG)** | **Thymine glycol (TG)**

3-Methyladenine (3-meA) | **1,N⁶-ethenoadenine (εA)** | **Abasic site (AP site)**

(b) Typical Nucleotide Excision Repair Substrates

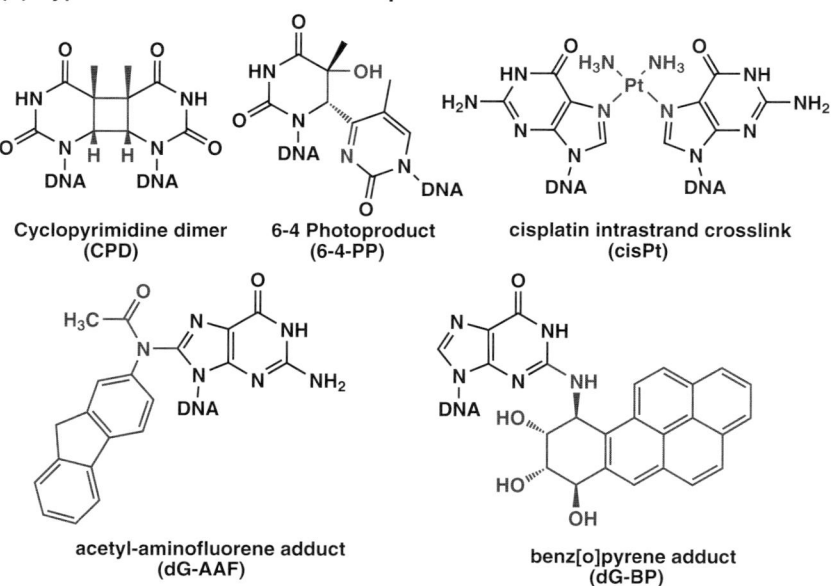

Cyclopyrimidine dimer (CPD) | **6-4 Photoproduct (6-4-PP)** | **cisplatin intrastrand crosslink (cisPt)**

acetyl-aminofluorene adduct (dG-AAF) | **benz[o]pyrene adduct (dG-BP)**

Figure 11.1 Representative BER and NER substrates. (a) Bases modified by hydrolytic deamination (U, Hyp), oxidation (8-oxo-G, TG), alkylation (3-meA, εA), and hydrolysis (AP site) are processed by BER. The atoms of the molecule that differ from native DNA are highlighted in gray. (b) Bulky DNA adducts formed by solar UV irradiation (CPD, 6-4 photoproduct), antitumor agents (cisplatin), and environmental agents (dG-AAF, dG-BP) are repaired by NER. The positions altered in the adducts are highlighted in dark gray.

DNA recognition mode in which the lesion is first recognized and verified in an ATP-dependent manner. Lesion removal then occurs by two latent endonucleases that are recruited to NER complexes in a subsequent step. In contrast to BER, damage recognition and incision activities do not reside in the same polypeptide, likely allowing NER to be a more adaptable system for lesion recognition and repair.

The physiological importance of BER and NER is underscored by the phenotypes displayed by mice and humans with deficiencies in the two repair systems. Mice with deficiencies in the BER genes *Ape1*, *Polβ*, *Xrcc1* and *Lig3α*, display an embryonic lethal phenotype [4], indicating that lesion repair by BER is essential for life. Interestingly, deficiencies in most DNA glycosylases do not display the same severe phenotypes [5]. This apparent discrepancy has been attributed to either redundancy among DNA glycosylases or the critical importance of abasic sites, which do not require DNA glycosylases for repair (see below). A notable exception is the MYH DNA glycosylase, which removes adenine bases that have been aberrantly introduced opposite 8-oxo-G residues during replication. Mutations in MYH are associated with hereditary colon cancer, similar to phenotypes resulting from defects in the mismatch repair pathway [6].

By contrast, deficiencies in GG-NER (see Section 11.3) lead to the inherited disorder xeroderma pigmentosum (XP), which can be viewed as the prototypical DNA repair disorder [7–9]. XP patients are unable to repair UV lesions, and as a consequence suffer from an extreme sensitivity to UV light and an over 1000-fold increased incidence of skin cancer. Severely affected XP patients additionally suffer from neurological abnormalities. The phenotypes of two additional symptoms associated with NER genes, Cockayne syndrome (CS) and trichothiodystrophy (TTD), are more complex as they are due to defects in transcription as well as defects in DNA repair [10].

11.2
BER

11.2.1
BER Overview – Short-Patch and Long-Patch BER

BER is generally believed to be the main "housekeeping" pathway and deals with lesions that occur as part of normal metabolism. Many of the lesions processed by BER occur with frequencies on the order of 100–10 000/human cell/day [11, 12], justifying the existence of dedicated enzymes that recognize these lesions with high specificity. Twelve human DNA glycosylases have been identified to date that recognize a small number of modified bases with unique substrate specificities. These enzymes can be grouped into two mechanistic classes: monofunctional glycosylases that generate an abasic site product, and bifunctional DNA glycosylases/AP lyases that form a transient covalent Schiff's base adduct with the abasic site, and cleave the phophodiester bond 3' and in some cases 5' to the abasic site using β- and δ-lyase activities.

The common products generated by the glycosylases and spontaneously generated abasic sites are then channeled into short-patch or long-patch BER (Figure 11.2) – two alternate pathways to complete the repair. In short-patch BER [13, 14] the abasic site is incised on its 5'-side by an AP endonuclease (APE1) and the resulting 3'-OH serves as a substrate for Pol β that covalently incorporates a nucleotide. The AP lyase activity of Pol β is then responsible for removing the remnants of the abasic site [15]. In the case of the NEIL1 and NEIL2 glycosylases that catalyze an AP lyase reaction 5' and 3' to the abasic site, AP endonuclease and the AP lyase activity of Pol β are not needed, but instead polynucleotide kinase (PNK) cleaves the 3'-phosphate group resulting from the lyase reaction (Figure 11.2, top right) [16]. In all cases the DNA ligase 3α/XRCC1 (excision repair cross-complementation group 1) complex is responsible for sealing the nick and completing the repair process.

Long-patch BER is believed to be the minor pathway and it might be of particular advantage if an abasic site is chemically altered so that the AP lyase reaction is no longer possible. In this pathway, the replicative Pols δ/ε, and associated factors proliferating cell nuclear antigen (PCNA) and replication factor C (RFC), catalyze strand displacement synthesis past the abasic site, generating a 5'-flap structure (Figure 11.2, right panel) [17, 18]. The abasic site containing the 5'-flap is then removed by FEN1 endonuclease, resulting in a nick that is sealed by DNA ligase 1.

11.2.2
Lesion Recognition by DNA Glycosylases

The first key step in BER is the recognition of the damaged base by a DNA glycosylase. The mechanism underlying the enzymatic recognition of a modified DNA residue (Figure 11.1) among the more than million-fold excess of nondamaged DNA is an intriguing question that has been the subject of intensive research [2, 19–21]. The human DNA glycosylases can be divided into four structural families that have been conserved throughout evolution. The largest family has a helix–hairpin–helix (HhH) motif as a defining feature [22], and includes OGG1, NTH1, MYH, and MBD4 proteins. These four enzymes do not share common substrates, however. The three uracil/thymine-specific glycosylases UNG, TDG, and SMUG form the second family, and these three proteins share a common structure of the glycosylase domain. NEIL1 and NEIL2 are members of a family of bifunctional glycosylase/AP lyases that remove a diverse set of oxidized bases from DNA. AAG appears to be an outlier, being the sole family member and is unique among human DNA glycosylases in that it processes structurally rather diverse lesions.

Despite the diversity of sequences and structures of the different families, all DNA glycosylases share key features of their mechanism to recognize and excise damaged bases, which will be illustrated here using the examples of members of three of the families – UNG [23–25], OGG1 [26], and AAG [27, 28]. All DNA glycosylases employ a nucleotide-flipping mechanism to extrude the target base into the active site pocket where catalysis takes place. A bulky aromatic or aliphatic

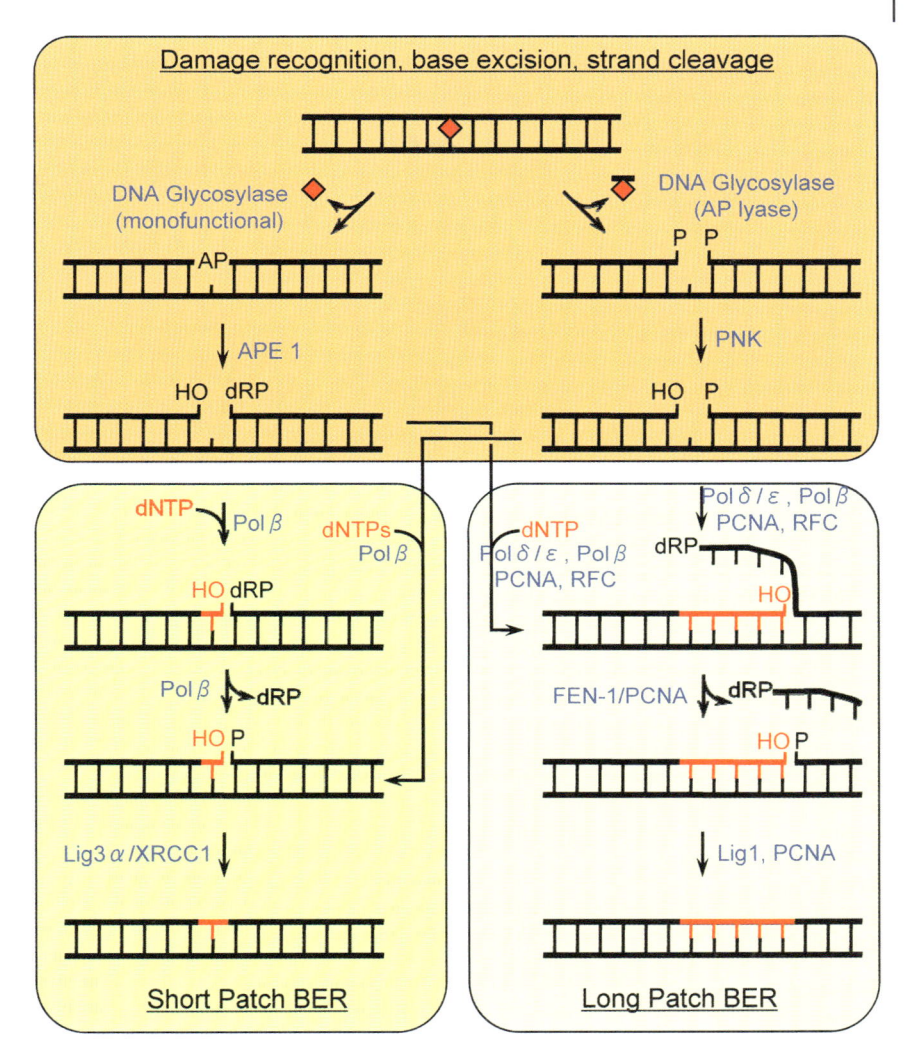

Figure 11.2 BER pathway. BER is initiated by DNA glycosylases that recognize lesions and hydrolyze the glycosidic bond linking the base to the sugar phosphate backbone. Monofunctional DNA glycosylases (top left) generate abasic site products that are cleaved at the 5'-phosphodiester bond by APE1. Bifunctional DNA glycosylases/AP lyases (top right) remove the abasic site and leave a 3'-phosphate group that is removed by PNK. The resulting gaps are then filled in and sealed either via short-patch BER (bottom left panel) by Pol β and DNA ligase 3α/XRCC1, or by long-path BER (bottom right panel) utilizing Pols δ/ε, PCNA, RFC, FEN1, and DNA ligase 1. Not shown are additional specialized BER pathways.

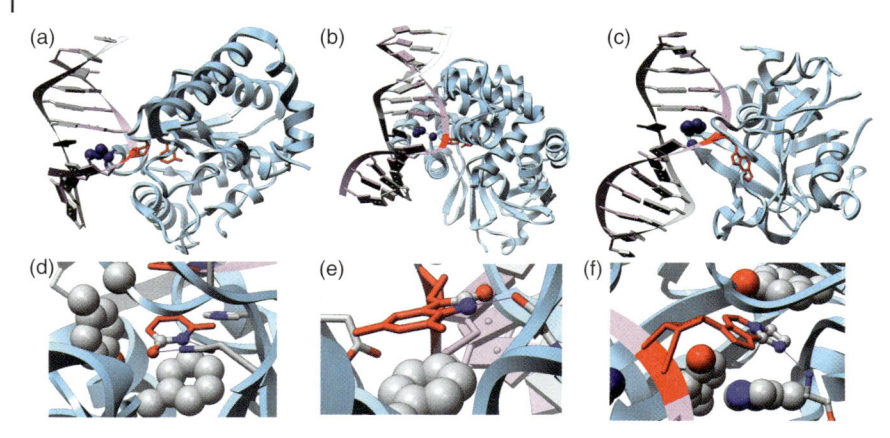

Figure 11.3 Damage recognition by the UDG, OGG1, and AAG glycosylases. (a–c) Overview of uracil-DNA glycosylase (UDG, a) 8-oxoguanine-DNA glycosylase (OGG1, b), and alkyladenine glycosylase complex (AAG, c) bound to substrates. Enzymes are colored light blue, the damaged DNA strand is in purple, and the undamaged strand is in gray color. The damaged nucleotide (red) is flipped-out into the enzyme active site and a residue of the enzyme (dark blue; Leu272 in UDG, Phe319 in OGG1, and Tyr127 in AAG) takes up the position of the base in the DNA helix. The base opposite the lesion (dark blue) is slightly displaced from the helix stack and the DNA is kinked at the lesion site. (d–f) Magnified view of the base-binding pockets of UDG, OGG1, and AAG. (d) UDG (light blue) bound to a uracil residue (red). N3 and O4 of U that distinguish it from C are shown in atom color, and hydrogen bonds to Asn204 are indicated by solid purple lines. Tyr 47 that packs against C5, excluding the binding of T, and Phe158 that packs against the U are represented in VDW atom color. (e) OGG1 (light blue) bound to 8-oxo-G (red)-containing DNA. The 7- and 8-positions of 8-oxo-G that distinguish it from G are shown in atom color. The hydrogen bond from the backbone carbonyl of Gly42 (atom color) to NH7 is indicated by a solid purple line. Phe319 packs against the base and is shown in VDW atom color. (f) AAG (light blue) bound to εA (red)-containing DNA. Tyr127, His136, and Tyr159 are represented by VDW atom color. The five-membered ring in εA linking the 1- and 6-positions is indicated in atom color. A solid purple line indicates the hydrogen bond from N6 of εA to the backbone amide of His136 (atom color). The figures were prepared using Chimera (http://www.cgl.ucsf.edu/chimera) using the structures PDB 1EMJ for UDG, PDB 1YQR for OGG1, and PDB 1EWN for AAG.

residue is inserted into the helix from the minor groove replacing the modified base (blue in Figure 11.3a–c). Together, these rearrangements in the DNA lead to a local kink in the DNA. Once moved out of the helix, the modified base is accommodated in a specific binding pocket. The lesion-binding pockets have some common features among the families of glycosylases, in particular aromatic amino acids that engage in coplanar stacking with the nucleobase (Phe158 in UDG, Phe319 in OGG1, and Tyr127 in AAG) (Figure 11.3d–f).

These three individual enzymes make a number of contacts that favor binding of the substrate and discriminate against binding of nondamaged bases. In UDG, which removes U from DNA, a bifurcated hydrogen bond from Asn204 to O4 and

NH3 of uracil discriminates against cytidine. The binding of thymine is excluded by the aromatic group of Tyr147, which packs against C5 and prevents the accommodation of the methyl group of T (Figure 11.3d) [25]. Purine bases are excluded from the binding pocket due to their size. The UDG binding pocket is therefore highly specific for uracil.

OGG1 removes oxidized purines, in particular 8-oxo-G, from DNA. The base-binding pocket of OGG1 discriminates between G and 8-oxoG based on a single hydrogen bond from the backbone carbonyl group of Gly42 to the NH group at the 7 position (Figure 11.3e) [26]. Other key residues in the binding pocket, including Gln315 that contacts the Watson–Crick face of the lesion, are also well suited to accommodate guanine.

AAG is special among DNA glycosylases in that it recognizes a fairly wide and structurally somewhat heterogeneous group of substrates that includes 3-methyladenine, hypoxanthine, and ethenoadenine (εA) [29]. The structure of a catalytically inactivated version of AAG bound to DNA shows a specific hydrogen bond from the backbone amide of His136 to the N6 position of εA (Figure 11.4f) [28]. This interaction suggests how AAG could favorably interact with εA, hypoxanthine, or 3-methyladenine over adenine. Adenine has two hydrogen atoms at N6 rather then just one like these three lesions and one of these hydrogen atoms which would clash with the backbone amide hydrogen.

The impressive number of available crystallographic structures of DNA glycosylases bound to their substrates have revealed how the glycosylases provide specificity for damaged bases in the flipped-out state, but do not provide any insight into how the enzymes might recognize the lesions while they are still intrahelical. Would DNA glycosylases flip out every base to check for damage or would they have a way of preselecting intrahelical damaged bases for recognition [30]? Innovative approaches have been used to address this question. Verdine *et al.* used an engineered disulfide cross-link bond between the enzyme and the DNA to force the enzyme not only to bind to lesioned bases, but also to nondamaged bases [31–33]. These studies revealed that a G residue can also be extruded from the helix. Intriguingly, however, it does not bind in the same pocket as 8-oxo-G. Instead G binds in a different, shallower binding pocket (termed the exo-site), which is located along the trajectory to a fully flipped-out state in the periphery of the active site. These studies of OGG1 are consistent with a mechanism whereby the flipping out of the damaged base occurs through successive intermediates until it is accommodated in the active site pocket. It is likely that the selectivity of OGG1 for 8-oxo-G and against G is achieved at various steps of the nucleotide flipping process that provides a much higher degree of selectivity than could be achieved by the single hydrogen bond in the active site. Such a model is indeed consistent with kinetic studies that revealed several discrete reaction intermediates along the reaction coordinate [34, 35].

Extensive kinetic and structural studies of uracil recognition by UDG have been published mainly by Stivers *et al.* revealing similarities, but also important differences in how UDG recognizes damaged sites as compared to OGG1 [21]. Both enzymes can bind native DNA bases in an exo-site [36]. In the case of UDG, a T

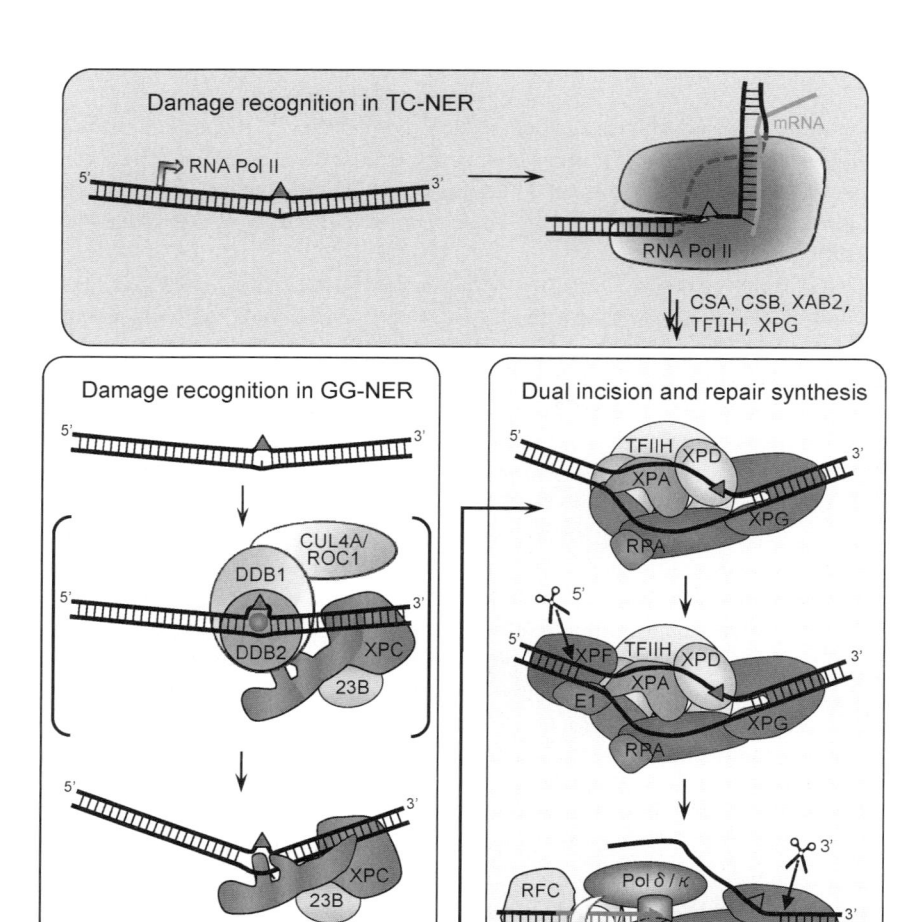

Figure 11.4 NER pathway. Bulky DNA lesions are repaired via GG-NER (left) or TC-NER. In GG-NER, strongly distorting lesions are directly recognized by XPC-RAD23B; less distorting lesions are initially bound by DDB2–DDB1, which forms a complex with the ubiquitin ligase CUL4A/ROC1. Ubiquitination of DDB2 and XPC then leads to the dissociation of DDB2 and facilitates binding of XPC-RAD23B. XPC-RAD23B binds to the undamaged strand of DNA, allowing for the recruitment of TFIIH and likely the verification of the lesion by the XPD subunit. This allows the formation of the preincision complex by recruitment of XPA, RPA, and XPA. The preincision complex may also be formed by stalling of an RNA polymerase at the site of the lesion in TC-NER, and the subsequent action of CSB, CSA, XAB2, TFIIH, and XPG. Recruitment of ERCC1-XPF to the preincision complex leads to incision 5′ to the lesion, initiation of repair synthesis by Pols δ and κ and associated factors, and 3′-incision by XPG. Completion of repair synthesis and sealing of the nick by DNA ligase 3α or DNA ligase I completes the process.

residue can be accommodated in a pocket that requires rotation by only about 30° out of the DNA helix compared to the rotation of close to 150° in OGG1. The T undergoes an *anti* to *syn* rotation about the glycosidic bond to bind in the exo-site. In this way the base binds the enzyme through the Watson–Crick pairing surface, allowing binding of T and U without differentiating between a hydrogen and methyl group at the 5-position. Further conformational rearrangements are required to fully extrude the U into the active site, providing selectivity against the methyl group in T in the fully flipped state (see above, Figure 11.3d). The strategy employed to obtain this transient complex was fundamentally different from the approach used to study OGG1. Nuclear magnetic resonance dynamic measurements have suggested that UDG substantially increases the lifetime of the open conformation of not only UA, but also TA base pairs [37]. This increase was found not to be due to an active role of the enzymes in opening base pairs, but rather due to a passive role in trapping spontaneously opened base pairs. To further study these issues, a DNA duplex containing a T residue paired with the adenine isostere 4-methylindole (4M) that does not form hydrogen bonds like the AT base pair was designed. The absence of hydrogen bonding predisposes the T in the 4MT base pair to assume an extrahelical position. This led to the formation of a stable complex with UDG that could be crystallized, yielding the structure of UDG with T bound within the exo-site [36, 38]. The studies of UDG collectively suggest a different model for the very early steps in damaged base recognition than for OGG1. Presently it is thought that UDG has no active role in extruding U from the helix, but rather traps bases that are already present in the extrahelical conformation. Further studies are required to show whether differences in OGG1 and UDG are intrinsic to the enzymes or are due to the different experimental approaches used.

11.2.3
Passing the Baton – Abasic Site Removal and Repair

The reaction product formed by DNA glycosylases, abasic sites (or breaks formed by bifunctional enzymes) are more dangerous to cells than most of the damaged base lesions due their labile and mutagenic nature. The same is true for additional intermediates that occur downstream that contain strand breaks (Figure 11.2). To avoid the persistence of such intermediates, multiple layers of coordination are in place in BER and are discussed here using short patch BER. The first level is intrinsic to the biochemical mechanisms underlying BER: both DNA glycosylases and APE1 have high affinity for their products, abasic sites and nicked abasic sites, respectively. This allows for the protection of these labile intermediates before the next enzyme in the pathway arrives. The DNA intermediates bound to DNA glycosylases, APE1, and Pol β all have similar bent conformations, facilitating the handover from one enzyme to the next [39, 40]. It has been shown that APE1 can actively displace the hOGG1 or TDG DNA glycosylases from DNA, further reducing the lifetime of unprotected abasic sites [41, 42]. The handover of intermediates is facilitated by a number of protein–protein interactions. For example interactions between APE1, Pol β and XRCC1 serve to coordinate short-patch BER [43, 44]. Many additional interactions between BER proteins have been found and these have been reviewed comprehensively elsewhere [45, 46]. A number of post-translational modifications of BER enzymes have recently been discovered. One of these, sumoylation of TDG, has been shown to increase the turnover rate of the enzyme, providing a possible regulatory mechanism for a handoff of the abasic site product to a subsequent enzyme in the pathway [47, 48].

11.3
NER

NER is the principal pathway in mammals that removes lesions, caused by environmental factors, from DNA (see Figure 11.1 for some common substrates). As the NER substrates are diverse in structure, the etiology behind recognition and repair is fundamentally different from the damage recognition strategy employed by BER described above. NER has therefore evolved to employ a strategy for lesion recognition that relies on general rather than specific recognition.

11.3.1
Subpathways of NER: Global Genome and Transcription-Coupled NER

NER operates through two subpathways: global genome (GG)-NER and transcription-coupled (TC)-NER. GG-NER is active throughout the genome, while TC-NER is responsible for the accelerated repair of the coding strand of actively transcribed genes. The two pathways differ in their mode of lesion recognition, but are thought to share the modes of excising lesions and carrying out repair synthesis (Figure 11.4). GG-NER is initiated by XPC-RAD23B, and UV-DDB (damaged DNA-binding protein) is also used for certain lesions and is additionally important for NER in the context of chromatin (see below). TC-NER, on the other hand, is triggered by the blocking of a transcribing RNA polymerase at the lesion site, which leads to the recruitment of several TC-NER-specific factors (in particular CSA and CSB). These two pathways are believed to share the same core NER factors following damage recognition, leading to the assembly of the full preincision complex, excision of a lesion-containing oligonucleotide of about 25–30 nucleotides in length, and the filling of the resulting gap using the undamaged strand as a template.

11.3.2
Damage Recognition in GG-NER

The GG-NER reaction has been reconstituted with purified protein factors and chromatin-free DNA substrates [49–51]. Consequently, we have a relatively good understanding of the roles of the various GG-NER proteins. XPC-RAD23B is the factor that initiates damage recognition *in vitro*, and it has been shown to be required for the recruitment of all subsequent factors to sites of active NER *in vitro*

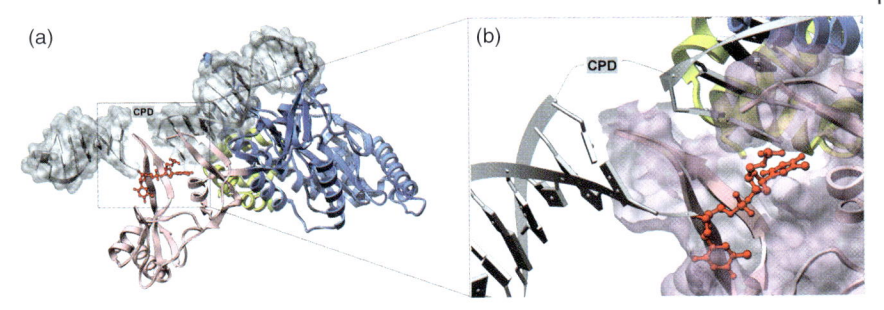

Figure 11.5 Rad4/XPC binds to the nondamaged strand of a CPD-containing DNA. (a) The TDG/BHD1 domains (blue) of Rad4/XPC bind to undamaged DNA, the BHD2/BHD3 (pink) encircle two nucleotides in the nondamaged strand of DNA. DNA is represented in gray with the two thymidines opposite the lesion in red. The position of the CPD lesion that is disordered in this structure is indicated. A fragment of RAD23 is shown in yellow. (b) Two β-hairpins of the BHD2/3 domain (VDW and pink) intercalate into the DNA at the lesion site and encircle two T residues (red) in the nondamaged DNA strand. Note that only native DNA bases, not bulky adducts will fit into the binding pocket. The figure was prepared using Chimera (http://www.cgl.ucsf.edu/chimera) using the structure PDB 2QSG.

and *in vivo* [52–55]. XPC-RAD23B binds specifically to helix-distorting lesions such as 6-4 pyrimidine–pyrimidone photoproducts, benzo[a]pyrene (B[a]P)-derived or AAF adducts, and also to thermodynamically destabilized sites such as base–base mismatches that do not contain chemical modifications, but do not form proper base pairs [56]. Biochemical studies have suggested that XPC-RAD23B has higher affinity to DNA with a single-stranded character, as might be found opposite a distorting lesion [57, 58]. Recent structural studies of the yeast homolog of XPC-RAD23B, Rad4-Rad23, have provided a mechanistic basis for how the distortion-specific DNA binding is achieved [59]. Rad4 binds DNA using two functionally distinct domains (Figure 11.5). A damage and sequence unspecific part of the protein, made up of transglutamase-homology domain (TGD) and a β-hairpin domain (BHD1), binds roughly one turn of a double helix on the 3′-side of the lesion. A pair of BHDs (BHD2/3) interacts specifically with the site of the DNA lesion. BHD2/3 does not, however, interact with the lesion directly, but instead tightly encircles two bases on the nondamaged strand of DNA. One of the hairpins encircles the DNA from the minor groove, the other from the major groove, and together they form a binding pocket for two native DNA bases. Bulky adducts would not fit into this binding pocket, directing the binding of XPC to the nondamaged single-stranded DNA region rather than the damaged strand of a helix-destabilizing lesion. This structure also suggests how XPC can bind single-stranded DNA regions as well as DNA mismatches. An interesting question for the future will be whether all lesions are bound by XPC-RAD23B in the same way. Footprinting analysis of XPC-RAD23B bound to various B[a]P-derived adducts suggest that there might be significant differences in the binding modes depending on the structures and stereochemical properties of the adducts [60].

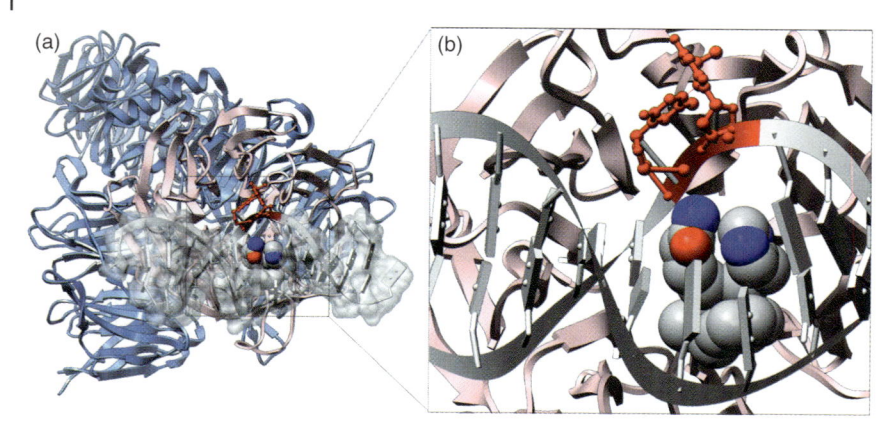

Figure 11.6 DDB2 uses a wedge to probe for lesions in DNA. (a) The UV-DDB complex, made up of DDB1 (light blue) and DDB2 (pink), binds DNA (gray) containing a 6-4 photoproduct (red) through the DDB2 subunit. The DNA-binding site in DDB2 is located on one site of β-propeller; DDB1 binds DDB2 on the opposite side. A wedge made up of residues F371, Q372, and H373 in DDB2 (shown in VDW atom color) inserts into the DNA helix at the lesion site. (b) Magnified view of DDB2 bound to DNA using the same colors as in (a). The wedge of DDB2 inserts into the DNA from the minor groove and displaces the two modified nucleotides of the 6-4 photoproduct into a shallow binding pocket in DDB2. The figure was prepared using Chimera (http://www.cgl.ucsf.edu/chimera) using the structure PDB 3EI1.

The binding mode of XPC-RAD23B does not explain the recognition of less distorting lesions, in particular the UV-induced cyclopyrimidine dimer (CPD). Indeed, CPDs and other lesions specifically require the UV-DDB factor for damage recognition upstream of XPC [61]. UV-DDB binds specifically to DNA lesions through its DDB2 subunit [62], while the DDB1 subunit forms a complex with the ubiquitin ligase CUL4A/ROC1 [63]. Recruitment of CUL4A/ROC1 to NER complexes leads to ubiquitination of DDB2 and XPC, weakening the binding of DDB2, but not of XPC to damaged sites [64]. This is thought to enable the handover of the lesion from DDB2 to XPC and this may be especially important in the context of chromatin. A recent structure of DDB1–DDB2 bound to 6-4 photoproducts provided insight into how DDB2 binds to weakly distorting lesions (Figure 11.6) [65]. DDB2 senses the damage through the insertion of a hairpin wedge made up of residues Phe371, Gln372, and His373 into the minor groove of DNA. This displaces the lesion into a shallow binding pocket. This binding pocket is well suited to recognize intrastrand cross-links such as CPDs or 6-4 photoproducts. Like XPC-RAD23B, UV-DDB exhibits only modestly higher affinity to UV lesions over nondamaged DNA and also binds to DNA distorted by mismatches rather than lesions [66, 67]. How the rather modest binding selectivity of the two factors enables the selective recognition in the context of genomic DNA remains to be elucidated.

11.3.3
Damage Verification and Lesion Demarcation in NER

Since NER only excises bulky lesions and not mismatches [56, 68] after initial recognition, a step is needed to verify that a chemical modification is present since XPC and UV-DDB can bind single-stranded DNA which therefore may include DNA mismatches. Although not yet experimentally proven, it is widely believed that the damage verification step is accomplished by the TFIIH transcription/repair factor that is recruited by direct interaction with XPC [69]. TFIIH is a 10-subunit factor consisting of the core complex (XPB, XPD, p62, p52, p44, p32, and p8/TTDA) and the cyclin-dependent kinase-activating (CAK) complex (cdk7, MAT1, and cyclin H). The core complex is needed for TFIIH to function properly in NER and it has recently been shown that dissociation of CAK from TFIIH is required for the progression through the NER pathway [70]. Following the recruitment of TFIIH the helicase subunits of TFIIH, XPB and XPD, generate a more open DNA structure around the lesion [52, 71, 72]. The available data is consistent with the model for damage verification by TFIIH: XPB uses its ATPase activity to pry open the DNA around the lesion [72, 73], providing an entry site for the XPD helicase. XPD then tracks along the DNA and separates the two strands until it stalls at the site of the lesion. Recent structural studies of archeal XPD homologs have revealed that single-stranded DNA passes through a narrow channel in the helicase-mediated opening reaction. Bulky DNA lesions are not likely to fit through this channel, thus bringing XPD to a halt [74–76]. Structural alterations such as mismatches that are bound by XPC-RAD23 but do not contain a chemical modification will not lead to a stalling of XPD, leading to the disengagement of TFIIH and XPC-RAD23.

The immobilization of TFIIH facilitates the recruitment of XPA, replication protein A (RPA), and XPG to sites of DNA damage, leading to the formation of a relatively stable complex [77]. XPC-RAD23 dissociates from the complex upon XPG binding. XPA, RPA, and XPG appear to assemble at NER sites independently of each other [54, 78], but an interaction with TFIIH seems necessary for the recruitment of XPA and XPG [79–81]. XPA has a DNA-binding domain that interacts with non-B-form structures [82] and thus likely has a role in verifying important structural features of the NER preincision complex, perhaps in concert with the single-stranded DNA-binding protein RPA – one of its interaction partners.

XPG is a structure-specific endonuclease with two discernable roles in NER. It makes the incision 3' to the lesion and has nuclease activity on substrates with single/double-stranded DNA junctions mimicking the open intermediate in the NER reaction [83, 84]. Prior to exerting its catalytic activity, it plays a structural role in assembling the preincision complex in NER. The presence, but not the catalytic activity of XPG is required for the assembly of the preincision complex [85, 86].

RPA, the trimeric single-stranded DNA-binding protein that is involved in many other aspects of DNA metabolism, is believed to bind to the nondamaged strand of DNA in NER [87]. The size of the excised fragment in NER is strikingly similar to the preferred binding size of RPA trimer on single-stranded DNA, which is 30 nucleotides.

11.3.4
Dual-Incision and Repair Synthesis in NER

The last factor to join the preincision complex is the second endonuclease, ERCC1-XPF [54, 55, 77]. A small region in XPA mediates the recruitment of ERCC1-XPF to sites of NER through interaction with the central domain of ERCC1 [88]. ERCC1 and XPF form an obligate heterodimer through their C-terminal HhH domains. While both subunits have a role in DNA binding, the nuclease active site is located in the XPF subunit [89].

The catalytic activities of the two endonucleases ERCC1-XPF and XPG are subject to tight control, so that DNA is not randomly cut and that the excision of the damage-containing oligonucleotide does not lead to the formation and exposure of single-stranded DNA gaps. Consequently, the two incision events occur in a near simultaneous manner [90–92] and the dual-incision and repair synthesis steps are subject to further coordination. One possible key factor in this process is RPA, which remains associated with the NER machinery well past the dual-incision stage [55, 60], possibly providing a platform for this transition. RPA is ideally suited to play this role, as it is believed that it is bound to the nondamaged complementary strand during NER, where it likely plays a role in directing the two incision reactions and protecting the single-stranded DNA [87].

An additional level of coordination is likely achieved through a temporal coordination of the two dual-incision reactions. Studies using catalytically inactive versions of ERCC1 and XPG have shown that 5′-incision by XPF requires the presence, but not catalytic activity of XPG [85, 86], while efficient 3′-incision by XPG only occurs following 5′-incision by XPF [71, 93]. Furthermore, repair synthesis can be initiated in the presence of catalytically inactive XPG, providing another possible mechanism ensuring that exposure of persistent single-stranded DNA intermediates is actively prevented [93].

Repair synthesis is executed by components of the replication machinery – the clamp loader RFC, the processivity factor PCNA, and the replicative Pols δ/ε [51, 94]. Recent studies have additionally implicated the translesion synthesis Pol κ in repair synthesis [95], but what role the different polymerases play in the process remains to be elucidated. Following the fill-in reaction, the nick in the DNA is sealed by XRCC1-DNA ligase 3α in all stages of the cell cycle and by DNA ligase 1 in proliferating cells [96], thus completing the NER process.

11.3.5
Damage Recognition in TC-NER

TC-NER is less well understood than GG-NER at the mechanistic level. TC-NER does not require the XPC-RAD23B and UV-DDB factors, but instead involves a

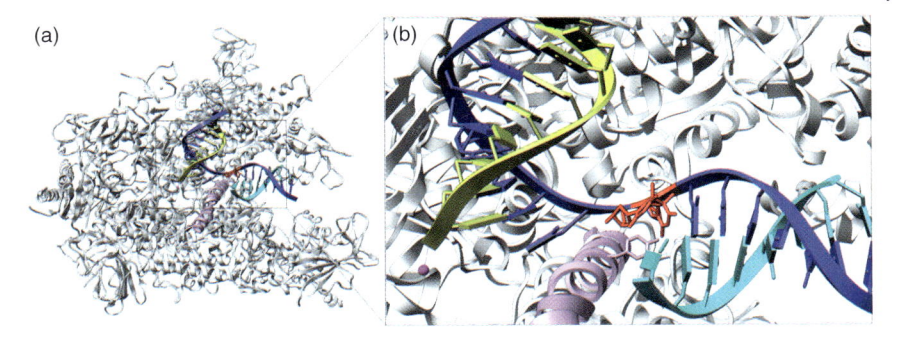

Figure 11.7 Structure of a transcribing RNA Pol II stalled at a CPD lesion. (a) RNA Pol II (gray) is shown with a template strand (dark blue), nontemplate strand (light blue), and growing mRNA strand (yellow). The translocation of the CPD lesion (red) on the template strand into the active site (active site metal is shown in magenta) is blocked by the bridge helix (purple). Tyr836, represented in licorice on the bridge helix, is likely to have a key role as a gatekeeper on the bridge helix that inhibits translocation of the damaged template strand. A large section of the enzyme was deleted for clarity. (b) Magnified view of the RNA Pol II stalled at a CPD lesion. The figure was prepared using Chimera (http://www.cgl.ucsf.edu/chimera) using the structure PDB 2JA5.

number of dedicated factors, including RNA Pol II, CSA, and CSB [97, 98]. It is believed that these factors together recognize DNA lesions in actively transcribed DNA strands, and subsequently channels them into the dual-incision and repair synthesis steps shared with the GG-NER pathway (Figure 11.4).

Recent structural studies have provided a molecular basis for how RNA polymerase is stalled at a lesion [99, 100]. A nondamaged DNA template strand passes through the RNA polymerase in an L-shaped form, with the active site located in the kink of the L (Figure 11.7). The growing mRNA chain (yellow) forms a duplex with the template strand as it exits the RNA polymerase, while the nontemplate strand (light blue) is separated from the template strand prior through the translocation of the template strand over the bridge helix (magenta) that serves as a gate to entering the active site where an NTP is incorporated into the mRNA chain. If a 1,2-cisplatin interstrand lesion is present in the template, the translocation over the bridge helix is blocked, leading to a stalling of the RNA polymerase [100]. With a CPD lesion in the template (red), translocation is very slow, but still possible. Following this translocation, an ATP is incorporated opposite the first T residue of the CPD and the misincorporation of UTP opposite the second T leads to stalling of the RNA polymerase [99].

The steps that follow the stalling of RNA polymerase are less well understood, but several available studies are consistent with a model whereby both TC-NER and GG-NER employ the same core reaction following damage recognition (Figure 11.4). *In vitro* studies using a reconstituted system demonstrated that stalling of RNA Pol II at a cisplatin lesion leads to the recruitment of TFIIH, and subsequently RPA, XPA, and finally XPG and ERCC1-XPF [101]. Use of a chromatin immunoprecipitation assay designed to study the arrival of NER factors in living cells following the stalling of RNA Pol II at a UV lesion confirmed the presence of those same core NER factors at TC-NER sites [102]. The recruitment of these factors was dependent on the CSB protein, while CSB and CSA were required for the recruitment of additional chromatin factors such as XAB2 and HMGN1. A different study proposed an early role for the XPG protein in TC-NER by direct association with RNA Pol II [103]. The stage is now set for more detailed studies of TC-NER.

11.4 Conclusions

Extensive biochemical and structural studies of the BER and NER pathways have led to a detailed mechanistic understanding of how lesions formed by endogenous and exogenous agents are recognized in our genomes. BER uses a damage-specific recognition strategy that can efficiently process lesions that occur with relatively high frequency. This is achieved by a number of enzymes that initiate the repair, DNA glycosylases, which recognize only one or a small set of lesions with high specificity. By contrast, NER employs a more general recognition strategy that involves three distinct steps to recognize minor (UV-DDB) and major (XPC-RAD23B) alterations in the DNA structure as well as the chemical modification induced by a lesion (TFIIH). The mechanisms for BER and NER discussed in this chapter are therefore fully consistent with evolutionary roles for BER to deal with specific frequent lesions, while the general damage recognition strategy in NER is capable of dealing with new lesions that arise in the environment, at the cost of a less-efficient repair mechanism.

References

1 Friedberg, E.C., Walker, G.C., Siede, W., Wood, R.D., Schultz, R.A., and Ellenberger, T. (2005) *DNA Repair and Mutagenesis*, 2nd edn, ASM Press, Washington, DC.

2 Yang, W. (2008) Structure and mechanism for DNA lesion recognition. *Cell Res.*, **18**, 184–197.

3 Gillet, L.C. and Schärer, O.D. (2006) Molecular mechanisms of mammalian global genome nucleotide excision repair. *Chem. Rev.*, **106**, 253–276.

4 Larsen, E., Meza, T.J., Kleppa, L., and Klungland, A. (2007) Organ and cell specificity of base excision repair mutants in mice. *Mutat. Res.*, **614**, 56–68.

5 Parsons, J.L. and Elder, R.H. (2003) DNA N-glycosylase deficient mice: a tale of redundancy. *Mutat. Res.*, **531**, 165–175.

6 David, S.S., O'Shea, V.L., and Kundu, S. (2007) Base-excision repair of oxidative DNA damage. *Nature*, **447**, 941–950.

7 Cleaver, J.E. (1968) Defective repair replication of DNA in xeroderma pigmentosum. *Nature*, **218**, 652–656.

8 Bootsma, D., Kraemer, K.H., Cleaver, J.E., and Hoeijmakers, J.H.J. (1998) Nucleotide excision repair syndromes: xeroderma pigmentosum, cockayne syndrome, and trichothiodystrophy, in *The Genetic Basis of Cancer* (eds B. Vogelstein and K.W. Kinzler), McGraw-Hill, New York, pp. 245–274.

9 Lehmann, A.R. (2003) DNA repair-deficient diseases, xeroderma pigmentosum, Cockayne syndrome and trichothiodystrophy. *Biochimie*, **85**, 1101–1111.
10 Kraemer, K.H., Patronas, N.J., Schiffmann, R., Brooks, B.P., Tamura, D., and DiGiovanna, J.J. (2007) Xeroderma pigmentosum, trichothiodystrophy and Cockayne syndrome: a complex genotype–phenotype relationship. *Neuroscience*, **145**, 1388–1396.
11 Kunkel, T.A. (1999) The high cost of living. American Association for Cancer Research Special Conference: Endogenous Sources of Mutations, Fort Myers, Florida, USA, 11–15 November 1998. *Trends Genet.*, **15**, 93–94.
12 Lindahl, T. (1993) Instability and decay of the primary structure of DNA. *Nature*, **362**, 709–715.
13 Dianov, G. and Lindahl, T. (1994) Reconstitution of the DNA base excision-repair *pathway. Curr. Biol.*, **4**, 1069–1076.
14 Kubota, Y., Nash, R.A., Klungland, A., Schar, P., Barnes, D.E., and Lindahl, T. (1996) Reconstitution of DNA base excision-repair with purified human proteins: interaction between DNA polymerase β and the XRCC1 protein. *EMBO J.*, **15**, 6662–6670.
15 Matsumoto, Y. and Kim, K. (1995) Excision of deoxyribose phosphate residues by DNA polymerase β during DNA repair. *Science*, **269**, 699–702.
16 Wiederhold, L., Leppard, J.B., Kedar, P., Karimi-Busheri, F., Rasouli-Nia, A., Weinfeld, M., Tomkinson, A.E., Izumi, T., Prasad, R., Wilson, S.H., Mitra, S., and Hazra, T.K. (2004) AP endonuclease-independent DNA base excision repair in human cells. *Mol. Cell*, **15**, 209–220.
17 Pascucci, B., Stucki, M., Jonsson, Z.O., Dogliotti, E., and Hübscher, U. (1999) Long patch base excision repair with purified human proteins. DNA ligase I as patch size mediator for DNA polymerases delta and epsilon. *J. Biol. Chem.*, **274**, 33696–33702.
18 Matsumoto, Y., Kim, K., Hurwitz, J., Gary, R., Levin, D.S., Tomkinson, A.E., and Park, M.S. (1999) Reconstitution of proliferating cell nuclear antigen-dependent repair of apurinic/apyrimidinic sites with purified human proteins. *J. Biol. Chem.*, **274**, 33703–33708.
19 Fromme, J.C., Banerjee, A., and Verdine, G.L. (2004) DNA glycosylase recognition and catalysis. *Curr. Opin. Struct. Biol.*, **14**, 43–49.
20 Hitomi, K., Iwai, S., and Tainer, J.A. (2007) The intricate structural chemistry of base excision repair machinery: implications for DNA damage recognition, removal, and repair. *DNA Repair*, **6**, 410–428.
21 Stivers, J.T. (2008) Extrahelical damaged base recognition by DNA glycosylase enzymes. *Chemistry*, **14**, 786–793.
22 Nash, H.M., Bruner, S.D., Schärer, O.D., Kawate, T., Addona, T.A., Spooner, E., Lane, W.S., and Verdine, G.L. (1996) Cloning of a yeast 8-oxoguanine DNA glycosylase reveals the existence of a base-excision DNA-repair protein superfamily. *Curr. Biol.*, **6**, 968–980.
23 Savva, R., McAuley-Hecht, K., Brown, T., and Pearl, L. (1995) The structural basis of specific base-excision repair by uracil-DNA glycosylase. *Nature*, **373**, 487–493.
24 Mol, C.D., Arvai, A.S., Slupphaug, G., Kavli, B., Alseth, I., Krokan, H.E., and Tainer, J.A. (1995) Crystal structure and mutational analysis of human uracil-DNA glycosylase: structural basis for specificity and catalysis. *Cell*, **80**, 869–878.
25 Parikh, S.S., Walcher, G., Jones, G.D., Slupphaug, G., Krokan, H.E., Blackburn, G.M., and Tainer, J.A. (2000) Uracil-DNA glycosylase-DNA substrate and product structures: conformational strain promotes catalytic efficiency by coupled stereoelectronic effects. *Proc. Natl. Acad. Sci. USA*, **97**, 5083–5088.
26 Bruner, S.D., Norman, D.P., and Verdine, G.L. (2000) Structural basis for recognition and repair of the endogenous mutagen 8-oxoguanine in DNA. *Nature*, **403**, 859–866.
27 Lau, A.Y., Schärer, O.D., Samson, L., Verdine, G.L., and Ellenberger, T. (1998) Crystal structure of a human

alkylbase-DNA repair enzyme complexed to DNA: mechanisms for nucleotide flipping and base excision. *Cell*, **95**, 249–258.

28 Lau, A.Y., Wyatt, M.D., Glassner, B.J., Samson, L.D., and Ellenberger, T. (2000) Molecular basis for discriminating between normal and damaged bases by the human alkyladenine glycosylase, AAG. *Proc. Natl. Acad. Sci. USA*, **97**, 13573–13578.

29 Wyatt, M.D., Allan, J.M., Lau, A.Y., Ellenberger, T.E., and Samson, L.D. (1999) 3-Methyladenine DNA glycosylases: structure, function, and biological importance. *Bioessays*, **21**, 668–676.

30 Verdine, G.L. and Bruner, S.D. (1997) How do DNA repair proteins locate damaged bases in the genome? *Chem. Biol.*, **4**, 329–334.

31 Banerjee, A., Santos, W.L., and Verdine, G.L. (2006) Structure of a DNA glycosylase searching for lesions. *Science*, **311**, 1153–1157.

32 Banerjee, A. and Verdine, G.L. (2006) A nucleobase lesion remodels the interaction of its normal neighbor in a DNA glycosylase complex. *Proc. Natl. Acad. Sci. USA*, **103**, 15020–15025.

33 Banerjee, A., Yang, W., Karplus, M., and Verdine, G.L. (2005) Structure of a repair enzyme interrogating undamaged DNA elucidates recognition of damaged DNA. *Nature*, **434**, 612–618.

34 Kuznetsov, N.A., Koval, V.V., Nevinsky, G.A., Douglas, K.T., Zharkov, D.O., and Fedorova, O.S. (2007) Kinetic conformational analysis of human 8-oxoguanine-DNA glycosylase. *J. Biol. Chem.*, **282**, 1029–1038.

35 Kuznetsov, N.A., Koval, V.V., Zharkov, D.O., Nevinsky, G.A., Douglas, K.T., and Fedorova, O.S. (2005) Kinetics of substrate recognition and cleavage by human 8-oxoguanine-DNA glycosylase. *Nucleic Acids Res.*, **33**, 3919–3931.

36 Parker, J.B., Bianchet, M.A., Krosky, D.J., Friedman, J.I., Amzel, L.M., and Stivers, J.T. (2007) Enzymatic capture of an extrahelical thymine in the search for uracil in DNA. *Nature*, **449**, 433–437.

37 Cao, C., Jiang, Y.L., Stivers, J.T., and Song, F. (2004) Dynamic opening of DNA during the enzymatic search for a damaged base. *Nat. Struct. Mol. Biol.*, **11**, 1230–1236.

38 Cao, C., Jiang, Y.L., Krosky, D.J., and Stivers, J.T. (2006) The catalytic power of uracil DNA glycosylase in the opening of thymine base pairs. *J. Am. Chem. Soc.*, **128**, 13034–13035.

39 Mol, C.D., Izumi, T., Mitra, S., and Tainer, J.A. (2000) DNA-bound structures and mutants reveal abasic DNA binding by APE1 and DNA repair coordination. *Nature*, **403**, 451–456.

40 Wilson, S.H. and Kunkel, T.A. (2000) Passing the baton in base excision repair. *Nat. Struct. Biol.*, **7**, 176–178.

41 Sidorenko, V.S., Nevinsky, G.A., and Zharkov, D.O. (2007) Mechanism of interaction between human 8-oxoguanine-DNA glycosylase and AP endonuclease. *DNA Repair*, **6**, 317–328.

42 Fitzgerald, M.E. and Drohat, A.C. (2008) Coordinating the initial steps of base excision repair. Apurinic/apyrimidinic endonuclease 1 actively stimulates thymine DNA glycosylase by disrupting the product complex. *J. Biol. Chem.*, **283**, 32680–32690.

43 Bennett, R.A., Wilson, D.M., 3rd, Wong, D., and Demple, B. (1997) Interaction of human apurinic endonuclease and DNA polymerase β in the base excision repair pathway. *Proc. Natl. Acad. Sci. USA*, **94**, 7166–7169.

44 Vidal, A.E., Boiteux, S., Hickson, I.D., and Radicella, J.P. (2001) XRCC1 coordinates the initial and late stages of DNA abasic site repair through protein–protein interactions. *EMBO J.*, **20**, 6530–6539.

45 Almeida, K.H. and Sobol, R.W. (2007) A unified view of base excision repair: lesion-dependent protein complexes regulated by post-translational modification. *DNA Repair*, **6**, 695–711.

46 Hegde, M.L., Hazra, T.K., and Mitra, S. (2008) Early steps in the DNA base excision/single-strand interruption repair pathway in mammalian cells. *Cell Res.*, **18**, 27–47.

47 Hardeland, U., Steinacher, R., Jiricny, J., and Schär, P. (2002) Modification of the human thymine-DNA glycosylase by ubiquitin-like protein facilitates

enzymatic turnover. *EMBO J.*, **21**, 1456–1464.

48 Steinacher, R. and Schär, P. (2005) Functionality of human thymine DNA glycosylase requires SUMO-regulated changes in protein conformation. *Curr. Biol.*, **15**, 616–623.

49 Aboussekhra, A., Biggerstaff, M., Shivji, M.K., Vilpo, J.A., Moncollin, V., Podust, V.N., Protic, M., Hübscher, U., Egly, J.M., and Wood, R.D. (1995) Mammalian DNA nucleotide excision repair reconstituted with purified protein components. *Cell*, **80**, 859–868.

50 Mu, D., Park, C.H., Matsunaga, T., Hsu, D.S., Reardon, J.T., and Sancar, A. (1995) Reconstitution of human DNA repair excision nuclease in a highly defined system. *J. Biol. Chem.*, **270**, 2415–2418.

51 Araujo, S.J., Tirode, F., Coin, F., Pospiech, H., Syvaoja, J.E., Stucki, M., Hubscher, U., Egly, J.M., and Wood, R.D. (2000) Nucleotide excision repair of DNA with recombinant human proteins: definition of the minimal set of factors, active forms of TFIIH, and modulation by CAK. *Genes Dev.*, **14**, 349–359.

52 Evans, E., Moggs, J.G., Hwang, J.R., Egly, J.M., and Wood, R.D. (1997) Mechanism of open complex and dual incision formation by human nucleotide excision repair factors. *EMBO J.*, **16**, 6559–6573.

53 Sugasawa, K., Ng, J.M., Masutani, C., Iwai, S., van der Spek, P.J., Eker, A.P., Hanaoka, F., Bootsma, D., and Hoeijmakers, J.H. (1998) Xeroderma pigmentosum group C protein complex is the initiator of global genome nucleotide excision repair. *Mol. Cell*, **2**, 223–232.

54 Volker, M., Mone, M.J., Karmakar, P., van Hoffen, A., Schul, W., Vermeulen, W., Hoeijmakers, J.H., van Driel, R., van Zeeland, A.A., and Mullenders, L.H. (2001) Sequential assembly of the nucleotide excision repair factors *in vivo*. *Mol. Cell*, **8**, 213–224.

55 Riedl, T., Hanaoka, F., and Egly, J.M. (2003) The comings and goings of nucleotide excision repair factors on damaged DNA. *EMBO J.*, **22**, 5293–5303.

56 Sugasawa, K., Okamoto, T., Shimizu, Y., Masutani, C., Iwai, S., and Hanaoka, F. (2001) A multistep damage recognition mechanism for global genomic nucleotide excision repair. *Genes Dev.*, **15**, 507–521.

57 Sugasawa, K., Shimizu, Y., Iwai, S., and Hanaoka, F. (2002) A molecular mechanism for DNA damage recognition by the xeroderma pigmentosum group C protein complex. *DNA Repair*, **1**, 95–107.

58 Maillard, O., Solyom, S., and Naegeli, H. (2007) An aromatic sensor with aversion to damaged strands confers versatility to DNA repair. *PLoS Biol.*, **5**, e79.

59 Min, J.H. and Pavletich, N.P. (2007) Recognition of DNA damage by the Rad4 nucleotide excision repair protein. *Nature*, **449**, 570–575.

60 Mocquet, V., Laine, J.P., Riedl, T., Yajin, Z., Lee, M.Y., and Egly, J.M. (2008) Sequential recruitment of the repair factors during NER: the role of XPG in initiating the resynthesis step. *EMBO J.*, **27**, 155–167.

61 Fitch, M.E., Nakajima, S., Yasui, A., and Ford, J.M. (2003) *In vivo* recruitment of XPC to UV-induced cyclobutane pyrimidine dimers by the DDB2 gene product. *J. Biol. Chem.*, **278**, 46906–46910.

62 Tang, J. and Chu, G. (2002) Xeroderma pigmentosum complementation group E and UV-damaged DNA binding protein. *DNA Repair*, **1**, 601–616.

63 Groisman, R., Polanowska, J., Kuraoka, I., Sawada, J., Saijo, M., Drapkin, R., Kisselev, A.F., Tanaka, K., and Nakatani, Y. (2003) The ubiquitin ligase activity in the DDB2 and CSA complexes is differentially regulated by the COP9 signalosome in response to DNA damage. *Cell*, **113**, 357–367.

64 Sugasawa, K., Okuda, Y., Saijo, M., Nishi, R., Matsuda, N., Chu, G., Mori, T., Iwai, S., Tanaka, K., Tanaka, K., and Hanaoka, F. (2005) UV-induced ubiquitylation of XPC protein mediated by UV-DDB–ubiquitin ligase complex. *Cell*, **121**, 387–400.

65 Scrima, A., Konickova, R., Czyzewski, B.K., Kawasaki, Y., Jeffrey, P.D.,

Groisman, R., Nakatani, Y., Iwai, S., Pavletich, N.P., and Thoma, N.H. (2008) Structural basis of UV DNA-damage recognition by the DDB1–DDB2 complex. *Cell*, **135**, 1213–1223.

66 Wittschieben, B.O., Iwai, S., and Wood, R.D. (2005) DDB1-DDB2 (xeroderma pigmentosum group E) protein complex recognizes a cyclobutane pyrimidine dimer, mismatches, apurinic/apyrimidinic sites, and compound lesions in DNA. *J. Biol. Chem.*, **280**, 39982–39989.

67 Kulaksiz, G., Reardon, J.T., and Sancar, A. (2005) Xeroderma pigmentosum complementation group E protein (XPE/DDB2): purification of various complexes of XPE and analyses of their damaged DNA binding and putative DNA repair properties. *Mol. Cell. Biol.*, **25**, 9784–9792.

68 Hess, M.T., Schwitter, U., Petretta, M., Giese, B., and Naegeli, H. (1997) Bipartite substrate discrimination by human nucleotide excision repair. *Proc. Natl. Acad. Sci. USA*, **94**, 6664–6669.

69 Yokoi, M., Masutani, C., Maekawa, T., Sugasawa, K., Ohkuma, Y., and Hanaoka, F. (2000) The xeroderma pigmentosum group C protein complex XPC-HR23B plays an important role in the recruitment of transcription factor IIH to damaged DNA. *J. Biol. Chem.*, **275**, 9870–9875.

70 Coin, F., Oksenych, V., Mocquet, V., Groh, S., Blattner, C., and Egly, J.M. (2008) Nucleotide excision repair driven by the dissociation of CAK from TFIIH. *Mol. Cell*, **31**, 9–20.

71 Tapias, A., Auriol, J., Forget, D., Enzlin, J.H., Scharer, O.D., Coin, F., Coulombe, B., and Egly, J.M. (2004) Ordered conformational changes in damaged DNA induced by nucleotide excision repair factors. *J. Biol. Chem.*, **279**, 19074–19083.

72 Coin, F., Oksenych, V., and Egly, J.M. (2007) Distinct roles for the XPB/p52 and XPD/p44 subcomplexes of TFIIH in damaged DNA opening during nucleotide excision repair. *Mol. Cell*, **26**, 245–256.

73 Fan, L., Arvai, A.S., Cooper, P.K., Iwai, S., Hanaoka, F., and Tainer, J.A. (2006) Conserved XPB core structure and motifs for DNA unwinding: implications for pathway selection of transcription or excision repair. *Mol. Cell*, **22**, 27–37.

74 Fan, L., Fuss, J.O., Cheng, Q.J., Arvai, A.S., Hammel, M., Roberts, V.A., Cooper, P.K., and Tainer, J.A. (2008) XPD helicase structures and activities: insights into the cancer and aging phenotypes from XPD mutations. *Cell*, **133**, 789–800.

75 Liu, H., Rudolf, J., Johnson, K.A., McMahon, S.A., Oke, M., Carter, L., McRobbie, A.M., Brown, S.E., Naismith, J.H., and White, M.F. (2008) Structure of the DNA repair helicase XPD. *Cell*, **133**, 801–812.

76 Wolski, S.C., Kuper, J., Hanzelmann, P., Truglio, J.J., Croteau, D.L., Van Houten, B., and Kisker, C. (2008) Crystal structure of the FeS cluster-containing nucleotide excision repair helicase XPD. *PLoS Biol.*, **6**, e149.

77 Wakasugi, M. and Sancar, A. (1998) Assembly, subunit composition, and footprint of human DNA repair excision nuclease. *Proc. Natl. Acad. Sci. USA*, **95**, 6669–6674.

78 Rademakers, S., Volker, M., Hoogstraten, D., Nigg, A.L., Mone, M.J., Van Zeeland, A.A., Hoeijmakers, J.H., Houtsmuller, A.B., and Vermeulen, W. (2003) Xeroderma pigmentosum group A protein loads as a separate factor onto DNA lesions. *Mol. Cell. Biol.*, **23**, 5755–5767.

79 Park, C.H., Mu, D., Reardon, J.T., and Sancar, A. (1995) The general transcription-repair factor TFIIH is recruited to the excision repair complex by the XPA protein independent of the TFIIE transcription factor. *J. Biol. Chem.*, **270**, 4896–4902.

80 Araujo, S.J., Nigg, E.A., and Wood, R.D. (2001) Strong functional interactions of TFIIH with XPC and XPG in human DNA nucleotide excision repair, without a preassembled repairosome. *Mol. Cell. Biol.*, **21**, 2281–2291.

81 Dunand-Sauthier, I., Hohl, M., Thorel, F., Jaquier-Gubler, P., Clarkson, S.G., and Schärer, O.D. (2005) The spacer region of XPG mediates recruitment to

nucleotide excision repair complexes and determines substrate specificity. *J. Biol. Chem.*, **280**, 7030–7037.

82 Missura, M., Buterin, T., Hindges, R., Hubscher, U., Kasparkova, J., Brabec, V., and Naegeli, H. (2001) Double-check probing of DNA bending and unwinding by XPA-RPA: an architectural function in DNA repair. *EMBO J.*, **20**, 3554–3564.

83 O'Donovan, A., Davies, A.A., Moggs, J.G., West, S.C., and Wood, R.D. (1994) XPG endonuclease makes the 3? incision in human DNA nucleotide excision repair. *Nature*, **371**, 432–435.

84 Evans, E., Fellows, J., Coffer, A., and Wood, R.D. (1997) Open complex formation around a lesion during nucleotide excision repair provides a structure for cleavage by human XPG protein. *EMBO J.*, **16**, 625–638.

85 Wakasugi, M., Reardon, J.T., and Sancar, A. (1997) The non-catalytic function of XPG protein during dual incision in human nucleotide excision repair. *J. Biol. Chem.*, **272**, 16030–16034.

86 Constantinou, A., Gunz, D., Evans, E., Lalle, P., Bates, P.A., Wood, R.D., and Clarkson, S.G. (1999) Conserved residues of human XPG protein important for nuclease activity and function in nucleotide excision repair. *J. Biol. Chem.*, **274**, 5637–5648.

87 de Laat, W.L., Appeldoorn, E., Sugasawa, K., Weterings, E., Jaspers, N.G., and Hoeijmakers, J.H. (1998) DNA-binding polarity of human replication protein A positions nucleases in nucleotide excision repair. *Genes Dev.*, **12**, 2598–2609.

88 Tsodikov, O.V., Ivanov, D., Orelli, B., Staresincic, L., Shoshani, I., Oberman, R., Schärer, O.D., Wagner, G., and Ellenberger, T. (2007) Structural basis for the recruitment of ERCC1-XPF to nucleotide excision repair complexes by XPA. *EMBO J.*, **26**, 4768–4776.

89 Enzlin, J.H. and Schärer, O.D. (2002) The active site of XPF-ERCC1 forms a highly conserved nuclease motif. *EMBO J.*, **21**, 2045–2053.

90 Matsunaga, T., Mu, D., Park, C.H., Reardon, J.T., and Sancar, A. (1995) Human DNA repair excision nuclease. Analysis of the roles of the subunits involved in dual incisions by using anti-XPG and anti-ERCC1 antibodies. *J. Biol. Chem.*, **270**, 20862–20869.

91 Mu, D., Hsu, D.S., and Sancar, A. (1996) Reaction mechanism of human DNA repair excision nuclease. *J. Biol. Chem.*, **271**, 8285–8294.

92 Moggs, J.G., Yarema, K.J., Essigmann, J.M., and Wood, R.D. (1996) Analysis of incision sites produced by human cell extracts and purified proteins during nucleotide excision repair of a 1,3-intrastrand d(GpTpG)-cisplatin adduct. *J. Biol. Chem.*, **271**, 7177–7186.

93 Staresincic, L., Fagbemi, A.F., Enzlin, J.H., Gourdin, A.M., Wijgers, N., Dunand-Sauthier, I., Giglia-Mari, G., Clarkson, S.G., Vermeulen, W., and Scharer, O.D. (2009) Coordination of dual incision and repair synthesis in human nucleotide excision repair. *EMBO J.*, **28**, 1111–1120.

94 Shivji, M.K., Podust, V.N., Hubscher, U., and Wood, R.D. (1995) Nucleotide excision repair DNA synthesis by DNA polymerase epsilon in the presence of PCNA, RFC, and RPA. *Biochemistry*, **34**, 5011–5017.

95 Ogi, T. and Lehmann, A.R. (2006) The Y-family DNA polymerase κ (pol κ) functions in mammalian nucleotide-excision repair. *Nat. Cell Biol.*, **8**, 640–642.

96 Moser, J., Kool, H., Giakzidis, I., Caldecott, K., Mullenders, L.H., and Fousteri, M.I. (2007) Sealing of chromosomal DNA nicks during nucleotide excision repair requires XRCC1 and DNA ligase III α in a cell-cycle-specific manner. *Mol. Cell*, **27**, 311–323.

97 Fousteri, M. and Mullenders, L.H. (2008) Transcription-coupled nucleotide excision repair in mammalian cells: molecular mechanisms and biological effects. *Cell Res.*, **18**, 73–84.

98 Hanawalt, P.C. and Spivak, G. (2008) Transcription-coupled DNA repair: two decades of progress and surprises. *Nat. Rev. Mol. Cell Biol.*, **9**, 958–970.

99 Brueckner, F., Hennecke, U., Carell, T., and Cramer, P. (2007) CPD damage recognition by transcribing RNA

polymerase II. *Science*, **315**, 859–862.

100 Damsma, G.E., Alt, A., Brueckner, F., Carell, T., and Cramer, P. (2007) Mechanism of transcriptional stalling at cisplatin-damaged DNA. *Nat. Struct. Mol. Biol.*, **14**, 1127–1133.

101 Laine, J.P. and Egly, J.M. (2006) Initiation of DNA repair mediated by a stalled RNA polymerase IIO. *EMBO J.*, **25**, 387–397.

102 Fousteri, M., Vermeulen, W., van Zeeland, A.A., and Mullenders, L.H. (2006) Cockayne syndrome A and B proteins differentially regulate recruitment of chromatin remodeling and repair factors to stalled RNA polymerase II *in vivo*. *Mol. Cell*, **23**, 471–482.

103 Sarker, A.H., Tsutakawa, S.E., Kostek, S., Ng, C., Shin, D.S., Peris, M., Campeau, E., Tainer, J.A., Nogales, E., and Cooper, P.K. (2005) Recognition of RNA polymerase II and transcription bubbles by XPG, CSB, and TFIIH: insights for transcription-coupled repair and Cockayne Syndrome. *Mol. Cell*, **20**, 187–198.

12
Recognition and Removal of Bulky DNA Lesions by the Nucleotide Excision Repair System

Yuqin Cai, Konstantin Kropachev, Marina Kolbanovskiy, Alexander Kolbanovskiy, Suse Broyde, Dinshaw J. Patel, and Nicholas E. Geacintov

12.1
Introduction

Nucleotide excision repair (NER) is one of the most important mammalian defense mechanisms against bulky DNA lesions. The importance of DNA repair is illustrated by several genetic human NER-deficiency syndromes [1], such as xeroderma pigmentosum (XP) [2, 3], Cockayne syndrome [4], and trichothiodystrophy [5]. These deficiencies in DNA repair capacities are due to mutant forms of proteins that compromise the overall efficiency of the normal repair process (see Chapter 11). While efficient NER is critically important for the survival of cells and for preventing mutations, a high repair capacity is undesirable in chemotherapeutic applications. For example, a common chemotherapeutic drug is cisplatin, which forms covalent adducts with DNA that inhibit DNA replication. The levels of these covalent DNA adducts are diminished by efficient DNA repair, thus inhibiting the effectiveness of the therapeutic regimens [6].

While NER can remove a large variety of DNA lesions, it is well established that the DNA repair capacity, or efficiency, can vary by two orders of magnitude or more, depending on the lesion [7]. Therefore, a fundamental knowledge of the structural factors of the DNA lesions that are critical to their recognition and processing by the mammalian NER apparatus is important for understanding the origins of chemical carcinogenesis and perhaps the design of more effective chemotherapeutic agents.

12.2
Overview of Mammalian NER

The two subpathways of NER, global genome (GG)-NER [8] and transcription-coupled (TC)-NER [9], employ a common set of proteins including XPC/HR23B (also known as XPC-RAD23B; Chapter 11), TFIIH, XPA, RPA (replication protein A), XPG, and ERCC1 (excision repair cross-complementation group 1)/XPF. The

GG-NER and TC-NER pathways are essentially identical except for differences in their mechanisms of lesion recognition. The GG-NER pathway relies on the XPC/HR23B protein complex to sense the presence of the DNA lesions, while TC-NER is activated by a stalled RNA polymerase during transcription [9]. Once a lesion is identified, GG-NER and TC-NER proceed in an essentially identical manner to excise the lesion. The GG-NER and TC-NER phenomena are considered in greater detail in Chapters 11 and 17. To summarize briefly, in the GG-NER pathway, the local distortions and destabilizations of the DNA structure associated with the lesion are recognized by the XPC/HR23B heteroprotein complex that binds tightly to the damaged region. This complex initiates the recruitment of other NER factors to the site of the lesion [10–13]. The multiprotein TFIIH complex is the first one to be recruited to this site [14–16]. The TFIIH complex contains the helicases XPB and XPD that induce the unwinding of a 20- to 25-nucleotide region around the lesion [17, 18]. The subsequent binding of the factors XPA and RPA stabilizes this bubble-like structure, and XPC/HR23B is released. The endonucleases XPG and XPF/ERCC1 then bind to the complex and incise the damaged strand on the 3'- and 5'-sides of the lesion, respectively, releasing a 24- to 32-nucleotide dual-incision fragment containing the damage – these fragments are the hallmark of successful mammalian NER incision activity (Figure 12.1) [13, 14, 17, 19]. The displaced oligonucleotide creates a gap that is filled by DNA synthesis, catalyzed by DNA Pol δ, proliferating cell nuclear antigen, and replication factor C. Finally, the nick in the repaired strand is sealed by DNA ligase I, thus completing the NER process.

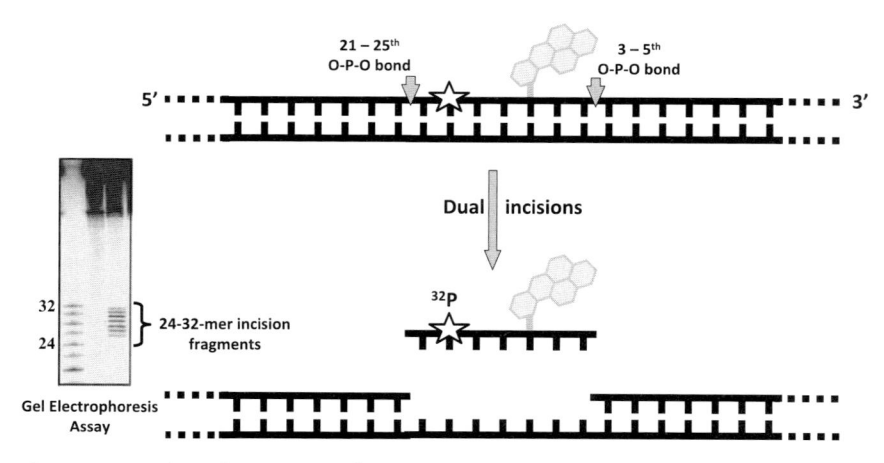

Figure 12.1 Outline of NER assay in human cell extracts. The hallmark of a successful mammalian NER dual-incision activity is the excision of internally ^{32}P-labeled 24- to 32-nucleotide fragments containing the lesion. The amount of these fragments, relative to the total (incised plus nonincised duplexes) is a relative measure of the efficiency of NER.

12.3
Prokaryotic NER

The NER pathway is also present in prokaryotic organisms, such as *Escherichia coli*, as well as in eukaryotes from yeast to mammals. Many of the basic steps of NER are evolutionarily conserved, including damage recognition and dual incisions to excise the lesion, helicase activity to displace the excised oligonucleotide and repair factors, and synthesis and ligation enzymes to seal the nick [19, 20]. Nevertheless, the biochemical features in prokaryotes and eukaryotes are distinct. NER in bacteria involves only three proteins to carry out the complete process of damage recognition and excision: UvrA, UvrB, and UvrC. Owing to its relative simplicity, the UvrABC system has been studied extensively, particularly in *E. coli*, and serves as a model system for NER [20, 21].

12.4
Recognition of Bulky Lesions by Mammalian NER Factors

Although mammalian NER recognizes and removes a wide range of chemically and conformationally diverse DNA lesions, it is well known that the efficiency of repair can vary over several orders of magnitude [7, 22, 23]. The NER system is well adapted to remove lesions of very different structural characteristics because an entire oligonucleotide fragment is removed rather than the lesion itself, as in base excision repair (Chapter 11). Nevertheless, the structures of the DNA lesions that elicit NER repair do affect the efficiency of this mechanism and the underlying reasons have been a subject of considerable interest [24]. It has been suggested that local changes in the properties of DNA duplexes caused by DNA lesions, including enhanced flexibility [25, 26], DNA bends and kinks [25, 27], oscillatory motions of the unmodified complementary strand [28–30], impaired base stacking [31, 32], and distorted Watson–Crick hydrogen-bonding [33], constitute important recognition signals for NER proteins [29, 30].

Much evidence now supports the hypothesis that the NER factors, especially XPC/HR23B, do not recognize the lesion itself, but the local distortions and destabilizations in the DNA that are associated with the lesions [13, 14, 23, 29, 34–36]. This concept is further supported by the X-ray crystallographic structure of a truncated form of Rad4/Rad23 (the yeast *Saccharomyces cerevisiae* homologs of the mammalian XPC and HR23B, respectively) in a complex with an oligonucleotide containing a cyclobutane pyrimidine dimer (CPD) lesion [37]. There are three β-hairpin domains in the Rad4 protein and one of these is inserted into the DNA helix, thus separating the CPD lesion from the unmodified strand. The CPD lesion is positioned in a disordered region of the crystal where it has no visible contacts with the protein [37]. In contrast, the two thymines in the complementary strand that were positioned opposite the CPD dimer in the duplex interact with some of the amino acids of Rad4. The strand separation around the site of the lesion in the XPC/HR23B–DNA complex comprises initially around 6 bp in the case of cisplatin [18] and bulky benzo[a]pyrene (B[a]P) diol epoxide-N^2-guanine adducts [38]. It is likely that this open structure is stabilized by the insertion of a β-hairpin as it is in the Rad4–DNA complex containing a CPD lesion [37]. From these observations, it can be concluded that the binding of Rad4/XPC to damaged DNA and the concomitant strand separation is favored by lesions that cause a local thermodynamic destabilization [24]. It has indeed been noted previously that lesions that strongly destabilize double-stranded DNA sequences are better substrates of mammalian NER than those that do not [22]. Furthermore, for certain bulky polycyclic aromatic hydrocarbons (PAH)-derived adenine lesions (see below) in double-stranded DNA, the NER efficiency in mammalian cell extracts is correlated with the melting points of the modified DNA duplexes [39–41]. However, the thermodynamic stabilities of lesion-containing duplexes is a crude indicator of NER only within families of structurally and conformationally similar lesion-containing DNA, since the DNA duplex melting points depend not only on the distortions caused by the lesions, but also on favorable interactions between the lesions and the DNA that tend to stabilize the duplexes [39, 40].

12.5
Bipartite Model of Mammalian NER and the Multipartite Model of Lesion Recognition

The local destabilizations that facilitate the binding of Rad4/XPC to the damaged DNA are due to interactions of the lesions themselves with nearby DNA residues; these interactions can result in a weakening of local Watson–Crick hydrogen-bonding and base-stacking interactions. The recognition of bulky lesions by XPC/HR23B and their subsequent processing by NER factors depend in a highly complex manner on the site of attachment of the lesion to the nucleotide, size, conformation, and topology of the bulky adducts, their interactions with neighboring DNA residues, and the structural distortions that they cause [39–43]. Moreover, the deviations of the B-DNA structural parameters from their normal values depend not only on the characteristics of the lesions, but also on the base sequence context in which the lesions are embedded. Since these DNA structural parameters are coupled, changes in one causes changes in others [44, 45]. For example, the loss of Watson–Crick hydrogen-bonding may be accompanied by changes in base-stacking interactions, and coupled changes in helicoidal parameters such as Twist, Roll, and Rise [46]. On the other hand, van der Waals and electrostatic interactions between the lesion and the DNA may cause stabilizing interactions that counteract the destabilizing ones. As lesions can produce changes in multiple DNA parameters, we have developed a multipartite recognition model of NER which evaluates the various disturbances utilizing computational methods (e.g., [26, 39, 41, 43, 47]). This multipartite model pertains to the initial recognition mechanism of NER in which the properties of the modified double strands elicit the initial step of GG-NER – the binding of XPC/HR23B to the lesion-containing site of the damaged

DNA duplex. Naegeli et al. have previously advanced a bipartite model of NER substrate discrimination that is initiated by the detection of disrupted Watson–Crick base-pairing followed by a lesion-sensing step that "verifies" the presence of a chemically altered nucleotide [29, 33]. The nature of the critically important verification step that leads to the dual incision is still not well understood [24]. The bipartite model is consistent with previous observatisons of Sugasawa et al. who found that XPC/HR23B binds to DNA that contains bubbles of several mismatched DNA bases in the absence of lesions or chemically modified nucleotides, but incisions occur only when a chemically modified base is also present [13, 35].

Our multipartite model is a more extended view of the first step of the bipartite hypothesis and is based on the well-established observations that disruption of Watson–Crick base-pairing cannot occur without changing some of the other structural DNA parameters that are affected by the presence of DNA lesions. Changes in properties such as hydrogen bonding, base stacking, base-pair opening, groove widths, phosphodiester backbone torsion angles, and bending are coupled [44, 45], and not mutually exclusive. Understanding the multipartite changes that lesions impose on the DNA structural parameters, and how these changes impact the recognition and subsequent dual incisions that characterize both prokaryotic and eukaryotic NER, is therefore highly pertinent to elucidating the signals that the NER machinery detects [40].

12.6
DNA Lesions Derived from the Reactions of PAH Diol Epoxides with DNA are Excellent Substrates for Probing the Mechanisms of NER

B[a]P and structurally related PAHs constitute a well-known class of environmental genotoxic pollutants [48, 49]. They are metabolized in mammalian cells to highly reactive intermediates [50] that bind covalently to DNA [51, 52]. B[a]P (Figure 12.2) is the best known and most widely studied PAH carcinogen [56]. Like other PAH compounds, it is metabolized to reactive diol epoxide intermediates and PAH diol epoxide derivatives (see Chapter 6). The diol epoxide derivatives are highly mutagenic and tumorigenic [50], and the formation of covalent DNA adducts has been linked to mutagenesis and tumorigenesis in mammalian cell [57–59] and animal [60–64] model systems. The most reactive diol epoxide derivative of B[a]P is the (+)-7R,8S,9S,10R enantiomer of r7,t8-dihydroxy-t9,10-epoxy-7,8,9,10-tetrahydro-B[a]P (r7,t8-dihydroxy-t9,10-epoxy-7,8,9,10-tetrahydrobenzo[a]pyrene*anti*-B[a]PDE); the numbering system of the metabolized ring is defined in Figure 12.2, while the notation *anti* designates the opposite orientations of the 9,10-oxirane group and the 7-OH group relative to the planar pyrene-like aromatic ring system (Figure 12.3). The major B[a]PDE product is the (+)-*anti*-B[a]PDE enantiomer, and only minor proportions of the mirror-image enantiomer, (–)-*anti*-B[a]PDE, are formed (for a recent review, see [65]). Both *anti*-B[a]PDE enantiomers form predominantly *trans* adducts, while smaller proportions of stereoisomeric *cis* adducts are formed by the addition of the exocyclic amino groups of guanine or

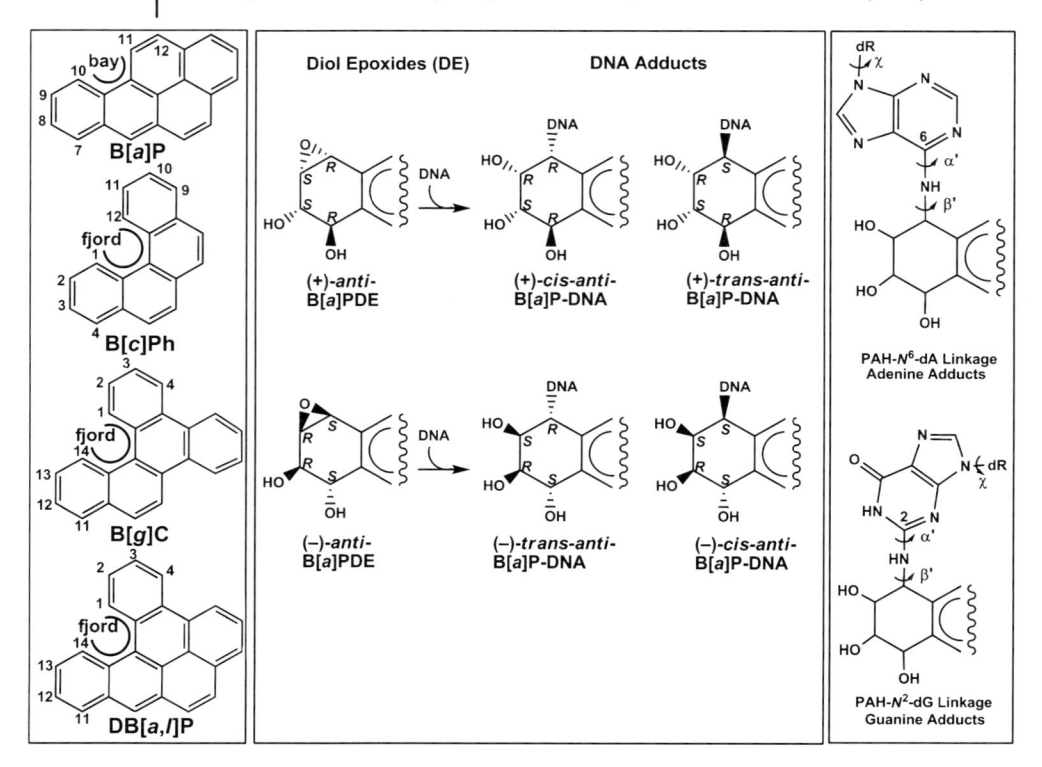

Figure 12.2 Chemical structures of the PAH compounds considered in this chapter, definitions of the bay and fjord region topologies, the stereochemical characteristics of the biologically relevant PAH diol epoxide metabolites, and stereochemical characteristics of PAH diol epoxide-DNA adducts. The aromatic rings metabolized by mammalian CYP 450 enzymes are the 7,8,9,10 ring of B[a]P, the 1,2,3,4 ring of B[c]Ph, and the 11,12,13,14 rings of B[g]C and DB[a,l]P. The absolute configurations of the PAH diol epoxides and DNA adducts are illustrated in the case of the (+)- and (−)-anti-B[a]PDE diol epoxides and the pairs of cis and trans covalent adducts formed from the reactions of these two diol epoxide enantiomers with either N^2-dG or N^6-dA. The torsion angles that determine the sterically allowed adduct conformations [53, 54] are shown by curved arrow. The absolute configurations of the stereochemically analogous fjord region diol epoxides of DB[a,l]PDE, B[c]Ph, and B[g]C are identical except that the (+) and (−) designations of optical rotation are reversed from those of the B[a]P diol epoxides with the same absolute configurations [55].

adenine (N^2-dG or N^6-dA, respectively) in DNA to the C10 carbon atom of (+)- or (−)-anti-B[a]PDE. Thus, four stereoisomeric adducts (the stereochemistry is defined in Figure 12.3) are formed when (±)-anti-B[a]PDE reacts with guanine in DNA [66]. The most abundant stable adduct derived from the reaction of (±)-anti-B[a]PDE with DNA in mammalian cells [67, 68] is the 10S (+)-trans-anti-B[a]P-N^2-dG adduct (Figures 12.2 and 12.3).

B[a]P is known as a bay region PAH; it has a relatively uncrowded region near the critical aliphatic C10 position that is linked to N^2-dG or N^6-dA in covalent DNA

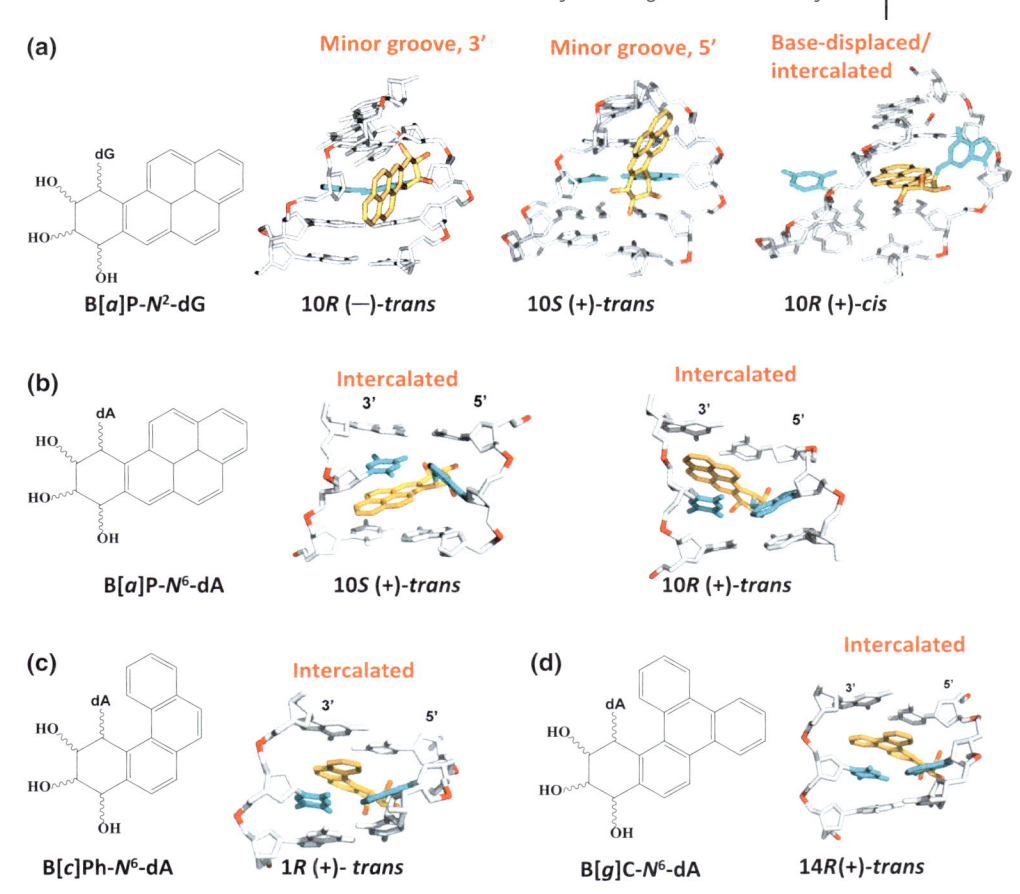

Figure 12.3 Chemical structures and conformations of PAH-DNA adducts (see text). The PAH ring systems are shown in yellow, and the modified bases and their partners are in cyan.

adducts (Figure 12.2). In contrast, the PAHs benzo[c]phenanthrene (B[c]Ph), benzo[g]chrysene (B[g]C), and dibenzo[a,l]pyrene (DB[a,l]P) have a sterically crowded region, called the fjord region, because of the proximities of the pairs of hydrogen atoms at C1 and C12 in B[c]Ph, and C14 and C1 in B[g]C and DB[a,l]P (Figure 12.2). The steric hindrance between these two protons distorts the aromatic ring in the fjord regions (the 1,2,3,4 rings in B[g]C and DB[a,l]P, and the 9,10,11,12 ring in B[c]Ph, Figure 12.2) so that they are no longer coplanar with the remaining aromatic ring systems in these molecules [69]. In the case of B[a]P, there is no steric hindrance between the protons at C10 and C11, and the aromatic ring system remains planar, as in other bay region PAH compounds. Like B[a]P, the fjord PAHs are also metabolically activated to stereoisomeric diol epoxides that are stereochemically analogous to the *anti*-B[a]PDEs except that the sign of the optical rotation is reversed; for example, the absolute configuration of the

(+)-7R,8S,9S,10R or (+)-*anti*-B[*a*]PDE enantiomer is the same as that of the (−)-11R,12S,13S,14R or (−)-*anti*-DB[*a,l*]PDE enantiomer (Figure 12.2). Furthermore, while the reaction of the B[*a*]PDEs with native DNA yield predominantly N^2-guanine adducts [66], the reactions of the fjord diol epoxides yield comparable or greater proportions of N^6-adenine adducts [70–73]. The fjord region PAHs, and/or their diol epoxides and DNA adducts (Figures 12.2 and 12.3), have attracted attention because of their high tumorigenic and genotoxic activities [60–64]. Among these compounds, DB[*a,l*]P has been the most extensively studied one because of its exceptional tumorigenic activity in animal model systems [74]. Luch estimated that (−)-*anti*-DB[*a,l*]PDE is approximately 60 times more mutagenic in V79 rodent cells (mutation frequencies) than the bay region (+)-*anti*-B[*a*]PDE with the identical absolute configuration [75]. Interestingly, in the case of the latter, dG→dT transversion base substitution mutations are dominant, while in the case of the fjord (−)-*anti*-DB[*a,l*]PDE diol epoxide, dA→dT transversions are more abundant [75–77]. These mutagenic specificities are consistent with the preferential reactions of the bay region (+)-*anti*-B[*a*]PDE with guanine and the preferred reactions of the fjord (−)-*anti*-DB[*a,l*]PDE with adenine in native DNA, as mentioned earlier.

These differences in mutagenic activities parallel the tumorigenic activities of these enantiomeric diol epoxides. This provides tantalizing clues that point to the effects of molecular structure on the biological activities of these two PAH metabolites that differ from one another by the single additional aromatic ring protruding into the fjord region of DB[*a,l*]PDE (Figure 12.2). The first line of cellular defense marshaled against bulky DNA adducts, such as the covalent bay and fjord region PAH diol epoxide-DNA adducts depicted in Figure 12.2, is the mammalian NER system. We have therefore sought to gain an understanding of the structural differences between the different stereoisomeric DNA adducts depicted in Figures 12.2 and 12.3, and how the conformations of these DNA lesions affects their recognition and removal by NER.

12.7
Multidisciplinary Approach Towards Investigating Structure–Function Relationships in the NER of Bulky PAH-DNA Adducts

In order to delineate the structural and dynamic properties of PAH diol epoxide-DNA adducts and their correlations to DNA repair susceptibility, a powerful array of complementary methods have been employed (e.g., [78]):

i) Site-specifically modified oligonucleotides (usually 11 or 12 nucleotides in length) containing single, stereochemically well-defined PAH diol epoxide-N^2-dG or -N^6-dA lesions (Figure 12.2) are synthesized; the structural features of these lesions in double-stranded DNA are then investigated using standard nuclear magnetic resonance (NMR) methods [53, 79]. The thermal stabilities of the duplexes are determined because they provide a crude measure of the extent of structural distortions in the modified double-stranded DNA duplexes

[39, 80], while gel electrophoresis methods are used to evaluate properties such as lesion-induced bending of double-stranded DNA [78, 81, 82] to distinguish between rigid and flexible bends.

ii) The modified 11 or 12mer oligonucleotides are embedded into 135mer DNA duplexes using ligation methods [38], and NER efficiencies are evaluated in human cell extracts as shown schematically in Figure 12.1 and described in greater detail elsewhere [38, 78].

iii) Based on sets of interproton distances and other NMR data, and employing structural refinement with molecular mechanics and molecular dynamics (MD) methods, three-dimensional models of the PAH diol epoxide-N^2-dG or -N^6-dA adducts in the modified duplexes are obtained that are consistent with experimental data as described elsewhere [79, 83–89].

iv) Molecular modeling and MD simulations with free energy analyses are employed to reveal the dynamic structural and thermodynamic properties of the DNA lesions in order to define the underlying structural origins of the experimental observations [47, 78]. We perform MD simulations utilizing the AMBER simulation package [90] with explicit solvent and counterions to provide an ensemble of many thousands of structures simulated at 300 K and 1 atm of pressure. These ensembles provide the basis for structural analyzes as well as for evaluating free energy differences between different conformations of the same lesion or a set of stereoisomeric lesions; the molecular mechanics Poisson–Boltzmann surface area (MM-PBSA) method [91] in the AMBER package is used for this purpose. The detailed structural information is then correlated with *in vitro* NER experiments in human HeLa cell extracts to assess the various structural distortions and destabilizations that can account for the differential NER dual-incision efficiencies.

12.8
Dependence of DNA Adduct Conformations and NER on PAH Topology and Stereochemistry

Extensive studies by NMR methods have shown that the conformations of PAH diol epoxide-N^2-dG and -N^6-dA adducts in double-stranded DNA depend markedly on the stereochemistry, PAH topology, and base sequence context. Earlier work has been summarized by us [53]. Since then, we have published the NMR solution structures of a number of other DNA adducts [79, 88, 89, 92]. The structural properties of various DNA adducts have been reviewed more recently by Lukin and de los Santos [93]. The basic structural motifs, based on our work and that of others [94–98] that are relevant to the NER studies described in this chapter, are summarized in Figure 12.3. Computational analyses have provided insights into the origins and stereochemistry dependence of these remarkably different adduct conformations [26, 42, 47, 54, 99]. In the following, we summarize the main features of these different conformations.

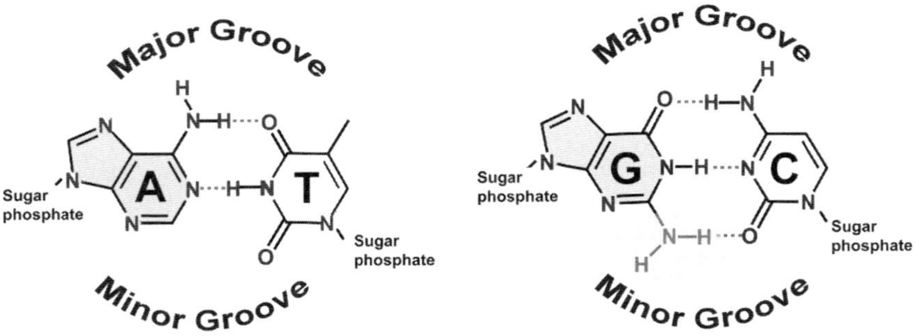

Figure 12.4 Watson–Crick base-pairing in B-DNA showing the locations of the exocyclic amino groups of guanine and adenine in the minor and major grooves, respectively.

There are fundamental structural differences between PAH-N^2-dG and -N^6-dA adducts that may be understood in terms of Figure 12.4. In B-DNA, the exocyclic amino group of guanine is positioned in the sterically hindered narrow minor groove. Since the site of attachment of the bulky PAH ring system in N^2-dG adducts is in the minor groove of normal B-DNA, the polycyclic aromatic residues are either accommodated in the minor groove as in the *trans-anti*-B[a]P-N^2-dG adducts or intercalated from the crowded minor groove side with base displacement as in the *cis-anti*-B[a]P-N^2-dG adducts (see below). A third structural family, intercalation from the minor groove, but with base pairing maintained, has also been observed in the case of the fjord region *trans-anti*-B[c]Ph-N^2-dG adducts [88], but discussion of these structures is beyond the scope of this chapter.

In the case of PAH-N^6-dA adducts, the site of attachment is positioned in the more spacious major groove of B-DNA (Figure 12.4) and all of the B[a]P-N^6-dA adducts studied to date have been found to be intercalated from the major groove without base displacement (see below). We note in passing that external conformations were observed in the case of major groove adducts derived from the reaction of the one-ringed styrene oxide with the exocyclic amino group of adenine in double-stranded DNA [100]. We now briefly summarize the main features of the PAH-DNA adducts depicted in Figure 12.3.

12.8.1
Guanine B[a]P Adducts (Figure 12.3a): Minor Groove and Base-Displaced/Intercalative Conformations

The NMR solution structures of all B[a]P-N^2-dG adducts described below were determined with the modified guanine residue (**G***) embedded in the 11mer duplex (5'-d(CCATCG*CTACC)) · (5'-d(GGTAGCGTAGG)). The bulky aromatic ring system of the B[a]P residues is bound to the exocyclic amino group of guanine in all cases. In the 10*S* (+)-*trans-anti*-B[a]P-N^2-dG adduct (10*S* absolute configuration at the C10 linkage site), the pyrenyl residue (Figure 12.3) points towards the 5'-end of the modified strand, while in the stereoisomeric 10*R* (−)-*trans-anti* adduct it is also positioned in the minor groove, but points towards the 3′-direction. All Watson–Crick base pairs are intact [84, 87]. The opposite orientations with respect to the modified base of these DNA adducts derived from a pair of enantiomeric PAH diol epoxides is a recurring theme that has been predicted [101] and experimentally observed in many cases [53, 93].

In the stereoisomeric 10R (+)-*cis-anti*-B[a]P-N^2-dG adduct the B[a]P residue is intercalated with the benzylic ring in the minor groove. The aromatic B[a]P ring system is intercalated between the neighboring base pairs, but the modified guanine is displaced into the minor groove, while the partner C residue is displaced into the major groove [83]. The structure of the stereoisomeric 10S (−)-*cis-anti*-B[a]P-N^2-dG adduct is not shown, but its benzylic ring is positioned in the major groove, and the modified guanine and partner C residues are displaced into the major groove [102].

All of these B[a]P-N^2-dG adducts destabilize the DNA duplexes in a manner that depends on the adduct stereochemistry [53]. Detailed thermodynamic studies of some of these adducts [103] and the effects of mismatched bases in the complementary strand opposite all four stereoisomeric B[a]P-N^2-dG lesions (**G***) in C**G***C and T**G***C sequence contexts have been published [104]. The melting points of unmodified and B[a]P-modified duplexes are defined as T_m – the temperature at which 50% of the duplexes are dissociated into single strands and are in equilibrium with one another [105]. For convenience, we define the melting points of the modified duplexes by T_m (adduct) and the T_m of the unmodified duplexes by T_m (um). The difference in melting points $\Delta T_m = T_m(\text{adduct}) - T_m$ (um) is −8 ((+)-*trans*-), −10 ((−)-*trans*-), −4 ((+)-*cis*-), and −5 °C ((−)-*cis-anti*-B[a]P-N^2-dG) [53].

12.8.2
Bay Region B[a]P-N^6-Adenine Adducts (Figure 12.3b): Distorting Intercalative Insertions from the Major Groove

All Watson–Crick base pairs are intact in the bay region 10R (+)-*trans*- and 10S (−)-*trans-anti*-B[a]P-N^6-dA adducts. The NMR structure of the 10R adduct is well established [92, 94–98, 106], but the conformation of the 10S adduct could not be determined in fully double-stranded DNA with Watson–Crick base pairs [95]. The structure of the 10S (−)-*trans-anti*-B[a]P-N^6-dA adduct shown in Figure 12.3(b) is therefore a simulated one based on information from experiments [42]. The B[a]P residues are oriented on opposite sides of the modified adenine residues, but are not quite parallel to neighboring base pairs; furthermore, a flanking base pair is distorted in the 10S case (Figure 12.3b), and the glycosidic torsion angle fluctuates between *anti* and *syn* domains only for this stereoisomeric adduct. The greater structural disorder in the 10S (−)-*trans*- compared to the 10R (+)-*trans-anti*-B[a]P-N^6-dA adduct is reflected in the melting points, T_m, of the 9-mer duplexes (5′-d(GGTCA*CGAG)) · (5′-d(CTCGGTGACC)) containing these *trans-anti*-B[a]P-N^6-dA lesions (**A***), which is considerably lower in the case of the 10S ($T_m = 16$ °C) than the 10R ($T_m = 28$ °C) duplex [95]. Both lesions significantly destabilize the modified duplexes, since the T_m of the unmodified 9mer duplex is 43 °C under the experimental conditions adopted by Schurter et al. [95]. Thus, the difference in melting points ΔT_m is −27 and −15 °C in the case of the 10S (−)-trans- and 10R (−)-trans-anti-B[a]P-N^6-dA adducts, respectively, and indicate that this pair of stereoisomeric lesions destabilize double-stranded DNA to different extents.

12.8.3
Fjord Region PAH N^6-Adenine Adducts (Figure 12.3c and d): Minimally Distorting Intercalation from the Major Groove

All Watson–Crick base pairs are intact in the fjord region 1R (+)-trans- and 1S (−)-trans-anti-B[c]Ph-N^6-dA adducts [85, 86] and a structurally related fjord region 14R (+)-trans-anti-B[g]C-N^6-dA adduct [89]. The approximately parallel orientation of the B[c]Ph residue with respect to the neighboring base pairs suggests that these intercalated structures (without base displacement) are less distorting than those derived from the bay region B[a]P diol epoxides [41]. These conclusions are supported by comparisons of the differences in the melting points in the identical sequence (5′-d(CTCTCA*CTTCC)) · (5′-d(GGAAGTGAGAG)) duplexes. For the 10R (+)-trans- and the 10S (−)-trans-anti-B[c]Ph-N^6-dA adducts (A*) in 11mer duplexes, ΔT_m is 0 °C in both cases (no destabilization). For the 14S (−)-trans-anti-B[g]C-N^6-dA adduct, ΔT_m = −4 °C and for the 14R (+)-trans-anti-B[g]C-N^6-dA adduct, ΔT_m = +9 °C. Similar differences are found for the 14S (−)-trans-anti-DB[a,l]P-N^6-dA (ΔT_m = −4 °C) and the 14R (+)-trans-anti-DB[a,l]P-N^6-dA adducts (ΔT_m = +7 °C). The positive ΔT_m values denote that the fjord PAH lesions stabilize the DNA duplexes, which is at first sight surprising, given the large bulk of the B[g]CDE and DB[a,l]PDE residues. In contrast, the large negative ΔT_m values detailed above indicate that the bay region trans-anti-B[a]P-N^6-dA adducts significantly destabilize DNA. This is remarkable because the sizes of the B[a]P-derived adducts are smaller by one aromatic ring than those derived from DB[a,l]P, but are otherwise identical. The origin of these differences between fjord and bay region PAH diol epoxide-derived adducts is attributed to differences in topology: the fjord region aromatic ring systems are nonplanar and have twist flexibility (the 1,2,3,4 aromatic rings in the DB[a,l]P and B[g]C rings, and the 9,10,11,12 aromatic ring on the B[c]Ph residue); however, the bay region aromatic ring systems are planar and rigid, and hence cannot stack as well within the duplex [39–41].

12.8.4
Dependence of NER Efficiencies on the Conformations of the Bay Region B[a]P-N^2-dG Adducts

The NER efficiencies in extracts from human HeLa cells of the (+)-trans-, (−)-trans-, and (+)-cis-anti-B[a]P-N^2-dG adducts in the identical sequence context (5′-d(CCAT CG*CTACC)) · (5′-d(GGTAGCGATGG)) embedded in 135mer duplexes are shown in Figure 12.5(a). The lengths of the incision products coincide with those of the marker oligonucleotides around 26–32 nucleotides in length (lane M in Figure 12.5a), which is consistent with dual excision products obtained by the mamma-

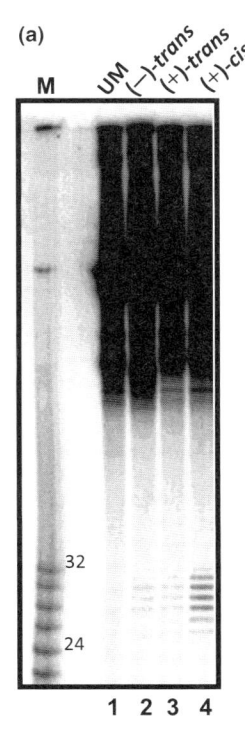

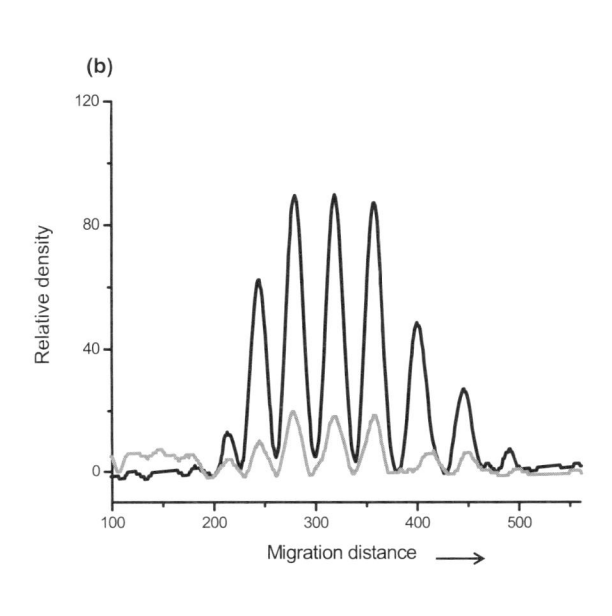

12.8 Dependence of DNA Adduct Conformations and NER on PAH Topology and Stereochemistry

Figure 12.5 (a) Autoradiographs of dual excision products: NER of (+)-*cis-I*, (+)-*trans-I*, and (−)-*trans-I* 135mer duplexes in the 5′-...CCATCG*CTACC...sequence context (CG*C-I in Figure 12.9) hybridized with a fully complementary strand with C opposite **G***. The duplexes were internally ^{32}P 5′-end-labeled at the first C shown here at the sixth phosphodiester bond on the 5′-side of the B[*a*]P-N^2-dG lesions (**G***). The dual-incision products were obtained after incubation of the 135mer duplexes containing different stereoisomeric **G*** lesions with nuclear extracts from human HeLa cells for 30 min at 30 °C (lanes 2–4). Lane 1 (UM): unmodified oligonucleotide 135mer duplex control after treatment with the nuclear extracts. Lane M: oligonucleotide size markers 32, 30, 28, 26, 24, etc., nucleotides long (from top to bottom). (b) Densitometric analysis of lanes 3 and 4 (the bottom and top traces, respectively, are shown only). After normalization with respect to the total radioactivity in each lane, quantitative analysis indicates that the ratios of dual-excision products of the (+)-*cis-I*, (+)-*trans-I*, and (−)-*trans-I* duplexes are 5 : 1 : 0.9, with error bars of ±20% using different extracts (eight experiments).

lian NER mechanism [107]. Densitometry tracings of lanes 2 (lower) and 4 (upper tracing) are shown in Figure 12.5(b). The areas under these individual curves in the range of 24–32 oligonucleotides were evaluated and divided by the total area of analogous traces from one end of the lane to the other that includes all of the non-NER products [107]. The amounts of dual-incision products are smaller by a factor of 5–7 in the case of the minor groove (+)-*trans*- and (−)-*trans-anti*-B[*a*]P-N^2-dG adducts than in the case of the base-displaced/intercalated (+)-*cis-anti*-B[*a*]P-N^2-dG adduct.

It is interesting to compare the NER efficiencies of the B[a]P-N^2-dG adducts to those of the familiar and widely studied UV-induced thymine–thymine (<TT>) photodimers. Irradiation of native DNA *in vitro* or in cells with UV-C light produces isomeric CPD dimers, predominantly the *cis-syn* form, which involves the covalent linking of two adjacent pyrimidines via a four-membered ring structure (Figure 12.6b). While the CPD dimers do not cause much distortion of double-

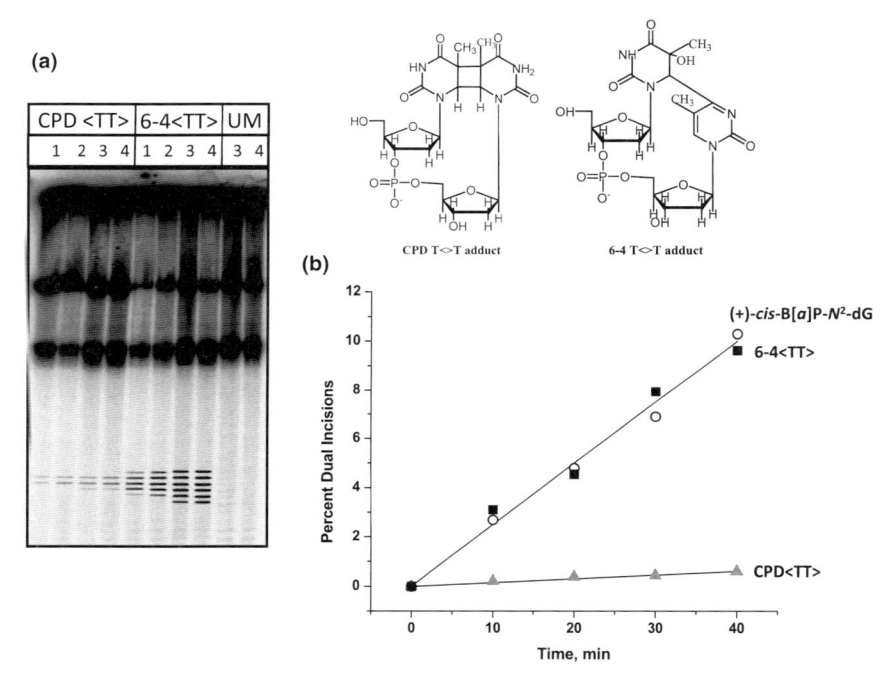

Figure 12.6 (a) NER results in HeLa cell extracts for 135mer duplexes containing the UV photodimer CPD <TT> or 6-4 <TT> duplexes embedded in the sequence context 5′-GCAAG<TT>GGAG-3′ hybridized with a fully complementary strand with ... AA ... opposite <TT>. The total radioactivity levels in each lane were comparable, to allow for a visual inspection of the differences in dual-incision efficiencies. Lanes 1, 2, 3, and 4 represent incubation times of 10, 20, 30, and 40 min, respectively. The lane marked <TT> represents parallel incubations with unmodified <TT> duplexes, while lane M contains unmodified oligonucleotide markers of 32, 30, 28, 26, and 24 nucleotides in length. UM lanes 3 and 4: 30 and 40 min incubation of unmodified 135mer control duplexes. (b) Time dependence of formation of dual-incision products. The average values are based on five independent experiments in the case of the two UV photodimers. Black squares: duplexes with 6-4 <TT> photodimers; triangles: duplexes with CPD <TT> photodimers. The (+)-*cis-I* duplexes (open circles, 10R (+)-*cis-anti*-B[a]P-N^2-dG adduct) were incubated under similar conditions and the dual-incision results were used as a reference standard in subsequent experiments to adjust for the variable NER activities of cell extracts prepared at different times. The straight lines are the least-squares fits to the data points. In all these independent experiments, the data points for the CPD <TT> and 6-4 <TT> were normalized to the 40 min value of the 10R (+)-*cis-anti*-B[a]P-N^2-dG adduct-containing CG*C duplex obtained in the same experiment.

stranded DNA, another product that involves a single covalent bond between the 6- and 4-positions of two adjacent thymines is also formed that causes significant distortion in the local DNA structure [108]. The human NER apparatus in cell extracts removes both lesions, but the 6-4 photoproduct is repaired at least 10 times more efficiently than the CPD lesion [109]. This difference is usually attributed to and is consistent with the greater distortion of the native DNA structure by the 6-4 photoproduct.

In order to compare the relative NER efficiencies of our PAH diol epoxide-DNA adducts we prepared site-specifically modified oligonucleotides containing either CPD or 6-4 <TT> photodimers using the same sequence and methods as described in the literature [110]. Typical results obtained in our laboratory are displayed in Figure 12.6(a). There are fewer NER dual-incision bands in the case of the CPD <TT> dimer than in the case of the 6-4 <TT> photoproduct and the latter are clearly more intense than the former. Densitometry analysis shows that the time course of dual incisions is linear at least up to 40 min and that the DNA strand bearing the 6-4 <TT> adduct is around 17 times faster than the strand with the CPD <TT> dimer (Figure 12.6c). The repair rate of the (+)-*cis-anti*-B[a]P-N^2-dG adduct and the 6-4 <TT> photoproduct coincide with one another (Figure 12.6c), as described in more detail elsewhere [78].

While all three B[a]P-N^2-dG adducts destabilize double-stranded DNA, the extent of destabilization is significantly smaller in the case of the (+)-*cis-anti* adduct ($\Delta T_m = -4\,°C$) than in the two *trans-anti*-B[a]P-N^2-dG adducts ($\Delta T_m = -8$ to $-10\,°C$). Nevertheless, the NER efficiency of the (+)-*cis-anti*-B[a]P-N^2-dG adduct is 5–7 times greater than that of either *trans-anti* adduct. The smaller extent of destabilization caused by the (+)-*cis-anti*-B[a]P-N^2-dG adduct is attributed to carcinogen base π–π stacking interactions that counteract the loss of Watson–Crick hydrogen-bonding and base–base stacking interactions in this base-displaced/intercalated conformation [38].

The studies of Mocquet *et al.* provide an important clue to the mechanism of discrimination of these stereoisomeric lesions by human NER factors [38]. A permanganate footprinting method [14, 17] was utilized to probe the extent of helix opening associated with the binding of the lesion-recognizing XPC/HR23B NER factor to the sites of the conformationally distinct and stereoisomeric B[a]P-derived DNA adducts in 135mer duplexes. The helix-opening patterns are remarkably different for these stereoisomeric DNA adducts; these patterns are well correlated with the efficiencies of dual excisions in both human cell extracts and when catalyzed by a set of reconstituted NER factors (XPC/HR23B, TFIIH, XPA, RPA, XPG, and XPF/ERCC1). The footprinting patterns showed that opening of the duplex was favored on the 5′-side of the lesion in the case of the minor groove 5′-oriented B[a]P residue in the (+)-*trans-anti*-B[a]P-N^2-dG adduct. In the case of the 3′-oriented B[a]P residue in the (−)-*trans-anti*-B[a]P-N^2-dG adduct, helix opening was favored on the 3′-side of the adduct. In both cases, the relative extent of opening was weak. The extent of opening of the double-helix is markedly more pronounced in the case of the (+)-*cis-anti* adduct on its 5′-side. The extent of opening is qualitatively correlated with the higher NER efficiency characterizing the *cis-anti* adducts [38].

Structural analyses with MD simulation methods based on the NMR solution structures [83, 84, 87] of this set of stereoisomeric adducts revealed that minor groove enlargement is a distinctive structural feature of these lesions. The patterns of minor groove enlargement determined by MD simulation methods are correlated with the helix-opening patterns caused by the binding of the NER recognition factor XPC/hHR23B to the three stereoisomeric B[a]P-N^2-dG duplexes and with the observed NER efficiencies [38]. Thus, minor groove enlargement appears to play a role in lesion recognition. However, the NER recognition signals for the base-displaced/intercalated (+)-cis-anti adduct structure include flipped-out base pairs, ruptured Watson–Crick hydrogen-bonding at the lesion site together with enlarged minor groove widths [38]. We note that base flipping is favored when the duplex is thermodynamically destabilized by the lesion; there are also changes in the local values of some of the structural parameters that can facilitate the extrusion of one or more bases out of the double helix. Such coordinated structural changes are essential features of the multipartite recognition of DNA lesions by NER factors [39, 40]. The ease of flipping a base out of the helix, as suggested by Naegeli et al. [36, 111], may be an important contributing factor to efficient NER; this hypothesis is supported by the observation that the cis-anti-B[a]P-N^2-dG adducts are excellent substrates for NER [107] because they adopt base-displaced intercalated confirmations with flipped-out bases [83, 102].

Removing the base C in the complementary strand opposite the (+)-cis-anti-B[a]P-N^2-dG adduct gives rise to the deletion duplex (DEL), (5′-d(CCATCG*CTACC)) · (5′-d(GGTAG–GATGG)), where the dash denotes the missing C in the complementary strand. In this DEL duplex, the aromatic B[a]P ring system is intercalated into a wedge-shaped site which is flanked on both sides by intact Watson–Crick dG · dC base pairs. The modified deoxyguanosine stacks over the minor groove face of the sugar ring of the 5′-flanking dC residue [112]. Interestingly, the DEL duplex is resistant to NER [107] as shown by comparisons of the time course of NER of the same (+)-cis-anti-B[a]P-N^2-dG adduct intercalated into either the full duplex (5′-d(CCATCG*CTACC)) · (5′-d(GGTAGCGATGG)) or the DEL duplex (Figure 12.7). There is a complete absence of dual-incision products in the latter case. These findings are consistent with the Rad4/CPD crystal structure [37] in which two unmodified bases opposite the CPD <TT> dimer in the complementary strand are apparently involved in anchoring the Rad4/XPC protein to the damaged DNA duplex [29, 36, 111]. However, an additional factor that could account for the resistance to NER of the DEL duplex is the unusual degree of thermal stabilization of the DEL duplex by the intercalated (+)-cis-anti-B[a]P aromatic ring system. The ΔT_m is +25 °C, the highest we have ever observed. The T_m (um) value of the unmodified 11 · 10mer duplex is 24 °C, which is 27 °C lower than the T_m value of the full, unmodified 11 · 11mer duplex (5′-d(CCATCG CTACC)) · (5′-d(GGTAGCGATGG)). The extraordinarily high T_m (mod) value of the DEL duplex is due to the strong π–π stacking interactions between the flanking base pairs and the intercalated aromatic ring system of the B[a]P residue. It is therefore plausible that the opening of the duplex by XPC/HR23B is inhibited by these strong B[a]P base-stacking interactions, while the absence of the partner

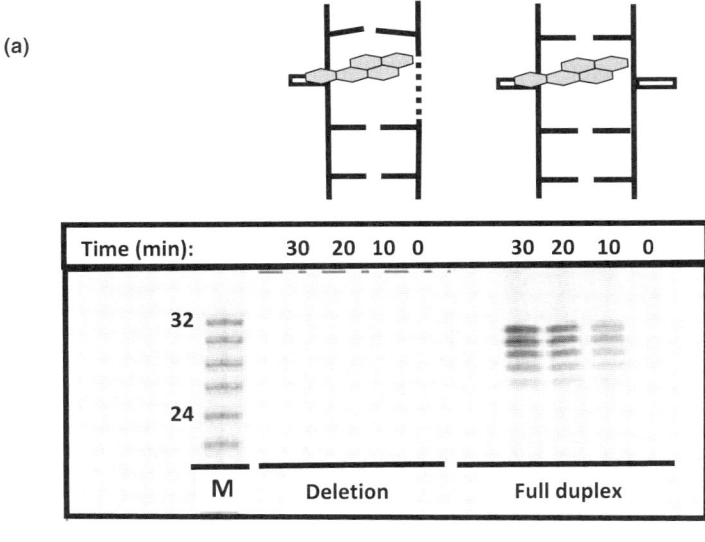

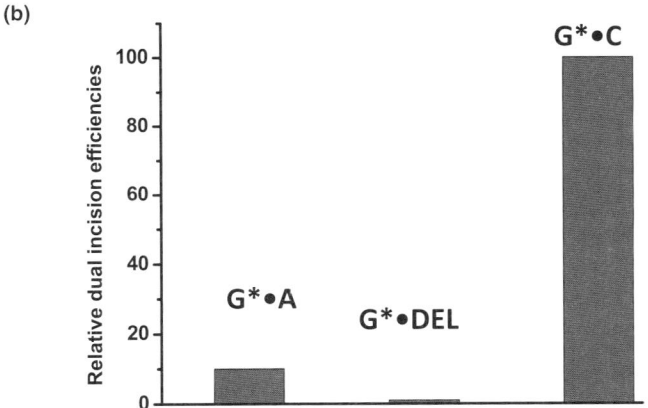

12.8 Dependence of DNA Adduct Conformations and NER on PAH Topology and Stereochemistry

Figure 12.7 (a) Schematic illustration of the time dependence of NER in nuclear cell extracts of 135·134mer deletion duplexes (G*·DEL) that lack the C opposite G* (the 10R (+)-cis-anti-B[a]P-N^2-dG adduct) and full 135·135mer full duplexes (G*·C) that have C in the complementary strand opposite G* (full (+)-cis-I 135mer duplexes). (b) Relative NER efficiencies for the mismatched duplex with A opposite G* in full 135·135mer duplexes (G*·A), the 135·134mer deletion, and fully complementary 135·135mer duplexes, all containing the 10R (+)-cis-anti-B[a]P-N^2-dG adduct.

base, needed to anchor the duplex to XPC, may also play a role. The NER in cell extracts of an A-mismatch duplex (A-MM) in which the A is substituted for the normal C opposite the (+)-cis-anti-B[a]P-N^2-dG adduct in the (5'-d(CCATCG*CTACC))·(5'-d(GGTAGAGATGG)) 11·11mer duplex, provides yet another interesting case. In this duplex, all nucleotides are present in the complementary strand, but the NER efficiency is around 10 times smaller (Figure 12.7b) than it is in the normal 11·11mer duplex (5'-d(CCATCG*CTACC))·(5'-d(GGTAGCGATGG)) [107]. In this connection, with the mismatched A the modified duplex has a T_m of 49.1 °C while the unmodified duplex containing the mismatch has a T_m of 49.8 °C [104]. Thus, the modified duplex is not destabilized as compared to the unmodified duplex with the mismatch, consistent with its weak repair susceptibility. However, structural information for the adduct with the mismatch is not available. Thus, it is clear that the important features in the lesion recognition mechanism that lead to successful NER still remain to be clarified in the case of the DEL and A-MM duplexes.

12.8.5
NER Efficiencies: Bay and Fjord Region PAH Diol Epoxide-N^6-dA Adducts

The relative NER efficiencies of the two stereoisomeric B[a]P-N^6-dA adducts are compared to those of the B[a]P-N^2-dG adducts in Figure 12.8. The sequence context was (5'-d(CCGGACA*AGAAGC))·(5'-d(GCTTCTTGTCCGG)), and the repair efficiencies are similar to those of the minor groove 10R (+)-*trans*- and 10S (−)-*trans-anti*-B[a]P-N^2-dG adducts. However, the 10S (−)-*trans* adduct is repaired more efficiently by a factor of around 2 than the 10R (+)-*trans* adduct; this difference is sequence-dependent since the same 10S adduct is repaired around 9 times more efficiently than the 10R adduct in the context (5'-d(CTCTCA*CTTCC))·(5'-d(GG AAGTGAGAG)) [43, 113]. The origin of this sequence-governed difference was analyzed by Yan *et al.* [43] who concluded that the higher NER efficiency of the 10S (−)-*trans-anti*-B[a]P-N^6-dA adduct is due to the existence of a sequence-specific

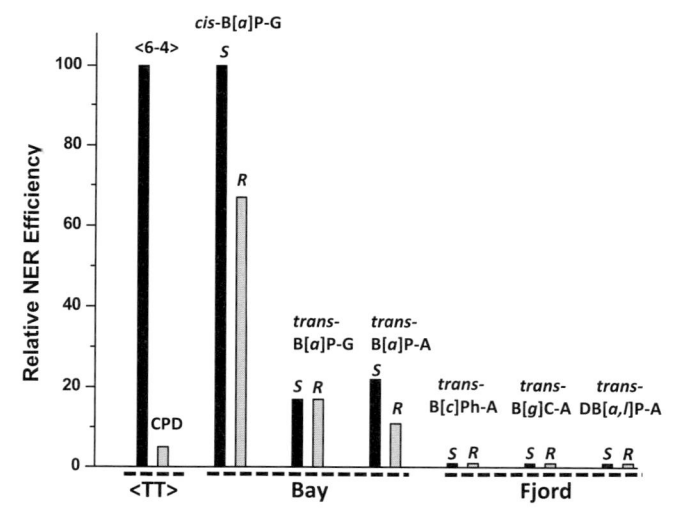

Figure 12.8 Relative NER efficiencies. The *cis*- and *trans-anti*-B[a]P-N^2-dG (B[a]P-G) data are from Hess *et al.* [107], and the B[a]P-N^6-dA (B[a]P-A) and the fjord region PAH-N^6-dA adducts (PAH-A) are from Buterin *et al.* [113].

hydrogen bond, accompanied by a significantly increased Roll and consequent bending in this adduct in the ... CA*C ... sequence context.

The higher NER efficiency of the 10S (+)-*trans-anti*-B[a]P-N^6-dA adduct compared to the 10R case may be understood in terms of the greater destabilization of the double-stranded helix and the structural factors that account for this destabilization. In the 10R (−)-*trans-anti*-B[a]P-N^6-dA adduct, the B[a]P aromatic ring system is intercalated on the 5′-side of the modified adenine base, while in the stereoisomeric 10S adduct it is positioned on the 3′-side. A greater degree of conformational disorder in the 10S (+)-*trans-anti* adduct is indicated by the NMR studies [95, 97] and by the T_m data, since ΔT_m is −27 °C for the 3′-oriented 10S (+)-*trans-anti* adduct and −15 °C for the stereoisomeric 10R (−)-*trans-anti* adduct. The key features that account for the differences in the duplex melting points T_m are the orientation of the B[a]P residues on the 5′- or 3′-side of the modified adenine. Owing to the right-handed helical twist of B-DNA, the interactions with neighboring DNA moieties are substantially different for the two stereoisomeric adducts; this leads to a significantly greater extent of steric crowding in the case of the 10S adduct [42, 114]. The structural underpinnings of the differences in the thermodynamic stabilities of the 10S (+)- and 10R (−)-*trans-anti*-B[a]P-N^6-dA adducts were investigated in detail by Yan *et al.* by MD simulations and MM-PBSA methods [42].

Owing to the effect of the helical twist, the 10S (+)-*trans-anti* adduct changes its glycosidic bond conformation χ (Figure 12.2) linking the base to the 2′-deoxyribose residue within its normal *anti* domain to values in the *syn* domain; furthermore, there is a wider distribution of χ glycosidic torsion angles in the 10S than in the 10R adduct. These facts suggest (i) a greater heterogeneity of conformations in the case of the 10S (+)-*trans-anti* adduct and (ii) a disruption of the Watson–Crick hydrogen bonds that accompanies a change of χ from the *anti* to the *syn* domain. These differences explain, in part, the observation of well-defined NMR spectra in the case of the 10R (−)-*trans-anti* adduct and the poorly resolved heterogeneous spectra of the 10S (+)-*trans-anti* adduct. The greater destabilization of the duplex by the 10S (+)-*trans-anti*-B[a]P-dA adduct (ΔT_m = −27 °C) relative to smaller destabilization by the 10R (−)-*trans-anti*-B[a]P-dA adduct (ΔT_m = −15 °C) is thus also explained. These conclusions are further supported by the computed free energy differences ($\Delta \Delta G$) between the two stereoisomeric adducts that are lower by around 13 kcal/mol in the case of the 10R (−)-*trans-anti*-B[a]P-dA adduct than in the case of the 10S adduct, following the trend of the ΔT_m values. The fjord PAH-adenine and the bay region B[a]P-adenine adducts share a common intercalative motif without displacement of any of the bases since base pairing is maintained even at the site of the lesions. However, in contrast to the bay region *trans-anti*-B[a]P-N^6-dA adducts, neither of the stereoisomeric fjord region *trans-anti*-B[c]Ph-N^6-dA adducts are repaired by the human NER apparatus in cell extracts [113]. Also, in contrast to the *trans-anti*-B[a]P-N^6-dA adducts, neither the 1S (−)-*trans* nor the 1R (+)-*trans-anti*-B[c]Ph-N^6-dA adducts destabilize double-stranded DNA since the ΔT_m = 0 °C [115]. The other fjord region adducts studied, *trans-anti*-B[g]C-N^6-dA and *trans-anti*-DB[a,l]P-N^6-dA, are also NER-resistant (Figure 12.8) [113]. Both types of 14R (+)-*trans-anti* adducts actually stabilize the modified DNA duplexes, while both of the 14S (−)-*trans-anti* adducts induce only relatively small or no destabilization of the duplexes [80]. Thus, in this family of conformationally related intercalated adducts, there is an approximate and qualitative correlation between the thermal stabilities of the modified duplexes and their susceptibilities to NER.

12.8.6
Why the *trans-anti*-B[c]Ph-N^6-dA and Related Fjord Region N^6-dA Adducts do not Destabilize DNA and are Resistant to NER

The availability of NMR structural data for both the 1R (+)-*trans*- and 1S (−)-*trans-anti*-B[c]Ph-N^6-dA adducts [85, 86] made possible an in-depth structural analysis of these DNA lesions by MD simulation and MM-PBSA methods [41]. The only other NMR structure of a fjord PAH-DNA adduct available up till now is the 14R (+)-*trans-anti*-B[g]C-N^6-dA adduct [88]. The structural DNA distortions induced by the fjord region *trans-anti*-B[c]Ph-dA adducts are minimal, and the free energy ($\Delta\Delta G$) differences between the 1S and 1R adducts are also negligibly small, in agreement with the observations that the T_m values are identical [115]. The lack of an effect on T_m is attributed to the nonplanar, flexible 9,10,11,12 aromatic rings in the fjord region of the B[c]Ph-N^6-dA and the 1,2,3,4 ring in the case of the B[g]C-N^6-dA aromatic residues (Figure 12.2) that allow for optimal stacking interactions with neighboring bases in the intercalation pockets. The excellent fit of the bulky aromatic ring systems between adjacent DNA base pairs without disruption of Watson–Crick hydrogen-bonding is consistent with the minimal destabilization of the DNA duplexes, as reflected in the T_m values. In contrast, in the case of the *trans-anti*-B[a]P-N^6-dA adduct, the bulky aromatic ring system is planar and rigid, and lacks the additional flexible ring of the three fjord region adducts; this ring most likely contributes to the stabilities of the modified duplexes by additional favorable stacking interactions with the bases, since the twist adjusts to optimize stacking [53].

In summary, our hypothesis is that the increased polycyclic aromatic ring–base stacking interactions in the fjord region PAH diol epoxide-N^6-dA adducts stabilize the duplexes and inhibit the extent of strand opening in XPC/HR23B complexes that is critically important for successful NER [38], possibly by preventing the flipping out of nucleotides from the complementary strand.

12.9
Dependence of NER of the 10S (+)-*trans-anti*-B[a]P-N^2-dG Adduct on Base Sequence Context

We first consider why base sequence effects can affect the thermodynamic stabilities of double-stranded DNA. The neighboring base–base stacking energies have been analyzed in detail (e.g., see [116, 117]). These interactions are base sequence dependent because of potential steric clashes and electrostatic effects between the exocyclic groups (-NH$_2$ and >C=O) of adjacent bases in the minor and major grooves in DNA [118]; these steric hindrance and electrostatic effects influence the local helical twist, and helicoidal parameters and thus produce distortions to ideal Watson–Crick base-pairing alignments. Since bending and weakened stacking coupled with transient unpairing of Watson–Crick hydrogen bonds may be recognized by the human NER apparatus, such disturbances could influence the NER rates of excision in a manner dependent on the bases flanking the lesions. Purine–purine steps such as GG or AA are generally the least flexible dinucleotide steps, and runs of Gs (e.g., GG or GGG) are stable and rigid. On the other hand, the 5'-TG dinucleotide step represents a step of unusual anisotropic flexibility [119, 120], while the 5'-GT step is less flexible than the TG step [121] and bends slightly into the minor groove [122]. The stacking enthalpies of the TG, TA, and AT dinucleotide steps are lower than those of the CG and GC steps. We found that the bending associated with *trans-anti*-B[a]P-N^2-dG adducts in oligonucleotides can be well accounted for in terms of such a dinucleotide step model [123].

12.9.1
Structural Characteristics of the Identical 10S (+)-*trans-anti*-B[a]P-N^2-dG Adduct in Different Sequence Contexts

The structural properties of B[a]P diol epoxide-derived DNA adducts depend not only on their stereochemical properties as described above, but also on the local base sequence contexts. In order to gain some insights into these phenomena, we focused on the major, biologically significant 10S (+)-*trans-anti*-B[a]P-N^2-dG adduct in the sequence contexts shown in Figure 12.9. We first discuss what is special

(a)
CG*C-I 5'-CCATCG*CTACC-3'
CG*C-II 5'-CACACG*CACAC-3'
TG*T 5'-CACATG*TACAC-3'
G6*G7 5'-CATGCG*GCCTAC-3'
G6G7* 5'-CATGCGG*CCTAC-3'
I6G7* 5'-CATGCIG*CCTAC-3'

(b) Guanine / Inosine structures

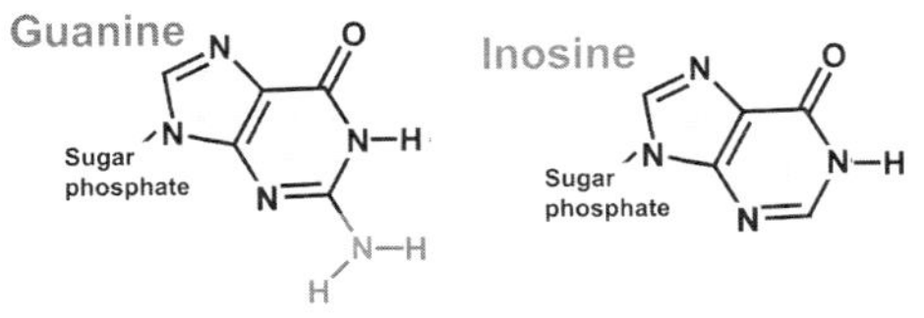

Figure 12.9 (a) Base sequence contexts investigated, where **G*** represents the 10S (+)-*trans-anti*-B[a]P-N^2-dG adduct in all sequences. The duplex I6**G7*** is identical to G6**G7*** except for the designated 2'-deoxyinosine (I). (b) Structures for guanine and inosine: the -NH$_2$ group in guanine is replaced by a hydrogen atom in inosine. For simplicity we utilize the term inosine, although formally, inosine is the nucleoside and hypoxanthine is the corresponding base.

and interesting about these sequences, and then discuss how the structural properties of the lesion in these different sequence contexts affect NER.

12.9.1.1 CG*C and TG*T Sequences

The sequence CG*C-I is the standard sequence that was used in many NMR structural studies [53]. The pairs of sequences CG*C-II and TG*T are identical, except for the C or T bases flanking the adducts G* on both sides. The CG*C and TG*T sequences are of unusual interest for several reasons. (i) The TG*T 11mer duplex has a lower thermal melting point than the 11mer CG*C duplex (by about 14–15 °C) [124] as expected from the thermodynamic properties of TA and CG base pairs [125, 126]. (ii) A single, well-defined adduct conformation is observed in CG*C duplexes [84], while the same 10S (+)-trans-anti-B[a]P-N^2-dG adduct is conformationally heterogeneous with partially ruptured base pairs on its 5′-side in the TG*T sequence context [127]. (iii) Gel electrophoresis studies show that this adduct induces a bend in the CG*C DNA duplex [81, 128] that is rigid [82], while the same lesion in the TG*T sequence context induces a highly flexible bend [82, 127]. MD simulations for the TG*T and CG*C-II duplexes provided molecular insights on these experimental observations [26]. Specifically, we observed that the TG*T duplex with the B[a]P rings 5′-directed in the minor groove manifests much greater overall dynamic mobility than the CG*C-II duplex. This is consistent with the conformational heterogeneity observed in the NMR studies [127]. The base pairs on the 5′-side of the lesion were very dynamic and exhibited episodic denaturation of one of the two Watson–Crick hydrogen bonds, in agreement with the partial rupturing of this base pair observed by the NMR methods [127]. Also the TG*T duplex showed somewhat increased and more dynamic Roll and untwisting compared to the CG*C-II duplex, consistent with the flexible bend observed only for the TG*T case. The overall greater dynamics for the TG*T sequence was also shown by greater mobility of the B[a]P ring system and greater variability in groove dimensions. By contrast, for the CG*C-II case, the presence of the lesion caused mainly just an enlarged minor groove and modest perturbation of Watson–Crick base-pairing without denaturation at the lesion site. The differences are accounted for by the intrinsically weaker stacking properties of TG compared to CG steps, which allowed for greater flexibility in the TG*T duplex. Furthermore, the weaker TA pair, with only two hydrogen bonds, compared to the CG pair, with three bonds, is also a contributing component to the enhanced flexibility. Moreover, the absence of guanine amino groups adjacent to the G* in the TG*T case allows greater mobility for the B[a]P ring system; when amino groups are present adjacent to the G* as in the CG*C-II case, they compete for space with the bulky B[a]P moiety, since both are present in the narrow minor groove, and thereby constrain the positioning of the B[a]P rings.

12.9.1.2 G6*G7, G6G7*, and I6G7* Sequences

These are all identical duplexes, except for the following differences: the G6G7* and G6*G7 duplexes are identical in composition and 10S (+)-trans-anti-B[a]P-N^2-dG adduct stereochemistry except that the lesions are positioned at either of the

12.9 Dependence of NER of the 10S (+)-trans-anti-B[a]P-N^2-dG Adduct on Base Sequence Context

(a)
G6*G7   5'- C1 A2 T3 G4 C5 G6* G7 C8 C9 T10 A11 C12 -3'
         3'- G24 T23 A22 C21 G20 C19 C18 G17 G16 A15 T14 G13 -5'

G6G7*   5'- C1 A2 T3 G4 C5 G6 G7* C8 C9 T10 A11 C12 -3'
         3'- G24 T23 A22 C21 G20 C19 C18 G17 G16 A15 T14 G13 -5'

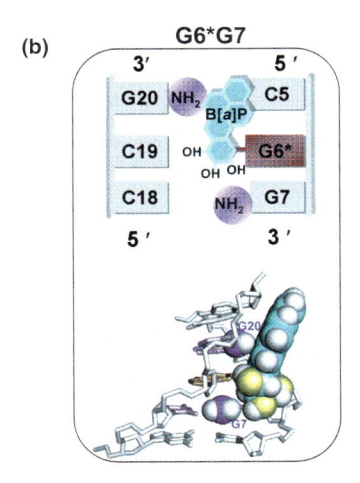

(b)

Figure 12.10 Steric competition from guanine amino groups adjacent to the bulky B[a]P residues of 10S (+)-trans-anti-B[a]P-N^2-dG lesions (**G6*** or **G7***) in the minor groove produces sequence-dependent conformational differences. (a) Base sequence and numbering scheme of **G6*G7** and G6**G7*** duplexes (the central 12mer sequence in the 135mer duplexes used in NER experiments is shown; the 12mer duplexes actually shown were used in NMR studies [79, 129]). The region including **G*** and its flanking bases is underlined to highlight the differences in the flanking bases in the otherwise chemically identical **G6*G7** and G6**G7*** duplexes. (b) Upper panel: schemes illustrating different sequence contexts and the relative positioning of the 10S (+)-trans-anti-B[a]P-N^2-dG lesion and guanine amino groups in the minor groove of B-DNA. Lower panel: representative structures of the lesions in the minor groove in the **G6*G7** and G6**G7*** duplexes, revealing orientational differences of the B[a]P aromatic rings (cyan) governed by guanine amino groups (violet) flanking the modified guanine on either the 5'- or 3'-side.

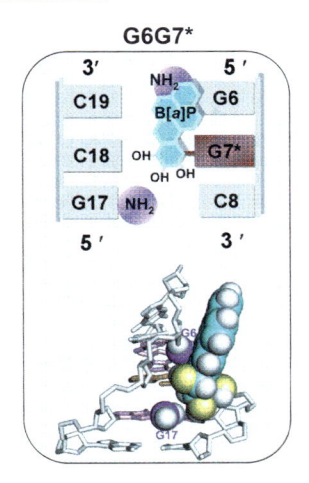

two adjacent G6 or G7 sites in … G6**G7***C8 … or … C5**G6***G7 … sequence contexts, respectively (Figure 12.9). Detailed NMR studies have shown that in both duplexes G6**G7*** and **G6*G7**, the bulky aromatic B[a]P residues are positioned in the minor groove on the 5'-side of **G*** [79]. Since the exocyclic amino groups of guanine are also positioned in the minor groove (Figure 12.4), we surmised that the bulky polycyclic aromatic B[a]P residue and the amino group of guanine must compete for the same space in the minor groove of the B-DNA duplex (Figure 12.10) [79, 129]. We also reasoned that the resolution of these steric hindrance effects might result in significant differences in the characteristics of the **G6*G7** and G6**G7*** duplexes. Indeed, these two sequences elicited our interest because the G6**G7*** duplex is characterized by an unusually slow electrophoretic mobility that is

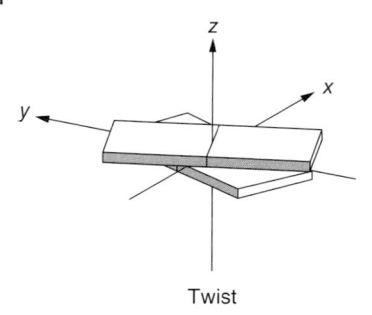

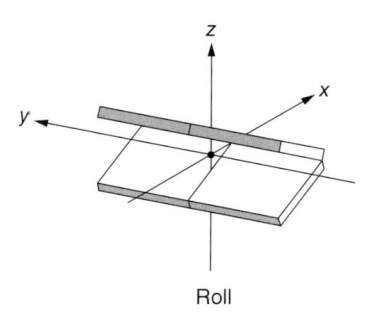

Figure 12.11 Definition of the helicoidal parameters Twist and Roll. With permission from [134].

significantly slower than that of the sequence isomer **G6*G7** duplex [78]. The NMR data combined with restrained MD simulations for these two duplexes revealed that the unusual electrophoretic properties of the G6**G7*** duplex are due to a large untwisting and increased Roll (Figure 12.11), a molecular manifestation of DNA bending [44, 45, 130], at the site of the lesion; this results from the steric crowding between the aromatic B[a]P residue and the exocyclic amino group of the flanking G6 residue, due to their competition for space in the minor groove [79]. This steric hindrance effect is resolved by the local distortion of the duplex that results in the observed prominent kink [78, 79]. Consistent with this conclusion, when G6 is replaced by inosine (I) (Figure 12.9) that is identical to guanine, but lacks the exocyclic amino group, the kink disappears since the electrophoretic mobility of the I6**G7*** duplex is the same as that of the **G6*G7** [78] or the C**G*C**-I duplexes [82 128]. The NMR characteristics of the I6**G7*** duplex have also been studied; multiple conformations were found in the I6**G7*** duplex, although two major conformers with the B[a]P residues positioned in the minor groove and oriented towards the 5′-direction of the modified strand were identified [129]. Thus, the I6**G7*** and T**G*T** duplexes [127] resemble one another since the replacement of a CG base pair by either an IC or an AT base pair destabilizes the local adduct structure, giving rise to multiple adduct conformations rather than a single dominant one as in the **G6*G7** and C**G*C** sequence contexts. The steric competition between the guanine amino groups and the B[a]P rings produce a different dis-

12.9 Dependence of NER of the 10S (+)-trans-anti-B[a]P-N²-dG Adduct on Base Sequence Context

tortion in the **G6***G7 duplex. Specifically, the NMR data revealed a significantly destabilized Watson–Crick base pair on the 5′-side of **G6***. However, this **G6***G7 duplex did not display the unusual gel mobility characteristic of a flexible bend that was seen in the G6**G7*** case.

To obtain further detailed molecular insights into these experimental findings, we carried out unrestrained MD simulations. Our simulations for the G6**G7*** duplex provided an explanation of the molecular origins of the unusual electrophoretic mobility of this duplex. Specifically, the results revealed that the enlarged Roll and untwisting (Figure 12.11) of the duplex at the G6–**G7*** step were highly flexible, consistent with the electrophoretic mobility studies that showed that the bend was flexible. In addition, the specific structural origin of this increased Roll and untwisting was uncovered from the MD simulations: steric hindrance between the B[a]P moiety on **G7*** and the amino group of G6 forces the local untwisting and increased Roll, which relieves the steric crowding (Figure 12.10). When this amino group is eliminated as in the case of I6**G7***, this phenomenon in the MD simulations was abolished, consistent with the observed lack of the anomalously slow gel mobility. The MD simulations also showed that the base pair involving I6, with only two hydrogen bonds, is more dynamic than the analogous base pair involving G6; in addition, there is episodic denaturation of one hydrogen bond at I6 in the I6**G7*** case. These structural properties can account for the observed conformational heterogeneity in the I6**G7*** duplex. The specific disturbance to the **G6***G7 duplex (i.e., destabilization of the base pair on the 5′-side of **G***) was clearly explained on a structural level from the MD simulations. We observed episodic denaturation of this base pair and determined the structural origin of this phenomenon. Specifically, the amino group of G7 crowds the B[a]P moiety on the OH-containing benzylic ring (Figure 12.10); this forces the B[a]P moiety to rotate and, in doing so, it intrudes between the Watson–Crick hydrogen bonds at the base pair on the 5′-side of **G6*** (C5:G20) (Figure 12.10). Overall, our series of studies with the same lesion in a number of sequence contexts revealed just how the steric hindrance between the B[a]P ring system and nearby guanine amino groups lead to different types of structural distortions and destabilizations that provide distinct recognition signals to the NER machinery.

Since the identical lesion positioned in different sequence contexts causes different kinds of structural DNA perturbations, it is of interest to compare the NER repair efficiencies in these different sequences. Comparing the T**G***T versus the C**G***C-I duplexes gives insights on the effects of the local stabilities surrounding **G***: these differ because TA base pairs flanking **G*** destabilize the duplexes more than CG flanking base pairs [124, 127]. In the case of the C**G***C-I and C**G***C-II duplexes, the flanking base pairs are the same, but the overall sequence context is different; therefore, comparing the NER of the same 10S (+)-*trans-anti*-B[a]P-N^2-dG adduct in these different C**G***C sequence contexts provides information about the effects of changing the more distant bases on NER efficiencies [135]. Finally, comparing the NER efficiencies of the **G6***G7, G6**G7***, and I6**G7*** sequences allows one to evaluate the effects of flexible kinks or bends and disrupted Watson–Crick hydrogen-bonding on NER efficiencies.

12.9.2
Hierarchies of Mammalian NER Recognition Signals

The relative NER efficiencies of the 10S (+)-*trans-anti*-B[a]P-N^2-dG adduct in different sequence contexts is summarized in Figure 12.12. The TG*T duplex is a better NER substrate than the CG*C-II and CG*C-I duplexes by factors of 1.5 ± 0.1 and 2.4 ± 0.2, respectively. The conformational heterogeneity and the flexible bend account for the greater NER efficiency of the TG*T compared to the CG*C-II duplex that is identical except for the bases flanking the lesion. These results also demonstrate that the sequence context beyond the flanking bases can play a role since the CG*C-II/CG*C-I NER efficiency ratio is around 1.6. The repair efficiency of the G6*G7 duplex is 2.4 ± 0.2 times greater than that of the G6G7* duplex. Taken together, the results for this series suggest that the perturbation of Watson–Crick hydrogen-bonding, as observed at the CG base pair flanking the modified G6*C base pair in the G6*G7 duplex is the most important signal for lesion recognition in this series. A flexible bend, as observed with G6G7* and TG*T, provides a less prominent recognition signal. The enlarged minor groove and modestly perturbed Watson–Crick base-pairing at the lesion site found for the CG*C sequences are the least disturbing and are weaker NER recognition signals in and of themselves, but can be correlated with other signals. Table 12.1 summarizes key structural disturbances that relate to relative NER efficiencies shown in Figure 12.12.

The sequence-dependent trend is conserved in the prokaryotic UvrABC system, in which the NER dual-incision efficiency is around 2.4-fold higher for the same 10S (+)-*trans-anti*-B[a]P-N^2-dG adduct in the TG*T case than the CG*C-II case [124]. Similarly, the G6*G7 duplex is incised more efficiently by a factor of around

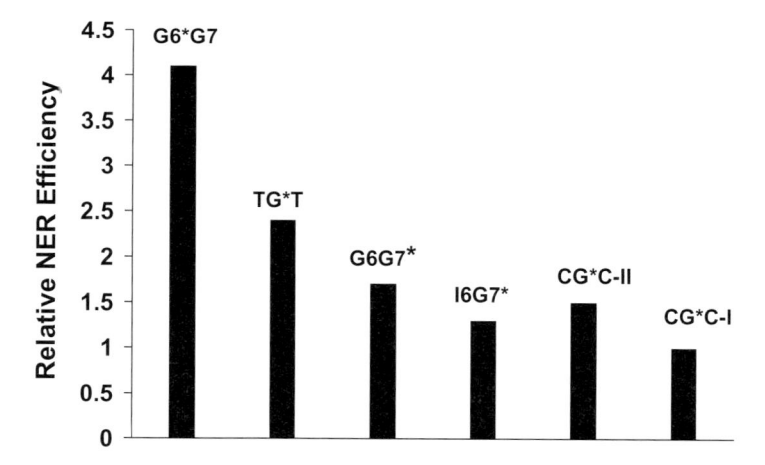

Figure 12.12 Relative NER efficiencies for the 10S (+)-*trans-anti*-B[a]P-N^2-dG adduct embedded in various sequence contexts in otherwise identical 135mer duplexes hybridized with their fully complementary strands.

Table 12.1 Key structural disturbances of B[a]P-dG adducts.

Sequence/adduct	Conformation	Prominent Watson–Crick hydrogen-bonding disturbance	Bend	Minor groove width
G6*G7 (+)-trans-	minor groove [79]	destabilized base pair 5′-flanking G* [47, 79]	rigid [78]	enlarged [47, 79]
TG*T (+)-trans-	heterogeneous minor groove [127]	partially destabilized base pair 5′-flanking G* [26, 127]	flexible [82, 127]	enlarged [26]
G6G7* (+)-trans-	minor groove [79]		flexible [78]	enlarged [47, 79]
I6G7* (+)-trans-	heterogeneous minor groove [129]	partially destabilized base pair 5′-flanking G* [79, 129]	rigid [78]	enlarged [47]
CG*C-II (+)-trans-	minor groove [26]		rigid [82]	enlarged [26]
CG*C-I (+)-trans-	minor groove [84]		rigid [131]	enlarged [38]
CG*C-I (−)-trans-	minor groove [87]		rigid [131]	enlarged [38]
CG*C-I (+)-cis-	base-displaced/ intercalated [112]	completely ruptured at G* [112]	rigid [131]	enlarged [38]

2 than the G6**G7*** duplex in the UvrABC system (Liu and Geacintov, in preparation).

12.10 Conclusions

The binding of the highly reactive metabolic diol epoxide intermediates of bay and fjord region PAHs to the exocyclic amino groups of adenine and guanine in DNA give rise to a variety of structurally different lesions that may locally distort and destabilize the DNA structure. The observed adduct conformations depend on the topology of the polycyclic aromatic ring systems and the stereochemical properties of the DNA adducts formed. The variable conformations that have been observed include intercalation with displacement of the modified and partner bases from the interior of the double helix, intercalation without base displacement, and external minor groove conformations. The conformations of the PAH diol epoxide-DNA adducts and the local structural DNA perturbations they cause depend not only on the adduct, but also on the surrounding base sequence context. These PAH-DNA adducts therefore represent ideal model substrates for probing the relationships between the structures of these DNA lesions, the local distortions and destabilization that they cause in the DNA duplexes, and the efficiencies of NER.

Correlations between the structural properties of the series of the bay region B[a]P-N^2-dG and -N^6-dA adducts, as well as selected fjord region PAH diol epoxide-N^6-dA adducts have provided new structural insights into the factors that govern the efficiencies of recognition and dual incisions in this family of DNA lesions. The highest NER efficiencies are observed in the case of B[a]P-derived guanine adducts with base-displaced intercalative conformations, where the modified guanine and partner base are displaced out of the double helix; the stereoisomeric adducts with minor groove external conformations are incised with efficiencies that are lower by a factor of 5 or more [38, 107]. The dual-incision efficiencies in the case of the base-displaced intercalated adducts were as high as those manifested by the 6-4 <TT> UV photoproduct—an excellent substrate of mammalian NER [78]. In sharp contrast, intercalative fjord region PAH-adenine adducts in DNA without base displacement are resistant to NER [113]. This resistance is attributed to the topology of the aromatic ring systems of fjord PAHs that have a nonplanar flexible aromatic ring in the fjord region; this allows for more extensive carcinogen base-stacking interactions that stabilize the site of the modified DNA duplex [39–41, 113]. On the other hand, intercalated bay region B[a]P-derived adducts with rigid aromatic ring systems are incised by the NER system with intermediate efficiencies because this rigid, extended aromatic ring system stacks less well within the DNA and is more distorting [39, 42]. In this family of intercalated DNA lesions, there is a qualitative correlation between NER efficiencies and the melting points of the duplexes that reflect the extent of destabilization of the DNA duplexes by the intercalated PAH residues [39, 40, 42, 43].

Further, but still incomplete insights into the distortions in the DNA duplexes induced by these bulky lesions were evident from an analysis of the effects of base sequence context on NER efficiencies [26, 43, 47, 78]. Destabilized Watson–Crick hydrogen-bonding and diminished base–base stacking interactions of the base pairs flanking the lesion on the 5'-side, provide a stronger recognition signal than a lesion-induced flexible kink or bend in the DNA duplex. An enlarged minor groove and modest perturbation of Watson–Crick base-pairing at the lesion site appears to be a lesser signal of lesion recognition in and of itself, but can be correlated with other signals. A multipartite model combining alterations of multiple DNA structural parameters and destabilizing features associated with the lesions can produce a relatively strong recognition signal. These results are consistent with the finding [38] that the binding of the initial lesion-recognizing NER factor XPC/hHR23B to the stereochemically different B[a]P-derived DNA adducts causes different extents and patterns of opening of the double helix that are correlated with NER dual-incision efficiencies. Our working model of recognition of bulky lesions by the NER apparatus involves local thermodynamic destabilization of the double helix near the site of the lesion that facilitates strand separation and helix opening; in turn, strand separation allows for the extrusion of the bases opposite the adduct [111] facilitating the insertion of the XPC/Rad4 β-hairpin into the helix and stabilizing the complex by the interaction of the flipped out bases with the amino acids of the Rad4 protein [37]. The formation of such a stable complex is dependent on the structure of the DNA lesions and the extent and type of DNA structural perturbations that they cause. As a result, these lesion-containing substrates are processed by the NER machinery with different efficiencies.

It is noteworthy that the NER proteins in both prokaryotic and eukaryotic systems share common structural features of β-hairpin insertion between the strands of the DNA duplex at the lesion site, as revealed by the structures of UvrB binding to damaged DNA [132, 133] and the yeast ortholog Rad4 binding to damaged DNA [37]. Therefore, in both systems, structures with more pronounced distortions and thermodynamic destabilization are likely to facilitate the β-hairpin insertion and thus produce a higher initial incision rate. Future detailed studies of the structural properties of different PAH-derived lesions with different conformations in selected base sequence contexts, and comparisons with patterns of helix opening by XPC/HR23B and dual-incision efficiencies, should provide further clues to the mechanisms underlying the recognition and excision of bulky DNA lesions by the mammalian NER apparatus.

There is still much to learn about the recognition of well-defined bulky lesions in double-stranded DNA. However, it is clear that the next challenging frontier in this field will be to determine whether similar or different rules of recognition and dual excision of these bulky DNA lesions apply in a nucleosomal and chromatin environment at the cellular level.

Acknowledgements

This research was supported by research grants from the NCI, NIH CA-099194 (N.E.G.), CA-28038 (S.B.), CA-75449 (S.B.), and CA-046533 (D.J.P).

References

1 Friedberg, E.C., Walker, G.C., Siede, W., Wood, R.D., Shchultz, R.A., and Ellenberger, T. (2005) *DNA Repair and Mutagenesis*, 2nd edn, ASM Press, Washington, DC.

2 Kraemer, K.H., Lee, M.M., and Scotto, J. (1984) DNA repair protects against cutaneous and internal neoplasia: evidence from xeroderma pigmentosum. *Carcinogenesis*, **5**, 511–514.

3 Kraemer, K.H., Lee, M.M., and Scotto, J. (1987) Xeroderma pigmentosum. Cutaneous, ocular, and neurologic abnormalities in 830 published cases. *Arch. Dermatol.*, **123**, 241–250.

4 Nance, M.A. and Berry, S.A. (1992) Cockayne syndrome: review of 140 cases. *Am. J. Med. Genet.*, **42**, 68–84.

5 Kraemer, K.H., Patronas, N.J., Schiffmann, R., Brooks, B.P., Tamura,

D., and DiGiovanna, J.J. (2007) Xeroderma pigmentosum, trichothiodystrophy and Cockayne syndrome: a complex genotype–phenotype relationship. *Neuroscience*, **145**, 1388–1396.

6 Martin, L.P., Hamilton, T.C., and Schilder, R.J. (2008) Platinum resistance: the role of DNA repair pathways. *Clin. Cancer Res.*, **14**, 1291–1295.

7 Gillet, L.C. and Scharer, O.D. (2006) Molecular mechanisms of mammalian global genome nucleotide excision repair. *Chem. Rev.*, **106**, 253–276.

8 Shuck, S.C., Short, E.A., and Turchi, J.J. (2008) Eukaryotic nucleotide excision repair: from understanding mechanisms to influencing biology. *Cell Res.*, **18**, 64–72.

9 Fousteri, M. and Mullenders, L.H. (2008) Transcription-coupled nucleotide excision repair in mammalian cells: molecular mechanisms and biological effects. *Cell Res.*, **18**, 73–84.

10 Wakasugi, M. and Sancar, A. (1998) Assembly, subunit composition, and footprint of human DNA repair excision nuclease. *Proc. Natl. Acad. Sci. USA*, **95**, 6669–6674.

11 Aboussekhra, A., Biggerstaff, M., Shivji, M.K., Vilpo, J.A., Moncollin, V., Podust, V.N., Protic, M., Hubscher, U., Egly, J.M., and Wood, R.D. (1995) Mammalian DNA nucleotide excision repair reconstituted with purified protein components. *Cell*, **80**, 859–868.

12 Sugasawa, K., Ng, J.M., Masutani, C., Iwai, S., van der Spek, P.J., Eker, A.P., Hanaoka, F., Bootsma, D., and Hoeijmakers, J.H. (1998) Xeroderma pigmentosum group C protein complex is the initiator of global genome nucleotide excision repair. *Mol. Cell*, **2**, 223–232.

13 Sugasawa, K., Okamoto, T., Shimizu, Y., Masutani, C., Iwai, S., and Hanaoka, F. (2001) A multistep damage recognition mechanism for global genomic nucleotide excision repair. *Genes Dev.*, **15**, 507–521.

14 Evans, E., Moggs, J.G., Hwang, J.R., Egly, J.M., and Wood, R.D. (1997) Mechanism of open complex and dual incision formation by human nucleotide excision repair factors. *EMBO J.*, **16**, 6559–6573.

15 Uchida, A., Sugasawa, K., Masutani, C., Dohmae, N., Araki, M., Yokoi, M., Ohkuma, Y., and Hanaoka, F. (2002) The carboxy-terminal domain of the XPC protein plays a crucial role in nucleotide excision repair through interactions with transcription factor IIH. *DNA Repair*, **1**, 449–461.

16 Yokoi, M., Masutani, C., Maekawa, T., Sugasawa, K., Ohkuma, Y., and Hanaoka, F. (2000) The xeroderma pigmentosum group C protein complex XPC-HR23B plays an important role in the recruitment of transcription factor IIH to damaged DNA. *J. Biol. Chem.*, **275**, 9870–9875.

17 Riedl, T., Hanaoka, F., and Egly, J.M. (2003) The comings and goings of nucleotide excision repair factors on damaged DNA. *EMBO J.*, **22**, 5293–5303.

18 Tapias, A., Auriol, J., Forget, D., Enzlin, J.H., Scharer, O.D., Coin, F., Coulombe, B., and Egly, J.M. (2004) Ordered conformational changes in damaged DNA induced by nucleotide excision repair factors. *J. Biol. Chem.*, **279**, 19074–19083.

19 Reardon, J.T. and Sancar, A. (2005) Nucleotide excision repair. *Prog. Nucleic Acid Res. Mol. Biol.*, **79**, 183–235.

20 Truglio, J.J., Croteau, D.L., Van Houten, B., and Kisker, C. (2006) Prokaryotic nucleotide excision repair: the UvrABC system. *Chem. Rev.*, **106**, 233–252.

21 Van Houten, B. and Snowden, A. (1993) Mechanism of action of the *Escherichia coli* UvrABC nuclease: clues to the damage recognition problem. *Bioessays*, **15**, 51–59.

22 Gunz, D., Hess, M.T., and Naegeli, H. (1996) Recognition of DNA adducts by human nucleotide excision repair. Evidence for a thermodynamic probing mechanism. *J. Biol. Chem.*, **271**, 25089–25098.

23 Wood, R.D. (1999) DNA damage recognition during nucleotide excision repair in mammalian cells. *Biochimie*, **81**, 39–44.

24 Scharer, O.D. (2008) A molecular basis for damage recognition in eukaryotic

nucleotide excision repair. *ChemBioChem*, **9**, 21–23.
25 Isaacs, R.J. and Spielmann, H.P. (2004) A model for initial DNA lesion recognition by NER and MMR based on local conformational flexibility. *DNA Repair*, **3**, 455–464.
26 Cai, Y., Patel, D.J., Geacintov, N.E., and Broyde, S. (2007) Dynamics of a benzo[a]pyrene-derived guanine DNA lesion in TGT and CGC sequence contexts: enhanced mobility in TGT explains conformational heterogeneity, flexible bending, and greater susceptibility to nucleotide excision repair. *J. Mol. Biol.*, **374**, 292–305.
27 Missura, M., Buterin, T., Hindges, R., Hubscher, U., Kasparkova, J., Brabec, V., and Naegeli, H. (2001) Double-check probing of DNA bending and unwinding by XPA-RPA: an architectural function in DNA repair. *EMBO J.*, **20**, 3554–3564.
28 Blagoev, K.B., Alexandrov, B.S., Goodwin, E.H., and Bishop, A.R. (2006) Ultra-violet light induced changes in DNA dynamics may enhance TT-dimer recognition. *DNA Repair*, **5**, 863–867.
29 Maillard, O., Camenisch, U., Blagoev, K.B., and Naegeli, H. (2008) Versatile protection from mutagenic DNA lesions conferred by bipartite recognition in nucleotide excision repair. *Mutat. Res.*, **658**, 271–286.
30 Maillard, O., Camenisch, U., Clement, F.C., Blagoev, K.B., and Naegeli, H. (2007) DNA repair triggered by sensors of helical dynamics. *Trends Biochem. Sci.*, **32**, 494–499.
31 Yang, W. (2006) Poor base stacking at DNA lesions may initiate recognition by many repair proteins. *DNA Repair*, **5**, 654–666.
32 Yang, W. (2008) Structure and mechanism for DNA lesion recognition. *Cell Res.*, **18**, 184–197.
33 Hess, M.T., Schwitter, U., Petretta, M., Giese, B., and Naegeli, H. (1997) Bipartite substrate discrimination by human nucleotide excision repair. *Proc. Natl. Acad. Sci. USA*, **94**, 6664–6669.
34 Fujiwara, Y., Masutani, C., Mizukoshi, T., Kondo, J., Hanaoka, F., and Iwai, S. (1999) Characterization of DNA recognition by the human UV-damaged DNA-binding protein. *J. Biol. Chem.*, **274**, 20027–20033.
35 Sugasawa, K., Shimizu, Y., Iwai, S., and Hanaoka, F. (2002) A molecular mechanism for DNA damage recognition by the xeroderma pigmentosum group C protein complex. *DNA Repair*, **1**, 95–107.
36 Clement, F.C., Camenisch, U., Fei, J., Kaczmarek, N., Mathieu, N., and Naegeli, H. (2010) Dynamic two-stage mechanism of versatile DNA damage recognition by xeroderma pigmentosum group C protein. *Mutat Res.*, **685**, 21–28.
37 Min, J.H. and Pavletich, N.P. (2007) Recognition of DNA damage by the Rad4 nucleotide excision repair protein. *Nature*, **449**, 570–575.
38 Mocquet, V., Kropachev, K., Kolbanovskiy, M., Kolbanovskiy, A., Tapias, A., Cai, Y., Broyde, S., Geacintov, N.E., and Egly, J.M. (2007) The human DNA repair factor XPC-HR23B distinguishes stereoisomeric benzo[a]pyrenyl-DNA lesions. *EMBO J.*, **26**, 2923–2932.
39 Geacintov, N.E., Broyde, S., Buterin, T., Naegeli, H., Wu, M., Yan, S., and Patel, D.J. (2002) Thermodynamic and structural factors in the removal of bulky DNA adducts by the nucleotide excision repair machinery. *Biopolymers*, **65**, 202–210.
40 Geacintov, N.E., Naegeli, H., Patel, D.J., and Broyde, S. (2006) Structural aspects of polycyclic aromatic carcinogen-damaged DNA and its recognition by NER proteins, in *DNA Damage and Recognition* (W. Siede, Y.W. Kow, and P.W. Doetsch), Taylor & Francis, London, pp. 263–296.
41 Wu, M., Yan, S., Patel, D.J., Geacintov, N.E., and Broyde, S. (2002) Relating repair susceptibility of carcinogen-damaged DNA with structural distortion and thermodynamic stability. *Nucleic Acids Res.*, **30**, 3422–3432.
42 Yan, S., Shapiro, R., Geacintov, N.E., and Broyde, S. (2001) Stereochemical, structural, and thermodynamic origins of stability differences between stereoisomeric benzo[a]pyrene diol epoxide deoxyadenosine adducts in a

DNA mutational hot spot sequence. *J. Am. Chem. Soc.*, **123**, 7054–7066.

43 Yan, S., Wu, M., Buterin, T., Naegeli, H., Geacintov, N.E., and Broyde, S. (2003) Role of base sequence context in conformational equilibria and nucleotide excision repair of benzo[a]pyrene diol epoxide-adenine adducts. *Biochemistry*, **42**, 2339–2354.

44 Gorin, A.A., Zhurkin, V.B., and Olson, W.K. (1995) B-DNA twisting correlates with base-pair morphology. *J. Mol. Biol.*, **247**, 34–48.

45 Olson, W.K., Gorin, A.A., Lu, X.J., Hock, L.M., and Zhurkin, V.B. (1998) DNA sequence-dependent deformability deduced from protein–DNA crystal complexes. *Proc. Natl. Acad. Sci. USA*, **95**, 11163–11168.

46 Dickerson, R.E. (1989) Definitions and nomenclature of nucleic acid structure components. *Nucleic Acids Res.*, **17**, 1797–1803.

47 Cai, Y., Patel, D.J., Geacintov, N.E., and Broyde, S. (2009) Differential nucleotide excision repair susceptibility of bulky DNA adducts in different sequence contexts: hierarchies of recognition signals. *J. Mol. Biol.*, **385**, 30–44.

48 Bostrom, C.E., Gerde, P., Hanberg, A., Jernstrom, B., Johansson, C., Kyrklund, T., Rannug, A., Tornqvist, M., Victorin, K., and Westerholm, R. (2002) Cancer risk assessment, indicators, and guidelines for polycyclic aromatic hydrocarbons in the ambient air. Environ. *Health Perspect.*, **110** (Suppl. 3), 451–488.

49 Luch, A. (ed.) (2005) *The Carcinogenic Effects of Polycyclic Aromatic Hydrocarbons*, Imperial College Press, London.

50 Conney, A.H. (1982) Induction of microsomal enzymes by foreign chemicals and carcinogenesis by polycyclic aromatic hydrocarbons: G. H. A. Clowes Memorial Lecture. *Cancer Res.*, **42**, 4875–4917.

51 Courter, L.A., Luch, A., Musafia-Jeknic, T., Arlt, V.M., Fischer, K., Bildfell, R., Pereira, C., Phillips, D.H., Poirier, M.C., and Baird, W.M. (2008) The influence of diesel exhaust on polycyclic aromatic hydrocarbon-induced DNA damage, gene expression, and tumor initiation in Sencar mice *in vivo*. *Cancer Lett.*, **265**, 135–147.

52 Luch, A. (2005) Nature and nurture – lessons from chemical carcinogenesis. *Nat. Rev. Cancer*, **5**, 113–125.

53 Geacintov, N.E., Cosman, M., Hingerty, B.E., Amin, S., Broyde, S., and Patel, D.J. (1997) NMR solution structures of stereoisometric covalent polycyclic aromatic carcinogen-DNA adduct: principles, patterns, and diversity. *Chem. Res. Toxicol.*, **10**, 111–146.

54 Xie, X.M., Geacintov, N.E., and Broyde, S. (1999) Stereochemical origin of opposite orientations in DNA adducts derived from enantiomeric *anti*-benzo[a]pyrene diol epoxides with different tumorigenic potentials. *Biochemistry*, **38**, 2956–2968.

55 Szeliga, J. and Dipple, A. (1998) DNA adduct formation by polycyclic aromatic hydrocarbon dihydrodiol epoxides. *Chem. Res. Toxicol.*, **11**, 1–11.

56 Phillips, D.H. (1983) Fifty years of benzo[a]pyrene. *Nature*, **303**, 468–472.

57 Nesnow, S., Ross, J.A., Mass, M.J., and Stoner, G.D. (1998) Mechanistic relationships between DNA adducts, oncogene mutations, and lung tumorigenesis in strain A mice. *Exp. Lung Res.*, **24**, 395–405.

58 Ross, J.A., Nelson, G.B., Wilson, K.H., Rabinowitz, J.R., Galati, A., Stoner, G.D., Nesnow, S., and Mass, M.J. (1995) Adenomas induced by polycyclic aromatic hydrocarbons in strain A/J mouse lung correlate with time-integrated DNA adduct levels. *Cancer Res.*, **55**, 1039–1044.

59 Ross, J.A. and Nesnow, S. (1999) Polycyclic aromatic hydrocarbons: correlations between DNA adducts and *ras* oncogene mutations. *Mutat. Res.*, **424**, 155–166.

60 Amin, S., Desai, D., Dai, W., Harvey, R.G., and Hecht, S.S. (1995) Tumorigenicity in newborn mice of fjord region and other sterically hindered diol epoxides of benzo[g]chrysene, dibenzo[a,l]pyrene (dibenzo[def,p]chrysene), 4H-cyclopenta[def]chrysene and fluoranthene. *Carcinogenesis*, **16**, 2813–2817.

61 Amin, S., Krzeminski, J., Rivenson, A., Kurtzke, C., Hecht, S.S., and El-Bayoumy, K. (1995) Mammary carcinogenicity in female CD rats of fjord region diol epoxides of benzo[c]phenanthrene, benzo[g]chrysene and dibenzo[a,l]pyrene. *Carcinogenesis*, **16**, 1971–1974.

62 Ronai, Z.A., Gradia, S., el-Bayoumy, K., Amin, S., and Hecht, S.S. (1994) Contrasting incidence of *ras* mutations in rat mammary and mouse skin tumors induced by *anti*-benzo[c]phenanthrene-3,4-diol-1,2-epoxide. *Carcinogenesis*, **15**, 2113–2116.

63 Hecht, S.S., el-Bayoumy, K., Rivenson, A., and Amin, S. (1994) Potent mammary carcinogenicity in female CD rats of a fjord region diol-epoxide of benzo[c]phenanthrene compared to a bay region diol-epoxide of benzo[a]pyrene. *Cancer Res.*, **54**, 21–24.

64 Cavalieri, E.L., Higginbotham, S., RamaKrishna, N.V., Devanesan, P.D., Todorovic, R., Rogan, E.G., and Salmasi, S. (1991) Comparative dose–response tumorigenicity studies of dibenzo[a,l]pyrene versus 7,12-dimethylbenz[a]anthracene, benzo[a]pyrene and two dibenzo[a,l]pyrene dihydrodiols in mouse skin and rat mammary gland. *Carcinogenesis*, **12**, 1939–1944.

65 Luch, A. and Baird, W.M. (2005) Metabolic activation and detoxification of polycyclic aromatic hydrocarbons, in *The Carcinogenic Effects of Polycyclic Aromatic Hydrocarbons* (ed. A. Luch), Imperial College Press, London, pp. 19–96.

66 Cheng, S.C., Hilton, B.D., Roman, J.M., and Dipple, A. (1989) DNA adducts from carcinogenic and noncarcinogenic enantiomers of benzo[a]pyrene dihydrodiol epoxide. *Chem. Res. Toxicol.*, **2**, 334–340.

67 Koreeda, M., Moore, P.D., Wislocki, P.G., Levin, W., Yagi, H., and Jerina, D.M. (1978) Binding of benzo[a]pyrene 7,8-diol-9,10-epoxides to DNA, RNA, and protein of mouse skin occurs with high stereoselectivity. *Science*, **199**, 778–781.

68 Weinstein, I.B., Jeffrey, A.M., Jennette, K.W., Blobstein, S.H., Harvey, R.G., Harris, C., Autrup, H., Kasai, H., and Nakanishi, K. (1976) Benzo[a]pyrene diol epoxides as intermediates in nucleic acid binding *in vitro* and *in vivo*. *Carcinogenesis*, **5**, 1421–1430.

69 Katz, A.K., Carrell, H.L., and Glusker, J.P. (1998) Dibenzo[a,l]pyrene (dibenzo[def,p]chrysene): fjord-region distortions. *Carcinogenesis*, **19**, 1641–1648.

70 Dipple, A., Pigott, M.A., Agarwal, S.K., Yagi, H., Sayer, J.M., and Jerina, D.M. (1987) Optically active benzo[c]phenanthrene diol epoxides bind extensively to adenine in DNA. *Nature*, **327**, 535–536.

71 Ralston, S.L., Seidel, A., Luch, A., Platt, K.L., and Baird, W.M. (1995) Stereoselective activation of dibenzo[a,l]pyrene to (−)-*anti* (11R,12S,13S,14R)- and (+)-*syn* (11S,12R,13S,14R)-11,12-diol-13,14-epoxides which bind extensively to deoxyadenosine residues of DNA in the human mammary carcinoma cell line MCF-7. *Carcinogenesis*, **16**, 2899–2907.

72 Jankowiak, R., Ariese, F., Hewer, A., Luch, A., Zamzow, D., Hughes, N.C., Phillips, D., Seidel, A., Platt, K.L., Oesch, F., *et al.* (1998) Structure, conformations, and repair of DNA adducts from dibenzo[a,l]pyrene: ^{32}P-postlabeling and fluorescence studies. *Chem. Res. Toxicol.*, **11**, 674–685.

73 Nesnow, S., Davis, C., Nelson, G., Ross, J.A., Allison, J., Adams, L., and King, L.C. (1997) Comparison of the morphological transforming activities of dibenzo[a,l]pyrene and benzo[a]pyrene in C3H10T1/2CL8 cells and characterization of the dibenzo[a,l]pyrene-DNA adducts. *Carcinogenesis*, **18**, 1973–1978.

74 Luch, A., Friesel, H., Seidel, A., and Platt, K.L. (1999) Tumor-initiating activity of the (+)-(S,S)- and (−)-(R,R)-enantiomers of *trans*-11,12-dihydroxy-11,12-dihydrodibenzo[a,l]pyrene in mouse skin. *Cancer Lett.*, **136**, 119–128.

75 Luch, A. (2009) On the impact of the molecule structure in chemical carcinogenesis. *EXS*, **99**, 151–179.

76 Mahadevan, B., Dashwood, W.M., Luch, A., Pecaj, A., Doehmer, J.,

Seidel, A., Pereira, C., and Baird, W.M. (2003) Mutations induced by (−)-*anti*-11R,12S-dihydrodiol 13S,14R-epoxide of dibenzo[*a,l*]pyrene in the coding region of the hypoxanthine phosphoribosyltransferase (*Hprt*) gene in Chinese hamster V79 cells. *Environ. Mol. Mutagen.*, **41**, 131–139.

77 Yoon, J.H., Besaratinia, A., Feng, Z., Tang, M.S., Amin, S., Luch, A., and Pfeifer, G.P. (2004) DNA damage, repair, and mutation induction by (+)-*syn* and (−)-*anti*-dibenzo[*a,l*]pyrene-11,12-diol-13,14-epoxides in mouse cells. *Cancer Res.*, **64**, 7321–7328.

78 Kropachev, K., Kolbanovskii, M., Cai, Y., Rodriguez, F., Kolbanovskii, A., Liu, Y., Zhang, L., Amin, S., Patel, D., Broyde, S., *et al.* (2009) The sequence dependence of human nucleotide excision repair efficiencies of benzo[*a*]pyrene-derived DNA lesions: insights into the structural factors that favor dual incisions. *J. Mol. Biol.*, **386**, 1193–1203.

79 Rodríguez, F.A., Cai, Y., Lin, C., Tang, Y., Kolbanovskiy, A., Amin, S., Patel, D.J., Broyde, S., and Geacintov, N.E. (2007) Exocyclic amino groups of flanking guanines govern sequence-dependent adduct conformations and local structural distortions for minor groove-aligned benzo[*a*]pyrenyl-guanine lesions in a GG mutation hotspot context. *Nucleic Acids Res.*, **35**, 1555–1568.

80 Ruan, Q., Kolbanovskiy, A., Zhuang, P., Chen, J., Krzeminski, J., Amin, S., and Geacintov, N.E. (2002) Synthesis and characterization of site-specific and stereoisomeric fjord dibenzo[*a,l*]pyrene diol epoxide-N^6)-adenine adducts: unusual thermal stabilization of modified DNA duplexes. *Chem. Res. Toxicol.*, **15**, 249–261.

81 Liu, T., Xu, J., Tsao, H., Li, B., Xu, R., Yang, C., Amin, S., Moriya, M., and Geacintov, N.E. (1996) Base sequence-dependent bends in site-specific benzo[*a*]pyrene diol epoxide-modified oligonucleotide duplexes. *Chem. Res. Toxicol.*, **9**, 255–261.

82 Tsao, H., Mao, B., Zhuang, P., Xu, R., Amin, S., and Geacintov, N.E. (1998) Sequence dependence and characteristics of bends induced by site-specific polynuclear aromatic carcinogen-deoxyguanosine lesions in oligonucleotides. *Biochemistry*, **37**, 4993–5000.

83 Cosman, M., de los Santos, C., Fiala, R., Hingerty, B.E., Ibanez, V., Luna, E., Harvey, R., Geacintov, N.E., Broyde, S., and Patel, D.J. (1993) Solution conformation of the (+)-*cis-anti*-[BP]dG adduct in a DNA duplex: intercalation of the covalently attached benzo[*a*]pyrenyl ring into the helix and displacement of the modified deoxyguanosine. *Biochemistry*, **32**, 4145–4155.

84 Cosman, M., de los Santos, C., Fiala, R., Hingerty, B.E., Singh, S.B., Ibanez, V., Margulis, L.A., Live, D., Geacintov, N.E., Broyde, S., *et al.* (1992) Solution conformation of the major adduct between the carcinogen (+)-*anti*-benzo[*a*]pyrene diol epoxide and DNA. *Proc. Natl. Acad. Sci. USA*, **89**, 1914–1918.

85 Cosman, M., Fiala, R., Hingerty, B.E., Laryea, A., Lee, H., Harvey, R.G., Amin, S., Geacintov, N.E., Broyde, S., and Patel, D. (1993) Solution conformation of the (+)-*trans-anti*-[BPh]dA adduct opposite dT in a DNA duplex: intercalation of the covalently attached benzo[*c*]phenanthrene to the 5′-side of the adduct site without disruption of the modified base pair. *Biochemistry*, **32**, 12488–12497.

86 Cosman, M., Laryea, A., Fiala, R., Hingerty, B.E., Amin, S., Geacintov, N.E., Broyde, S., and Patel, D.J. (1995) Solution conformation of the (−)-*trans-anti*-benzo[*c*]phenanthrene-dA ([BPh]dA) adduct opposite dT in a DNA duplex: intercalation of the covalently attached benzo[*c*]phenanthrenyl ring to the 3′-side of the adduct site and comparison with the (+)-*trans-anti*-[BPh]dA opposite dT stereoisomer. *Biochemistry*, **34**, 1295–1307.

87 de los Santos, C., Cosman, M., Hingerty, B.E., Ibanez, V., Margulis, L.A., Geacintov, N.E., Broyde, S., and Patel, D.J. (1992) Influence of benzo[*a*]pyrene diol epoxide chirality on solution conformations of DNA covalent adducts: the (−)-*trans-anti*-[BP]G · C

adduct structure and comparison with the (+)-*trans-anti*-[BP]G · C enantiomer. *Biochemistry*, **31**, 5245–5252.

88 Lin, C.H., Huang, X., Kolbanovskii, A., Hingerty, B.E., Amin, S., Broyde, S., Geacintov, N.E., and Patel, D.J. (2001) Molecular topology of polycyclic aromatic carcinogens determines DNA adduct conformation: a link to tumorigenic activity. *J. Mol. Biol.*, **306**, 1059–1080.

89 Suri, A.K., Mao, B., Amin, S., Geacintov, N.E., and Patel, D.J. (1999) Solution conformation of the (+)-*trans-anti*-benzo[g]chrysene-dA adduct opposite dT in a DNA duplex. *J. Mol. Biol.*, **292**, 289–307.

90 Case, D.A., Darden, T.A., Cheatham, I.T.E., Simmerling, C.L., Wang, J., Duke, R.E., Luo, R., Crowley, M., Walker, R., Zhang, W., *et al.* (2008) *AMBER 9, University of California*, San Francisco, CA.

91 Kollman, P.A., Massova, I., Reyes, C., Kuhn, B., Huo, S., Chong, L., Lee, M., Lee, T., Duan, Y., Wang, W., *et al.* (2000) Calculating structures and free energies of complex molecules: combining molecular mechanics and continuum models. *Acc. Chem. Res.*, **33**, 889–897.

92 Mao, B., Gu, Z., Gorin, A., Chen, J., Hingerty, B.E., Amin, S., Broyde, S., Geacintov, N.E., and Patel, D.J. (1999) Solution structure of the (+)-*cis-anti*-benzo[a]pyrene-dA ([BP]dA) adduct opposite dT in a DNA duplex. *Biochemistry*, **38**, 10831–10842.

93 Lukin, M. and de Los Santos, C. (2006) NMR structures of damaged DNA. *Chem. Rev.*, **106**, 607–686.

94 Yeh, H.J., Sayer, J.M., Liu, X., Altieri, A.S., Byrd, R.A., Lakshman, M.K., Yagi, H., Schurter, E.J., Gorenstein, D.G., and Jerina, D.M. (1995) NMR solution structure of a nonanucleotide duplex with a dG mismatch opposite a 10S adduct derived from *trans* addition of a deoxyadenosine N^6-amino group to (+)-(7R,8S,9S,10R)-7,8-dihydroxy-9,10-epoxy-7,8,9,10-tetrahydrobenzo[a]pyrene: an unusual *syn* glycosidic torsion angle at the modified dA. *Biochemistry*, **34**, 13570–13581.

95 Schurter, E.J., Yeh, H.J., Sayer, J.M., Lakshman, M.K., Yagi, H., Jerina, D.M., and Gorenstein, D.G. (1995) NMR solution structure of a nonanucleotide duplex with a dG mismatch opposite a 10R adduct derived from *trans* addition of a deoxyadenosine N^6-amino group to (−)-(7S,8R,9R,10S)-7,8-dihydroxy-9,10-epoxy-7,8,9,10-tetrahydrobenzo[a]pyrene. *Biochemistry*, **34**, 1364–1375.

96 Schurter, E.J., Sayer, J.M., Oh-hara, T., Yeh, H.J., Yagi, H., Luxon, B.A., Jerina, D.M., and Gorenstein, D.G. (1995) Nuclear magnetic resonance solution structure of an undecanucleotide duplex with a complementary thymidine base opposite a 10R adduct derived from *trans* addition of a deoxyadenosine N^6-amino group to (−)-(7R,8S,9R,10S)-7,8-dihydroxy-9,10-epoxy-7,8,9,10-tetrahydrobenzo[a]pyrene. *Biochemistry*, **34**, 9009–9020.

97 Volk, D.E., Rice, J.S., Luxon, B.A., Yeh, H.J., Liang, C., Xie, G., Sayer, J.M., Jerina, D.M., and Gorenstein, D.G. (2000) NMR evidence for *syn–anti* interconversion of a *trans* opened (10R)-dA adduct of benzo[a]pyrene (7S,8R)-diol (9R,10S)-epoxide in a DNA duplex. *Biochemistry*, **39**, 14040–14053.

98 Zegar, I.S., Kim, S.J., Johansen, T.N., Horton, P.J., Harris, C.M., Harris, T.M., and Stone, M.P. (1996) Adduction of the human N-*ras* codon 61 sequence with (−)-(7S,8R,9R,10S)-7,8-dihydroxy-9,10-epoxy-7,8,9,10-tetrahydrobenzo[a]pyrene: structural refinement of the intercalated *SRSR*(61,2) (−)-(7S,8R,9S,10R)-N^6-[10-(7,8,9,10-tetrahydrobenzo[a]pyrenyl)]-2′-deoxyadenosyl adduct from ^{1}H NMR. *Biochemistry*, **35**, 6212–6224.

99 Xie, X.M., Geacintov, N.E., and Broyde, S. (1999) Origins of conformational differences between *cis* and *trans* DNA adducts derived from enantiomeric *anti*-benzo[a]pyrene diol epoxides. *Chem. Res. Toxicol.*, **12**, 597–609.

100 Feng, B., Voehler, M., Zhou, L., Passarelli, M., Harris, C.M., Harris, T.M., and Stone, M.P. (1996) Major groove (S)-α-(N^6-adenyl)styrene oxide adducts in an oligodeoxynucleotide containing the human N-*ras* codon 61 sequence: conformations of the S(61,2)

and S(61,3) sequence isomers from ^{1}H NMR. *Biochemistry*, **35**, 7316–7329.

101 Singh, S.B., Hingerty, B.E., Singh, U.C., Greenberg, J.P., Geacintov, N.E., and Broyde, S. (1991) Structures of the (+)- and (−)-*trans*-7,8-dihydroxy-anti-9,10-epoxy-7,8,9,10-tetrahydrobenzo[a]pyrene adducts to guanine-N^2 in a duplex dodecamer. *Cancer Res.*, **51**, 3482–3492.

102 Cosman, M., Hingerty, B.E., Luneva, N., Amin, S., Geacintov, N.E., Broyde, S., and Patel, D.J. (1996) Solution conformation of the (−)-*cis-anti*-benzo[a]pyrenyl-dG adduct opposite dC in a DNA duplex: intercalation of the covalently attached BP ring into the helix with base displacement of the modified deoxyguanosine into the major groove. *Biochemistry*, **35**, 9850–9863.

103 Marky, L.A., Rentzeperis, D., Luneva, N.P., Cosman, M., Geacintoy, N.E., and Kupke, D.W. (1996) Differential hydration thermodynamics of stereoisomeric DNA-benzo[a]pyrene adducts derived from diol epoxide enantiomers with different tumorigenic potentials. *J. Am. Chem. Soc.*, **118**, 3804–3810.

104 Arghavani, M.B., SantaLucia, J., Jr., and Romano, L.J. (1998) Effect of mismatched complementary strands and 5′-change in sequence context on the thermodynamics and structure of benzo[a]pyrene-modified oligonucleotides. *Biochemistry*, **37**, 8575–8583.

105 Marky, L.A. and Breslauer, K.J. (1987) Calculating thermodynamic data for transitions of any molecularity from equilibrium melting curves. *Biopolymers*, **26**, 1601–1620.

106 Zegar, I.S., Chary, P., Jabil, R.J., Tamura, P.J., Johansen, T.N., Lloyd, R.S., Harris, C.M., Harris, T.M., and Stone, M.P. (1998) Multiple conformations of an intercalated (−)-(7S,8R,9S,10R)-N^6-[10-(7,8,9,10-tetrahydrobenzo[a]pyrenyl)]-2′-deoxyadenosyl adduct in the N-*ras* codon 61 sequence. *Biochemistry*, **37**, 16516–16528.

107 Hess, M.T., Gunz, D., Luneva, N., Geacintov, N.E., and Naegeli, H. (1997) Base pair conformation-dependent excision of benzo[a]pyrene diol epoxide-guanine adducts by human nucleotide excision repair enzymes. *Mol. Cell. Biol.*, **17**, 7069–7076.

108 Friedberg, E.C., Walker, G.C., and Siede, W. (1995) *DNA Repair and Mutagenesis*, ASM Press, Washington, DC.

109 Szymkowski, D.E., Lawrence, C.W., and Wood, R.D. (1993) Repair by human cell extracts of single (6-4) and cyclobutane thymine–thymine photoproducts in DNA. *Proc. Natl. Acad. Sci. USA*, **90**, 9823–9827.

110 Banerjee, S.K., Christensen, R.B., Lawrence, C.W., and LeClerc, J.E. (1988) Frequency and spectrum of mutations produced by a single *cis-syn* thymine–thymine cyclobutane dimer in a single-stranded vector. *Proc. Natl. Acad. Sci. USA*, **85**, 8141–8145.

111 Buterin, T., Meyer, C., Giese, B., and Naegeli, H. (2005) DNA quality control by conformational readout on the undamaged strand of the double helix. *Chem. Biol.*, **12**, 913–922.

112 Cosman, M., Fiala, R., Hingerty, B.E., Amin, S., Geacintov, N.E., Broyde, S., and Patel, D.J. (1994) Solution conformation of the (+)-*cis-anti*-[BP]dG adduct opposite a deletion site in a DNA duplex: intercalation of the covalently attached benzo[a]pyrene into the helix with base displacement of the modified deoxyguanosine into the minor groove. *Biochemistry*, **33**, 11518–11527.

113 Buterin, T., Hess, M.T., Luneva, N., Geacintov, N.E., Amin, S., Kroth, H., Seidel, A., and Naegeli, H. (2000) Unrepaired fjord region polycyclic aromatic hydrocarbon-DNA adducts in *ras* codon 61 mutational hot spots. *Cancer Res.*, **60**, 1849–1856.

114 Schwartz, J.L., Rice, J.S., Luxon, B.A., Sayer, J.M., Xie, G., Yeh, H.J., Liu, X., Jerina, D.M., and Gorenstein, D.G. (1997) Solution structure of the minor conformer of a DNA duplex containing a dG mismatch opposite a benzo[a]pyrene diol epoxide/dA adduct: glycosidic rotation from *syn* to *anti* at the modified deoxyadenosine. *Biochemistry*, **36**, 11069–11076.

115 Laryea, A., Cosman, M., Lin, J.M., Liu, T., Agarwal, R., Smirnov, S., Amin, S., Harvey, R.G., Dipple, A., and Geacintov, N.E. (1995) Direct synthesis and characterization of site-specific adenosyl adducts derived from the binding of a 3,4-dihydroxy-1,2-epoxybenzo[c]phenanthrene stereoisomer to an 11-mer oligodeoxyribonucleotide. *Chem. Res. Toxicol.*, **8**, 444–454.

116 Packer, M.J., Dauncey, M.P., and Hunter, C.A. (2000) Sequence-dependent DNA structure: tetranucleotide conformational maps. *J. Mol. Biol.*, **295**, 85–103.

117 Packer, M.J., Dauncey, M.P., and Hunter, C.A. (2000) Sequence-dependent DNA structure: dinucleotide conformational maps. *J. Mol. Biol.*, **295**, 71–83.

118 Calladine, C.R. (1982) Mechanics of sequence-dependent stacking of bases in B-DNA. *J. Mol. Biol.*, **161**, 343–352.

119 Lyubchenko, Y.L., Shlyakhtenko, L.S., Appella, E., and Harrington, R.E. (1993) CA runs increase DNA flexibility in the complex of lambda Cro protein with the OR3 site. *Biochemistry*, **32**, 4121–4127.

120 Young, M.A., Ravishanker, G., Beveridge, D.L., and Berman, H.M. (1995) Analysis of local helix bending in crystal structures of DNA oligonucleotides and DNA–protein complexes. *Biophys. J.*, **68**, 2454–2468.

121 Sarai, A., Mazur, J., Nussinov, R., and Jernigan, R.L. (1989) Sequence dependence of DNA conformational flexibility. *Biochemistry*, **28**, 7842–7849.

122 Bolshoy, A., McNamara, P., Harrington, R.E., and Trifonov, E.N. (1991) Curved DNA without A–A: experimental estimation of all 16 DNA wedge angles. *Proc. Natl. Acad. Sci. USA*, **88**, 2312–2316.

123 Ruan, Q., Zhuang, P., Li, S., Perlow, R., Srinivasan, A.R., Lu, X.J., Broyde, S., Olson, W.K., and Geacintov, N.E. (2001) Base sequence effects in bending induced by bulky carcinogen-DNA adducts: experimental and computational analysis. *Biochemistry*, **40**, 10458–10472.

124 Ruan, Q., Liu, T., Kolbanovskiy, A., Liu, Y., Ren, J., Skorvaga, M., Zou, Y., Lader, J., Malkani, B., Amin, S., *et al.* (2007) Sequence context–and temperature-dependent nucleotide excision repair of a benzo[*a*]pyrene diol epoxide-guanine DNA adduct catalyzed by thermophilic UvrABC proteins. *Biochemistry*, **46**, 7006–7015.

125 Breslauer, K.J., Frank, R., Blocker, H., and Marky, L.A. (1986) Predicting DNA duplex stability from the base sequence. *Proc. Natl. Acad. Sci. USA*, **83**, 3746–3750.

126 SantaLucia, J., Jr., Allawi, H.T., and Seneviratne, P.A. (1996) Improved nearest-neighbor parameters for predicting DNA duplex stability. *Biochemistry*, **35**, 3555–3562.

127 Xu, R., Mao, B., Amin, S., and Geacintov, N.E. (1998) Bending and circularization of site-specific and stereoisomeric carcinogen–DNA adducts. *Biochemistry*, **37**, 769–778.

128 Suh, M., Ariese, F., Small, G.J., Jankowiak, R., Liu, T.M., and Geacintov, N.E. (1995) Conformational studies of the (+)-*trans*, (−)-*trans*, (+)-*cis*, and (−)-*cis* adducts of anti-benzo[*a*]pyrene diolepoxide to N^2-dG in duplex oligonucleotides using polyacrylamide gel electrophoresis and low-temperature fluorescence spectroscopy. *Biophys. Chem.*, **56**, 281–296.

129 Rodríguez, F.A. (2007) Nuclear magnetic resonance solution structure of covalent polycyclic aromatic carcinogen–DNA adducts: influence of base sequence context and carcinogen topology. PhD Thesis. New York University, New York.

130 Dickerson, R.E. (1998) DNA bending: the prevalence of kinkiness and the virtues of normality. *Nucleic Acids Res.*, **26**, 1906–1926.

131 Xu, R. (1994) Studies of characteristics of DNA–polynuclear aromatic carcinogen adducts at atomic and molecular resolution levels. PhD Thesis. New York University, New York.

132 Truglio, J.J., Karakas, E., Rhau, B., Wang, H., DellaVecchia, M.J., Van Houten, B., and Kisker, C. (2006)

Structural basis for DNA recognition and processing by UvrB. *Nat. Struct. Mol. Biol.*, **13**, 360–364.

133 Jia, L., Kropachev, K., Ding, S., Van Houten, B., Geacintov, N.E., and Broyde, S. (2009) Exploring damage recognition models in prokaryotic nucleotide excision repair with a benzo[a]pyrene-derived lesion in UvrB. *Biochemistry*, **48**, 8948–8957.

134 Lu, X.-J. and Olson, W.K. (2003) *Nucleic Acid Res.*, **31**, 5108–5121.

135 Cai et al. (2010) *Journal of Molecular Biology*, in press.

13
Impact of Chemical Adducts on Translesion Synthesis in Replicative and Bypass DNA Polymerases: From Structure to Function

Robert L. Eoff, Martin Egli, and F. Peter Guengerich

13.1
Introduction

DNA polymerases can be organized into at least seven families based on sequence and structural similarities: A, B, C, D, X, Y, and reverse transcriptases [1, 2]. The first DNA polymerase to be isolated and shown to possess DNA synthesis capabilities was DNA Pol I from *Escherichia coli* [3, 4]. Since Arthur Kornberg's initial studies, countless researchers have focused upon the biological roles and catalytic properties of DNA polymerases [5–9]. Substrate selection by polymerases involves several distinct steps and determines the accuracy of DNA replication. Checks upon base-pair geometry, hydrogen-bonding patterns, metal-ion coordination, and conformational changes all contribute to the efficiency of each nucleotidyl transfer reaction [9–11]. Subsequent work by Kornberg and others revealed that some DNA polymerases possess 3′–5′ exonuclease activity, which serves as a means of proofreading the accuracy of DNA synthesis [12–17]. The kinetic benefit toward accurate DNA synthesis is 10- to 100-fold when proofreading activity is present, which results in some polymerases making a mistake once in every 10^9 incorporation events [11]. Accessory proteins may be important in the stabilization of complexes that undergo proofreading [18–20], and some polymerases (e.g., Pol III from *E. coli*) rely upon subunits within the holoenzyme complex for exonuclease activity [14]. The Y-family DNA polymerase Dpo4 from the crenarchaeote *Sulfolobus solfataricus* can remove incorporated nucleotides through pyrophosphorolysis [21, 22] and the X-family member Pol β has been shown to possess 5′-deoxyribose phosphate lyase activity [23]. Each one of these mechanisms converges upon finding the most thermodynamically stable form for the DNA double helix [24].

The large number of genomic sequencing projects has led to the discovery of many new polymerases from various organisms. *E. coli* possess five DNA polymerases [25], Pols I–V (I and II are both A-family members and are thought to play roles in assisting replication/repair). DNA Pol III is a member of the C-family and is the major replicative polymerase in *E. coli* [26, 27]. The final two polymerases in *E. coli*, Pols IV and V, are both Y-family members that are thought to help bypass DNA damage and facilitate adaptive mutation [28]. The absolute number of

The Chemical Biology of DNA Damage. Edited by Nicholas E. Geacintov and Suse Broyde
© 2010 WILEY-VCH Verlag GmbH & Co. KGaA, Weinheim
ISBN: 978-3-527-32295-4

polymerases in Archaea and eukaryotes is less certain. The crenarchaeote *S. solfataricus* has at least four DNA polymerases (three B-family members and the model Y-family member Dpo4) [29]. Some other extremophiles (e.g., *Methanosarcina mazei* (strain Gol) and *Methanosaeta thermophila*) possess putative homologs of the X-family Pol β. Eukaryotes vary in their number of polymerases. Humans possess at least 19 enzymes that catalyze some type of nucleotidyl transfer [30]: Pols α, δ, ε, and ζ are B-family members; mitochondrial Pol γ and Pols ν and θ are members of the A-family; Pols β, λ, and μ are X-family members; Pols η, ι, κ, and REV1 comprise the Y-family; the remaining enzymes include Pols σ1, σ2, ϕ, terminal deoxynucleotidyl transferase, and telomerase (the sole human reverse transcriptase).

The domain organization and structural features of DNA polymerases are largely conserved across species and families (Figure 13.1) [31, 32]. The modular domain organization has been likened to a right hand with fingers, thumb, and palm subdomains. The palm domain universally contains the active-site aspartate residues that coordinate two divalent metal ions. The thumb and finger domains, which differ substantially across polymerase families, are more closely associated with nucleotide selection and DNA binding/translocation, respectively. The little-finger (or palm-associated) domain is unique to the Y-family, and is involved in DNA binding and the lesion bypass properties of these enzymes.

The catalytic mechanisms of DNA polymerases have been studied in great detail [9, 33–35]. From these analyses many inferences have been made regarding the accuracy of nucleotide incorporation and the relevance of *in vitro* fidelity toward the biological functions of different polymerases. All polymerases require a single-stranded DNA template. A few notable deviations from the typical primer/template substrate include telomerase, which carries an RNA template in its active site, and A-family Pol θ, which possesses ATPase-driven helicase activity such that it can catalyze polymerization on nicked double-stranded DNA by displacing a strand ahead of DNA synthesis [36]. The minimal kinetic description of polymerase catalysis includes binding of DNA, divalent metal ions, and the incoming dNTP [37]. Generally a noncovalent step is then included in the cycle just prior to the phosphoryl transfer step (i.e., phosphodiester bond formation, or "chemistry"). The noncovalent step has often been referred to as the "rate-limiting" step in the polymerase catalytic cycle (at least for correct base pairs) [9]. The physical basis of the noncovalent step has been a topic of some debate and is often attributed to a conformational transition from an "open" to "closed" state. Indeed, several crystal structures have shown large conformational changes when comparing "binary" (no dNTP) to ternary (with incoming dNTP) complexes [38, 39]. However, subsequent studies have cast doubt upon the view that the large conformational changes observed in some crystal structures limits the rate of catalysis [10, 40]. Nevertheless, these changes are undoubtedly important mechanistic features of polymerase activity.

Substrate selection by polymerases determines, to a large extent, the accuracy of genomic replication. The efficiency of nucleotidyl transfer is dependent upon several interrelated factors. The two-metal-ion-dependent mechanism of polymerase catalysis appears to be universally conserved in most respects, with

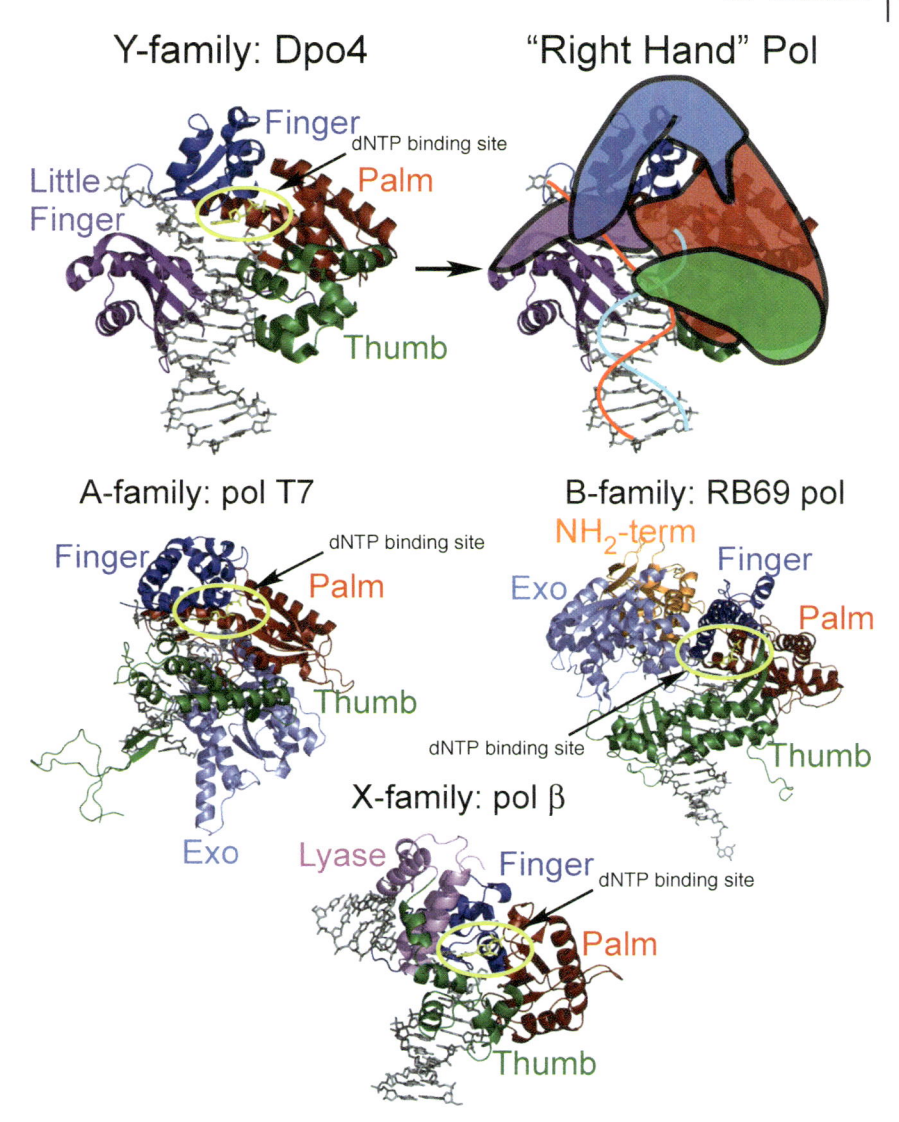

Figure 13.1 General structural features for replicative and translesion polymerases. A structure of Dpo4 from *S. solfotaricus* is shown to highlight general structural features associated DNA polymerases. The overall topology of polymerase subdomains has been likened to a "right hand," as indicated in cartoon form. Three other polymerase structures are shown to illustrate the "right-handedness" of A-, B-, and X-family members Pol T7, RB69 gp43, and Pol β, respectively (PDB 1T7P, 1IG9, and 1BPY). All of the structures are oriented with the incoming dNTP (yellow) as a point of reference.

deprotonation of the 3′-hydroxyl group being an important initial step in the catalytic cycle, but the exact mechanism used to facilitate nucleophilic attack upon the α-phosphate may differ among enzymes [41, 42]. Transfer of the 3′-hydroxyl proton and the subsequent formation of the transition state intermediate is subject to the dynamics of a given polymerase active site and the physicochemical characteristics of the nascent base pair. Deviation from the normal Watson–Crick pattern is often the result of chemical modification to some portion of either the template strand or the pool of nucleotide triphosphate molecules.

Polymerase fidelity is a measure of how frequently the enzyme makes a "mistake" (i.e., inserts a dNTP that does not result in a canonical Watson–Crick pair). Often steady-state (multiple catalytic turnovers) analysis is used to compare dNTP insertion activities. While steady-state kinetic analysis does often correlate well with insertion frequencies observed in sequencing-based cellular analyses (e.g., shuttle vectors or M13 replication assays), it cannot discern mechanistic features that define the fidelity being measured. Transient-state kinetic analysis can provide a direct measure of distinct steps in the catalytic cycle and therefore is a more informative measure of polymerase fidelity. Both approaches involve providing a single nucleotide to the polymerase and then measuring the kinetic constants that define incorporation. The catalytic efficiency of polymerase activity is defined as the rate constant defining nucleotide incorporation (k_{cat} or k_{pol}) divided by the dependence of the rate upon dNTP concentration (K_m or K_d). These terms are frequently used to understand the effect of covalent DNA modifications upon polymerase action. Misinsertion frequency ($f = (k_{pol}/K_d)_{incorrect}/(k_{pol}/K_d)_{correct}$, or more precisely $f = (k_{pol}/K_d)_{incorrect}/[(k_{pol}/K_d)_{incorrect} + (k_{pol}/K_d)_{correct}]$) is often used as quantitative measure of polymerase fidelity for specific lesions. Recently, a mass spectrometry-based method has been developed for analyzing *in vitro* polymerase fidelity in the presence of all four dNTPs, as opposed to single nucleotide comparisons [43]. We will summarize what is currently known concerning how DNA polymerases perform catalysis with modified DNA substrates in an effort to highlight mechanisms that balance accurate replication and mutagenesis.

13.2
Bypass of Abasic Sites

Loss of the purine/pyrimidine moiety from the DNA backbone is probably the most frequent form of DNA damage encountered in living cells and results in the generation of noninstructional abasic sites within the genetic code (estimated at 10^4/cell/day) [44, 45]. Abasic sites may be produced by spontaneous depurination (e.g., hydrolysis or nucleophilic attack upon the N7 position of guanine by reactive chemicals), by UV or γ-irradiation, or through the action of lesion specific DNA glycosylases (e.g., uracil DNA glycosylase, 7,8-dihydro-8-oxo-deoxyguanosine (8-oxo-G) DNA glycosylase, etc.) during enzymatic repair of modified bases [46]. Naturally occurring abasic sites exists in equilibrium between the ring-closed α- and β-hemiacetals (99%) and ring-opened aldehyde or hydrated aldehyde (less than

1%). Oxidized abasic sites, resulting in a lactone moiety, can be formed upon C4′ proton abstraction by hydroxyl radicals and/or antitumor agents such as bleomycin [47].

If an abasic site is encountered by the replication machinery then the DNA polymerase bypassing the lesion must either select a nucleotide from the dNTP pool without instruction from the template or must rely upon the bases adjacent to the abasic site for guidance. In general, abasic sites and abasic site analogs inhibit catalysis by DNA polymerases [46, 48–51]. Most polymerases balance insertion of adenosine (the "A-rule") with the generation of −1 frameshift mutations [51]. *In vitro* experiments with *E. coli* polymerases have shown that the A-family member Pol II can bypass abasic sites more efficiently than either Pol I or III at physiologic salt concentrations [49]. Catalysis by all three enzymes is substantially reduced. Exonuclease-deficient mutants of Pols I, II, and III display an increased ability to bypass abasic lesions, consistent with the idea that insertion opposite the lesion leads to a futile cycle of insertion/excision when "proofreading" is active. The action of the Y-family polymerase Pol V (UmuD′$_2$C complex) is the most efficient means of bypassing abasic sites in *E. coli*, with insertion of dATP being the most favored product and accessory proteins (e.g., RecA recombinase) increasing the efficiency of Pol V catalysis around 3000-fold [52]. The DinB homolog Pol IV is rather inefficient at bypass of abasic lesions, although inclusion of the β,γ complex (clamp and clamp loader, respectively) increases DNA synthesis efficiency around 15 000-fold [52].

Eukaryotic bypass of abasic lesions is more complex. Studies with yeast polymerases suggest that bypass of abasic moieties can result from the action of several enzymes, including Pols δ and ζ and REV1 [53, 54]. Similar to prokaryotic enzymes, the addition of accessory factors can stimulate eukaryotic polymerase bypass of abasic moieties [55]. In the proposed model, the replicative polymerase (Pol δ or ϵ) inserts dATP opposite an abasic site but cannot extend beyond the lesion. The C-terminal portion of REV1 may then serve to recruit Pol ζ, which can extend from the newly inserted dATP with reasonable efficiency. Consistent with a role for replicative polymerases in catalysis opposite abasic sites, inhibition of the B-family polymerases with aphidicolin decreases bypass around 4-fold in cell culture assays [56]. The Y-family Pol η is strongly inhibited by abasic sites and genetic studies indicate little involvement in the bypass of abasic sites [53, 54, 56, 57]. Any contribution from the eukaryotic DinB homolog Pol κ is probably also negligible because Pol κ is inefficient at both insertion and extension steps opposite abasic lesions [58, 59]. Some studies are suggestive of a role for the Y-family member REV1 in the accurate bypass of natural abasic sites [60, 61]. Involvement of REV1 in the bypass of abasic sites is an attractive model since guanine is most susceptible to spontaneous depurination events [46, 62, 63] and the strict deoxycytidyl transferase activity of REV1 would prove the most "accurate" means of reconstituting the information lost at the abasic site.

Significant insight into the mechanism of nucleotide selection opposite abasic lesions has been achieved from structural studies. Crystal structures of the B-family polymerase from bacteriophage RB69 and the Y-family Dpo4 from

S. solfataricus have increased our understanding of the "A-rule" and mechanistic determinants of abasic bypass [64–66]. Both of these polymerases are inefficient at bypass of abasic sites. RB69 can insert dATP opposite an abasic site but extension is completely inhibited. Dpo4 catalysis is inhibited several hundred fold when it encounters the abasic tetrahydrofuran (THF) analog, and shows little kinetic difference between following the "A-rule" and skipping to the next template base to generate a −1 frameshift [67, 68]. A series of crystal structures with Dpo4 in complex with template DNA containing a THF moiety showed that the abasic site can be accommodated in an extrahelical or intrahelical conformation [65]. The extrahelical conformation is easily accommodated in the gap between the finger and little-finger domains. Neither the structural nor mechanistic basis for the "A-rule" is apparent from the crystal structures of Dpo4 (see Figure 13.2).

Dpo4 bypass of THF abasic analogues

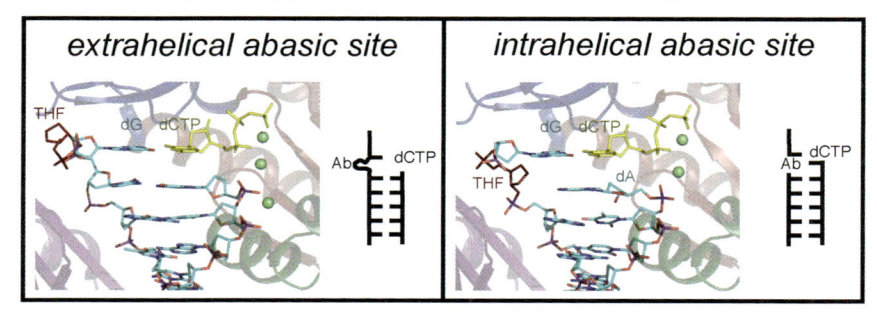

extrahelical abasic site *intrahelical abasic site*

RB69 gp43 bypass of THF abasic analogues

intrahelical abasic site

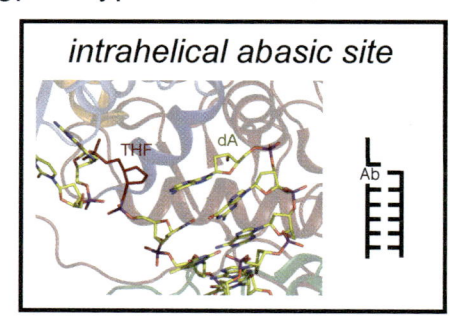

Figure 13.2 Comparison of bypass of abasic site analogs by different polymerases. Substrates containing the abasic analog THF can adopt at least two conformations during Dpo4-catalyzed bypass of the lesion, namely extrahelical (PDB 1S0N) and intrahelical (PDB 1S10) conformations. The intrahelical orientation represents what occurs following insertion of dATP ("A-rule"), while the extrahelical conformation represents generation of a −1 frameshift deletion. The replicative polymerase gp43 from bacteriophage RB69 (PDB 2P5O) bypasses a THF moiety using an intrahelical "A-rule" that relies primarily upon the influence of base-stacking interactions.

A series of very informative crystal structures with the model B-family DNA polymerase from bacteriophage RB69 have been solved [64, 66]. The initial work revealed a very unusual crystal structure that possesses four distinct molecules in the asymmetric unit [64]. Insertion of dATP opposite the THF moiety results in a distortion of the template DNA near the lesion and RB69 fails to translocate past the abasic site, consistent with previous studies showing that replicative polymerases can incorporate dATP opposite abasic sites but then fail to extend from the incorporated base. Further mechanistic elucidation of the A-rule was achieved by comparing dATP incorporation opposite the THF to what is observed with 5-nitro-1-indolyl-2'-deoxyriboside-5'-triphosphate (5-NITP). B-family polymerases incorporate the non-natural purine analog 5-NITP opposite the THF moiety around 1000-fold more efficiently than dATP [69]. The 5-NITP base analog can pair with all natural bases and is known to stabilize DNA, presumably because of the superior base stacking capacity conferred by the nitro moiety. In a ternary complex with RB69 and THF-modified DNA, 5-NITP does not distort the phosphate backbone as strongly as the dATP opposite the THF ring, but extension of the 5-NITP:THF complex is much slower than the dATP:THF complex. These results suggest that base stacking is the most important feature of nucleotide selection when RB69 is faced with a noninstructional lesion and that local distortion of the DNA backbone is a less influential factor in the "A-rule".

13.3
Lesions Generated by Oxidative Damage to DNA

Oxidative damage to DNA is thought to be a contributing factor to cellular dysfunction [70–73]. Reactive oxygen species (ROS) can be generated from environmental toxins, uncoupling the electron transport chain, ionizing radiation, and/or metal-catalyzed reactions [74–76]. In some instances ROS are generated by endogenous mechanisms, as means of defense (e.g., cytotoxic macrophage generation of superoxide/nitric oxide) and/or signaling [77, 78]. These oxidants can react with DNA directly to generate adducts such as 8-oxo-G and N-(2-deoxy-D-pentofuranosyl)-N-(2,6-diamino-4-hydroxy-5-formamidopyrimidine) (Fapy-dG) [79, 80]. The pyrimidine glycols 5,6-dihydro-5,6-dihydroxythymine (Tg) and 5,6-dihydro-5,6-dihydroxycytosine (Cg) are also common forms of oxidative damage [79, 80], although only Tg is stable under physiological conditions as Cg undergoes deamination to form uracil derivatives [81].

Polymerase bypass of the mutagenic lesion 8-oxo-G has been studied in considerable detail. All DNA polymerases studied to date insert either dCTP or dATP opposite 8-oxo-G [82–90]. Relatively modest decreases in catalytic efficiency have been observed for some polymerases (e.g., around 30-fold decrease for *E. coli* Pols I and II (exo$^-$), around 10-fold for calf thymus Pol δ), whereas the catalytic efficiency of the model B-family polymerase from bacteriophage T7 (Pol T7) was inhibited around 270-fold as judged by pre-steady-state measurements [83, 85, 86, 91, 92]. Only the Y-family polymerases Pol η (from *Saccharomyces cerevisiae*) and

Dpo4 (from *S. solfataricus*) show unaltered or enhanced efficiency when bypassing 8-oxo-G [90, 93]. Both of these enzymes bypass 8-oxo-G in a highly accurate manner, with a 20-fold preference for dCTP over dATP. The eukaryotic DinB homolog Pol κ is very error-prone when catalyzing nucleotide incorporation opposite 8-oxo-G, showing much greater preference for dATP insertion [58].

The ratio of dCTP to dATP incorporated opposite 8-oxo-G varies for other polymerases. Some polymerases, such as *E. coli* Pol I (Klenow fragment (KF) exo⁻) and Pol II and bacteriophage Pol T7⁻, insert dCTP and dATP with almost equal efficiency, although all three enzymes can extend a dATP:8-oxo-G pair with greater efficiency than a dCTP:8-oxo-G pair [85, 86, 92]. Mitochondrial Pols γ and β both favor dATP incorporation slightly, with upstream mutations occurring when Pol β performs short gap-filling reactions opposite 8-oxo-G [94, 95]. The DNA polymerase I fragment from the thermostable organism *Bacillus stearothermophilus* (*Bacillus* fragment (BF)) and HIV-1 reverse transcriptase both preferentially inserted dATP opposite 8-oxo-G [85, 86, 96].

The mechanism of miscoding by 8-oxo-G has been studied from several perspectives. In order to alleviate a steric clash between the C8 oxygen and the O4′ atom of deoxyribose, 8-oxo-G adopts a *syn* conformation, as opposed to the normal *anti* configuration observed in duplex DNA, allowing dATP to form two stable hydrogen bonds with the Hoogsteen face of 8-oxo-G (Figure 13.3) [97]. Hoogsteen-type pairing between A and 8-oxo-G has been observed for oligonucleotides in isolation (e.g., nuclear magnetic resonance (NMR) studies) and in the active sites of several DNA polymerases, including Pol T7⁻, BF, Dpo4, Pol β, and RB69 [82, 90, 96, 98, 99]. The dATP:8-oxo-G pair is geometrically similar to a dTTP:dATP pair and, as such, can evade the proofreading activity of some polymerases. In order to incorporate dCTP opposite 8-oxo-G a polymerase must somehow overcome the thermodynamic barrier that favors the *syn* configuration of the lesion. At least two polymerases, Dpo4 and Pol η, have mechanisms that favor accurate bypass of 8-oxo-G. A recent report provides evidence that Pol η plays an important role during accurate bypass of 8-oxo-G in human cells [100]. In the case of Dpo4, stabilization of the *anti* configuration is achieved through an electrostatic contact between the C8 oxygen and a charged side-chain in the little finger [84, 90]. Superimposition of the ternary Dpo4 · 8-oxo-G · dCTP structure with yeast Pol η reveals that both enzymes possess a positive center (Arg332 for Dpo4 and Lys498 for yeast Pol η) that can contact the C8 oxygen (Figure 13.3). By way of comparison, the positive center is moved further from 8-oxo-G in Pol κ (Arg507) and could explain the more error-prone nature of Pol κ-catalyzed bypass of 8-oxo-G (Figure 13.3).

Less is known concerning the Fapy-dG lesion. Like 8-oxo-G, Fapy-dG is mutagenic in simian kidney cells, resulting in mainly G → T transitions. However, the lesion appears to be only weakly mutagenic in bacteria [101]. *E. coli* DNA Pol I (KF exo⁻) strongly preferred incorporation of dCTP opposite Fapy-dG, and unlike studies with 8-oxo-G, extension of the Fapy-dG:dATP pair was also inhibited [101]. Structural studies with DNA polymerases and Fapy-dG are lacking. Molecular modeling with Pol β suggests that pairing modes may be similar for Fapy-dG and

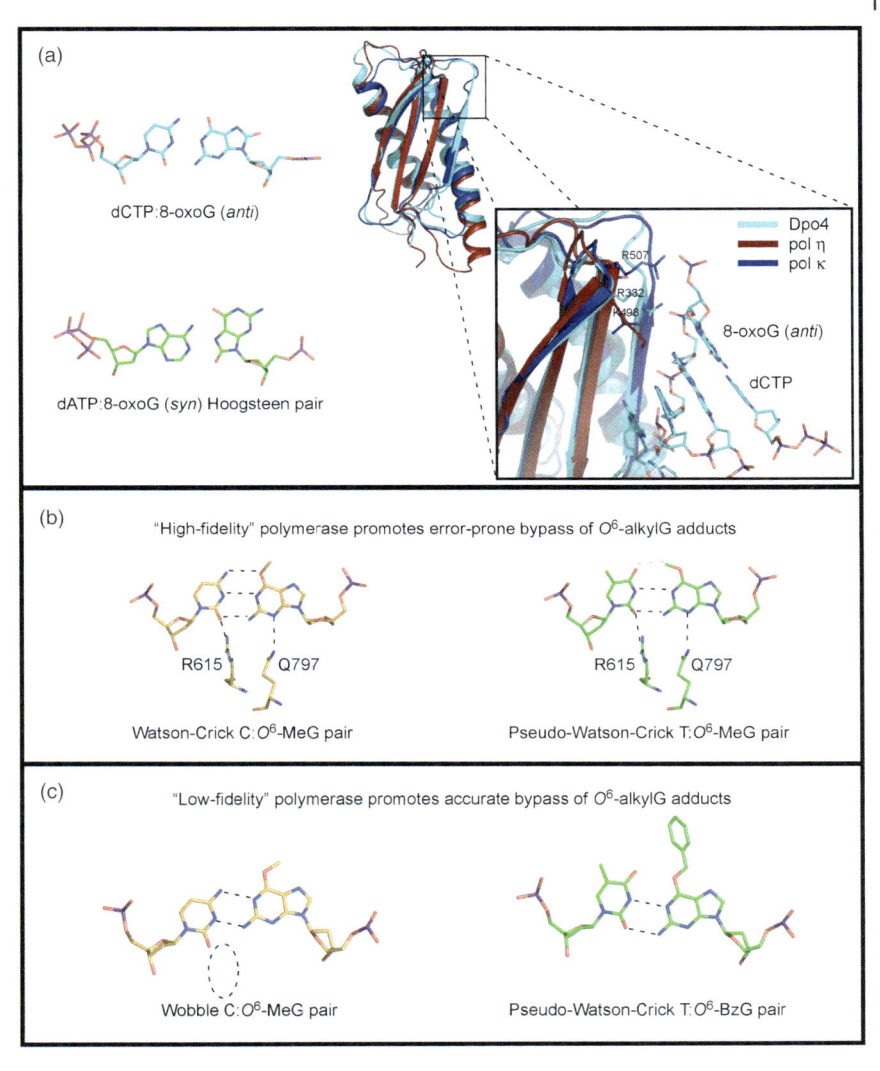

13.3 Lesions Generated by Oxidative Damage to DNA

Figure 13.3 Polymerase bypass of the small nonblocking lesions 8-oxo-G and O^6-MeG. (a) The Y-family polymerase Dpo4 stabilizes the *anti* conformation of 8-oxo-G during accurate bypass of the lesion (PDB 2C2E). A positive center in the little-finger domain of Dpo4 and (based on structural superimposition) Pol η facilitates accurate bypass of 8-oxo-G (superimposing PDB 2C2E and 1JIH). Pol κ, on the other hand, does not have a homologous residue in contact with 8-oxo-G (PDB 2OH2). Insertion of dATP opposite 8-oxo-G proceeds through a Hoogsteen base pair for Dpo4 and Pol T7. Replicative polymerases adopt similar conformations, but are generally more prone to misincorporation of dATP opposite 8-oxo-G. (b) The highly mutagenic O^6-MeG is forced into Watson–Crick geometry when a model replicative polymerase, BF, bypasses the lesion. (c) The bypass polymerase Dpo4 tolerates a wobble pair during accurate synthesis. Both enzymes utilize a pseudo-Watson–Crick geometry during misincorporation of dTTP opposite O^6-alkylG lesions.

8-oxo-G, and that base stacking with neighboring base pairs may influence mutagenicity [102].

The pyrimidine adduct Tg is formed at levels of around 400 lesions/cell/day [103]. Tg is not particularly mutagenic, but it does produce significant distortion to the double helix because the addition of hydroxyl groups to the 5- and 6-positions of thymidine results in a nonplanar (i.e., nonaromatic) ring [104]. Tg is a strong block to polymerase catalysis; both KF exo⁻ and the bacteriophage T4 polymerase gp43 can insert dATP opposite Tg, but extension is strongly impeded [105]. Likewise, Y-family Pol η can incorporate dATP opposite Tg with around 60-fold decrease in catalytic efficiency, but extension of the dATP:Tg pair is inhibited around 300-fold [106, 107]. The low-fidelity B-family Pol ζ shows little to no kinetic inhibition at either the insertion and extension steps of Tg bypass (around 7- and 2-fold, respectively), and genetic studies with mice suggest a role for Pol ζ in bypass of Tg *in vivo* [106]. Pol κ exhibits a 20- to 50-fold decrease in catalytic efficiency when inserting dATP opposite Tg, with slightly less efficient incorporation when the 5*R* stereoisomers represent a greater proportion of the isomeric mixture [108]. During extension of dATP:Tg pairs, Pol κ is inhibited around 230-fold, similar to what is observed with Pol η. Given the tissue and developmentally specific nature of Pol κ expression, as well as its induction in response to agents that induce oxidative damage, it is quite possible that both Pols κ and ζ are important during bypass of Tg *in vivo*, but Pol ζ is by far the most efficient extender of dATP:Tg pairs *in vitro*.

The structural basis for the ability of Tg to block DNA polymerase activity was elucidated by solving the crystal structure of the RB69 polymerase in complex with Tg-modified DNA [109]. Tg is intrahelical in the binary complex, in contrast to what was observed with Tg-modified oligonucleotides [110], and forms a Watson–Crick pair with the terminal dATP of the primer. Extension of a dATP:Tg pair is impaired because the C5 methyl group prevents stacking of the base to the 5′-side of the lesion by protruding axially from the ring of the damaged base. While Tg is blocking in most contexts, bypass can occur more readily when a cytosine is positioned 5′ of Tg [111, 112]. The reason for enhanced bypass is not entirely clear, but may stem from the ability of cytosine to better accommodate the axial methyl group.

13.4
Exocyclic DNA Adduct Bypass

Reactions between the purine/pyrimidine ring systems and various bis-electrophiles can result in the extension of the normal ring system. These exocyclic DNA adducts differ in ring size, planarity, and substitution. They often block the Watson–Crick side of the base, masking the coding potential for the damaged site. The etheno-dG adduct $1,N^2$-etheno-dG ($1,N^2$-ε-dG) is mutagenic in multiple cell lines from both prokaryotic and eukaryotic organisms, producing a wide range of products that included base substitutions, deletions, and rearrangements [113–

116]. Polymerase bypass of 1,N^2-ε-dG is error-prone, regardless of which polymerase is being studied, with a general preference for inserting purines when the lesion is encountered [43, 117, 118]. The lesion is also quite blocking to several polymerases, reducing the catalytic efficiency considerably. Human Pol η can insert dGTP and dATP opposite 1,N^2-ε-dG with an efficiency value ($k_{cat}/K_{M,dNTP}$) that is higher than Pol ι [117]. However, the actual decrease in catalytic efficiency that occurs relative to unmodified DNA is smallest for Pol ι, which is consistent with the argument that Pol ι is important for the insertion step opposite 1,N^2-ε-dG [117, 119]. It is difficult to establish which polymerase is most important for *in vivo* bypass, but a general conclusion is that 1,N^2-ε-dG presents a strong block to all polymerase families.

Crystal structures of Dpo4 in complex with 1,N^2-ε-dG-modified DNA show that the modified base is stacked between adjacent bases during and after generation of a −1 frameshift deletion, with only slight buckling of the base pair to the 3′-side of the lesion [43]. Dpo4 catalysis is strongly impeded by 1,N^2-ε-dG (100- to 1000-fold), although bypass does occur more readily than with an enzyme such as Pol T7⁻ [118]. *In vitro* insertion/extension products include insertion of dATP and −1 and −2 frameshift deletions and appear to be somewhat dependent upon the sequence context. The sequence dependence observed for Dpo4-catalyzed bypass of 1,N^2-ε-dG might be due to the interconversion between *syn* and *anti* conformations observed for modified oligonucleotides when dTTP is 5′ of 1,N^2-ε-dG [120, 121].

Attack upon DNA to generate base propenals and the reaction of DNA with lipid peroxidation byproducts, such as acrolein, crotonaldehyde, malondialdehyde, and 4-hydroxynonenal, can generate secondary forms of damage including exocyclic dG adducts such as 3-(2′-deoxy-β-D-*erythro*-pentofuranosyl)-pyrimido[1,2-α]purin-10(3H)-one (M_1dG) and γ-hydroxy-1,N^2-propano-2′-deoxyguanosine (HOPdG) adducts [73, 122]. These exocyclic DNA adducts can exist in equilibrium between ring-opened and ring-closed forms. For example, the major malondialdehyde-derived deoxyguanosine adduct can exist in either the ring-closed (M_1dG) or the ring-opened form (N^2-(3-oxo-1-propenyl)-dG (N^2-OPdG)). Ring opening to from N^2-OPdG is stabilized in duplex DNA when cytosine is placed opposite the lesion [123]. Normal Watson–Crick base-pairing is observed between dC and N^2-OPdG when the oligonucleotides are in isolation. A crystal structure with a dCTP:M_1dG pair in the active site of Dpo4 captured a nonproductive complex in which the dCTP was flipped down into the growing minor groove and M_1dG exists in the ring-closed form, well-stacked between the adjacent bases [214]. Dpo4 catalysis is strongly impeded by M_1dG (200- to 2000-fold as judged by steady-state analysis), with a slight kinetic preference for dATP incorporation. Analysis of insertion and extension products revealed highly error-prone bypass of M_1dG (52–83% nonaccurate bypass). Products included base substitutions, frameshifts, and untargeted mutations. Similar to the results observed with 1,N^2-ε-dG [43], more error-prone bypass was observed when dTTP was 5′ of M_1dG. In contrast to the permanently ring-closed 1,N^2-ε-dG (but see [124]), Dpo4 can perform a substantial amount of dCTP incorporation opposite M_1dG, perhaps indicating that accurate synthesis can occur when M_1dG converts to the ring-opened N^2-OPdG.

A study with human Pol η also showed mutagenic bypass of M_1dG by insertion of dATP and generation of −1 frameshifts [215]. The level of Pol η inhibition during bypass of M_1dG (around 8-fold for dATP insertion) was not as strong as that observed with Dpo4. Pol η shows a preference for dATP insertion over generation of −1 frameshifts both kinetically and as judged by analysis of full-length incorporation products by liquid chromatography/tandem mass spectrometry.

Similar to M_1dG, the acrolein-derived exocyclic deoxyguanosine adduct γ-HOPdG can exist in either a ring-closed or a ring-opened forms [125]. Pol η appears to bypass γ-HOPdG more accurately and efficiently when the ring-opened form is present, because the ring-closed structural analog $1,N^2$-propanodeoxyguanosine (PdG) presents a strong block to Pol η catalysis [126]. The mutation frequency observed during replication of site-specifically modified vectors containing γ-HOPdG in human fibroblasts and xeroderma pigmentosum type V (XPV)-derived cells indicate that Pol η is not the only polymerase responsible for bypass of the lesion [126]. The Y-family member REV1 can bypass the permanently ring-closed PdG adduct by ejecting the modified template residue from the active site and using Arg324 to pair with the incoming dCTP [127]. Presumably, REV1 should bypass ring-closed γ-HOPdG in an accurate manner, so REV1-mediated bypass of γ-HOPdG would not account for the mutations observed in Pol η-deficient XPV cells. Steady-state analysis indicates that Pol ι can bypass γ-HOPdG in a highly accurate and efficient manner, while Pol κ appears to be blocked by the lesion [119, 128]. Dpo4-catalyzed bypass of PdG appears to be similar to that observed with $1,N^2$-ε-dG, based on crystal structures of the enzyme in complex with PdG-modified DNA [129]. Permanently ring-closed exocyclic adducts can be considered strong blocks to replication and result in error-prone bypass, while those adducts that can ring open result in higher levels of accurate synthesis by Y-family polymerases.

$1,N^6$-ε-dA and $3,N^4$-ε-dC had been investigated earlier than $1,N^2$-ε-dG in oligonucleotides [130] and using site-specific mutagenesis [131], although no reports of bypass properties with individual DNA polymerases have been published. Another etheno adduct of interest is $N^2,3$-ε-dG, which exists in DNA treated with certain bifunctional electrophiles [132]. The only evidence of misincorporation involves a study with mispairing of $N^2,3$-ε-dGTP [133]. A problem with this latter adduct is the inherent instability of the glycosidic bond [134] and we are unaware of efforts to address details of misincorporation mechanisms.

13.5
Alkylated DNA

The addition of alkyl or aryl (or aralkyl) groups to DNA occurs mostly as a result of exposure to exogenous compounds that occur in cigarette smoke, chemotherapeutic agents, and industrial chemicals [135, 136]. All four bases are subject to alkylation, but the N7 position of guanine is by far the most susceptible to attack

[137]. However, alkylation at the N7 position weakens the glycosidic bond and this often leads to depurination (i.e., formation of an abasic site). Nevertheless, the possibility that N7-alkylG adducts could be directly miscoding (at some level) has not been tested carefully and cannot be dismissed. The alkylation changes the pK_a of the pyrimidine ring atoms of the purine and could induce mispairs via minor tautomers [138].

Modification at the O6 position is not as frequent as attack at N7 but O^6-alkylG is one of the most mutagenic forms of DNA damage [139]. One of the best-studied forms of alkylated DNA is O^6-methylguanine (O^6-MeG). DNA polymerase bypass of O^6-MeG has been the subject of many studies because it is observed in many organisms (including human DNA), is highly mutagenic, and is a principal basis of cell toxicity in chemotherapeutic regimes [140, 141]. Alkylation at the O6 position of guanine is principally recognized and repaired by O^6-alkylguanine DNA-alkyltransferase (AGT; also called methyl guanine methyl transferase) in prokaryotes and eukaryotes [142–144]. Epigenetic silencing of the human AGT promoter region has been implicated in the formation of neoplasms in the colon, further illustrating the transformative potential of O^6-alkylguanine lesions [145]. The most common mutation observed in cells is a C → T transversion [146–148], and most polymerases tend to insert a mixture of dCTP and dTTP *in vitro* [149–152].

Examination of the chemical structure of O^6-MeG provides a rationale for its mutagenic potential (Figure 13.3), in that alkylation at the O6 position alters the hydrogen bonding capability of guanine by deprotonating the N1 position and provides a potential steric block in the form of the alkyl group. Crystallographic and NMR studies of oligonucleotides containing O^6-MeG suggested that the O^6-MeG:dTTP pair adopts a pseudo-Watson–Crick geometry, while the O^6-MeG:dCTP pair was found to adopt three configurations including a "wobble" pair, a bifurcated pairing, and a protonated Watson–Crick-type pairing [153–156]. Compared with unmodified bases, where the energetic cost of mispairing is around 5 kcal/mol, the thermodynamic difference between dCTP and dTTP:O^6-MeG pairs is predicted to be small (below 1 kcal/mol) [157]. Two studies with model DNA polymerases have revealed different modes of O^6-MeG bypass. In one study the model B-family polymerase BF was crystallized in complex with O^6-MeG and either a dCTP or dTTP pair [158]. The second study solved the crystal structure of the Y-family member Dpo4, also in complex with dC or dT:O^6-MeG pairs [150]. The replicative polymerase BF is strongly inhibited by the O^6-MeG lesion (around 10^4- to 10^5-fold for dTTP and dCTP insertion, respectively). Both dCTP and dTTP were found to pair in the Watson–Crick-type mode when constrained by the BF active site (Figure 13.3). BF is found in an "open" (i.e., noncatalytic) conformation when the dC:O^6-MeG and dT:O^6-MeG pairs are moved into the postinsertion register. Notably, the O4 position of thymine forms a weak electrostatic interaction with the protons of the methyl group, as evidenced by convincing electron density maps. The dCTP:O^6-MeG pair adopts the wobble mode once it exits the constraints of the BF active site and moves into the −10 position (with 0 position being insertion of dNTP).

The extreme level of BF inhibition is in contrast to steady-state analysis of human Pol δ. The degree of inhibition by O^6-MeG observed for both the replicative Pol δ (with the sliding clamp) and three recombinant Y-family polymerases (η, ι, and κ) was found to be relatively modest (10- to 100-fold) [149]. With one exception, all of the human enzymes tested in the aforementioned study incorporated dCTP and dTTP equally well opposite O^6-MeG. The exception to this pattern was human Pol ι, which performed insertion of dTTP opposite O^6-MeG 10 times better than dCTP opposite the lesion and with greater efficiency than dCTP opposite G (around 2-fold). It is entirely possible that inclusion of accessory proteins and/or post-translational modifications to the Y-family members could influence the catalytic efficiency of O^6-MeG bypass. *In vitro* studies with all polymerases studied to date verify the mutagenic potential of the O^6-MeG adduct.

Dpo4 utilizes a means of bypassing O^6-MeG that is mechanistically distinct from BF. In the postinsertion context, a wobble dCTP : O^6-MeG pair was observed in the Dpo4 active site (Figure 13.3). Dpo4 is also more accurate at O^6-MeG than other polymerases, inserting dCTP around 70% of the time. Transient-state kinetic analysis of Dpo4-catalyzed bypass of O^6-MeG indicated that the enzyme is inhibited around 14-fold during accurate bypass of the lesion and steady-state analysis indicates a around 3-fold kinetic preference for dCTP over dTTP. Similar kinetic and structural results were obtained when Dpo4-catalyzed bypass of the larger O^6-benzylguanine (O^6-BzG) was studied [159]. Increasing the size of the O^6-alkylG group from methyl to benzyl decreased the catalytic efficiency of Dpo4 5- to 10-fold. Similar to O^6-MeG, the dCTP : O^6-BzG pair adopts a wobble conformation in the Dpo4 active site. A crystal structure of Dpo4 in complex with a dTTP : O^6-BzG pair showed that the mispair is in a pseudo-Watson–Crick geometry, suggesting a similar mode of O^6-alkylG bypass for Dpo4 and BF during misincorporation of dTTP opposite the lesion. Overall, the major difference between bypass of O^6-alkylG by BF and Dpo4 appears to be the relaxed active site of Dpo4, which allows more facile and accurate bypass of the lesion.

Work with Pol T7⁻ and HIV-1 reverse transcriptase revealed that these enzymes form nonproductive complexes with O^6-MeG-modified substrates [151, 152]. Transient-state kinetic experiments showed that transition from a nonproductive to a productive complex can occur in a single binding event, without dissociation of the polymerase · DNA complex. In contrast, Dpo4 catalysis opposite O^6-MeG is slower, which drives kinetic partitioning between k_{pol} and k_{off} toward dissociation and results in a reduced burst amplitude for product formation. The different effect of O^6-MeG upon the catalytic activities of Pol T7⁻ and Dpo4 highlights an important mechanistic difference between polymerase families. While conformational changes do occur during Dpo4 catalysis, they do not appear to be large changes and therefore the free-energy landscape defining different conformational states should be relatively uniform. The open and flexible active site of Dpo4 is tolerant of the conformational adjustments required for catalysis to occur during bypass of the dCTP : O^6-MeG wobble pair. Therefore, Dpo4 is less likely to stall when it encounters O^6-MeG. The more solvent-exposed active site of Dpo4 most likely contributes to higher than normal fidelity opposite O^6-alkylG by allowing the O6:O4 repulsion between O^6-alkylG and dTTP to drive the binding equilibrium of an incoming dTTP towards dissociation. In contrast, the large structural changes observed for enzymes such as BF and Pol T7⁻ should create more pronounced alterations in the free-energy landscape, making it difficult (in a thermodynamic sense) to escape from a free-energy minimum that may not be catalytically competent (i.e., the nonproductive state). In other words, both enzymes probably enter nonproductive states during bypass of O^6-alkylG lesions, but these catalytically incompetent states are more problematic for replicative enzymes.

13.6
Polycyclic Aromatic Hydrocarbons and the Effect of Adduct Size upon Polymerase Catalysis

There are numerous types of bulky DNA lesions. Many of these lesions arise from exposure to exogenous chemicals such as automotive exhaust, cigarette smoke, and environmental contaminants and from cooked foods [135, 139]. The aromatic hydrocarbon benzo[a]pyrene (B[a]P) is one of the most studied carcinogenic compounds and has served as a prototype for the other polycyclic aromatic hydrocarbons [160]. Activation of B[a]P to epoxide intermediates during cytochrome P450-catalyzed detoxication yields the (+)-7R,8S,9S,10R enantiomer of the B[a]P diol epoxide, r7,t8-dihydroxy-t9,10-epoxy-7,8,9,10-dihydroxy-B[a]P (r7,t8-dihydroxy-t9,10-epoxy-7,8,9,10-dihydroxybenzo[a]pyreneB[a]PDE), which can react with DNA to form N^2-deoxyguanosine adducts, with a major adduct being (+)-*trans-anti*-B[a]PDE-N^2-dG (B[a]P-dG) [161, 162]. B[a]P adducts, in general, present strong blocks to most polymerases tested, including both the high-fidelity and translesion enzymes [163–169].

A crystal structure of the model replicative polymerase BF in complex with B[a]P-dG was solved and reveals several features that explain the blocking nature of the lesion [167]. When placed in a postinsertion complex opposite dC, B[a]P-dG forms a relatively normal Watson–Crick geometry. However, placement of the B[a]P moiety into the minor groove blocks binding of the next dNTP and disrupts coordination between Asp830 and the 3′-hydroxyl group at the primer terminus, effectively preventing extension of the dC:B[a]P-dG pair (Figure 13.4). Furthermore, minor groove contacts between the template base and BF that normally contribute to selection of Watson–Crick geometry are blocked by the B[a]P ring system. Finally, the normally underwound A-form duplex observed near the active site of BF is widened further from position −1 (postinsertion site) to −4 (four residues into the duplex), which is in contrast to the B-form duplex observed for B[a]P-dG-modified oligonucleotides in isolation [161, 170].

The Y-family polymerase Dpo4 has been crystallized in complex with two B[a]P adducts, B[a]P-dG [171] and an N^6-B[a]PDE-adenine adduct B[a]P-dA [172]. In

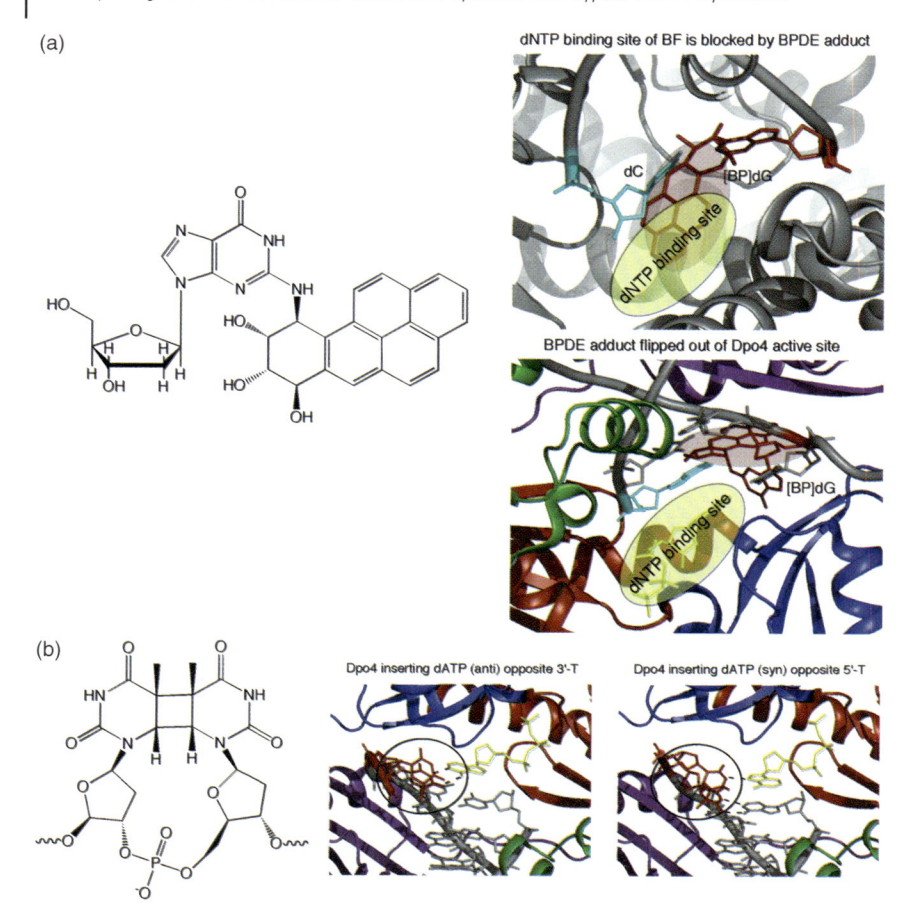

Figure 13.4 Polymerase bypass of bulky lesions. (a) Chemical structure of B[a]P-dG. The dNTP binding site (yellow circle) of the model replicative polymerase BF is blocked by the B[a]P-dG adduct (red circle, labeled [BP]dG) (PDB 1XC9). Dpo4 can flip the B[a]P-dG adduct (red circle) out of the polymerase active site, which allows the incoming dNTP to bind (PDB 2IA6). (b) Y-family polymerase Dpo4 can accommodate a CPD in its active site. Correct insertion of dATP opposite the 3′-T of the CPD occurs through normal Watson–Crick geometry (PDB 1RYR). Adenine is flipped into the *syn* orientation during bypass of the 5′-T (PDB 1RYS). Replicative polymerases cannot accommodate two bases of a CPD in their active site.

contrast to what is observed for BF, the crystal structure of Dpo4 bound to B[a]P-dG-modified DNA reveals two orientations. B[a]P-dG adopts a *syn* orientation, and intercalates the B[a]P moiety between the base pair at the primer/template junction and the nascent base pair to the 5′-side of the lesion, which is effectively a nonproductive, or blocking, conformation because the α-phosphate of the incoming dNTP is 9.0 Å removed from the 3′-hydroxyl group at the primer terminus.

This conformation also results in overwinding of the DNA helix and buckling of the primer/template base pair. A second conformation results in the B[a]P-dG group being "flipped" into a cleft between the finger and little-finger domains of Dpo4 (Figure 13.4). Moving the B[a]P-dG group into the gap between domains minimizes the distortion to the DNA helix and allows the incoming dATP to pair with dTTP to the 5′-side of B[a]P-dG with a distance of 3.9 Å between the 3′-hydroxyl group and the α-phosphate of dATP. This second conformation represents one mode of generating a −1 frameshift mutation, but it also indicates that base substitutions would be noninstructional when B[a]P-dG is flipped out of the Dpo4 active site.

Similar to B[a]P-dG, the B[a]P-dA adduct is found in two conformations when in complex with Dpo4 [172]. In one complex, B[a]P-dA is intercalated between base pairs, blocking nucleotidyl transfer. In the second conformation, B[a]P-dA is in the solvent-exposed major groove. Stabilization of B[a]P-dA in the major groove shifts the purine moiety toward the major groove and is more favorable for pairing the exocyclic amino groups of either dCTP or dATP with the N1 position of the modified adenine ring. Both Dpo4 and Pol κ insert dATP most frequently opposite B[a]P-dA, consistent with the crystal structures [173, 174].

The B[a]P adducts represent very large, bulky additions to DNA bases and, as such, the fact that they are generally blocking to polymerase catalysis is not surprising. Adduct size and steric interference with polymerase catalysis has been studied using a series of N^2-dG adducts of increasing size [175–179]. Simple addition of a methyl or ethyl group to the N2 position of guanine to form N^2-methylguanine or N^2-ethylguanine, respectively, can strongly impede catalysis by some polymerases (e.g., N^2-methylguanine inhibits Pol T7⁻ and HIV-1 reverse transcriptase around 400- and 2000-fold, respectively) [176]. Bulkier modifications further decrease catalytic efficiency to the point of effectively blocking replicative polymerases [176, 180]. Human Pol κ is perhaps the most efficient polymerase for bypassing bulky minor groove adducts. The catalytic efficiency and fidelity of nucleotide incorporation parameters are only reduced slightly even when bypassing bulky N^2-CH$_2$-anthracenyl-dG and N^2-B[a]P-dG, and in some cases (e.g., N^2-isobutyl-dG and N^2-benzyl-dG) the efficiency of bypass is slightly better than unmodified substrates [175]. Pol η shows a slightly greater sensitivity to bulk at the N2 position of guanine compared to Pol κ, with increasing loss of fidelity with adduct size [177]. Catalysis by Pols ι and κ is inhibited by N^2-dG adducts to a similar extent (i.e., the efficiency is decreased maximally only around 60-fold), but Pol ι is much more error-prone [178].

In general, most polymerases do not accommodate bulky adducts efficiently, although specific examples of efficient bypass, such as that observed with Pol κ and N^2-dG adducts, do exist. Pol κ may be important during accurate bypass of bulky minor groove adducts in that diets high in cooked meats have been associated with colorectal adenoma (a precursor to colon cancer) [181–183], and 2- to 4-fold downregulation of Pol κ expression (through downregulation of the cAMP response element binding protein and/or inhibition of deacetylation pathways) has been observed in colorectal biopsies of tumors [184].

13.7
Cyclobutane Pyrimidine Dimers and UV Photoproducts

Exposure to UV radiation is the most important environmental factor contributing to skin cancer. Cyclobutane pyrimidine dimers (CPDs) are the most abundant lesion found in whole skin following exposure to UV-A radiation (320–400 nm) [185]. UV-B radiation (290–320 nm) results primarily in the formation of pyrimidine–pyrimidone (6-4) photoproducts (6-4 lesions) [186]. CPDs are bypassed accurately and efficiently by Y-family Pol η [187] and loss of Pol η activity *in vivo* results in XPV [188] – a syndrome that is characterized by an increased propensity to develop skin cancer. Replication in XPV cells is defective and hypermutable following UV irradiation and treatment with some DNA-damaging agents [189–195]. Other polymerases may delay onset of skin cancer in the absence of Pol η [196], but Pol η is clearly the most important factor in bypass of UV damage. Both yeast and human Pol η can bypass CPDs accurately and with high efficiency by inserting dATP opposite both 3′- and 5′-thymines of the dimer, with only a slight reduction in dATP-binding affinity for each step [187]. The crystal structure of Dpo4 in complex with a *cis-syn* CPD has shown that this Y-family polymerase can accommodate both thymines of the CPD in its active site (Figure 13.4) [197]. For Dpo4, insertion of dATP opposite the first (3′) thymine occurs through normal Watson–Crick geometry, but the second adenine adopts a *syn* orientation to form a Hoogsteen pair with the 5′-dTTP of the CPD. Whether Pol η uses a similar means of bypassing CPDs is not known. The replicative Pol T7 cannot accommodate the 3′-T in a CPD, flipping the lesion out of the active site and failing to adopt a closed conformation in the crystal structure [198]. The unfavorable equilibrium for forming a nascent base pair with the 3′-T in a *cis-syn* thymine dimer is reflective of the restrictive nature of the Pol T7 active site, which is an important feature of high-fidelity enzymes.

The 6-4 lesions generated by exposure to UV-B radiation strongly perturb duplex DNA [199]. Both yeast and human Pol η can insert dGTP opposite the 3′-T in a 6-4 lesion, albeit 50- to 100-fold less efficiently than with an undamaged template [200]. The low-fidelity B-family member Pol ζ can apparently insert dATP opposite the 5′-T of a 6-4 lesion with very little reduction in catalytic efficiency [200]. In contrast to Pol η, Pol ι prefers to accurately insert dATP opposite the 3′-T of a 6-4 lesion [201]. Therefore, accurate bypass of 6-4 lesions may be achieved by the combined action of Pols ι and ζ.

13.8
Inter- and Intrastrand DNA Cross-Links

Fusion of the purine/pyrimidine rings either as intra- or interstrand cross-links can occur upon exposure to ionizing radiation, chemotherapeutic agents (e.g., cisplatin), or photosensitization reactions (psoralens), or from endogenous generation of bisfunctional enals mentioned earlier (e.g., acrolein, crotonaldehyde, and

4-hydroxynonenal) [122, 139]. Interstrand cross-links present an obvious block to the replication fork because they prevent strand separation by helicases and the subsequent generation of single-stranded DNA template. Recognition and repair of cross-linked DNA is not completely understood, and a summary of the relevant literature is beyond the scope of this chapter. Polymerase bypass of some cross-linked DNA substrates has been investigated. For example, Pol κ can apparently bypass a model acrolein-derived N^2-dG–N^2-dG cross-link better than Pol ζ or REV1 [202]. E. coli Pol IV can incorporate dCTP opposite N^2-dG–N^2-dG cross-links *in vitro* and Pol IV-deficient cells show a decreased ability to replicate a cross-link-containing vector [203]. Still, much remains unknown concerning translesion synthesis past cross-linked DNA.

In bacteria, interstrand cross-links can be repaired through a combination of nucleotide excision repair and homologous recombination-dependent pathways [204]. The A-family member DNA Pol I plays an important role in repair of interstrand cross-links in E. coli [205]. Recently, two human A-family members have been identified, Pols θ and ν [36, 206, 207]. Mutations in the coding region for the mouse ortholog of these enzymes (*mus308*) can result in an increased sensitivity to cross-linking agents [208]. The *mus308* gene is predicted to encode a polypeptide that possesses both polymerase and helicase activities. The N-terminal portion of Pol θ encodes seven helicase motifs and the C-terminal region encodes an A-family DNA polymerase [36]. There is substantial evidence to support the idea that Pol θ plays an important, if not dominant role in somatic hypermutation [209]. The ability of a polymerase to catalyze strand displacement ahead of DNA synthesis makes Pol θ an obvious candidate for bypass of unrepaired interstrand cross-links, but kinetic and structural analysis of either Pol θ– or ν-mediated bypass of interstrand cross-links is lacking.

Recently two crystal structures of yeast Pol η bound to an intrastrand cisplatin-derived cross-link between the N7 positions of two adjacent guanines were reported (Pt-GG) [210]. These were the first structures to contain a ternary complex of Pol η, and they provide some insight regarding both translesion DNA synthesis opposite Pt-GG cross-links and basic mechanistic questions regarding catalysis by this Y-family member. The domain and substrate orientation observed in the Pol η structures is similar to that observed with other Y-family polymerases. In both structures, the purine rings of the Pt-GG adduct are perpendicular to one another. Pol η qualitatively prefers to insert dCTP opposite the first guanine residue in the cross-link, and the crystal structures revealed normal Watson–Crick base-pairing between an incoming dCTP and the 3′-guanine. The second incorporation appears to be less faithful and less efficient. The orientation of the Pt-GG moiety changes little during bypass of the second guanine, which severely compromises any ability of Pol η to utilize Watson–Crick geometry. Instead the exocyclic amino group of an incoming dATP forms a single hydrogen bond with the O6 atom of the 5′-dG in the Pt-GG cross-link, providing a rationale for the qualitative decrease in fidelity during insertion opposite the second guanine.

Analysis of the Pol η apo and ternary structures reveals some important features of Y-family catalysis. Binding of DNA and the incoming dNTP results in the thumb and little-finger domains drawing closer to one another (i.e., like a hand grasping a rope). Two different primer/template orientations were observed in the asymmetric unit of the Pol η crystal. One molecule contained DNA in the "pre-elongation" state. In the pre-elongation mode, the 3′-hydroxyl group is located 8.5 Å away from the α-phosphate of the incoming dCTP. The second molecule in the asymmetric unit adopts what is termed the "elongation" state, in which the primer/template DNA has rotated in the binding cleft between the thumb and little finger. This rotation moves the 3′-hydroxyl group to within 5 Å of the α-phosphate. Conformational changes in the thumb and little-finger domains most likely drive the transition from the pre-elongation to the elongation state and may reflect some aspect of the "induced-fit" mechanism derived from kinetic studies with Y-family members. Contacts between the incoming dNTP and the finger domain (specifically the highly conserved Arg67 and the γ-phosphate moiety) are also important features of nucleotide selection for Y-family DNA polymerases. Like other polymerases, Pol η discriminates against ribonucleotides through a so-called "steric-gate" residue (Phe35 for Pol η; Tyr12 for Dpo4), with a hydrogen bond between the amide proton of the steric gate and the 3′-hydroxyl group contributing to dNTP stabilization. The relative contribution of these features towards the catalytic properties of individual Y-family members is still a topic of investigation.

13.9
Conclusions

The past decade has been one of remarkable progress in the interactions of carcinogen-modified DNA with DNA polymerases. In the late 1990s there was a generally good understanding of the basic enzymology of replicative DNA polymerases and some serious kinetic analysis of these DNA polymerases with adducted DNA. However, the translesion DNA polymerases had not been characterized as such and no structures of DNA polymerases bound to carcinogen-adducted DNA were known. The field was dominated by NMR structures of oligonucleotides containing DNA adducts and inferences about how these might relate to biological events.

Today we have a large inventory of translesion DNA polymerases, as well as many interesting replicative ones. We have come from no crystal structures of polymerases with DNA adducts to many. These show the versatility of DNA polymerases in the interactions with adducts, including Watson–Crick, reverse wobble, and Hoogsteen pairing, and even transient pairing of dCTP with an amino acid (Arg324 in REV1). Some of these structures are consistent with the NMR structures obtained with "sealed-in" adducts in oligonucleotides, but others are not predicted. The roles of some individual amino acid residues in modulating DNA base interactions have been defined (e.g., Dpo4 Arg332 with 8-oxo-G [84, 90, 99]).

Much more information is available from kinetic analyses, particularly since it has become possible to begin relating this information to structure. Earlier concepts about equilibria of multiple conformations have been validated by structural studies [40, 152, 159, 172]. Indeed, we have begun to realize that the kinetics (i.e., function) are probably even more complicated than previously thought due to the possibility of more than two conformations of ternary (polymerase · DNA · dNTP) complexes. Progress is being made in identifying rate-controlling steps in the DNA polymerase cycle and the perturbations of the kinetics by DNA adducts.

Despite the progress in the enzymology of DNA polymerases, there are many unanswered questions and future needs. One area is the extension of structural work from the prototypic DNA polymerases to more human enzymes. Further, some of the polymerases work with accessory proteins (e.g., proliferating cell nuclear antigen) and there is only very limited structural work on the complexes [211–213]. With the more relevant eukaryotic DNA polymerases, more kinetic investigation is in order. With all of the DNA polymerases available in a cell, the question of selectivity and trafficking at blocked DNA forks is a complex one. What controls the recruitment of a particular DNA polymerase for some replication events and why does it leave after doing its transient work? How much of an effect does simple "mass action" (i.e., multiple molecules of a polymerase in the vicinity of damage) have upon lesion bypass? Finally, there are many biological questions about which of the translesion and other specialized DNA polymerases are most important in particular cells and tissues, and how downstream DNA damage recognition systems are linked.

References

1 Ito, J. and Braithwaite, D.K. (1991) Compilation and alignment of DNA polymerase sequences. *Nucleic Acids Res.*, **19**, 4045–4057.

2 Ohmori, H., Friedberg, E.C., Fuchs, R.P., Goodman, M.F., Hanaoka, F., Hinkle, D., Kunkel, T.A., Lawrence, C.W., et al. (2001) The Y-family of DNA polymerases. *Mol. Cell*, **8**, 7–8.

3 Bessman, M.J., Lehman, I.R., Simms, E.S., and Kornberg, A. (1958) Enzymatic synthesis of deoxyribonucleic acid. II. General properties of the reaction. *J. Biol. Chem.*, **233**, 171–177.

4 Lehman, I.R., Bessman, M.J., Simms, E.S., and Kornberg, A. (1958) Enzymatic synthesis of deoxyribonucleic acid. I. Preparation of substrates and partial purification of an enzyme from *Escherichia coli*. *J. Biol. Chem.*, **233**, 163–170.

5 Burgers, P.M. (1998) Eukaryotic DNA polymerases in DNA replication and DNA repair. *Chromosoma*, **107**, 218–227.

6 Friedberg, E.C., Fischhaber, P.L., and Kisker, C. (2001) Error-prone DNA polymerases: novel structures and the benefits of infidelity. *Cell*, **107**, 9–12.

7 Friedberg, E.C., Lehmann, A.R., and Fuchs, R.P. (2005) Trading places: how do DNA polymerases switch during translesion DNA synthesis? *Mol. Cell*, **18**, 499–505.

8 Guengerich, F.P. (2006) Interactions of carcinogen-bound DNA with individual DNA polymerases. *Chem. Rev.*, **106**, 420–452.

9 Johnson, K.A. (1993) Conformational coupling in DNA polymerase fidelity. *Annu. Rev. Biochem.*, **62**, 685–713.

10 Joyce, C.M. and Benkovic, S.J. (2004) DNA polymerase fidelity: kinetics, structure, and checkpoints. *Biochemistry*, **43**, 14317–14324.

11 Kunkel, T.A. (2004) DNA replication fidelity. *J. Biol. Chem.*, **279**, 16895–16898.

12 Lehman, I.R. and Richardson, C.C. (1964) The deoxyribonucleases of

Escherichia coli. IV. An exonuclease activity present in purified preparations of deoxyribonucleic acid polymerase. *J. Biol. Chem.*, **239**, 233–241.

13 Livingston, D.M. and Richardson, C.C. (1975) Deoxyribonucleic acid polymerase III of *Escherichia coli*. Characterization of associated exonuclease activities. *J. Biol. Chem.*, **250**, 470–478.

14 Maki, H. and Kornberg, A. (1987) Proofreading by DNA polymerase III of *Escherichia coli* depends on cooperative interaction of the polymerase and exonuclease subunits. *Proc. Natl. Acad. Sci. USA*, **84**, 4389–4392.

15 Richardson, C.C. and Kornberg, A.A. (1964) Deoxyribonucleic acid phosphatase-exonuclease from *Escherichia coli*. I. Purification of the enzyme and characterization of the phosphatase activity. *J. Biol. Chem.*, **239**, 242–250.

16 Richardson, C.C., Lehman, I.R., and Kornberg, A.A. (1964) Deoxyribonucleic acid phosphatase-exonuclease from *Escherichia coli*. II. Characterization of the exonuclease activity. *J. Biol. Chem.*, **239**, 251–258.

17 Tabor, S. and Richardson, C.C. (1989) Selective inactivation of the exonuclease activity of bacteriophage T7 DNA polymerase by *in vitro* mutagenesis. *J. Biol. Chem.*, **264**, 6447–6458.

18 Bedinger, P. and Alberts, B.M. (1983) The 3′–5′ proofreading exonuclease of bacteriophage T4 DNA polymerase is stimulated by other T4 DNA replication proteins. *J. Biol. Chem.*, **258**, 9649–9656.

19 Capson, T.L., Peliska, J.A., Kaboord, B.F., Frey, M.W., Lively, C., Dahlberg, M., and Benkovic, S.J. (1992) Kinetic characterization of the polymerase and exonuclease activities of the gene 43 protein of bacteriophage T4. *Biochemistry*, **31**, 10984–10994.

20 Venkatesan, M. and Nossal, N.G. (1982) Bacteriophage T4 gene 44/62 and gene 45 polymerase accessory proteins stimulate hydrolysis of duplex DNA by T4 DNA polymerase. *J. Biol. Chem.*, **257**, 12435–12443.

21 Irimia, A., Zang, H., Loukachevitch, L.V., Eoff, R.L., Guengerich, F.P., and Egli, M. (2006) Calcium is a cofactor of polymerization but inhibits pyrophosphorolysis by the *Sulfolobus solfataricus* DNA polymerase Dpo4. *Biochemistry*, **45**, 5949–5956.

22 Vaisman, A., Ling, H., Woodgate, R., and Yang, W. (2005) Fidelity of Dpo4: effect of metal ions, nucleotide selection and pyrophosphorolysis. *EMBO J.*, **24**, 2957–2967.

23 Prasad, R., Beard, W.A., Strauss, P.R., and Wilson, S.H. (1998) Human DNA polymerase β deoxyribose phosphate lyase. Substrate specificity and catalytic mechanism. *J. Biol. Chem.*, **273**, 15263–15270.

24 SantaLucia, J., Jr. (1998) A unified view of polymer, dumbbell, and oligonucleotide DNA nearest-neighbor thermodynamics. *Proc. Natl. Acad. Sci. USA*, **95**, 1460–1465.

25 Blattner, F.R., Plunkett, G., 3rd, Bloch, C.A., Perna, N.T., Burland, V., Riley, M., Collado-Vides, J., Glasner, J.D., et al. (1997) The complete genome sequence of *Escherichia coli* K-12. *Science*, **277**, 1453–1474.

26 Kelman, Z. and O'Donnell, M. (1995) DNA polymerase III holoenzyme: structure and function of a chromosomal replicating machine. *Annu. Rev. Biochem.*, **64**, 171–200.

27 McHenry, C.S. (1988) DNA polymerase III holoenzyme of *Escherichia coli*. *Annu. Rev. Biochem.*, **57**, 519–550.

28 Goodman, M.F. (2002) Error-prone repair DNA polymerases in prokaryotes and eukaryotes. *Annu. Rev. Biochem.*, **71**, 17–50.

29 She, Q., Singh, R.K., Confalonieri, F., Zivanovic, Y., Allard, G., Awayez, M.J., Chan-Weiher, C.C., Clausen, I.G., et al. (2001) The complete genome of the crenarchaeon *Sulfolobus solfataricus* P2. *Proc. Natl. Acad. Sci. USA*, **98**, 7835–7840.

30 Loeb, L.A. and Monnat, R.J., Jr. (2008) DNA polymerases and human disease. *Nat. Rev. Genet.*, **9**, 594–604.

31 Steitz, T.A. (1999) DNA polymerases: structural diversity and common mechanisms. *J. Biol. Chem.*, **274**, 17395–17398.

32 Yang, W. and Woodgate, R. (2007) What a difference a decade makes: insights into translesion DNA synthesis. *Proc. Natl. Acad. Sci. USA*, **104**, 15591–15598.

33 Patel, P.H. and Loeb, L.A. (2001) Getting a grip on how DNA polymerases function. *Nat. Struct. Biol.*, **8**, 656–659.

34 Steitz, T.A. (1998) A mechanism for all polymerases. *Nature*, **391**, 231–232.

35 von Hippel, P.H., Fairfield, F.R., and Dolejsi, M.K. (1994) On the processivity of polymerases. *Ann. NY Acad. Sci.*, **726**, 118–131.

36 Seki, M., Marini, F., and Wood, R.D. (2003) POLQ (Pol θ), a DNA polymerase and DNA-dependent ATPase in human cells. *Nucleic Acids Res.*, **31**, 6117–6126.

37 Patel, S.S., Wong, I., and Johnson, K.A. (1991) Pre-steady-state kinetic analysis of processive DNA replication including complete characterization of an exonuclease-deficient mutant. *Biochemistry*, **30**, 511–525.

38 Doublié, S., Sawaya, M.R., and Ellenberger, T. (1999) An open and closed case for all polymerases. *Structure*, **7**, R31–R35.

39 Doublié, S., Tabor, S., Long, A.M., Richardson, C.C., and Ellenberger, T. (1998) Crystal structure of a bacteriophage T7 DNA replication complex at 2.2 A resolution. *Nature*, **391**, 251–258.

40 Tsai, Y.C. and Johnson, K.A. (2006) A new paradigm for DNA polymerase specificity. *Biochemistry*, **45**, 9675–9687.

41 Lin, P., Pedersen, L.C., Batra, V.K., Beard, W.A., Wilson, S.H., and Pedersen, L.G. (2006) Energy analysis of chemistry for correct insertion by DNA polymerase β. *Proc. Natl. Acad. Sci. USA*, **103**, 13294–13299.

42 Wang, L., Yu, X., Hu, P., Broyde, S., and Zhang, Y. (2007) A water-mediated and substrate-assisted catalytic mechanism for *Sulfolobus solfataricus* DNA polymerase IV. *J. Am. Chem. Soc.*, **129**, 4731–4737.

43 Zang, H., Goodenough, A.K., Choi, J.Y., Irimia, A., Loukachevitch, L.V., Kozekov, I.D., Angel, K.C., Rizzo, C.J., et al. (2005) DNA adduct bypass polymerization by *Sulfolobus solfataricus* DNA polymerase Dpo4: analysis and crystal structures of multiple base pair substitution and frameshift products with the adduct 1,N^2-ethenoguanine. *J. Biol. Chem.*, **280**, 29750–29764.

44 Lindahl, T. and Andersson, A. (1972) Rate of chain breakage at apurinic sites in double-stranded deoxyribonucleic acid. *Biochemistry*, **11**, 3618–3623.

45 Nakamura, J., Walker, V.E., Upton, P.B., Chiang, S.Y., Kow, Y.W., and Swenberg, J.A. (1998) Highly sensitive apurinic/apyrimidinic site assay can detect spontaneous and chemically induced depurination under physiological conditions. *Cancer Res.*, **58**, 222–225.

46 Loeb, L.A. and Preston, B.D. (1986) Mutagenesis by apurinic/apyrimidinic sites. *Annu. Rev. Genet.*, **20**, 201–230.

47 Xue, L. and Greenberg, M.M. (2007) Use of fluorescence sensors to determine that 2-deoxyribonolactone is the major alkali-labile deoxyribose lesion produced in oxidatively damaged DNA. *Angew. Chem. Int. Ed.*, **46**, 561–564.

48 Kokoska, R.J., McCulloch, S.D., and Kunkel, T.A. (2003) The efficiency and specificity of apurinic/apyrimidinic site bypass by human DNA polymerase η and *Sulfolobus solfataricus* Dpo4. *J. Biol. Chem.*, **278**, 50537–50545.

49 Paz-Elizur, T., Takeshita, M., Goodman, M., O'Donnell, M., and Livneh, Z. (1996) Mechanism of translesion DNA synthesis by DNA polymerase II. Comparison to DNA polymerases I and III core. *J. Biol. Chem.*, **271**, 24662–24669.

50 Randall, S.K., Eritja, R., Kaplan, B.E., Petruska, J., and Goodman, M.F. (1987) Nucleotide insertion kinetics opposite abasic lesions in DNA. *J. Biol. Chem.*, **262**, 6864–6870.

51 Shibutani, S., Takeshita, M., and Grollman, A.P. (1997) Translesional synthesis on DNA templates containing a single abasic site. A mechanistic study of the "A rule". *J. Biol. Chem.*, **272**, 13916–13922.

52 Tang, M., Pham, P., Shen, X., Taylor, J.S., O'Donnell, M., Woodgate, R., and Goodman, M.F. (2000) Roles of *E. coli* DNA polymerases IV and V in lesion-targeted and untargeted SOS mutagenesis. *Nature*, **404**, 1014–1018.

53 Gibbs, P.E., McDonald, J., Woodgate, R., and Lawrence, C.W. (2005) The relative roles *in vivo* of *Saccharomyces cerevisiae* Pol η, Pol ζ, Rev1 protein and

Pol32 in the bypass and mutation induction of an abasic site, T-T (6-4) photoadduct and T-T *cis-syn* cyclobutane dimer. *Genetics*, **169**, 575–582.
54 Haracska, L., Unk, I., Johnson, R.E., Johansson, E., Burgers, P.M., Prakash, S., and Prakash, L. (2001) Roles of yeast DNA polymerases δ and ζ and of Rev1 in the bypass of abasic sites. *Genes Dev.*, **15**, 945–954.
55 Mozzherin, D.J., Shibutani, S., Tan, C.K., Downey, K.M., and Fisher, P.A. (1997) Proliferating cell nuclear antigen promotes DNA synthesis past template lesions by mammalian DNA polymerase δ. *Proc. Natl. Acad. Sci. USA*, **94**, 6126–6131.
56 Avkin, S., Adar, S., Blander, G., and Livneh, Z. (2002) Quantitative measurement of translesion replication in human cells: evidence for bypass of abasic sites by a replicative DNA polymerase. *Proc. Natl. Acad. Sci. USA*, **99**, 3764–3769.
57 Haracska, L., Washington, M.T., Prakash, S., and Prakash, L. (2001) Inefficient bypass of an abasic site by DNA polymerase η. *J. Biol. Chem.*, **276**, 6861–6866.
58 Haracska, L., Prakash, L., and Prakash, S. (2002) Role of human DNA polymerase κ as an extender in translesion synthesis. *Proc. Natl. Acad. Sci. USA*, **99**, 16000–16005.
59 Wolfle, W.T., Washington, M.T., Prakash, L., and Prakash, S. (2003) Human DNA polymerase κ uses template–primer misalignment as a novel means for extending mispaired termini and for generating single-base deletions. *Genes Dev.*, **17**, 2191–2199.
60 Otsuka, C., Loakes, D., and Negishi, K. (2002) The role of deoxycytidyl transferase activity of yeast Rev1 protein in the bypass of abasic sites. *Nucleic Acids Res. Suppl.*, **2**, 87–88.
61 Zhao, B., Xie, Z., Shen, H., and Wang, Z. (2004) Role of DNA polymerase η in the bypass of abasic sites in yeast cells. *Nucleic Acids Res.*, **32**, 3984–3994.
62 Greer, S. and Zamenhof, S. (1962) Studies on depurination of DNA by heat. *J. Mol. Biol.*, **4**, 123–141.
63 Suzuki, T., Ohsumi, S., and Makino, K. (1994) Mechanistic studies on depurination and apurinic site chain breakage in oligodeoxyribonucleotides. *Nucleic Acids Res.*, **22**, 4997–5003.
64 Hogg, M., Wallace, S.S., and Doublie, S. (2004) Crystallographic snapshots of a replicative DNA polymerase encountering an abasic site. *EMBO J.*, **23**, 1483–1493.
65 Ling, H., Boudsocq, F., Woodgate, R., and Yang, W. (2004) Snapshots of replication through an abasic lesion; structural basis for base substitutions and frameshifts. *Mol. Cell*, **13**, 751–762.
66 Zahn, K.E., Belrhali, H., Wallace, S.S., and Doublié, S. (2007) Caught bending the A-rule: crystal structures of translesion DNA synthesis with a non-natural nucleotide. *Biochemistry*, **46**, 10551–10561.
67 Fiala, K.A., Hypes, C.D., and Suo, Z. (2007) Mechanism of abasic lesion bypass catalyzed by a Y-family DNA polymerase. *J. Biol. Chem.*, **282**, 8188–8198.
68 Fiala, K.A. and Suo, Z. (2007) Sloppy bypass of an abasic lesion catalyzed by a Y-family DNA polymerase. *J. Biol. Chem.*, **282**, 8199–8206.
69 Reineks, E.Z. and Berdis, A.J. (2004) Evaluating the contribution of base stacking during translesion DNA replication. *Biochemistry*, **43**, 393–404.
70 Ames, B.N. (1982) Mutagens, carcinogens, and anti-carcinogens. *Basic Life Sci.*, **21**, 499–508.
71 Fraga, C.G., Shigenaga, M.K., Park, J.W., Degan, P., and Ames, B.N. (1990) Oxidative damage to DNA during aging: 8-hydroxy-2′-deoxyguanosine in rat organ DNA and urine. *Proc. Natl. Acad. Sci. USA*, **87**, 4533–4537.
72 Marnett, L.J. (2000) Oxyradicals and DNA damage. *Carcinogenesis*, **21**, 361–370.
73 Marnett, L.J. (2002) Oxy radicals, lipid peroxidation and DNA damage. *Toxicology*, **181–182**, 219–222.
74 Ames, B.N. (1983) Dietary carcinogens and anticarcinogens. Oxygen radicals and degenerative diseases. *Science*, **221**, 1256–1264.
75 Valko, M., Morris, H., and Cronin, M.T. (2005) Metals, toxicity and oxidative stress. *Curr. Med. Chem.*, **12**, 1161–1208.

76 Wallace, K.B. and Starkov, A.A. (2000) Mitochondrial targets of drug toxicity. *Annu. Rev. Pharmacol. Toxicol.*, **40**, 353–388.

77 MacMicking, J., Xie, Q.W., and Nathan, C. (1997) Nitric oxide and macrophage function. *Annu. Rev. Immunol.*, **15**, 323–350.

78 Nathan, C. and Shiloh, M.U. (2000) Reactive oxygen and nitrogen intermediates in the relationship between mammalian hosts and microbial pathogens. *Proc. Natl. Acad. Sci. USA*, **97**, 8841–8848.

79 Aruoma, O.I., Halliwell, B., and Dizdaroglu, M. (1989) Iron ion-dependent modification of bases in DNA by the superoxide radical-generating system hypoxanthine/xanthine oxidase. *J. Biol. Chem.*, **264**, 13024–13028.

80 Aruoma, O.I., Halliwell, B., Gajewski, E., and Dizdaroglu, M. (1989) Damage to the bases in DNA induced by hydrogen peroxide and ferric ion chelates. *J. Biol. Chem.*, **264**, 20509–20512.

81 Demple, B. and Harrison, L. (1994) Repair of oxidative damage to DNA: enzymology and biology. *Annu. Rev. Biochem.*, **63**, 915–948.

82 Brieba, L.G., Eichman, B.F., Kokoska, R.J., Doublié, S., Kunkel, T.A., and Ellenberger, T. (2004) Structural basis for the dual coding potential of 8-oxoguanosine by a high-fidelity DNA polymerase. *EMBO J.*, **23**, 3452–3461.

83 Einolf, H.J. and Guengerich, F.P. (2001) Fidelity of nucleotide insertion at 8-oxo-7,8-dihydroguanine by mammalian DNA polymerase δ. Steady-state and pre-steady-state kinetic analysis. *J. Biol. Chem.*, **276**, 3764–3771.

84 Eoff, R.L., Irimia, A., Angel, K.C., Egli, M., and Guengerich, F.P. (2007) Hydrogen bonding of 7,8-dihydro-8-oxodeoxyguanosine with a charged residue in the little finger domain determines miscoding events in *Sulfolobus solfataricus* DNA polymerase Dpo4. *J. Biol. Chem.*, **282**, 19831–19843.

85 Furge, L.L. and Guengerich, F.P. (1997) Analysis of nucleotide insertion and extension at 8-oxo-7,8-dihydroguanine by replicative T7 polymerase exo$^-$ and human immunodeficiency virus-1 reverse transcriptase using steady-state and pre-steady-state kinetics. *Biochemistry*, **36**, 6475–6487.

86 Furge, L.L. and Guengerich, F.P. (1998) Pre-steady-state kinetics of nucleotide insertion following 8-oxo-7,8-dihydroguanine base pair mismatches by bacteriophage T7 DNA polymerase exo$^-$. *Biochemistry*, **37**, 3567–3574.

87 Haracska, L., Prakash, S., and Prakash, L. (2003) Yeast DNA polymerase ζ is an efficient extender of primer ends opposite from 7,8-dihydro-8-oxoguanine and O^6-methylguanine. *Mol. Cell. Biol.*, **23**, 1453–1459.

88 Haracska, L., Yu, S.L., Johnson, R.E., Prakash, L., and Prakash, S. (2000) Efficient and accurate replication in the presence of 7,8-dihydro-8-oxoguanine by DNA polymerase η. *Nat. Genet.*, **25**, 458–461.

89 Shibutani, S., Takeshita, M., and Grollman, A.P. (1991) Insertion of specific bases during DNA synthesis past the oxidation-damaged base 8-oxodG. *Nature*, **349**, 431–434.

90 Zang, H., Irimia, A., Choi, J.Y., Angel, K.C., Loukachevitch, L.V., Egli, M., and Guengerich, F.P. (2006) Efficient and high fidelity incorporation of dCTP opposite 7,8-dihydro-8-oxodeoxyguanosine by *Sulfolobus solfataricus* DNA polymerase Dpo4. *J. Biol. Chem.*, **281**, 2358–2372.

91 Einolf, H.J., Schnetz-Boutaud, N., and Guengerich, F.P. (1998) Steady-state and pre-steady-state kinetic analysis of 8-oxo-7,8-dihydroguanosine triphosphate incorporation and extension by replicative and repair DNA polymerases. *Biochemistry*, **37**, 13300–13312.

92 Lowe, L.G. and Guengerich, F.P. (1996) Steady-state and pre-steady-state kinetic analysis of dNTP insertion opposite 8-oxo-7,8-dihydroguanine by *Escherichia coli* polymerases I exo$^-$ and II exo$^-$. *Biochemistry*, **35**, 9840–9849.

93 Carlson, K.D. and Washington, M.T. (2005) Mechanism of efficient and accurate nucleotide incorporation opposite 7,8-dihydro-8-oxoguanine by *Saccharomyces cerevisiae* DNA polymerase η. *Mol. Cell. Biol.*, **25**, 2169–2176.

94 Efrati, E., Tocco, G., Eritja, R., Wilson, S.H., and Goodman, M.F. (1997) Abasic

translesion synthesis by DNA polymerase β violates the "A-rule". Novel types of nucleotide incorporation by human DNA polymerase β at an abasic lesion in different sequence contexts. *J. Biol. Chem.*, **272**, 2559–2569.

95 Pinz, K.G., Shibutani, S., and Bogenhagen, D.F. (1995) Action of mitochondrial DNA polymerase γ at sites of base loss or oxidative damage. *J. Biol. Chem.*, **270**, 9202–9206.

96 Hsu, G.W., Ober, M., Carell, T., and Beese, L.S. (2004) Error-prone replication of oxidatively damaged DNA by a high-fidelity DNA polymerase. *Nature*, **431**, 217–221.

97 McAuley-Hecht, K.E., Leonard, G.A., Gibson, N.J., Thomson, J.B., Watson, W.P., Hunter, W.N., and Brown, T. (1994) Crystal structure of a DNA duplex containing 8-hydroxydeoxyguanine-adenine base pairs. *Biochemistry*, **33**, 10266–10270.

98 Freisinger, E., Grollman, A.P., Miller, H., and Kisker, C. (2004) Lesion (in)tolerance reveals insights into DNA replication fidelity. *EMBO J.*, **23**, 1494–1505.

99 Rechkoblit, O., Malinina, L., Cheng, Y., Kuryavyi, V., Broyde, S., Geacintov, N.E., and Patel, D.J. (2006) Stepwise translocation of Dpo4 polymerase during error-free bypass of an oxoG lesion. *PLoS Biol.*, **4**, e11.

100 Lee, D.H. and Pfeifer, G.P. (2008) Translesion synthesis of 7,8-dihydro-8-oxo-2′-deoxyguanosine by DNA polymerase η *in vivo*. *Mutat. Res.*, **641**, 19–26.

101 Patro, J.N., Wiederholt, C.J., Jiang, Y.L., Delaney, J.C., Essigmann, J.M., and Greenberg, M.M. (2007) Studies on the replication of the ring opened formamidopyrimidine, Fapy · dG in *Escherichia coli*. *Biochemistry*, **46**, 10202–10212.

102 Kalam, M.A., Haraguchi, K., Chandani, S., Loechler, E.L., Moriya, M., Greenberg, M.M., and Basu, A.K. (2006) Genetic effects of oxidative DNA damages: comparative mutagenesis of the imidazole ring-opened formamidopyrimidines (Fapy lesions) and 8-oxo-purines in simian kidney cells. *Nucleic Acids Res.*, **34**, 2305–2315.

103 Cathcart, R., Schwiers, E., Saul, R.L., and Ames, B.N. (1984) Thymine glycol and thymidine glycol in human and rat urine: a possible assay for oxidative DNA damage. *Proc. Natl. Acad. Sci. USA*, **81**, 5633–5637.

104 Flippen, J.L. (1973) Crystal and molecular structures of reaction products from γ-irradiation of thymine and cytosine. *cis*-Thymine glycol and *trans*-1-cabamoylimidazolidone-4,5-diol. *Acta Crystallogr. B*, **29**, 1756–1762.

105 Clark, J.M. and Beardsley, G.P. (1987) Functional effects of *cis*-thymine glycol lesions on DNA synthesis *in vitro*. *Biochemistry*, **26**, 5398–5403.

106 Johnson, R.E., Yu, S.L., Prakash, S., and Prakash, L. (2003) Yeast DNA polymerase ζ is essential for error-free replication past thymine glycol. *Genes Dev.*, **17**, 77–87.

107 Kusumoto, R., Masutani, C., Iwai, S., and Hanaoka, F. (2002) Translesion synthesis by human DNA polymerase η across thymine glycol lesions. *Biochemistry*, **41**, 6090–6099.

108 Fischhaber, P.L., Gerlach, V.L., Feaver, W.J., Hatahet, Z., Wallace, S.S., and Friedberg, E.C. (2002) Human DNA polymerase κ bypasses and extends beyond thymine glycols during translesion synthesis *in vitro*, preferentially incorporating correct nucleotides. *J. Biol. Chem.*, **277**, 37604–37611.

109 Aller, P., Rould, M.A., Hogg, M., Wallace, S.S., and Doublié, S. (2007) A structural rationale for stalling of a replicative DNA polymerase at the most common oxidative thymine lesion, thymine glycol. *Proc. Natl. Acad. Sci. USA*, **104**, 814–818.

110 Kung, H.C. and Bolton, P.H. (1997) Structure of a duplex DNA containing a thymine glycol residue in solution. *J. Biol. Chem.*, **272**, 9227–9236.

111 Clark, J.M. and Beardsley, G.P. (1989) Template length, sequence context, and 3′–5′ exonuclease activity modulate replicative bypass of thymine glycol lesions *in vitro*. *Biochemistry*, **28**, 775–779.

112 Hayes, R.C. and LeClerc, J.E. (1986) Sequence dependence for bypass of thymine glycols in DNA by DNA polymerase I. *Nucleic Acids Res.*, **14**, 1045–1061.

113 Akasaka, S. and Guengerich, F.P. (1999) Mutagenicity of site-specifically located 1,N^2-ethenoguanine in Chinese hamster ovary cell chromosomal DNA. *Chem. Res. Toxicol.*, **12**, 501–507.

114 Guengerich, F.P., Langouet, S., Mican, A.N., Akasaka, S., Muller, M., and Persmark, M. (1999) Formation of etheno adducts and their effects on DNA polymerases. *IARC Sci. Publ.*, **150**, 137–145.

115 Moriya, M., Zhang, W., Johnson, F., and Grollman, A.P. (1994) Mutagenic potency of exocyclic DNA adducts: marked differences between *Escherichia coli* and simian kidney cells. *Proc. Natl. Acad. Sci. USA*, **91**, 11899–11903.

116 Pandya, G.A. and Moriya, M. (1996) 1,N^6-ethenodeoxyadenosine, a DNA adduct highly mutagenic in mammalian cells. *Biochemistry*, **35**, 11487–11492.

117 Choi, J.Y., Zang, H., Angel, K.C., Kozekov, I.D., Goodenough, A.K., Rizzo, C.J., and Guengerich, F.P. (2006) Translesion synthesis across 1,N^2-ethenoguanine by human DNA polymerases. *Chem. Res. Toxicol.*, **19**, 879–886.

118 Langouet, S., Mican, A.N., Muller, M., Fink, S.P., Marnett, L.J., Muhle, S.A., and Guengerich, F.P. (1998) Misincorporation of nucleotides opposite five-membered exocyclic ring guanine derivatives by *Escherichia coli* polymerases *in vitro* and *in vivo*: 1,N^2-ethenoguanine, 5,6,7,9-tetrahydro-9-oxoimidazo[1,2-*a*]purine, and 5,6,7,9-tetrahydro-7-hydroxy-9-oxoimidazo[1,2-*a*]purine. *Biochemistry*, **37**, 5184–5193.

119 Wolfle, W.T., Johnson, R.E., Minko, I.G., Lloyd, R.S., Prakash, S., and Prakash, L. (2005) Human DNA polymerase ι promotes replication through a ring-closed minor-groove adduct that adopts a *syn* conformation in DNA. *Mol. Cell Biol.*, **25**, 8748–8754.

120 Shanmugam, G., Goodenough, A.K., Kozekov, I.D., Guengerich, F.P., Rizzo, C.J., and Stone, M.P. (2007) Structure of the 1,N^2-etheno-2′-deoxyguanosine adduct in duplex DNA at pH 8.6. *Chem. Res. Toxicol.*, **20**, 1601–1611.

121 Shanmugam, G., Kozekov, I.D., Guengerich, F.P., Rizzo, C.J., and Stone, M.P. (2008) Structure of the 1,N^2-ethenodeoxyguanosine adduct opposite cytosine in duplex DNA: Hoogsteen base pairing at pH 5.2. *Chem. Res. Toxicol.*, **21**, 1795–1805.

122 Stone, M.P., Cho, Y.J., Huang, H., Kim, H.Y., Kozekov, I.D., Kozekova, A., Wang, H., Minko, I.G., *et al.* (2008) Interstrand DNA cross-links induced by α,β-unsaturated aldehydes derived from lipid peroxidation and environmental sources. *Acc. Chem. Res.*, **41**, 793–804.

123 Mao, H., Schnetz-Boutaud, N.C., Weisenseel, J.P., Marnett, L.J., and Stone, M.P. (1999) Duplex DNA catalyzes the chemical rearrangement of a malondialdehyde deoxyguanosine adduct. *Proc. Natl. Acad. Sci. USA*, **96**, 6615–6620.

124 Guengerich, F.P., Persmark, M., and Humphreys, W.G. (1993) Formation of 1,N^2- and N^2,3-ethenoguanine from 2-halooxiranes: isotopic labeling studies and isolation of a hemiaminal derivative of N^2-(2-oxoethyl)guanine. *Chem. Res. Toxicol.*, **6**, 635–648.

125 de los Santos, C., Zaliznyak, T., and Johnson, F. (2001) NMR characterization of a DNA duplex containing the major acrolein-derived deoxyguanosine adduct γ-OH-1,-N^2-propano-2′-deoxyguanosine. *J. Biol. Chem.*, **276**, 9077–9082.

126 Minko, I.G., Washington, M.T., Kanuri, M., Prakash, L., Prakash, S., and Lloyd, R.S. (2003) Translesion synthesis past acrolein-derived DNA adduct, γ-hydroxypropanodeoxyguanosine, by yeast and human DNA polymerase η. *J. Biol. Chem.*, **278**, 784–790.

127 Nair, D.T., Johnson, R.E., Prakash, L., Prakash, S., and Aggarwal, A.K. (2008) Protein-template-directed synthesis across an acrolein-derived DNA adduct by yeast Rev1 DNA polymerase. *Structure*, **16**, 239–245.

128 Washington, M.T., Minko, I.G., Johnson, R.E., Wolfle, W.T., Harris, T.M., Lloyd, R.S., Prakash, S., and Prakash, L. (2004) Efficient and error-free replication past a minor-groove DNA adduct by the sequential action of human DNA polymerases ι and κ. *Mol. Cell. Biol.*, **24**, 5687–5693.

129 Wang, Y., Musser, S.K., Saleh, S., Marnett, L.J., Egli, M., and Stone, M.P.

(2008) Insertion of dNTPs opposite the 1,N^2-propanodeoxyguanosine adduct by *Sulfolobus solfataricus* P2 DNA polymerase IV. *Biochemistry*, **47**, 7322–7334.

130 Basu, A.K., Niedernhofer, L.J., and Essigmann, J.M. (1987) Deoxyhexanucleotide containing a vinyl chloride induced DNA lesion, 1,N^6-ethenoadenine: synthesis, physical characterization, and incorporation into a duplex bacteriophage M13 genome as part of an amber codon. *Biochemistry*, **26**, 5626–5635.

131 Basu, A.K., Wood, M.L., Niedernhofer, L.J., Ramos, L.A., and Essigmann, J.M. (1993) Mutagenic and genotoxic effects of three vinyl chloride-induced DNA lesions: 1,N^6-ethenoadenine, 3,N^4-ethenocytosine, and 4-amino-5-(imidazol-2-yl) imidazole. *Biochemistry*, **32**, 12793–12801.

132 Müller, M., Belas, F.J., Blair, I.A., and Guengerich, F.P. (1997) Analysis of 1,N^2-ethenoguanine and 5,6,7,9-tetrahydro-7-hydroxy-9-oxoimidazo[1,2-α]purine in DNA treated with 2-chlorooxirane by high performance liquid chromatography/electrospray mass spectrometry and comparison of amounts to other DNA adducts. *Chem. Res. Toxicol.*, **10**, 242–247.

133 Cheng, K.C., Preston, B.D., Cahill, D.S., Dosanjh, M.K., Singer, B., and Loeb, L.A. (1991) The vinyl chloride DNA derivative N^2,3-ethenoguanine produces G → A transitions in *Escherichia coli*. *Proc. Natl. Acad. Sci. USA*, **88**, 9974–9978.

134 Kusmierek, J.T., Folkman, W., and Singer, B. (1989) Synthesis of N^2,3-ethenodeoxyguanosine, N^2,3-ethenodeoxyguanosine 5′-phosphate, and N^2,3-ethenodeoxyguanosine 5′-triphosphate. Stability of the glycosyl bond in the monomer and in poly(dG,epsilon dG–dC). *Chem. Res. Toxicol.*, **2**, 230–233.

135 Lawley, P.D. (1984) Carcinogenesis by alkylating agents, in *Chemical Carcinogens*, 2nd edn (ed. C.E. Searle), ACS Monograph 182, American Chemical Society, Washington, DC, pp. 325–484.

136 Loveless, A. (1969) Possible relevance of O^6 alkylation of deoxyguanosine to the mutagenicity and carcinogenicity of nitrosamines and nitrosamides. *Nature*, **223**, 206–207.

137 Singer, B. and Kusmierek, J.T. (1982) Chemical mutagenesis. *Annu. Rev. Biochem.*, **51**, 655–693.

138 Lawley, P.D. and Brookes, P. (1961) Acidic dissociation of 7:9-dialkylguanines and its possible relation to mutagenic properties of alkylating agents. *Nature*, **192**, 1081–1082.

139 Friedberg, E.C., Walker, G.C., Siede, W., Wood, R.D., Shultz, R.A., and Ellenberger, T. (2006) *DNA Repair and Mutagenesis*, 2nd edn, ASM Press, Washington, DC.

140 Margison, G.P., Santibanez Koref, M.F., and Povey, A.C. (2002) Mechanisms of carcinogenicity/chemotherapy by O^6-methylguanine. *Mutagenesis*, **17**, 483–487.

141 Wyatt, M.D. and Pittman, D.L. (2006) Methylating agents and DNA repair responses: methylated bases and sources of strand breaks. *Chem. Res. Toxicol.*, **19**, 1580–1594.

142 Grafstrom, R.C., Pegg, A.E., Trump, B.F., and Harris, C.C. (1984) O^6-alkylguanine-DNA alkyltransferase activity in normal human tissues and cells. *Cancer Res.*, **44**, 2855–2857.

143 Pegg, A.E., Roberfroid, M., von Bahr, C., Foote, R.S., Mitra, S., Bresil, H., Likhachev, A., and Montesano, R. (1982) Removal of O^6-methylguanine from DNA by human liver fractions. *Proc. Natl. Acad. Sci. USA*, **79**, 5162–5165.

144 Pegg, A.E., Scicchitano, D., Morimoto, K., and Dolan, M.E. (1987) Specificity of O^6-alkylguanine-DNA alkyltransferase. *IARC Sci. Publ.*, **84**, 30–34.

145 Shen, L., Kondo, Y., Rosner, G.L., Xiao, L., Hernandez, N.S., Vilaythong, J., Houlihan, P.S., Krouse, R.S., *et al.* (2005) MGMT promoter methylation and field defect in sporadic colorectal cancer. *J. Natl. Cancer Inst.*, **97**, 1330–1338.

146 Essigmann, J.M., Loechler, E.L., and Green, C.L. (1986) Mutagenesis and repair of O^6-substituted guanines. *IARC Sci. Publ.*, **70**, 393–399.

147 Essigmann, J.M., Loechler, E.L., and Green, C.L. (1986) Genetic toxicology of

O^6 methylguanine. *Prog. Clin. Biol. Res.*, **209A**, 433–440.

148 Loechler, E.L., Green, C.L., and Essigmann, J.M. (1984) *In vivo* mutagenesis by O^6-methylguanine built into a unique site in a viral genome. *Proc. Natl. Acad. Sci. USA*, **81**, 6271–6275.

149 Choi, J.Y., Chowdhury, G., Zang, H., Angel, K.C., Vu, C.C., Peterson, L.A., and Guengerich, F.P. (2006) Translesion synthesis across O^6-alkylguanine DNA adducts by recombinant human DNA polymerases. *J. Biol. Chem.*, **281**, 38244–38256.

150 Eoff, R.L., Irimia, A., Egli, M., and Guengerich, F.P. (2007) *Sulfolobus solfataricus* DNA polymerase Dpo4 is partially inhibited by "wobble" pairing between O^6-methylguanine and cytosine, but accurate bypass is preferred. *J. Biol. Chem.*, **282**, 1456–1467.

151 Woodside, A.M. and Guengerich, F.P. (2002) Effect of the O^6 substituent on misincorporation kinetics catalyzed by DNA polymerases at O^6-methylguanine and O^6-benzylguanine. *Biochemistry*, **41**, 1027–1038.

152 Woodside, A.M. and Guengerich, F.P. (2002) Misincorporation and stalling at O^6-methylguanine and O^6-benzylguanine: evidence for inactive polymerase complexes. *Biochemistry*, **41**, 1039–1050.

153 Ginell, S.L., Kuzmich, S., Jones, R.A., and Berman, H.M. (1990) Crystal and molecular structure of a DNA duplex containing the carcinogenic lesion O^6-methylguanine. *Biochemistry*, **29**, 10461–10465.

154 Patel, D.J., Shapiro, L., Kozlowski, S.A., Gaffney, B.L., and Jones, R.A. (1986) Structural studies of the O^6MeG · C interaction in the d(C-G-C-G-A-A-T-T-C-O^6MeG-C-G) duplex. *Biochemistry*, **25**, 1027–1036.

155 Patel, D.J., Shapiro, L., Kozlowski, S.A., Gaffney, B.L., and Jones, R.A. (1986) Structural studies of the O^6MeG · T interaction in the d(C-G-T-G-A-A-T-T-C-O^6MeG-C-G) duplex. *Biochemistry*, **25**, 1036–1042.

156 Sriram, M., van der Marel, G.A., Roelen, H.L., van Boom, J.H., and Wang, A.H. (1992) Structural consequences of a carcinogenic alkylation lesion on DNA: effect of O^6-ethylguanine on the molecular structure of the d(CGC[e6G]AATTCGCG)-netropsin complex. *Biochemistry*, **31**, 11823–11834.

157 Gaffney, B.L. and Jones, R.A. (1989) Thermodynamic comparison of the base pairs formed by the carcinogenic lesion O^6-methylguanine with reference both to Watson–Crick pairs and to mismatched pairs. *Biochemistry*, **28**, 5881–5889.

158 Warren, J.J., Forsberg, L.J., and Beese, L.S. (2006) The structural basis for the mutagenicity of O^6-methyl-guanine lesions. *Proc. Natl. Acad. Sci. USA*, **103**, 19701–19706.

159 Eoff, R.L., Angel, K.C., Egli, M., and Guengerich, F.P. (2007) Molecular basis of selectivity of nucleoside triphosphate incorporation opposite O^6-benzylguanine by *Sulfolobus solfataricus* DNA polymerase Dpo4: steady-state and pre-steady-state kinetics and X-ray crystallography of correct and incorrect pairing. *J. Biol. Chem.*, **282**, 13573–13584.

160 Phillips, D.H. (1983) Fifty years of benzo[a]pyrene. *Nature*, **303**, 468–472.

161 Cosman, M., de los Santos, C., Fiala, R., Hingerty, B.E., Singh, S.B., Ibanez, V., Margulis, L.A., Live, D., *et al.* (1992) Solution conformation of the major adduct between the carcinogen (+)-*anti*-benzo[a]pyrene diol epoxide and DNA. *Proc. Natl. Acad. Sci. USA*, **89**, 1914–1918.

162 Graslund, A. and Jernstrom, B. (1989) DNA–carcinogen interaction: covalent DNA-adducts of benzo[a]pyrene 7,8-dihydrodiol 9,10-epoxides studied by biochemical and biophysical techniques. *Q. Rev. Biophys.*, **22**, 1–37.

163 Alekseyev, Y.O., Dzantiev, L., and Romano, L.J. (2001) Effects of benzo[a]pyrene DNA adducts on *Escherichia coli* DNA polymerase I (Klenow fragment) primer–template interactions: evidence for inhibition of the catalytically active ternary complex formation. *Biochemistry*, **40**, 2282–2290.

164 Alekseyev, Y.O. and Romano, L.J. (2000) *In vitro* replication of primer-templates containing benzo[a]pyrene adducts by exonuclease-deficient *Escherichia coli* DNA polymerase I (Klenow fragment):

effect of sequence context on lesion bypass. *Biochemistry*, **39**, 10431–10438.
165 Alekseyev, Y.O. and Romano, L.J. (2002) Effects of benzo[a]pyrene adduct stereochemistry on downstream DNA replication *in vitro*: evidence for different adduct conformations within the active site of DNA polymerase I (Klenow fragment). *Biochemistry*, **41**, 4467–4479.
166 Hruszkewycz, A.M., Canella, K.A., Peltonen, K., Kotrappa, L., and Dipple, A. (1992) DNA polymerase action on benzo[a]pyrene-DNA adducts. *Carcinogenesis*, **13**, 2347–2352.
167 Hsu, G.W., Huang, X., Luneva, N.P., Geacintov, N.E., and Beese, L.S. (2005) Structure of a high fidelity DNA polymerase bound to a benzo[a]pyrene adduct that blocks replication. *J. Biol. Chem.*, **280**, 3764–3770.
168 Rechkoblit, O., Zhang, Y., Guo, D., Wang, Z., Amin, S., Krzeminsky, J., Louneva, N., and Geacintov, N.E. (2002) *trans*-Lesion synthesis past bulky benzo[a]pyrene diol epoxide N^2-dG and N^6-dA lesions catalyzed by DNA bypass polymerases. *J. Biol. Chem.*, **277**, 30488–30494.
169 Shibutani, S., Margulis, L.A., Geacintov, N.E., and Grollman, A.P. (1993) Translesional synthesis on a DNA template containing a single stereoisomer of dG-(+)- or dG-(–)-*anti*-BPDE (7,8-dihydroxy-*anti*-9,10-epoxy-7,8,9,10-tetrahydrobenzo[a]pyrene). *Biochemistry*, **32**, 7531–7541.
170 Feng, B., Gorin, A., Hingerty, B.E., Geacintov, N.E., Broyde, S., and Patel, D.J. (1997) Structural alignment of the (+)-*trans-anti*-benzo[a]pyrene-dG adduct positioned opposite dC at a DNA template–primer junction. *Biochemistry*, **36**, 13769–13779.
171 Bauer, J., Xing, G., Yagi, H., Sayer, J.M., Jerina, D.M., and Ling, H. (2007) A structural gap in Dpo4 supports mutagenic bypass of a major benzo[a]pyrene dG adduct in DNA through template misalignment. *Proc. Natl. Acad. Sci. USA*, **104**, 14905–14910.
172 Ling, H., Sayer, J.M., Plosky, B.S., Yagi, H., Boudsocq, F., Woodgate, R., Jerina, D.M., and Yang, W. (2004) Crystal structure of a benzo[a]pyrene diol epoxide adduct in a ternary complex with a DNA polymerase. *Proc. Natl. Acad. Sci. USA*, **101**, 2265–2269.
173 Ogi, T., Shinkai, Y., Tanaka, K., and Ohmori, H. (2002) Pol κ protects mammalian cells against the lethal and mutagenic effects of benzo[a]pyrene. *Proc. Natl. Acad. Sci. USA*, **99**, 15548–15553.
174 Suzuki, N., Ohashi, E., Kolbanovskiy, A., Geacintov, N.E., Grollman, A.P., Ohmori, H., and Shibutani, S. (2002) Translesion synthesis by human DNA polymerase κ on a DNA template containing a single stereoisomer of dG-(+)- or dG-(–)-*anti*-N^2-BPDE (7,8-dihydroxy-*anti*-9,10-epoxy-7,8,9,10-tetrahydrobenzo[a]pyrene). *Biochemistry*, **41**, 6100–6106.
175 Choi, J.Y., Angel, K.C., and Guengerich, F.P. (2006) Translesion synthesis across bulky N^2-alkyl guanine DNA adducts by human DNA polymerase κ. *J. Biol. Chem.*, **281**, 21062–21072.
176 Choi, J.Y. and Guengerich, F.P. (2004) Analysis of the effect of bulk at N^2-alkylguanine DNA adducts on catalytic efficiency and fidelity of the processive DNA polymerases bacteriophage T7 exonuclease⁻ and HIV-1 reverse transcriptase. *J. Biol. Chem.*, **279**, 19217–19229.
177 Choi, J.Y. and Guengerich, F.P. (2005) Adduct size limits efficient and error-free bypass across bulky N^2-guanine DNA lesions by human DNA polymerase η. *J. Mol. Biol.*, **352**, 72–90.
178 Choi, J.Y. and Guengerich, F.P. (2006) Kinetic evidence for inefficient and error-prone bypass across bulky N^2-guanine DNA adducts by human DNA polymerase ι. *J. Biol. Chem.*, **281**, 12315–12324.
179 Choi, J.Y. and Guengerich, F.P. (2008) Kinetic analysis of translesion synthesis opposite bulky N^2- and O^6-alkylguanine DNA adducts by human DNA polymerase REV1. *J. Biol. Chem.*, **283**, 23645–23655.
180 Zang, H., Harris, T.M., and Guengerich, F.P. (2005) Kinetics of nucleotide incorporation opposite DNA bulky guanine N2 adducts by processive bacteriophage T7 DNA polymerase (exonuclease⁻) and HIV-1 reverse

transcriptase. *J. Biol. Chem.*, **280**, 1165–1178.

181 Cross, A.J., Pollock, J.R., and Bingham, S.A. (2002) Red meat and colorectal cancer risk: the effect of dietary iron and haem on endogenous N-nitrosation. *IARC Sci. Publ.*, **156**, 205–206.

182 Cross, A.J. and Sinha, R. (2004) Meat-related mutagens/carcinogens in the etiology of colorectal cancer. *Environ. Mol. Mutagen.*, **44**, 44–55.

183 Sinha, R., Peters, U., Cross, A.J., Kulldorff, M., Weissfeld, J.L., Pinsky, P.F., Rothman, N., and Hayes, R.B. (2005) Meat, meat cooking methods and preservation, and risk for colorectal adenoma. *Cancer Res.*, **65**, 8034–8041.

184 Lemee, F., Bavoux, C., Pillaire, M.J., Bieth, A., Machado, C.R., Pena, S.D., Guimbaud, R., Selves, J., et al. (2007) Characterization of promoter regulatory elements involved in downexpression of the DNA polymerase κ in colorectal cancer. *Oncogene*, **26**, 3387–3394.

185 Mouret, S., Baudouin, C., Charveron, M., Favier, A., Cadet, J., and Douki, T. (2006) Cyclobutane pyrimidine dimers are predominant DNA lesions in whole human skin exposed to UVA radiation. *Proc. Natl. Acad. Sci. USA*, **103**, 13765–13770.

186 Douki, T. and Cadet, J. (2001) Individual determination of the yield of the main UV-induced dimeric pyrimidine photoproducts in DNA suggests a high mutagenicity of CC photolesions. *Biochemistry*, **40**, 2495–2501.

187 Washington, M.T., Prakash, L., and Prakash, S. (2003) Mechanism of nucleotide incorporation opposite a thymine–thymine dimer by yeast DNA polymerase η. *Proc. Natl. Acad. Sci. USA*, **100**, 12093–12098.

188 Masutani, C., Kusumoto, R., Yamada, A., Dohmae, N., Yokoi, M., Yuasa, M., Araki, M., Iwai, S., et al. (1999) The XPV (xeroderma pigmentosum variant) gene encodes human DNA polymerase η. *Nature*, **399**, 700–704.

189 Lehmann, A.R., Kirk-Bell, S., Arlett, C.F., Paterson, M.C., Lohman, P.H., de Weerd-Kastelein, E.A., and Bootsma, D. (1975) Xeroderma pigmentosum cells with normal levels of excision repair have a defect in DNA synthesis after UV-irradiation. *Proc. Natl. Acad. Sci. USA*, **72**, 219–223.

190 Limoli, C.L., Giedzinski, E., Bonner, W.M., and Cleaver, J.E. (2002) UV-induced replication arrest in the xeroderma pigmentosum variant leads to DNA double-strand breaks, γ-H2AX formation, and Mre11 relocalization. *Proc. Natl. Acad. Sci. USA*, **99**, 233–238.

191 Lin, Q., Clark, A.B., McCulloch, S.D., Yuan, T., Bronson, R.T., Kunkel, T.A., and Kucherlapati, R. (2006) Increased susceptibility to UV-induced skin carcinogenesis in polymerase η-deficient mice. *Cancer Res.*, **66**, 87–94.

192 Maher, V.M., Ouellette, L.M., Curren, R.D., and McCormick, J.J. (1976) Frequency of ultraviolet light-induced mutations is higher in xeroderma pigmentosum variant cells than in normal human cells. *Nature*, **261**, 593–595.

193 Misra, R.R. and Vos, J.M. (1993) Defective replication of psoralen adducts detected at the gene-specific level in xeroderma pigmentosum variant cells. *Mol. Cell. Biol.*, **13**, 1002–1012.

194 Wang, Y.C., Maher, V.M., and McCormick, J.J. (1991) Xeroderma pigmentosum variant cells are less likely than normal cells to incorporate dAMP opposite photoproducts during replication of UV-irradiated plasmids. *Proc. Natl. Acad. Sci. USA*, **88**, 7810–7814.

195 Waters, H.L., Seetharam, S., Seidman, M.M., and Kraemer, K.H. (1993) Ultraviolet hypermutability of a shuttle vector propagated in xeroderma pigmentosum variant cells. *J. Invest. Dermatol.*, **101**, 744–748.

196 Dumstorf, C.A., Clark, A.B., Lin, Q., Kissling, G.E., Yuan, T., Kucherlapati, R., McGregor, W.G., and Kunkel, T.A. (2006) Participation of mouse DNA polymerase ι in strand-biased mutagenic bypass of UV photoproducts and suppression of skin cancer. *Proc. Natl. Acad. Sci. USA*, **103**, 18083–18088.

197 Ling, H., Boudsocq, F., Plosky, B.S., Woodgate, R., and Yang, W. (2003) Replication of a *cis-syn* thymine dimer at atomic resolution. *Nature*, **424**, 1083–1087.

198 Li, Y., Dutta, S., Doublié, S., Bdour, H.M., Taylor, J.S., and Ellenberger, T. (2004) Nucleotide insertion opposite a *cis-syn* thymine dimer by a replicative DNA polymerase from bacteriophage T7. *Nat. Struct. Mol. Biol.*, **11**, 784–790.

199 Taylor, J.S., Garrett, D.S., and Cohrs, M.P. (1988) Solution-state structure of the Dewar pyrimidinone photoproduct of thymidylyl-(3′ → 5′)-thymidine. *Biochemistry*, **27**, 7206–7215.

200 Johnson, R.E., Haracska, L., Prakash, S., and Prakash, L. (2001) Role of DNA polymerase ζ in the bypass of a (6-4) TT photoproduct. *Mol. Cell Biol.*, **21**, 3558–3563.

201 Johnson, R.E., Washington, M.T., Haracska, L., Prakash, S., and Prakash, L. (2000) Eukaryotic polymerases ι and ζ act sequentially to bypass DNA lesions. *Nature*, **406**, 1015–1019.

202 Minko, I.G., Harbut, M.B., Kozekov, I.D., Kozekova, A., Jakobs, P.M., Olson, S.B., Moses, R.E., Harris, T.M., et al. (2008) Role for DNA polymerase κ in the processing of N^2–N^2-guanine interstrand cross-links. *J. Biol. Chem.*, **283**, 17075–17082.

203 Kumari, A., Minko, I.G., Harbut, M.B., Finkel, S.E., Goodman, M.F., and Lloyd, R.S. (2008) Replication bypass of interstrand cross-link intermediates by *Escherichia coli* DNA polymerase IV. *J. Biol. Chem.*, **283**, 27433–27437.

204 Cole, R.S. (1973) Repair of DNA containing interstrand crosslinks in *Escherichia coli*: sequential excision and recombination. *Proc. Natl. Acad. Sci. USA*, **70**, 1064–1068.

205 van Houten, B. (1990) Nucleotide excision repair in *Escherichia coli*. *Microbiol. Rev.*, **54**, 18–51.

206 Marini, F., Kim, N., Schuffert, A., and Wood, R.D. (2003) POLN, a nuclear PolA family DNA polymerase homologous to the DNA cross-link sensitivity protein Mus308. *J. Biol. Chem.*, **278**, 32014–32019.

207 Marini, F. and Wood, R.D. (2002) A human DNA helicase homologous to the DNA cross-link sensitivity protein Mus308. *J. Biol. Chem.*, **277**, 8716–8723.

208 Boyd, J.B., Sakaguchi, K., and Harris, P.V. (1990) Mus308 mutants of *Drosophila* exhibit hypersensitivity to DNA cross-linking agents and are defective in a deoxyribonuclease. *Genetics*, **125**, 813–819.

209 Zan, H., Shima, N., Xu, Z., Al-Qahtani, A., Evinger Iii, A.J., Zhong, Y., Schimenti, J.C., and Casali, P. (2005) The translesion DNA polymerase θ plays a dominant role in immunoglobulin gene somatic hypermutation. *EMBO J.*, **24**, 3757–3769.

210 Alt, A., Lammens, K., Chiocchini, C., Lammens, A., Pieck, J.C., Kuch, D., Hopfner, K.P., and Carell, T. (2007) Bypass of DNA lesions generated during anticancer treatment with cisplatin by DNA polymerase η. *Science*, **318**, 967–970.

211 Bruning, J.B. and Shamoo, Y. (2004) Structural and thermodynamic analysis of human PCNA with peptides derived from DNA polymerase-δ p66 subunit and flap endonuclease-1. *Structure*, **12**, 2209–2219.

212 Bunting, K.A., Roe, S.M., and Pearl, L.H. (2003) Structural basis for recruitment of translesion DNA polymerase Pol IV/DinB to the β-clamp. *EMBO J.*, **22**, 5883–5892.

213 Shamoo, Y. and Steitz, T.A. (1999) Building a replisome from interacting pieces: sliding clamp complexed to a peptide from DNA polymerase and a polymerase editing complex. *Cell*, **99**, 155–166.

214 Eoff, R.L., Stafford, J.B., Szekely, J., Rizzo, C.J., Egli, M., Guengerich, F.P., and Marnett, L.J. (2009) Structural and functional analysis of *Sulfolobus solfataricus* Y-family DNA polymerase Dpo4-catalyzed bypass of the malondialdehyde–deoxyguanosine adduct. *Biochemistry*, **48**, 7079–7088.

215 Stafford, J.B., Eoff, R.L., Kozekova, A., Rizzo, C.J., Guengerich, F.P., and Marnett, L.J. (2009) Translesion DNA synthesis by human DNA polymerase η on templates containing a pyrimidopurinone deoxyguanosine adduct, 3-(2′-deoxy-β-D-*erythro*-pentofuranosyl) pyrimido-[1,2-*a*]purin-10(3*H*) one. *Biochemistry*, **48**, 471–480.

14
Elucidating Structure–Function Relationships in Bulky DNA Lesions: From Solution Structures to Polymerases

Suse Broyde, Lihua Wang, Dinshaw J. Patel, and Nicholas E. Geacintov

14.1
Introduction

In the present chapter we provide an overview of the development of new powerful approaches, based on a combination of experimental investigations and computer simulations, for gaining an understanding of the mechanisms of bypass of bulky DNA lesions by polymerases. Simulations complement experiment and can provide information that can be difficult to gain experimentally. Examples include dynamic and thermodynamic information, enzyme mechanisms, and where experimental difficulties, such as growing sufficiently well-diffracting crystals of DNA polymerase complexes for high-resolution structural studies, are encountered. Here, we focus on the structures of DNA lesions and their impact on DNA polymerases during the replication process. For an introduction to molecular mechanics (MM), molecular dynamics (MD), and quantum mechanical (QM) simulation methods utilized for biomacromolecules, we refer the reader to a useful website: http://cmm.cit.nih.gov/intro_simulation.

14.2
Benzo[a]pyrene-Derived DNA Lesions as a Useful Model

The chemical procarcinogen benzo[a]pyrene (B[a]P) has long been a model substance for the investigation of the biological consequences arising from exposure to such environmental pollutants [1, 2]. This substance is a product of fossil fuel combustion, and is therefore present in cooked foods and tobacco smoke, and is generally widespread in the environment [1, 3, 4]. Upon ingestion or inhalation, B[a]P is metabolically converted to substances that are highly reactive and capable of forming covalent adducts/lesions with biological macromolecules including DNA. While the purpose of this metabolic activity is to produce partially water-soluble derivatives so that these hydrophobic compounds may be more easily excreted, this process produces chemically reactive and harmful intermediates that can ultimately lead to cancer. In the case of B[a]P and other polycyclic aromatic

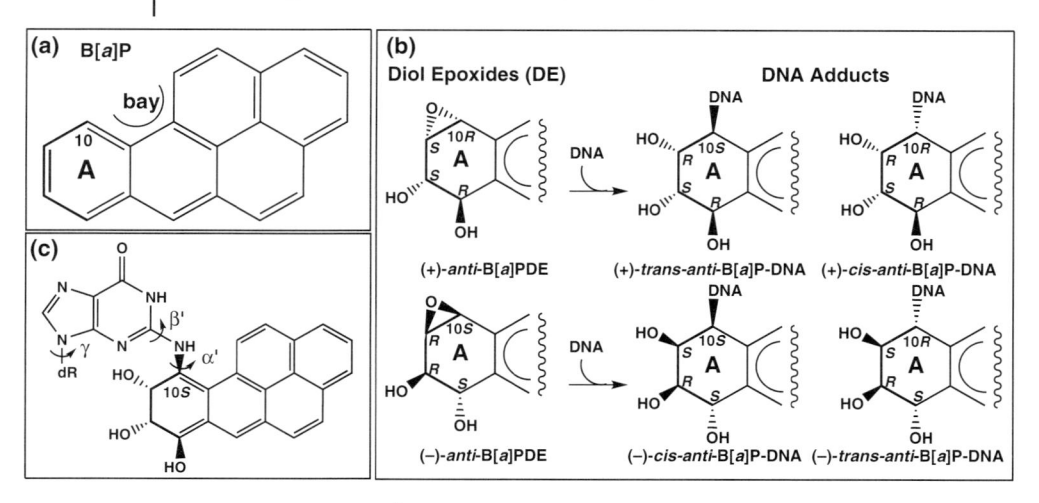

Figure 14.1 Structures of (a) B[a]P, (b) its diol epoxide metabolites and DNA adducts, and (c) 10S-(+)-trans-anti-B[a]PDE-N^2-dG.

chemicals, several metabolic activation pathways have been delineated [5–7], and Chapter 6 by Penning provides a detailed discussion of this topic. Of particular interest for studies of structure–function relationships are the pair of stereoisomeric DNA lesions derived through the diol epoxide metabolic activation pathway, namely the (+)- and (−)-*trans-anti*-B[a]PDE-N^2-dGuo adducts [5] (Figure 14.1). The former is derived from the reaction of the ultimate carcinogenic metabolite of B[a]P, (+)-*anti*-B[a]PDE, with the exocyclic amino group of guanine in DNA, while the latter lesion results from the reaction of the enantiomer (mirror image) and non-tumorigenic (−)-*anti*-B[a]PDE metabolite [8]. The (+)-*trans-anti*-B[a]PDE-N^2-dG adduct is the major product of the tumorigenic (+)-*anti*-B[a]PDE metabolite and hence has been the subject of intense interest for many years [9–11]. The (−)-*trans-anti*-B[a]PDE-N^2-dG stereoisomer, derived from the nontumorigenic (−)-*anti*-B[a]PDE enantiomer, presents a fascinating counterpoint for gaining insight into the important question: why does a given lesion have the potential for initiating cancer while another structurally similar one is significantly less active or inactive? Figure 14.1 also shows the stereochemical characteristics of other DNA adducts derived from (+)- and (−)-*anti*-B[a]PDE, including those derived from the less-efficient reactions of these two enantiomers with adenine in DNA [9–11].

The current understanding of the initiation phase of chemical carcinogenesis by B[a]P-derived DNA adducts [1, 3, 4], which involves the mutagenic bypass of these lesions by DNA polymerases, is reviewed in Chapter 15 by Chandani and Loechler. Such mutations, when occurring in cell cycle control genes, can ultimately set in motion the unrestrained replicative process that is the hallmark of tumorigenesis. The (+)-*trans-anti*-B[a]PDE-N^2-dG lesion is highly mutagenic in both mammalian and prokaryotic systems [12–24]. A goal of our research has been to utilize this stereoisomeric pair of DNA adducts as a model system for opening the door to understanding critical differences in biological outcomes derived from the tumorigenic (+)-*anti*-B[*a*]PDE metabolite *vis-à-vis* its nontumorigenic (−)-*anti*-B[*a*]PDE enantiomer. Combined computational and experimental studies, beginning with damaged DNA itself and eventually progressing to investigations with DNA polymerases are described here. Chapter 13 by Eoff *et al.* presents a detailed overview of translesion synthesis in DNA polymerases from mainly crystallographic and kinetic perspectives.

14.3
Computational Elucidation of the Structural Properties of B[*a*]P-Derived DNA Lesions in Solution

The most intriguing question for many years was how the structures of the (+)- and (−)-*trans-anti-trans*-B[*a*]PDE-N^2-dG adducts differed in duplex DNA in aqueous environments. This subject was the focus of extensive spectroscopic investigations over a number of years. A Site II-type structure [25–27] with the aromatic ring system of the B[*a*]PDE adducts exposed to solvent had been delineated by spectroscopic methods, but the difference between the (+) and the (−) stereoisomeric adducts remained elusive. A computational prediction provided novel and fundamental insights [28] that has since found wide applicability to stereoisomeric pairs of DNA lesions [29–33] derived from enantiomeric pairs of diol epoxides originating from a number of different polycyclic aromatic hydrocarbon (PAH) carcinogens [31, 34, 35]. Specifically, the computations predicted that the (+)- and (−)-*trans-anti*-B[*a*]PDE-N^2-dG adducts would both be positioned in the minor groove, but with opposite orientations relative to the 5′ → 3′ polarity of double-stranded B-form DNA [28]. Specifically, the B[*a*]P aromatic rings are 5′-directed along the modified strand in the case of the (+) stereoisomer, while they are 3′-directed in the (−) stereoisomer. The results of high-resolution nuclear magnetic resonance (NMR) solution studies of the (+)- and (−)-*trans-anti-trans*-B[*a*]PDE-N^2-dG adducts provided the essential and critical experimental evidence for the opposite orientations of this adduct pair in solution [34, 36, 37]. A crystal structure of the (+)-*anti-trans*-B[*a*]PDE-N^2-dG adduct at the nucleoside level was structurally nearly identical to the adduct structure in duplex DNA [38]. More broadly, computations and NMR solution structures [35, 39] have shown that such stereoisomer pairs adopt opposite orientations in the DNA duplex, regardless of the specific position assumed by the aromatic rings; in some cases these are intercalated into the DNA through various possible modes, assuming solvent-shielded positions (experimentally termed Site II [25, 27, 40]). NMR solution studies [35, 39] of numerous stereoisomeric adduct pairs derived from various PAHs have shown that the opposite orientation phenomenon is observed without exception to date as far as we are aware, irrespective of the specific ring structure of the PAH, the base to which the adduct is linked, or the specific conformation adopted with respect to the DNA duplex. Representative examples of these conformational families are illustrated in Figure 14.2. The chemical/

(a) *anti*-BP-N^2-dG, minor groove

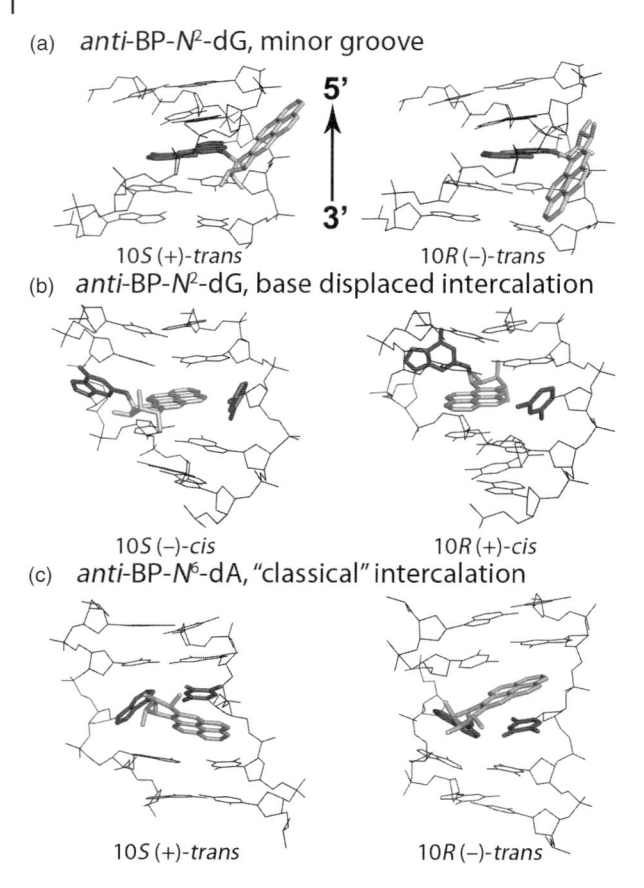

10S (+)-*trans*   10R (−)-*trans*

(b) *anti*-BP-N^2-dG, base displaced intercalation

10S (−)-*cis*   10R (+)-*cis*

(c) *anti*-BP-N^6-dA, "classical" intercalation

10S (+)-*trans*   10R (−)-*trans*

Figure 14.2 Conformational themes for B[*a*]P adducts in duplex DNA in solution. The 5′ → 3′ directionality of all modified strands is shown in (a). (a) and (b) are structures from [34, 36, 41, 42], and (c) are structures from [43].

structural/energetic origins of this opposite orientation phenomenon was defined through a comprehensive investigation utilizing MM calculations, with our representative (+)- and (−)-*trans-anti-trans*-B[*a*]PDE-N^2-dG adducts [33] as well as other stereoisomeric pairs derived from B[*a*]P and benzo[*c*]phenanthrene [29–32]. Specifically, the explanation came from studying simply the damaged nucleosides, containing just a B[*a*]PDE damaged base and the deoxyribose sugar attached to it [33]. The two flexible torsions, α′ and β′ (Figure 14.1), connecting the B[*a*]PDE moiety and the base, and the flexible torsion χ connecting the base to the sugar, govern the orientation of the B[*a*]PDE to the DNA. We constructed 373 248 different conformations for each nucleoside adduct by stepping over α′, β′, and χ at 5° intervals in combination (known as a grid search), and computed the MM energy for each case. This enabled us to obtain remarkable potential energy maps for each stereoisomer that revealed the near mirror image properties of the pair of adducts at the nucleoside level (Figure 14.3a): the map for the (+) stereoisomer can be essentially converted into that of the (−) stereoisomer by inverting α' and β' (i.e., α' becomes $-\alpha'$ and β' becomes $-\beta'$). The prominent importance of the linkage torsion angle β' in governing the opposite orientation of the B[a]P rings was dramatically displayed, with the (+) stereoisomer favoring the −90° domain and the (−)stereoisomer favoring the +90° domain. This directly produces the opposite orientations of the stereoisomer pairs. The opposite orientations stem from simple steric hindrance differences between the hydroxyl groups at carbons 8 and 9 of the B[a]P puckered cyclohexene-like benzylic ring and the guanine base. The differing stereochemistries at these hydroxyl groups (Figure 14.1) produce steric crowding when the (+) stereoisomer adopts the β' orientation that is preferred by the (−) stereoisomer and vice versa. The NMR solution structures of the lesion-containing duplexes [34, 36], as well as the crystal structure of the damaged nucleoside [38], adopted structures that are consistent with the computed low-energy domains in each stereoisomer.

From the biological perspective, these opposite orientational preferences would be expected to produce differential processing of the lesions by DNA and RNA polymerases (see Chapter 17 by Dreij *et al*.). Indeed, experimental observations of such differences have been reported (e.g., [17, 45–50]). Of particular interest is whether the β' orientations that are favored energetically and observed in solution remain prevalent when the lesions are bound to polymerase and repair enzymes. In this connection, crystal structures of the (+)-*trans-anti*-B[a]PDE-N^2-dG adduct in the Y-family lesion bypass DNA polymerase IV (Dpo4) (see also Chapter 15 by Chandani and Loechler) from the archaeon *Sulfolobus solfataricus* reveal that the favored β' domain is preserved in the enzyme in this case [51]. Overall, a survey of crystal structures of DNA damaged by polycyclic carcinogenic chemicals shows that structures observed in DNA duplexes in solution by high-resolution NMR methods are often observed in polymerases [37]; however, the preferred β' domain can be overridden by strong lesion–polymerase interactions, as was manifested in a crystal structure of a B[a]PDE-derived adenine lesion [52].

14.4
DNA Polymerase Structure–Function Relationships Elucidated with B[a]P-Derived Lesions

We now address a representative series of experimental investigations with DNA polymerases, and the interplay between the experimental observations and the computational simulations that yielded insights into the mechanisms of DNA lesion–polymerase interactions at the molecular level. In this connection, we first discuss briefly the current views of the processing of bulky lesions, such as those derived from B[a]PDE, by polymerases. Chapter 13 by Eoff *et al*. provides further detailed insights on the interactions of other DNA lesions with polymerases. It is now well understood that the progress of high-fidelity polymerases is mostly

336 *14 Elucidating Structure–Function Relationships in Bulky DNA Lesions*

(a)

(b)

(c)

(d)

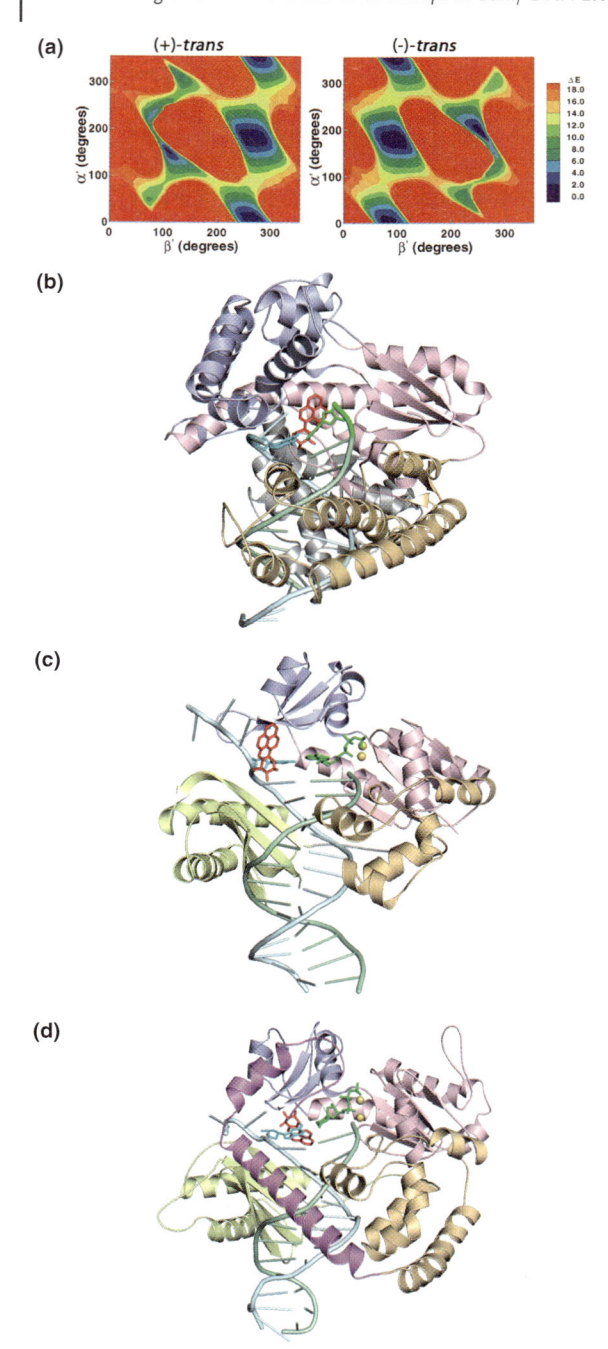

Figure 14.3 (a) α', β' van der Waals energy component maps to 20 kcal/mol (adapted with permission from [33], Figure 3, Copyright 1999, American Chemical Society). Left: (+)-*trans-anti*-B[a]PDE-N^2-dG adduct; right: (−)-*trans-anti*-B[a]PDE-N^2-dG adduct. The glycosidic bond χ is in the *anti* domain (230°) for both cases. (b) Crystal structure of BF binary complex with (+)-*trans-anti*-B[a]PDE-N^2-dG as the templating base (PDB ID:1XC9, [68]). The B[a]P moiety is wedged between the polymerase palm domain and the nascent base pair, thus breaking critical interactions between the polymerase minor groove scanning track and the nascent base pair, and causing polymerase blockage. (c) Dpo4 ternary complex structure from an MD simulation [44] with (+)-*trans-anti*-B[a]PDE-N^2-dG as the templating base and a mismatched incoming dATP. The modified guanine adopts the *syn* conformation and the B[a]P moiety is comfortably positioned on the major groove side of the nascent base pair. This conformation is not feasible in Pol κ (see (d)) because the B[a]P rings would collide with the N-clasp of Pol κ. (d) Pol κ ternary complex structure with (+)-*trans-anti*-B[a]PDE-N^2-dG Watson–Crick paired with an incoming dCTP [87]. The B[a]P moiety is well accommodated in the open space on the minor groove side, thus supporting error-free bypass. All polymerase views are into the major groove side of the nascent base pair, with the minor groove side in back. Color code: light orange, thumb; light pink, palm; light blue, fingers; gray, BF exonuclease domain; light yellow, little-finger domain; purple, Pol κ N-clasp; light cyan, template strand; light green, primer strand; red, the B[a]P moiety; cyan, the B[a]P damaged guanine; green, base paired with the damaged template; golden spheres, Mg^{2+}.

blocked when they encounter a bulky lesion [53–55]. One subsequent result of this blocking event is known as polymerase switch – the high-fidelity and processive polymerase, which synthesizes large quantities of DNA without dissociating from the template, is replaced by one or more specialized lesion bypass polymerases [56–58]. These distributive polymerases catalyze addition of sufficient numbers of nucleotide residues so that the distortion engendered by the lesion is transited [59–62]. They are mainly from the Y-family of DNA polymerases, are low-fidelity on unmodified DNA, and are frequently error-prone in bypassing DNA lesions although they may be specialized for error-free bypass in certain cases [61, 63]. Structurally, the high-fidelity and the bypass polymerases share common features of being shaped like a hand with palm, finger, and thumb domains, and utilize a triad of conserved carboxylate amino acid residues and a pair of Mg^{2+} ions in facilitating the catalytic mechanism of the nucleotidyl transfer reaction [64]. However, the high-fidelity polymerases utilize an induced-fit mechanism with finger closing upon entry of the incoming dNTP that is the Watson–Crick partner to the template; this creates a tight-fitting active-site region with multiple amino acid–DNA interactions on the minor groove side of the nascent duplex, whose function is to probe for rare misincorporation events following reaction. The major groove side of the nascent duplex is, however, open. By contrast, Y-family bypass polymerases, which possess an additional little-finger domain, do not appear to utilize an open/closing induced-fit mechanism [61, 65] and lack the minor groove scanning track. Consequently, the nascent duplex is somewhat solvent exposed on the minor groove side, as well as being open on the major groove side. Reviews of DNA polymerase structures and their relations to the processing of DNA lesions have been published [53, 60, 61, 66].

A representative high-fidelity DNA polymerase that has been extensively characterized structurally through X-ray crystallographic analyses is the A-family *Bacillus* fragment (BF) from *Bacillus stearothermophilus* [67]. The crystal structures showed replicative stages where the DNA in the polymerase existed as a binary complex containing primer and template DNA strands, but without incoming dNTP, and in a reaction-ready ternary complex with dNTP also present. Experimental primer extension studies have been carried out with this polymerase and utilizing a DNA template containing the (+)-*trans-anti*-B[a]PDE-N^2-dG adduct [55]. In these *in vitro* experiments, nucleotide incorporation opposite the lesion and the extension of the nascent DNA primer strand beyond the lesion were investigated. The overall findings were that the lesion primarily blocked the progress of the polymerase at the site of the lesion and that a small amount of lesion bypass with full-length extension of the primer strand was observable. Purines rather than pyrimidines were preferentially inserted opposite the lesion and nucleotide insertion was significantly more facile than primer strand extension beyond the lesion. Computer modeling with MD simulations carrying the adducted templating base through an entire replicative cycle provided structural understanding of these functional observations. The computer simulations were based on BF crystal structures, and utilized computer graphics and extensive conformational searches to model the lesion and its possible placement within the polymerase in a milieu of solvent and counterions. The subsequent MD simulations provided ensembles of thousands of structures and their fluctuations in time. Specifically, the blockage of the polymerase stems from the placement of the B[a]PDE aromatic ring system into the BF minor groove scanning track, thus disrupting the structure as well as the normal function of the enzyme. This had been previously demonstrated in a crystal structure of BF containing this same lesion [68] (Figure 14.3b). The B[a]PDE residue is positioned in the minor groove of the nascent double strand because it is linked to the exocyclic amino group of guanine; in normal B-DNA this amino group is located in the minor groove of B-form DNA (Figure 14.4b). In order to account for the experimentally observed low efficiency of translesion bypass by BF [55], the bulky aromatic B[a]PDE ring system must be displaced into the major groove side of the nascent duplex in the large open pocket of the polymerase; this necessitates a rotation of the glycosidic torsion χ of the damaged guanine from its normal B-DNA *anti* conformation by around 180° into the abnormal *syn* domain where it is incapable of Watson–Crick hydrogen-bonding with an incoming dNTP (Figure 14.4a). However, the polymerase structure itself is much less distorted by the lesion. In the *syn* conformation, the modified guanine presents its pyrimidine-like side to the incoming dNTP; this explains on a structural basis the preferential incorporation of purines rather than pyrimidines opposite the (+)-*trans-anti*-B[a]PDE-N^2-dG lesion: the *syn*-guanine mimics the appearance of a pyrimidine in the active site, leading to selection of a purine partner to form a nascent base pair that is most Watson–Crick-like in shape. Modeling and MD simulations have shown that the high-fidelity DNA polymerase from T7 phage [69] adopts a similar strategy for bypassing the same lesion with low efficiency.

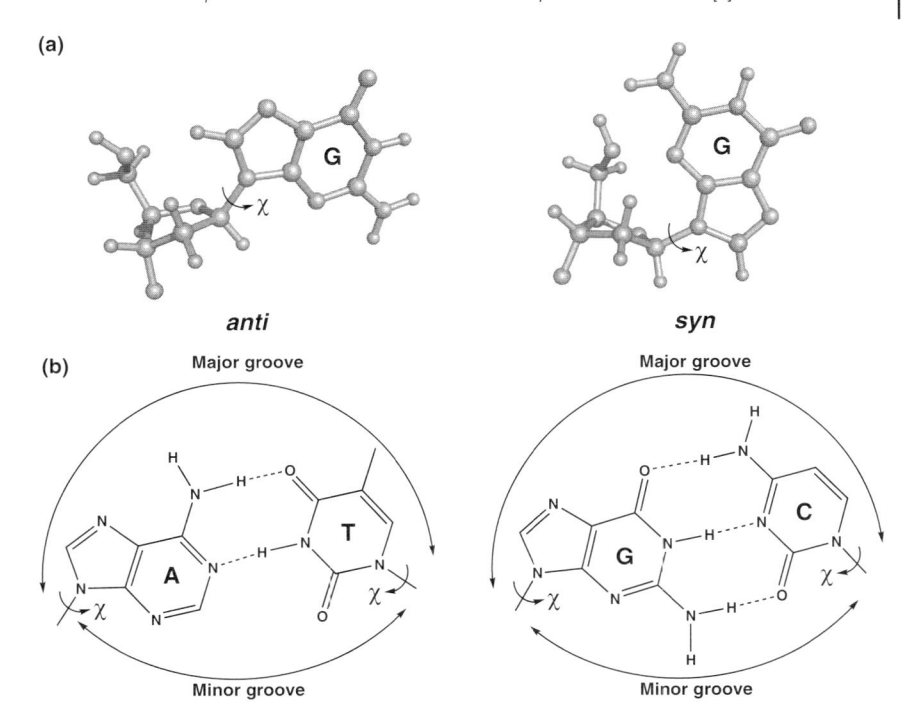

14.4 DNA Polymerase Structure–Function Relationships Elucidated with B[a]P-Derived Lesions

Figure 14.4 (a) Relative to the sugar moiety, the base can adopt two main orientations, defined by torsion angle χ, the common *anti* conformation and the less common *syn* conformation. (b) Since the two glycosidic bonds in a base pair in B-DNA are not diametrically opposite each other, each base pair has a larger major groove side and a smaller minor groove side.

Among the Y-family of lesion bypass polymerases, Dpo4 has been extensively studied by X-ray crystallography both without and with different lesions (reviewed in Chapter 13 by Eoff *et al.*), including those derived from B[a]P metabolites [51, 52]. The first published crystal structure of this enzyme revealed both structural similarities and differences with high-fidelity polymerases [70]. Both families of polymerases share similar features of the active site which includes the palm/finger/thumb domains, the three conserved carboxylate amino acid residues, and the two metal ions that play critical roles in catalyzing the nucleotidyl transfer reaction. However, there are also key structural differences that govern the functional properties of Dpo4: in translesion synthesis it functions as a low-fidelity polymerase that is prone to frameshift (deletion and insertion) mutations, both with unmodified DNA and in the presence of bulky lesions [51, 52, 71–74]. However, it bypasses the small 7,8-dihydro-8-oxoguanine lesion derived from reactive oxygen species (see Chapter 3 by Cadet *et al.*) in an error-free manner [65]. Dpo4 has a spacious and solvent-accessible active site region that can accommodate two templating nucleotides, rather than one as in A-family replicative polymerases, so that one template base can be easily skipped to cause deletion mutations [70]. Furthermore, Dpo4 does not have a minor groove scanning track so there is open space on both the major and minor groove side of the evolving DNA duplex. It does not utilize an open/closing induced-fit mechanism upon entry of the dNTP [53, 61, 65] and relies for fidelity of replication on Watson–Crick hydrogen-bonding [75]. Additionally, Dpo4 contains the flexible little-finger domain not found in high-fidelity DNA polymerases. Other Y-family bypass polymerases share many of these features but also differ significantly on a case-by-case basis [61].

Experimental primer extension studies with the $(+)$-*trans-anti*-B[*a*]PDE-N^2-dG adduct utilizing Dpo4 revealed stark contrasts with the A-family polymerase BF in the way it processes this DNA lesion [71, 73, 74]. In Dpo4, this lesion is bypassed to a significant extent although at a much slower rate than an unmodified guanine at the same position. The major primer extension product contains a −1 deletion. Full-length extension products are also observed, but the nature of the nucleotide inserted opposite the lesion during bypass depends on the sequence context of the lesion. In the case of the AG*C sequence context (G* denotes the damaged base), nucleotide insertion is promiscuous [44]. This observation was explained by molecular modeling and MD simulation studies that demonstrated that Dpo4 has the capacity to house the bulky B[*a*]PDE ring system on either the major or the minor groove open spaces within the polymerase. When the lesion is on the major groove side, the damaged guanine adopts the *syn* orientation which permits various mismatch hydrogen-bonding schemes between the guanine residue of G* and the incoming dNTPs (Figure 14.4b); when positioned on the minor groove side, the G* is in the *anti* glycosidic torsion conformation and the modified guanine residue can form three Watson–Crick hydrogen bonds with incoming dCTP. However, in the case of the CG*G sequence context, the incorporation of dGTP is predominant, while dATP or dCTP are favored in the TG*G and GG*G sequence contexts, respectively [71, 73, 74]. The modeling and dynamics simulations suggested an unusual slippage mechanism. Usual slippage patterns with bulky lesions feature a G* that is skipped (Figure 14.5a) [52]. However, in sequences containing G*G, the unmodified G rather than G* may be skipped, the selected incoming dNTP Watson–Crick pairs with the base on the 5′-side of the G* (Figure 14.5b), and further extension from this Watson–Crick pair is relatively facile. Slippage mechanisms for insertion and deletion mutations were first proposed by Streisinger [76], and have been extensively discussed by Kunkel [77, 78] and Ripley [79].

Overall, the Dpo4 structure with its open minor and major groove sides allows both *syn*- and *anti*-G* conformations, which explains promiscuous nucleotide incorporation; its slippage-prone nature permits sequence-governed nucleotide selectivity because of its propensity for −1 deletion mutations. By contrast, in the relatively constrained active site of the A-family polymerase BF that can accommodate only one template nucleotide at a time, slippage is unlikely because the modified guanine residue of G* remains in the *anti* glycosidic conformation; this places the bulky B[*a*]P ring system into the minor groove so that the essential

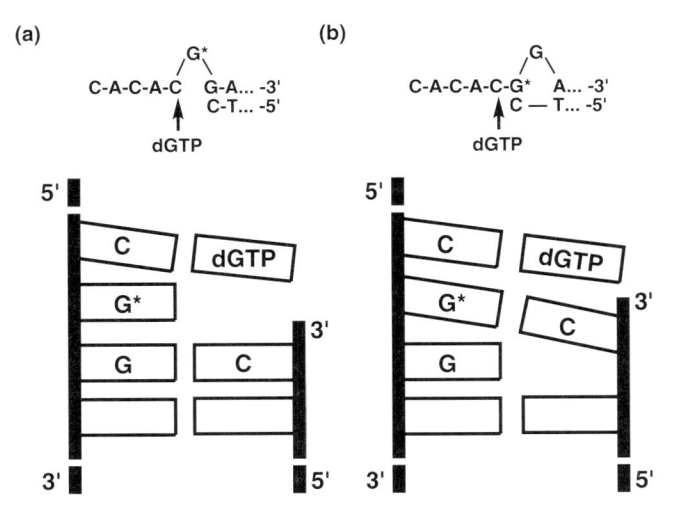

14.4 DNA Polymerase Structure–Function Relationships Elucidated with B[a]P-Derived Lesions | 341

Figure 14.5 Cartoons showing the (a) G*-skipped and (b) G-skipped slippage patterns in the ternary complex models (adapted with permission from [71], Figure 2, Copyright 2008, American Chemical Society).

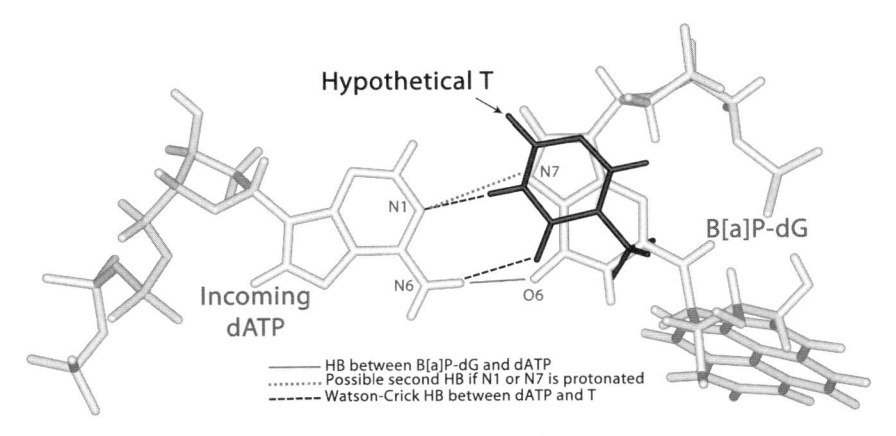

Figure 14.6 The shape of the *syn*-(+)-*trans-anti*-B[a]PDE-N^2-dG:dATP mismatch pair mimics that of a normal *anti*-T:dATP Watson–Crick base pair (adapted with permission from [69], Figure 9, Copyright 2001, Elsevier). The B[a]P-modified guanine is shown in gray and the hypothetical thymine is shown in black. HB, hydrogen bond.

minor groove scanning track interactions are disrupted [68]. Primer extension is thus nearly completely blocked by the *anti*-G* conformation, which is consistent with experimental observations [55]. The limited extension observed is explained by the preferred misincorporation of purines opposite the *syn*-G* conformation (Figure 14.6) and the resulting G* · purine mismatches are elongated with very poor efficiencies.

The human homolog of the Y-family lesion bypass polymerase Dpo4 is Pol κ. A crystal structure of this polymerase in a ternary complex with primer/template

DNA and dNTP has revealed strong overall structural similarities with Dpo4, as well as a very striking difference [80]. Specifically, Pol κ contains a unique N-terminal domain termed the N-clasp; in the presence of this domain the DNA primer/template in the active-site region is completely encircled by the enzyme. Primer extension experiments utilizing the (+)-*trans-anti*-B[a]PDE-N^2-dG adduct revealed nearly exclusive error-free bypass of the lesion in Pol κ [47, 81, 82]. *In vitro*, the bypass of this lesion catalyzed by Pol κ is significantly slowed relative to the bypass of a guanine residue at the same template position, although the bypass efficiency of G* is strongly dependent on the bases flanking the lesion [83]. Moreover, Pol κ bypasses a variety of other N^2-dG adducts in an essentially error-free manner [84–86]. The high-fidelity bypass of the bulky B[a]PDE-N^2-dG lesion by Pol κ is in sharp contrast to the error-prone bypass of the same adduct by Dpo4. Therefore, this pair of polymerases represents useful model systems for gaining a basic understanding of why a given polymerase can bypass a lesion without causing mutations, while another one generates replication errors. Our simulation studies provided new insights and a plausible explanation of these differences [87]. The key point is that the unique Pol κ N-clasp sterically interferes with the formation of the *syn* glycosidic conformation of the (+)-*trans-anti*-B[a]PDE-N^2-dG lesion, while the *anti* → *syn* conformational change is feasible in Dpo4 (Figure 14.3c). Indeed, the N-clasp of Pol κ obstructs the major groove region of the nascent base pair (Figure 14.3d); however, in Dpo4 the *syn* orientation of G* is accommodated in the major groove region which is open to the solvent environment. Furthermore, a Watson–Crick paired *anti* conformation of the (+)-*trans-anti*-B[a]PDE-N^2-dG lesion opposite dCTP in Pol κ was shown on the basis of energetic considerations to be most favorable (Figure 14.3d) [87]. Interestingly, this structure was stabilized by a stacking interaction between the B[a]PDE rings and a specific phenylalanine in Pol κ, while in Dpo4 the analogous residue is a nonaromatic lysine. Similar Watson–Crick paired structures involving other aromatic lesions may explain the propensity of Pol κ to bypass N^2-dG adducts in an error-free manner [84–86].

In contrast to the error-free bypass of the (+)-*trans-anti*-B[a]PDE-N^2-dG adduct by Pol κ, the stereochemically identical adduct to adenine, the (+)-*trans-anti*-B[a]PDE-N^6-dAde adduct, completely blocks the progression of Pol κ in primer extension experiments *in vitro* [47]. The polymerase is stalled at the step that involves nucleotide insertion opposite the lesion. The key difference between the N^6-dA and the N^2-dG adducts resides in the positioning of the linkage site of the adducted base: in B-DNA, whether the glycosidic bond conformation is *syn* or *anti*, the (+)-*trans-anti*-B[a]PDE-N^6-dA adduct is placed on the major groove side of the nascent base pair; however, for the (+)-*trans-anti*-B[a]PDE-N^2-dG adduct, the *anti* conformation positions the bulky B[a]PDE ring system on the minor groove side, while in the *syn* conformation it is on the major groove side (Figure 14.4b). In Pol κ, with the N-clasp on the major groove side of the active site region, the (+)-*trans-anti*-B[a]PDE-N^6-dA adduct is obstructed from entering the templating position, thus accounting for the observed blockage of the polymerase Pol κ. Moreover, the observed slowing of Pol κ by other lesions that are necessarily

14.5
Mechanism of the Nucleotidyl Transfer Reaction

positioned on the major groove side of the nascent base pair, such as bulky O^6-alkylated guanine lesions [88], might be explainable in a similar manner.

While significant understanding of DNA polymerase structures has been acquired over the last decade via X-ray crystal structure studies, augmented with computer simulations, the actual mechanism of the nucleotidyl transfer reaction remains a research frontier and is approachable experimentally only indirectly. Based on crystal structures, a two-metal-ion (usually Mg^{2+}) mechanism involving three conserved carboxylate-containing amino acid residues (usually Asp and Glu) was proposed [64]. Overall, in this mechanism, the attack of the 3′-O on the α-phosphate of the dNTP is facilitated by the lowered affinity of the primer terminus 3′-OH for its hydrogen due to its liganding with the catalytic Mg^{2+} ion, termed MgA. The expected pentacovalent transition state is stabilized by both metal ions and the nucleotide-binding Mg^{2+} (termed MgB) aids in the leaving of the pyrophosphate (Figure 14.7). These metal ions are essential for catalysis in the DNA synthesis reactions. Still, details concerning the reaction mechanism remained unresolved and have been the focus of recent QM/MM simulations [89–91]. The issues include determination of the base to which the proton of the 3′-OH is initially transferred, characterization of the transition states and intermediates, and the elucidation of the exact roles of the metal ions and the conserved amino acid residues. Generally this approach utilizes computationally costly *ab initio* quantum mechanical treatments of the active-site region of the polymerase that contains the primer/template DNA and the incoming dNTP. The remainder of the polymerase system, including solvent and counterions, is treated with computationally more affordable classical MM/MD. Broadly speaking, the nucleotidyl transfer reaction is driven through chemically plausible reaction mechanisms, and potential energy or free-energy profiles are obtained. Mechanisms with the lowest barriers are most likely and these barriers can be compared with experimental kinetic data for the rate-limiting step. A number of different treatments have been developed for the QM/MM method that focus particularly on the thorny problem of how to best handle the QM/MM interface [92]. These kinds of QM/MM calculations for defining mechanism of enzyme reactions are reviewed by Hu and Yang [92].

In the case of Dpo4, the pseudobond approach [93–95] has been utilized for treating the QM/MM interface to determine a reaction mechanisms (Figure 14.7); its free-energy profile is the lowest of the eight investigated and agrees well with the upper limit set by the measured rate-limiting DNA polymerization step in this enzyme [96]. The mechanism takes advantage of a water molecule observable in the crystal structure that is conserved in many DNA polymerase crystal structures: the H of the 3′-OH group is first transferred to the dNTP α-phosphate via this water molecule, and is then relayed to the dNTP γ-phosphate group via a

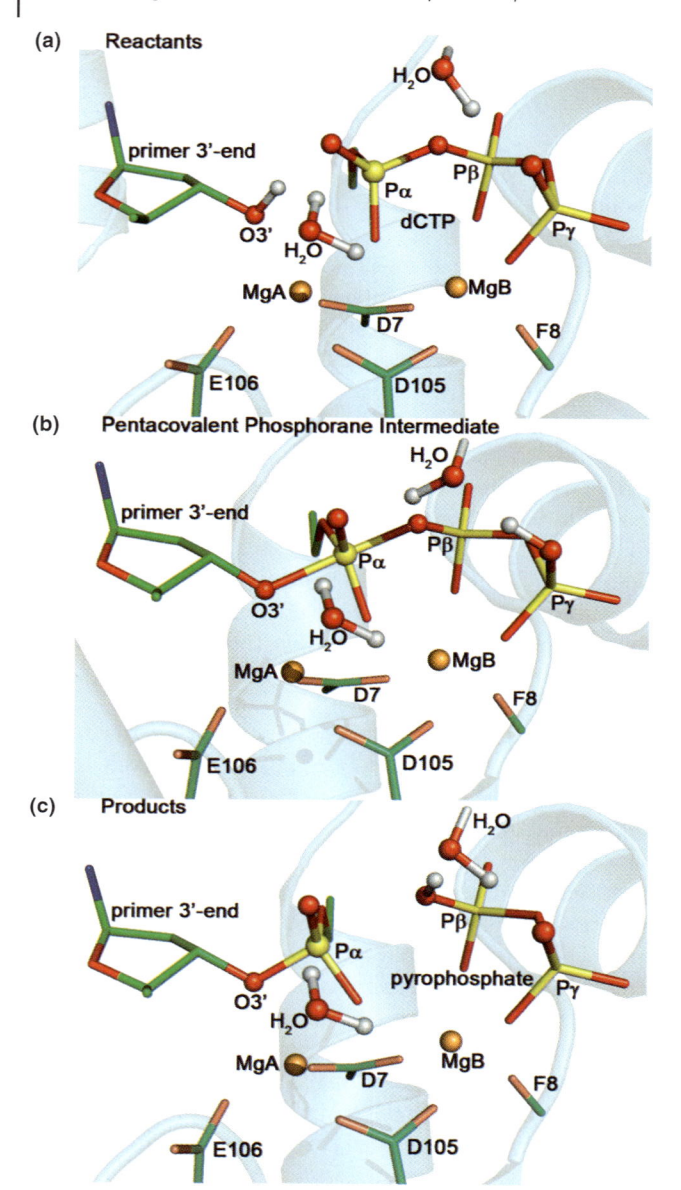

Figure 14.7 Critical structures determined for nucleotidyl transfer catalyzed by Dpo4 (adapted with permission from [96], Figure 2, Copyright 2007, American Chemical Society). (a) Reactants (ground state). (b) Metastable pentacovalent phosphorane intermediate. (c) Products. For clarity, the base of the primer 3'-end and the nucleoside of the incoming dNTP are not shown.

second, solvent water molecule; next, the deprotonated 3′-O attacks the dNTP α-P, producing a pentacovalent phosphorane intermediate; subsequently, as the pyrophosphate leaves, the proton on the γ-phosphate is transferred to the α–β bridging oxygen to neutralize the evolving negative charge. The conserved amino acid residues and the octahedrally coordinated Mg^{2+} ions serve to organize the active site in reaction-ready geometry prior to the chemical steps and stabilize the transition state. A related water-mediated and substrate-assisted mechanism has recently been shown to be feasible, as well in the high-fidelity replicative DNA polymerase from phage T7 [97]. One important insight gained from these studies for the Dpo4 and T7 DNA polymerases is that, for a correct incoming dNTP, the initial proton transfer relay may be rate limiting relative to the formation of the new P–O bond and the leaving of the pyrophosphate. Experimental studies have shown that the nucleotidyl transfer reaction entails two proton transfers in the rate-limiting step, presumably loss of 3′-OH of the primer terminus and protonation of the leaving pyrophosphate [98, 99]. The specific path that leads to the ultimate deprotonation of primer 3′-end and protonation of the pyrophosphate leaving group may vary for different polymerases [89–91, 100–103], and may involve multiple amino acid residues and water molecules. However, all amino acid residues, as well as the polymerase active site organization, need to be restored to their prechemistry state for each round of reaction. Future studies in this area with damaged templating bases are needed to gain insights on how polymerase mechanisms are impacted by lesions. Furthermore, for this frontier technology designed to uncover enzyme reaction mechanisms, innovative advances such as the recently developed implementation of MD to the QM regime of the QM/MM methodology (QM/MM/MD) promise more powerful capabilities in the near future [104].

14.6
Conclusions and Future Perspectives

With the current growing appreciation of cellular processes as molecular chemical machines, it is clear that more ample understanding, beyond lesion processing by polymerases, is needed. In the cell, there is a dynamic interplay between replication, transcription, and various repair mechanisms [105–108]. How these various processes are called into play in the face of DNA lesions is currently the focus of research in areas that span structural, molecular, cellular, and systems biology, with computational approaches playing increasingly vital roles [109]. Recent advances in these areas promise exciting future developments.

Acknowledgements

This work is supported by NIH grants CA28038 and CA75449 to S.B., CA99194 to N.E.G., and CA46533 to D.J.P. Molecular images were made with PyMOL (DeLano Scientific; www.pymol.org).

References

1 Luch, A. (2005) Nature and nurture – lessons from chemical carcinogenesis. *Nat. Rev. Cancer*, **5**, 113–125.
2 Phillips, D.H. (1983) Fifty years of benzo(a)pyrene. *Nature*, **303**, 468–472.
3 Clapp, R.W., Jacobs, M.M., and Loechler, E.L. (2008) Environmental and occupational causes of cancer: new evidence 2005–2007. *Rev. Environ. Health*, **23**, 1–37.
4 Luch, A. (2006) The mode of action of organic carcinogens on cellular structures. *EXS*, 65–95.
5 Conney, A.H. (1982) Induction of microsomal enzymes by foreign chemicals and carcinogenesis by polycyclic aromatic hydrocarbons: G. H. A. Clowes Memorial Lecture. *Cancer Res.*, **42**, 4875–4917.
6 Cavalieri, E.L. and Rogan, E.G. (1995) Central role of radical cations in metabolic activation of polycyclic aromatic hydrocarbons. *Xenobiotica*, **25**, 677–688.
7 Burczynski, M.E., Harvey, R.G., and Penning, T.M. (1998) Expression and characterization of four recombinant human dihydrodiol dehydrogenase isoforms: oxidation of *trans*-7,8-dihydroxy-7,8-dihydrobenzo[a]pyrene to the activated o-quinone metabolite benzo[a]pyrene-7,8-dione. *Biochemistry*, **37**, 6781–6790.
8 Buening, M.K., Wislocki, P.G., Levin, W., Yagi, H., Thakker, D.R., Akagi, H., Koreeda, M., Jerina, D.M., and Conney, A.H. (1978) Tumorigenicity of the optical enantiomers of the diastereomeric benzo[a]pyrene 7,8-diol-9,10-epoxides in newborn mice: exceptional activity of (+)-7β,8α-dihydroxy-9α,10α-epoxy-7,8,9,10-tetrahydrobenzo[a]pyrene. *Proc. Natl. Acad. Sci. USA*, **75**, 5358–5361.
9 Szeliga, J. and Dipple, A. (1998) DNA adduct formation by polycyclic aromatic hydrocarbon dihydrodiol epoxides. *Chem. Res. Toxicol.*, **11**, 1–11.
10 Cheng, S.C., Hilton, B.D., Roman, J.M., and Dipple, A. (1989) DNA adducts from carcinogenic and noncarcinogenic enantiomers of benzo[a]pyrene dihydrodiol epoxide. *Chem. Res. Toxicol.*, **2**, 334–340.
11 Meehan, T. and Straub, K. (1979) Double-stranded DNA steroselectively binds benzo(a)pyrene diol epoxides. *Nature*, **277**, 410–412.
12 Alekseyev, Y.O. and Romano, L.J. (2000) In vitro replication of primer-templates containing benzo[a]pyrene adducts by exonuclease-deficient *Escherichia coli* DNA polymerase I (Klenow fragment): effect of sequence context on lesion bypass. *Biochemistry*, **39**, 10431–10438.
13 Hanrahan, C.J., Bacolod, M.D., Vyas, R.R., Liu, T., Geacintov, N.E., Loechler, E.L., and Basu, A.K. (1997) Sequence specific mutagenesis of the major (+)-*anti*-benzo[a]pyrene diol epoxide-DNA adduct at a mutational hot spot *in vitro* and in *Escherichia coli* cells. *Chem. Res. Toxicol.*, **10**, 369–377.
14 Zhuang, P., Kolbanovskiy, A., Amin, S., and Geacintov, N.E. (2001) Base sequence dependence of *in vitro* translesional DNA replication past a bulky lesion catalyzed by the exo⁻ Klenow fragment of Pol I. *Biochemistry*, **40**, 6660–6669.
15 Seo, K.Y., Nagalingam, A., Tiffany, M., and Loechler, E.L. (2005) Mutagenesis studies with four stereoisomeric N^2-dG benzo[a]pyrene adducts in the identical 5′-CGC sequence used in NMR studies: G → T mutations dominate in each case. *Mutagenesis*, **20**, 441–448.
16 Seo, K.Y., Nagalingam, A., Miri, S., Yin, J., Chandani, S., Kolbanovskiy, A., Shastry, A., and Loechler, E.L. (2006) Mirror image stereoisomers of the major benzo[a]pyrene N^2-dG adduct are bypassed by different lesion-bypass DNA polymerases in *E. coli*. *DNA Repair*, **5**, 515–522.
17 Shukla, R., Jelinsky, S., Liu, T., Geacintov, N.E., and Loechler, E.L. (1997) How stereochemistry affects mutagenesis by N^2-deoxyguanosine adducts of 7,8-dihydroxy-9,10-epoxy-7,8,9,10-tetrahydrobenzo[a]pyrene:

configuration of the adduct bond is more important than those of the hydroxyl groups. *Biochemistry*, **36**, 13263–13269.

18 Rodriguez, H. and Loechler, E.L. (1995) Are base substitution and frameshift mutagenesis pathways interrelated? An analysis based upon studies of the frequencies and specificities of mutations induced by the (+)-*anti* diol epoxide of benzo[a]pyrene. *Mutat. Res.*, **326**, 29–37.

19 Moriya, M., Spiegel, S., Fernandes, A., Amin, S., Liu, T., Geacintov, N., and Grollman, A.P. (1996) Fidelity of translesional synthesis past benzo[a]pyrene diol epoxide-2′-deoxyguanosine DNA adducts: marked effects of host cell, sequence context, and chirality. *Biochemistry*, **35**, 16646–16651.

20 Jelinsky, S.A., Liu, T., Geacintov, N.E., and Loechler, E.L. (1995) The major, N^2-Gua adduct of the (+)-*anti*-benzo[a]pyrene diol epoxide is capable of inducing G → A and G → C, in addition to G → T, mutations. *Biochemistry*, **34**, 13545–13553.

21 Shen, X., Sayer, J.M., Kroth, H., Ponten, I., O'Donnell, M., Woodgate, R., Jerina, D.M., and Goodman, M.F. (2002) Efficiency and accuracy of SOS-induced DNA polymerases replicating benzo[a]pyrene-7,8-diol 9,10-epoxide A and G adducts. *J. Biol. Chem.*, **277**, 5265–5274.

22 Fernandes, A., Liu, T., Amin, S., Geacintov, N.E., Grollman, A.P., and Moriya, M. (1998) Mutagenic potential of stereoisomeric bay region (+)- and (−)-*cis-anti*-benzo[a]pyrenediolepoxide-N^2-2′-deoxyguanosine adducts in *Escherichia coli* and simian kidney cells. *Biochemistry*, **37**, 10164–10172.

23 Zhao, B., Wang, J., Geacintov, N.E., and Wang, Z. (2006) Polη, Polζ and Rev1 together are required for G to T transversion mutations induced by the (+)- and (−)-*trans-anti*-BPDE-N^2-dG DNA adducts in yeast cells. *Nucleic Acids Res.*, **34**, 417–425.

24 Avkin, S., Goldsmith, M., Velasco-Miguel, S., Geacintov, N., Friedberg, E.C., and Livneh, Z. (2004) Quantitative analysis of translesion DNA synthesis across a benzo[a]pyrene-guanine adduct in mammalian cells: the role of DNA polymerase κ. *J. Biol. Chem.*, **279**, 53298–53305.

25 Geacintov, N.E., Gagliano, A.G., Ibanez, V., and Harvey, R.G. (1982) Spectroscopic characterizations and comparisons of the structures of the covalent adducts derived from the reactions of 7,8-dihydroxy-7,8,9,10-tetrahydrobenzo[a]pyrene-9,10-oxide, and the 9,10-epoxides of 7,8,9,10-tetrahydrobenzo[a]pyrene and 9,10,11,12-tetrahydrobenzo[e]pyrene with DNA. *Carcinogenesis*, **3**, 247–253.

26 Geacintov, N.E., Yoshida, H., Ibanez, V., Jacobs, S.A., and Harvey, R.G. (1984) Conformations of adducts and kinetics of binding to DNA of the optically pure enantiomers of anti-benzo(a)pyrene diol epoxide. *Biochem. Biophys. Res. Commun.*, **122**, 33–39.

27 Harvey, R.G. and Geacintov, N.E. (1988) Intercalation and binding of carcinogenic hydrocarbon metabolites to nucleic acids. *Acc. Chem. Res.*, **21**, 66–73.

28 Singh, S.B., Hingerty, B.E., Singh, U.C., Greenberg, J.P., Geacintov, N.E., and Broyde, S. (1991) Structures of the (+)- and (−)-*trans*-7,8-dihydroxy-*anti*-9,10-epoxy-7,8,9,10-tetrahydrobenzo(a)pyrene adducts to guanine-N^2 in a duplex dodecamer. *Cancer Res.*, **51**, 3482–3492.

29 Tan, J., Geacintov, N.E., and Broyde, S. (2000) Conformational determinants of structures in stereoisomeric *cis*-opened *anti*-benzo[a]pyrene diol epoxide adducts to adenine in DNA. *Chem. Res. Toxicol.*, **13**, 811–822.

30 Tan, J., Geacintov, N.E., and Broyde, S. (2000) Principles governing conformations in stereoisomeric adducts of bay region benzo[a]pyrene diol epoxides to adenine in DNA: steric and hydrophobic effects are dominant. *J. Am. Chem. Soc.*, **122**, 3021–3032.

31 Wu, M., Yan, S.F., Tan, J., Patel, D.J., Geacintov, N.E., and Broyde, S. (2004) Conformational searches elucidate effects of stereochemistry on structures of deoxyadenosine covalently bound to tumorigenic metabolites of benzo[c]phenanthrene. *Front. Biosci.*, **9**, 2807–2818.

32 Xie, X.M., Geacintov, N.E., and Broyde, S. (1999) Origins of conformational differences between *cis* and *trans* DNA adducts derived from enantiomeric *anti*-benzo[a]pyrene diol epoxides. *Chem. Res. Toxicol.*, **12**, 597–609.

33 Xie, X.M., Geacintov, N.E., and Broyde, S. (1999) Stereochemical origin of opposite orientations in DNA adducts derived from enantiomeric *anti*-benzo[a]pyrene diol epoxides with different tumorigenic potentials. *Biochemistry*, **38**, 2956–2968.

34 de los Santos, C., Cosman, M., Hingerty, B.E., Ibanez, V., Margulis, L.A., Geacintov, N.E., Broyde, S., and Patel, D.J. (1992) Influence of benzo[a]pyrene diol epoxide chirality on solution conformations of DNA covalent adducts: the (−)-*trans-anti*-[BP]G · C adduct structure and comparison with the (+)-*trans-anti*-[BP]G · C enantiomer. *Biochemistry*, **31**, 5245–5252.

35 Geacintov, N.E., Cosman, M., Hingerty, B.E., Amin, S., Broyde, S., and Patel, D.J. (1997) NMR solution structures of stereoisometric covalent polycyclic aromatic carcinogen-DNA adduct: principles, patterns, and diversity. *Chem. Res. Toxicol.*, **10**, 111–146.

36 Cosman, M., de los Santos, C., Fiala, R., Hingerty, B.E., Singh, S.B., Ibanez, V., Margulis, L.A., Live, D., Geacintov, N.E., Broyde, S., *et al.* (1992) Solution conformation of the major adduct between the carcinogen (+)-*anti*-benzo[a]pyrene diol epoxide and DNA. *Proc. Natl. Acad. Sci. USA*, **89**, 1914–1918.

37 Broyde, S., Wang, L., Zhang, L., Rechkoblit, O., Geacintov, N.E., and Patel, D.J. (2008) DNA adduct structure–function relationships: comparing solution with polymerase structures. *Chem. Res. Toxicol.*, **21**, 45–52.

38 Karle, I.L., Yagi, H., Sayer, J.M., and Jerina, D.M. (2004) Crystal and molecular structure of a benzo[a]pyrene 7,8-diol 9,10-epoxide N^2-deoxyguanosine adduct: absolute configuration and conformation. *Proc. Natl. Acad. Sci. USA*, **101**, 1433–1438.

39 Lukin, M. and de Los Santos, C. (2006) NMR structures of damaged DNA. *Chem. Rev.*, **106**, 607–686.

40 Geacintov, N.E. (1988) Mechanisms of reaction of polycyclic aromatic epoxide derivatives with nucleic acids, in *Polycyclic Aromatic Hydrocarbon Carcinogenesis: Structure–Activity Relationships*, vol. II (eds S.K. Yang and B.D. Silverman), CRC Press, Boca Raton, FL, pp. 181–206.

41 Cosman, M., de los Santos, C., Fiala, R., Hingerty, B.E., Ibanez, V., Luna, E., Harvey, R., Geacintov, N.E., Broyde, S., and Patel, D.J. (1993) Solution conformation of the (+)-*cis-anti*-[BP]dG adduct in a DNA duplex: intercalation of the covalently attached benzo[a]pyrenyl ring into the helix and displacement of the modified deoxyguanosine. *Biochemistry*, **32**, 4145–4155.

42 Cosman, M., Hingerty, B.E., Luneva, N., Amin, S., Geacintov, N.E., Broyde, S., and Patel, D.J. (1996) Solution conformation of the (−)-*cis-anti*-benzo[a]pyrenyl-dG adduct opposite dC in a DNA duplex: intercalation of the covalently attached BP ring into the helix with base displacement of the modified deoxyguanosine into the major groove. *Biochemistry*, **35**, 9850–9863.

43 Yan, S., Shapiro, R., Geacintov, N.E., and Broyde, S. (2001) Stereochemical, structural, and thermodynamic origins of stability differences between stereoisomeric benzo[a]pyrene diol epoxide deoxyadenosine adducts in a DNA mutational hot spot sequence. *J. Am. Chem. Soc.*, **123**, 7054–7066.

44 Perlow-Poehnelt, R.A., Likhterov, I., Scicchitano, D.A., Geacintov, N.E., and Broyde, S. (2004) The spacious active site of a Y-family DNA polymerase facilitates promiscuous nucleotide incorporation opposite a bulky carcinogen-DNA adduct: elucidating the structure–function relationship through experimental and computational approaches. *J. Biol. Chem.*, **279**, 36951–36961.

45 Chary, P. and Lloyd, R.S. (1995) *In vitro* replication by prokaryotic and eukaryotic polymerases on DNA templates containing site-specific and stereospecific benzo[a]pyrene-7,8-dihydrodiol-9,10-epoxide adducts. *Nucleic Acids Res.*, **23**, 1398–1405.

46 Shibutani, S., Margulis, L.A., Geacintov, N.E., and Grollman, A.P. (1993) Translesional synthesis on a DNA template containing a single stereoisomer of dG-(+)- or dG-(−)-anti-BPDE (7,8-dihydroxy-*anti*-9,10-epoxy-7,8,9,10-tetrahydrobenzo[*a*]pyrene). *Biochemistry*, **32**, 7531–7541.

47 Rechkoblit, O., Zhang, Y., Guo, D., Wang, Z., Amin, S., Krzeminsky, J., Louneva, N., and Geacintov, N.E. (2002) *trans*-Lesion synthesis past bulky benzo[*a*]pyrene diol epoxide N^2-dG and N^6-dA lesions catalyzed by DNA bypass polymerases. *J. Biol. Chem.*, **277**, 30488–30494.

48 Mocquet, V., Kropachev, K., Kolbanovskiy, M., Kolbanovskiy, A., Tapias, A., Cai, Y., Broyde, S., Geacintov, N.E., and Egly, J.M. (2007) The human DNA repair factor XPC-HR23B distinguishes stereoisomeric benzo[*a*]pyrenyl-DNA lesions. *EMBO J.*, **26**, 2923–2932.

49 Schinecker, T.M., Perlow, R.A., Broyde, S., Geacintov, N.E., and Scicchitano, D.A. (2003) Human RNA polymerase II is partially blocked by DNA adducts derived from tumorigenic benzo[*c*]phenanthrene diol epoxides: relating biological consequences to conformational preferences. *Nucleic Acids Res.*, **31**, 6004–6015.

50 Perlow, R.A., Kolbanovskii, A., Hingerty, B.E., Geacintov, N.E., Broyde, S., and Scicchitano, D.A. (2002) DNA adducts from a tumorigenic metabolite of benzo[*a*]pyrene block human RNA polymerase II elongation in a sequence- and stereochemistry-dependent manner. *J. Mol. Biol.*, **321**, 29–47.

51 Bauer, J., Xing, G., Yagi, H., Sayer, J.M., Jerina, D.M., and Ling, H. (2007) A structural gap in Dpo4 supports mutagenic bypass of a major benzo[*a*]pyrene dG adduct in DNA through template misalignment. *Proc. Natl. Acad. Sci. USA*, **104**, 14905–14910.

52 Ling, H., Sayer, J.M., Plosky, B.S., Yagi, H., Boudsocq, F., Woodgate, R., Jerina, D.M., and Yang, W. (2004) Crystal structure of a benzo[*a*]pyrene diol epoxide adduct in a ternary complex with a DNA polymerase. *Proc. Natl. Acad. Sci. USA*, **101**, 2265–2269.

53 Broyde, S., Wang, L., Rechkoblit, O., Geacintov, N.E., and Patel, D.J. (2008) Lesion processing: high-fidelity versus lesion-bypass DNA polymerases. *Trends Biochem. Sci.*, **33**, 209–219.

54 Choi, J.Y. and Guengerich, F.P. (2004) Analysis of the effect of bulk at N^2-alkylguanine DNA adducts on catalytic efficiency and fidelity of the processive DNA polymerases bacteriophage T7 exonuclease⁻ and HIV-1 reverse transcriptase. *J. Biol. Chem.*, **279**, 19217–19229.

55 Xu, P., Oum, L., Beese, L.S., Geacintov, N.E., and Broyde, S. (2007) Following an environmental carcinogen N^2-dG adduct through replication: elucidating blockage and bypass in a high-fidelity DNA polymerase. *Nucleic Acids Res.*, **35**, 4275–4288.

56 Pages, V. and Fuchs, R.P. (2002) How DNA lesions are turned into mutations within cells? *Oncogene*, **21**, 8957–8966.

57 Friedberg, E.C., Lehmann, A.R., and Fuchs, R.P. (2005) Trading places: how do DNA polymerases switch during translesion DNA synthesis? *Mol. Cell*, **18**, 499–505.

58 Lehmann, A.R., Niimi, A., Ogi, T., Brown, S., Sabbioneda, S., Wing, J.F., Kannouche, P.L., and Green, C.M. (2007) Translesion synthesis: Y-family polymerases and the polymerase switch. *DNA Repair*, **6**, 891–899.

59 Friedberg, E.C., Wagner, R., and Radman, M. (2002) Specialized DNA polymerases, cellular survival, and the genesis of mutations. *Science*, **296**, 1627–1630.

60 Prakash, S., Johnson, R.E., and Prakash, L. (2005) Eukaryotic translesion synthesis DNA polymerases: specificity of structure and function. *Annu. Rev. Biochem.*, **74**, 317–353.

61 Yang, W. and Woodgate, R. (2007) What a difference a decade makes: insights into translesion DNA synthesis. *Proc. Natl. Acad. Sci. USA*, **104**, 15591–15598.

62 Goodman, M.F. (2002) Error-prone repair DNA polymerases in prokaryotes and eukaryotes. *Annu. Rev. Biochem.*, **71**, 17–50.

63 Waters, L.S., Minesinger, B.K., Wiltrout, M.E., D'Souza, S., Woodruff, R.V., and Walker, G.C. (2009) Eukaryotic translesion polymerases and their roles and regulation in DNA damage tolerance. *Microbiol. Mol. Biol. Rev.*, **73**, 134–154.

64 Steitz, T.A. (1998) A mechanism for all polymerases. *Nature*, **391**, 231–232.

65 Rechkoblit, O., Malinina, L., Cheng, Y., Kuryavyi, V., Broyde, S., Geacintov, N.E., and Patel, D.J. (2006) Stepwise translocation of Dpo4 polymerase during error-free bypass of an oxoG lesion. *PLoS Biol.*, **4**, e11.

66 Rothwell, P.J. and Waksman, G. (2005) Structure and mechanism of DNA polymerases. *Adv. Protein Chem.*, **71**, 401–440.

67 Johnson, S.J., Taylor, J.S., and Beese, L.S. (2003) Processive DNA synthesis observed in a polymerase crystal suggests a mechanism for the prevention of frameshift mutations. *Proc. Natl. Acad. Sci. USA*, **100**, 3895–3900.

68 Hsu, G.W., Huang, X., Luneva, N.P., Geacintov, N.E., and Beese, L.S. (2005) Structure of a high fidelity DNA polymerase bound to a benzo[a]pyrene adduct that blocks replication. *J. Biol. Chem.*, **280**, 3764–3770.

69 Perlow, R.A. and Broyde, S. (2001) Evading the proofreading machinery of a replicative DNA polymerase: induction of a mutation by an environmental carcinogen. *J. Mol. Biol.*, **309**, 519–536.

70 Ling, H., Boudsocq, F., Woodgate, R., and Yang, W. (2001) Crystal structure of a Y-family DNA polymerase in action: a mechanism for error-prone and lesion-bypass replication. *Cell*, **107**, 91–102.

71 Xu, P., Oum, L., Geacintov, N.E., and Broyde, S. (2008) Nucleotide selectivity opposite a benzo[a]pyrene-derived N^2-dG adduct in a Y-family DNA polymerase: a 5′-slippage mechanism. *Biochemistry*, **47**, 2701–2709.

72 Boudsocq, F., Iwai, S., Hanaoka, F., and Woodgate, R. (2001) *Sulfolobus solfataricus* P2 DNA polymerase IV (Dpo4): an archaeal DinB-like DNA polymerase with lesion-bypass properties akin to eukaryotic polh. *Nucleic Acids Res.*, **29**, 4607–4616.

73 Oum, L. (2007) Base-sequence and temperature effects on *in vitro* studies of translesion synthesis past *anti*-BPDE-N^2-dG adducts, PhD Thesis, New York University.

74 Xu, P., Oum, L., Geacintov, N.E., and Broyde, S. (2009) Visualizing sequence-governed nucleotide selectivities and mutagenic consequences through a replicative cycle: processing of a bulky carcinogen N^2-dG lesion in a Y-family polymerase. *Biochemistry*, **48**, 4677–4690.

75 Mizukami, S., Kim, T.W., Helquist, S.A., and Kool, E.T. (2006) Varying DNA base-pair size in subangstrom increments: evidence for a loose, not large, active site in low-fidelity Dpo4 polymerase. *Biochemistry*, **45**, 2772–2778.

76 Streisinger, G., Okada, Y., Emrich, J., Newton, J., Tsugita, A., Terzaghi, E., and Inouye, M. (1966) Frameshift mutations and the genetic code. *Cold Spring Harb. Symp. Quant. Biol.*, **31**, 77–84.

77 Kunkel, T.A. (1990) Misalignment-mediated DNA synthesis errors. *Biochemistry*, **29**, 8003–8011.

78 Kunkel, T.A. (1992) DNA replication fidelity. *J. Biol. Chem.*, **267**, 18251–18254.

79 Ripley, L.S. (1990) Frameshift mutation: determinants of specificity. *Annu. Rev. Genet.*, **24**, 189–213.

80 Lone, S., Townson, S.A., Uljon, S.N., Johnson, R.E., Brahma, A., Nair, D.T., Prakash, S., Prakash, L., and Aggarwal, A.K. (2007) Human DNA polymerase κ encircles DNA: implications for mismatch extension and lesion bypass. *Mol. Cell*, **25**, 601–614.

81 Suzuki, N., Ohashi, E., Kolbanovskiy, A., Geacintov, N.E., Grollman, A.P., Ohmori, H., and Shibutani, S. (2002) Translesion synthesis by human DNA polymerase κ on a DNA template containing a single stereoisomer of dG-(+)- or dG-(−)-*anti*-N^2-BPDE (7,8-dihydroxy-*anti*-9,10-epoxy-7,8,9,10-tetrahydrobenzo[a]pyrene). *Biochemistry*, **41**, 6100–6106.

82 Zhang, Y., Wu, X., Guo, D., Rechkoblit, O., and Wang, Z. (2002) Activities of human DNA polymerase κ in response to the major benzo[a]pyrene DNA adduct: error-free lesion bypass and extension synthesis from opposite the lesion. *DNA Repair*, **1**, 559–569.

83 Huang, X., Kolbanovskiy, A., Wu, X., Zhang, Y., Wang, Z., Zhuang, P., Amin, S., and Geacintov, N.E. (2003) Effects of base sequence context on translesion synthesis past a bulky (+)-*trans-anti*-B[a]P-N^2-dG lesion catalyzed by the Y-family polymerase pol κ. *Biochemistry*, **42**, 2456–2466.

84 Choi, J.Y., Angel, K.C., and Guengerich, F.P. (2006) Translesion synthesis across bulky N^2-alkyl guanine DNA adducts by human DNA polymerase κ. *J. Biol. Chem.*, **281**, 21062–21072.

85 Poon, K., Itoh, S., Suzuki, N., Laxmi, Y.R., Yoshizawa, I., and Shibutani, S. (2008) Miscoding properties of 6α- and 6β-diastereoisomers of the N^2-(estradiol-6-yl)-2′-deoxyguanosine DNA adduct by Y-family human DNA polymerases. *Biochemistry*, **47**, 6695–6701.

86 Yasui, M., Dong, H., Bonala, R.R., Suzuki, N., Ohmori, H., Hanaoka, F., Johnson, F., Grollman, A.P., and Shibutani, S. (2004) Mutagenic properties of 3-(deoxyguanosin-N^2-yl)-2-acetylaminofluorene, a persistent acetylaminofluorene-derived DNA adduct in mammalian cells. *Biochemistry*, **43**, 15005–15013.

87 Jia, L., Geacintov, N.E., and Broyde, S. (2008) The N-clasp of human DNA polymerase κ promotes blockage or error-free bypass of adenine- or guanine-benzo[a]pyrenyl lesions. *Nucleic Acids Res.*, **36**, 6571–6584.

88 Choi, J.Y., Chowdhury, G., Zang, H., Angel, K.C., Vu, C.C., Peterson, L.A., and Guengerich, F.P. (2006) Translesion synthesis across O^6-alkylguanine DNA adducts by recombinant human DNA polymerases. *J. Biol. Chem.*, **281**, 38244–38256.

89 Wang, Y. and Schlick, T. (2008) Quantum mechanics/molecular mechanics investigation of the chemical reaction in Dpo4 reveals water-dependent pathways and requirements for active site reorganization. *J. Am. Chem. Soc.*, **130**, 13240–13250.

90 Lin, P., Pedersen, L.C., Batra, V.K., Beard, W.A., Wilson, S.H., and Pedersen, L.G. (2006) Energy analysis of chemistry for correct insertion by DNA polymerase β. *Proc. Natl. Acad. Sci. USA*, **103**, 13294–13299.

91 Florian, J., Goodman, M.F., and Warshel, A. (2005) Computer simulations of protein functions: searching for the molecular origin of the replication fidelity of DNA polymerases. *Proc. Natl. Acad. Sci. USA*, **102**, 6819–6824.

92 Hu, H. and Yang, W. (2008) Free energies of chemical reactions in solution and in enzymes with *ab initio* quantum mechanics/molecular mechanics methods. *Annu. Rev. Phys. Chem.*, **59**, 573–601.

93 Zhang, Y., Liu, H., and Yang, W. (2000) Free energy calculation on enzyme reactions with an efficient iterative procedure to determine minimum energy paths on a combined *ab initio* QM/MM potential energy surface. *J. Chem. Phys.*, **112**, 3483–3492.

94 Zhang, Y., Lee, T., and Yang, W. (1999) A pseudobond approach to combining quantum mechanical and molecular mechanical methods. *J. Chem. Phys.*, **110**, 46–54.

95 Zhang, Y. (2005) Improved pseudobonds for combined *ab initio* quantum mechanical/molecular mechanical methods. *J. Chem. Phys.*, **122**, 024114.

96 Wang, L., Yu, X., Hu, P., Broyde, S., and Zhang, Y. (2007) A water-mediated and substrate-assisted catalytic mechanism for *Sulfolobus solfataricus* DNA polymerase IV. *J. Am. Chem. Soc.*, **129**, 4731–4737.

97 Wang, L., Broyde, S., and Zhang, Y. (2009) Polymerase-tailored variations in the water-mediated and substrate-assisted mechanism for nucleotidyl transfer: insights from a study of T7 DNA polymerase. *J. Mol. Biol.*, **389**, 787–796.

98 Castro, C., Smidansky, E., Maksimchuk, K.R., Arnold, J.J., Korneeva, V.S., Gotte, M., Konigsberg, W., and Cameron, C.E. (2007) Two proton

transfers in the transition state for nucleotidyl transfer catalyzed by RNA- and DNA-dependent RNA and DNA polymerases. *Proc. Natl. Acad. Sci. USA*, **104**, 4267–4272.

99 Castro, C., Smidansky, E.D., Arnold, J.J., Maksimchuk, K.R., Moustafa, I., Uchida, A., Gotte, M., Konigsberg, W., and Cameron, C.E. (2009) Nucleic acid polymerases use a general acid for nucleotidyl transfer. *Nat. Struct. Mol. Biol.*, **16**, 212–218.

100 Florian, J., Goodman, M.F., and Warshel, A. (2003) Computer simulation of the chemical catalysis of DNA polymerases: discriminating between alternative nucleotide insertion mechanisms for T7 DNA polymerase. *J. Am. Chem. Soc.*, **125**, 8163–8177.

101 Radhakrishnan, R. and Schlick, T. (2006) Correct and incorrect nucleotide incorporation pathways in DNA polymerase b. *Biochem. Biophys. Res. Commun.*, **350**, 521–529.

102 Bojin, M.D. and Schlick, T. (2007) A quantum mechanical investigation of possible mechanisms for the nucleotidyl transfer reaction catalyzed by DNA polymerase β. *J. Phys. Chem. B*, **111**, 11244–11252.

103 Cisneros, G.A., Perera, L., Garcia-Diaz, M., Bebenek, K., Kunkel, T.A., and Pedersen, L.G. (2008) Catalytic mechanism of human DNA polymerase l with Mg^{2+} and Mn^{2+} from *ab initio* quantum mechanical/molecular mechanical studies. *DNA Repair*, **7**, 1824–1834.

104 Wang, S., Hu, P., and Zhang, Y. (2007) *Ab initio* quantum mechanical/molecular mechanical molecular dynamics simulation of enzyme catalysis: the case of histone lysine methyltransferase SET7/9. *J. Phys. Chem. B*, **111**, 3758–3764.

105 de Bruin, R.A., Kalashnikova, T.I., Aslanian, A., Wohlschlegel, J., Chahwan, C., Yates, J.R., 3rd, Russell, P., and Wittenberg, C. (2008) DNA replication checkpoint promotes G_1–S transcription by inactivating the MBF repressor Nrm1. *Proc. Natl. Acad. Sci. USA*, **105**, 11230–11235.

106 de Bruin, R.A. and Wittenberg, C. (2009) All eukaryotes: before turning off G_1–S transcription, please check your DNA. *Cell Cycle*, **8**, 214–217.

107 Guo, C., Kosarek-Stancel, J.N., Tang, T.S., and Friedberg, E.C. (2009) Y-family DNA polymerases in mammalian cells. *Cell Mol. Life Sci.*, **66**, 2363–2381.

108 Schneider, R. and Grosschedl, R. (2007) Dynamics and interplay of nuclear architecture, genome organization, and gene expression. *Genes Dev.*, **21**, 3027–3043.

109 Rooney, J.P., George, A.D., Patil, A., Begley, U., Bessette, E., Zappala, M.R., Huang, X., Conklin, D.S., Cunningham, R.P., and Begley, T.J. (2009) Systems based mapping demonstrates that recovery from alkylation damage requires DNA repair, RNA processing, and translation associated networks. *Genomics*, **93**, 42–51.

15
Translesion Synthesis and Mutagenic Pathways in *Escherichia coli* Cells

Sushil Chandani and Edward L. Loechler

15.1
Introduction

In this chapter we consider how the study of carcinogen-induced mutagenesis in the model system *Escherichia coli* has helped illuminate mutagenesis pathways in general. Carcinogens are usually mutagens, which makes sense given that tumor cells have mutations in key growth control genes that then do not operate properly, thus allowing cells to grow in an improperly regulated fashion, ultimately into a tumor [1, 2]. The study of biological processes is complicated, especially in complex systems, such as humans, and so model systems are often useful in teasing out the fundamentals of important processes that can then be pursued in higher organisms. Mutagenesis is one process that appears to be fundamentally similar in all cells and insights about mutagenesis often were first revealed in model systems, notably in *E. coli* (see below).

The steps leading to mutagenesis vary depending on the carcinogen, but the paradigm in Figure 15.1, which utilizes one particularly well-studied chemical carcinogen (benzo[a]pyrene (B[a]P)), illustrates many of the typical steps [3–5]. Steps in the horizontal direction lead toward carcinogenicity and include: metabolic activation (Step 1), reaction with DNA (Step 2), adduct mutagenesis (Step 3), and tumorigenesis (Step 4). Steps in the vertical direction lead to diminished carcinogenicity and include: metabolic detoxification (Step 5), carcinogen deactivation (Step 6), and DNA repair (Step 7). Diminished carcinogenicity is also associated with other cellular processes, such as delaying the cell cycle and apoptosis [1].

Although this chapter considers various aspects of the mutagenesis in *E. coli*, the focus is on what is known about the process by which adducts/lesions are bypassed by DNA polymerases.

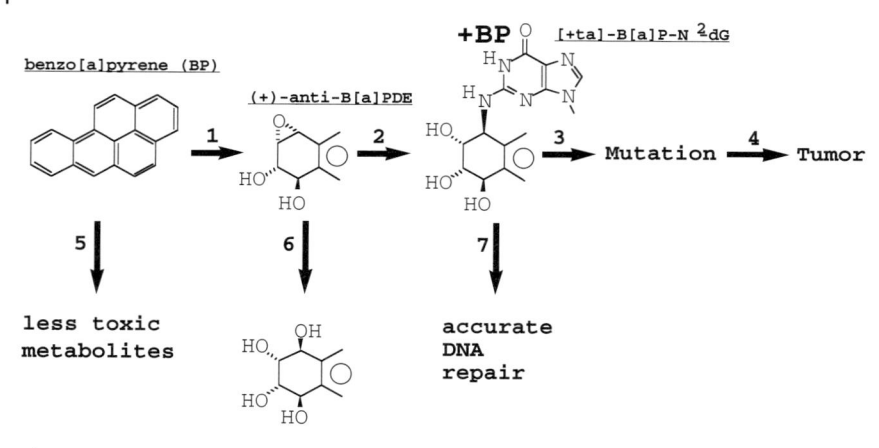

Figure 15.1 Mutagenesis/carcinogenesis paradigm with B[a]P.

15.2
Mutagenesis in *E. coli* has Illuminated Our Understanding of Mutagenesis in General

Cancer is the second leading cause of death in the United States with around 1 450 000 new cases and around 565 000 deaths expected for 2008. (Such estimates can be found at the websites for the American Cancer Society (www.cancer.org) and the National Cancer Institute (www.cancer.gov)). Approximately 60–90% of all cancers are now generally believed to be due to environmental factors (i.e., chemicals, radiation, and viruses), to which humans are exposed in food, water, or air [2]. Thus, it is of interest to understand the steps that lead from carcinogen exposure to mutagenesis to cellular transformation. We briefly discuss a few examples where insights in *E. coli* have helped forward our general understanding of mutagenesis.

Perhaps the earliest example of how *E. coli* would prove useful as a model system for mutagenesis studies is that the first DNA polymerase to be isolated was the *E. coli* DNA polymerase Pol I [6], which provided the foundation from which DNA polymerases were identified in other cells, including human cells [7–10]. DNA repair is remarkably well conserved from *E. coli* to humans [10]. Nucleotide excision repair (NER) was first discovered in *E. coli* [11–13], which provided the foundation both for the investigation of a functionally analogous NER process in higher organisms [10] and for the discovery that humans with the disease xeroderma pigmentosum usually have a defect in NER [14]. The RER (replication error) phenomenon associated with tumor cells from hereditary nonpolyposis colorectal cancer (HNPCC) individuals was reminiscent of the behavior of *E. coli* deficient in mismatch repair, which led to the discovery that HNPCC is due to a deficiency in the human orthologs of the *E. coli* mismatch repair proteins (see [15] and references therein). Base excision repair (BER) and recombinational repair in higher organisms are also similar in *E. coli* [10].

DNA polymerases that perform translesion synthesis of mutagenic/carcinogenic adducts/lesions are also strikingly similar in cells as disparate as *E. coli* and humans. Cells possess many DNA polymerases; for example, human cells, yeast (*Saccharomyces cerevisiae*), and *E. coli* have at least 15, eight, and five, respectively [7–9]. If not removed by DNA repair, mutagen/carcinogen adducts/lesions usually block replicative DNA polymerases. To overcome such potentially lethal blockage, cells have DNA polymerases that perform translesion synthesis past DNA lesions/adducts [7–9, 16–21].

Most translesion synthesis DNA polymerases are in the Y-family [7–9, 16–21], where humans have three members (Pols η, ι, and κ), yeast has one (Pol η), and *E. coli* has two (Pols IV and V). Y-family DNA polymerases have a conserved approximately 350-amino-acid core, which includes the polymerase active site (e.g., [22–32]). As with all DNA polymerases, Y-family members resemble a right hand with thumb, palm, and fingers domains, although their "stubby" fingers and thumb result in more solvent-accessible surface around the template/dNTP-binding pocket [19] – presumably to accommodate the bypass of bulky and/or deforming DNA adducts/lesions, which protrude into these open spaces. Y-family DNA polymerases grip DNA with an additional domain [22, 23, 27], usually called the "little finger." As outlined in Section 15.7, steps in the mechanism of Y-family DNA polymerases have been proposed for both protein structural changes based on a series of X-ray structures [31, 32] and for chemical catalysis based on theoretical studies [33].

The study of *E. coli*'s Y-family DNA polymerases may provide insights about Y-family DNA polymerases in general. Human Pol κ was originally discovered because its sequence closely resembles *E. coli* Pol IV [34–36] and dNTP insertion opposite a variety of adducts/lesions is remarkably similar for the Pol IV/κ pair (Table 15.1), suggesting they are functional orthologs (discussed in [37]). This notion was reinforced when the identical mutation in a conserved residue in the active site of Pols IV and κ had a similar effect on lesion bypass versus normal

Table 15.1 Dominant dNTP insertions opposite various DNA adducts/lesions by *E. coli* Pols IV and V, and human Pols κ and η.

Lesion	Pol V	Pol η	Pol IV	Pol κ
[+ta]-BP-N^2-dG	A/C	A > G	C	C
AAF-C8-dG	C	C	C/T	C/T
AF-C8-dG	–	–	C	C
TT CPD	AA	AA	n	n
T(6-4)T	AG	nG	n	n
AP site	A	A	n	A*

Dominant dNTP insertion using purified DNA polymerases, where "n" indicates "no" or low activity, "A*" indicates bypass by an unusual mechanism, and "–" indicates data unavailable. Data as reviewed in [37].

replication both *in vitro* and in cells [38]. *E. coli* Pol V and human Pol η are also functional orthologs, based on their similarity of dNTP insertion opposite a variety of adducts/lesions (Table 15.1 [37]). Cases have been made that the IV/κ class is present in cells to bypass endogenously generated N^2-dG adducts and the V/η class is present to bypass UV-induced photoproducts, as discussed below.

Some B-family DNA polymerases can also be involved in translesion synthesis, such as Pol II in *E. coli* and Pol ζ, which is present in most eukaryotes [7–9]. B-family translesion synthesis DNA polymerases are involved in a DNA repair pathway of some interstrand DNA cross-links [39–42] and in translesion synthesis of some adduct/lesions (see below).

Thus, cellular processes that affect mutagenesis (e.g., DNA repair and translesion synthesis DNA polymerases) are remarkably similar in organisms as disparate as *E. coli* and human cells. This probably explains why mutagenesis with a variety of agents is remarkably similar. p53 mutations in human colon cancer are often GC → AT mutations at CpG sites [43–49] and are associated with 5-methylcytosines, which are located at CpG sequences in many eukaryotes. The mutagenic mechanism is generally thought to involve deamination of 5-methylcytosine, as first described in *E. coli* (at *dcm* sites) over 25 years ago [50], and has been pursued extensively by a number of workers ([51] and references therein). Evidence would suggest that a similar pathway of mutagenesis and DNA repair is present in eukaryotes [10]. p53 mutations in certain human skin cancers are often GC → AT in 5′-PyC-3′ sequences [43–49, 52–55]. This pattern is reminiscent of UV light mutagenesis in *E. coli* (e.g., [56, 57]), and based on these and other results a role for UV light in skin cancer has been established [43–46, 52–55]. There is a strong preference for GC → TA mutations at codon 249 in p53 in human liver tumors associated with aflatoxin exposures [43–49, 58, 59]; it is well known that aflatoxins can induce GC → TA mutations (discussed in [60]). There are also other suggestive examples. Many p53 mutations in certain human lung cancers are GC → TA mutations, reminiscent of bulky agent mutagenesis [43–46, 61–67], where an argument for B[*a*]P involvement has been made [61–67], although arguments have been made for a nitrosamine [68].

15.3
Why Does *E. coli* have Three Translesion Synthesis DNA Polymerases [126, 127]?

E. coli's translesion synthesis polymerases are Pols II (*polB* gene, 783 amino acids, 90 kDa), IV (*dinB* gene, 351 amino acids, 39.5 kDa) and V, which is composed of one subunit of UmuC (*umuC* gene, 422 amino acids, 47.7 kDa) and two subunits of UmuD′ (see above), which is derived from UmuD (*umuD* gene, 139 amino acids, 15 kDa) following autodigestive removal of its 24 N-terminal amino acids, when stimulated by RecA* [8, 9, 16–21, 69–72]. Pols II, IV and V are each induced as part of the SOS response, which is triggered by DNA damage and leads to the induction of around 40 proteins that help *E. coli* cope with the damage [70]. The basal and SOS-induced levels are different for each, where the values in parentheses are monomers per cell (uninduced/induced): Pols II (around 40/280), IV (around 250/2500), and V (around 15/200) [73–75]. It may be that each translesion synthesis DNA polymerase is present in E. coli principally to overcome the cellular problems presented by a lesion commonly encountered in cells, as discussed next.

Although Pol V replicates undamaged templates with relatively low fidelity (10^{-3} to 10^{-4}) [76], one striking quality is Pol V's ability to accurately bypass UV photoproducts (e.g., inserting dATP opposite thymine–thymine (TT) cyclobutane pyrimidine dimers (CPDs) [76]). Analysis of insertion tendencies opposite a variety of adducts/lesions led to the observation that Pol V seems to have two insertion modes: (i) correct dNTP insertion and (ii) default dATP insertion [37]. UV light is a frequently encountered form of DNA damage for which a translesion synthesis DNA polymerase might be important and since TT CPDs are the major UV lesion [77], a default dATP insertion mode might help minimize UV mutagenesis. However, the utilization of this second mode in other circumstances may have drawbacks. For example, UV mutagenesis also depends on the *umuD/C* genes, implying that Pol V is required for UV mutagenesis, where C → T mutations in 5′-PyC sequences predominate, which also implies dATP insertion (discussed in [78]). Pol V is involved in other mutagenesis pathways; for example, it inserts dATP opposite +BP in the G → T mutational pathway ([79], discussed below). In fact, the preferential mutagenic insertion of dATP opposite a variety of DNA lesions in E. coli has been called the "A-rule" ([60, 80] and references therein) and it seems likely that this is attributable to Pol V's tendency to insert dATP [37]. Based on lesion-bypass specificity, E. coli Pol V appears to be the functional ortholog of human Pol η [37], which is almost certainly responsible for correct bypass of UV lesions in human cells and minimizes UV light mutagenesis that leads to skin cancer [81–86].

While Pols II and IV appear to operate as simple monomers, Pol V is more complex [70–72]. On its own, the polymerase subunit UmuC either misfolds or aggregates and is found in inclusion bodies [70–72], and although UmuC copurifies with UmuD′, the yield is invariably low [70–72]. Evidence suggests that for proper functioning in E. coli, Pol V requires the β-clamp and RecA, which forms a filament on single-stranded DNA downstream of the lesion [70–72]. The RecA monomer closest to the lesion site is thought to contact Pol V and stimulate bypass, which can be thought of as a role for *cis*-RecA [70–72]. Recent evidence suggests that *trans*-RecA is also required [87]. RecFOR is necessary to remove single-stranded DNA-binding protein that initially coats single-stranded DNA downstream of the lesion and to load the *cis*-RecA filament [88]. Finally, $UmuD_2C$ (not $UmuD'_2C$) is thought to slow down DNA replication in response to DNA damage, thus allowing additional time for lesion removal, which is considered a DNA damage checkpoint analogous to what happens in eukaryotic cells [89].

Pol IV replicates undamaged DNA only around 5-fold less accurately than the catalytic α-subunit of Pol III [76]. It is prone to making −1 frameshift mutations in homopolymeric runs of six or more GC base pairs; base substitutions also result (18–30%) [71, 90]. However, Pol IV's most striking quality is its ability to accurately bypass a variety of N^2-dG adducts [38, 91–95]. Methylglyoxal is produced nonenzymatically from various cellular trioses and forms N^2-(1-carboxyethyl)-2'-dG as its major stable adduct, which is bypassed accurately by Pol IV [94]. Oxidative metabolism forms reactive oxygen species that generate lipid peroxidation products that give exocyclic adducts, some of which can ring-open to N^2-dG adducts in double-stranded DNA [96] and might be bypassed by Pol IV, although this has not been investigated experimentally. These observations have led several groups to speculate that the cellular rationale for the genesis of the IV/κ class of Y-family DNA polymerases is the accurate bypass of N^2-dG adducts derived from various endogenous mechanisms [38, 94].

No analogous story *vis-á-vis* adducts/lesions has yet emerged to provide a rationale for the presence of Pol II in cells, although one possibility is the involvement of Pol II in an accurate DNA repair pathway for interstrand cross-links [39], which has been proposed to include the following steps. NER makes a nick on either side of the interstrand cross-linked lesion in one DNA strand (Figure 15.2, Step 1) to generate a gap with the other half of the cross-link still present in the second strand (Figure 15.2, Step 2). The lesion in the second strand is then bypassed by DNA polymerase II (Figure 15.2, Step 3), which, for example, accurately inserts dCTP opposite the N^7-dG adduct derived from a nitrogen mustard interstrand cross-link. Repair is completed by another round of NER (Figure 15.2, Step 4). An analogous pathway involving Pol ζ exists in yeast and other eukaryotic cells, which, in fact, has been studied more extensively [40–42]. As discussed below, Pol II functions in other translesion synthesis pathways.

The translesion synthesis DNA polymerases also confer selective advantage on *E. coli* during long periods in stationary phase – the "growth advantage in stationary phase" (GASP) phenotype [97]. Finally, Pol IV is particularly elevated in stationary phase (around 7500/cell) and is implicated in adaptive mutagenesis [98].

15.4
Overview of the Steps Leading to Translesion Synthesis

An overarching model for the steps in mutagenic bypass of adduct/lesions in *E. coli* has emerged. Replicative Pol III stalls at many adducts. For example, in the case of the N-2-acetylaminofluorene (AAF) adduct, AAF-C8-dG, the polymerase activity and the 3'–5' exonuclease activity of Pol III compete to give an around 10 : 1 ratio of [L−1] : [L0] primers, as determined *in vitro* [72]. A translesion synthesis DNA polymerase probably helps dissociate a stalled Pol III from the lesion site (see below). Pol III reinitiates replication downstream of the adduct/lesion at a primosome assembly site in a process called "replication restart," either on the lagging strand using the normal lagging strand machinery (i.e., PriA/B/C, DnaB/C/T, and primase) or on the leading strand, whose details are being worked out [99, 100]. This leaves a single-strand gap between the lesion site and the site where DNA synthesis was reinitiated. This gap is either filled via recombination or via the action of translesion synthesis DNA polymerases [10, 99, 100].

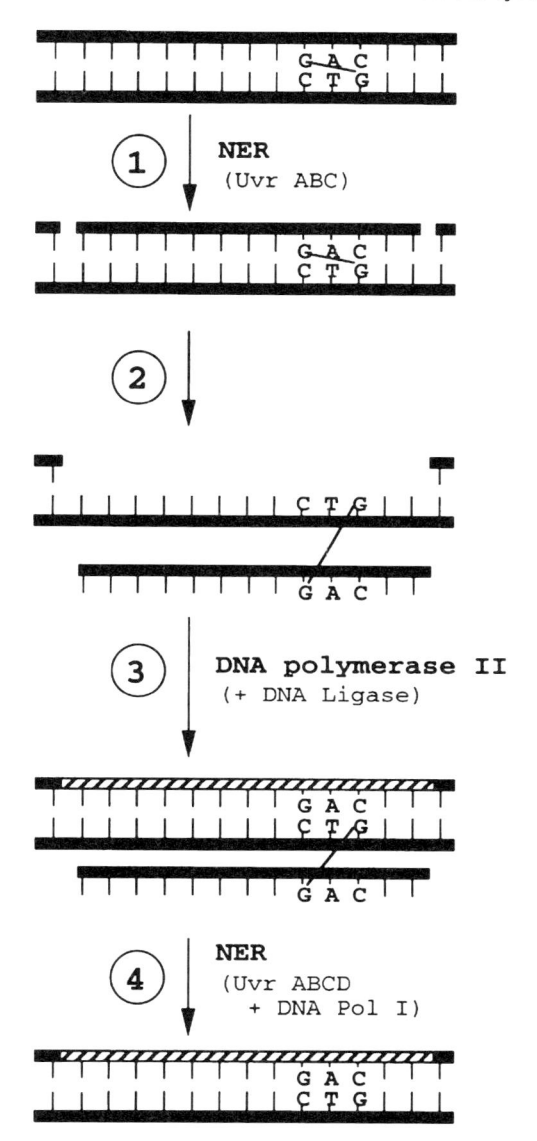

Figure 15.2 The NER/Pol II pathway for the repair of a nitrogen mustard interstrand cross-link. Repair of the top strand is proposed to involve NER (Step 1), followed by Pol II filling of the gap, including bypass of the lesion and ligation (Steps 2 and 3), and finally NER of the resulting monoadduct in the bottom strand (Step 4).

A recent study showed that Pol IV binds the β-clamp to help release a stalled Pol III from the same β-clamp, leaving Pol IV/β-clamp at the site of the lesion [101]. This process is rapid ($t < 15\,\text{s}$). Presumably, this mechanism operates for each translesion synthesis DNA polymerase (II, IV and V), all of which have β-clamp binding sites (consensus: QLxLx) that are required for them to be active in *E. coli* [102]. An X-ray crystal structure shows that the underlined amino acids QLVLGL at the C-terminus of Pol IV form the main interactions with a pocket in the β-clamp [103]. The α-subunit of Pol III and the δ-subunit of the γ-complex also bind to the same site on the β-clamp. *In vitro* studies show that the β-clamp stimulates both polymerase activity and processivity of translesion synthesis DNA polymerases: the addition of β-clamp *in vitro* increases Pol IV activity around 2000-fold and processivity from 1 to around 400 nucleotides, and also increases Pol V activity around 100-fold and processivity from 1–2 to around 18 nucleotides [71].

What factors affect the choice about which translesion synthesis DNA polymerase will insert opposite a particular lesion? Several lines of evidence suggest that *E. coli* has a hierarchy for the replication of normal, unadducted DNA when Pol III is inactivated: Pol II > IV > V [104]. (The assays did not permit an assessment of Pol I.) Since this order (III > II > IV > V) does not reflect the relative concentration of these DNA polymerases in cells (see above), another mechanism for decision making was suggested, such as relative DNA polymerase affinity for the β-clamp. This order does reflect relative fidelity of these DNA polymerases and would be a sensible order for *E. coli* to allow translesion synthesis DNA polymerases to initially sample adducts/lesions prior to a decision about which polymerase will perform translesion synthesis. However, the ultimate decision is probably predominantly controlled by which translesion synthesis DNA polymerase is most efficient at bypassing a particular adduct/lesion biochemically.

After insertion opposite the lesion, additional extension synthesis by a translesion synthesis DNA polymerase is required or else Pol III's proofreading 3′–5′ exonuclease activity will remove the inserted nucleotides back to the site of the lesion [69, 72]. The amount of extension required for Pol III to resume normal synthesis is pathway-dependent, being [L + 4] for the AAF-C8-dG nonmutagenic pathway with Pol V and [L + 3] for the AAF-C8-dG −2 frameshift pathway with Pol V [69, 72].

15.5
Case Studies: AAF-C8-dG and N^2-dG Adducts, Such as +BP

More is known about translesion synthesis of the major adduct of AAF and N^2-dG adducts in *E. coli* than any other adducts/lesions, as outlined in this section.

AAF was originally developed as a potential pesticide, but it was abandoned when it was found to be a potent rat carcinogen [105]. Following activation, AAF principally binds at C8-dG, as do most aromatic amine mutagens/carcinogens, where AAF and AAF-C8-dG have frequently been used as models to probe the mutagenic and carcinogenic mechanisms of aromatic amines [69]. In *E. coli*, AAF has a major mutational hotspot in 5′-CG_1CG_2 sequences in which it induces −2 frameshift mutations [69]. AAF-C8-dG at G_2 (but not G_1) causes a −2 frameshift mutation in a Pol II-dependent process or causes no mutation in a Pol V-dependent process [69]. The current model is that AAF-C8-dG at a replication fork exists in two different conformations [69, 72]. In one conformation the adducted dG moiety is in a −2 slipped mutagenic intermediate form, which Pol II uses for insertion, and then in the presence of β-clamp an additional three extension steps (to L + 3) are accomplished, at which point replication can be successfully continued by Pol III [72]. From a nonslipped intermediate, Pol V inserts dCTP opposite AAF-C8-dG and then extends by adding four more dNTPs (to L + 4), after which Pol III can successfully continue replication [72]. A modest alteration in sugar pucker is required before AAF-C8-dG can Watson–Crick base-pair with dCTP [106]. These two pathways are followed approximately equally in cells, although by manipulating the concentration of Pol II versus IV, the ratio −2 frameshift:no mutation can be modulated, suggesting that the two conformations interconvert [69]. *In vitro* in 5′-CG$_1$CG$_2$ sequences Pol II also does translesion synthesis to give a bypass product that should ultimately yield a −1 frameshift mutation, which is not, however, observed *in vivo*; recent *in vitro* studies suggest that Pol II cannot extend far enough from the −1 frameshift intermediate and, thus, the 3′–5′ exonuclease activity of Pol III degrades the intermediates in the −1 frameshift pathway [72].

B[*a*]P is a well-studied DNA-damaging agent that is a potent mutagen/carcinogen and an example of a polycyclic aromatic hydrocarbon (PAH) – a class of ubiquitous environmental substances produced by incomplete combustion [107, 108]. PAHs, in general, and B[*a*]P, in particular, induce the kinds of mutations thought to be relevant to carcinogenesis and may be important in human cancer [61–67]. B[*a*]P mutational spectra were established with the major metabolite that reacts with DNA (i.e., ((+)-7β,8α-dihydroxy-7,8-dihydro-9α,10α-oxo-benzo[*a*]pyrene or *anti*-B[*a*]PDE), in *E. coli* [109], yeast [110, 111], and mammalian (CHO) cells [112]. Mutagenesis has also been studied with [+ta]-B[*a*]P-N^2-dG (+BP, Figure 15.1), the major adduct of (+)-*anti*-B[*a*]PDE, and G → T mutations predominate in most cases ([113] and references therein).

Pols IV and V of *E. coli* are both involved in translesion synthesis with B[*a*]P-N^2-dG adducts, although they play very different roles (Figure 15.3). In studies with purified proteins, Pol IV inserted dCTP (greater than 99%) opposite both +BP and its stereoisomer −BP ([−ta]-B[*a*]P-N^2-dG) in a 5′-CGA sequence, while Pol V inserted dATP (greater than 99%) [95]; these two adducts are derived from the mirror-image metabolites, (+)- and (−)-*anti*-B[*a*]PDE, of B[*a*]P (Chapter 6). This tendency to insert dCTP is evident in *E. coli*. Pol IV is required in the nonmutagenic pathway with +BP [38, 91–93], −BP [93], and other N^2-dG adducts [38, 94]. An amino acid change (F12I) at the conserved "steric gate" (which excludes rNTPs) decreases dCTP insertion *in vitro* opposite several N^2-dG adducts and similarly decreases translesion synthesis *in vivo*, which argues that Pol IV does dCTP insertion *in vivo* [38]. In the nonmutagenic pathway Pol V is required in addition to Pol IV with +BP [91–93]. Why are two DNA polymerases required for nonmutagenic translesion synthesis with +BP: certain lesions need one DNA polymerase for insertion and a second for extension [114, 115]. Thus, if Pol IV does dCTP insertion [38, 91–95], then Pol V must do extension, which is reasonable given kinetic findings with purified proteins, where Pol V can be up to around 1500 times better than Pol IV at the

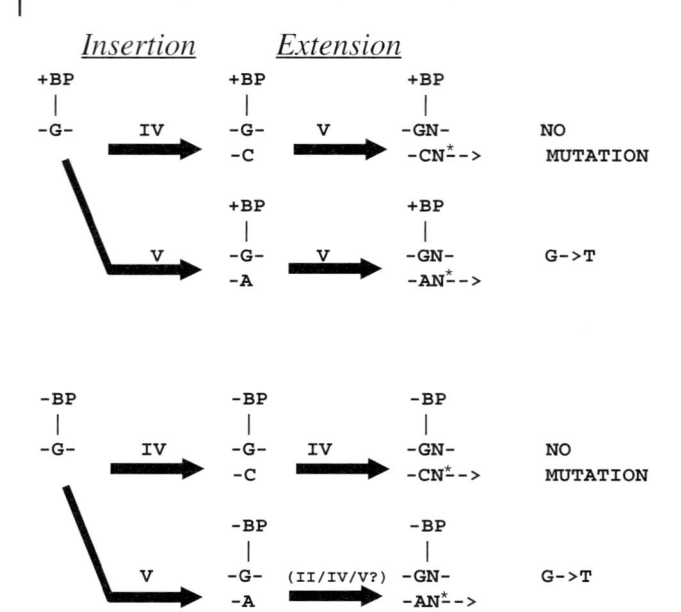

```
        Insertion            Extension
+BP              +BP              +BP
 |                |                |
-G-      IV      -G-      V      -GN-           NO
                 -C              -CN*->       MUTATION

                 +BP              +BP
                  |                |
         V       -G-      V      -GN-          G->T
                 -A              -AN*->


-BP              -BP              -BP
 |                |                |
-G-      IV      -G-     IV      -GN-           NO
                 -C              -CN*->       MUTATION

                 -BP              -BP
                  |                |
         V       -G-  (II/IV/V?)  -GN-          G->T
                 -A              -AN*->
```

Figure 15.3 Current model for insertion and extension of +BP and −BP adducts by Pols IV and V in the nonmutagenic and G → T mutagenic pathway in *E. coli* (see text).

step directly following adduct-GC formation (i.e., extension) in the case of +BP compared to −BP (discussed in greater detail in [93]). Regarding the nonmutagenic pathway with −BP, only Pol IV is required for efficient translesion synthesis [93], suggesting it does both insertion and extension. Pol V is required in the G → T pathway for +BP, while Pols II and IV are not, implying that Pol V must do insertion and extension [79]. We have unpublished findings that Pol V is required for the G → T pathway with −BP, but we have not yet investigated whether Pols II or IV might have a role (unpublished). Finally, random mutagenesis studies with (+)-*anti*-B[*a*]PDE suggest that most G → T mutations with B[*a*]P adducts require SOS-inducible Y-family DNA polymerases, although a minor non-SOS-inducible G → T pathway does exist (discussed in [79]), which has been studied with +BP in a 5′-G<u>G</u>A sequence [91, 92]; translesion synthesis was proposed to require Pol III.

15.6
Structure–Function Analysis of Y-Family Pols IV and V of *E. coli*

Table 15.1 [37] shows that dNTP insertion opposite a variety of adducts/lesions, including +BP, is remarkably similar for the Pol IV/κ pair, suggesting they are functional orthologs, and is remarkably similar for the Pol V/η pair, also suggesting they are functional orthologs. There must be structural reasons for the insertion preferences of these DNA polymerases, although the key elements are not obvious, given that in alignments, for example, UmuC(V) shares only 20% amino acid identity with its functional ortholog human Pol η, which is about the same as the 21% identity that it shares with its nonfunctional ortholog human Pol κ [37]. The extent of this dilemma is further revealed by the fact that human Pol η is no more identical to yeast Pol η (24%) than it is to human Pol κ (24%). Nevertheless, a careful examination of Y-family DNA polymerase structure is beginning to reveal likely key structural features.

Regarding structure, the absence of X-ray crystal structures for UmuC (the polymerase subunit of Pol V), Pol IV, and human Pol η prompted us to build models taking a homology modeling approach [37]. Analysis of X-ray structures, modeled structures, and sequence alignment suggests that Y-family DNA polymerases lend themselves to accurate homology modeling [37, 116] in the thumb, palm, and fingers domain, although alignment of the little-finger domain is less clear. Fortunately, an X-ray structure exists for Pol IV's little finger domain [103]. An X-ray crystal structure does exist for human Pol κ with DNA [30].

In the rest of this chapter we reflect on what we have observed in our models of Pol IV and UmuC(V), and compare this to the X-ray structures for other Y-family DNA polymerases. Our laboratory has focused on the relationship between DNA polymerase structure and function for B[a]P-N^2-dG adducts, which will be described below, although much of this analysis is unpublished [116].

To form an adduct-dG:dCTP base pair, the B[a]P moiety must be in the developing minor groove, since the adduction site (N^2-dG) is in the minor groove in a Watson–Crick base pair. On the minor groove side, Y-family DNA polymerases have an opening (or gap) next to the active site between the fingers and little-finger domains. This opening looks like an elliptical hole of varying sizes in Dpo4 [23–26], Dbh [22], human Pol ι [29], and in our models of Pol IV and UmuC(V) (see below), while it looks like a slot in human Pol κ [30]. The character of this opening can be analyzed based on a simple analogy to a "chimney," where a cluster of nearby amino acids can be thought of as a "flue" that either plug the chimney, leaving a small opening, or do not plug the chimney, leaving a large opening. Next to the flue is a single amino acid, which can be thought of as a "flue-handle" that controls whether the flue is open or closed. Nearest the active site, three regions of the protein contribute to the chimney as shown in Figure 15.4(a) for our model of Pol IV: an upper lip (turquoise), a lower lip (dark blue), and a left lip (blue) [37]. It is not unreasonable to think that chimney opening size and shape might influence dNTP insertional mechanism given that the bulky B[a]P moiety (red, Figure 15.4a) must fit in it.

Our models reveal that Pol IV has a large chimney opening (Figure 15.4a), which can accommodate the pyrene and allows +BP to pair with dCTP, when dCTP adopts the canonical shape observed in all other families of DNA polymerases. In contrast, our molecular models suggest that UmuC(V) has a small chimney opening (Figure 15.4b), which forces +BP downward in the active site into a position where catalysis is less likely to be facile.

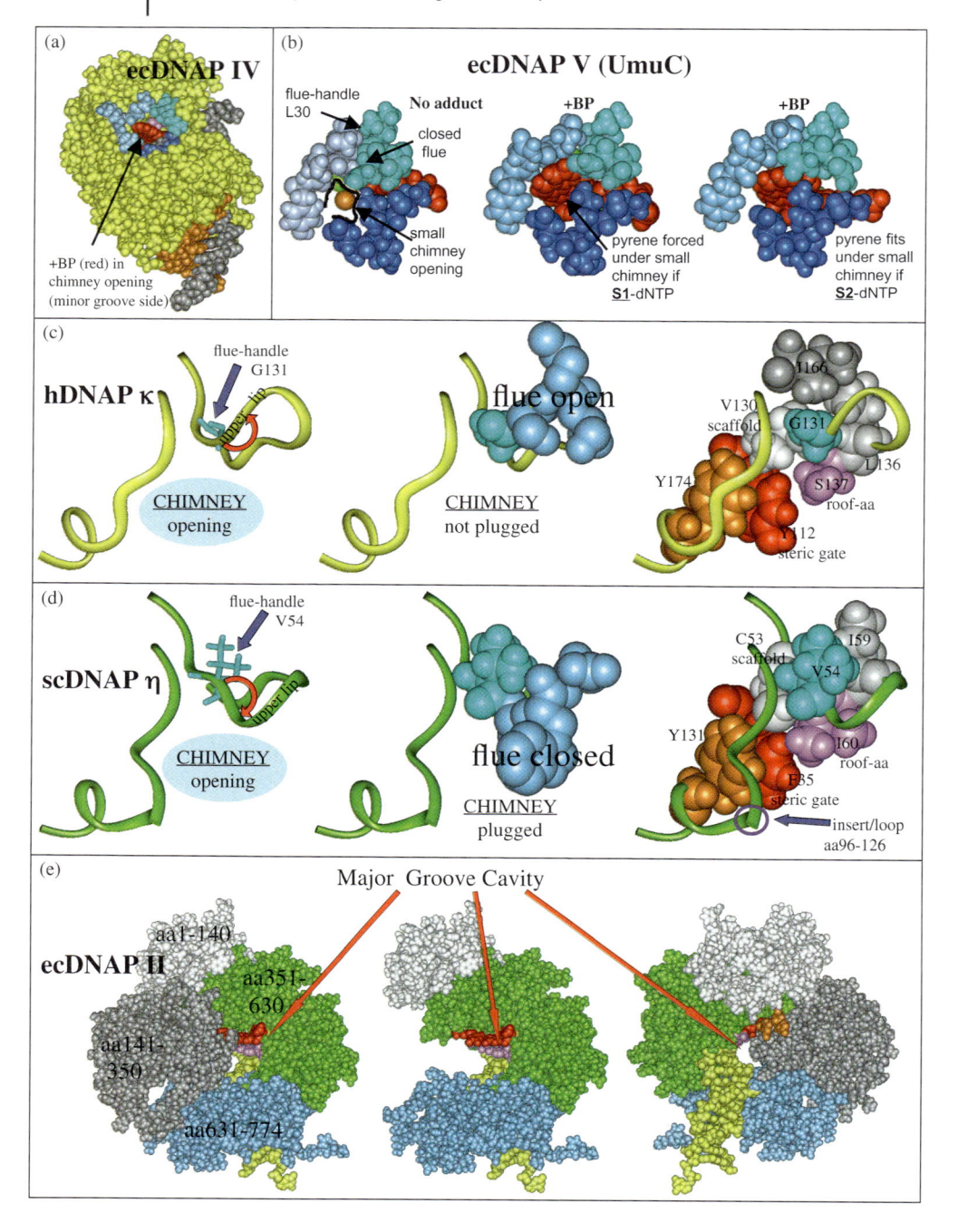

364 | *15 Translesion Synthesis and Mutagenic Pathways in Escherichia coli Cells*

(a) **ecDNAP IV**
+BP (red) in chimney opening (minor groove side)

(b) **ecDNAP V (UmuC)**
No adduct — flue-handle L30, closed flue, small chimney opening
+BP — pyrene forced under small chimney if **S1**-dNTP
+BP — pyrene fits under small chimney if **S2**-dNTP

(c) **hDNAP κ**
flue-handle G131
CHIMNEY opening
flue open
CHIMNEY not plugged
I166, V130 scaffold, G131, L136, Y174, S137 roof-aa, F112 steric gate

(d) **scDNAP η**
flue-handle V54
CHIMNEY opening
flue closed
CHIMNEY plugged
C53 scaffold, I59, V54, Y131, I60 roof-aa, F35 steric gate, insert/loop aa96-126

(e) **ecDNAP II**
Major Groove Cavity
aa1-140, aa141-350, aa351-630, aa631-774

Figure 15.4 Structures of regions of various DNA polymerases (DNAPs): *E. coli* Pol IV (a), *E. coli* UmuC(V) (b), human Pol κ (c), yeast Pol η (d), and *E. coli* Pol II (e). (a) View from the minor groove side of Pol IV (yellow), showing the "chimney" opening (cleft or hole), which is encircled by the upper lip (turquoise, amino acids 32–35), left lip (blue, amino acids 73–76), and lower lip (dark blue, amino acids 244–247), along with the pyrene moiety of +BP (red) emerging from the chimney opening, which is large. In Pol IV, the dG moiety of +BP can base-pair comfortably with the canonical S1-dCTP shape Figure 15.6. (Neither the dG moiety of +BP nor the dCTP are visible.) The template (gray) and primer (brown) are also shown. (b) Models of UmuC(V) with no adduct (left) or +BP in the canonical, "chair-like" S1-dNTP shape (center) or +BP in the noncanonical, "goat-tail-like" S2-dNTP shape (right). The chimney opening is small (left) and the pyrene moiety of +BP does not fit into the chimney in the case of the S1-dNCP arrangement (center). In contrast, the pyrene moiety of +BP fits under the chimney opening in the case of the S2-dCTP shape, because it sits lower down in the active site compared to S1-dNTP. (c) Regions of human Pol κ. Y-family DNA polymerases in the IV/κ class have a glycine "flue-handle," such as G131 (turquoise, left) in Pol κ, which leads to upward curvature of the protein backbone in the chimney upper lip (red arrow, left) and results in the "flue" amino acids (S132/M133, blue, center) pointing away from the chimney, giving a large opening. The right structure shows V130 (white) that serves as a scaffold to organize the chimney's upper lip and left lip (yellow ribbons), along with the roof-amino acid (S137, purple), the steric gate (Y112, red), and a conserved tyrosine (Y174, brown), which stacks on the backbone of the left lip and helps orient it. V130 forms a rectangle with the G131 flue-handle, L136 (gray), and the S137 roof (pink) upon which I166 stacks (dark gray). (d) Regions in yeast Polη. Y-family DNA polymerases in the V/η class have a bulky "flue-handle," such as V54 in Pol η (turquoise, left), which causes downward curvature of the chimney upper lip (red arrow, center) and results in the "flue" amino acids (Q33/W34, blue, center) plugging the chimney, giving a small opening. The structure on the right shows scaffold C53 (white) organizing the chimney's upper lip and left lip (green ribbons in panel c), along with the roof-amino acid (I60, purple), the steric gate (F35, red), and a conserved tyrosine (Y131, brown), which stacks on the backbone of the left lip and orients it. C53 forms a rectangle with the V54 flue-handle (turquoise), I59 (gray), and the I60 roof (pink). Pol η has a large insert/loop (amino acids 96–126) in the left lip, which is represented as a discontinuity. X-ray coordinates are from PDB 2OH2 for human Pol κ [30] and from PDB 1JIH for yeast Pol η [28], where hydrogens were added using insightII. (e) Preliminary model of Pol II, which was developed from the X-ray crystal structure of Pol II [118] and the B-family Pol Rpb69 ([119]; PDB 2P5O, 2OZS). The left-hand structure shows the thumb domain (blue, amino acids 631–774), palm domain (green, amino acids 351–630), and finger domain (white, amino acids −1 to 141 and gray, amino acids 141–350), along with duplex DNA (yellow), the template-dT:dATP base pair (red), and the L + 1 base pair (pink). The center structure is the same as the left-hand structure with a portion of the finger domain removed (amino acids 141–350). The right-hand structure is the same as the left-hand structure except Pol II is rotated (counter-clockwise from the top) and single-stranded DNA is visible (brown). The palm domain (green) and finger domain (blue) wrap around the double-stranded DNA by contacting the minor groove, thus leading the major groove open to solvent. Although amino acids 141–350 cover the major groove of the active site, they do not contact the DNA, thus leaving a cavity.

We consider why chimney opening size might be different in Pol IV versus UmuC(V), before assessing how chimney size might affect dNTP insertion. We also consider why in cells Pol IV might insert dCTP, while Pol V might insert dATP. Finally, we reflect on how Pol IV might be similar to Pol κ, as well as how UmuC(V) might be similar to Pol η.

15.6.1
Structural Basis for a Large versus Small Chimney Opening

What structural difference(s) in Pol IV versus UmuC(V) might result in a large versus a small chimney opening and is this structural difference(s) conserved in other Y-family DNA polymerases in the IV/κ versus V/η class? The chimney has three sides (Figure 15.4a): the upper lip (numbered amino acids 32–35 in Pol IV, turquoise) and left lip (amino acids 73–76, blue) are in the finger domain, while the lower lip (amino acids 244–247, dark blue) is in the little-finger domain. The amino acids associated with the chimney lips can be assigned reliably based on conserved amino acids. For example, the regions surrounding the chimney upper lip for the IV/κ and V/η classes (as shown in Figure 15.5) have 22 and 18 conserved amino acids, respectively, and, of these, seven are identical between the sets (boxed in gray in Figure 15.5). The assignment of amino acids to the left lip and lower lip are similarly anchored by the alignment of nearby conserved amino acids.

The chimney upper lip (turquoise, Figure 15.4a) is closest to the active site and principally defines whether the chimney can accommodate the bulky B[a]P moiety. The first amino acid in the upper lip of Pol IV is glycine (G32) and all IV/κ class members listed in Figure 15.5(a) have a glycine at this position, often embedded in a VGS sequence. The one X-ray structure available for this class, human Pol κ [30], shows that this glycine (G131, turquoise, Figure 15.4c) is followed by upward curvature of the chimney upper lip (red arrow, Figure 15.4c). This glycine can be thought of as a "flue-handle" whose backbone properties allow this upward curvature (discussed below), with the consequence being that the R-groups on the next several amino acids ("flue"; S132/R133, blue in Figure 15.4c) are turned away from the chimney opening, which remains open. Our models of Pol IV also have this upward curvature (Figure 15.4a) with an open flue, which depends on an analogous flue-handle glycine (G32).

In contrast, all V/η class members have bulky "flue-handles," which is usually valine in a VVQ sequence (Figure 15.5b). The one X-ray structure in the V/η class, yeast Pol η [28], shows that its bulky V54 flue-handle (turquoise, Figure 15.4d) is associated with downward curvature of the chimney upper lip (red arrow), which forces the "flue" (Q55/Y56, blue in Figure 15.4d) to plug the chimney. In UmuC(V) the sequence is slightly different (VLSN), although the outcome is the same – the bulky L30 flue-handle causes downward curvature and an asparagine (N32) plugs the chimney giving a closed flue (Figure 15.4b).

Glycine has greater flexibility in the ϕ/φ angles it can adopt compared to other amino acids, which appears to be the fundamental conformational reason why a glycine versus a nonglycine flue-handle gives different curvature of the chimney

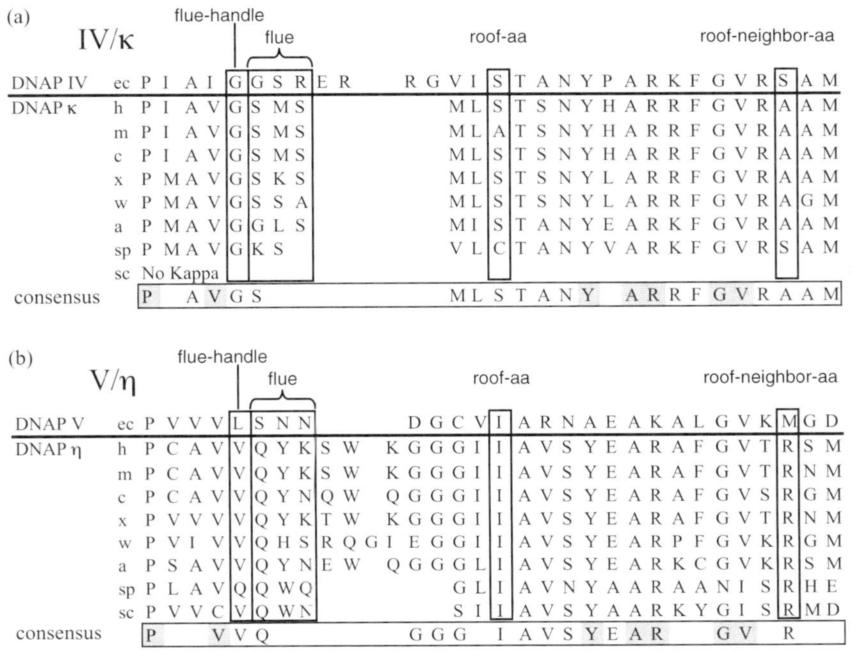

15.6 Structure–Function Analysis of Y-Family Pols IV and V of E. coli

(a)

IV/κ

				flue-handle / flue						roof-aa									roof-neighbor-aa	
DNAP IV	ec	P I A I	G	G S R	E R		R G V I	S	T A N Y P A R K F G V R	S	A M
DNAP κ	h	P I A V	G	S M S			M L	S	T S N Y H A R R F G V R	A	A M
	m	P I A V	G	S M S			M L	A	T S N Y H A R R F G V R	A	A M
	c	P I A V	G	S M S			M L	S	T S N Y H A R R F G V R	A	A M
	x	P M A V	G	S K S			M L	S	T S N Y L A R R F G V R	A	A M
	w	P M A V	G	S S A			M L	S	T S N Y L A R R F G V R	A	G M
	a	P M A V	G	G L S			M I	S	T A N Y E A R K F G V R	A	A M
	sp	P M A V	G	K S			V L	C	T A N Y V A R K F G V R	S	A M
	sc	No Kappa									
consensus		P A V G S					M L S T A N Y A R R F G V R A A M				

(b)

V/η

			flue-handle / flue			roof-aa			roof-neighbor-aa	
DNAP V	ec	P V V V	L	S N N		D G C V	I	A R N A E A K A L G V K	M	G D
DNAP η	h	P C A V	V	Q Y K	S W	K G G G	I	A V S Y E A R A F G V T	R	S M
	m	P C A V	V	Q Y K	S W	K G G G	I	A V S Y E A R A F G V T	R	N M
	c	P C A V	V	Q Y N	Q W	Q G G G	I	A V S Y E A R A F G V S	R	G M
	x	P V V V	V	Q Y K	T W	K G G G	I	A V S Y E A R A F G V T	R	N M
	w	P V I V	V	Q H S	R Q G I	E G G I	I	A V S Y E A R P F G V K	R	G M
	a	P S A V	V	Q Y N	E W	Q G G G L	I	A V S Y E A R K C G V K	R	S M
	sp	P L A V	Q	Q W Q		G L	I	A V N Y A A R A A N I S	R	H E
	sc	P V V C	V	Q W N		S I	I	A V S Y A A R K Y G I S	R	M D
consensus		P V V Q				G G G	I A V S Y E A R G V R				

Figure 15.5 Amino acid alignment of several key regions of representative Y-family DNA polymerases (DNAPs) in the (a) IV/κ and (b) V/η class. The R-groups of the final three amino acids ("flue") of the chimney's upper lip either do not plug (IV/κ class) or plug (V/η class) the chimney, depending upon whether the first amino acid in the upper lip ("flue-handle") is glycine (flue open, IV/κ class) or is a bulky amino acid (often valine, flue closed, V/η class). The lowest line in each panel shows conserved residues, with residues conserved (75% or greater) between the IV/κ class and the V/η class highlighted in gray. Abbreviations: aa, amino acid; ss, *Sulfolobus solfataricus* (hyperthermophilic acidophilic sulfur-metabolizing archeon); h, human (*Homo sapiens*); m, mouse (*Mus musculus*); c, chicken (*Gallus gallus*); x, *Xenopus laevis* (frog); w, worm (*Caenorhabditis elegans*); a, *Arabidopsis thaliana* (plant); sp, *Saccharomyces pombe* (fission yeast); sc, *Saccharomyces cerevisiae* (budding yeast).

upper lip. The ϕ/φ angles for the flue-handles in yeast DNA Pol η (V54, $-113°/157°$), human Pol η (V37, $-130°/145°$), UmuC(V) (L30, $-128°/173°$), and Dpo4 (C31, $-118°/54°$) are all clustered in the region expected for extended (β) secondary structure (i.e., ϕ angles: $-60°$ to $-150°$; φ angles: $90°$ to $170°$), with the consequence being that this combination of ϕ/φ angles leads to downward curvature of the chimney's upper lip, causing the flue to plug the chimney and the chimney opening to be small. In contrast, the ϕ/φ angles for the glycine flue-handles are outside the extended (β) region for human Pol κ (G131, $172°/-136°$) and Pol IV (G32, $178°/-160°$), which results in upward curvature of the chimney's upper lip, the opening of the flue and the large chimney opening.

Glycines are *not* excluded from adopting the φ/φ angles observed in the extended (β) region, so in principle IV/κ class DNA polymerases could have a closed flue leading to a small chimney opening as in the V/η class. The glycine flue-handles prefer the "up" orientation, because they are stabilized, for example, by a R34/E25 salt bridge in our model of Pol IV and by a M133/I166 hydrophobic interaction for Pol κ (not shown).

15.6.2
Roof-Amino Acids and Roof-Neighbor-Amino Acids

Another key difference between the IV/κ and V/η class is the roof-amino acid (Figure 15.4c and d, pink), which is a positionally conserved residue that lies above the nucleobase of the dNTP, as seen in the active site of Dpo4 [23–26], yeast Pol η [28], human Pol ι [29], and human Pol κ [30]. The roof-amino acid is next to the chimney (Section 15.6.3). The roof-amino acid is isoleucine for all members of the V/η class listed in Figure 15.5(b) and is nonbulky (usually serine) for all IV/κ class members (Figure 15.5a).

The amino acid after the roof-amino acid also correlates: alanine follows when the roof-amino acid is isoleucine (i.e., UmuC: I38/A39, yeast Pol η: I60/A61, and human Pol η: I48/A49), while threonine follows a nonbulky alanine or serine roof-amino acid (i.e., Dpo4: A44/T45, Dbh: A44/T45, Pol IV: S41/T42, and human Pol κ: S137/T138). This pattern is observed in all examples listed in Figure 15.5. In the yeast Pol η X-ray structure [28] I60/A61 forms a hydrophobic layer above the nucleobase of the dNTP, which is also the case in our model of UmuC(V). In X-ray structures the threonine methyls in Dpo4 (T45) and in human Pol κ (T138) sit near the roof-amino acid (A44 and S137, respectively), while the hydroxyl of the threonine forms a hydrogen bond with a nonbonded oxygen on Pβ of the dNTP.

Another amino acid sits beside the roof-amino acid and can also be in contact with nucleobase of the dNTP (designated "roof-neighbor" in Figure 15.5). For the V/η class the roof-neighbor-amino acid is always bulky (arginine or methionine) and for the IV/κ class the roof-neighbor-amino acid is nonbulky (usually alanine).

It is not unreasonable to imagine that the roof-amino acid, as well as the two amino acids it contacts could influence dNTP insertion patterns, although the details remain to be worked out.

15.6.3
Interconnected Architecture of the Chimney and Roof Regions [128]

To understand the architecture of the chimney/roof region of Y-family DNA polymerases, it is useful to focus on a bulky aliphatic amino acid, usually a valine (e.g., V29/UmuC(V), I30/Pol IV, V37/yeast Pol η, and V130/human Pol κ, Figure 15.5), that plays a scaffolding role as revealed in X-ray structures [22–32] and in our models [37, 116, 119]. Using human Pol κ [30] as an example (Figure 15.4c), this scaffolding valine (V130, white) begins a loop that ends with the roof-amino acid, with which it forms a backbone hydrogen bond (scaffold-C=O : HN-roof). This type of hydrogen bond is also observed in X-ray structures from Dpo4 [23–26], human Pol κ [30] and yeast Pol η [28]. The flue-handle (G131) is adjacent to scaffold-V130, which also contacts the bulky amino acid L136 (Figure 15.4c, gray) next to roof-S137. Thus, the base of this loop is anchored by a square of four amino acids (V130/G131/L136/S137), above which sits the R-group of I166 (dark gray, Figure 15.4c) giving a rectangular pyramid of amino acids that orient the flue-handle/flue/roof-amino acid region. In our model of Pol IV, the rectangle includes I31/G32/I41/S42 with L71 sitting above it. This region in yeast Pol η (Figure 15.4d) looks similar, although only the rectangle exists (C53/V54/I59/I60), because the bulky R-group of flue-handle/V54 serves the role of the fifth amino acid (i.e., the R-group of flue-handle-V54 takes the place of I166 in human Pol κ). In our model of UmuC(V), the rectangle involves V29/L30/V37/I38.

Scaffold-V130 in human Pol κ (white, Figure 15.4c) also helps organize the steric gate (Y12, red), which face-stacks with Y174, a highly conserved tyrosine whose other aromatic face contacts the backbone of the left lip of the chimney (i.e., amino acids 168–171 in human Pol κ), thus helping to orient it. The yeast Pol η structure is similar (Figure 15.4d) except the chimney left lip has an insert/loop (amino acids 96–126, which is indicated by the circle). In spite of this insert/loop, the chimney left lip is similarly positioned in the vicinity of the upper lip. (Of the DNA polymerases shown in Figure 15.5, only Pols η from yeast have an insert at this position.)

Amino acid structure in the upper lip/left lip/roof region is similar in all Y-family DNA polymerases, for which X-ray structures [22–32] or modeled structures [37, 116, 119] exist.

15.6.4
dCTP Insertion by Pol IV

Pol IV can pair dCTP with +BP, importantly, because the bulky pyrene can be accommodated in Pol IV's large chimney opening (Figure 15.4a). For phospho-ester bond formation to occur the distance between primer-3'-O and Pα-dCTP must be reaction ready and can be compared to the closest possible distance, a van der Waals' contact (around 3.5 Å). In our models of +BP in Pol IV [116], the distance between primer-3'-O and Pα-dCTP was around 3.7 Å, which comes close to van der Waals' contact, and, thus, can be thought of as being "reaction ready." The no-adduct control had a similar primer-3'-O–Pα-dCTP distance (around 3.7 Å).

In contrast, Pol V does not give a satisfactory structure when +BP is paired with dCTP (Figure 15.4b, center), because Pol V's small chimney opening forces the bulky pyrene moiety downward. Asparagine at position 31 is the main problem, as its side-chain plugs the UmuC(V) chimney leading to clashes with +BP. In the unadducted structure (Figure 15.4b, left), N31 adopts its lowest energy rotational conformer with respect to the Cα–Cβ bond The presence of +BP causes a rotation about the Cα–Cβ bond (Figure 15.4b, center); however, no other rotations can get N31 farther out of the way. Consequently, UmuC(V)'s small chimney forces template BP-dG and its paired dCTP to move downward such that the primer 3′-O–Pα-dCTP distance is elongated to a non-reaction-ready distance of around 5.0 Å.

These observations provide a reasonable rationale for why Pol IV preferentially does cellular dCTP insertion in cells, since a reasonable adduct-dG : dCTP structure emerges with near reaction-ready distances between primer-3′-O and Pα-dCTP.

15.6.5
How Does UmuC(V) Insert dATP?

If UmuC(V)'s small chimney is likely to enforce a non-reaction-ready distance between the primer-3′-O and Pα-dNTP, then how does UmuC(V) insert a dATP opposite +BP? Recently, we offered a hypothesis [116]. X-ray structures from all families of DNA polymerases show a canonical dNTP shape that has been called "chair-like" (dNTPS1-dNTP, Figure 15.6, white) [25]. In some X-ray structures of the best-studied Y-family DNA polymerase, Dpo4, the dNTP also adopts a second noncanonical "goat-tail-like" shape [25]. The goat-tail-like shape (dNTPS2-dNTP, Figure 15.6, gray) lies lower down in the active site than the canonical S1-dNTP and +BP paired with the S2-dNTP shape allows the pyrene to lie comfortably under the small chimney opening of UmuC(V) (Figure 15.4b, right), thus allowing the primer-3′-O–Pα-dNTP distance to be reaction-ready (around 3.8 Å). Furthermore, the adducted dG can Hoogsteen pair with *syn*-adenine using the S2-dATP shape, which is in marked contrast to the situation with S1-dATP, where *syn*-adenine can have steric clashes with atoms in the deoxyribose and the α-phosphate. Adduct-dG : *syn*-dATP pairing in UmuC(V) gave near-reaction-ready structures with primer-3′-O–Pα-dNTP distances of around 3.6 Å. We emphasize that this is merely a hypothesis, although the S2-dNTP shape does have accompanying protein components that should allow phosphodiester bond making and breaking [116, 119].

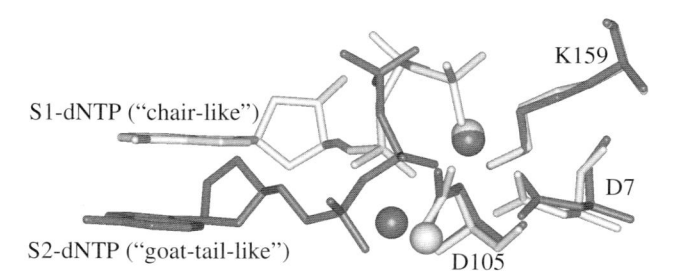

Figure 15.6 Side view of a dNTP in the "chair-like" shape S1-dNTP (white) versus the "goat-tail-like" shape S2-dNTP (gray). Key amino acids (only D7, D105, and K159 are shown) from a Dpo4 structure adopting the S1-dNTP shape were superimposed on the same amino acids in a Dpo4 structure adopting the S2-dNTP. Spheres are divalent cations (S1-dNTP/white and S2-dNTP/gray). X-ray coordinates are from PDB 1SOM-B for Dpo4/S1-dNT and PDB 1RYS-A for Dpo4/S2-dNTP.

Other G:A mispairings are also possible. In principle, *anti*-dATP can pair with *syn*-guanine in adduct-dG, which requires the pyrene moiety to be in the major groove. *Anti*-dATP can also pair with *anti*-guanine in adduct-dG in an elongated mispair. Thus, there are scenarios other than the one involving our *syn*-dATP:*anti*-adduct-dG hypothesis.

15.6.6
A Cautionary Note about Dpo4

Dpo4 is by far the best-studied Y-family DNA polymerase, both structurally and biochemically. Based on biochemical and X-ray findings [26], Dpo4 insertion opposite +BP was proposed to follow a "dislocation" or "templated" pathway. Dislocation/templated insertion ([120, 121] and references therein) involves DNA polymerase stalling at an adduct, slippage to the next 5'-base along the template, and its use to direct incorporation (e.g., dATP insertion opposite the 5'-T in a 5'-TG sequence context), whereupon the newly incorporated dA slips back to form an adduct-G:A mispair, from which extension yields the mispair that ultimately gives a G → T mutation. Dpo4 preferentially inserted dCTP, dTTP, dATP, and dGTP opposite +BP in 5'- GG, 5'-AG, 5'-TG, and 5'-CG sequences, respectively [26], which is consistent with a dislocation/templated mechanism.

Although the dislocation/templated mechanism is attractive for Dpo4, considerable evidence both *in vitro* and *in vivo* suggests that neither Pol IV nor V follow a dislocation/templated mechanism with +BP (discussed extensively in [116], as well as in [79, 93, 113]).

Why might Dpo4 be different than Pols IV and V? Both the roof-amino acid and roof-neighbor-amino acid for Dpo4 are nonbulky (A44/A57), while its bulky flue-handle (C31) causes downward curvature of the chimney upper lip, leading to a closed flue (V32) and a small chimney opening (based on structures from references 125 and other Dpo4 structures). Thus, Dpo4 is a hybrid with a roof similar to the IV/κ class and a chimney similar to the V/η class. In fact, Dpo4's chimney is exceedingly blocked, both by its V32 flue, which is inserted deeper into the chimney than, for example, the N32 flue of UmuC(V), and by M76, which is the second amino acid in Dpo4's left lip and also plugs the chimney [116]. (Pol IV and UmuC(V) have nonbulky G74 and S72, respectively, in the equivalent position to Dpo4's M76.) The excessively plugged chimney of Dpo4 forces the pyrene moiety of +BP so far from the active site that pairing with either S1-dCTP or S2-dCTP is impossible; consequently, an entirely different structure is observed with both the pyrene and the dG moieties of +BP being extrahelical, resulting in no pairing with its complementary dC [26].

This analysis suggests a reason for caution when applying conclusions from Dpo4 to other Y-family DNA polymerases, especially those purely in the IV/κ or V/η class. Perhaps Dpo4 evolved its hybrid roof/chimney structure to bypass a unique set of lesions encountered by a thermophilic bacterium. Alternatively, perhaps the structure of Dpo4 at physiologically relevant elevated temperatures is different than at the temperature at which it was crystallized (room temperature) and assayed (37 °C), and this affects its structure and behavior.

15.6.7
Why is Pol IV Efficient at Extension with −BP, but Inefficient with +BP?

In the nonmutagenic translesion synthesis pathway in *E. coli*, Pol IV inserts dCTP opposite +BP, but Pol IV is not efficient at the next step, extension, which requires a switch to Pol V (Figure 15.3). In models of an extension structure, Pol IV has a leucine (L102) facing into the minor groove that clashes with +BP (Figure 15.7a). L102 forces a repositioning of both +BP and its complementary dC, whose primer 3′-OH moves away from the Pα of the dATP to a non-reaction-ready distance of around 4.9 Å, after molecular dynamics (our unpublished findings). No such steric clash was observed in the extension model of UmuC(V), which has I100 in the equivalent position, and after molecular dynamics, the primer-3′-OH–Pα distance is near-reaction-ready at around 3.4 Å (Figure 15.7b). A possible role for leucine versus isoleucine at this position was investigated further by making mutations *in silico* − L102I-Pol IV gave a reaction-ready distance of around 3.7 Å, while I100L-UmuC(V) gave a non-reaction-ready distance of around 5.2 Å. Preliminary experimental studies (J. Yin and G. Sholder, personal communication) support the view that isoleucine is better than leucine at this extension-amino acid position both in Pol IV and UmuC(V). If we assume that −BP faces the opposite direction in the minor groove compared to +BP (as it does in duplex DNA [122]), then there is no clash between leucine and −BP (Figure 15.7c), which may rationalize why Pol IV is able to do extension with −BP. In our models virtually all DNA polymerases in the IV/κ class have leucine in the extension-amino acid position, while DNA polymerases in the V/η class have isoleucine (data not shown), so this amino acid is conserved.

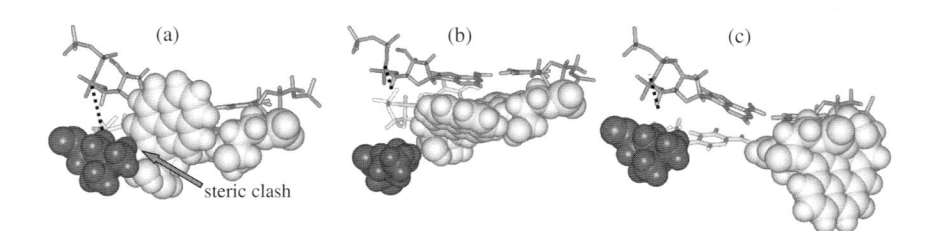

Figure 15.7 Clash between +BP and the extension-amino acid L102 in Pol IV (a); there is no clash between +BP and the extension-amino acid I100 in UmuC(V) (b) or between −BP and L102 in Pol IV (c). The extension-amino acid is black. +BP/−BP (white) are in the L + 1 position and dATP:template-dT (gray stick figures) is in the L0 position. The primer-3′-OH–Pα distance is indicated by dotted lines.

15.7
Y-Family DNA Polymerase Mechanistic Steps

Steps in the mechanism of Y-family DNA polymerases have been proposed for protein structural changes during catalysis based on a series of X-ray structures for Dpo4 [31, 32] and the most dramatic steps include the following. Upon DNA binding to Apo-Dpo4, the thumb/palm/finger domains do not change their structure dramatically. However, the little-finger domain acts like a door, which is open in Apo-Dpo4 and then rotates around 130° to close around DNA; in particular, it binds in the minor groove in the duplex region from about L + 3 to L + 8 [32]. This motion is facilitated by the fact that the little finger is connected to the rest of the protein by a simple 10-amino-acid tether. Once binary-Dpo4 is formed, the palm, finger and little-finger domains translate around 3.3 Å along the helix as the next template base slides into the active site, thus opening the space into which the complementary dNTP can bind to give ternary-Dpo4 [31]; the thumb domain, however, does not move in this step, but rather moves either before, during, or after the subsequent covalent reaction step. Kinetic studies reveal that Y-family DNA polymerases have a rate-determining conformational change before dNTP incorporation [123, 124], although it is unclear which of these protein conformational changes is key. A variety of subtler changes in Dpo4 structure also accompany these steps [31, 32].

The steps in covalent catalysis by Dpo4 have been explored using a combination of molecular modeling/dynamics and *ab initio* quantum mechanical/molecular mechanical minimizations; a novel water-mediated and substrate-assisted mechanism was proposed [33]. In the first step, a water molecule in the active site serves as a conduit to deprotonate the primer 3′-OH and protonate an oxygen on the α-phosphate of the dNTP. In the second step, a second water molecule in the active site serves as a conduit to deprotonate the oxygen on the α-phosphate of the dNTP and to protonate an oxygen on the γ-phosphate. Following these two steps the deprotonated 3′-O$^-$ of the primer is a stronger nucleophile and attacks the α-phosphate, while the second water molecule serves as a conduit again – this time to deprotonate the γ-phosphate of the dNTP and to protonate the β-phosphate, which is on the pyrophosphate leaving group, thus facilitating its removal.

15.8
Structure of B-Family Pol II of *E. coli*

We have built a preliminary model of *E. coli* Pol II, by taking components of the X-ray structure of Pol II [117], which has no DNA, is misshapen in regions, and is missing some amino acids, and adapting them onto the Rb69 DNA polymerase [118], which is the only B-Family member with DNA in its X-ray structures. Figure 15.4(e) shows our preliminary model following initial refinement. The thumb (blue), palm (green), and finger (white and gray) domains surround DNA (yellow, red, and pink).

Pol II inserts and extends the −2 frameshift intermediate of AAF-C8-dG, which must have two looped-out nucleotides as well as the AAF moiety protruding into the major groove. Data suggests that Pol II is involved in the bypass of a nitrogen mustard interstrand cross-link [39–42], which must have a large oligonucleotide protruding into the major groove during translesion synthesis.

If the protein surrounds the DNA, then how can Pol II accommodate such bulky features in its major groove? In fact, Pol II has a protein dome that leaves a large open cavity on the major groove side of its active site. Figure 15.4(e, center) shows Pol II with amino acids 141–350 removed. The palm domain (green) cradles the DNA by binding on the minor groove side, including in the vicinity of the dNTP/template base pair (red) and the L + 1 base pair (pink). The thumb domain (blue) continues to contact DNA as it coils around the minor groove. Amino acids 141–350 sit over the major groove, but this domain is concave and leaves a cavity when it covers the major groove, as seen in the side view of Figure 15.4(e, right), which shows more of the exposed major groove.

Although Y-family DNA polymerases are open to solvent on their major groove side, the solvent exposed DNA surface inside the cavity for Pol II ($411\,\text{Å}^2$, when considering, for example, the template:dNTP base pair plus the L + 1 base pair) is actually larger than with either Pol IV ($231\,\text{Å}^2$) or UmuC(V) ($135\,\text{Å}^2$). Though more work is needed, the large cavity and solvent exposed region on the major groove side may explain how Pol II is able to perform translesion synthesis on lesions having bulky protrusions into the major groove.

An improved X-ray structure of Pol II has recently been published [129].

References

1 Weinberg, R.A. (2007) *The Biology of Cancer*, Garland Science, New York.
2 Luch, A. (2006) Nature and nurture – lessons from chemical carcinogenesis. *Nat. Rev. Cancer*, **5**, 113–125.
3 Singer, B. and Grunberger, D. (1983) *Molecular Biology of Mutagens and Carcinogens*, Plenum Press, New York.
4 Balmain, A., Brown, R., and Harris, C.C. (eds) (2000) Twentieth Anniversary Special Issue. *Carcinogenesis*, **21**, 339–531.
5 Conney, A.H. (1982) Induction of microsomal enzymes by foreign chemicals and carcinogens by polycyclic aromatic hydrocarbons: G. H. A. Clowes Memorial Lecture. *Cancer Res.*, **42**, 4875–4917.
6 Kornberg, A. (1960) Biologic synthesis of deoxyribonucleic acid. *Science*, **131**, 1503–1508.
7 McCulloch, S.D. and Kunkel, T.A. (2008) The fidelity of DNA synthesis by eukaryotic replicative and translesion synthesis polymerases. *Cell Res.*, **18**, 148–161.
8 Bebenek, K. and Kunkel, T.A. (2004) Functions of DNA polymerases. *Adv. Protein Chem.*, **69**, 137–165.
9 Rothwell, P.J. and Waksman, G. (2005) Structure and mechanism of DNA polymerases. *Adv. Protein Chem.*, **71**, 401–440.
10 Friedberg, E.C., Walker, G.W., Siede, W., Wood, R.D., Schultz, R.A., and Ellenberger, T. (2006) *DNA Repair and Mutagenesis*, 2nd edn, ASM Press, Washington, DC.
11 Howard-Flanders, P., Boyce, R.P., and Theriot, L. (1966) Three loci in *Escherichia coli* K-12 that control the excision of pyrimidine dimers and certain other mutagen products

from DNA. *Genetics*, **53**, 1119–1136.

12 van de Putte, P., van Sluis, C.A., van Dillewijn, J., and Rorsch, A. (1965) The location of genes controlling radiation sensitivity in *Escherichia coli*. *Mutat. Res.*, **2**, 97–110.

13 Mattern, I.E., van Winden, M.P., and Rorsch, A. (1965) The range of action of genes controlling radiation sensitivity in *Escherichia coli*. *Mutat. Res.*, **2**, 111–131.

14 Cleaver, J.E. (1968) Defective repair replication of DNA in xeroderma pigmentosum. *Nature*, **218**, 652–656.

15 Modrich, P. (1995) Mismatch repair, genetic stability, and cancer. *Science*, **266**, 1959–1960.

16 Yang, W. and Woodgate, R. (2007) What a difference a decade makes: insights into translesion DNA synthesis. *Proc. Natl. Acad. Sci. USA*, **104**, 15591–15598.

17 Ohmori, H., Friedberg, E.C., Fuchs, R.P., Goodman, M.F., Hanaoka, F., Hinkle, D., Kunkel, T.A., Lawrence, C.W., Livneh, Z., Nohmi, T., Prakash, L., Prakash, S., Todo, T., Walker, G.C., Wang, Z., and Woodgate, R. (2001) The Y-family of DNA polymerases. *Mol. Cell*, **8**, 7–8.

18 Nohmi, T. (2006) Environmental stress and lesion-bypass DNA polymerases. *Annu. Rev. Microbiol.*, **60**, 231–253.

19 Yang, W. (2003) Damage repair DNA polymerases. *Curr. Opin. Struct. Biol.*, **13**, 23–30.

20 Prakash, S., Johnson, R.E., and Prakash, L. (2005) Eukaryotic translesion synthesis DNA polymerases: specificity of structure and function. *Annu. Rev. Biochem.*, **74**, 317–353.

21 Jarosz, D.F., Beuning, P.J., Cohen, S.E., and Walker, G.C. (2007) Y-family DNA polymerases in *Escherichia coli*. *Trends Microbiol.*, **15**, 70–77.

22 Zhou, B.L., Pata, J.D., and Steitz, T.A. (2001) Crystal structure of a DinB lesion bypass DNA polymerase catalytic fragment reveals a classic polymerase catalytic domain. *Mol. Cell*, **8**, 427–437.

23 Ling, H., Boudsocq, F., Woodgate, R., and Yang, W. (2001) Crystal structure of a Y-family DNA polymerase in action: a mechanism for error-prone and lesion-bypass replication. *Cell*, **107**, 91–102.

24 Ling, H., Boudsocq, F., Plosky, B.S., Woodgate, R., and Yang, W. (2004) Replication of a *cis-syn* thymine dimer at atomic resolution. *Nature*, **424**, 1083–1087.

25 Vaisman, A., Ling, H., Woodgate, R., and Yang, W. (2005) Fidelity of Dpo4: effect of metal ions, nucleotide selection and pyrophosphorolysis. *EMBO J.*, **25**, 2957–2967.

26 Bauer, J., Xing, G., Yagi, H., Sayer, J.M., Jerina, D.M., and Ling, H. (2007) A structural gap in Dpo4 supports mutagenic bypass of a major benzo[a]pyrene dG adduct in DNA through template misalignment. *Proc. Natl. Acad. Sci. USA*, **104**, 14905–14910.

27 Trincao, J., Johnson, R.E., Escalante, C.R., Prakash, S., Prakash, L., and Aggarwal, A.K. (2001) Structure of the catalytic core of *S. cerevisiae* DNA polymerase η: implications for translesion DNA synthesis. *Mol. Cell*, **8**, 417–426.

28 Alt, A., Lammens, K., Chiocchini, C., Lammens, A., Pieck, J.C., Kuch, D., Hopfner, K.P., and Carell, T. (2007) Bypass of DNA lesions generated during anticancer treatment with cisplatin by DNA polymerase η. *Science*, **318**, 967–970.

29 Nair, D.T., Johnson, R.E., Prakash, S., Prakash, L., and Aggarwal, A.K. (2004) Replication by human DNA polymerase-ι occurs by Hoogsteen base-pairing. *Nature*, **430**, 377–380.

30 Lone, S., Townson, S.A., Uljon, S.N., Johnson, R.E., Brahma, A., Nair, D.T., Prakash, S., Prakash, L., and Aggarwal, A.K. (2007) Human DNA polymerase κ encircles DNA: implications for mismatch extension and lesion bypass. *Mol. Cell*, **23**, 601–614.

31 Rechkoblit, O., Malinina, L., Cheng, Y., Kuryavyi, V., Broyde, S., Geacintov, N.E., and Patel, D.J. (2006) Stepwise translocation of Dpo4 polymerase during error-free bypass of an oxoG lesion. *PLoS Biol.*, **4**, 25–42.

32 Wong, J.H., Fiala, K.A., Suo, Z., and Ling, H. (2008) Snapshots of a Y-family DNA polymerase in replication:

substrate-induced conformational transitions and implications for fidelity of Dpo4. *J. Mol. Biol.*, **379**, 317–330.

33 Wang, L., Yu, X., Hu, P., Broyde, S., and Zhang, Y. (2007) A water-mediated and substrate-assisted catalytic mechanism for *Sulfolobus solfataricus* DNA polymerase IV. *J. Am. Chem. Soc.*, **129**, 4731–4737.

34 Johnson, R.E., Prakash, S., and Prakash, L. (2000) The human *DINB1* gene encodes the DNA polymerase Polθ. *Proc. Natl. Acad. Sci. USA*, **97**, 3838–3843.

35 Ohashi, E., Ogi, T., Kusumoto, R., Iwai, S., Masutani, C., Hanaoka, F., and Ohmori, H. (2000) Error-prone bypass of certain DNA lesions by the human DNA polymerase κ. *Genes Dev.*, **14**, 1589–1594.

36 Zhang, Y., Yuan, F., Wu, X., Wang, M., Rechkoblit, O., Taylor, J.S., Geacintov, N.E., and Wang, Z. (2000) Error-free and error-prone lesion bypass by human DNA polymerase κ *in vitro*. *Nucleic Acids Res.*, **28**, 4138–4146.

37 Lee, C.H., Chandani, S., and Loechler, E.L. (2006) Homology modeling of four lesion-bypass DNA polymerases: structure and lesion bypass findings suggest that *E. coli* pol IV and human Pol κ are orthologs, and *E. coli* pol V and human Pol η are orthologs. *J. Mol. Graph. Model.*, **25**, 87–102.

38 Jarosz, D.F., Godoy, V.G., Delaney, J.C., Essigmann, J.M., and Walker, G.C. (2006) A single amino acid governs enhanced activity of DinB DNA polymerases on damaged templates. *Nature*, **439**, 225–228.

39 Berardini, M., Mackay, W., and Loechler, E.L. (1997) A site-specific study of a plasmid containing single nitrogen mustard interstrand cross-link: evidence for a second, recombination-independent pathway for the DNA repair of interstrand cross-links. *Biochemistry*, **36**, 3506–3513.

40 Sarkar, S., Davies, A.A., Ulrich, H.D., and McHugh, P.J. (2006) DNA interstrand crosslink repair during G_1 involves nucleotide excision repair and DNA polymerase ζ. *EMBO J.*, **25**, 1285–1294.

41 McHugh, P.J. and Sarkar, S. (2006) DNA interstrand cross-link repair in the cell cycle: a critical role for polymerase ζ in G_1 phase. *Cell Cycle*, **5**, 1044–1047.

42 Lehoczky, P., McHugh, P.J., and Chovanex, M. (2007) DNA interstrand cross-link repair in *Saccharomyces cerevisiae*. *FEMS Microbiol. Rev.*, **31**, 109–133.

43 Hollstein, M., Sidransky, D., Vogelstein, B., and Harris, C.C. (1991) p53 mutations in human cancers. *Science*, **253**, 49–53.

44 Vogelstein, B. and Kinzler, K.W. (1992) Carcinogens leave fingerprints. *Nature*, **355**, 209–210.

45 Greenblatt, M.S., Bennett, W.P., Hollstein, M., and Harris, C.C. (1994) Perspectives in cancer research: mutations in the p53 tumor suppressor gene: clues to cancer etiology and molecular pathogenesis. *Cancer Res.*, **54**, 4855–4878.

46 Nagase, H. and Nakamura, Y. (1993) Mutations of the APC (adenomatous polyposis coli) gene. *Hum. Mutat.*, **2**, 425–434.

47 Miyaki, M., *et al.* (1994) Characteristics of somatic mutation of the adenomatous polyposis coli gene in colorectal tumors. *Cancer Res.*, **54**, 3011–3020.

48 Hussain, S.P., Hollstein, M.H., and Harris, C.C. (2000) p53 tumor suppressor gene: at the crossroads of molecular carcinogenesis, molecular epidemiology, and human risk assessment. *Ann. NY Acad. Sci.*, **919**, 79–85.

49 Olivier, M., Eeles, R., Hollstein, M., Khan, M.A., Harris, C.C., and Hainaut, P. (2002) The IARC TP53 database: new online mutation analysis and recommendations to users. *Hum. Mutat.*, **19**, 607–614.

50 Duncan, B.K. and Miller, J.H. (1980) Mutagenic deamination of cytosine residues in DNA. *Nature*, **287**, 560–561.

51 Lieb, M. (1995) Very short patch repair of T:G mismatches *in vivo*: importance of context and accessory proteins. *J. Bacteriol.*, **177**, 660–666.

52 Brash, D.E., *et al.* (1991) A role for sunlight in skin cancer: UV-induced p53 mutations in squamous cell carcinoma. *Proc. Natl. Acad. Sci. USA*, **88**, 10124.

53 Dumaz, N., Stary, A., Soussi, T., Daya-Grosjean, L., and Sarasin, A. (1994) Can we predict solar ultraviolet radiation as the causal event in human tumors by analyzing the mutation spectra of the p53 gene? *Mutat. Res.*, **307**, 375–386.

54 Daya-Grosjean, L., Dumaz, N., and Sarasin, A. (1995) The specificity of p53 mutation spectra in sunlight induced human cancers. *J. Photochem. Photobiol.*, **B 28**, 115–124.

55 D'errico, M., Calcagnile, A., and Dogliotti, E. (1996) Genetic alterations in skin cancer. *Ann. Ist. Super. Sanita*, **32**, 53–63.

56 Miller, J.H. (1983) Mutational specificity in bacteria. *Annu. Rev. Genet.*, **17**, 215–238.

57 Miller, J.H. (1980) The *lacI* gene, in *The Operon*, 2nd edn (eds J.H. Miller and W.S. Reznikoff), Cold Spring Harbor Laboratory Press, Cold Spring Harbor, NY, pp. 31–88.

58 Hsieh, D.P. and Atkinson, D.N. (1995) Recent aflatoxin exposure and mutation at codon 249 of the human p53 gene: lack of association. *Food Addit. Contam.*, **12**, 421–424.

59 Takeshima, Y., Seyama, T., Bennett, W.P., Akiyama, M., Tokuoka, S., Inai, K., Mabuchi, K., Land, C.E., and Harris, C.C. (1993) p53 mutations in lung cancers from non-smoking atomic-bomb survivors. *Lancet*, **342**, 1520–1521.

60 Loechler, E.L. (1994) Mechanism by which aflatoxins and other bulky carcinogens induce mutations, in *The Toxicology of Aflatoxins: Human Health, Veterinary, and Agricultural Significance* (ed. D.L. Eaton and J.D. Groopman), Academic Press, Orlando, FL, pp. 149–178.

61 Denissenko, M.F., Pao, A., Tang, M., and Pfeifer, G.P. (1996) Preferential formation of benzo[a]pyrene adducts at lung cancer mutational hotspots in p53. *Science*, **274**, 430–432.

62 Denissenko, M.F., Pao, A., Pfeifer, G.P., and Tang, M.-S. (1998) Slow repair of bulky DNA adducts along the nontranscribed strand of the human p53 gene may explain the strand bias of transversion mutations in cancers. *Oncogene*, **16**, 1241–1247.

63 Tang, M.S., Zheng, J.B., Denissenko, M.F., Pfeifer, G.P., and Zheng, Y. (1999) Use of UvrABC nuclease to quantify benzo[a]pyrene diol epoxide-DNA adduct formation at methylated versus unmethylated CpG sites in the p53 gene. *Carcinogenesis*, **20**, 1085–1089.

64 Pfeifer, G.P. and Denissenko, M.F. (1998) Formation and repair of DNA lesions in the p53 gene: relation to cancer mutations? *Environ. Mol. Mutagen.*, **31**, 197–205.

65 Pfeifer, G.P., Denissenko, M.F., Olivier, M., Tretyakova, N., Hecht, S.S., and Hainaut, P. (2002) Tobacco smoke carcinogens, DNA damage and p53 mutations in smoking-associated cancers. *Oncogene*, **21**, 7435–7451.

66 Smith, L.E., Denissenko, M.F., Bennett, W.P., Li, H., Amin, S., Tang, M., and Pfeifer, G.P. (2000) Targeting of lung cancer mutational hotspots by polycyclic aromatic hydrocarbons. *J. Natl. Cancer Inst.*, **92**, 803–811.

67 Pfeifer, G.P. and Hainaut, P. (2003) On the origin of G → T transversions in lung cancer. *Mutat. Res.*, **526**, 39–43.

68 Hoffmann, D. and Hoffmann, I. (1997) The changing cigarette, 1950–1995. *J. Toxicol. Environ. Health*, **50**, 307–364.

69 Fuchs, R.P. and Fujii, S. (2007) Translesion synthesis in *Escherichia coli*: lessons from the *Nar*I mutation hot spot. *DNA Repair*, **6**, 1032–1041.

70 Schlacher, K. and Goodman, M.F. (2000) Lessons from 50 years of SOS DNA-damage-induced mutagenesis. *Nat. Rev. Mol. Cell Biol.*, **8**, 587–594.

71 Fuchs, R.P., Fujii, S., and Wagner, J. (2004) Properties and functions of *Escherichia coli*: Pol IV and Pol V. *Adv. Protein Chem.*, **69**, 230–264.

72 Fujii, S. and Fuchs, R.P. (2007) Interplay among replicative and specialized DNA polymerases determines failure or success of translesion synthesis pathways. *J. Mol. Biol.*, **372**, 883–893.

73 Qiu, A. and Goodman, M.F. (1997) The *Escherichia coli polB* locus is identical to *dinA*, the structural gene for DNA polymerase II. Characterization of Pol

II purified from a *polB* mutant. *J. Biol. Chem.*, **272**, 8611–8617.

74 Kim, S.R., Matsui, K., Yamada, M., Gruz, P., and Nohmi, T. (2001) Roles of chromosomal and episomal *dinB* genes encoding DNA pol IV in targeted and untargeted mutagenesis in *Escherichia coli*. *Mol. Genet. Genomics*, **266**, 207–215.

75 Woodgate, R. and Ennis, D.G. (1991) Levels of chromosomally encoded Umu proteins and requirements for *in vivo* UmuD cleavage. *Mol. Gen. Genet*, **229**, 10–16.

76 Tang, M., Pham, P., Shen, X., Taylor, J.S., O'Donnell, M., Woodgate, R., and Goodman, M.F. (2000) Roles of *E. coli* DNA polymerases IV and V in lesion-targeted and untargeted SOS mutagenesis. *Nature*, **404**, 1014–1018.

77 Brash, D.E. and Haseltine, W.A. (1982) UV-induced mutation hotspots occur at DNA damage hotspots. *Nature*, **298**, 189–192.

78 Goodman, M.F. (2002) Error-prone repair DNA polymerases in prokaryotes and eukaryotes. *Annu. Rev. Biochem.*, **71**, 17–50.

79 Yin, J., Seo, K.-Y., and Loechler, E.L. (2004) A role for DNA polymerase V in G-to-T mutagenesis from the major benzo[a]pyrene N^2-dG adduct when studied in a 5′-TGT sequence in *Escherichia coli*. *DNA Repair*, **3**, 323–334.

80 Strauss, B.S. (2002) The "A" rule revisited: polymerases as determinants of mutational specificity. *DNA Repair*, **1**, 125–135.

81 Masutani, C., Kusumoto, R., Yamada, A., Dohmae, N., Yokoi, M., Yuasa, M., Araki, M., Iwai, S., Takio, K., and Hanaoka, F. (1999) The XPV (xeroderma pigmentosum variant) gene encodes human DNA polymerase η. *Nature*, **399**, 700–704.

82 Johnson, R.E., Prakash, S., and Prakash, L. (1999) Efficient bypass of a thymine–thymine dimer by yeast DNA polymerase, Polη. *Science*, **283**, 1001–1004.

83 Johnson, R.E., Kondratick, C.M., Prakash, S., and Prakash, L. (1999) hRAD30 mutations in the variant form of xeroderma pigmentosum. *Science*, **285**, 263–265.

84 Washington, M.T., Johnson, R.E., Prakash, S., and Prakash, L. (1999) Fidelity and processivity of *Saccharomyces cerevisiae* DNA polymerase η. *J. Biol. Chem.*, **274**, 36835–36838.

85 Washington, M.T., Johnson, R.E., Prakash, S., and Prakash, L. (2000) Accuracy of thymine–thymine dimer bypass by *Saccharomyces cerevisiae* DNA polymerase η. *Proc. Natl. Acad. Sci. USA*, **97**, 3094–3099.

86 Johnson, R.E., Washington, M.T., Prakash, S., and Prakash, L. (2000) Fidelity of human DNA polymerase η. *J. Biol. Chem.*, **275**, 7447–7450.

87 Schlacher, K., Cox, M.M., Woodgate, R., and Goodman, M.F. (2006) RecA acts in *trans* to allow replication of damaged DNA by DNA polymerase V. *Nature*, **442**, 883–887.

88 Fujii, S., Isogawa, A., and Fuchs, R.P. (2006) RecFOR proteins are essential for Pol V-mediated translesion synthesis and mutagenesis. *EMBO J.*, **25**, 5754–5763.

89 Opperman, T., Murli, S., Smith, B.T., and Walker, G.C. (1999) A model for a umuDC-dependent prokaryotic DNA damage checkpoint. *Proc. Natl. Acad. Sci. USA*, **96**, 9218–9223.

90 Kobayashi, S., Valentine, M.R., Pham, P., O'Donnell, M., and Goodman, M.F. (2002) Fidelity of *Escherichia coli* DNA polymerase IV. Preferential generation of small deletion mutations by dNTP-stabilized misalignment. *J. Biol. Chem.*, **277**, 34198–34207.

91 Lenne-Samuel, N., Janel-Bintz, R., Kolbanovskiy, A., Geacintov, N.E., and Fuchs, R.P. (2000) The processing of a benzo(a)pyrene adduct into a frameshift or a base substitution mutation requires a different set of genes in *Escherichia coli*. *Mol. Microbiol.*, **38**, 299–307.

92 Napolitano, R., Janel-Bintz, R., Wagner, J., and Fuchs, R.P. (2000) All three SOS-inducible DNA polymerases (Pol II, Pol IV and Pol V) are involved in induced mutagenesis. *EMBO J.*, **19**, 6259–6265.

93 Seo, K.-Y., Nagalingam, A., Miri, S., Yin, J., Kolbanovskiy, A., Shastry, A., and Loechler, E.L. (2006) Mirror image Stereoisomers of the major

benzo[a]pyrene N^2-dG adduct are bypassed by different lesion-bypass DNA polymerases in *E. coli*. *DNA Repair*, **5**, 515–522.

94 Yuan, B., Cao, H., Jiang, Y., Hong, H., and Wang, Y. (2008) Efficient and accurate bypass of N^2-(1-carboxyethyl)-2′-deoxyguanosine by DinB DNA polymerase *in vitro* and *in vivo*. *Proc. Natl. Acad. Sci. USA*, **105**, 8679–8684.

95 Shen, X., Sayer, J.M., Kroth, H., Ponten, I., O'Donnell, M., Woodgate, R., Jerina, D.M., and Goodman, M.F. (2002) Efficiency and accuracy of SOS-induced DNA polymerases replicating benzo[a]pyrene-7,8-diol 9,10-epoxide A and G adducts. *J. Biol. Chem.*, **277**, 5265–5674.

96 Marnett, L.J. (2000) Oxyradicals and DNA damage. *Carcinogenesis*, **21**, 361–370.

97 Yeiser, B., Pepper, E.D., Goodman, M.F., and Finkel, S.E. (2002) SOS-induced DNA polymerases enhance long-term survival and evolutionary fitness. *Proc. Natl. Acad. Sci. USA*, **99**, 8737–8741.

98 Tompkins, J.D., Nelson, J.L., Leugers, S.L., Stumpf, J.D., and Foster, P.L. (2003) Error-prone polymerase, DNA polymerase IV, is responsible for transient hypermutation during adaptive mutation in *Escherichia coli*. *J. Bacteriol.*, **185**, 3469–3472.

99 Heller, R.C., and Marians, K.J. (2006) Replisome assembly and the direct restart of stalled replication forks. *Nat. Rev. Mol. Cell Biol.*, **7**, 932–943.

100 Langston, L.D. and O'Donnell, M. (2006) DNA replication: keep moving and don't mind the gap. *Mol. Cell*, **23**, 155–160.

101 Furukohri, A., Goodman, M.F., and Maki, H. (2008) A dynamic polymerase exchange with *Escherichia coli* DNA polymerase IV replacing DNA polymerase III on the sliding clamp. *J. Biol. Chem.*, **283**, 11260–11269.

102 Becherel, O.J., Fuchs, R.P.P., and Wagner, J. (2002) Pivotal role of the β-clamp in translesion DNA synthesis and mutagenesis in *E. coli* cells. *DNA Repair*, **4**, 703–708.

103 Bunting, K.A., Roe, S.M., and Pearl, L.H. (2003) Structural basis for recruitment of translesion DNA polymerase Pol IV/DinB to the β-clamp. *EMBO J.*, **22**, 5883–5892.

104 Delmas, S. and Matic, I. (2006) Interplay between replication and recombination in *Escherichia coli*: impact of the alternative DNA polymerases. *Proc. Natl. Acad. Sci. USA*, **103**, 4564–4569.

105 Heflich, R.H. and Neft, R.E. (1994) Genetic toxicity of 2-acetylaminofluorene, 2-aminofluorene and some of their metabolites and model metabolites. *Mutat. Res.*, **318**, 73–114.

106 Wang, L. and Broyde, S. (2006) A new anti conformation for N-(deoxyguanosin-8-yl)-2-acetylaminofluorene (AAF-dG) allows Watson–Crick pairing in the *Sulfolobus solfataricus* P2 DNA polymerase IV (Dpo4). *Nucleic Acids Res.*, **34**, 785–795.

107 Harvey, R.G. (1997) *Polycyclic Aromatic Hydrocarbons: Chemistry and Cancer*, John Wiley & Sons, Inc., New York.

108 Dipple, A. (1985) Polycyclic aromatic hydrocarbon carcinogens, in *Polycyclic Aromatic Hydrocarbons and Carcinogenesis* (ed. R.G. Harvey), American Chemical Society, Washington, DC, pp. 1–17.

109 Rodriguez, H. and Loechler, E.L. (1993) Mutagenesis by the (+)-*anti*-diol epoxide of benzo[a]pyrene: what controls mutagenic specificity? *Biochemistry*, **32**, 373–383.

110 Xie, Z., Braithwaite, E., Guo, D., Zhao, B., Geacintov, N.E., and Wang, Z. (2003) Mutagenesis of benzo[a]pyrene diol epoxide in yeast: requirement for DNA polymerase ζ and involvement of DNA polymerase η. *Biochemistry*, **42**, 11253–11262.

111 Yoon, J.H., Lee, C.S., and Pfeifer, G.P. (2003) Simulated sunlight and benzo[a]pyrene diol epoxide induced mutagenesis in the human p53 gene evaluated by the yeast functional assay: lack of correspondence to tumor mutation spectra. *Carcinogenesis*, **24**, 113–119.

112 Schiltz, M., Cui, X.X., Lu, Y.P., Yagi, H., Jerina, D.M., Zdzienicka, M.Z.,

Chang, R.L., Conney, A.H., and Wei, S.J. (1999) Characterization of the mutational profile of (+)-7R,8S-dihydroxy-9S,10R-epoxy-7,8,9,10-tetrahydrobenzo[a]pyrene at the hypoxanthine (guanine) phosphoribosyltransferase gene in repair-deficient Chinese hamster V-H1 cells. *Carcinogenesis*, **20**, 2279–2286.

113 Seo, K.-Y., Nagalingam, A., and Loechler, E.L. (2005) Mutagenesis studies on four stereoisomeric N^2-dG benzo[a]pyrene adducts in the identical 5'-CGC sequence used in NMR studies: although adduct conformation differs, mutagenesis outcome does not as G → T mutations dominate in each case. *Mutagenesis*, **20**, 441–448.

114 Johnson, R.E., Washington, M.T., Haracska, L., Prakash, S., and Prakash, L. (2000) Eukaryotic polymerases ι and ζ act sequentially to bypass DNA lesions. *Nature*, **406**, 1015–1019.

115 Yuan, F., Zhang, Y., Rajpal, D.K., Wu, X., Guo, D., Wang, M., Taylor, J.S., and Wang, Z. (2000) Specificity of DNA lesion bypass by the yeast DNA polymerase η. *J. Biol. Chem.*, **275**, 8233–8239.

116 Chandani, S. and Loechler, E.L. (2009) Y-family DNA polymerases may use two different dNTP shapes for insertion: a hypothesis and its implications. *J. Mol. Graph. Mod.*, **27**, 759–769.

117 Anderson, W.F., Prince, D.B., Yu, H., McEntee, K., and Goodman, M.F. (1994) Crystallization of DNA polymerase II from *Escherichia coli*. *J. Mol. Biol.*, **238**, 120–122.

118 Hogg, M., Wallace, S.S., and Doublie, S. (2004) Crystallographic snapshots of a replicative DNA polymerase encountering an abasic site. *EMBO J.*, **23**, 1483–1493.

119 Chandani, S., Lee, C.H., and Loechler, E.L. (2007) Molecular modeling benzo[a]pyrene N^2-dG adducts in two partially overlapping active sites of the Y-family DNA polymerase Dpo4. *J. Mol. Graph. Model.*, **25**, 658–670.

120 Papanicolaou, C. and Ripley, L.S. (1991) An *in vitro* approach to identifying specificity determinants of mutagenesis mediated by DNA misalignments. *J. Mol. Biol.*, **221**, 805–821.

121 Kunkel, T.A. and Soni, A. (1988) Mutagenesis by transient misalignment. *J. Biol. Chem.*, **263**, 14784–14789.

122 Geacintov, N.E., Cosman, M., Hingerty, B.E., Amin, S., Broyde, B., and Patel, D.J. (1997) NMR solution structures of stereoisomeric polycyclic aromatic carcinogen-DNA adducts: principles, patterns and diversity. *Chem. Res. Toxicol.*, **10**, 111–146.

123 Washington, M.T., Prakash, L., and Prakash, S. (2001) Yeast DNA polymerase η utilizes an induced-fit mechanism of nucleotide incorporation. *Cell*, **107**, 917–927.

124 Fiala, K.A. and Suo, Z. (2004) Mechanism of DNA polymerization catalyzed by *Sulfolobus solfataricus* P2 DNA polymerase IV. *Biochemistry*, **43**, 2116–2125.

125 Chandani, S. and Loechler, E.L. (2009) Y-Family DNA polymerases may use two different dNTP shapes for insertion: a hypothesis and its implications. *J. Mol. Graph. Mod.*, **27**, 759–769.

126 Jiang, Q., Karata, K., Woodgate, R., Cox, M.M., and Goodman, M.F. (2009) The active form of DNA polymerase V is UmuD'(2)C-RecA-ATP. *Nature*, **460**, 359–363.

127 Räschle, M., Knipscheer, P., Enoiu, M., Angelov, T., Sun, J., Griffith, J.D., Ellenberger, T.E., Schärer, O.D., and Walter, J.C. (2008) Mechanism of replication-coupled DNA interstrand crosslink repair. *Cell*, **134**, 969–980.

128 Seo, K.Y., Yin, J., Donthamsetti, P., Chandani, S., Lee, C.H., and Loechler, E.L. (2009) Amino acid architecture that influences dNTP insertion efficiency in Y-family DNA polymerase V of E. coli. *J Mol Biol*, **392**, 270–282.

129 Wang, F. and Yang, W. (2009) Structural insight into translesion synthesis by DNA Pol II. *Cell.*, **139**, 1279–1289.

16
Insight into the Molecular Mechanism of Translesion DNA Synthesis in Human Cells using Probes with Chemically Defined DNA Lesions

Zvi Livneh

16.1
Introduction

DNA damage in inevitable and is inherent to the biochemistry of cells. Most of the damage is believed to be endogenous, originating from spontaneous decay of DNA, such as depurination and deamination, and from byproducts of metabolism, such as oxidative reactive species. Beyond that, there are DNA-damaging agents in our environment such as sunlight, in the air that we breathe, and in the food that we consume. The most conservative estimates are that approximately 20 000 DNA lesions are formed each day in a human cell [1, 2]. Since DNA lesions interfere with the most fundamental aspects of DNA functions such as replication and transcription, they must be dealt with for the cell to be able to function. This poses a significant challenge to the cell, as DNA lesions make up a very large variety of chemical structures. The large numbers of DNA lesions and their chemical diversity imply that no single solution is likely to overcome the problem. Indeed, organisms have evolved a series of DNA repair mechanisms, which operate using several strategies, and are directed to different types of DNA damage and physiological situations. Most DNA repair mechanisms function to eliminate the damaged DNA part and restore the original DNA sequence. This is usually done by an excision repair strategy, in which the damaged nucleotide is recognized and cut out from DNA along with several neighboring nucleotides, and the gap thus formed is filled in by DNA synthesis using the intact complementary strand as a template [1]. However, DNA replication often encounters unrepaired DNA lesions. This can lead to the arrest of replication forks or the formation of postreplication gaps. The common denominator among these structures is the presence of a modified nucleotide in a region of single-stranded DNA. This situation precludes repair by excision repair, since the complementary strand is missing. Under such condition the greater threat to the cell is the presence of a persisting single-stranded DNA that can be easily broken and transform a point damage to a chromosomal damage, which is much more severe. Essentially all organisms have evolved two major strategies to deal with this situation and convert the single-stranded DNA region into double-stranded DNA, without removing the damaged nucleotide. These mechanisms are termed DNA damage tolerance mechanisms and they operate in two main strategies: translesion DNA synthesis (TLS) and homology-dependent repair (HDR) [1, 3]. HDR can operate when a homologous DNA is present, such as a sister chromatid generated by replication. The single-stranded DNA region can then be converted into double-stranded DNA by either physical transfer of the complementary strand from the sister chromatid (homologous recombination repair), or by copying the complementary strand from the sister chromatid, (a template switch mechanism). This process is inherently error-free. TLS is the process in which the single-stranded DNA segment is converted into double-stranded DNA by DNA synthesis, carried out by specialized low-fidelity DNA polymerases, which are capable of replicating across DNA lesions. Due to the miscoding nature of most DNA lesions, TLS is inherently mutagenic. It is responsible for most mutations caused by agents such as sunlight and tobacco smoke. Given the central role of mutations in disease and evolution, the mechanisms of TLS have drawn significant attention [4–8].

16.2
Overview of TLS

TLS presents a fundamental enigma. It is a process that is inherently mutagenic and yet it functions to protect humans against DNA damage. Since mutations are a major cause of cancer, the mode of operation of TLS and how it regulates mutation rates are of great interest. The biological significance of TLS is illustrated by the hereditary disease xeroderma pigmentosum variant (XPV). Most xeroderma pigmentosum (XP) patients exhibit sunlight sensitivity and extreme predisposition to sunlight-induced skin cancer, caused by a deficiency in nucleotide excision repair (NER). XPV patients, on the other hand, while manifesting clinical features of XP, have no defect in NER. Instead, they lack a TLS DNA polymerase, Pol η – the product of the *POLH* (*XPV*) gene [9, 10]. This polymerase is specialized to replicate across UV light-induced cyclobutane pyrimidine dimers (CPDs) with high efficiency and relatively accurately [11, 12]. In its absence, another polymerase performs the TLS reaction, but with lower efficiency – hence the UV sensitivity and higher error-frequency, and hence the hypermutability [13, 14] and cancer predisposition [1]. This highlights the role of TLS as an inherently mutagenic process that protects humans against cancer. One of the big surprises in the TLS field was the discovery of many specialized DNA polymerases [15]. There are five specialized DNA polymerase for which a role in TLS in mammalian cells was demonstrated. Four of these, Pol η, Pol κ, Pol ι and REV1 belong to the Y-family, whereas the fifth, Pol ζ, belongs to the B-family [16]. Five additional DNA polymerases belong to other polymerase families and may also be involved in TLS [17]. These 10 specialized DNA polymerase share a low fidelity and the ability to replicate across DNA lesions. Together with the "classical" five DNA polymerases that were known prior to 1999, this adds up to a total of 15 DNA polymerases that may be involved in TLS. The multiplicity of TLS polymerases suggested that their activity must be carefully regulated to prevent an escalation in the rates of mutations. Indeed, it has been shown that in the yeast *Saccharomyces cerevisiae*, TLS is regulated by monoubiquitination of proliferating cell nuclear antigen (PCNA) and a similar mechanism occurs in mammalian cells [5, 18–20]. PCNA is a ring-shaped homotrimer, which functions as a sliding DNA clamp [21]. It tethers the replicative polymerases to DNA, therefore endowing them with high processivity. In addition, PCNA is also a regulatory protein involved in other DNA transaction and it also interacts with TLS polymerases. Upon treatment with DNA-damaging agents, such as UV light, PCNA becomes monoubiquitinated and binds several TLS polymerases via their ubiquitin-binding domain (Figure 16.1). It has been hypothesized that this function serves to recruit TLS polymerases to the damaged site in DNA [22], thereby enabling the bypass of the lesion (Figure 16.1). This hypothesis has not yet been directly proven and an alternative indirect role for PCNA ubiquitination has been suggested [23, 24]. Monoubiquitination of PCNA is subject to regulation by the tumor suppressor p53 [25], primarily through its target protein p21 [25, 26] – a cell-cycle inhibitor known to interact with PCNA. Both p53 and p21 were shown also to regulate TLS itself, limiting its extent, but increasing its accuracy [8, 25].

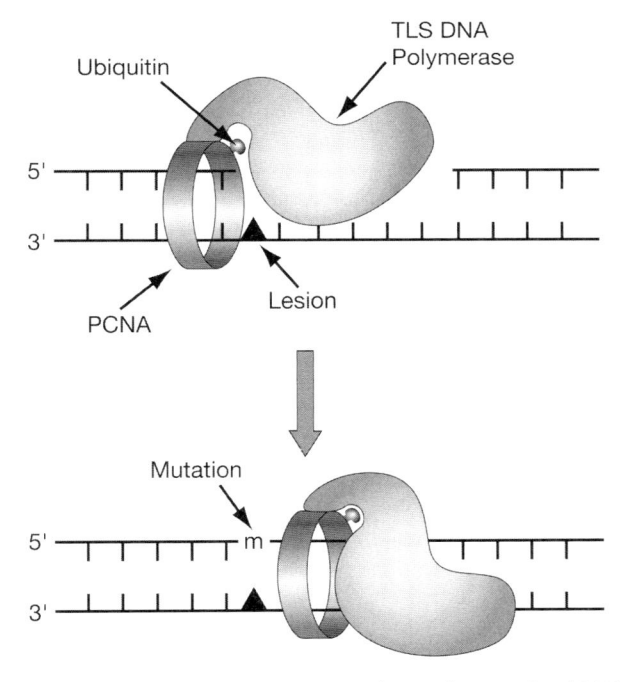

Figure 16.1 Model of TLS across DNA lesions by specialized DNA polymerases. PCNA, the sliding DNA clamp, becomes monoubiquitinated in response to exposure to genotoxic agents. Binding of a TLS polymerase to the ubiquitinated PCNA via both ubiquitin- and PCNA-binding sites helps to recruit it to the damage site, and enables TLS. See text for details.

16.3
Plasmid Model Systems with Defined Lesions for Studying TLS

A fundamental question in TLS is the DNA damage specificity of individual TLS polymerases. This requires the ability to assay TLS of defined lesions, in cells in which specific TLS polymerases were either knocked-out or their expression silenced. Unfortunately, most DNA-damaging agents such as alkylating agents, radiation, oxidative agents, and so on, produce a multitude of DNA lesions, which precludes an unequivocal assignment of specific lesions to their cognate polymerases. The ability to use DNA substrates with single chemically defined lesions provides a powerful approach to address this problem. This is typically done with plasmids, which are convenient for engineering and manipulation. Indeed, over the years, plasmids with site-specific lesions were used to study TLS in *Escherichia coli*, *S. cerevisiae*, chicken DT40 cells, and mammalian cells, both *in vitro* and *in vivo* [27–34]. Here, we shall concentrate on the effort to elucidate the molecular mechanism of TLS in mammalian cells.

Plasmid-based systems can employ plasmids that can replicate and propagate in mammalian cells or plasmids that are unable to replicate. Each approach has its advantages and disadvantages. Using replicating plasmids enables one to study TLS during replication, but it complicates the ability to quantify the absolute extents of TLS when comparing plasmids with and without a lesion. Moreover, plasmid replication may be fundamentally different than chromosomal replication. Another problem is that the plasmid needs to be fully double-stranded and therefore the lesion is subject not only to TLS, but also to excision repair, which eliminates the damage and reduces the amount of substrate available for TLS [34, 35]. The other approach is to use nonreplicating plasmids carrying gaps opposite lesions [25, 36–38]. The advantage of this approach is that TLS can be quantified on an absolute scale relative to DNA synthesis in the absence of a lesion. Moreover, due to the presence of the gap opposite the lesion, the only way to fill in the gap is by TLS. Excision repair is impossible because there is no complementary strand and homology-dependent repair is impossible due to the lack of a homologous donor. The disadvantage of the system is that TLS is assayed uncoupled from replication, although it may be visualized as a model system for TLS in postreplication gaps. It should be realized that currently it is not clear whether chromosomal TLS occurs at arrested forks, in gaps behind the fork, or both; however there is increasing evidence suggesting the importance of gaps behind the fork [39–42].

16.4
Gap-Lesion Plasmid Assay for Mammalian TLS

The key element in the gap-lesion plasmid assay for mammalian TLS is a plasmid carrying a chemically defined, single lesion opposite a gap (Figure 16.2). The construction of the gap-lesion plasmid involves building a gapped duplex oligonucleotide carrying the lesion and ligating it to a linearized vector plasmid (Figure

16.4 Gap-Lesion Plasmid Assay for Mammalian TLS

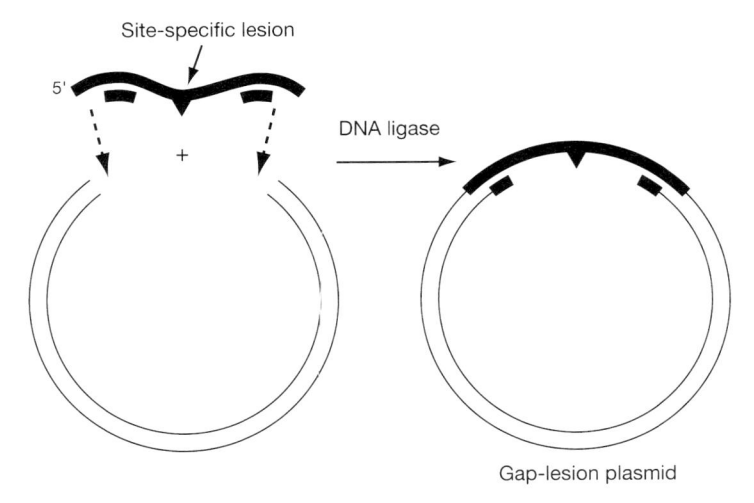

Figure 16.2 Outline of the construction of a gap-lesion plasmid. A duplex oligonucleotide is prepared with a gap opposite a defined site-specific lesion, and with termini complementary of those generated by cleavage of a plasmid with two restriction enzymes. Ligation of the two forms the gap-lesion plasmid, which can be purified by gel electrophoresis. A key feature of the method is the use of restriction enzymes that cleave at non-palindromic sequences.

16.2). To improve the yield of the desired ligation product, the vector plasmid is cleaved with restriction nucleases that cleave at nonpalindromic sequences and the gapped duplex is engineered to have compatible ends [43, 44]. Construction usually starts with a short (e.g., 12mer) oligonucleotide, carrying a site-specific lesion. This oligonucleotide can be prepared by complete chemical synthesis using a damaged nucleotide building block or by reacting with a chemical reagent and isolating the desired product carrying a site-specific lesion. The short oligonucleotide is then extended to be 50–60 nucleotides long by ligating it to two additional short oligos, using a complementary scaffold [37, 38]. The long lesion-carrying oligonucleotide is then annealed to two complementary short nucleotides, such that a duplex oligonucleotide with a gap of 20 nucleotides or so is generated opposite the lesion, and with termini which are compatible with the cleaved vector (Figure 16.2). The vector and the gapped oligonucleotide are then ligated, and the gap-lesion plasmid is fractionated by gel electrophoresis and extracted from the gel by electroelution. A similar gapped plasmid is prepared with a gapped duplex oligonucleotide with a control template without a lesion. To facilitate discrimination between the two, the gap-lesion plasmid carries a gene conferring kanamycin resistance (Kan^R), whereas the control gapped plasmid carries a gene conferring chloramphenicol resistance (Cm^R).

To assay TLS, mammalian cultured cells are transfected with a mixture of the gap-lesion plasmid and the control gapped plasmid, along with an intact plasmid (e.g., pUC18), which serves as a carrier to increase transfection yields (Figure 16.3). After allowing time for TLS in the mammalian cells, the plasmids are extracted under mild alkaline conditions, followed by renaturation. Under such conditions

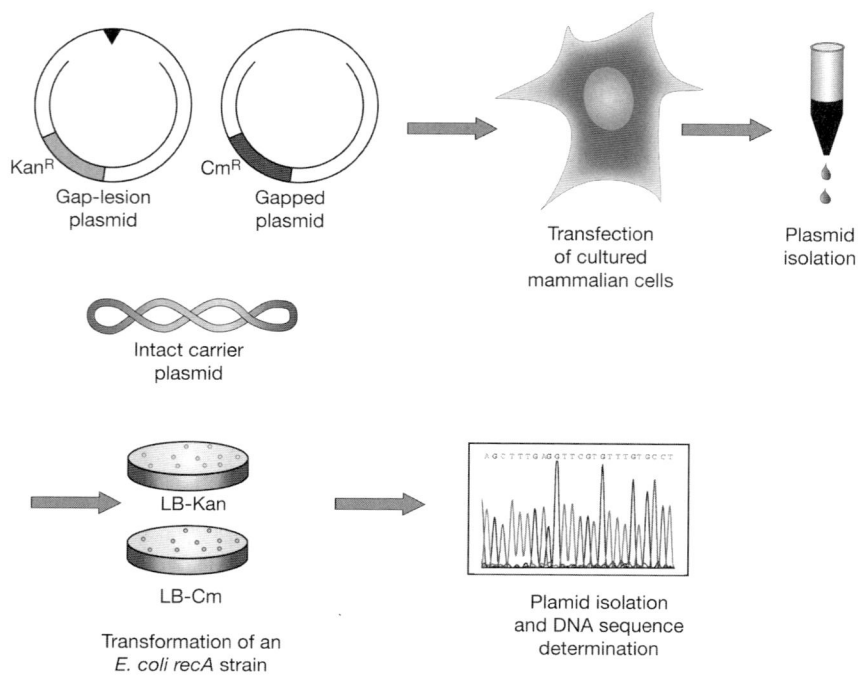

Figure 16.3 Outline of the quantitative mammalian TLS assay. Cultured cells are co-transfected with a mixture containing the gap-lesion plasmid, a control gapped plasmid, and a carrier intact plasmid. After allowing time for gap filling via TLS, the plasmids are extracted from the cells under alkaline conditions followed by renaturation, such that only plasmids that have been fully filled in and ligated remain intact. These are used to transform TLS-deficient *E. coli recA* cells, which are then plated in parallel on LB plates containing kanamycine (kan), to select for descendants of the gap-lesion plasmid, and LB containing chloramphenicol (cm), to select for descendants of the control gapped plasmid. The ratio of kanR/cmR colonies provides an estimate of the efficiency of TLS in the mammalian cells. Plasmids are extracted from the colonies, and subjected to DNA sequence analysis at the region of the original gap-lesion.

plasmids that contain nicks or gaps are denatured, and only plasmids in which the gap was fully filled in and ligated remain intact. The analysis of the fully repaired plasmids is done by using them to transform an *E. coli recA* indicator strain, which is TLS-deficient. This enables isolation and amplification of a single gapped plasmid that had undergone gap filling. During this process, *E. coli* cells usually acquire a single plasmid each, which propagates and endows antibiotic resistance on its host, leading to the formation of an antibiotic-resistant colony. The *E. coli recA* cells transformed with the plasmid mixture are plated in parallel on LB plates containing kanamycin, to select for descendants of the gap-lesion plasmids that have been filled in, and LB plates containing chloramphenicol, to select for descendants of gapped plasmids without a lesion that were filled in (Figure 16.3). The ratio of Kan^R/Cm^R colonies provides an estimate for the extent of TLS. Individual colonies are then picked, their plasmid content isolated, and subjected to DNA sequence analysis across the original gap-lesion region. Each such plasmid represents a single TLS event and provides information at a high single-nucleotide resolution. It should be noted that plasmids become chromatinized in mammalian cells, which was verified also for our gapped plasmids [38]. Many cell types can be used in the assay, including human, mouse, and chicken cell lines, as well as primary cells. In addition, a variety of transfection methods can be used. An extensive series of control experiments have demonstrated that the results obtained reflect TLS events that occur in the mammalian cells.

16.5 Some Lesions are Bypassed Most Effectively and Most Accurately by Specific Cognate TLS DNA Polymerases

The best-studied example of efficient and relatively accurate TLS is the bypass of a thymine–thymine (TT) CPD by Pol η. This was clearly indicated by using the purified polymerase and oligonucleotide substrates carrying a site-specific cis-syn TT CPD [9, 11, 12, 45]. The UV hypermutability of XPV Pol η-deficient cells compared to normal Pol η-proficient cells strongly supports this explanation [13]. We have used the gap-lesion plasmid system to assay and compare TLS across site-specific TT CPD in normal cells and XPV cells. This was done in three pairs of Pol η-proficient and Pol η-deficient human cells: primary fibroblasts from a normal individual and from an XPV patient, SV40-immortalized fibroblasts from a normal individual and an XPV patient, and the human Burkitt lymphoma cell line BL2 and a $POLH^{-/-}$ derivative of the latter [46]. In all three pairs, the extent of TLS across a site-specific TT CPD was 2- to 4-fold lower in the Pol η-deficient cells than in their normal counterparts [14]. In contrast, there was a marginal or no effect on TLS across a site-specific TT 6-4 photoproduct – the other main UV-induced lesion in DNA. DNA sequence analysis of TLS events revealed even a more dramatic effect – in all three cell pairs, TLS across the TT CPD was between 8- and 16-fold more mutagenic in Pol η-deficient cells as compared to normal cells [14]. TLS across the TT 6-4 photoproduct was similarly and highly mutagenic in both cell types. These results, obtained with the gap-lesion plasmid system, are fully consistent with the major role of Pol η in efficient and relatively accurate TLS across CPDs in human cells and with the UV hypermutability of XPV cells. Moreover, they suggest that this UV hypermutability is due to mutagenic TLS at CPDs, carried out by DNA polymerases other than polη. Moreover, when Pol η is missing, TLS across a TT CPD is less efficient and more mutagenic than in cells that have Pol η. This fact clearly indicates that Pol η performs TLS across TT CPDs more efficiently and more accurately than any of the other 14 DNA polymerases present in the cell.

The analysis described above can be extended to other types of DNA damage and other TLS polymerases. This can be done using cells from mice in which specific polymerase genes were knocked-out or human cells in which the expression of specific polymerase genes was knocked-down using RNA interference methods. Using such an approach, it was found that Pol κ is involved in the relatively accurate bypass of a benzo[a]pyrene (B[a]P)-guanine adduct (B[a]P-G), a major tobacco smoke DNA-damage agent [37], and Pol η was involved in TLS across the intrastrand cisplatin-guanine–guanine (GG) adduct, formed by the chemotherapeutic drug cisplatin [38]. TLS across an abasic site did not require Pol η nor Pol κ, but was aphidicolin-sensitive [36], perhaps representing the involvement of Pol δ, which was shown to be able to bypass AP sites *in vitro* [47, 48] and to be involved in TLS across this site *in vivo* in the yeast *S. cerevisiae* [49].

16.6
Pivotal Role for Pol ζ in TLS Across a Wide Variety of DNA Lesions

The role of Pol ζ in TLS was addressed by assaying its involvement in TLS across a broad range of DNA lesions using the gap-lesion plasmid assay. This included seven lesions: a TT CPD and a TT 6-4 photoproduct, the two main UV photoproducts in DNA; an abasic site, representing one of the most common spontaneous lesions in DNA; and cisplatin-GG, B[*a*]P-G, and 4-hydroxyequilenin-C – an adduct formed in DNA by metabolites of equilin and equilenin (Chapter 8), which are widely used in estrogen replacement therapy (Figure 16.4). In addition, an artificial lesion in the form of a hydrocarbon insert into DNA was used, consisting of a chain of 12 methylene residues, $[\text{-}(CH_2)_{12}\text{-}]$, embedded into the backbone of the single-stranded region of the gapped plasmid (M12; Figure 16.4). This is an extreme form of DNA damage, since it is in fact a non-DNA segment, lacking all DNA features. We use it to challenge the cells in order to examine the limits of the TLS system. The cells used were p53-deficient mouse embryo fibroblasts (MEFs) in which the *Rev3L* gene, encoding the catalytic subunit of Pol ζ, was knocked-out [50]. The results were quite remarkable – TLS of all lesions, except for the TT CPD, was significantly reduced in cells deficient in Pol ζ [38]. Thus, despite the presence of a multiplicity of DNA polymerases, TLS is eventually funneled to Pol ζ, therefore generating a bottleneck, with the apparent lack of a backup. This may explain the observation that Pol ζ is essential in mammals, as the attempts to knockout *Rev3L* in mice led to embryonic lethality [51]. Of course, being a very large protein, Pol ζ may be involved in processes other than TLS, which might be essential for life. DNA sequence analysis of individual plasmid isolates revealed that the residual TLS in the absence of Pol ζ was more accurate than in the Pol ζ-proficient parental mouse cells [38].

16.7
Knocking-Down the Expression of TLS Polymerases using Small Interfering RNA Provides a useful Tool for the Analysis of TLS using the Gapped Plasmid Assay

The ability to knock down the expression of specific genes using small interfering RNA (siRNA) revolutionized the ability to manipulate mammalian cells and provided a powerful tool for the analysis of various cellular processes. We adapted our assay to be used with cells in which the expression of specific DNA polymerases

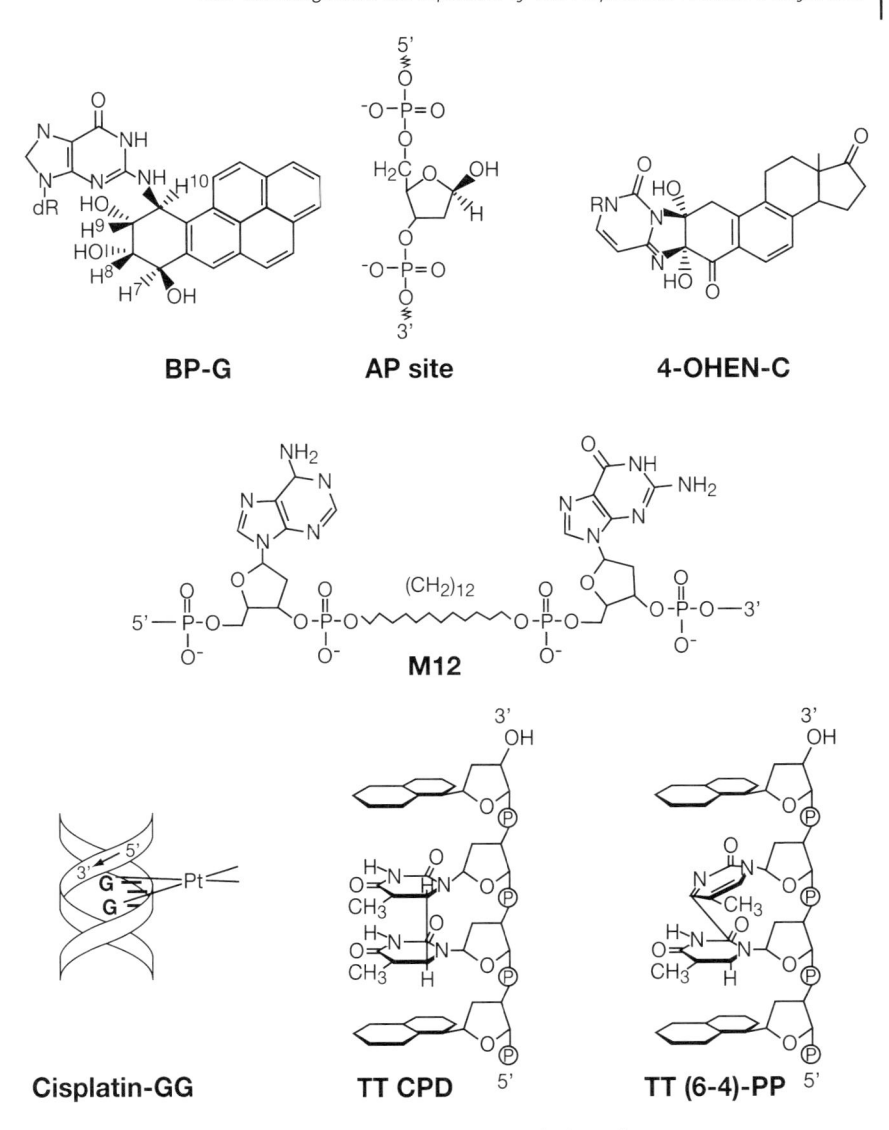

Figure 16.4 DNA lesions used in TLS studies with gapped plasmids. BP-G, B[a]P-guanine adduct; AP site, apurinic/apyrimidinic site; 4-OHEN-C, 4-hydroxyequilenin-C adduct; M12, dodecamethylene; Cisplatin-GG, cis-diamminedichloroplatinum(II)-GG intrastrand crosslink; TT CPD, TT cyclobutane pyrimidine dimmer; TT (6-4) PP, TT 6-4 photoproduct.

was knocked-down using siRNA. The experimental outline involves transfecting cells with siRNA against a specific DNA polymerase gene and after 2–3 days the cells are transfected again with the plasmid mixture to assay TLS (Figure 16.5). The two main concerns in this approach are: (i) the extent of the knockdown and (ii) the specificity of the effects observed, due to the possibility of off-target action of the siRNA. The extent of the knockdown can be optimized by the choice of the

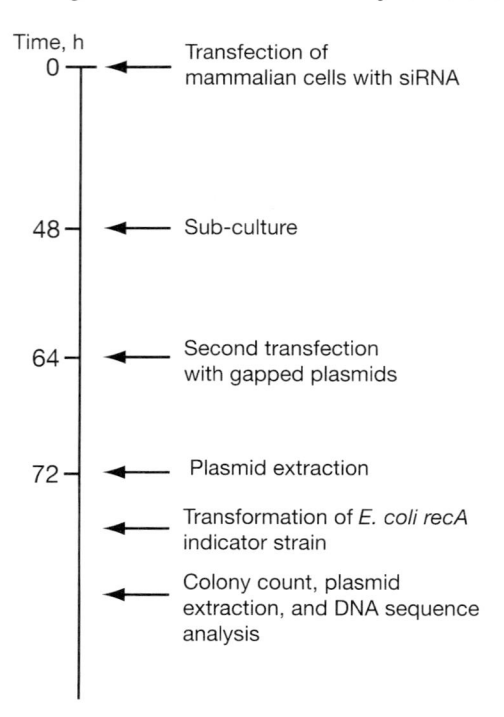

```
Time, h
  0  ──◄── Transfection of
           mammalian cells with siRNA



 48  ──◄── Sub-culture



 64  ──◄── Second transfection
           with gapped plasmids


 72  ──◄── Plasmid extraction

       ──◄── Transformation of E. coli recA
             indicator strain

       ──◄── Colony count, plasmid
             extraction, and DNA sequence
             analysis
```

Figure 16.5 Timeline of the gapped plasmid TLS assay in cells treated with siRNA. See text for details.

transfection reagent (to ensure maximal transfection efficiency of above 90%) and the choice of siRNA. Levels of expression should be carefully monitored, preferably with antibodies to the target protein. In the absence of an appropriate antibody, when only levels of mRNA can be measured (by reverse transcription- or quantitative-PCR), it is best to allow extra time before the assay to ensure maximal depletion of pre-exisiting target protein. Typically siRNA effects can last as long as 72 or 96 h and, if needed, a boost of siRNA can be used to extend the duration of the knockdown. The recently developed siRNA derivatives that can be added to the growing cells are a potentially powerful addition to extend the effects of siRNA knockdown.

The specificity problem can be addressed by: (i) the use of a control nonrelevant siRNA, (ii) the use of two or three different siRNAs to the same gene, (iii) comparison to cells carrying a knockout in the target gene and (iv) complementation with the ectopically expressed target protein. This possibility requires extra engineering to avoid knockdown of the expression of the complementing gene. The flexibility of the siRNA approach allows more than one gene to be knocked down, thereby essentially enabling quite robust genetic analysis of genes involved in various cellular processes.

16.8
Evidence that TLS Occurs by Two-Polymerase Mechanisms, in Combinations that Determine the Accuracy of the Process

There is currently no way to measure TLS in mammalian chromosomes. It is possible to measure chromosomal-induced mutagenesis that is caused by TLS. However, this method assays the integrated effect of error-prone TLS and various DNA repair mechanisms, and is blind to error-free TLS, or at most relates to it as any other error-free DNA repair. This is an important point since most TLS across several lesions, such as B[a]P-G and cisplatin-GG, is error-free [38]. Still, if reducing the expression of a TLS polymerase results in a change in induced mutagenesis, this is consistent with a role in TLS. Similarly, if cells in which a particular TLS polymerase was silenced exhibit decreased viability upon exposure to a genotoxic agent, this is also consistent with the involvement of that polymerase in TLS. Of course, this is not definitive proof, since TLS polymerases may be involved in processes other than TLS, which affect viability or mutagenesis. As far as the involvement of TLS polymerases in chromosomal TLS across B[a]P-induced lesions it was reported that $PolK^{-/-}$ MEFs are sensitive to such compounds, and exhibit also reduced mutagenesis [52]. In addition, B[a]P diol epoxide-induced chromosomal mutagenesis was reported to be reduced in cells in which the expression of $REV3L$ was suppressed [53], consistent with a role of Pol ζ in error-prone TLS of UV lesions. Thus, there is evidence consistent with the involvement of both Pol κ and Pol ζ in TLS across B[a]P-induced lesions, consistent with our results on the TLS across a site-specific B[a]P-G adduct. As for cisplatin-GG, it was reported that cells lacking Pol η are more sensitive to cisplatin than Pol η-proficient cells [54, 55] and, similarly, an increased sensitivity was reported for cells in which the expression of Pol ζ was suppressed [56, 57], suggesting involvement of both Pol η and Pol ζ in TLS across cisplatin-induced adducts, again consistent with our results that both Pol η and Pol ζ are involved in TLS across a site-specific intrastrand cisplatin-GG adduct.

The question arises of whether the two polymerases act in the same reaction or in competing pathways. A convenient approach to address this question is an epistasis analysis performed with siRNA. Thus, if Pol κ and Pol ζ are the only polymerases that function in the bypass of a B[a]P-G adduct, and they act in two parallel reactions, silencing the expression of both should yield an effect larger than silencing each polymerase alone. On the other hand, if they act in the same pathway, knocking out both, should give an effect similar to knocking out one of them. The situation may become more complex if additional polymerases are involved. For this analysis, the gap-lesion plasmid assay system, in conjunction with siRNA silencing of specific polymerases, offers a clear advantage, as both the extent of TLS and its accuracy can be assayed in the same cell in a quantitative way.

Such experiments were performed in the p53-proficient human osteosarcoma cell line U2OS using the gap-lesion plasmid with a site-specific cisplatin-GG [38].

Silencing the expression of *POLH* caused a 2-fold reduction in TLS, suggesting that Pol η is involved in TLS across at least half of the cisplatin-GG lesions. Silencing the expression of Pol ζ caused a greater 5-fold reduction of TLS, suggesting that that Pol ζ is involved in the bypass of at least 80% of the cisplatin-GG lesions. Knocking down both, had an effect similar to knocking down Pol ζ alone [38]. These results indicate that at least some of the cisplatin-GG lesions are bypassed in a reaction that involves the action of both Pol η and Pol ζ. In addition, they suggest that an additional polymerase might be involved, since the effect of Pol η was not as big as that of Pol ζ. We examined whether the additional polymerase involved might be Pol κ. Knocking down the expression of *POLK* alone had a modest effect of reducing TLS by 34%. However, knocking down both *POLH* and *POLK* caused a strong 4.5-fold decrease in TLS, similar to the effect observed with *REV3L*. This indicates that Pol ζ cooperates with *POLH* and with *POLK* in two polymerase TLS reactions to bypass the cisplatin-GG lesion. How about fidelity? cisplatin-GG is not a highly mutagenic lesion and in U2OS cells it is bypassed with an error frequency of 18%. Knocking down the expression of *POLH* or *REV3L* did not significantly change the mutation frequency. However, knocking down the expression of *POLK* caused a strong 5.7-fold decrease in mutagenicity, down to an error frequency of 3.2%, indicating that *POLK* is responsible for mutagenic TLS across cisplatin-GG. Thus, TLS by Pol η and Pol ζ carries out primarily accurate TLS, whereas Pol κ and Pol ζ carry out primarily mutagenic TLS [38]. The involvement of Pol κ in TLS across cisplatin-GG was unexpected, since purified Pol κ was reported to be unable to bypass a cisplatin-GG lesion. Presumably accessory proteins present in the cell (PCNA, replication protein A, or others?) enable Pol κ to perform the TLS reaction, albeit in a mutagenic manner. This highlights the potential differences between *in vivo* and *in vitro* TLS, underscoring the need to examine the TLS of specific DNA lesions *in vivo*.

Two-polymerase mechanisms were found using a similar approach also for B[*a*]P-G. In this case Pol κ was found to act with Pol ζ to perform accurate TLS across B[*a*]P-G, whereas Pol η acted with Pol ζ to perform mutagenic TLS. Interestingly, in this case whereas knocking down the expression of *REV3L* caused a strong 5.5-fold decrease in the extent of TLS, knocking down both *POLK* and *POLH* caused only a 2.4-fold decrease in TLS, similar to the effect of *POLK* alone [38]. This suggests that an additional polymerase is involved in TLS across B[*a*]P-G in human cells, acting with Pol ζ in an error-free manner, similarly to the Pol κ and Pol ζ combination. The identity of that polymerase is yet unknown.

Based on experiments with purified DNA polymerases, it was suggested that TLS can operate in two-polymerase mechanisms, in which one polymerase, the inserter, incorporates a nucleotide opposite the lesion and a second polymerase, the extender, continues synthesis past the lesion [58]. Moreover, based on the properties of the *S. cerevisiae* Pol ζ and the human Pol κ, the two were suggested to act as extenders [59]. Our results are generally consistent with this hypothesis, and in fact provide evidence for the action of two-polymerase mechanisms in

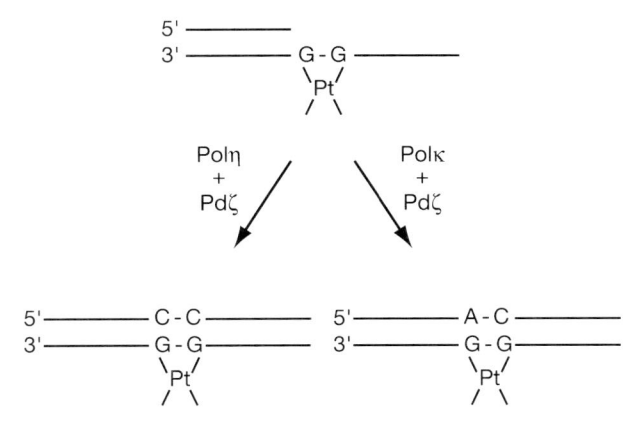

Figure 16.6 Model for two-polymerase bypass of a cisplatin-GG adduct. Polη cooperates with polζ to carryout accurate TLS across cisplarin-GG is the main mechanism for bypassing this lesion. Polκ and polζ cooperate in a minor reaction to perform mutagenic TLS across this lesion.

human cells. The requirement for Pol ζ in the two-polymerase reactions is consistent with a function of a general extender, similar to the yeast Pol ζ. Thus, the simplest explanation for our results is that Pol η or Pol κ act as inserters, whereas Pol ζ acts a universal extender (Figure 16.6). Moreover, we have also shown that different combinations of polymerases dictate whether the TLS reaction will be accurate or mutagenic (Figure 16.6).

16.9
Conclusions

The use of plasmid model systems with site-specific defined DNA lesions has proven to be a powerful tool in elucidating the mechanism of TLS. Further studies are needed to elucidate the identity of DNA polymerases involved in TLS across other significant DNA lesions. Moreover, such systems can be also useful to elucidate regulatory aspects of TLS. The big challenge for the future is to harness the power of chemically defined lesions to the development of novel TLS assays in mammalian chromosomes.

Acknowledgements

Z.L. is the incumbent of the Maxwell Ellis Professorial Chair in Biomedical Research. This work was supported by grants from the Flight Attendant Medical Research Institute, Florida, USA, the Israel Science Foundation (no. 564/04), and the M.D. Moross Institute for Cancer Research at the Weizmann Institute of Science.

References

1 Friedberg, E.C., Walker, G.C., Siede, W., Wood, R.D., Schultz, R.A., and Ellenberger, T. (2006) *DNA Repair and Mutagenesis*, 2nd edn, ASM Press, Washington, DC.

2 Lindahl, T. (1993) Instability and decay of the primary structure of DNA. *Nature*, **362**, 709–715.

3 Friedberg, E.C. (2005) Suffering in silence: the tolerance of DNA damage. *Nat. Rev. Mol. Cell. Biol.*, **6**, 943–953.

4 Prakash, S., Johnson, R.E., and Prakash, L. (2005) Eukaryotic translesion synthesis DNA polymerases: specificity of structure and function. *Annu. Rev. Biochem.*, **74**, 317–353.

5 Lehmann, A.R., Niimi, A., Ogi, T., Brown, S., Sabbioneda, S., Wing, J.F., Kannouche, P.L., and Green, C.M. (2007) Translesion synthesis: Y-family polymerases and the polymerase switch. *DNA Repair*, **6**, 891–899.

6 Yang, W. and Woodgate, R. (2007) What a difference a decade makes: insights into translesion synthesis. *Proc. Natl. Acad. Sci. USA*, **104**, 15591–15598.

7 Livneh, Z. (2001) DNA damage control by novel DNA polymerases: translesion replication and mutagenesis. *J. Biol. Chem.*, **276**, 25639–25642.

8 Livneh, Z. (2006) Keeping mammalian mutation load in check. Regulation of the activity of error-prone DNA polymerases by p53 and p21. *Cell Cycle*, **5**, 1918–1922.

9 Masutani, C., Kusumoto, R., Yamada, A., Dohmae, N., Yokoi, M., Yuasa, M., Araki, M., Iwai, S., Takio, K., and Hanaoka, F. (1999) The XPV (xeroderma pigmentosum variant) gene encodes human DNA polymerase eta. *Nature*, **399**, 700–704.

10 Johnson, R.E., Kondratick, C.M., Prakash, S., and Prakash, L. (1999) hRAD30 mutations in the variant form of xeroderma pigmentosum. *Science*, **285**, 263–265.

11 Masutani, C., Araki, M., Yamada, A., Kusumoto, R., Nogimori, T., Maekawa, T., Iwai, S., and Hanaoka, F. (1999) Xeroderma pigmentosum variant (XP-V) correcting protein from HeLa cells has a thymine dimer bypass DNA polymerase activity. *EMBO J.*, **18**, 3491–3501.

12 Johnson, R.E., Washington, M.T., Prakash, S., and Prakash, L. (2000) Fidelity of human DNA polymerase η. *J. Biol. Chem.*, **275**, 7447–7450.

13 Maher, V.M., Ouellette, L.M., Curren, R.D., and McCormick, J.J. (1976) Frequency of ultraviolet light-induced mutations is higher in xeroderma pigmentosum variant cells than in normal human cells. *Nature*, **261**, 593–595.

14 Hendel, A., Ziv, O., Gueranger, Q., Geacintov, N., and Livneh, Z. (2008) Reduced fidelity and increased efficiency of translesion DNA synthesis across a TT cyclobutane pyrimidine dimer, but not a TT 6-4 photoproduct, in human cells lacking DNA polymerase eta. *DNA Repair*, **7**, 1636–1646.

15 Ohmori, H., Friedberg, E.C., Fuchs, R.P.P., Goodman, M.F., Hanaoka, F., Hinkle, D., Kunkel, T.A., Lawrence, C.W., Livneh, Z., Nohmi, T., Prakash, L., Prakash, S., Todo, T., Walker, G.C., Wang, Z., and Woodgate, R. (2001) The Y-family of DNA polymerases. *Mol. Cell*, **8**, 7–8.

16 Nelson, J.R., Lawrence, C.W., and Hinkle, D.C. (1996) Thymine–thymine dimer bypass by yeast DNA polymerase ζ. *Science*, **272**, 1646–1649.

17 Sweasy, J.B., Lauper, J.M., and Eckert, K.A. (2006) DNA polymerases and human disease. *Radiat. Res.*, **166**, 693–714.

18 Hoege, C., Pfander, B., Moldovan, G.L., Pyrowolakis, G., and Jentsch, S. (2002) RAD6-dependent DNA repair is linked to modification of PCNA by ubiquitin and SUMO. *Nature*, **419**, 135–141.

19 Stelter, P. and Ulrich, H.D. (2003) Control of spontaneous and damage-induced mutagenesis by SUMO and ubiquitin conjugation. *Nature*, **425**, 188–191.

20 Lee, K.Y. and Myung, K. (2008) PCNA modifications for regulation of post-replication repair pathways. *Mol. Cell*, **26**, 5–11.

21 Moldovan, G.L., Pfander, B., and Jentsch, S. (2007) PCNA, the maestro of the replication fork. *Cell*, **129**, 665–679.

22 Bienko, M., Green, C.M., Crosetto, N., Rudolf, F., Zapart, G., Coull, B., Kannouche, P., Wilder, G., Peter, M., Lehmann, A.R., Hofmann, K., and Dikic, I. (2005) Ubiquitin-binding domains in Y-family polymerases regulate translesion synthesis. *Science*, **310**, 1821–1824.

23 Zhuang, Z., Johnson, R.E., Haracska, L., Prakash, L., Prakash, S., and Benkovic, S.J. (2008) Regulation of polymerase exchange between Polη and Polδ by monoubiquitination of PCNA and the movement of DNA polymerase holoenzyme. *Proc. Natl. Acad. Sci. USA*, **105**, 5361–5366.

24 Acharya, N., Yoon, J.H., Gali, H., Unk, I., Haracska, L., Johnson, R.E., Hurwitz, J., Prakash, L., and Prakash, S. (2008) Roles of PCNA-binding and ubiquitin-binding domains in human DNA polymerase η in translesion DNA synthesis. *Proc. Natl. Acad. Sci. USA*, **105**, 17724–17729.

25 Avkin, S., Sevilya, Z., Toube, L., Geacintov, N.E., Chaney, S.G., Oren, M., and Livneh, Z. (2006) p53 and p21 regulate error-prone DNA repair to yield a lower mutation load. *Mol. Cell*, **22**, 407–413.

26 Soria, G., Podhajcer, O., and Gottifredi, V. (2006) P21$^{Cip1/WAF1}$ downregulation is required for efficient PCNA ubiquitination after UV irradiation. *Oncogene*, **25**, 2829–2838.

27 Loechler, E.L., Green, C.L., and Essigmann, J.M. (1984) *In vivo* mutagenesis by O^6-methylguanine built into a unique site in a viral genome. *Proc. Natl. Acad. Sci. USA*, **81**, 6271–6275.

28 Singer, B. and Essigmann, J.M. (1991) Site-specific mutagenesis: retrospective and prospective. *Carcinogenesis*, **12**, 949–955.

29 LeClerc, J.E., Borden, A., and Lawrence, C.W. (1991) The thymine–thymine pyrimidine–pyrimidone (6-4) ultraviolet light photoproduct is highly mutagenic and specifically induces 3′ thymine-to-cytosine transitions in *Escherichia coli*. *Proc. Natl. Acad. Sci. USA*, **88**, 9685–9689.

30 Napolitano, R., Janel-Bintz, R., Wagner, J., and Fuchs, R.P.P. (2000) All three SOS-inducible DNA polymerases (Pol II, Pol IV, and Pol V) are involved in induced mutagenesis. *EMBO J.*, **19**, 6259–6265.

31 Shibutani, S., Takeshita, M., and Grollman, A.P. (1991) Insertion of specific bases during DNA synthesis past the oxidation-damaged base 8-oxodG. *Nature*, **349**, 431–434.

32 Gibbs, P.E., McDonald, J., Woodgate, R., and Lawrence, C.W. (2005) The relative roles *in vivo* of *Saccharomyces cerevisiae* Pol η, Pol ζ, Rev1 protein and Pol32 in the bypass and mutation induction of an abasic site, T–T (6-4) photoadduct and T–T *cis-syn* cyclobutyl dimer. *Genetics*, **169**, 575–582.

33 Reuven, N.B., Arad, G., Maor-Shoshani, A., and Livneh, Z. (1999) The mutagenesis protein UmuC is a DNA polymerase activated by UmuD′, RecA and SSB, and specialized for translesion replication. *J. Biol. Chem.*, **274**, 31763–31766.

34 Szüts, D., Marcus, A.P., Himoto, M., Iwai, S., and Sale, J.E. (2008) REV1 restrains DNA polymerase ζ to ensure frame fidelity during translesion synthesis of UV photoproducts *in vivo*. *Nucleic Acids Res.*, **36**, 6767–6780.

35 Moriya, M., Zhang, W., Johnson, F., and Grollman, A.P. (1994) Mutagenic potency of exocyclic DNA adducts: marked differences between *Escherichia coli* and simian kidney cells. *Proc. Natl. Acad. Sci. USA*, **91**, 11899–11903.

36 Avkin, S., Adar, S., Blander, G., and Livneh, Z. (2002) Quantitative measurement of translesion replication in human cells: evidence for bypass of abasic sites by a replicative DNA polymerase. *Proc. Natl. Acad. Sci. USA*, **99**, 3764–3769.

37 Avkin, S., Goldsmith, M., Velasco-Miguel, S., Geacintov, N., Friedberg, E.C., and Livneh, Z. (2004) Quantitative analysis of translesion DNA synthesis across a benzo[*a*]pyrene-guanine adduct in mammalian cells. The role of DNA polymerase κ. *J. Biol. Chem.*, **279**, 53298–53305.

38 Shachar, S., Ziv, O., Avkin, S., Adar, S., Wittschieben, J., Reisner, T., Chaney, S.G., Friedberg, E.C., Wang, Z., Carell,

T., Geacintov, N., and Livneh, Z. (2009) Two-polymerase mechanisms dictate error-free and error-prone translesion DNA synthesis in mammals. *EMBO J.*, **28**, 383–393.

39 Lopes, M., Foiani, M., and Sogo, J.M. (2006) Multiple mechanisms control chromosome integrity after replication fork uncoupling and restart at irreparable UV lesions. *Mol. Cell*, **21**, 15–27.

40 Mojas, N., Lopes, M., and Jiricny, J. (2007) Mismatch repair-dependent processing of methylation damage gives rise to persistent single-stranded gaps in newly replicated DNA. *Genes Dev.*, **21**, 3342–3355.

41 Lehmann, A.R. and Fuchs, R.P. (2006) Gaps and forks in DNA replication: rediscovering old models. *DNA Repair*, **5**, 1595–1498.

42 Niimi, A., Brown, S., Sabbioneda, S., Kannouche, P.L., Scott, A., Yasui, A., Green, C.M., and Lehmann, A.R. (2008) Regulation of proliferating cell nuclear antigen ubiquitination in mammalian cells. *Proc. Natl. Acad. Sci. USA*, **105**, 16125–16130.

43 Tomer, G., Reuven, N.B., and Livneh, Z. (1998) The β subunit sliding DNA clamp is responsible for unassisted mutagenic translesion replication by DNA polymerase III holoenzyme. *Proc. Natl. Acad. Sci. USA*, **95**, 14106–14111.

44 Tomer, G. and Livneh, Z. (1999) Analysis of unassisted translesion replication by the DNA polymerase III holoenzyme. *Biochemistry*, **38**, 5948–5958.

45 Johnson, R.E., Prakash, S., and Prakash, L. (1999) Efficient bypass of a thymine–thymine dimer by yeast DNA polymerase, Pol η. *Science*, **283**, 1001–1004.

46 Gueranger, Q., Stary, A., Aoufouchi, S., Faili, A., Sarasin, A., Reynaud, C.A., and Weill, J.C. (2008) Role of human DNA polymerases η, ι, and ζ in UV resistance and UV-induced mutagenesis. *DNA Repair*, **7**, 1551–1562.

47 Mozzherin, D.J., Shibutani, S., Tan, C.K., Downey, K.M., and Fisher, P.A. (1997) Proliferating cell nuclear antigen promotes DNA synthesis past template lesions by mammalian DNA polymerase δ. *Proc. Natl. Acad. Sci. USA*, **94**, 6126–6131.

48 Daube, S.S., Tomer, G., and Livneh, Z. (2000) Translesion replication by DNA polymerase δ depends on processivity accessory proteins and differs in specificity from DNA polymerase β. *Biochemistry*, **39**, 348–355.

49 Haracska, L., Unk, I., Johnson, R.E., Johansson, E., Burgers, P.M., Prakash, S., and Prakash, L. (2001) Roles of yeast DNA polymerases δ and ζ and of Rev1 in the bypass of abasic sites. *Genes Dev.*, **15**, 945–954.

50 Wittschieben, J., Shivji, M.K., Lalani, E., Jacobs, M.A., Marini, F., Gearhart, P.J., Rosewell, I., Stamp, G., and Wood, R.D. (2000) Disruption of the developmentally regulated Rev3l gene causes embryonic lethality. *Curr. Biol.*, **10**, 1217–1220.

51 Gan, G.N., Wittschieben, J.P., Wittschieben, B.O., and Wood, R.D. (2008) DNA polymerase ζ (pol ζ) in higher eukaryotes. *Cell Res.*, **18**, 174–183.

52 Ogi, T., Shinkai, Y., Tanaka, K., and Ohmori, H. (2002) Pol κ protects mammalian cells against the lethal and mutagenic effects of benzo[a]pyrene. *Proc. Natl. Acad. Sci. USA*, **99**, 15548–15553.

53 Li, Z., Zhang, H., McManus, T.P., McCormick, J.J., Lawrence, C.W., and Maher, V.M. (2002) hREV3 is essential for error-prone translesion synthesis past UV or benzo[a]pyrene diol epoxide-induced DNA lesions in human fibroblasts. *Mutat. Res.*, **510**, 71–80.

54 Bassett, E., King, N.M., Bryant, M.F., Hector, S., Pendyala, L., Chaney, S.G., and Cordeiro-Stone, M. (2004) The role of DNA polymerase η in translesion synthesis past platinum-DNA adducts in human fibroblasts. *Cancer Res.*, **64**, 6469–6475.

55 Albertella, M.R., Green, C.M., Lehmann, A.R., and O'Connor, M.J. (2005) A role for polymerase η in the cellular response to cisplatin-induced damage. *Cancer Res.*, **65**, 9799–9806.

56 Wu, F., Lin, X., Okuda, T., and Howell, S.B. (2004) DNA polymerase ζ regulates cisplatin cytotoxicity, mutagenicity, and the rate of development of cisplatin resistance. *Cancer Res.*, **64**, 8029–8035.

57 Zander, L. and Bemark, M. (2004) Immortalized mouse cell lines that lack a functional Rev3 gene are hypersensitive to UV irradiation and cisplatin treatment. *DNA Repair*, **3**, 743–752.

58 Johnson, R.E., Washington, M.T., Haracska, L., Prakash, S., and Prakash, L. (2000) Eukaryotic polymerases ι and ζ act sequentially to bypass DNA lesions. *Nature*, **406**, 1015–1019.

59 Prakash, S. and Prakash, L. (2002) Translesion DNA synthesis in eukaryotes: a one- or two-polymerase affair. *Genes Dev.*, **16**, 1872–1883.

17
DNA Damage and Transcription Elongation: Consequences and RNA Integrity

Kristian Dreij, John A. Burns, Alexandra Dimitri, Lana Nirenstein, Taissia Noujnykh, and David A. Scicchitano

17.1
Introduction

DNA is continuously exposed to a wide range of agents that can alter its fundamental structure. Such damage to the genome is unavoidable since many of the offending agents are ubiquitous, originating from endogenous and exogenous sources. Endogenous DNA-damaging agents often arise from metabolic processes that can generate reactive oxygen species (ROS) or methylating agents, among others. Exogenous DNA-damaging agents include different types of radiation and chemicals that are often found among environmental pollutants. Some chemicals can react with DNA directly, whereas others are metabolized in cells to intermediate compounds that damage DNA. When one considers the wide range of chemical species that can react with DNA, it should come as no surprise that modifications to the double helix have been observed at most of its chemical constituents, including the nitrogenous bases, sugars, and phosphate groups. Furthermore, the primary damage that occurs can result in the formation of unstable intermediates that give rise to secondary genomic alterations, including DNA strand breaks, abasic sites, and inter- and intrastrand DNA cross-links.

DNA damage interferes with the fundamental cellular processes that use DNA as a template – replication and transcription. DNA damage that is present during replication can block the progress of nascent DNA synthesis and also decrease fidelity, leading to collapsed replication forks, chromosomal aberrations, and mutations [1–3]. In an analogous way, DNA damage that is present in the transcription units of active genes can stall or block RNA elongation, or induce base misincorporation events in growing RNA – an event that has been called transcriptional mutagenesis by Paul Doetsch, who pioneered research in this area [4, 5]. These perturbations to DNA and RNA synthesis exact severe consequences, triggering events like apoptosis or inducing transformation [1]. In fact, errors induced by DNA damage during RNA synthesis can lead to compromised gene expression, affecting the spatial and temporal information flow in a cell, which can lead to growth defects during development [6].

The Chemical Biology of DNA Damage. Edited by Nicholas E. Geacintov and Suse Broyde
© 2010 WILEY-VCH Verlag GmbH & Co. KGaA, Weinheim
ISBN: 978-3-527-32295-4

Over the last several decades, biochemical and molecular biological methods have been developed to study the effects of DNA damage on elongating RNA polymerases during transcription and to investigate the potentially detrimental downstream cellular events that the damage can elicit. This chapter focuses on RNA polymerase behavior at DNA lesions, providing an overview of the adducts that interfere with transcription, their effects on transcript integrity, and, when available, potential mechanisms that explain their action.

17.2
DNA Repair

The ever-present nature of DNA-damaging agents implies that accumulated alterations to the genome and the resulting synthesis of aberrant DNA and RNA should pose serious threats to living systems. However, cells have evolved mechanisms to minimize these threats. For example, the detrimental effects of DNA damage on replication blockage can be nullified in part by recombination or by a process known as translesion DNA synthesis, which allows a specialized group of DNA polymerases to perform strand extension past lesions positioned in the template strand (reviewed in [7, 8]). Also of great importance in guarding against the consequences of DNA damage are an evolutionarily conserved set of DNA repair systems that exist in prokaryotes and eukaryotes. These include nucleotide excision repair (NER), base excision repair (BER), mismatch repair, and direct reversal mechanisms that clear DNA of damage and, for the most part, restore the original genetic information. (See Chapters 5 and 10–12 for further discussion of some of these pathways.)

NER is perhaps the most versatile DNA repair pathway in that it recognizes and removes many different structural alterations from DNA. It is often divided into two general subpathways based on the mechanism by which a DNA lesion is recognized and verified: global genomic (GG)-NER, which operates throughout the genome (see Chapter 11), and transcription-coupled (TC)-NER, which removes DNA lesions from actively transcribed sequences. Isabel Mellon, Graciela Spivak, and Philip Hanawalt originally characterized TC-NER when they discovered that the transcribed strand of the actively expressed housekeeping gene that encodes dihydrofolate reductase in human cells is cleared of cyclobutane pyrimidine dimers (CPDs) faster than its complementary, nontranscribed strand or the genome overall [9]. Similar results were observed for the transcribed strand of the actively expressed *lac* operon in *Escherichia coli* [10]. Since then, TC-NER has been documented in cells from many organisms and it operates on a variety of different DNA adducts (see [11, 12] and references therein).

In *E. coli*, GG-NER is executed by the UvrABC excinuclease and it is responsible for repairing a wide array of helix-distorting lesions (reviewed in [13]). Damage recognition is carried out by the $UvrA_2UvrB$ complex that binds to DNA, and moves along the helix scanning for damage. When damage is encountered, $UvrA_2$ dissociates and UvrC binds to UvrB at the site of damage, creating an endonuclease. This UvrC protein activity nicks the damaged DNA strand 3′ and 5′ to the lesion. An additional protein, UvrD, which acts as a helicase, enhances the release of a segment of DNA about 12 bases in length that contains the site of damage; the UvrD protein also displaces UvrC. Repair is complete following synthesis of DNA by DNA Pol I and ligation. (See Chapter 12, Section 3 for details.)

TC-NER in *E. coli* occurs when an elongating RNA polymerase complex that stalls at a site of damage in DNA is recognized by a protein that is the product of the mutation frequency decline (*mfd*) gene [14–17]. The Mfd protein displaces the transcription complex, remains bound at the damaged site, and then recruits UvrA$_2$UvrB. Hence, the Mfd protein is a *bona fide* transcription–DNA repair coupling factor and TC-NER in *E. coli* requires the same proteins as GG-NER, with the principal difference between the two pathways lying at the level of damage recognition.

In eukaryotes, GG-NER also removes a wide array of DNA lesions from the genome and lesion recognition is carried out by XPC/HR23B (as detailed in Chapter 12). The XPC protein is essential in GG-NER and is among the proteins that are defective in the disease xeroderma pigmentosum (XP). XP patients are prone to cancer and sun sensitivity due to defective DNA repair and the associated accumulation of DNA damage and mutations. The XPC/HR23B heterodimer detects and binds to a wide range of DNA lesions, and damage recognition requires both base-pair disruption and actual modification to the DNA [18–20]. In fact, there are certain "bulky" DNA adducts that are poorly repaired by NER because they minimally distort the double helix, making them poor substrates for XPC/HR23B [21–25]. Following damage recognition, there is a sequential assembly of additional proteins, including the XP proteins, which unwind the DNA in the vicinity of the damage and make incisions on the 5′- and 3′-sides of the lesion, generating an oligodeoxynucleotide that contains the damage. Following removal of this region, DNA Pol δ/ε then fills in the resulting gap and ligase seals the newly synthesized DNA to regenerate an undamaged parental strand.

GG-NER and TC-NER overlap in eukaryotes, with DNA damage recognition acting as the principal difference between them. Cells that lack functional XPC carry out TC-NER, but not GG-NER. It is believed that RNA Pol II that stalls at a DNA lesion stimulates TC-NER by replacing XPC in the recruitment of the additional XP proteins, but the mechanistic details are not well understood. TC-NER also requires the CSA and CSB proteins that are mutated in patients with the disease Cockayne syndrome (CS) [26–28]. However, the precise roles of CSA, CSB, and additional elongation factors in this repair pathway are also not well understood, although the CSB protein may operate by helping RNA polymerase bypass certain lesions [29]. Since it is thought that stalled RNA polymerases at sites of DNA damage stimulate TC-NER, understanding how DNA lesions affect RNA polymerases during transcription elongation is critical for understanding their removal from actively expressed genes [30–32].

17.3
Transcription Elongation and DNA Damage

As with DNA polymerases, the DNA template-dependent action of RNA polymerases requires an undamaged template. When an RNA polymerase encounters DNA damage during transcription elongation, the lesion can affect RNA synthesis in several ways: it can block RNA polymerase progression (resulting in truncated transcripts), slow RNA polymerase translocation (perturbing temporal features of RNA synthesis), and induce incorporation of incorrect bases into nascent RNA (potentially altering the transcript's function via transcriptional mutagenesis). The absence or incorrect synthesis of mRNA sequences at a given point during cell growth can contribute to changes in cell physiology.

The behavior of RNA polymerases at sites of DNA damage can best be understood by using site-specific, modified DNA templates in which the precise location, base linkage site, and stereochemistry of the lesion is known. Studies using this approach have collectively shown that the capacity of a DNA lesion to block transcription is a function of several interrelated variables, which have been summarized in prior articles and still hold true [11, 12, 33, 34]. These parameters include the structure of the specific RNA polymerase's active site, the size and shape of the DNA adduct, the stereochemistry of the adduct, the particular base incorporated into the growing transcript at the site of the damage, and the local DNA sequence.

17.4
RNA Polymerases: A Brief Overview

Cells depend on regulated gene expression for their growth, proliferation, and survival. DNA-dependent RNA polymerases act in the first step of gene expression by executing transcription, converting the information stored in DNA into RNA in a tightly regulated process. Typically, transcription is divided into several stages: initiation, elongation, and termination.

The first stage in transcription, initiation, involves assembly of a preinitiation RNA polymerase complex at the gene's promoter with subsequent unwinding of the duplex around the +1 start of transcription and addition of the first few ribonucleoside triphosphates via the formation of phosphodiester bonds. The RNA polymerase reads the DNA template strand, which is referred to as the transcribed or noncoding strand, obeying base-pairing rules as it catalyzes the synthesis of RNA. The second stage, elongation, results in the addition of the remaining bases to the transcript, again via phosphodiester bond formation. The third and final stage, termination, results in the release of the full-length RNA, which may undergo significant post-transcriptional modification, especially in eukaryotic cells.

The behavior of an RNA polymerase at DNA damage can be best understood in light of structural information about the specific RNA polymerase being studied. As Table 17.1 illustrates, studies using site-specifically damaged DNA have employed bacteriophage, prokaryotic, and eukaryotic RNA polymerases.

Table 17.1 The effect of DNA lesions on transcription elongation and TC-NER.

Lesion	Bypass/arrest	TC-NER	References
Bacteriophage			
abasic sites	stall with 50–100% bypass[a]		[35–37]
strand breaks	strong block to bypass[a]		[35, 38]
gaps	significant bypass		[39, 40]
8-oxoguanine	95–100% bypass		[41, 42]
thymine glycol	50% bypass		[43, 44]
5,6-dihydrouracil	pause with 100% bypass		[4]
2-deoxyribonolactone	block		[45]
5-guanidino-4-nitroimidazole	87% bypass		[46]
pyrimido[1,2-α]purin-10(3H)one	60% bypass		[47]
uracil	100% bypass		[35]
O^6-methylguanine	96% bypass		[48]
N^2-ethyl-2'-dG	block		[49]
1,N^2-ethenoguanine	60% bypass		[50]
cyclobutane pyrimidine dimer	stall with 45% bypass		[51–53]
6-4 photoproducts	stall with some bypass		[52]
4'-hydroxymethyl-4,5',8-trimethylpsoralen			
1,1'-d(TA/AT)	block		[54]
mono dT adduct	block		[54]
cisplatin			
cis-1,2-d(GG)	~10% bypass		[55, 56]
cis-1,3-d(GTG)	up to 30% bypass		[55, 56]
benzo[a]pyrene diol epoxide-N^2-dG			
(+)-*trans-anti*	~5% bypass		[57]
(−)-*trans-anti*	~10% bypass		[57]
(+)-*cis-anti*	~20% bypass		[57]
(−)-*cis-anti*	~40% bypass		[57]
benzo[a]pyrene diol epoxide-N^6-dA			
(+)-*trans-anti*	18% bypass		[58]
(−)-*trans-anti*	32% bypass		[58]

Table 17.1 *Continued*

Lesion	Bypass/arrest	TC-NER	References
dibenzo[a,l]pyrene diol epoxide-N^2-dG			
(−)-*trans-anti*	25% bypass		this chapter
benzo[c]phenanthrene-N^6-dA			
(+)-*trans-anti*	~20% bypass		[59]
(−)-*trans-anti*	~10% bypass		[59]
2-aminofluorene-C8-dG	70% bypass		[41]
2-acetylaminofluorene-C8-dG	20% bypass		[41]
4-hydroxyequilenin			
dA	8% bypass		this chapter
dC	2% bypass		this chapter
Prokaryotic			
abasic sites	stall with bypass		[35, 60]
strand breaks	block		[35, 60]
gaps	stall with bypass		[39]
8-oxoguanine	stall with 90–100% bypass	+	[60–62]
5,6-dihydrouracil	80% bypass		[63]
uracil	100% bypass		[35, 61]
O^6-methylguanine	100% bypass		[61]
N^2-ethyl-2'-dG	block		[49]
cyclobutane pyrimidine dimer	block	+	[14, 64]
4'-hydroxymethyl-4,5',8-trimethylpsoralen			
1,1'-d(TA/AT)	block		[65]
mono dT adduct	block		[65]
cisplatin			
cis-1,2-d(GG)	block		[66, 67]
cis-1,2-d(AG)	block		[66, 67]
cis/trans-1,3-d(GTG)	block		[68]
cis-1,1'-d(GC/GC)	block		[68]
mono dG adduct	some bypass		[68]
Eukaryotic			
abasic sites	10% bypass		[37]
strand breaks	10–35% bypass[a]	−	[69, 70]
8-oxoguanine	stall with 80–100% bypass[a]		[29, 42, 69, 71–73]

Table 17.1 Continued

Lesion	Bypass/arrest	TC-NER	References
thymine glycol	stall with 50–100% bypass[a]		[29, 44, 69, 73]
5-hydroxyuracil	40–60% bypass		[29]
2-deoxyribonolactone	<10% bypass		[45]
5-guanidino-4-nitroimidazole	9% bypass		[46]
8,5′-cyclo-2′-dA	Some bypass		[74]
pyrimido[1,2-α]purin-10(3H)one	30% bypass		[47]
uracil	100% bypass		[72]
O^6-methylguanine	40% bypass		[48]
N^2-ethyl-2′-dG	block		[49]
1,N^2-ethenoguanine	block		[50]
cyclobutane pyrimidine dimer	block	+	[9, 51, 53, 75–79]
6-4 photoproducts	block	+	[78, 80]
4′-hydroxymethyl-4,5′,8-trimethylpsoralen			
1,1′-d(TA/AT)	block	+	[81, 82]
mono dT adduct	block	–	[81, 82]
cisplatin			
cis-1,2-d(GG)	~10–100% bypass[a]	+	[55, 79, 83]
cis-1,3-d(GTG)	up to 20% bypass	+	[55, 79, 83, 84]
benzo[a]pyrene diol epoxide-N^2-dG		±	[85, 86]
(+)-*trans*-anti	block		[87]
(–)-*trans*-anti	block		[87]
dibenzo[a,l]pyrene diol epoxide-N^2-dG			
(–)-*trans*-anti	4% bypass		this chapter
benzo[c]phenanthrene-N^6-dA		+	[88, 89]
(+)-*trans*-anti	9% bypass		[90]
(–)-*trans*-anti	16% bypass		[90]
2-aminofluorene-C8-dG	pause with bypass	–	[91, 92]
2-acetylaminofluorene-C8-dG	block		[91]
4-hydroxyequilenin			
dA	block		this chapter
dC	block		this chapter

a) Variable because of differences in sequence context and/or transcription system used.

Among the bacteriophage RNA polymerases, T7 RNA polymerase has been widely used to study the effects of DNA damages on transcription, although some results using SP6 and T3 RNA polymerases have been published. Bacteriophage RNA polymerases are able to recognize and initiate transcription specifically and exclusively from short promoter sequences, often without the need for any auxiliary factors [93]. The organization of a transcribing T7 RNA polymerase has been solved based on data from several crystal structures, including ternary DNA:RNA:RNA polymerase complexes [71, 94–96]. During transcription elongation, T7 RNA polymerase assumes a conformation that has been described as a cupped right hand. The elongating polymerase rapidly reads the template DNA, which crosses the palm of the cupped-hand structure with a relatively open active site [96].

In contrast to the single-subunit RNA polymerase found in bacteriophages, the model prokaryotic RNA polymerase from *E. coli* is a multisubunit enzyme. This polymerase has a five-subunit core that forms a constricted, tunnel-shaped catalytic site [97]. Prokaryotic RNA polymerases require an additional subunit, σ, for promoter-specific initiation of transcription [98, 99].

Eukaryotes have three RNA polymerases rather than one, named RNA Pols I, II, and III. Each is responsible for the synthesis of specific RNA molecules: RNA Pol I transcribes genes encoding rRNA, RNA Pol II transcribes genes that encode proteins and some other genes that encode regulatory RNA molecules, and RNA Pol III transcribes genes that encode rRNA, tRNA, and other small RNAs.

RNA Pol II has been the most widely studied eukaryotic RNA polymerase in relation to transcription elongation past DNA damage. The features of RNA Pol II transcription are based on the available crystal structures from the yeast enzyme complex [100, 101]. The eukaryotic and prokaryotic RNA polymerases share the conserved tunnel-forming core that the transcribed DNA passes through during transcription. For eukaryotic RNA Pol II, accessory proteins known as general transcription factors are needed for the core RNA polymerase to locate a promoter. An understanding of one of these, TFIIH, is of particular interest in the context of both transcription and DNA damage. TFIIH is a multisubunit complex, and two of the peptides within it are helicases encoded by the *XPB* and *XPD* genes [102]. TFIIH opens the double helix, both for transcription initiation and at sites of DNA damage for purposes of repair, creating overlap between transcription and DNA repair [103–105].

During transcription elongation, the RNA polymerase complex translocates, allowing for addition of the incoming complementary nucleotide, which enters via a ribonucleotide entry channel. Elongation factors such as elongin, which enhances the catalytic rate of RNA Pol II, and TFIIS, which stimulates RNA polymerase endonuclease activity, play important roles in modulating transcription (reviewed in [106, 107]). The numerous transcription elongation factors found in the cell are not always associated with the elongation complex; more than likely, each is recruited depending on when it is needed for a specific task. The role of elongation factors in TC-NER is an area of ongoing investigation.

17.5
RNA Polymerase Elongation Past DNA Damage

The data in Table 17.1 hint at the potential complexity by which a particular lesion in DNA can block transcription elongation. Strand breaks, small modifications such as oxidized or methylated bases, distorting cross-linked adducts like thymine dimers, and so-called "bulky" adducts permit or impede transcription bypass to varying degrees. Explanations for these results that simply attempt to correlate adduct size with the ability of a lesion to block transcription are not satisfactory. While Table 17.1 shows that bulky adducts can block transcription to a greater extent than smaller lesions, the latter can impede RNA synthesis nonetheless. Furthermore, experiments using different stereoisomeric configurations of adducts derived from polycyclic aromatic hydrocarbons (PAHs) show that the three-dimensional organization of the lesion influences the degree of bypass as does the specific position at which the RNA polymerase stalls [57–59, 87, 90]. In essence, most structural perturbations that affect the DNA template strand within a transcription unit compromise RNA polymerase progression and alter transcript integrity, and the relationship between blocks and pauses to transcript elongation and the quality of the RNA product are clearly inseparable.

17.5.1
Abasic Sites, Single-Strand Nicks, and Gaps

Abasic sites occur in DNA when the glycosidic bond between a 2'-deoxyribose and a nitrogenous base is cleaved, leaving the DNA backbone intact. These sites, which lack information for proper base-pairing during replication and transcription, arise from spontaneous hydrolysis, which occurs to a measurable extent for purines, or as an intermediate during the removal of some DNA adducts during repair. Transcription elongation past abasic sites has been documented and all RNA polymerases studied pause to varying extents, with abasic sites posing a very strong block to mammalian RNA Pol II elongation [35–37, 60]. When bacteriophage RNA polymerases, *E. coli* RNA polymerase, or RNA Pol II bypasses an abasic site, each favors incorporation of adenine into the growing transcript [35, 108]. Hence, there is a strong possibility that transcriptional mutagenesis would result at any abasic site that was generated from the removal of a base other than thymine.

DNA single-strand breaks often result from oxidative damage or during DNA repair. They can be simple nicks in the backbone with no associated loss of a base, or they can occur at sites where the nitrogenous base is absent; hence, this class of DNA lesions can differ with respect to the chemistry of the 3'- and 5'-termini at the break site, depending on how they are generated. The structural details of single-strand DNA breaks play important roles in how specific RNA polymerases behave at such sites.

Bacteriophage SP6 RNA polymerase is blocked at single-strand breaks that form at an abasic site where the 2'-deoxyribose is present on either the 3'- or 5'-side of the nick, but bacteriophage T7 RNA polymerase bypasses abasic strand breaks when the sugar is linked to the 5'-terminus as long as a 3'-OH is present [35, 38]. For both T7 and SP6 RNA polymerases, 5'- and 3'-phosphate groups located on both sides of the break pose very strong blocks to elongation. These results led to the idea that there is a distance constraint that must be met between adjacent bases on DNA templates for T7 RNA polymerase to translocate successfully; breaks in the DNA backbone that do not meet these constraints block elongation by that polymerase [35, 38]. Surprisingly, T7 RNA polymerase and SP6 RNA polymerase are capable of bypassing large gaps in the DNA template, up to 24 bases for T7 RNA polymerase and 19 bases for SP6 RNA polymerase, and the resulting transcripts do not contain the bases missing from the template. Furthermore, there is an inverse correlation between bypass efficiency and the length of the gap [39, 40].

Single-strand breaks block *E. coli* RNA polymerase elongation, regardless of the nature of the nicks' termini [35–37, 60]. However, this polymerase is capable of bypassing a single-base gap on the transcribed strand at low efficiency [39]. When the elongating complex bypasses such a gap, the bacterial RNA polymerase omits the base and makes a transcript that is one nucleotide shorter than that encoded in the control, undamaged template. In cells, such deletions in the resulting RNA could dramatically affect the transcript. For example, mRNA in which such an alteration occurred in the translated region would more than likely encode a nonsense protein.

Single-strand DNA breaks generated at abasic sites also pose strong blocks to RNA Pol II transcription elongation. Experiments with human RNA Pol II showed over 80% blockage at the various types of nicks in the template. There was no evidence for eventual bypass, with the exception of a one-base gap with a 5'-phosphate and 3'-OH that paused elongation and eventually permitted some translocation and continued elongation [37, 69]. These data indicate that persistent single-strand breaks pose fundamental problems for both replication and transcription in mammalian cells by blocking both DNA and RNA synthesis, respectively.

17.5.2
Oxidative DNA Damage

The ubiquitous ROS that arise during metabolism or from exposure to ionizing radiation can form a variety of lesions in DNA, including 8-oxoguanine (8-oxo-G), thymine glycol, 5-hydroxyuracil, and 5,6-dihydrouracil, among others (Figure 17.1). Oxidative DNA damage is implicated in carcinogenesis and neurodegenerative diseases such as Alzheimer's disease. In addition, there is strong evidence for the role of oxidative DNA damage in the aging process (see [109, 110] and references therein).

8-Oxo-G is found in DNA even in the absence of exposure to specific oxidizing agents, but it also is produced under oxidative stress conditions (Section 2.1). Bacteriophage RNA polymerases bypasses 8-oxo-G with high efficiency, misincorporating adenine and cytosine opposite the adduct [41, 42]. *E. coli* RNA polymerase and RNA Pol II bypass 8-oxo-G with only some detectable pausing or arrest [29,

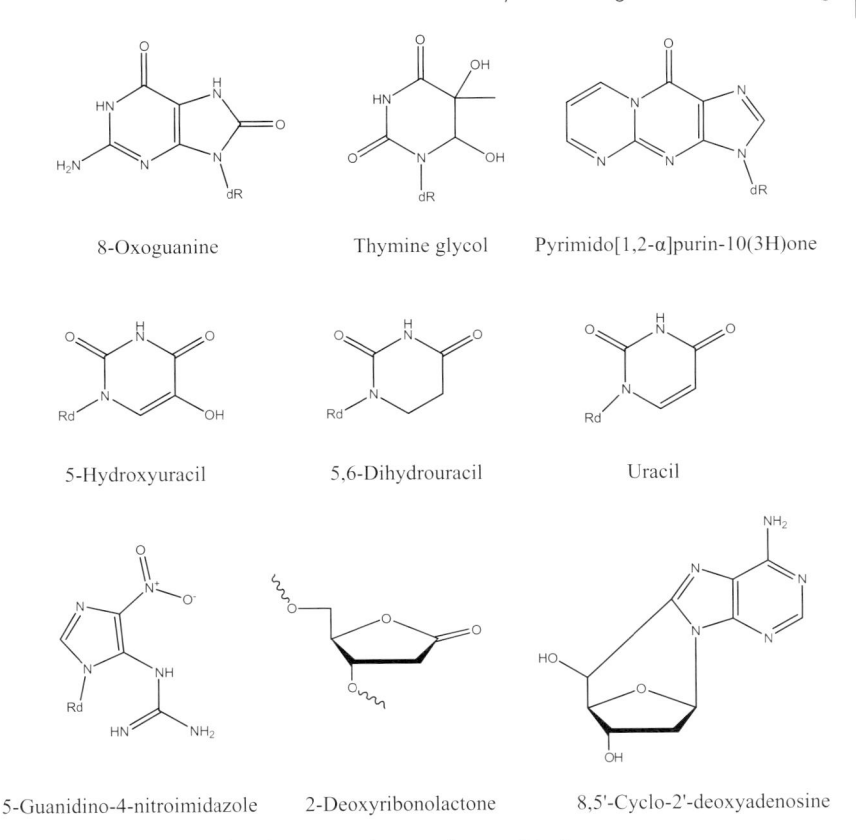

Figure 17.1 Structures of oxidative DNA lesions discussed in the text.

42, 60, 61, 71–73]. *E. coli* RNA polymerase incorporates either adenine or cytosine into nascent transcripts opposite the lesion, but RNA Pol II primarily incorporates cytosine, with about 8% of the transcripts incorporating adenine when transcription is initiated from an adenovirus promoter [29, 61]. These data suggest that 8-oxo-G may contribute to transcriptional mutagenesis. Importantly, elongation factors such as TFIIS enhance RNA Pol II bypass of 8-oxo-G, demonstrating their potential role in modulating RNA synthesis past DNA lesions as well as their ability to impact transcript integrity [29, 73].

Thymine glycol is a common thymine lesion found following exposure to oxidizing agents and this lesion poses an efficient block to T7 RNA polymerase progression, permitting only 50% bypass [43, 44]. However, the effects of thymine glycol on RNA Pol II transcription are variable, with some studies reporting a complete, efficient bypass of this lesion, while others report that it significantly blocks progression [29, 44, 69, 73]. These variations may be due to the assays used, with factors such as the presence or absence of a promoter, or the use of a defined transcription system rather than nuclear extracts, affecting the results. In addition, differences in stereochemistry of thymine glycol, which were not determined in these studies, could affect the extent of RNA polymerase bypass. The available information on sequencing RNA transcripts resulting from the bypass of thymine glycol during transcription past thymine glycol lesions is also limited. Transcripts generated by RNA Pol II exhibit no base misincorporation events, suggesting that thymine glycol does not contribute to transcriptional mutagenesis [29].

Transcription past 5-hydroxyuracil, which is derived from the oxidation of cytosine, has been examined using RNA Pol II [29]. The lesion poses a block or pause site, resulting in 40 and 60% bypass in a defined biochemical system and nuclear extract, respectively. Sequencing of the resulting full-length transcripts indicated that the lesion directed adenine incorporation opposite the 5-hydroxyuracil lesion in over 95% of the RNA transcripts, suggesting that the lesion should contribute significantly to transcriptional mutagenesis.

5,6-Dihydrouracil is an abundant, nondistorting DNA lesion produced by ionizing radiation under anoxic conditions. It is formed via the deamination of cytosine and the addition of hydrogen to C5 and C6 carbon atoms, thus eliminating the C5–C6 double bond producing a partially saturated and puckered pyrimidine. Both SP6 and T7 RNA polymerases transiently stall at this lesion, but eventual bypass does occur with the insertion of adenine opposite the lesion [4]. *E. coli* RNA polymerase bypasses 5,6-dihydrouracil with 80% efficiency and also tends to insert adenine [63]. Since 5,6-dihydrouracil is derived from cytosine, the incorporation of adenine opposite the lesion rather than guanine implies that the lesion should cause changes in the sequence of the transcribed RNA, thus contributing to transcriptional mutagenesis [4, 63].

Oxidation of abasic sites can also occur producing damage in DNA such as 2-deoxyribonolactone. Transcription elongation past 2-deoxyribonolactone using either bacteriophage T7 or T3 RNA polymerase is completely blocked, while the extent of stalling of the mammalian RNA Pol II is over 90% [45]. The addition of TFIIS results in transcript cleavage, suggesting that the arrested polymerase complexes are stable.

5′,8-cyclo-2′-deoxyadenosine is a free radical-induced lesion formed in considerable amounts in human cells after exposure to ionizing radiation and may be a factor in the neurodegenerative defects observed in XP patients [111]. Using host cell reactivation assays, it was shown that 5′,8-cyclo-2′-deoxyadenosine is a strong block to gene expression in human cells and that the lesion is repaired by NER [112]. Furthermore, transcription past this adduct by RNA Pol II results in a high level of transcriptional mutagenesis, with about 40% of the full-length transcripts containing misincorporations or nucleotide deletions at the site of the lesion [74]. Interestingly, the same study shows that the presence of CSB protein increases the probability of deletion transcript formation, which would be consistent with CSB facilitating the bypass of damage by RNA polymerase.

ROS present in cellular environments can also cause DNA damage (Chapters 2–5). For example, guanine radicals can react with NO_2 (Chapter 4) which generates ring-opened guanine lesions such as 5-guanidino-4-nitroimidazole [113]. This modified base is a ring-opened, conformationally flexible guanine lesion that stalls the elongation of T7 RNA polymerase, which is characterized by a rapid synthesis up to the site of the lesion, followed by slow bypass and the eventual formation of full-length transcripts. In some cases, the resulting RNA sequences contain the expected cytosine residue opposite the adduct, but adenine and guanine incorporation and single-base deletions are also observed. Computer modeling suggests that the lesion's torsional flexibility permits it to adopt conformations in which the ring-opened purine collides with amino acids in the phage polymerase active site, causing the complex to pause during elongation until the structure assumes a conformation in which steric hindrance is relieved and bypass can occur [46]. Furthermore, the modeling studies suggest that the base incorporated is a function of hydrogen bonding between the 5-guanidino-4-nitroimidazole and the incoming nucleotide, with the formation of three hydrogen bonds favoring CTP, two favoring ATP, and one favoring GTP incorporation [46].

5-Guanidino-4-nitroimidazole poses a strong block to RNA Pol II. In fact, the lesion completely blocks bypass during single rounds of transcription and permits about 14% bypass during multiple rounds [46]. Transcription stopped immediately prior to the adduct, and computer modeling suggests that the majority of the *syn* and *anti* conformations that the lesion can produce steric collisions between it and amino acids in the transcription complex active site. In fact, only one collision-free, *anti* conformation could be found, and it permitted three hydrogen bonds to form between the lesion and an incoming CTP, which offers an explanation for why full-length transcripts produced contained the cytosine at the position opposite the ring-opened purine.

Pyrimido[1,2-α]purin-10(*3H*)-one forms in DNA when malondialdehyde, a genotoxic byproduct of lipid peroxidation reacts with deoxyguanosine in DNA. This lesion is highly mutagenic in bacteria and mammalian cells, inducing both frameshifts and base substitutions [114, 115]. Pyrimido[1,2-α]purin-10(*3H*)-one was one of the first endogenous DNA lesions found to be repaired by NER (Chapters 5 and 9) [116]. Pyrimido[1,2-α]purin-10(*3H*)-one imposes strong blocks to translocation by both T7 RNA polymerase and mammalian RNA Pol II. Furthermore, addition of TFIIS to reactions with RNA Pol II results in transcript cleavage, suggesting that the arrested polymerase complexes are stable. This stability of the ternary complex suggests that pyrimido[1,2-α]purin-10(*3H*)-one should be a target for TC-NER [47].

Uracil can also be produced in DNA, primarily through deamination of cytosine. While this is not an oxidation, the impact of uracil on transcription is important and needs to be mentioned. Transcription past uracil is not associated with pauses or blocks for phage, prokaryotic, or eukaryotic RNA polymerases [35, 61, 72]. However, the lesion induces transcriptional mutagenesis as shown by Viswanathan *et al.* (1999) [5]. In nondividing *E. coli*, these investigators demonstrated that uracil in DNA could encode for adenine at the level of transcription, resulting in alterations to a reporter protein. Furthermore, they showed that the transcriptional mutagenesis of uracil inversely correlates with the ability of the cell to remove the base from DNA via BER. This salient piece of work demonstrated the generation of mutant proteins at the level of transcription.

Damaged bases in DNA that arise from ROS or by the deamination of cytosine to uracil are usually removed by BER; however, there is no evidence for the existence of TC-BER that would be analogous to TC-NER [11, 32]. Some oxidized bases may be repaired by TC-NER when they interfere with transcription elongation. In any case, oxidized bases that alter base incorporation events in nascent RNA could contribute significantly to the pool of altered proteins in cells. In fact, 8-oxo-G has been shown to be subject to TC-NER in *E. coli* and to induce altered proteins via transcriptional mutagenesis [62]. Furthermore, in a recent publication, Saxowsky *et al.* report that 8-oxo-G induces a change in the activity of the Ras protein in mammalian cells via transcriptional mutagenesis and that downstream phosphorylation events are dependent on Ras [117].

17.5.3
Alkylated Bases in DNA

DNA is susceptible to alkylation damage by exogenous compounds, like methyl nitrosourea and ethyl methanesulfonate, and endogenous compounds, like *S*-adenosylmethionine. Alkylation damage to DNA is often associated with tumor formation, which correlates with the high level of mutagenicity associated with many of the alkylating chemical agents or their metabolites [118].

In the case of methylating agents, the two most abundant lesions formed are 7-methylguanine and 3-methyladenine. 7-Methylguanine is considered to be a relatively innocuous adduct, but 3-methyladenine is cytotoxic [119]. Each of these *N*-methylpurines has an inherently unstable glycosidic bond and depurination at these sites in DNA occurs quite readily, resulting in the formation of noninstructional abasic sites (see Section 17.5.1) [120, 121]. O^6-Methylguanine is also formed to a significant extent by many methylating agents, particularly those that tend to be highly carcinogenic (Figure 17.2). It is stable in DNA and mutagenic during replication, producing GC → AT transitions. T7 RNA polymerase transcription past O^6-methylguanine results primarily in full-length transcripts, with a minimal pause occurring [48]. About 95% of the full-length transcripts contain a pyrimidine opposite the site of template damage, with an equal distribution between uracil and cytosine. The remaining full-length transcripts contain adenine opposite the

O^6-Methylguanine   1,N^2-Ethenoguanine N^2-Ethyl-2'-deoxyguanine

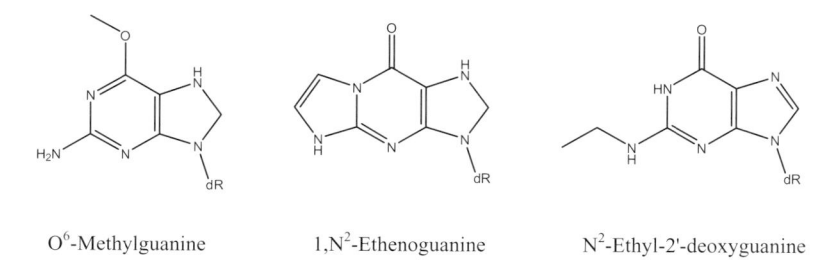

Figure 17.2 Structures of alkylated DNA lesions discussed in the text.

lesion. Computer modeling offered explanations for these results. When the adduct is in an *anti* conformation and the methyl group adopts a *proximal* conformation relative to the N1 atom, the lesion permitted incoming CTP to form Watson–Crick base pairs or UTP to form pseudo-Watson–Crick base pairs, thus extending the nascent transcript with the correct base in the former case or the incorrect base in the latter. The minimal amounts of ATP incorporated could be explained by the lesion adopting a *syn* conformation with the methyl group taking on a *distal* orientation relative to the N1 atom [48].

E. coli RNA polymerase also bypasses O^6-methylguanine with high efficiency, often incorporating uracil opposite the lesion [61]. No structural models have been offered to explain *E. coli* RNA polymerase behavior at the adduct.

For RNA Pol II, O^6-methylguanine produces both truncated and full-length transcripts, both during single and multiple rounds of transcriptions, suggesting that the lesion poses a block or pause during elongation, with full-length transcripts accounting for approximately 40% of the total [48]. Sequencing data show that the truncated transcripts are of the correct sequence and that the polymerase pauses immediately prior to the lesion. Full-length RNA contains the correct cytosine in about 75% of the transcripts while the remaining 25% contain uracil. Computer modeling suggests that the conformation of the O^6-methylguanine is critical for bypass [48]. Favorable adduct orientations have been found that can allow the enzyme's active site to accommodate the lesion with no unfavorable collisions between the O^6-methylguanine and any of the amino acids. Computer modeling shows that rotation of the methyl group can result in unfavorable interactions, causing the polymerase to stall until a permissive conformation is adopted. Similar to T7 RNA polymerase, RNA Pol II can incorporate either CTP or UTP when the adduct is in the *anti-proximal* conformation.

Ethylation of DNA can also occur. N^2-Ethyl-2′-deoxyguanine is among the most well-studied DNA lesions formed from the reaction of acetaldehyde with DNA (Figure 17.2). The most significant source of human exposure to acetaldehyde is via the metabolism of ethanol and consequently elevated levels of N^2-ethyl-2′-deoxyguanine have been observed after alcohol consumption [122]. N^2-Ethyl-2′-deoxyguanine poses a strong block to all RNA polymerases studied, including T7 RNA polymerase, *E. coli* RNA polymerase, and yeast and mammalian RNA Pol II [49].

Vinyl chloride and byproducts of lipid peroxidation can react with purines in DNA, forming stable three-ring molecules (Figure 17.2). Transcription past one of these, $1,N^2$-ethenoguanine, has been studied. The lesion poses a complete block to transcription elongation by human RNA Pol II, but permits T7 RNA polymerase bypass of about 40%, where the full-length transcripts often contained a purine or a single-base deletion opposite the adduct [50]. In essence, the exocyclic ring creates a moiety that has no hydrogen-bonding sites for typical Watson–Crick base-pairing, but its size allows it to enter easily into the active site of RNA polymerases. Hence, the lesion prevents translocation due to its inability to interact with an incoming nucleotide.

17.5.4
Intrastrand and Interstrand DNA Cross-links

UV radiation and some chemical agents damage DNA by creating intrastrand or interstrand DNA cross-links (Figure 17.3). Intrastrand cross-links, which occur when covalent bonds form between two bases on the same strand of DNA, tend to distort the double helix. Interstrand cross-links, which occur when covalent bonds form between two bases on opposite strands of the double helix, are particularly problematic since they prevent the melting at the point of the cross-link and pose absolute blocks to replication and transcription.

UV radiation induces DNA damage in part by causing adjacent pyrimidines in the same strand to react with one another, forming intrastrand cross-links. The major DNA lesions produced are CPDs and 6-4 photoproducts [123]. Transcription bypass has been investigated at site-specific thymine–thymine (TT) CPDs, which are known to be the major DNA photoproducts formed *in vivo*, and at 6-4 photoproducts, which are also helix-distorting lesions.

Studies using T7 RNA polymerase show that TT dimers block transcription, but to different extents depending on the specific isomer [51–53]. For example, TT CPDs in the *cis-syn* configuration pose less of a block to bypass than do the TT 6-4 photoproducts. Furthermore, either lesion only interferes with transcription elongation when it is present on the transcribed strand; lesions on the nontranscribed

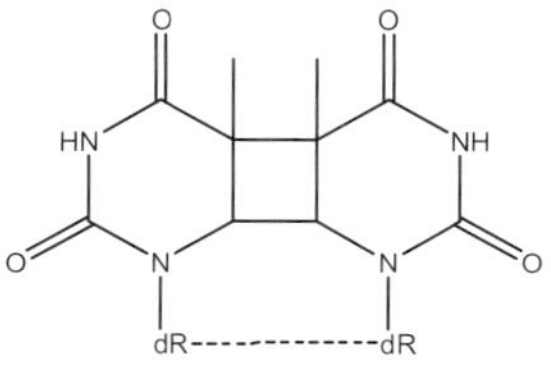

Thymidine cyclobutane dimer

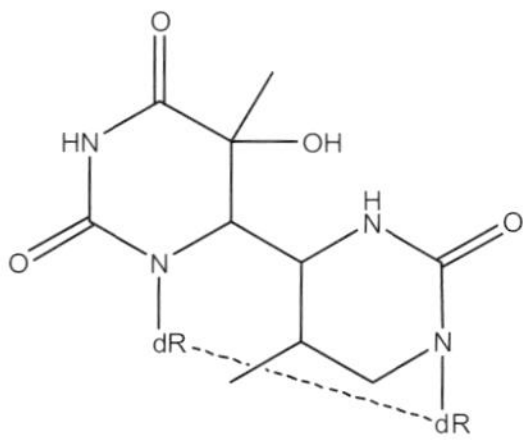

Thymidine 6-4 dimer

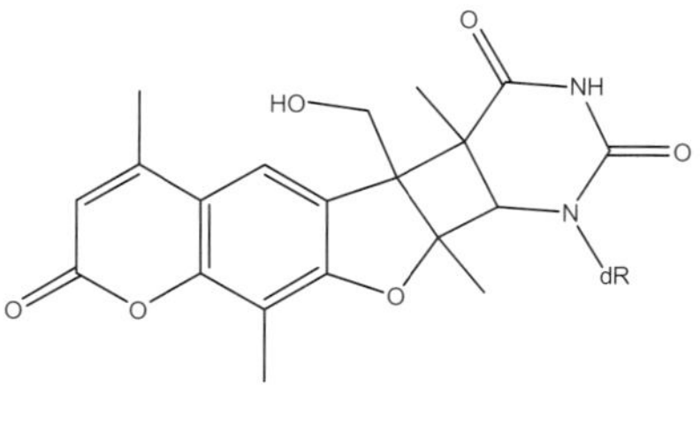

Psoralen thymidine monoadduct

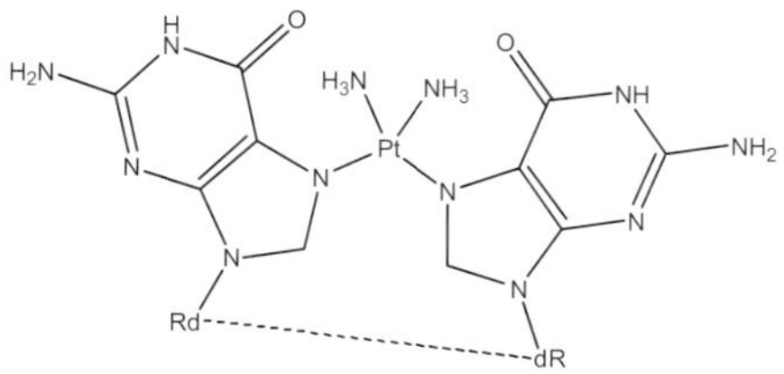

Cisplatin 1,2-d(GpG) crosslink

Figure 17.3 Structures of cross-linking DNA lesions discussed in the text.

strand do not impede RNA synthesis. These results are among those that show the importance of how a DNA lesion's specific stereoisomeric configuration can affect transcription.

E. coli RNA polymerase elongation is completely blocked by a TT dimer on the transcribed strand of DNA [64]. Approximately 90% of the elongating transcription complexes stall at the 3'-T; the other 10% stall at the 5'-T. As with phage RNA polymerase, TT dimers present on the nontranscribed strand exert no effect on transcription elongation. Once an RNA polymerase molecule has stalled at a TT CPD, the bacterial protein Mfd, which is the transcription–DNA repair coupling factor in *E. coli*, displaces the polymerase, permitting repair to occur through the UvrABC excinuclease [14]. As mentioned earlier, the initial observations of TC-NER were made by studying the repair of damage induced by UV radiation. The fact that TT dimers impede transcription elongation is consistent with the model that transcription-blocking DNA damage is subject to TC-NER.

Mammalian RNA Pol II transcription is completely blocked by TT dimers present on the transcribed strand, both when the lesion is a CPD or 6-4 photoproduct [51, 53, 75–78]. As with other RNA polymerases, these lesions do not affect RNA Pol II elongation when they are located on the nontranscribed strand. The elongation factor TFIIS modulates stalling by causing the polymerase to reverse translocate, inducing the polymerase to cleave the transcript [75, 77]. While this backtracking event may allow for repair of the DNA template or give the polymerase another opportunity to elongate past the lesion, there is evidence that TC-NER can remove a lesion even in the presence of a stalled RNA polymerase complex [51, 84, 124].

Kwei *et al.* reported that RNA Pol II stalls at TT CPDs and 6-4 photoproducts after inserting one or two nucleotides opposite the lesion [78]. For TT dimers, adenine is first inserted opposite the 3'-T in the dimer, but the next base is almost always an incorrect uracil. In contrast, 6-4 photoproducts direct incorporation of adenine at both positions. Experiments describing a crystal structure of yeast RNA Pol II attempting to elongate past a TT dimer provide a mechanistic reason for the observed base incorporation events [125]. The yeast RNA Pol II active site can accommodate the dimer's 3'-T in the same way as an undamaged template, but the template is bent at the dimer's 5'-T, which takes on a wobble orientation that can form two base pairs with an incoming UTP. Following incorporation of UMP, the resulting mismatch in the DNA:RNA poses a strong barrier to translocation; hence, dimer-directed misincorporation of uracil opposite the 5'-T leads to an absolute block to transcription elongation. When the polymerase is artificially provided with an RNA extension template that loads into the polymerase with a TT dimer that is correctly paired with two adenines, the polymerase can extend past the lesion to make full-length RNA. This suggests further that it is the mispair that prevents transcription elongation [125].

Psoralen is among the agents that forms interstrand DNA cross-links. It is a photosensitizing agent found in plants and has been used in a therapeutic regimen along with UV-A to treat psoriasis and other skin disorders. Psoralen can intercalate into DNA, absorb UV-A, and subsequently react with DNA, principally at thymine, to form a monoadduct. Subsequent photoactivation of the monoadduct via absorption of additional UV radiation can result in an additional reaction with a pyrimidine in the complementary DNA strand, forming a covalent cross-link that thwarts replication and transcription (reviewed in [126]).

Psoralen interstrand cross-links pose strong blocks to T7 RNA polymerase, *E. coli* RNA polymerase, and RNA Pol II [54, 65, 81], which is to be expected since the double helix cannot unwind in the vicinity of the adduct. Psoralen monoadducts on the transcribed strand of DNA pose blocks to transcription as well. Historically, psoralen adducts are of great interest since they were among the first chemically induced lesions to be investigated for TC-NER. Interestingly, psoralen interstrand cross-links are subject to TC-NER, but the monoadducts are not [82].

Intrastrand and interstrand DNA cross-links are also formed in DNA by another important class of chemotherapeutic drugs – the cisplatins. The majority of the adducts formed *in vivo* and *in vitro* are 1,2-intrastrand d(GpG) and d(ApG) cross-links that represent 80–90% of the bound platinum. In addition 1,3-intrastrand d(GpTpG) cross-links and a small percentage of interstrand cross-links and monofunctional adducts are formed [127].

Cisplatin 1,2-intrastrand d(GpG) cross-links and 1,3-intrastrand d(GpTpG) cross-links pose strong blocks to transcription by T7 RNA polymerase, *E. coli* RNA polymerase, and RNA Pol II [55, 56, 66–68, 83, 84]. Similar results have been reported using bacteriophage T3 RNA polymerase [83]. Furthermore, TFIIS induces transcript shortening for stalled RNA Pol II complexes *in vitro* [55], but in cells, the stalled RNA Pol II complexes are subject to ubiquitylation-mediated degradation [128]. In agreement with their distorting properties and ability to block transcription, cisplatin-induced DNA adducts are repaired by GG-NER and TC-NER [79, 129, 130].

The mechanism of RNA Pol II stalling at cisplatin lesions has begun to emerge with the use of crystallography and RNA extension assays using d(GpG) intrastrand cross-links. In these experiments, the polymerase stalls following addition of AMP opposite the 3′-G in the cross-link. However, the addition of AMP occurs opposite noninstructional, abasic sites, leading to the notion that the cisplatin intrastrand cross-link might not enter the active site and that the incorporation of AMP occurs without a template [108].

Clearly interstrand and intrastrand cross-links pose significant hurdles to RNA polymerase progression. These lesions also support several of the basic observations about transcription past sites of DNA damage, particularly the notion that the specific bases incorporated into the nascent RNA impact the ability of the complex to translocate.

17.5.5
"Bulky" DNA Adducts

A wide range of chemicals, especially those found among environmental pollutants, can react with DNA or be metabolized to reactive intermediates, producing so-called bulky chemical DNA damage. While the precise definition of bulky is

17.5 RNA Polymerase Elongation Past DNA Damage | 417

inherently unclear, the compounds that tend to fall in this category often contain ring systems. Among the chemicals that produce bulky DNA damage are the PAHs, aromatic amines, and intermediates in the biological metabolism of steroids (Figure 17.4).

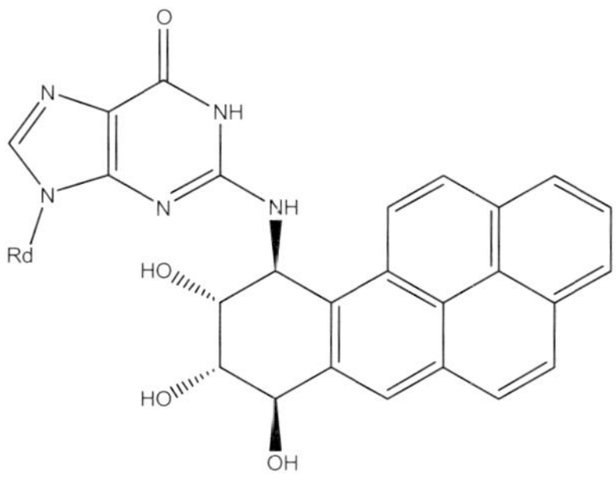

trans-anti-N^2-BPDE-dG

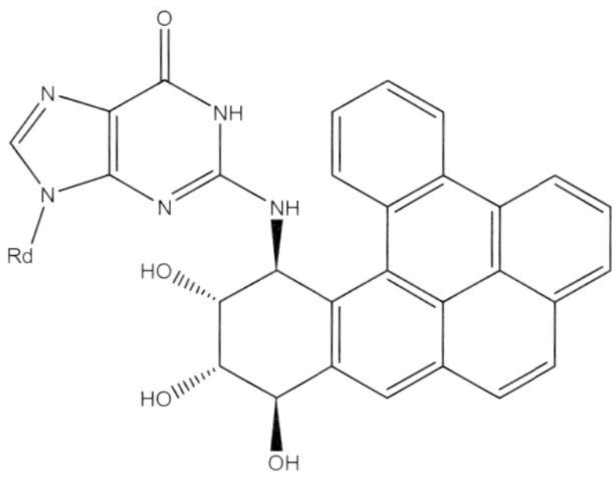

trans-anti-N^2-DBPDE-dG

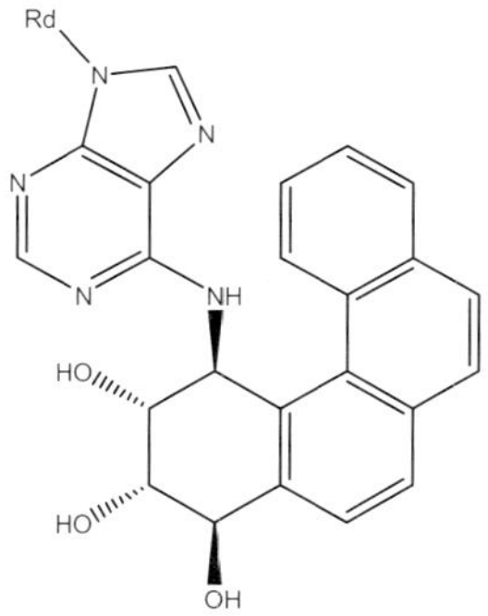

trans-anti-N^6-BPhDE-dA

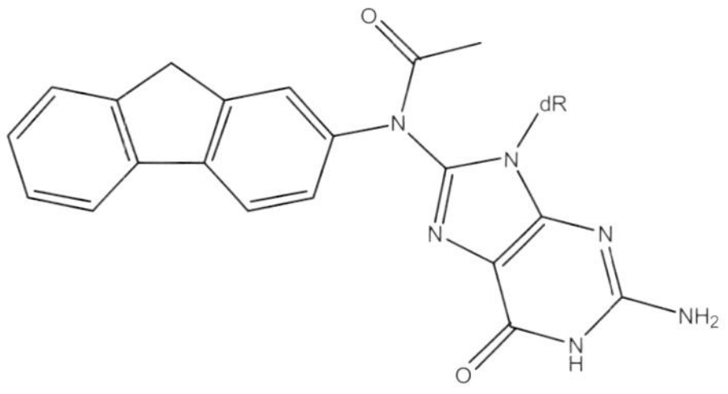

N-(deoxyguanosin-8-yl)-2-acetylaminofluorene

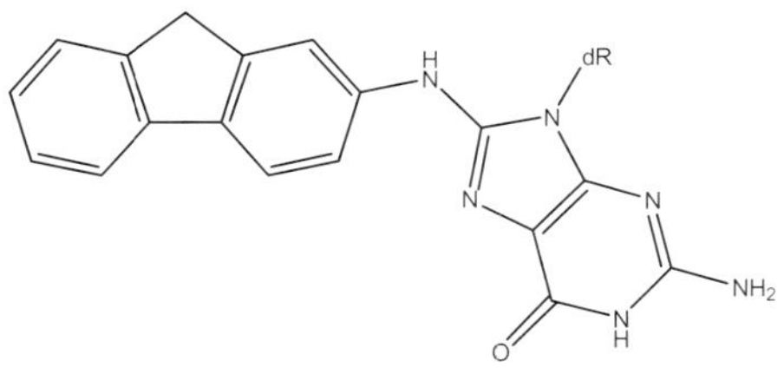

N-(deoxyguanosin-8-yl)-2-aminofluorene

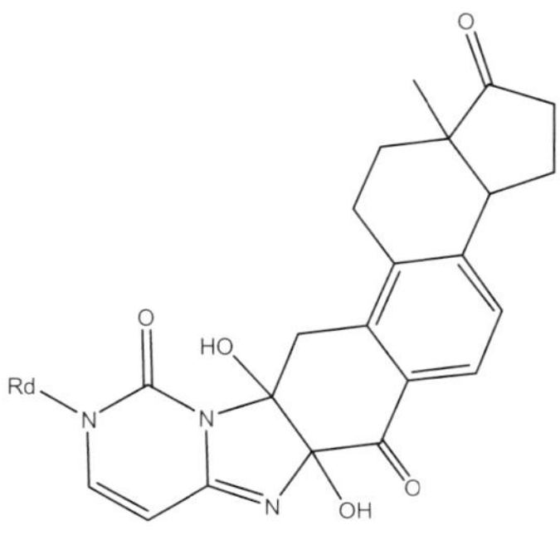

4-Hydroxyequilenin cytidine adduct

Figure 17.4 Structures of bulky DNA lesions discussed in the text.

PAHs are ubiquitous environmental pollutants known to be mutagenic and carcinogenic in mammalian cells [131, 132]. PAHs require metabolic activation that results in diol epoxide formation via reactions that are catalyzed by epoxide hydrolase and the CYP450 family of enzymes [133]. Diol epoxides are highly reactive, particularly toward purines in DNA, forming guanine and adenine adducts that exist as *cis* and *trans* stereoisomers [134]. Transcription past adducts derived from diol epoxides of benzo[*a*]pyrene (benzo[*a*]pyrene diol epoxide (B[*a*]PDE)), benzo[*c*]phenanthrene (benzo[*c*]phenanthrene diol epoxide (B[*c*]PhDE)), and dibenzo[*a,l*]pyrene (dibenzo[*a,l*]pyrene diol epoxide (B[*a,l*]PDE)) has been studied.

T7 RNA polymerase and RNA Pol II transcription past *anti*-N^2-B[*a*]PDE-dG has been investigated. The work provides evidence that such lesions impede elongation in a stereochemical-dependent manner. For T7 RNA polymerase, (+)- and (−)-*trans-anti*-N^2-B[*a*]PDE-dG adducts block transcription to a much greater extent than do the stereoisomeric (+)- and (−)-*cis* adducts. However, in each of these adducts, a significant ladder of truncated transcripts is observed, suggesting that there is no single, discrete stopping point or that the polymerase can slowly bypass the adduct [57]. In all cases, the shortened, truncated transcripts end with an incorrect base, usually adenine, while full-length transcripts exhibit no misincorporations [135]. B[*a*]PDE adducts linked to the N^6-position of adenine also interfere with T7 RNA polymerase transcription *in vitro*, with (−)-*trans-anti*-N^6-B[*a*]PDE-dA posing less of a block than (+)-*trans-anti*-N^6-B[*a*]PDE-dA [58]. In all cases, base misincorporation events are observed in full-length transcripts, including insertion of adenine or guanine opposite the lesion, or deletion of bases from the transcript. In contrast to the results for T7 RNA polymerase, (+)- or (−)-*trans-anti*-N^2-B[*a*]PDE-dG adducts completely block the progression of RNA Pol II in HeLa nuclear extracts [87].

Transcription past (+)- or (−)-*trans-anti*-N^6-B[*c*]PhDE-dA adducts by T7 RNA polymerase and RNA Pol II has also been examined. T7 RNA polymerase is able to bypass each stereoisomer when either is located on the template strand, and a significant ladder of truncated transcripts is observed [59]. Furthermore, the (−)-*trans-anti*-N^6-B[*c*]PhDE-dA poses a stronger block to the phage polymerase when compared to the (+)-*trans-anti*-N^6-B[*c*]PhDE-dA. In addition, the terminal base incorporated at the 3′-end opposite the adduct is more often a guanine than the expected uracil. In contrast, transcription past the two N^6-B[*c*]PhDE-dA stereoisomers by RNA Pol II resulted in a small percentage of read-through, with less than 20% bypass [90]. While no data are available for the base incorporated at the 3′-end of the truncated N^6-B[*c*]PhDE-dA transcripts, all truncated RNA molecules appear to be the same length.

T7 RNA polymerase transcription past (−)-*trans-anti*-N^2-DB[*a,l*]PDE-dG results in partial blockage with about 25% bypass (Figure 17.5). While full-length transcripts exhibit no base misincorporations, the majority of truncated transcripts either end opposite the lesion with an incorrect base, usually an adenine, or one base before the lesion. In contrast, (−)-*trans-anti*-N^2-DB[*a,l*]PDE-dG posed a strong block to human RNA Pol II, permitting little read-through at about 4% (Figure 17.5). In the case of N^2-DB[*a,l*]PDE-dG, the majority of the shortened transcripts

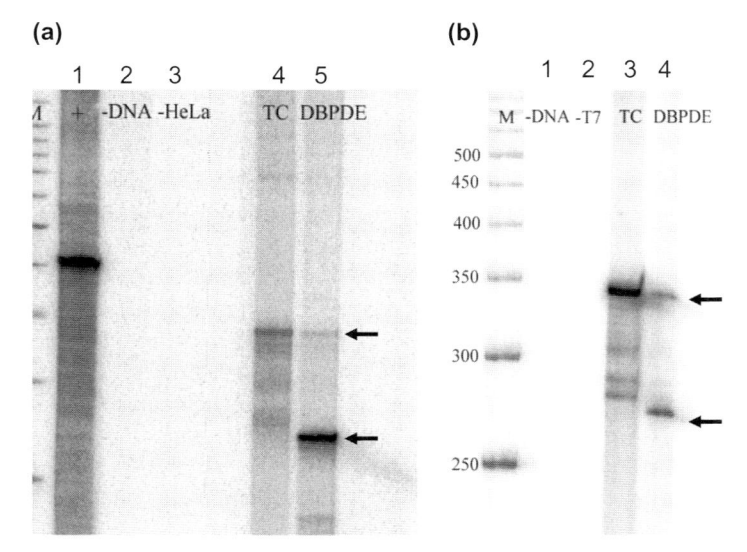

Figure 17.5 Transcription past DB[a,l] PDE-dG adducts. Template preparation and reaction conditions have been published [136]. For RNA Pol II, these were performed as follows. Briefly, HeLa nuclear extracts, DNA template, and 10 mM each of ATP, GTP, and UTP, and 0.4 µM [α-^{32}P]CTP (25 Ci/mmol) were added to the transcription buffer. The reaction was incubated at 30 °C for 60 min and quenched by the addition of HeLa extract stop solution. RNA was extracted, precipitated, isolated, and resolved using 7% denaturing polyacrylamide gel electrophoresis. Image Quant software (Molecular Dynamics) was used to visualize and quantify the radiolabeled transcripts. For T7 RNA polymerase, template and enzyme were mixed in transcription-optimized buffer containing 2.5 mM each of ATP, GTP, and UTP, and 200 µM [α-^{32}P]CTP (50 Ci/mmol). The reaction was incubated at 37 °C for 1 h. The same procedures for the extraction, precipitation, and resolving the RNA were employed as with RNA Pol II. The upper arrows indicate the full-length transcript; the lower arrows indicate the truncated transcripts. RNA Pol II A: 1, unmodified DNA control template supplied with the HeLa extract; 2, no template; 3, no RNA Pol II; 4, unmodified DNA template; 5, DB[a,l]PDE-dG DNA. T7 RNA polymerase B: 1, no template; 2, no T7 RNA polymerase; 3, unmodified DNA template; 4, DB[a,l]PDE-dG adduct.

end one to three bases before the lesion and full-length transcripts exhibit no base misincorporations. Initial sequencing of the truncated RNA suggests that RNA Pol II inserts either adenine or cytosine opposite the adduct.

The differences between the behavior of T7 RNA polymerase and RNA Pol II at PAH-derived DNA adducts can be explained by the structural features of the two different enzymes and their active sites as briefly described above. T7 RNA polymerase has a more open, flexible active site when compared to the more tunnel-like, constricted active site of RNA Pol II. Computer modeling using the crystal structures of T7 RNA polymerase and yeast RNA Pol II suggest that the open active site of the phage polymerase permits larger bulky adducts to enter the site, whereas PAH-derived lesions fit less readily into the eukaryotic polymerase active site with significant collisions between the aromatic moieties and the amino acid side-chains. However, there is another important model that has been generated and it is consistent with models obtained for transcription past O^6-methylguanine – a given adduct can potentially adopt different conformations in the active site, some yielding to bypass and others to blocking or stalling. For example, molecular modeling of the N^6-B[c]PhDE-dA lesions on the transcribed strand suggest that each diastereomer can exist in two orientations within the yeast RNA Pol II active site – one that permits nucleotide incorporation and subsequent transcription elongation, and another that blocks the RNA polymerase's nucleotide entry channel, thus preventing base incorporation with concomitant stalling [90]. For (+)- and (−)-*trans-anti*-N^2-B[a]PDE-dG, computer modeling suggests that the pyrenyl group of (+)- and (−)-*trans-anti*-N^2-B[a]PDE-dG adopts different conformations, which may explain the observed differences. The (+)-*trans-anti*-N^2-B[a]PDE-dG adduct predominantly blocks transcription before a base is incorporated at the adduct because the pyrenyl group is positioned in the minor groove, perhaps preventing the polymerase from clamping the damaged guanine and the incoming nucleotide. However, the (−)-adduct adopts a carcinogen/base-stacked conformation that may permit incorporation of a base opposite to it in the transcript [87]. The structural features of DB[a,l]PDE adducts in a DNA duplex or RNA:DNA heteroduplex have not been studied in detail, but the size of the lesion could easily account for the observed limited bypass during transcription. In all, the data from studying PAH-DNA adducts and their effects on transcription further support the principle that bypass of DNA lesions is influenced by stereochemistry and size of the lesion, carcinogen–base linkage position, conformational preference, and the RNA polymerase under investigation.

The synthetic aromatic amines N^2-aminofluorene and N^2-acetylaminofluorene were originally produced as insecticides, but their observed carcinogenicity precluded their use in the environment. Nonetheless, their tumorigenic properties led to their use as model compounds to investigate metabolic activation and its role in chemical mutagenesis. N^2-Aminofluorene and N^2-acetylaminofluorene are metabolized to reactive intermediates that produce mainly N-(deoxyguanosin-8-yl)-2-aminofluorene and N-(deoxyguanosin-8-yl)-2-acetylaminofluorene, respectively [137]. The fluorene group of N^2-aminofluorene-derived adducts can remain outside the double helix or it may insert into the helix. In contrast, the fluorene of N^2-acetylaminofluorene-derived lesions has a much greater propensity to insert itself between adjacent base pairs within the double helix; hence, this lesion is more distorting – a consequence of the presence of the bulky acetyl group [138–142]. The structural properties of these DNA adducts are detailed in Chapter 10. N-(Deoxyguanosin-8-yl)-2-aminofluorene or N-(deoxyguanosin-8-yl)-2-acetylaminofluorene adducts on the transcribed strand of DNA impede transcription by T7 RNA polymerase and RNA Pol II. For the phage RNA polymerase, N-(deoxyguanosin-8-yl)-2-acetylaminofluorene prematurely terminates more than half of the transcripts, while N-(deoxyguanosin-8-yl)-2-aminofluorene blocks progression to a slightly lesser extent of about one-third [41]. Furthermore, the time-course data indicate that the elongating polymerase quickly dissociates once the lesion is reached, and lesion bypass does not lead to base misincorporation during T7 RNA polymerase transcription. For mammalian RNA Pol II, *N*-(deoxyguanosin-8-yl)-2-aminofluorene poses a weak pause site with some bypass; in contrast, the *N*-(deoxyguanosin-8-yl)-2-acetylaminofluorene completely blocks transcriptional elongation [91]. Interestingly, TFIIS does not induce transcript cleavage by RNA Pol II at these adducts, which is consistent with rapid dissociation of the ternary complex and release of the truncated RNA. It is important to note that N^2-aminofluorene adducts do not appear to be subject to TC-NER even though they block RNA polymerase progression to a reasonable extent [92]. This further supports the model in which a stable, stalled RNA polymerase elongation complex is essential for inducing TC-NER.

As mentioned earlier, steroid metabolism can generate reactive species that can damage DNA. In fact, the two principal estrogen molecules used in hormone therapy for women are equilin and equilenin, which are harvested from the urine of pregnant mares [143]. Both are structurally very similar to the endogenous human estrogen estradiol. Humans metabolize equine estrogens to catechols that are subsequently oxidized to produce *o*-quinones that are reactive toward all DNA bases other bases than thymine [144]. The fact that these metabolites can damage DNA led to the hypothesis that the lesions formed may well interfere with transcription. This has been tested using adenine and cytosine adducts derived from 4-hydroxyequilenin. T7 RNA polymerase bypasses 4-hydroxyequilenin-dA and 4-hydroxyequilenin-dC 8 and 2% of the time, respectively, making them among the strongest blocks to a phage polymerase. In contrast, these lesions pose a complete block to human RNA Pol II with no detectable bypass (Figure 17.6). These data support the notion that these lesions should not contribute to transcriptional mutagenesis and should be subject to TC-NER.

17.6
Conclusions

Lesions in transcription units pose three fundamental problems for the cell: (i) they can permit an RNA polymerase to bypass with limited or no detectable pausing, but alter the nascent RNA, (ii) they can pause or transiently stall elongation, slowing transcription, or (iii) they can pose absolute blocks to elongation. In all these cases, the role of elongation factors, DNA repair, and documented or putative transcription–DNA repair coupling factors may well modulate these outcomes. Figure 17.7 illustrates these possibilities.

During transcription elongation, the DNA duplex is unwound, forming a transcription bubble in which RNA polymerase has access to the template strand. RNA Pol II reads the transcribed strand 3′ → 5′, with rates up to 70 bases/s [145]. The 7- to 9-bp DNA–RNA hybrid within the active site and the nascent RNA leaving the polymerase are in red (Figure 17.7a). Some lesions in DNA, which are illustrated as black triangles, are bypassed without detectable pausing; this might lead

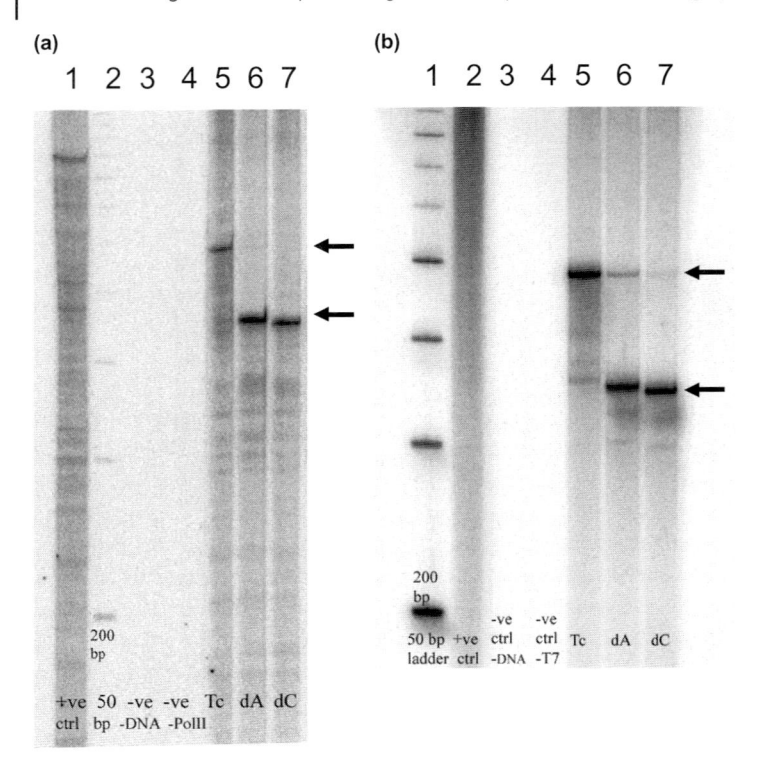

Figure 17.6 Transcription past 4-hydroxyequilenin-dA and 4-hydroxyequilenin-dC adducts. Transcription reactions were carried out as described in the legend for Figure 17.5. The upper arrows indicate the full-length transcript; the lower arrows indicate the truncated transcripts. RNA Pol II A: 1, size markers; 2, control template supplied with the HeLa extract; 3, no DNA; 4, no RNA Pol II; 5, unmodified DNA template; 6, 4-hydroxyequilenin-dA; 7, 4-hydroxyequilenin-dC. T7 RNA polymerase B: 1, size markers; 2, control template; 3, no DNA; 4, no T7 RNA polymerase; 5, unmodified template; 6, 4-hydroxyequilenin-dA; 7, 4-hydroxyequilenin-dC.

to a misincorporation event, resulting in the production of altered transcripts (transcriptional mutagenesis) [4] (Figure 17.7b). As Table 17.1 illustrates, there are many *in vitro* examples of RNA polymerases' ability to bypass lesions in a DNA template, often at the expense of fidelity and resulting in transcriptional mutagenesis (Table 17.2). The *in vivo* relevance of these studies is only beginning to emerge [5, 74]. The recent report from the laboratory of Paul Doetsch showing that damage in a transcription unit encoding Ras can alter the protein's structure due to changes directly made to the RNA during transcription past the lesion supports the physiological importance of the phenomenon and the need for further research in this area [117].

In contrast to lesion bypass – and depending on the nature of the lesion, its sequence context, and the structural properties of the polymerase active site –

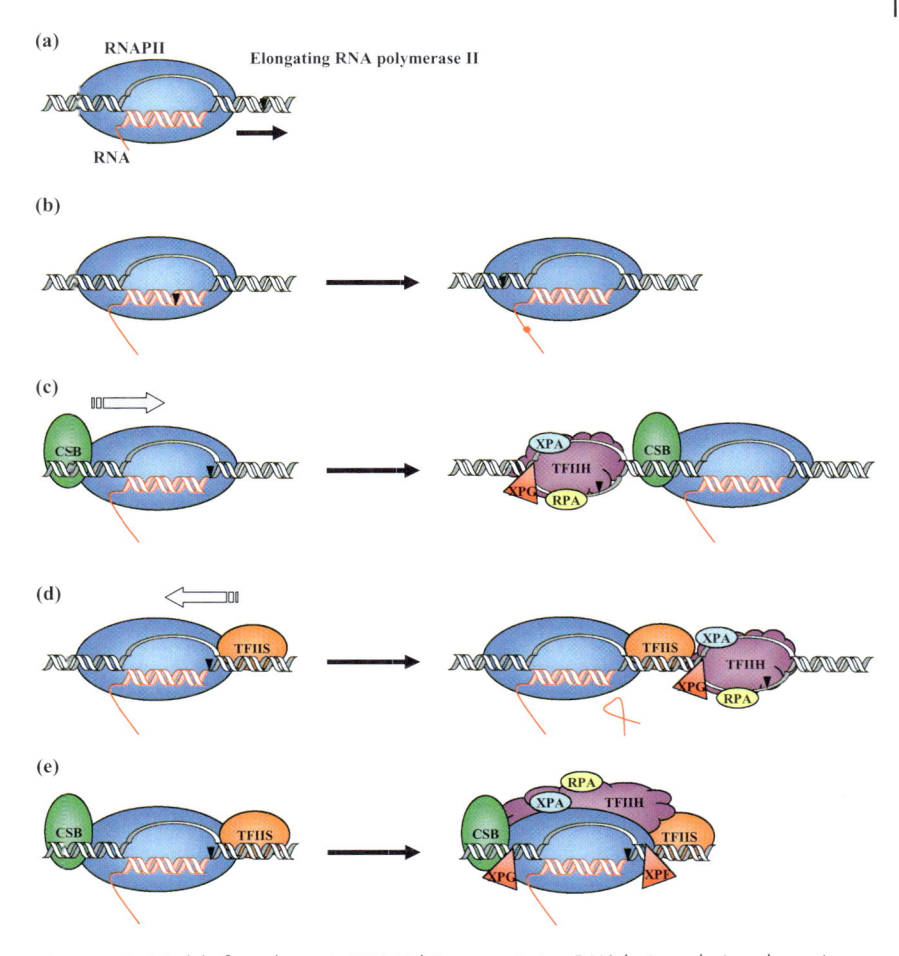

Figure 17.7 Models for eukaryotic RNA Pol II encountering DNA lesions during elongation. The models are based on research from publications cited in this chapter. For details, see Section 17.6.

transcription elongation can pause or terminate at DNA damage. Accordingly, a number of potential models for rescuing transcription and/or initiating TC-NER have been proposed (Figure 17.7c–e). Transiently stalled RNA polymerase can be rescued by CSB-induced lesion bypass [29]. Aberrant CSB could pose problems for the cell by perturbing temporal features of RNA synthesis. The displacement of the RNA Pol II also provides access to a lesion so that it can be removed by DNA repair processes. Furthermore, through the cooperative interaction of CSB and XPG with the stalled polymerase, the preincision NER core factors can be recruited [31, 124]. These include the TFIIH complex that contains the DNA helicases XPB and XPD, and the endonuclease XPF, which together with XPG is responsible for the 5'- and 3'-incision events, respectively. In addition, XPA

Table 17.2 DNA lesions effect on transcript integrity.

Lesion	Transcript effects	References
Bacteriophage		
abasic sites	full-length transcriptional mutagenesis misincorporation of A	[35]
strand breaks	shortened transcripts	[38]
gaps	shortened transcripts	[39, 40]
8-oxoguanine	full-length incorporation of C = A	[41]
thymine glycol	no data	
5,6-dihydrouracil	full-length transcriptional mutagenesis misincorporation of A	[4]
2-deoxyribonolactone	no data	
5-guanidino-4-nitroimidazole	full-length incorporation of C > A, truncated −1 base of lesion	[46]
pyrimido[1,2-α]purin-10(3H)one	no data	
uracil	full-length transcriptional mutagenesis misincorporation of A	[4, 35]
O^6-methylguanine	full-length incorporation of C = U	[48]
N^2-ethyl-2'-dG	truncated before lesion	[49]
1,N^2-ethenoguanine	full-length transcriptional mutagenesis misincorporation, A > G, majority truncated −1 base of lesion	[50]
cyclobutane pyrimidine dimer	no data	
6-4 photoproducts	no data	
4'-hydroxymethyl-4,5',8-trimethylpsoralen		
1,1'-d(TA/AT)	majority truncated −1 base of lesion with correct insertion	[54]
mono dT adduct	majority truncated −1 base of lesion with correct insertion	[54]
cisplatin		
cis-1,2-d(GG)	truncated at lesion with U misincorporation	[56]
cis-1,3-d(GTG)	truncated at lesion with U misincorporation	[56]
benzo[a]pyrene diol epoxide-N^2-dG		
(+)-trans-anti	no full-length data, truncated at lesion with A misincorporation	[135]

Table 17.2 Continued

Lesion	Transcript effects	References
(−)-trans-anti	full-length no transcriptional mutagenesis, truncated at lesion with A misincorporation	[135]
(+)-cis-anti	full-length no transcriptional mutagenesis, truncated at lesion with A misincorporation	[135]
(−)-cis-anti	full-length no transcriptional mutagenesis, truncated at lesion with A misincorporation	[135]
benzo[a]pyrene diol epoxide-N^6-dA		
(+)-trans-anti	full-length transcriptional mutagenesis misincorporation A > G, majority truncated at lesion	[58]
(−)-trans-anti	full-length transcriptional mutagenesis misincorporation A > G, majority truncated at lesion	[58]
dibenzo[a,l]pyrene diol epoxide-N^2-dG		
(−)-trans-anti	full-length no transcriptional mutagenesis, majority truncated −1 base of lesion	this chapter
benzo[c]phenanthrene-N^6-dA		
(+)-trans-anti	no full-length data, truncated ±1 base of lesion with G misincorporation	[59]
(−)-trans-anti	no full-length data, truncated ±1 base of lesion with G misincorporation	[59]
2-aminofluorene-C8-dG	no transcriptional mutagenesis	[41]
2-acetylaminofluorene-C8-dG	no transcriptional mutagenesis	[41]
4-hydroxyequilenin		
dA	no data	
dC	no data	
Prokaryotic		
abasic sites	full-length transcriptional mutagenesis misincorporation of A	[35]
strand breaks	no data	
gaps	shortened transcripts	[39]
8-oxoguanine	full-length incorporation of A = C	[61]

Table 17.2 Continued

Lesion	Transcript effects	References
5,6-dihydrouracil	full-length transcriptional mutagenesis incorporation of A	[63]
uracil	full-length transcriptional mutagenesis misincorporation of A	[35, 61]
O^6-methylguanine	full-length transcriptional mutagenesis misincorporation of U	[61]
N^2-ethyl-2′-dG	majority truncated −1 base of lesion	[49]
cyclobutane pyrimidine dimer	majority truncated at 3′-dT	[64]
4′-hydroxymethyl-4,5′,8-trimethylpsoralen		
1,1′-d(TA/AT)	majority truncated −1 base of lesion with correct insertion	[65]
mono dT adduct	majority truncated −1 base of lesion with correct insertion	[65]
cisplatin		
cis-1,2-d(GG)	no data	
cis-1,2-d(AG)	no data	
cis/*trans*-1,3-d(GTG)	no data	
cis-1,1′-d(GC/GC)	no data	
mono dG adduct	no data	
Eukaryotic		
abasic sites	misincorporation of A	[108]
strand breaks	no data	
pyrimido[1,2-α]purin-10(3H)one	no data	
8-oxoguanine	full-length and truncated incorporation of A ≥ C	[29, 72, 73]
thymine glycol	full-length no transcriptional mutagenesis, truncated at lesion with correct insertion	[29]
5-hydroxyuracil	full-length transcriptional mutagenesis misincorporation of A	[29]
2-deoxyribonolactone	no data	
8,5′-cyclo-2′-dA	full-length transcriptional mutagenesis misincorporation of A	[74]
5-guanidino-4-nitroimidazole	full-length no transcriptional mutagenesis, majority truncated −1 base of lesion correct insertion	[46]

Table 17.2 Continued

Lesion	Transcript effects	References
uracil	full-length transcriptional mutagenesis with misincorporation of A or G	[72]
O^6-methylguanine	full-length transcriptional mutagenesis with C > U, truncated 1 base before lesion with G > C	[48]
N^2-ethyl-2′-dG	majority truncated −1 base of lesion	[49]
1,N^2-ethenoguanine	majority truncated −1 base of lesion	[50]
cyclobutane pyrimidine dimer	majority truncated at 5′ dT with U misincorporation	[78, 125]
6-4 photoproducts	incorporation of A at both positions	[78]
4′-hydroxymethyl-4,5′,8-trimethylpsoralen		
1,1′-d(TA/AT)	truncated 1 or 2 bases before the lesion	[81]
mono dT adduct	truncated 1 or 2 bases before the lesion	[81]
cisplatin		
cis-1,2-d(GG)	no full-length data, truncated at lesion with misincorporation of A	[108]
cis-1,3-d(GTG)	no data	
benzo[a]pyrene diol epoxide-N^2-dG		
(+)-trans-anti	no data	
(−)-trans-anti	no data	
dibenzo[a,l]pyrene diol epoxide-N^2-dG		
(−)-trans-anti	majority truncated −1 base of lesion	this chapter
benzo[c]phenanthrene-N^6-dA		
(+)-trans-anti	full-length no transcriptional mutagenesis	[34]
(−)-trans-anti	full-length no transcriptional mutagenesis	[34]
2-aminofluorene-C8-dG	no data	
2-acetylaminofluorene-C8-dG	no data	
4-hydroxyequilenin		
dA	no data	
dC	no data	

and the single-strand binding protein replication protein A are recruited [32, 146] (Figure 17.7c). In the cases where DNA damage acts as a strong block that leads to an arrested RNA Pol II complex, the transcription factor TFIIS can induce reverse translocation [147]. TFIIS also induces a partial degradation of the nascent transcript and allows the 3′-end of the RNA to become realigned in the active site after the polymerase is positioned upstream from the lesion [30, 148]. Hence, the action of TFIIS could move the RNA polymerase complex away from the lesion, making the damage accessible to the TC-NER machinery and also allowing the RNA polymerase to resume transcription (Figure 17.7d). TC-NER can also be performed without the disassociation of RNA Pol II [51, 84, 124, 149]. The helicases in the TFIIH complex could further unwind the transcription bubble, permitting dual incision to occur [124, 125]. This would induce removal of the RNA polymerase along with the lesion and the nascent transcript.

The biochemical data, especially the structural information gleaned from crystallography, and the computer models support the original observations proposed several years ago – a complex combination of factors affect a DNA lesion's ability to interfere with transcription elongation, including the structure of the specific RNA polymerase's active site, the size and shape of the DNA adduct, the stereochemistry of the adduct, the particular base incorporated into the growing transcript at the site of the damage, and the local DNA sequence. In fact, the title of a recent commentary in the *Proceedings of the National Academy of Sciences* entitled "RNA polymerase: the most specific damage recognition protein in cellular responses to DNA damage?" asks an important question as the title implies [150]. As illustrated in this chapter, most RNA polymerases are sensitive to alterations to the transcribed strand, resulting in pausing or blocks to elongation, which for RNA Pol II can trigger repair of the lesion, ubiquitination of the polymerase and its subsequent destruction, or activation of p53 that can bring about apoptosis. Further work to elucidate structural details of RNA polymerases stalled at DNA lesions as well as the physiological relevance of transcriptional mutagenesis should provide mechanistic insights into the effects of DNA damage on gene expression, its repair, and perhaps most important, the relevance to human health.

Acknowledgements

We thank Professors Nicholas Geacintov and Suse Broyde for their continued collaboration in this work. We thank Laura Cartularo for critical reading of the manuscript. The work was supported by NIH ES010581 to D.A.S., and NIH 2R01 CA75449 and 5R01 CA28038 to S.B. J.A.B. and A.D. were supported in part by graduate fellowships from New York University. L.N. was supported in part by an "Administrative Supplement to Support High School Student and College Undergraduate Research Experiences" from the NIEHS and HHMI grant 52003738.

References

1. Friedberg, E.C., Walker, G.C., and Siede, W. (1995) *DNA Repair and Mutagenesis*, ASM Press, Washington, DC.
2. Kunkel, T.A. and Bebenek, K. (2000) DNA replication fidelity. *Annu. Rev. Biochem.*, **69**, 497–529.
3. Walker, G.C. (1984) Mutagenesis and inducible responses to deoxyribonucleic acid damage in *Escherichia coli*. *Microbiol. Rev.*, **48**, 60–93.
4. Liu, J., Zhou, W., and Doetsch, P.W. (1995) RNA polymerase bypass at sites of dihydrouracil: implications for transcriptional mutagenesis. *Mol. Cell. Biol.*, **15**, 6729–6735.
5. Viswanathan, A., You, H.J., and Doetsch, P.W. (1999) Phenotypic change caused by transcriptional bypass of uracil in nondividing cells. *Science*, **284**, 159–162.
6. Smith, J.M. and Koopman, P.A. (2004) The ins and outs of transcriptional control: nucleocytoplasmic shuttling in development and disease. *Trends Genet.*, **20**, 4–8.
7. Friedberg, E.C. (2005) Suffering in silence: the tolerance of DNA damage. *Nat. Rev. Mol. Cell Biol.*, **6**, 943–953.
8. Lehmann, A.R. (2006) Translesion synthesis in mammalian cells. *Exp. Cell Res.*, **312**, 2673–2676.
9. Mellon, I., Bohr, V.A., Smith, C.A., and Hanawalt, P.C. (1986) Preferential DNA repair of an active gene in human cells. *Proc. Natl. Acad. Sci. USA*, **83**, 8878–8882.
10. Mellon, I. and Hanawalt, P.C. (1989) Induction of the *Escherichia coli* lactose operon selectively increases repair of its transcribed DNA strand. *Nature*, **342**, 95–98.
11. Scicchitano, D.A., Olesnicky, E.C., and Dimitri, A. (2004) Transcription and DNA adducts: what happens when the message gets cut off? *DNA Repair*, **3**, 1537–1548.
12. Tornaletti, S. (2005) Transcription arrest at DNA damage sites. *Mutat. Res.*, **577**, 131–145.
13. Truglio, J.J., Croteau, D.L., Van Houten, B., and Kisker, C. (2006) Prokaryotic nucleotide excision repair: the UvrABC system. *Chem. Rev.*, **106**, 233–252.
14. Selby, C.P. and Sancar, A. (1993) Molecular mechanism of transcription-repair coupling. *Science*, **260**, 53–58.
15. Selby, C.P. and Sancar, A. (1995) Structure and function of transcription–repair coupling factor. I. Structural domains and binding properties. *J. Biol. Chem.*, **270**, 4882–4889.
16. Selby, C.P. and Sancar, A. (1995) Structure and function of transcription–repair coupling factor. II. Catalytic properties. *J. Biol. Chem.*, **270**, 4890–4895.
17. Park, J.S., Marr, M.T., and Roberts, J.W. (2002) E. coli transcription repair coupling factor (Mfd protein) rescues arrested complexes by promoting forward translocation. *Cell*, **109**, 757–767.
18. Sugasawa, K., Ng, J.M., Masutani, C., Iwai, S., van der Spek, P.J., Eker, A.P., Hanaoka, F., Bootsma, D., *et al.* (1998) Xeroderma pigmentosum group C protein complex is the initiator of global genome nucleotide excision repair. *Mol. Cell*, **2**, 223–232.
19. Sugasawa, K., Okamoto, T., Shimizu, Y., Masutani, C., Iwai, S., and Hanaoka, F. (2001) A multistep damage recognition mechanism for global genomic nucleotide excision repair. *Genes Dev.*, **15**, 507–521.
20. Hess, M.T., Schwitter, U., Petretta, M., Giese, B., and Naegeli, H. (1997) Bipartite substrate discrimination by human nucleotide excision repair. *Proc. Natl. Acad. Sci. USA*, **94**, 6664–6669.
21. Luch, A., Kudla, K., Seidel, A., Doehmer, J., Greim, H., and Baird, W.M. (1999) The level of DNA modification by (+)-*syn*-(11S,12R,13S,14R)- and (−)-anti-(11R,12S,13S,14R)-dihydrodiol epoxides of dibenzo[*a,l*]pyrene determined the effect on the proteins p53 and p21^{WAF1} in the human mammary carcinoma cell line MCF-7. *Carcinogenesis*, **20**, 859–865.
22. Buterin, T., Hess, M.T., Luneva, N., Geacintov, N.E., Amin, S., Kroth, H.,

Seidel, A., and Naegeli, H. (2000) Unrepaired fjord region polycyclic aromatic hydrocarbon-DNA adducts in *ras* codon 61 mutational hot spots. *Cancer Res.*, **60**, 1849–1856.

23 Wu, M., Yan, S., Patel, D.J., Geacintov, N.E., and Broyde, S. (2002) Relating repair susceptibility of carcinogen-damaged DNA with structural distortion and thermodynamic stability. *Nucleic Acids Res.*, **30**, 3422–3432.

24 Lloyd, D.R. and Hanawalt, P.C. (2002) p53 controls global nucleotide excision repair of low levels of structurally diverse benzo(g)chrysene-DNA adducts in human fibroblasts. *Cancer Res.*, **62**, 5288–5294.

25 Dreij, K., Seidel, A., and Jernström, B. (2005) Differential removal of DNA adducts derived from anti-diol epoxides of dibenzo[a,l]pyrene and benzo[a]pyrene in human cells. *Chem. Res. Toxicol.*, **18**, 655–664.

26 Venema, J., Mullenders, L., Natarajan, A., Zeeland, A., and Mayne, L. (1990) The genetic defect in Cockayne syndrome is associated with a defect in repair of UV-induced DNA damage in transcriptionally active DNA. *Proc. Natl. Acad. Sci. USA*, **87**, 4707–4711.

27 Troelstra, C., van Gool, A., de Wit, J., Vermeulen, W., Bootsma, D., and Hoeijmakers, J.H. (1992) ERCC6, a member of a subfamily of putative helicases, is involved in Cockayne's syndrome and preferential repair of active genes. *Cell*, **71**, 939–953.

28 van Hoffen, A., Natarajan, A.T., Mayne, L.V., van Zeeland, A.A., Mullenders, L.H., and Venema, J. (1993) Deficient repair of the transcribed strand of active genes in Cockayne's syndrome cells. *Nucleic Acids Res.*, **21**, 5890–5895.

29 Charlet-Berguerand, N., Feuerhahn, S., Kong, S.E., Ziserman, H., Conaway, J.W., Conaway, R., and Egly, J.M. (2006) RNA polymerase II bypass of oxidative DNA damage is regulated by transcription elongation factors. *EMBO J.*, **25**, 5481–5491.

30 Svejstrup, J.Q. (2007) Contending with transcriptional arrest during RNAPII transcript elongation. *Trends Biochem. Sci.*, **32**, 165–171.

31 Fousteri, M. and Mullenders, L.H. (2008) Transcription-coupled nucleotide excision repair in mammalian cells: molecular mechanisms and biological effects. *Cell Res.*, **18**, 73–84.

32 Hanawalt, P.C. and Spivak, G. (2008) Transcription-coupled DNA repair: two decades of progress and surprises. *Nat. Rev. Mol. Cell Biol.*, **9**, 958–970.

33 Tornaletti, S. and Hanawalt, P.C. (1999) Effect of DNA lesions on transcription elongation. *Biochimie*, **81**, 139–146.

34 Scicchitano, D.A. (2005) Transcription past DNA adducts derived from polycyclic aromatic hydrocarbons. *Mutat. Res.*, **577**, 146–154.

35 Zhou, W. and Doetsch, P.W. (1993) Effects of abasic sites and DNA single-strand breaks on prokaryotic RNA polymerases. *Proc. Natl. Acad. Sci. USA*, **90**, 6601–6605.

36 Sanchez, G. and Mamet-Bratley, M.D. (1994) Transcription by T7 RNA polymerase of DNA containing abasic sites. *Environ. Mol. Mutagen.*, **23**, 32–36.

37 Tornaletti, S., Maeda, L.S., and Hanawalt, P.C. (2006) Transcription arrest at an abasic site in the transcribed strand of template DNA. *Chem. Res. Toxicol.*, **19**, 1215–1220.

38 Zhou, W. and Doetsch, P.W. (1994) Transcription bypass or blockage at single-strand breaks on the DNA template strand: effect of different 3' and 5' flanking groups on the T7 RNA polymerase elongation complex. *Biochemistry*, **33**, 14926–14934.

39 Liu, J. and Doetsch, P.W. (1996) Template strand gap bypass is a general property of prokaryotic RNA polymerases: implications for elongation mechanisms. *Biochemistry*, **35**, 14999–15008.

40 Zhou, W., Reines, D., and Doetsch, P.W. (1995) T7 RNA polymerase bypass of large gaps on the template strand reveals a critical role of the nontemplate strand in elongation. *Cell*, **82**, 577–585.

41 Chen, Y.H. and Bogenhagen, D.F. (1993) Effects of DNA lesions on transcription elongation by T7 RNA polymerase. *J. Biol. Chem.*, **268**, 5849–5855.

42 Tornaletti, S., Maeda, L.S., Kolodner, R.D., and Hanawalt, P.C. (2004) Effect

of 8-oxoguanine on transcription elongation by T7 RNA polymerase and mammalian RNA polymerase II. *DNA Repair*, **3**, 483–494.

43 Hatahet, Z., Purmal, A.A., and Wallace, S.S. (1994) Oxidative DNA lesions as blocks to *in vitro* transcription by phage T7 RNA polymerase. *Ann. NY Acad. Sci.*, **726**, 346–348.

44 Tornaletti, S., Maeda, L.S., Lloyd, D.R., Reines, D., and Hanawalt, P.C. (2001) Effect of thymine glycol on transcription elongation by T7 RNA polymerase and mammalian RNA polymerase II. *J. Biol. Chem.*, **276**, 45367–45371.

45 Wang, Y., Sheppard, T.L., Tornaletti, S., Maeda, L.S., and Hanawalt, P.C. (2006) Transcriptional inhibition by an oxidized abasic site in DNA. *Chem. Res. Toxicol.*, **19**, 234–241.

46 Dimitri, A., Jia, L., Shafirovich, V., Geacintov, N.E., Broyde, S., and Scicchitano, D.A. (2008) Transcription of DNA containing the 5-guanidino-4-nitroimidazole lesion by human RNA polymerase II and bacteriophage T7 RNA polymerase. *DNA Repair*, **7**, 1276–1288.

47 Cline, S.D., Riggins, J.N., Tornaletti, S., Marnett, L.J., and Hanawalt, P.C. (2004) Malondialdehyde adducts in DNA arrest transcription by T7 RNA polymerase and mammalian RNA polymerase II. *Proc. Natl. Acad. Sci. USA*, **101**, 7275–7280.

48 Dimitri, A., Burns, J.A., Broyde, S., and Scicchitano, D.A. (2008) Transcription elongation past O^6-methylguanine by human RNA polymerase II and bacteriophage T7 RNA polymerase. *Nucleic Acids Res.*, **36**, 6459–6471.

49 Cheng, T.F., Hu, X., Gnatt, A., and Brooks, P.J. (2008) Differential blocking effects of the acetaldehyde-derived DNA lesion N^2-ethyl-2'-deoxyguanosine on transcription by multisubunit and single subunit RNA polymerases. *J. Biol. Chem.*, **283**, 27820–27828.

50 Dimitri, A., Goodenough, A.K., Guengerich, F.P., Broyde, S., and Scicchitano, D.A. (2008) Transcription processing at $1,N^2$-ethenoguanine by human RNA polymerase II and bacteriophage T7 RNA polymerase. *J. Mol. Biol.*, **375**, 353–366.

51 Selby, C.P., Drapkin, R., Reinberg, D., and Sancar, A. (1997) RNA polymerase II stalled at a thymine dimer: footprint and effect on excision repair. *Nucleic Acids Res.*, **25**, 787–793.

52 Smith, C.A., Baeten, J., and Taylor, J.S. (1998) The ability of a variety of polymerases to synthesize past site-specific *cis-syn*, *trans-syn*-II, (6-4), and Dewar photoproducts of thymidylyl-(3' → 5')-thymidine. *J. Biol. Chem.*, **273**, 21933–21940.

53 Kalogeraki, V.S., Tornaletti, S., and Hanawalt, P.C. (2003) Transcription arrest at a lesion in the transcribed DNA strand *in vitro* is not affected by a nearby lesion in the opposite strand. *J. Biol. Chem.*, **278**, 19558–19564.

54 Shi, Y.B., Gamper, H., and Hearst, J.E. (1988) Interaction of T7 RNA polymerase with DNA in an elongation complex arrested at a specific psoralen adduct site. *J. Biol. Chem.*, **263**, 527–534.

55 Tornaletti, S., Patrick, S.M., Turchi, J.J., and Hanawalt, P.C. (2003) Behavior of T7 RNA polymerase and mammalian RNA polymerase II at site-specific cisplatin adducts in the template DNA. *J. Biol. Chem.*, **278**, 35791–35797.

56 Jung, Y. and Lippard, S.J. (2003) Multiple states of stalled T7 RNA polymerase at DNA lesions generated by platinum anticancer agents. *J. Biol. Chem.*, **278**, 52084–52092.

57 Choi, D.J., Marino-Alessandri, D.J., Geacintov, N.E., and Scicchitano, D.A. (1994) Site-specific benzo[*a*]pyrene diol epoxide-DNA adducts inhibit transcription elongation by bacteriophage T7 RNA polymerase. *Biochemistry*, **33**, 780–787.

58 Remington, K.M., Bennett, S.E., Harris, C.M., Harris, T.M., and Bebenek, K. (1998) Highly mutagenic bypass synthesis by T7 RNA polymerase of site-specific benzo[*a*]pyrene diol epoxide-adducted template DNA. *J. Biol. Chem.*, **273**, 13170–13176.

59 Roth, R.B., Amin, S., Geacintov, N.E., and Scicchitano, D.A. (2001) Bacteriophage T7 RNA polymerase transcription elongation is inhibited by site-specific, stereospecific benzo[*c*]phenanthrene diol epoxide DNA lesions. *Biochemistry*, **40**, 5200–5207.

60 Smith, A.J. and Savery, N.J. (2008) Effects of the bacterial transcription–repair coupling factor during transcription of DNA containing non-bulky lesions. *DNA Repair*, **7**, 1670–1679.

61 Viswanathan, A. and Doetsch, P.W. (1998) Effects of nonbulky DNA base damages on *Escherichia coli* RNA polymerase-mediated elongation and promoter clearance. *J. Biol. Chem.*, **273**, 21276–21281.

62 Bregeon, D., Doddridge, Z.A., You, H.J., Weiss, B., and Doetsch, P.W. (2003) Transcriptional mutagenesis induced by uracil and 8-oxoguanine in *Escherichia coli*. *Mol. Cell*, **12**, 959–970.

63 Liu, J. and Doetsch, P.W. (1998) *Escherichia coli* RNA and DNA polymerase bypass of dihydrouracil: mutagenic potential via transcription and replication. *Nucleic Acids Res.*, **26**, 1707–1712.

64 Selby, C.P. and Sancar, A. (1990) Transcription preferentially inhibits nucleotide excision repair of the template DNA strand *in vitro*. *J. Biol. Chem.*, **265**, 21330–21336.

65 Shi, Y.B., Gamper, H., and Hearst, J.E. (1987) The effects of covalent additions of a psoralen on transcription by *E. coli* RNA polymerase. *Nucleic Acids Res.*, **15**, 6843–6854.

66 Corda, Y., Anin, M.F., Leng, M., and Job, D. (1992) RNA polymerases react differently at d(ApG) and d(GpG) adducts in DNA modified by cis-diamminedichloroplatinum(II). *Biochemistry*, **31**, 1904–1908.

67 Corda, Y., Job, C., Anin, M.F., Leng, M., and Job, D. (1991) Transcription by eucaryotic and procaryotic RNA polymerases of DNA modified at a d(GG) or a d(AG) site by the antitumor drug cis-diamminedichloroplatinum(II). *Biochemistry*, **30**, 222–230.

68 Corda, Y., Job, C., Anin, M.F., Leng, M., and Job, D. (1993) Spectrum of DNA–platinum adduct recognition by prokaryotic and eukaryotic DNA-dependent RNA polymerases. *Biochemistry*, **32**, 8582–8588.

69 Kathe, S.D., Shen, G.P., and Wallace, S.S. (2004) Single-stranded breaks in DNA but not oxidative DNA base damages block transcriptional elongation by RNA polymerase II in HeLa cell nuclear extracts. *J. Biol. Chem.*, **279**, 18511–18520.

70 Ljungman, M. (1999) Repair of radiation-induced DNA strand breaks does not occur preferentially in transcriptionally active DNA. *Radiat. Res.*, **152**, 444–449.

71 Larsen, E., Kwon, K., Coin, F., Egly, J.M., and Klungland, A. (2004) Transcription activities at 8-oxoG lesions in DNA. *DNA Repair*, **3**, 1457–1468.

72 Kuraoka, I., Endou, M., Yamaguchi, Y., Wada, T., Handa, H., and Tanaka, K. (2003) Effects of endogenous DNA base lesions on transcription elongation by mammalian RNA polymerase II. Implications for transcription-coupled DNA repair and transcriptional mutagenesis. *J. Biol. Chem.*, **278**, 7294–7299.

73 Kuraoka, I., Suzuki, K., Ito, S., Hayashida, M., Kwei, J.S., Ikegami, T., Handa, H., Nakabeppu, Y., et al. (2007) RNA polymerase II bypasses 8-oxoguanine in the presence of transcription elongation factor TFIIS. *DNA Repair*, **6**, 841–851.

74 Marietta, C. and Brooks, P.J. (2007) Transcriptional bypass of bulky DNA lesions causes new mutant RNA transcripts in human cells. *EMBO Rep.*, **8**, 388–393.

75 Donahue, B.A., Yin, S., Taylor, J.S., Reines, D., and Hanawalt, P.C. (1994) Transcript cleavage by RNA polymerase II arrested by a cyclobutane pyrimidine dimer in the DNA template. *Proc. Natl. Acad. Sci. USA*, **91**, 8502–8506.

76 Tornaletti, S., Donahue, B.A., Reines, D., and Hanawalt, P.C. (1997) Nucleotide sequence context effect of a cyclobutane pyrimidine dimer upon RNA polymerase II transcription. *J. Biol. Chem.*, **272**, 31719–31724.

77 Tornaletti, S., Reines, D., and Hanawalt, P.C. (1999) Structural characterization of RNA polymerase II complexes arrested by a cyclobutane pyrimidine dimer in the transcribed strand of template DNA. *J. Biol. Chem.*, **274**, 24124–24130.

78 Mei Kwei, J.S., Kuraoka, I., Horibata, K., Ubukata, M., Kobatake, E., Iwai, S., Handa, H., and Tanaka, K. (2004) Blockage of RNA polymerase II at a cyclobutane pyrimidine dimer and 6-4 photoproduct. *Biochem. Biophys. Res. Commun.*, **320**, 1133–1138.

79 May, A., Nairn, R.S., Okumoto, D.S., Wassermann, K., Stevnsner, T., Jones, J.C., and Bohr, V.A. (1993) Repair of individual DNA strands in the hamster dihydrofolate reductase gene after treatment with ultraviolet light, alkylating agents, and cisplatin. *J. Biol. Chem.*, **268**, 1650–1657.

80 Vreeswijk, M.P., van Hoffen, A., Westland, B.E., Vrieling, H., van Zeeland, A.A., and Mullenders, L.H. (1994) Analysis of repair of cyclobutane pyrimidine dimers and pyrimidine 6-4 pyrimidone photoproducts in transcriptionally active and inactive genes in Chinese hamster cells. *J. Biol. Chem.*, **269**, 31858–31863.

81 Wang, Z. and Rana, T.M. (1997) DNA damage-dependent transcriptional arrest and termination of RNA polymerase II elongation complexes in DNA template containing HIV-1 promoter. *Proc. Natl. Acad. Sci. USA*, **94**, 6688–6693.

82 Islas, A.L., Baker, F.J., and Hanawalt, P.C. (1994) Transcription-coupled repair of psoralen cross-links but not monoadducts in Chinese hamster ovary cells. *Biochemistry*, **33**, 10794–10799.

83 Cullinane, C., Mazur, S.J., Essigmann, J.M., Phillips, D.R., and Bohr, V.A. (1999) Inhibition of RNA polymerase II transcription in human cell extracts by cisplatin DNA damage. *Biochemistry*, **38**, 6204–6212.

84 Tremeau-Bravard, A., Riedl, T., Egly, J.M., and Dahmus, M.E. (2004) Fate of RNA polymerase II stalled at a cisplatin lesion. *J. Biol. Chem.*, **279**, 7751–7759.

85 Tang, M.S., Pao, A., and Zhang, X.S. (1994) Repair of benzo(a)pyrene diol epoxide- and UV-induced DNA damage in dihydrofolate reductase and adenine phosphoribosyltransferase genes of CHO cells. *J. Biol. Chem.*, **269**, 12749–12754.

86 Chen, R.H., Maher, V.M., and McCormick, J.J. (1990) Effect of excision repair by diploid human fibroblasts on the kinds and locations of mutations induced by ($\pm$)-7β,8α-dihydroxy-9α,10α-epoxy-7,8,9,10-tetrahydrobenzo[a]pyrene in the coding region of the HPRT gene. *Proc. Natl. Acad. Sci. USA*, **87**, 8680–8684.

87 Perlow, R.A., Kolbanovskii, A., Hingerty, B.E., Geacintov, N.E., Broyde, S., and Scicchitano, D.A. (2002) DNA adducts from a tumorigenic metabolite of benzo[a]pyrene block human RNA polymerase II elongation in a sequence- and stereochemistry-dependent manner. *J. Mol. Biol.*, **321**, 29–47.

88 Carothers, A.M., Zhen, W., Mucha, J., Zhang, Y.J., Santella, R.M., Grunberger, D., and Bohr, V.A. (1992) DNA strand-specific repair of ($\pm$)-3α,4β-dihydroxy-1α,2α-epoxy-1,2,3,4-tetrahydrobenzo[c]phenanthrene adducts in the hamster dihydrofolate reductase gene. *Proc. Natl. Acad. Sci. USA*, **89**, 11925–11929.

89 Carothers, A.M., Mucha, J., and Grunberger, D. (1991) DNA strand-specific mutations induced by ($\pm$)-3α,4β-dihydroxy-1α,2α-epoxy-1,2,3,4-tetrahydrobenzo[c]phenanthrene in the dihydrofolate reductase gene. *Proc. Natl. Acad. Sci. USA*, **88**, 5749–5753.

90 Schinecker, T.M., Perlow, R.A., Broyde, S., Geacintov, N.E., and Scicchitano, D.A. (2003) Human RNA polymerase II is partially blocked by DNA adducts derived from tumorigenic benzo[c]phenanthrene diol epoxides: relating biological consequences to conformational preferences. *Nucleic Acids Res.*, **31**, 6004–6015.

91 Donahue, B.A., Fuchs, R.P., Reines, D., and Hanawalt, P.C. (1996) Effects of aminofluorene and acetylaminofluorene DNA adducts on transcriptional elongation by RNA polymerase II. *J. Biol. Chem.*, **271**, 10588–10594.

92 Tang, M.S., Bohr, V.A., Zhang, X.S., Pierce, J., and Hanawalt, P.C. (1989) Quantification of aminofluorene adduct formation and repair in defined DNA sequences in mammalian cells using

the UVRABC nuclease. *J. Biol. Chem.*, **264**, 14455–14462.
93 Cheetham, G.M. and Steitz, T.A. (2000) Insights into transcription: structure and function of single-subunit DNA-dependent RNA polymerases. *Curr. Opin. Struct. Biol.*, **10**, 117–123.
94 Yin, Y.W. and Steitz, T.A. (2002) Structural basis for the transition from initiation to elongation transcription in T7 RNA polymerase. *Science*, **298**, 1387–1395.
95 Temiakov, D., Patlan, V., Anikin, M., McAllister, W.T., Yokoyama, S., and Vassylyev, D.G. (2004) Structural basis for substrate selection by T7 RNA polymerase. *Cell*, **116**, 381–391.
96 Durniak, K.J., Bailey, S., and Steitz, T.A. (2008) The structure of a transcribing T7 RNA polymerase in transition from initiation to elongation. *Science*, **322**, 553–557.
97 Cramer, P. (2002) Multisubunit RNA polymerases. *Curr. Opin. Struct. Biol.*, **12**, 89–97.
98 Mukherjee, K. and Chatterji, D. (1997) Studies on the omega subunit of *Escherichia coli* RNA polymerase – its role in the recovery of denatured enzyme activity. *Eur. J. Biochem.*, **247**, 884–889.
99 Minakhin, L., Bhagat, S., Brunning, A., Campbell, E.A., Darst, S.A., Ebright, R.H., and Severinov, K. (2001) Bacterial RNA polymerase subunit omega and eukaryotic RNA polymerase subunit RPB6 are sequence, structural, and functional homologs and promote RNA polymerase assembly. *Proc. Natl. Acad. Sci. USA*, **98**, 892–897.
100 Cramer, P., Bushnell, D.A., and Kornberg, R.D. (2001) Structural basis of transcription: RNA polymerase II at 2.8 Ångstrom resolution. *Science*, **292**, 1863–1876.
101 Wang, D., Bushnell, D.A., Westover, K.D., Kaplan, C.D., and Kornberg, R.D. (2006) Structural basis of transcription: role of the trigger loop in substrate specificity and catalysis. *Cell*, **127**, 941–954.
102 Schaeffer, L., Roy, R., Humbert, S., Moncollin, V., Vermeulen, W., Hoeijmakers, J.H., Chambon, P., and Egly, J.M. (1993) DNA repair helicase: a component of BTF2 (TFIIH) basic transcription factor. *Science*, **260**, 58–63.
103 Wang, Z., Svejstrup, J.Q., Feaver, W.J., Wu, X., Kornberg, R.D., and Friedberg, E.C. (1994) Transcription factor b (TFIIH) is required during nucleotide-excision repair in yeast. *Nature*, **368**, 74–76.
104 Drapkin, R., Reardon, J.T., Ansari, A., Huang, J.C., Zawel, L., Ahn, K., Sancar, A., and Reinberg, D. (1994) Dual role of TFIIH in DNA excision repair and in transcription by RNA polymerase II. *Nature*, **368**, 769–772.
105 Coin, F., Oksenych, V., and Egly, J.M. (2007) Distinct roles for the XPB/p52 and XPD/p44 subcomplexes of TFIIH in damaged DNA opening during nucleotide excision repair. *Mol. Cell*, **26**, 245–256.
106 Sims, R.J., 3rd, Belotserkovskaya, R., and Reinberg, D. (2004) Elongation by RNA polymerase II: the short and long of it. *Genes Dev.*, **18**, 2437–2468.
107 Shilatifard, A. (2004) Transcriptional elongation control by RNA polymerase II: a new frontier. *Biochim. Biophys. Acta*, **1677**, 79–86.
108 Damsma, G.E., Alt, A., Brueckner, F., Carell, T., and Cramer, P. (2007) Mechanism of transcriptional stalling at cisplatin-damaged DNA. *Nat. Struct. Mol. Biol.*, **14**, 1127–1133.
109 Cadet, J., Bellon, S., Douki, T., Frelon, S., Gasparutto, D., Muller, E., Pouget, J.P., Ravanat, J.L., et al. (2004) Radiation-induced DNA damage: formation, measurement, and biochemical features. *J. Environ. Pathol. Toxicol. Oncol.*, **23**, 33–43.
110 Valko, M., Leibfritz, D., Moncol, J., Cronin, M.T., Mazur, M., and Telser, J. (2007) Free radicals and antioxidants in normal physiological functions and human disease. *Int. J. Biochem. Cell Biol.*, **39**, 44–84.
111 Brooks, P.J. (2008) The 8,5′-cyclopurine-2′-deoxynucleosides: candidate neurodegenerative DNA lesions in xeroderma pigmentosum, and unique probes of transcription and nucleotide excision repair. *DNA Repair*, **7**, 1168–1179.

112 Brooks, P.J., Wise, D.S., Berry, D.A., Kosmoski, J.V., Smerdon, M.J., Somers, R.L., Mackie, H., Spoonde, A.Y., et al. (2000) The oxidative DNA lesion 8,5′-(S)-cyclo-2′-deoxyadenosine is repaired by the nucleotide excision repair pathway and blocks gene expression in mammalian cells. *J. Biol. Chem.*, **275**, 22355–22362.

113 Gu, F., Stillwell, W.G., Wishnok, J.S., Shallop, A.J., Jones, R.A., and Tannenbaum, S.R. (2002) Peroxynitrite-induced reactions of synthetic oligo 2′-deoxynucleotides and DNA containing guanine: formation and stability of a 5-guanidino-4-nitroimidazole lesion. *Biochemistry*, **41**, 7508–7518.

114 Burcham, P.C. and Marnett, L.J. (1994) Site-specific mutagenesis by a propanodeoxyguanosine adduct carried on an M13 genome. *J. Biol. Chem.*, **269**, 28844–28850.

115 VanderVeen, L.A., Hashim, M.F., Shyr, Y., and Marnett, L.J. (2003) Induction of frameshift and base pair substitution mutations by the major DNA adduct of the endogenous carcinogen malondialdehyde. *Proc. Natl. Acad. Sci. USA*, **100**, 14247–14252.

116 Johnson, K.A., Fink, S.P., and Marnett, L.J. (1997) Repair of propanodeoxyguanosine by nucleotide excision repair *in vivo* and *in vitro*. *J. Biol. Chem.*, **272**, 11434–11438.

117 Saxowsky, T.T., Meadows, K.L., Klungland, A., and Doetsch, P.W. (2008) 8-Oxoguanine-mediated transcriptional mutagenesis causes Ras activation in mammalian cells. *Proc. Natl. Acad. Sci. USA*, **105**, 18877–18882.

118 Pegg, A.E. (2000) Repair of O^6-alkylguanine by alkyltransferases. *Mutat. Res.*, **462**, 83–100.

119 Fronza, G. and Gold, B. (2004) The biological effects of N3-methyladenine. *J. Cell Biochem.*, **91**, 250–257.

120 Vodicka, P. and Hemminki, K. (1988) Depurination and imidazole ring-opening in nucleosides and DNA alkylated by styrene oxide. *Chem. Biol. Interact.*, **68**, 117–126.

121 Hemminki, K., Peltonen, K., and Vodicka, P. (1989) Depurination from DNA of 7-methylguanine, 7-(2-aminoethyl)-guanine and ring-opened 7-methylguanines. *Chem. Biol. Interact.*, **70**, 289–303.

122 Matsuda, T., Yabushita, H., Kanaly, R.A., Shibutani, S., and Yokoyama, A. (2006) Increased DNA damage in ALDH2-deficient alcoholics. *Chem. Res. Toxicol.*, **19**, 1374–1378.

123 Cadet, J., Sage, E., and Douki, T. (2005) Ultraviolet radiation-mediated damage to cellular DNA. *Mutat. Res.*, **571**, 3–17.

124 Sarker, A.H., Tsutakawa, S.E., Kostek, S., Ng, C., Shin, D.S., Peris, M., Campeau, E., Tainer, J.A., et al. (2005) Recognition of RNA polymerase II and transcription bubbles by XPG, CSB, and TFIIH: insights for transcription-coupled repair and Cockayne syndrome. *Mol. Cell*, **20**, 187–198.

125 Brueckner, F., Hennecke, U., Carell, T., and Cramer, P. (2007) CPD damage recognition by transcribing RNA polymerase II. *Science*, **315**, 859–862.

126 Cimino, G.D., Gamper, H.B., Isaacs, S.T., and Hearst, J.E. (1985) Psoralens as photoactive probes of nucleic acid structure and function: organic chemistry, photochemistry, and biochemistry. *Annu. Rev. Biochem.*, **54**, 1151–1193.

127 Eastman, A. (1986) Reevaluation of interaction of *cis*-dichloro(ethylenediamine)platinum(II) with DNA. *Biochemistry*, **25**, 3912–3915.

128 Jung, Y. and Lippard, S.J. (2006) RNA polymerase II blockage by cisplatin-damaged DNA. Stability and polyubiquitylation of stalled polymerase. *J. Biol. Chem.*, **281**, 1361–1370.

129 Jones, J.C., Zhen, W.P., Reed, E., Parker, R.J., Sancar, A., and Bohr, V.A. (1991) Gene-specific formation and repair of cisplatin intrastrand adducts and interstrand cross-links in Chinese hamster ovary cells. *J. Biol. Chem.*, **266**, 7101–7107.

130 Zhen, W., Link, C.J., Jr., O'Connor, P.M., Reed, E., Parker, R., Howell, S.B., and Bohr, V.A. (1992) Increased gene-specific repair of cisplatin interstrand cross-links in cisplatin-resistant human ovarian cancer cell lines. *Mol. Cell. Biol.*, **12**, 3689–3698.

131 Dipple, A. (1985) Polycyclic aromatic hydrocarbons: an introduction, in *Polycyclic Hydrocarbons and Carcinogenesis* (ed. R.G. Harvey), ACS Symposium Series 283, American Chemical Society, Washington, DC, pp. 1–17.

132 Thakker, D.R., Yagi, H., Levin, W., Wood, A.W., Conney, A.H., and Jerina, D.M. (1985) Polycyclic aromatic hydrocarbons: metabolic activation to ultimate carcinogens, in *Bioactivation of Foreign Compounds* (ed. M.W. Anders), Academic Press, London, pp. 177–242.

133 Sims, P. and Grover, P.L. (1974) Epoxides in polycyclic aromatic hydrocarbon metabolism and carcinogenesis. *Adv. Cancer Res.*, **20**, 165–274.

134 Szeliga, J. and Dipple, A. (1998) DNA adduct formation by polycyclic aromatic hydrocarbon dihydrodiol epoxides. *Chem. Res. Toxicol.*, **11**, 1–11.

135 Choi, D.J., Roth, R.B., Liu, T., Geacintov, N.E., and Scicchitano, D.A. (1996) Incorrect base insertion and prematurely terminated transcripts during T7 RNA polymerase transcription elongation past benzo[a]pyrenediol epoxide-modified DNA. *J. Mol. Biol.*, **264**, 213–219.

136 Perlow, R.A., Schinecker, T.M., Kim, S.J., Geacintov, N.E., and Scicchitano, D.A. (2003) Construction and purification of site-specifically modified DNA templates for transcription assays. *Nucleic Acids Res.*, **31**, e40.

137 Heflich, R.H. and Neft, R.E. (1994) Genetic toxicity of 2-acetylaminofluorene, 2-aminofluorene and some of their metabolites and model metabolites. *Mutat. Res.*, **318**, 73–114.

138 Mao, B., Hingerty, B.E., Broyde, S., and Patel, D.J. (1998) Solution structure of the aminofluorene [AF]-external conformer of the *anti*-[AF]-C8-dG adduct opposite dC in a DNA duplex. *Biochemistry*, **37**, 95–106.

139 Mao, B., Hingerty, B.E., Broyde, S., and Patel, D.J. (1998) Solution structure of the aminofluorene [AF]-intercalated conformer of the *syn*-[AF]-C8-dG adduct opposite dC in a DNA duplex. *Biochemistry*, **37**, 81–94.

140 Mao, B., Gu, Z., Gorin, A., Hingerty, B.E., Broyde, S., and Patel, D.J. (1997) Solution structure of the aminofluorene-stacked conformer of the *syn*-[AF]-C8-dG adduct positioned at a template–primer junction. *Biochemistry*, **36**, 14491–14501.

141 Mao, B., Gorin, A., Gu, Z., Hingerty, B.E., Broyde, S., and Patel, D.J. (1997) Solution structure of the aminofluorene-intercalated conformer of the *syn*-[AF]-C8-dG adduct opposite a −2 deletion site in the *Nar*I hot spot sequence context. *Biochemistry*, **36**, 14479–14490.

142 Patel, D.J., Mao, B., Gu, Z., Hingerty, B.E., Gorin, A., Basu, A.K., and Broyde, S. (1998) Nuclear magnetic resonance solution structures of covalent aromatic amine-DNA adducts and their mutagenic relevance. *Chem. Res. Toxicol.*, **11**, 391–407.

143 Hersh, A.L., Stefanick, M.L., and Stafford, R.S. (2004) National use of postmenopausal hormone therapy: annual trends and response to recent evidence. *J. Am. Med. Assoc.*, **291**, 47–53.

144 Kolbanovskiy, A., Kuzmin, V., Shastry, A., Kolbanovskaya, M., Chen, D., Chang, M., Bolton, J.L., and Geacintov, N.E. (2005) Base selectivity and effects of sequence and DNA secondary structure on the formation of covalent adducts derived from the equine estrogen metabolite 4-hydroxyequilenin. *Chem. Res. Toxicol.*, **18**, 1737–1747.

145 Darzacq, X., Shav-Tal, Y., de Turris, V., Brody, Y., Shenoy, S.M., Phair, R.D., and Singer, R.H. (2007) In vivo dynamics of RNA polymerase II transcription. *Nat. Struct. Mol. Biol.*, **14**, 796–806.

146 Laine, J.P. and Egly, J.M. (2006) When transcription and repair meet: a complex system. *Trends Genet.*, **22**, 430–436.

147 Mote, J., Jr, Ghanouni, P., and Reines, D.A. (1994) DNA minor groove-binding ligand both potentiates and arrests transcription by RNA polymerase II. Elongation factor SII enables readthrough at arrest sites. *J. Mol. Biol.*, **236**, 725–737.

148 Reines, D. and Mote, J., Jr. (1993) Elongation factor SII-dependent transcription by RNA polymerase II through a sequence-specific DNA-binding protein. *Proc. Natl. Acad. Sci. USA*, **90**, 1917–1921.

149 Laine, J.P. and Egly, J.M. (2006) Initiation of DNA repair mediated by a stalled RNA polymerase IIO. *EMBO J.*, **25**, 387–397.

150 Lindsey-Boltz, L.A. and Sancar, A. (2007) RNA polymerase: the most specific damage recognition protein in cellular responses to DNA damage? *Proc. Natl. Acad. Sci. USA*, **104**, 13213–13214.

Index

a

A-family DNA polymerase 300
A-rule 303f., 357
AAG glycosylase 242ff.
abasic site (AP site) 240, 302, 388, 407
– bypass 302f.
– removal and repair 247
ABP-DNA adduct formation 171
N^2-acetylaminofluorene
  (N-acetyl-2-aminofluorene, AAF) 157, 420
– AAF-C8-dG 358ff.
– adduct 358
– dG-AAF 240
N'-acetyl-dG-C8-Bz adduct 172
N-acetyl-HONH-Bz 164
N'-acetyl-N-hydroxyaminobenzidine
  (N-acetyl-HONH-Bz) metabolite 160
N-acetylation 160
O-acetylation 160
N-acetylcysteine 193
N-acetyltransferase (NAT) 159
– NAT1 and NAT2 160
acrolein 36, 107ff.
– adduct 203
– reaction of β-substituted acroleins with DNA base 109
adduct
– chemical 299ff.
– conformation 225
– linkage 164, 205ff., 232
adduct–DNA groove interaction 227
adenine (Ade) 62
– one-electron oxidation reaction 64
adenine lesion 264
adenosine (Ado) 146
aflatoxin 7
AKR 141ff.
aldehyde
– α,β-unsaturated 203ff.

aldehyde oxidase 125
aldehyde-reactive probe (ARP)-based
  enzyme-linked immunosorbent assay
  (ELISA) 136
AlkA 122
AlkB 122f.
alkylating agent 10
O^6-alkylguanine alkyltransferase (AGT) 10ff., 311
alkylnitrosamine 10
2-amino-5-[2-deoxy-β-D-erythro-
  pentofuranosyl)amino]-4H-imidazol-4-one 62
2-amino-3,4-dimethylimidazo[4,5-f]quinoline (MeIQ) 165
2-amino-3,4-dimethylimidazo[4,5-f]
  quinoxaline (MeIQx) 159ff.
2-amino-1-methyl-6-phenylimidazo[4,5-b]
  pyridine (PhIP) 159ff.
– dG adduct 221
– DNA adduct 172
2-amino-3-methyl-9H-pyrido[2,3-b]indole
  (MeAαC) 170
2-amino-6-methyldiprido[1,2-a: 3',2'-d]
  imidazole 168
2-amino-3-methylimidazo[4,5-f]quinoline
  (IQ) 161ff., 220
2-amino-1-naphthol 169
2-amino-9H-pyrido[2,3-b]indole (AαC) 161
2-amino-3,4,8-trimethylimidazo[4,5-f]
  quinoxaline (4,8-DiMeIQx) 170
4-aminobiphenyl (ABP) 157ff.
– ABP-DNA adduct formation 171
2-aminofluorene (AF) 157, 217, 420
aminoimidazoarene (AIA) 158
2-aminopurine (2-AP) 86, 224
aniline 157
AP endonuclease (APE1) 242ff.
AP lyase 242

The Chemical Biology of DNA Damage. Edited by Nicholas E. Geacintov and Suse Broyde
© 2010 WILEY-VCH Verlag GmbH & Co. KGaA, Weinheim
ISBN: 978-3-527-32295-4

APC gene   168
arachidonic acid-dependent peroxidase   169
Aristolochia clematitis   7
aristolochic acid   7
aromatic amine (AA)   157ff.
aromatic amine DNA adduct   161ff., 217ff.
– conformational motif   219
– detection   168
– S/B/W ratios   232
– structure–function characteristics   217ff.
Aspergillus species   7

b

B[a]P, *see* benzo[a]pyrene
B-DNA duplex   227
B-family DNA polymerase   300ff., 356
– structure   373
bacterial mutagenesis   164
Balkan endemic nephropathy (BEN)   7
base
– alkylated   310, 412
– secondary oxidation reaction   70
base damage
– indirect   35
base excision repair (BER)   121, 239ff.
– long-patch   241ff.
– mechanism   4ff., 239ff.
– protein   224
– short-patch   241ff.
– substrate   240
base oxidation product   23ff.
base sequence context   224, 280
benzidine (Bz)   157
benzo[g]chrysene (B[g]C)   267
– *trans-anti*-B[g]C-N^6-dA adduct   280
– 14R (+)-*trans-anti*-B[g]C-N^6-dA adduct   272
– fjord region   267
benzo[c]phenanthrene (B[c]Ph)   267
– 1R (+)-*trans*-B[c]Ph-N^6-dA adduct   272ff.
– 1S (−)-*trans-anti*-B[c]Ph-N^6-dA adduct   272ff.
– 10R (+)-*trans-anti*-B[c]Ph-N^6-dA adduct   272
– 10S (−)-*trans-anti*-B[c]Ph-N^6-dA adduct   272
– (−)-*trans-anti*-N^6-B[c]PhDE-dA   418
– (+)-*trans-anti*-N^6-B[c]PhDE-dA   418
– diol epoxide (B[c]PhDE)   418
– fjord region   267, 280
benzo[a]pyrene (B[a]P)   6ff., 134, 282ff., 315, 331
– 10S (+)-*trans-anti*-B[a]P adduct   232
– B[a]P-6-N7-Ade   136
– B[a]P-N^6-adenine adduct   271

– B[a]P-dA adduct   315
– B[a]P-N^6-dA adduct   278
– 10S (+)-*trans-anti*-B[a]P-N^6-dA adduct   232, 271ff., 418
– 10R (−)-*trans-anti*-B[a]P-N^6-dA adduct   279, 418
– B[a]P-*trans*-4,5-dihydrodiol   145
– B[a]P-*trans*-7,8-dihydrodiol   145
– B[a]P-1,6-dione   137ff.
– B[a]P-3–6-dione   137ff.
– B[a]P-6,12-dione   137ff.
– B[a]P-7,8-dione   141ff.
– B[a]P-7,8-dione-N^2-dGuo adduct   143
– B[a]P-6-N7-Gua   136
– B[a]P-N^2-dG adduct   271ff.
– 10R (+)-*cis-anti*-B[a]P-N^2-dG adduct   270, 272ff.
– 10R (−)-*trans-anti*-B[a]P-N^2-dG adduct   272ff.
– 10S (+)-*trans-anti*-B[a]P-N^2-dG adduct   266ff., 280ff.
– B[a]P-guanine adduct (B[a]P-G)   387f.
– (−)-*anti*-B[a]PDE   266, 332
– (+)-*anti*-B[a]PDE   138ff., 332
– (±)-*anti*-B[a]PDE   145, 266
– (+)-*syn*-B[a]PDE   138
– 10S (+)-*trans-anti*-B[a]PDE-N^6-dAde adduct   342
– 10R (−)-*trans-anti*-N^2-B[a]PDE-dG adduct   418ff.
– 10S (+)-*trans-anti*-N^2-B[a]PDE-dG adduct   418ff.
– 10S (+)-*anti*-B[a]PDE-N^2-dGuo adduct   140
– 10R (−)-*trans-anti*-B[a]PDE-N^2-dGuo adduct   332ff.
– 10S (+)-*trans-anti*-B[a]PDE-N^2-dGuo adduct   138ff., 332ff.
– 10S (+)-*trans*-B[a]PDE-N^2-dG adduct   225
– bay region PAH   266
– diol epoxide (B[a]PDE)   138ff., 225, 266, 332
– diol epoxide metabolite   332
– DNA adduct   332
– DNA lesion   331ff.
– dG-BP adduct   240
– 7,8,9,10-tetrahydroxytetrahydrobenzo[a]pyrene (B[a]PT)   87
– Watson–Crick base pair   271
benzo[a]pyrene diol epoxide (B[a]PDE)   138ff., 225, 266, 332
– metabolite   332
O^6-benzylguanine (O^6-BzG)   312
BER, *see* base excision repair

bleomycin   67
butadiene   10
bypass
– abasic site   302f.
– lesion   387, 422
bypass DNA polymerase   120, 232, 299ff., 316, 337ff., 355

c

C-family DNA polymerase   299
cancer
– environmentally related   6
cancer epidemiology   157
cancerogenesis
– chemical mechanism   188
– hormonal mechanism   187
carbinolamine   205
– Watson–Crick base-pairing   209
carbonatotetrammine Co(III) complex   87
catechol-O-methyl transferase (COMT) inhibitor   145, 193
charge transfer
– location of G oxidation products in DNA   25
– metal complex   86
checkpoint effector kinase (Chk)
– Chk1 and Chk2   9f.
chemical adduct   299ff.
chemotherapy   8
8-chloro-2′-deoxyadenosine   65
5-chloro-2′-deoxycytidine   65
8-chloro-2′-deoxyguanosine   65
chloroacetaldehyde   10, 114
8-chloroadenine   65
5-chlorocytosine   65
8-chloroguanine   65
1-chlorooxirane   115
cisplatin   8, 388
– 1,2-d(GpG) crosslink   414
– 1,2-intrastrand d(GpG) cross-link   416
– 1,3-intrastrand d(GpTpG) cross-link   416
– cisplatin-GG   388ff.
– intrastrand crosslink (cisPt)   240
Cockayne syndrome (CS)   241, 401
– CSA protein   254, 401
– CSB protein   254, 401, 423
COMT, see catechol-O-methyl transferase
conformation
– base-displaced stacked (S)   219
– major groove B-type (B)   219
– minor groove wedge (W)   219
conformational dynamics   224
conformational heterogeneity   221ff.
– translesion synthesis   227

conformational stability
– sequence effect   230
crotonaldehyde   36, 107, 204
5′,8-cyclo-2′-deoxyadenosine   69, 410
– (5′S)-5′,8-cyclo-2′-deoxyadenosine   69
5′,8-cyclo-2′-deoxyguanosine   69
cyclobutane pyrimidine dimer (CPD)   4, 63, 240, 263, 316, 357, 414
cytochrome P450   159, 188
– CYP1B1   192
5-(cytosilyl)methyl radical   66
cytosine adduct   67
– hydrogen abstraction at C4′   67

d

N^6-dA adduct   217
dATP insertion   370
DB[a,l]P, see dibenzo[a,l]pyrene
dCTP insertion   369
DDB2
– ubiquitination   250
dehydroguanidinohydantoin (DGh, Gh$_{ox}$)   24
– lesion   97ff.
deletion duplex (DEL)   276
N-(2-deoxy-D-pentofuranosyl)-N-(2,6-diamino-4-hydroxy-5-formamidopyrimidine) (Fapy-dG)   24, 60, 225, 305f.
N-(2-deoxy-β-D-erythro-pentofuranosyl) formamide   59
6-(2-deoxy-β-D-erythro-pentofuranosyl)-2-hydroxy-3(3-hydroxy-2-oxopropyl)-2,6-dihydroimidazo[1,2-c]pyrimidin-5(3H)-one   68
N^1-(2-deoxy-β-D-erythro-pentofuranosyl)-5-hydroxy-5-methylhydantoin   59
3-(2′-deoxy-β-D-erythro-pentofuranosyl)-pyrimido[1,2-α]purin-10(3H)-one (M₁dG)   110ff., 309f.
3-(2-deoxy-β-D-erythro-pentofuranosyl)-5,6,7,8-tetrahydro-6-hydroxypyrimido[1,2-α] purin-10(3H)-one (α-OH-PdG)   114, 203
3-(2-deoxy-β-D-erythro-pentofuranosyl)-5,6,7,8-tetrahydro-8-hydroxypyrimido[1,2-α]purin-10(3H)-one (γ-OH-PdG)   114, 203, 309f.
– interstrand DNA cross-linking   205
2′-deoxyadenosine (dA)   110
N-(deoxyadenosin-8-yl)-ABP (dA-C8-ABP) adduct   167f.
5-(deoxyadenosin-N^6-yl)-IQ (dA-N^6-IQ)   170
5-(deoxyadenosin-N^6-yl)-MeIQx (dA-N^6-MeIQx)   170
1-(deoxyadenosin-N^6-yl)-2-NA (dA-N^6-NA)   169

2′-deoxycytidine (dC)   110
2′-deoxyguanosine (dG)   110
1,N^2-deoxyguanosine adduct   203ff.
3-(deoxyguanosin-N^2-yl)-AAF (dG-N^2-AAF)   164f.
3-(deoxyguanosin-N^2-yl)-ABP (dG-N^2-ABP)   169
N-(deoxyguanosin-N^2-yl)-ABP (dG-N^2-N^4-ABP)   168
N-(deoxyguanosin-8-yl)-ABP (dG-C8-ABP) adduct   168
N-(deoxyguanosin-8-yl)-2-acetylaminofluorene   217ff., 420
N-(deoxyguanosin-8-yl)-N-acetylbenzidine   170
N-(deoxyguanosin-8-yl)-N'-acetylbenzidine (N'-acetyl-dG-C8-Bz)   170
N^4-(deoxyguanosin-N^2-yl)-2-amino-1,4-naphthoquinoneimine (dG-N^2-NAQI)   169
N-(deoxyguanosin-8-yl)-2-aminofluorene   420
N-(deoxyguanosin-N^2-yl)-4-azobiphenyl   169
N-(deoxyguanosin-8-yl)-benzidine   170
5-(deoxyguanosin-N^2-yl)-IQ (dG-N^2-IQ)   164
N-(deoxyguanosin-8-yl)-IQ (dG-C8-IQ)   164
N-(deoxyguanosin-8-yl)-2-NA (dG-C8-NA)   169
1-(deoxyguanosin-N^2-yl)-2-NA (dG-N^2-NA)   169
N-(deoxyguanosin-8-yl)-PhIP (dG-C8-PhIP)   162
2-deoxyribonolactone   410
2-deoxyribose
– hydroxyl radical-mediated oxidation   67
2′-deoxyribose   108
– 4′-autoxidation   109
– oxidant   34
– oxidation   30ff.
– peroxidation   107
N^2-dG adduct   217, 360
dG-C8-AA   163f.
dG-C8-AAF   217ff., 227, 420
dG-N^2-AAF adduct   217ff.
dG-C8-ABP   168ff., 220
dG-C8-AF   217ff.
dG-C8–1-aminopyrene (AP) adduct   220
dG-C8-HAA   163f.
dG-C8-IQ adduct   231
dG-C8-MeIQx   172
2,2-diamino-4-[(2-deoxy-β-D-erythro-pentofuranosyl)amino]-5(2H)-oxazolone   62

4,6-diamino-5-formamidopyrimidine   62
N-2,6-diamino-4-hydroxy-5-formamidopyrimidine (Fapy-G)   24, 60
2,5-diamino-4H-imidazolone (Iz) lesion   95ff.
dibenzo[a,l]pyrene (DB[a,l]P)   135
– trans-anti-DB[a,l]P-N^6-dA   279
– (–)-anti-DB[a,l]PDE   268
– (–)-trans-anti-N^6-DB[a,l]PDE-dG   418
– diol epoxide (DB[a,l]PDE)   418
– fjord region   267
11R,12R-dihydro-dihydroxy-DB[a,l]P   147
5,6-dihydro-5,6-dihydroxycytosine (Cg)   305
5,6-dihydro-5,6-dihydroxythymine (Tg)   305
7,8-dihydro-8-oxoguanine, see 8-oxo-G
trans-dihydrodiol   141
5,6-dihydrouracil   408
(–)-7R,8R-dihydroxy-dihydro-B[a]P   137
11R,12R-dihydroxy-11,12-dihydro-DB[a,l]P   141
(+)-7β,8α-dihydroxy-7,8-dihydro-9α,10α-oxo-B[a]P   137, 361
7R,8S-dihydroxy-7,8-dihydro-9S,10R-oxo-B[a]P (anti-B[a]PDE)   137ff.
– (–)-anti-B[a]PDE   266
– (+)-anti-B[a]PDE   138ff., 266
– (±)-anti-B[a]PDE   87, 145, 266
5,6-dihydroxy-5,6-dihydrothymidine   64
anti-r7,t8-dihydroxy-t9,10-epoxy-7,8,9,10-tetrahydrobenzo[a]pyrene ((±)-anti-B[a]PDE)   87, 145, 266
r7,t8-dihydroxy-t9,10-epoxy-7,8,9,10-tetrahydro-B[a]P
– (+)-7R,8S,9S,10R enantiomer of r7,t8-dihydroxy-t9,10-epoxy-7,8,9,10-tetrahydrobenzo[a]pyrene (anti-B[a]PDE)   265
– (+)-7R,8S,9S,10R enantiomer of r7,t8-dihydroxy-t9,10-epoxy-7,8,9,10-dihydrobenzo[a]pyrene ((+)-anti-B[a]PDE)   138ff., 266, 313
– notation   265f.
– numbering system   265f.
7,12-dimethylbenz[a]anthracene (DMBA)   135, 192
dimethylsulfate   10
diol epoxide   137
– B[a]P   332
diol epoxide DNA adduct   138
9,12-dioxo-10-dodecenoate   117
9,12-dioxo-10-dodecenoic acid   107
2,6-dioxo-M_1G   124
disease
– DNA damage   5

DNA
- alkylated 310, 412
- 2′-deoxyribose oxidation 30ff.
- cellular 53
- destabilization 280
- ethylation 413
- γ-irradiation 24
- positive hole migration in cellular DNA 65
DNA adduct 33, 110ff., 189
- aromatic amine (AA) 157ff.
- B[a]P 332
- biological effect 162
- bulky 416
- conformation 269
- diol epoxide 138
- effect of size upon polymerase catalysis 313
- exocyclic 125, 308
- heterocyclic aromatic amine (HAA) 157ff.
- interstrand DNA cross-linking 205ff.
- isolated 53
- PAH o-quinone 142
- radical cation 135
DNA adduct bypass
- exocyclic 308
DNA base
- alkylated 310, 412
- reaction of β-substituted acroleins 109
- reaction of MDA 109
DNA cross-link
- interstrand 203ff., 316, 414
- intrastrand 316, 414
DNA damage 4ff., 399ff.
- assessment 123
- biology 3ff.
- chemistry 3ff.
- disease 5
- free radical reaction 81ff.
- inflammation 21ff.
- oxidative 53ff., 188, 305, 408
- oxidative stress 105ff.
- protection 192
- recognition in GG-NER 248
- repair 121
- verification 251
DNA damage response (DDR) system 3
- cellular 9
DNA deamination 26
- analytical method 29
DNA duplex
- destabilization 271ff.
- fully complementary 219

B-DNA duplex 227
DNA glycosylase 242, 302
- lesion recognition 242
DNA lesion 12
- adenine 264
- bulky 261ff., 331ff.
- bypass 387, 422
- chemically defined 381ff.
- demarcation in NER 251
- DNA polymerase complex 231
- multipartite model of recognition 264
- oxidative 144
- oxidative damage 305
- PAH 131ff.
- PAH diol epoxide 265
- PAH o-quinone 144
- polymerase ζ 388
- recognition 261ff.
- removal 261ff.
- repair 10
- structure–function relationship 333ff.
- TC-NER 403ff.
- transcript integrity 424
- transcription elongation 403ff.
DNA ligase 3α/XRCC1 (excision repair cross-complementation group 1) complex 242
DNA nitration 23
DNA oxidation 23
DNA polymerase 120, 232, 299ff., 335ff.
- DNA lesion complex 231
- Pol β 241ff., 299ff.
- Pol ζ 317, 358, 388ff.
- Pol η 316, 356ff., 382ff.
- Pol θ 317
- Pol ι 363
- Pol κ 315, 341f., 362ff., 392
- Pol ν 317
- Pol I 354
- Pol II 357ff.
- Pol III 358ff.
- Pol IV 355ff., 366ff.
- Pol V 356ff.
- structure–function relationship 335
DNA radical reaction 86
DNA repair 400f.
DNA sequence
- context 224, 280
- effect on the conformational stability 230
- interstrand DNA cross-linking 210
- mutagenesis 224
DNA strand break 10, 30, 57
DODE (9,12-dioxo-10(E)-dodecenoic acid) 36

double-strand break
– repair   10
Dpo4   120, 232, 299ff., 341, 371ff.
dual-incision   252

e

electron
– hydrated   95
endonuclease   242ff., 252, 262
endoperoxide
– 4,8-endoperoxide   56
– bicyclic   107
enzyme
– genetic polymorphism   159
– metabolic activation   159
– peroxidative   169
enzyme-linked immunosorbent assay (ELISA)
– aldehyde-reactive probe (ARP)-based   136
4,5-epoxy-2-decenal   115
equilenin   187
equilenin catechol (4-OHEN)   189ff.
– 4-OHEN-o-quinone   191
equine estrogen   191
equine estrogen hormone replacement therapy   185ff.
ERCC1-XPF   212, 252f.
Escherichia coli
– mutagenic pathway   353ff.
– translesion synthesis   353ff.
estradiol   188ff.
estrogen   8
estrogen carcinogenesis
– mechanism   187
estrogen quinoid   189
estrogen receptor (ER)   193
estrogen receptor-mediated signaling pathway   187
etheno adduct   114
– 1,N^6-ε-dA   36, 115ff.
– 3,N^4-ε-dC   36, 115ff.
– 1,N^2-ε-dG   36, 115ff., 308f.
– N^2,3-ε-dG   115ff.
– lipid peroxidation   35
1,N^6-etheno-2′-deoxyadenosine (1,N^6-ε-dA)   36, 115ff.
3,N^4-etheno-2′-deoxycytidine (3,N^4-ε-dC)   36, 115ff.
1,N^2-etheno-2′-deoxyguanosine (1,N^2-ε-dG)   36, 115ff., 308f.
1,N^6-ethenoadenine (1,N^6-ε-dA)   36, 115ff., 240
1, N^2-ethenoguanine (1,N^2-ε-dG)   115ff., 308f., 413
N^2-ethyl-2′-deoxyguanine   413

f

Fapy (2,6-diamino-4-hydroxy-5-formamidopyrimidine)-dG   24, 60, 225, 305f.
fluorescence line narrowing spectrometry (FLNS)   137
food mutagen   157
formaldehyde   10
5-formyl-2′-deoxyuridine   64
free radical
– biological implication   99
– environmental consideration   88
– formation of DNA damage   81ff.
– lifetime   88
– mechanism   82
– reaction   89
– reaction with nucleic acid   83
– studying reaction   84
– type   83

g

G, *see* guanine
G[8–5]C adduct   66
G[8–5m]T adduct   66
gap   407
gap-lesion plasmid assay for mammalian TLS   384ff.
genetic polymorphism   159
genotoxic estrogen pathway   185ff.
genotoxicant   159
genotoxicity   188
GG-NER, *see* nucleotide excision repair
glutathione S-transferases (GST)   170, 192
glyoxal   33
5-guanidino-4-nitroimidazole (NIm)   95, 410f.
guanine (G)   60
– nitrosative deamination   29
– one-electron oxidation reaction   64
– oxidation product   23ff.
– oxidative degradation pathway   61f.
– singlet oxygen oxidation   55f.
guanine B[*a*]P adduct   270
– minor groove and base-displaced/intercalative conformation   270
guanine radical   91ff.
– oxyl radical   93
– reaction with nucleophile   91

h

halogenation
– HOCl acid-mediated   65
– pyrimidine and purine base   65
heptanone-1,N^2-ε-Gua   125

heptenal 36
heterocyclic aromatic amine (HAA) 157ff.
heterocyclic aromatic amine DNA adduct 161ff.
– detection 168
4-HNE (4-hydroxy-2(E)-nonenal) 36, 107ff., 146, 205
– interstrand DNA cross-linking 207f.
8-HO-dG· radical 91
5-HO-8-oxo-G 98
hormone replacement therapy (HRT) 185
9-HPODE (9-hydroperoxyoctadecadienoic acid) 107
13-HPODE (13-hydroperoxy-(9Z),(11E)-octadecadienoic acid) 36, 107
5-HPETE (5-hydroperoxyeicosatetraenoic acid) 107
15-HPETE (15-hydroperoxyeicosatetraenoic acid) 107
hydrated electron 95
hydrogen abstraction at C4'
– cytosine adduct 67
hydrogen atom abstraction 90
– intramolecular 90
hydrogen atom abstraction at C5'
– purine 5',8-cyclonucleoside 68
4-hydroperoxy-2-nonenal 107ff.
5-(hydroperoxymethyl)-2'-deoxyuridine 59
N-hydroxy-AA 159ff.
– DNA 161
N-hydroxy-AIA 162
N-hydroxy-2-amino-1-methyl-6-phenylmidazo[4,5-b]pyridine (HONH-PhIP) 161
N-hydroxy-4-aminobiphenyl (HONH-ABP) 164ff.
N-hydroxy-2-aminofluorene (HONH-AF) 164
N-hydroxy-2-aminonaphthalene (HNOH-NA) 169
8-hydroxy-7,8-dihydro-7-yl radical 61
8-hydroxy-7,8-dihydroadenyl radical 62
8-hydroxy-7,8-dihydroguanyl radical 64
6-hydroxy-5,6-dihydrothym-5-yl radical 64
N-hydroxy-HAA 159ff.
– DNA 161
N-hydroxy-2-naphthylamine (HONH-2-NA) 164
4-hydroxy-2-nonenal 107ff., 146
4-hydroxy-2-nonenal-dG adduct 121
4-hydroxy-8-oxo-4,8-dihydro-2'-deoxyguanosine 57
5-hydroxy-8-oxo-7,8-dihydro-2'-deoxyguanosine 56, 70

1,N^2-6-hydroxy-propanodeoxyguanosine (α-OH-PdG) 114, 203
1,N^2-8-hydroxy-propanodeoxyguanosine (γ-OH-PdG) 114, 203, 309f.
γ-hydroxy-1,N^2-propano-2'-deoxyguanosine (γ-HOPdG) 309f.
β-hydroxyacrolein 112
β-hydroxyalkyl radical 90
5-hydroxycytosine 70
4-hydroxyequilenin-C 388
2-hydroxyestradiol 190
4-hydroxyestradiol 190
2-hydroxyestradiol quinone methide 189
4-hydroxyestradiol-o-quinone 189
α-hydroxyhexyl-γ-OH-PdG 114
hydroxyl radical 53ff.
– 2-deoxyribose oxidation 67
hydroxyl radical reaction 58
5-(hydroxymethyl)-2'-deoxyuridine 64
hydroxynonenal (HNE) 36
– 4(R)-hydroxynonenal 203f.
– 4(S)-hydroxynonenal 203f.
hydroxysteroid SULT (SULT2) 161
5-hydroxyuracil 70, 408
hypoxanthine (Hyp) 27, 240
hypoxanthine phosphoribosyltransferase (hprt) 165

i

imine 205
inflammation 21ff.
– biomarker 25
inflammatory response 5
intercalative insertion 271f.
interstrand DNA cross-linking 203ff., 316, 414
– biological significance 212
– α-CH3-γ-OH-PdG adduct derived from crotonaldehyde 207
– DNA sequence 210
– 4-HNE 207f.
– γ-OH-PdG adduct 205
– stereochemistry 210
intrastrand DNA cross-link 316, 414
ionizing radiation 8
α-irradiation 34
γ-irradiation
– DNA 24, 34

l

laser flash photolysis 84
lesion, *see also* DNA lesion
– 6-4 lesion 316
– bypass 387, 422

– unrepaired   12
lesion recognition
– DNA glycosylase   242
– multipartite model   264
lipid hydroperoxide   107
lipid peroxidation   105ff.
– etheno adduct   35
lysine N-formylation   34

m

M_1dA   111
M_1dC   111
M_1dG   110ff., 309f.
– stability   112
M_1G adduct   33ff.
major groove
– intercalative insertion   271f.
malondialdehyde (MDA)   37, 107ff., 146
– reaction with DNA base   109
– MDA-dG adduct   110ff., 146
MeIQx, see 2-amino-3,4-dimethylimidazo[4,5-f] quinoxaline
metabolic activation
– PAH   134ff.
methyl guanine methyl transferase (MGMT), see O^6-alkylguanine DNA alkyltransferase
5-methyl-C   27
3-methyladenine (3-meA)   240, 412
3-methyladenine glycosylase (AlkA)   122
7-methylguanine   412
O^6-methylguanine (O^6-MeG)   311f., 412f.
4-methylindole (4M)   247
methylnitrosourea   10
misinsertion frequency   302
mismatch excision repair   11
molecular mechanism
– translesion DNA synthesis   381ff.
mutagenesis
– bacterial   164
– DNA sequence context   224
– $Escherichia\ coli$ 354
– mammalian   165
– transgenic rodent   166
mutagenic pathway
– $Escherichia\ coli$ 353ff.
mutagenicity
– peroxidation-derived adduct   117

n

2-naphthylamine (2-NA)   157
NEIL1 glycosylase   242
NEIL2 glycosylase   242
NER, see nucleotide excision repair

5-nitro-1-indolyl-2′-deoxyriboside-5′-triphosphate (5-NITP)   305
8-nitroguanine (8-nitro-G)   95ff.
nitrosoperoxycarbonate   63
nongenomic pathway   187
NONOate   28
NQO1 activity   193
nucleobase
– one-electron oxidation   63
nucleotide excision repair (NER)   121, 139, 225, 239ff., 354
– bipartite model   264
– damage recognition   248ff.
– damage verification   251
– DNA lesion   403ff.
– dual-incision   252
– efficiency   272ff.
– global genomic (GG-NER)   11, 139, 248ff., 262
– lesion demarcation   251
– mammalian   261ff., 286
– mechanism   239ff., 265
– Pol II pathway   359
– prokaryotic   263
– recognition   261ff., 286
– removal of bulky DNA lesion   261ff.
– repair synthesis   252
– structure–function relationship   268
– subpathway   248
– substrate   240
– system   11ff., 117
– transcription-coupled (TC-NER)   11, 139, 248ff., 262, 403ff., 421
nucleotidyl transfer reaction   343

o

OGG1 glycosylase   242ff.
4-OHEN   189ff.
4,17β-OHEN   194
4-OHEN-o-quinone   191
oncogene
– genetic alteration   167
one-electron oxidation of nucleobase   63
– adenine   64
– guanine   64
OPdA   111ff.
OPdC   111ff.
OPdG   112f.
Ox, see oxazalone
oxanine (O)   27ff.
oxazalone (Ox, Oz)   23f., 96ff.
5′-oxidation   33f.
N-oxidation   159

oxidative degradation pathway
– guanine   61
oxidative DNA damage   53ff., 188, 305, 408
oxidative stress
– biomarker   25
– DNA damage   105ff.
8-oxo-dA   189
8-oxo-7,8-dihydro-2′-deoxyadenosine   62
8-oxo-7,8-dihydro-2′-deoxyguanosine   54ff.
8-oxo-7,8-dihydroguanine   70
8-oxo-dG   189, 225
8-oxo-dGuo   144ff.
8-oxo-G (8-oxoguanine, 7,8-dihydro-8-oxoguanine)   23f., 97ff., 239, 305f., 408
– lesion   95, 339
– oxidation   97
– radical cation 8-oxo-G$^{·+}$ 98
8-oxo-G$^{·+}$ 98
8-oxo-G(-H)$^{·}$ 98
8-oxo-Gua   145
2-oxo-ε-Gua   125
2-oxo-heptanone-ε-Gua   125
6-oxo-M$_1$dG   124
4-oxo-2-nonenal   107ff., 146
N^6-(3-oxo-1-propenyl)-dA (OPdA, M$_1$dA)   111ff.
N^4-(3-oxo-1-propenyl)-dC (OPdC, M$_1$dC)   111ff.
N^2-(3-oxo-1-propenyl)-dG (N^2-OPdG)   309
oxopropenyl derivative   111
N^2-(3-oxopropyl)-dG lesion   205
oxyl radical   93
– guanine radical   93
Oz, see oxazalone

p

p21   383
p53   132ff., 167, 356, 383
P450 reductase   188
PAH, see polycyclic aromatic hydrocarbon
PAH o-quinone   144
– DNA lesion   144
– pathway   146
PAH o-quinone DNA adduct   142
– covalent   142f.
– 1,4- or 1,6-Michael addition   143
PAH-dA substrate   227
PdG (propanodeoxyguanosine)   120ff.
pentenal   36
peroxidase
– arachidonic acid-dependent   169
peroxidation-derived adduct
– mutagenicity   117

peroxidative enzyme   169
phenol SULT (SULT1)   161
PhIP, see 2-amino-1-methyl-6-phenylimidazo[4,5-b]pyridine
5′-(2-phosphoryl-1,4-dioxobutane)   33
photodissociation   86
photoionization   86
6-4 photoproduct (6-4-PP)   240, 414
polycyclic aromatic hydrocarbon (PAH)   6ff., 131ff., 217, 313
– N^6-adenine adduct   272
– bay region   138, 266, 279
– bulky PAH-DNA adduct   268
– carcinogen   132
– diol epoxide-N^6-dA adduct   278ff.
– diol epoxide-N^6-dA lesion   268
– diol epoxide-N^2-dG lesion   268
– DNA lesion   131ff., 265
– effect of adduct size upon polymerase catalysis   313
– fjord region   138, 267ff.
– metabolic pathway   131ff.
– methylated bay region   138
– NER efficiency   278
– o-quinone   141ff.
– stereochemistry   269
– topology   269
polymerase
– chimney opening   366
– Pol β   241ff., 299ff.
– Pol ζ   317, 358, 388ff.
– Pol η   316, 356ff., 382ff.
– Pol θ   317
– Pol ι   363
– Pol κ   315, 341f., 362ff., 392f.
– Pol ν   317
– Pol I   354
– Pol II   357ff.
– Pol III   358ff.
– Pol IV   355ff., 366ff.
– Pol V   356ff.
– roof region   368
– structure–function analysis   362
polymerase catalysis
– effect of DNA adduct size   313
polyunsaturated fatty acid (PUFA)   89, 105ff.
positive hole migration
– cellular DNA   65
Premarin®   186
procarcinogen   131
proliferating cell nuclear antigen (PCNA)   383

propano adduct   114
prostaglandin H synthase (PHS)   169
psoralen   415
psoralen thymidine monoadduct   414
purine base
– HOCl acid-mediated halogenation   65
purine 5′,8-cyclonucleoside   68f.
– hydrogen atom abstraction at C5′   68
pyrimidine base
– HOCl acid-mediated halogenation   65
– oxidized   57
pyrimidine–pyrimidone 6-4 photoproduct   63, 316
pyrimido[1,2-α]purin-10(3H)-one   411
pyrimidopurinone   205
pyrolytic HAA   158

q
quantum mechanical/molecular mechanics (QM/MM) simulation   343
quantum mechanical/molecular mechanics/molecular dynamics (QM/MM/MD)   345
o-quinone   188
quinone reductase (QR)   192

r
Rad4/XPC   264
α-radiation   34
γ-radiation   24, 34
radical
– free   81ff.
– mechanism of product formation   91
– reaction kinetics   85
– type of reaction   85
radical cation DNA adduct   135
radical cation pathway   134
K-*ras*   132
reactive nitrogen species (RNS)   4f., 21ff.
– indirect base damage   35
– inflammation   22
reactive oxygen species (ROS)   4ff., 21, 56, 108, 144
RecA   357
repair   400f.
– DNA damage   121
repair synthesis   252
replication error (RER) phenomenon   354
replication protein A (RPA)   251ff.
replicative DNA polymerase   299ff.
REV1   300ff.
ring-opening
– hydrolytic   112
RNA integrity   399ff.

RNA polymerase   253, 402ff., 419
– elongation   407
– Pol I   406
– Pol II   406ff., 418f.
– Pol III   406
ROS, *see* reactive oxygen species

s
Salmonella typhimurium   164
sequence, *see* DNA sequence
sequence effect
– conformational stability   230
single base damage   55
single-strand break
– repair   10
single-strand nick   407
singlet oxygen oxidation
– guanine   55
slipped mutagenic intermediate (SMI)   221ff.
– sequence effect on the conformational stability   230
spiroiminodihydantoin (Sp) lesion   23, 97ff.
stereochemistry
– DNA adduct conformation   269
– interstrand DNA cross-linking   210
strand break   30, 57
structure
– DNA lesion–DNA polymerase complex   231
structure–function relationship   203ff., 268
– bulky DNA lesion   331ff.
substance
– environmental cancer-causing   6
sulfotransferase (SULT)   159
– hydroxysteroid SULT (SULT2)   161
– phenol SULT (SULT1)   161

t
tandem base lesion   66
TC-NER, *see* nucleotide excision repair
tetrahydropyrimidopurinone ring   114
7,8,9,10-tetrahydroxytetrahydrobenzo[*a*]pyrene (B[*a*]PT)   87
TFIIH transcription/repair factor   251ff., 262, 406, 423
TFIIS   406ff., 421ff.
thermal/thermodynamic probing   226
thymine   27, 58
– hydroxyl radical-mediated oxidation   60
thymine glycol (TG)   239, 408
thymine–thymine (TT)   414
– CPD   387f., 414
– 6-4 photoproduct   388, 414

tobacco smoke   157
topology
– DNA adduct conformation   269
transcript integrity   424
transcription elongation   399ff.
– DNA lesion   403ff.
translesion bypass   12
translesion DNA polymerase   117, 387
– small interfering RNA (siRNA)   388
translesion DNA synthesis (TLS)   299ff., 353ff., 381ff.
– conformational heterogeneity   227
– *Escherichia coli*   353ff.
– gap-lesion plasmid assay for mammalian TLS   384
– molecular mechanism   381ff.
– study   384
trichothiodystrophy (TTD)   241
2,4,6-trioxo[1,3,5]triazinane-1-carboxamidine (CAC)   24
tumor suppressor gene
– genetic alteration   167
two-photon ionization   86ff.
two-photon mechanism   87
two-photon-induced ionization mechanism   63

u

ubiquitination
– DDB2   250
– PCNA   383
UDP-glucuronosyltransferase   170
UmuC(V)   363ff.
uracil (U)   27, 239f.
uracil DNA glycosylase   224
5-(uracilyl)methyl radical   66
UV photoproduct   316, 357
UV-DDB factor   250

UvrABC
– protein   225
– system   286

v

vinyl chloride   10

w

Watson–Crick base-pairing   209ff., 227ff.
– *10R* (+)-*trans*- and *10S* (–)-*trans-anti*-B[*a*]P-N^6-dA adduct   271
– carbinolamine   209

x

X-family DNA polymerase   299
xanthine   27ff.
xanthine oxidoreductase   125
xeroderma pigmentosum (XP)   241
xeroderma pigmentosum variant (XPV)   382ff.
XPA   251ff.
XPB   423
XPC-RAD23B (XPC/HR23B)   248ff., 263f., 275f.
XPD   251, 423
XPF   423
XPF/ERCC1 (xeroderma pigmentosum group F/excision repair cross-complementation group 1) complex   212, 252
XPG   251ff., 423
XRCC1   248

y

Y-family (bypass) DNA polymerase   120, 232, 299ff., 316, 337ff., 355ff.
– mechanistic step   373
– structure–function analysis   362